DNA Repair

Errol C. Friedberg
Stanford University

W. H. FREEMAN AND COMPANY
New York

Library of Congress Cataloging in Publication Data

Friedberg, Errol C.
DNA repair.

Includes bibliographies and indexes.
1. Deoxyribonucleic acid repair. I. Title. II. Title:
D.N.A. repair.
QH467.F75 1984 574.87'3282 84-18668
ISBN 0-7167-1674-7

Printed in the United States of America

1 2 3 4 5 6 7 8 9 0 HA 3 2 1 0 8 9 8 7 6 5

Produced by Ron Newcomer & Associates, San Francisco

To my sons,
Malcom Bradley
Andrew Seth

Contents

Preface

We live in an environment that poses continual threats to our genetic material. Ionizing radiations, ultraviolet light from the sun and a multitude of chemical agents (some manmade) cause alterations of DNA that would soon render our planet barren, were it not for constant cellular monitoring and repair of most defects as they occur. In addition, cellular DNA is subject to spontaneous damage that includes loss of bases, chemical alteration of bases and changes in base sequence due to replicative and recombinational infidelity. The repair of DNA must be regarded along with replication and recombination as an essential transaction of the genetic material in all life forms.

The study of DNA damage and the biological responses to such damage has undergone massive expansion during the past 30 years. Once the domain of a relatively small group of photobiologists interested in the responses of living cells to ultraviolet radiation, the field has since attracted the converging attention of biologists, chemists, physicists, biochemists, ecologists, radiobiologists, oncologists and dermatologists, among others. Investigators in these and other disciplines have brought to the field a wide range of new techniques and a wealth of new information.

Much of the recent excitement in this field has derived from the evident relevance of DNA repair to human health. Damage to DNA has been clearly implicated in cancer and there have been suggestions that it may be a component in the biology of aging as well. Many spontaneous abortions and some congenital defects may represent the adverse consequences of unrepaired or misrepaired DNA damage in embryonic cells. Interest and concern of the general public has increased as the list of known carcinogens in our environment expands. Is there a threshold level of exposure, below which these chemical agents or radiations may be considered "safe"? How dangerous is a chest X ray? Are some combinations of carcinogens much more potent than the health threat of the individual chemicals? The answers to these and to many other related questions require a detailed understanding of the cellular responses to DNA damage.

Within the past decade there have been a number of well-attended international conferences devoted to the subject of DNA damage and its repair. Although the development of the field can be followed through the published proceedings of these meetings, the literature for the most part is still dispersed in scientific journals. No comprehensive treatment of the subject of DNA repair has yet appeared in textbook form. Thus, courses on DNA repair and mutagenesis have relied on current review articles and journal articles as source material. This book was conceived with the specific goal of providing a comprehensive survey of DNA repair and other cellular responses to DNA damage. It is aimed primarily at the advanced undergraduate or graduate student who already possesses a solid understanding of the fundamental molecular biology of nucleic acids and of proteins, but who wants a deeper understanding of how living cells respond to genomic damage. I have thus attempted to organize into one logical scheme many of the diverse inputs from the different specialities that impinge on DNA repair, and I have provided features that I hope will be of value as a single reference source to students, teachers and investigators.

The book begins with a discussion of spontaneous DNA damage and damage from interactions with environmental, physical and chemical agents. Chapters 2 through 6 deal with DNA repair by the reversal of base damage (e.g., photoreactivation of pyrimidine dimers) and with excision of damaged or inappropriate bases from DNA. The latter topic is considered first in terms of the biochemical and enzymatic events that lead to the *incision* of DNA at sites of damage and then in terms of postincision events. Chapters 7 and 8 consider cellular responses that facilitate the *tolerance* of persisting damage in genetic material and Chapter 9 considers the rapidly expanding area of DNA repair in human cells and its relation to human disease. Throughout the text, I have deliberately confined my definition of DNA repair to those cellular responses associated with the reversal or removal of the *primary* damage to DNA and have largely avoided the general usage of terms such as "postreplication repair" and "inducible repair." These terms describe cellular responses associated with the *tolerance* of DNA base damage that do not typically involve *removal* of primary damage to the genome.

Although much of our understanding of and many of the models for DNA repair have derived from the well-studied bacterium *Escherichia coli*, I have included other biological systems as appropriate, for particular analysis. A deliberate major omission from this book is a comprehensive treatment of mutagenesis. This omission was made in the interests of providing a text of manageable proportions, with the realization that the topic of mutagenesis deserves a treatise of its own. Whenever possible, I have striven to provide an accurate assessment of what we know for sure about cellular responses to DNA damage, but I have also attempted to highlight unresolved problems and areas of controversy that merit further study. I have included an extensive bibliography in the hope that the book will serve its intended primary function—as an *initial* source of information. However, I am well aware that the bibliography is inevitably incomplete and apologize for any major omissions.

This work was conceived through a long association with my colleague and friend Philip Hanawalt. Indeed, it was originally intended as a coauthored text, and certainly the overall scope of the book, the selection of chapter topics and the sequence of chapters was a joint effort. Phil reluctantly was forced to yield to numerous other academic priorities and had I known then how long (and sometimes painful) this conception was going to be, I might have done the same.

I have benefited greatly in this enterprise from my interactions with a stimulating group of colleagues, both at Stanford and internationally in this thriving area of molecular biology. I am especially indebted to the numerous colleagues who generously gave of their time for critical reading and valuable discussion. Particular thanks go to Michele Calos, Janet Chenevert, Louie Naumovski and Charles Allen Smith, who read the entire manuscript, to Tom Lindahl and Eric Radany, who read large segments of the manuscript, and to Tom Bonura, Jim Cleaver, Ursula Ehmann, Phil Hanawalt, Bob Lehman, Dave Mount, Glenn Pure, Miroslav Radman, Gordon Robinson, Stan Rupert, Roger Schultz, Erling Seeberg, Bill Weiss, Harold Werbin, Evelyn Witkin and Elizabeth Yang, who each read one or more chapters.

It is my hope that the illustrations will be a major attribute of this book. In this regard I am deeply appreciative of the stimulating collaboration of Charlene Levering at Stanford University and gratefully acknowledge her staff for their skillful art work and consummate patience. While on the topic of patience, no words can adequately express my thanks to Margaret Beers, Jean Oberlindacher and Gina Johnson, who cheerfully typed and helped proof countless drafts of this manuscript. Jean also provided invaluable assistance in the preparation of the author index, while the subject index reflects the diligent efforts of my wife and colleague Anne Bowcock. I am most grateful to Neil Patterson and his staff at W. H. Freeman and Company for their commitment to this publication. Special thanks are due to Karen McDermott for her skillful editing, to Ron Newcomer for his outstanding job of production and to Linda Chaput and Alex Standefer for their constant and cooperative supervision of all phases of the translation of a manuscript into a book. Finally, I thank my many friends for their constant support and encouragement and the Department of Physical and Occupational Therapy at Stanford University Hospital for frequent relief from the physical toll of writer's neck, a malady discovered during this odyssey.

It is probably a safe assumption that no textbook is perfect—certainly I am aware of some of the limitations and shortcomings in this one. Please inform me of any that you uncover; they will certainly be addressed should any future editions materialize.

Errol C. Friedberg
September 1984

1

DNA Damage

1-1 Introduction

Once it was recognized that DNA is the informationally active chemical component of essentially all genetic material (with the notable exception of RNA viruses), it was assumed that this macromolecule must be extraordinarily stable in order to maintain the high degree of fidelity required of a master blueprint. It has been something of a surprise to learn that the primary structure of DNA is in fact quite dynamic and subject to constant change. In recent years the phenomenon of gene transposition has become a well-established phenomenon in prokaryotes (1–4), and there is increasing evidence for the existence of transposable elements in eukaryotes, including humans (1–4). The translocation of large nucleotide sequences is thought to constitute an important mechanism by which gene expression can be altered in living cells and is not considered here as an example of DNA damage.

In addition to these "macro" changes, DNA is also subject to alteration in the chemistry or sequence of individual nucleotides (5–13). Many of these changes arise as a consequence of errors introduced during DNA transactions such as replication, recombination and repair itself. Other base alterations arise from the inherent instability of specific chemical bonds that constitute the normal chemistry of nucleotides under physiological conditions of temperature and pH. Finally, the DNA of living cells is highly reactive to a large variety of chemical compounds and a smaller number of physical agents, many of which are present in any environment. Some of the chemicals are products of the metabolism or decomposition of the other living forms with which many organisms exist in intimate proximity. Others, particularly in recent decades, are manmade and possibly contribute significantly to the genetic insult faced by individuals living in highly industrialized communities. It should be appreciated, however, that in general the magnitude and real biological significance of this latter source of damage are difficult to assess, and hence the issue of industrial genetic toxicity remains highly controversial.

Each of these modifications of the molecular structure of genetic material is appropriately considered as DNA damage. Collectively, this multitude constitutes the substrate for the manifold DNA repair processes and other cellular responses that are considered in this book. For convenience, DNA damage can be grouped into two major classes, referred to as *spontaneous* and *environmental*. However, as we shall see, in some cases the actual chemical changes in DNA that occur "spontaneously" are indistinguishable from those brought about through interaction with certain environmental agents. In fact, the term "spontaneous" may sometimes merely imply that a particular environmental culprit has not yet been identified. The latter category comprises a very extensive list of known and potential agents, and it is not the intention of this text to treat each of these exhaustively. Rather, a number of specific physical and chemical agents have been selected about which there is substantial information concerning the mechanism of their interaction with DNA and the biological consequences of such interactions. Although all the primary components of DNA (bases, sugars and phosphodiester linkages) are subject to damage, this chapter focuses very largely on the nitrogenous bases, since these are the informational elements of the genetic blueprint encoded in DNA.

1-2 Spontaneous Damage to DNA

DNA Replication

The chief source of DNA damage arising during normal DNA metabolism is probably the mispairing of bases during the synthesis of DNA. Such mispairings can in theory arise from the synthetic events

TABLE 1-1
Mechanisms for maintaining genetic stability associated with DNA replication in *E. coli*

Mechanism	Cumulative Error Frequency
Base pairing	$\sim 10^{-1}$ to 10^{-2}
DNA polymerase actions (including base selection and $3' \rightarrow 5'$ proofreading exonuclease)	$\sim 10^{-5}$ to 10^{-6}
Accessory proteins (e.g., single-strand binding protein)	$\sim 10^{-7}$
Post-replicative mismatch correction	$\sim 10^{-10}$

After M. Radman et al., ref. 17.

associated with a *subset* of nuclear DNA during repair and recombination; however, *all* the bases in a genome must transact the process of replication prior to cell division, and thus semiconservative DNA synthesis probably accounts for the bulk of such lesions.

A number of parameters are known to affect replicational fidelity in *Escherichia coli* (14–19), and a summary of these is provided in Table 1-1. Without the intervention of any cellular factors, the difference in free energy for the stable pairing of a complementary base relative to that for a noncomplementary base during DNA synthesis is only 2 to 3 kcal/mol (the equivalent of a single hydrogen bond) (17, 18). In the absence of other influences, this translates into a potential error frequency of from 1 to 10 percent per nucleotide (14–18). However, measurements of the spontaneous mutation frequency show that the error frequency in newly replicated DNA in *E. coli* is in fact six to nine orders of magnitude less than that predicted from the energetic considerations given above (14–18). A significant contribution to this fidelity stems from the action of specific components of the replication machinery—the complex of DNA polymerases and accessory proteins that constitute the "replisome" required for normal semiconservative DNA synthesis in vivo (19). The combined effects of base selection and proofreading of newly inserted nucleotides by certain DNA polymerases, together with the enhancement of replication fidelity contributed by accessory factors such as single-strand-DNA binding protein, increase the accuracy of DNA synthesis by from three to six orders of magnitude (14–18) (Table 1-1).

Not all the determinants of semiconservative DNA replication that contribute to fidelity are known at the present time, but a variety of biological systems provide interesting clues. Thus, for example,

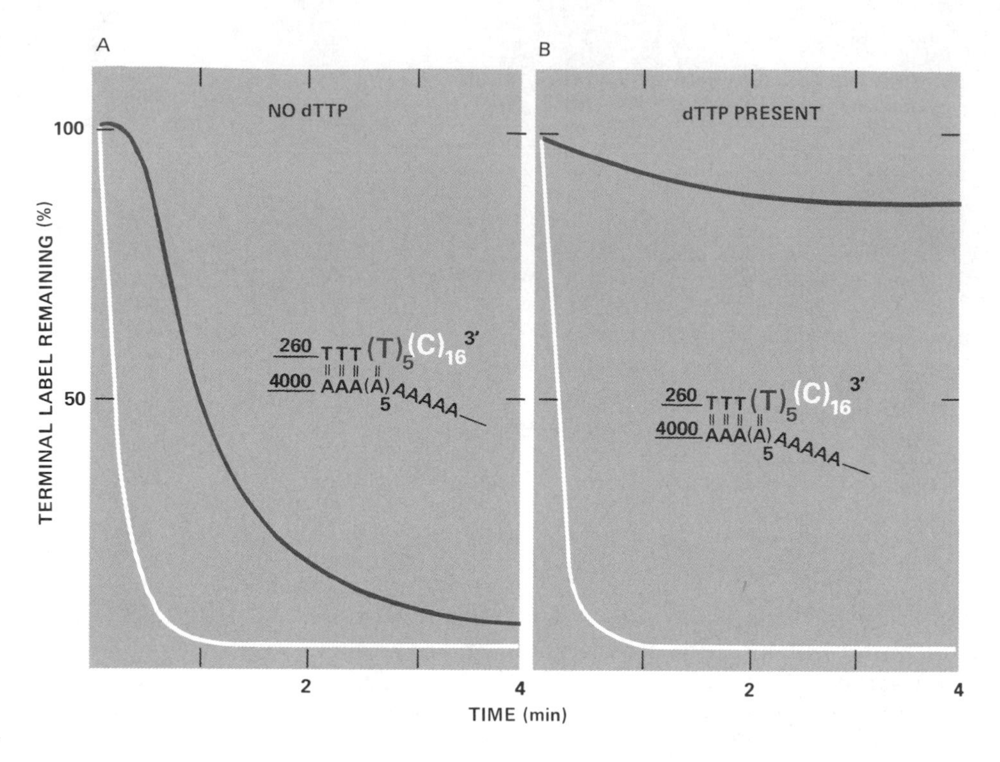

FIGURE 1-1
The 3′ → 5′ proofreading function of *E. coli* DNA polymerase I. When C is mispaired with A in a synthetic polymer, the mispaired nucleotide is excised both in the absence (A) or presence (B) of the appropriate triphosphate (dTTP). In the absence of the triphosphate, terminal fraying of the polymer is associated with excision of some dTMP as well. (From A. Kornberg, ref. 19.)

those DNA polymerases that contain associated 3′ → 5′ exonucleases can use this catalytic function to "edit out" (proofread) 3′-terminal mispaired nucleotides (14–19). The phenomenon of proofreading by DNA polymerase-associated 3′ → 5′ exonuclease activity was first demonstrated with purified DNA polymerase I of *E. coli* in vitro (20) (Fig. 1-1). The phenomenon has also been observed with the DNA polymerase of bacteriophage T4, an enzyme known to be required for the semiconservative replication of the DNA of this phage (19). Gene *43* of phage T4 is the structural gene that codes for the DNA polymerase core or holoenzyme (21), and mutations in this gene profoundly affect the overall mutation frequency of phage T4. Certain mutations in this gene (mutators) create a markedly enhanced mutation frequency in the phages (22, 23), whereas others significantly lower the mutation frequency (antimutators) (24) (Fig. 1-2). The enzyme from those phages that behave as mutators exhibits reduced 3′ → 5′ exonuclease relative to polymerase activity, whereas DNA polymerases from the antimutators have relatively enhanced 3′ → 5′

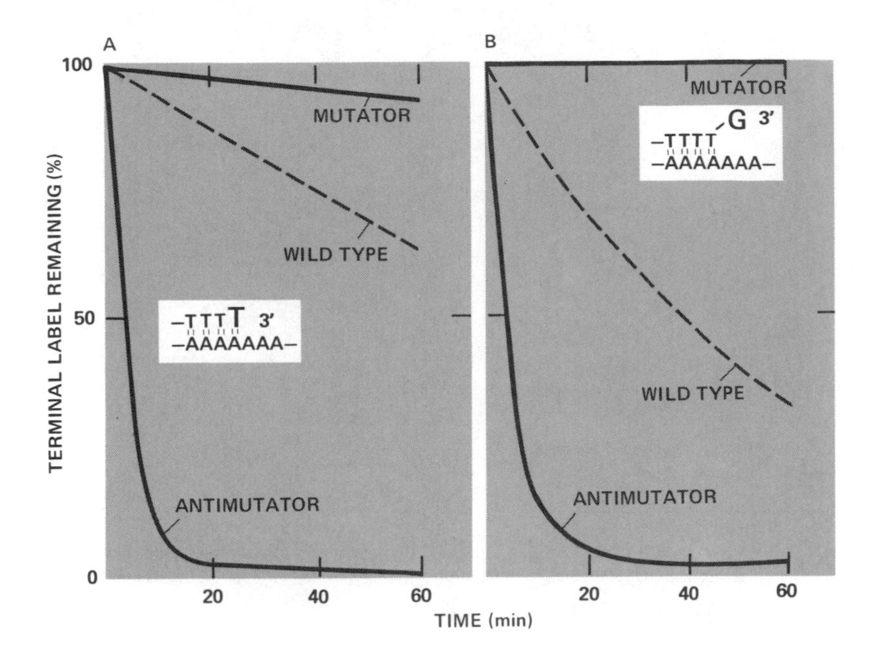

FIGURE 1-2
The 3′ → 5′ exonuclease activity of phage T4 mutator and antimutator DNA polymerases. The antimutator enzyme removes both base-paired (A) and mismatched (B) terminal nucleotides much more rapidly than the wild-type enzyme does, whereas the mutator enzyme does so much more slowly. (From A. Kornberg, ref. 19.)

exonuclease activity (25, 26). These observations have led to models that explain incorporation errors in DNA synthesis as a consequence of the interplay of polymerase specificity and exonuclease proofreading activity (27–29).

For many years, measurements of replicative fidelity in vitro utilized as templates homopolymers or copolymers missing one of the four nucleotides in order to quantitate the incorporation of a noncomplementary nucleotide (18). However, the polymers may be subject to artifactual effects such as complementary-strand slippage and unnatural base-stacking interactions that confer conformations not present in natural DNA (18). A sensitive alternative assay has been developed in which the error rate of DNA synthesis in vitro is quantitated by measuring the frequency of reversion of replicated phage ϕX174 DNA containing a defined *amber* mutation(s) to wild-type (18, 30) (Fig. 1-3). In this assay ϕX174 template DNA containing an amber mutation is replicated in vitro with a selected polymerase of interest and the resulting product is used for the transfection of *E. coli* spheroplasts. The titer of the progeny phage is then measured on bacterial indicators either permissive or nonpermissive for the amber mutation.

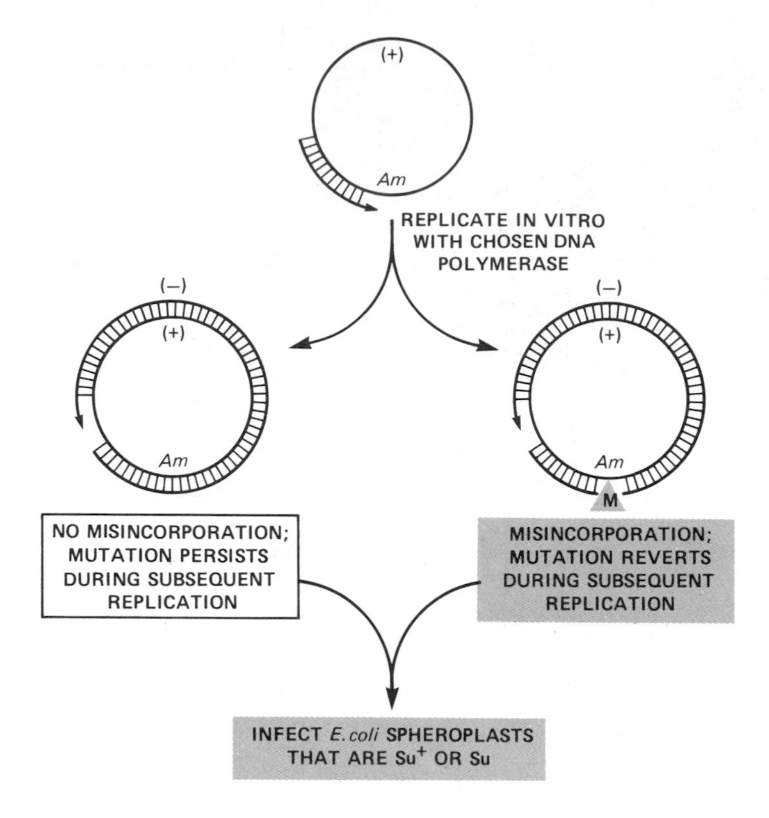

FIGURE 1-3
The fidelity of a DNA polymerase can be determined in vitro. ϕX174 template DNA (+ strand) containing an *amber* mutation (*Am*) is annealed to a primer and replicated with a given DNA polymerase. The replicated molecules are then introduced into *E. coli* spheroplasts that are either suppressors or not of the *amber* mutation. If replication results in an appropriate misincorporation opposite to the site of the *amber* mutation (right), the phenotype reverts to wild-type and these molecules yield phage progeny on both permissive and nonpermissive indicator bacteria. If replication past the *amber* mutation is accurate (left), the *amber* phenotype persists and viral progeny will only be obtained on the permissive (suppressor) host. Measurement of the reversion frequency can thus be converted into an error rate for in vitro DNA synthesis.

When deoxyribonucleoside (1-thio) triphosphates containing a sulfur atom in place of an oxygen atom on the phosphorus are used as substrates for replication of the ϕX174 template DNA, the analogue is incorporated as a thiomonophosphate at rates similar to those of the corresponding unmodified nucleosidetriphosphate (31). However, the phosphorothioate diester bond is not hydrolyzed by the $3' \rightarrow 5'$ exonuclease of either DNA polymerase I of *E. coli* or T4 DNA polymerase (Fig. 1-4) (31). Thus, reversion of amber mutations can be enhanced by incorporation of the analogue as a mispaired

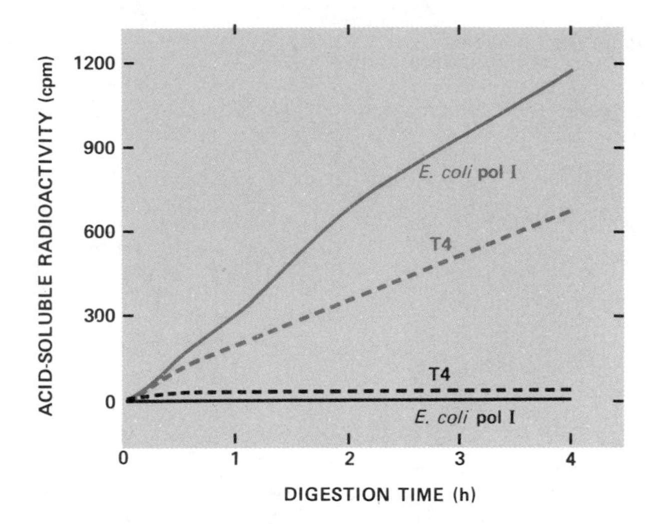

FIGURE 1-4
ϕX174 DNA containing phosphorothioate (black lines) is resistant to digestion by the $3' \rightarrow 5'$ exonuclease functions of *E. coli* DNA polymerase I and of phage T4 DNA polymerase. Unsubstituted DNA is readily degraded by these enzymes (blue lines). (From T. A. Kunkel et al., ref. 31.)

base that cannot be edited out. With this approach it can be inferred that the proofreading functions of the *E. coli* and phage T4 DNA polymerases increase the fidelity of replication by factors of 20-fold and 500-fold respectively (31).

An interesting hypothesis has been suggested relating replicative fidelity to the use of RNA primers for DNA synthesis by some organisms (19). At or near the beginning of a DNA chain, proofreading and other possible fidelity-promoting mechanisms may not function as efficiently as during subsequent chain elongation. However, since the RNA primers are eventually excised, this provides a possible mechanism for eliminating the adjacent stretches of the newly synthesized genome that may contain an unusually high level of errors.

A number of parameters affecting the replicative fidelity of various DNA polymerases in vitro have been explored. For example, replacement of Mg^{2+} by divalent metal ions such as Mn^{2+}, Co^{2+} and Be^{2+} reduces the replicative fidelity of a number of DNA polymerases (32) (Table 1-2). Alterations in the relative proportions of the different deoxyribonucleoside triphosphates used as precursors for DNA synthesis (33–35), the addition of deoxyribonucleoside monophosphates (36) and the presence of certain accessory proteins such as single-strand DNA binding protein (37) also affect the fidelity of DNA replication in the ϕX174 system. These accessory proteins may

TABLE 1-2
Effect of metals on replicative fidelity of AMV DNA polymerase

Compound Tested	Maximal Error Frequency (ratio)	Metal Conc. (mM)
$AgNO_3$	1.85	0.03
$BeCl_2$	15.0	10.0
$Cd(C_2H_3O_2)_2$	2.22	0.24
$CdCl_2$	1.35	0.04
$CoCl_2$	8.37	4.00
$CrCl_2$	3.70	0.64
CrO_3	3.83	16.0
$MnCl_2$	3.75	10.0
$NiCl_2$	1.92	8.0
$PbCl_2$	1.48	4.0

The metals shown in the table have all been implicated as mutagens and/or carcinogens. Fidelity of replication by AMV DNA polymerase was determined by measuring the frequency of misincorporation using polymer templates. The maximal error frequency values quoted are the ratio of the highest error frequency during titration with a given compound to that determined without the compound added. The metal concentration that yielded the largest observed change in error rate is also shown. From M. D. Sirover and L. A. Loeb, ref. 32.

be particularly important for those DNA polymerases that do not contain associated $3' \rightarrow 5'$ exonuclease activities. Such enzymes are characteristic of the majority of eukaryotic cells so far investigated and also of the retroviruses (19, 38).

When considering semiconservative DNA synthesis as a source of DNA damage, it should be noted that under certain circumstances replication of DNA occurs on parental template strands containing unrepaired base damage, which can act as miscoding lesions. Details of the cellular responses to damage in or near replication forks in DNA are considered in Chapters 7 and 8. Finally, it was pointed out earlier that the multitude of factors that determine replicative fidelity reduce the *error* frequency in newly replicated DNA to about 10^{-6} to 10^{-9} per nucleotide. However, the actual *mutation* frequency per nucleotide is even lower than this by about another three orders of magnitude (39). This reduction reflects the *repair* of mispaired bases in DNA (Table 1-1), a topic that is discussed in Sections 5-7 and 6-6.

Damage Arising during DNA Repair and DNA Rearrangement

Chapters 5 and 6 deal with the biochemical mechanisms involved in the replacement of nucleotides associated with the removal of damaged bases from DNA. Since this replacement process does not result in the generation of entire new daughter DNA strands as in classical semiconservative DNA synthesis, it is often referred to as nonsemiconservative DNA synthesis or *repair synthesis* of DNA (40). In some instances stretches of DNA as long as 1500 nucleotides may be replaced by these mechanisms (41); hence the potential also exists for replicative infidelity by the DNA polymerase(s) involved in the repair synthesis of DNA. Alterations in nucleotide sequence during various modes of DNA rearrangement, including homologous and nonhomologous recombination, are poorly understood at the present time; however, it is known that these processes can generate errors, and these too constitute sources of DNA damage (42).

Spontaneous Alterations in the Chemistry of DNA Bases

Tautomeric Shifts

Each of the common bases in DNA can spontaneously undergo a transient rearrangement of bonding, termed a tautomeric shift, to form a structural isomer (tautomer) of the base (43). Formation of the tautomer of any base alters its base-pairing properties. For example, although the N atoms of the purine and pyrimidine rings are usually in the amino (NH_2) form, on rare occasions they shift to the imino (NH) tautomeric form (43). When either cytosine or adenine is in the latter configuration, it can mispair with the other through two available H bonds (Fig. 1-5). Similarly, the oxygen on the C_6 atoms of guanine or thymine may shift from the usual keto (C═O) to the infrequent enol (C—OH) configuration, allowing for anomalous base pairing between them by three available H bonds (43) (Fig. 1-5). Thus, if any base in a template strand exists in its rare tautomeric form during DNA replication, misincorporation in the daughter strand can result.

Deamination of Bases

Three of the four bases normally present in DNA (cytosine, adenine and guanine) contain exocyclic amino groups. The loss of these groups (deamination) occurs spontaneously in pH- and temperature-dependent reactions of DNA (12, 44) and results in the conversion of the affected bases into uracil, hypoxanthine and xanthine respectively (Fig. 1-6). Some of these products of deamination are potentially mutagenic since during semiconservative synthesis of DNA they are miscoding lesions that result in altered base pairs in the genome (8, 44) (Fig. 1-7).

Deamination of Cytosine

The deamination of cytosine into uracil is one of the ways by which the latter base, normally confined to RNA, can occur in DNA. The formation of uracil in DNA by the deamination of cytosine is biolog-

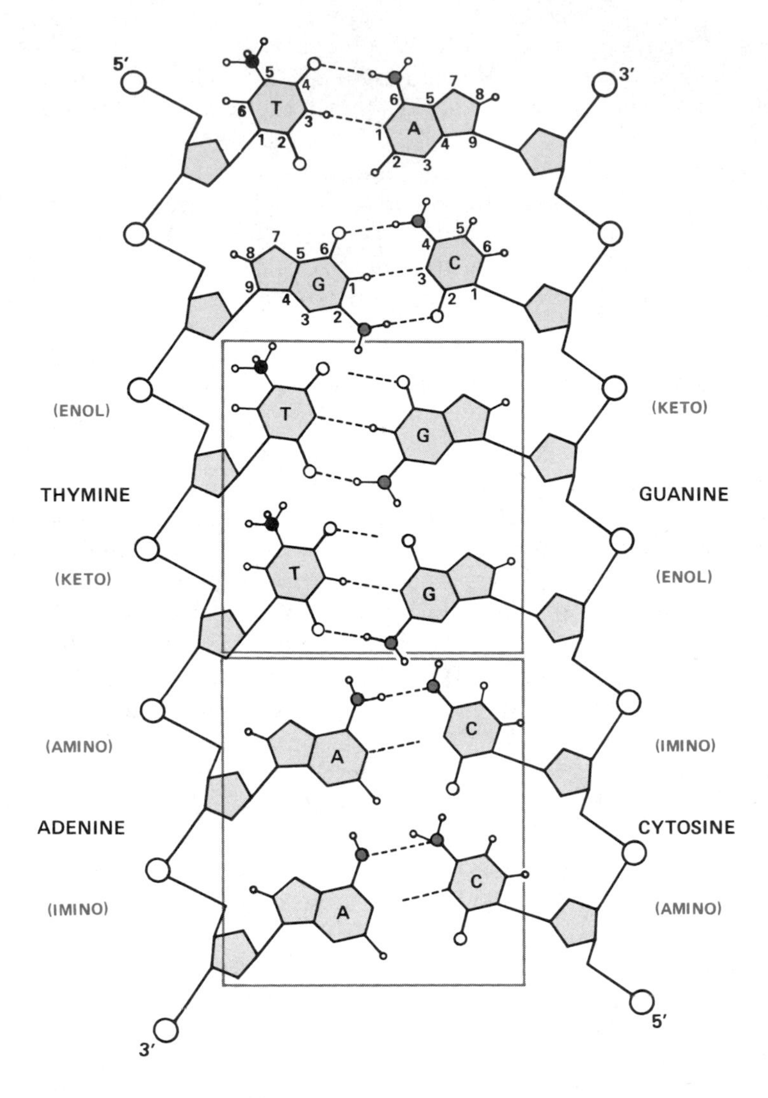

FIGURE 1-5
Anomalous base pairing involving rare tautomeric forms of the bases. When thymine or guanine is in the rare enol form, they can pair with each other. Similarly, when adenine or cytosine is in the rare imino form, they can pair. (After J. Drake, ref. 42.)

ically significant. For example, strains of *E. coli* that are defective in the removal (repair) of uracil from DNA (see Chapter 3) have an increased spontaneous mutation rate (45), and G·C → A·T base-pair transitions have been found at selected sites in such mutants (46).

Two chemical mechanisms have been proposed for the deamination of cytosine in solution at neutral pH (12, 47) (Fig. 1-8). One

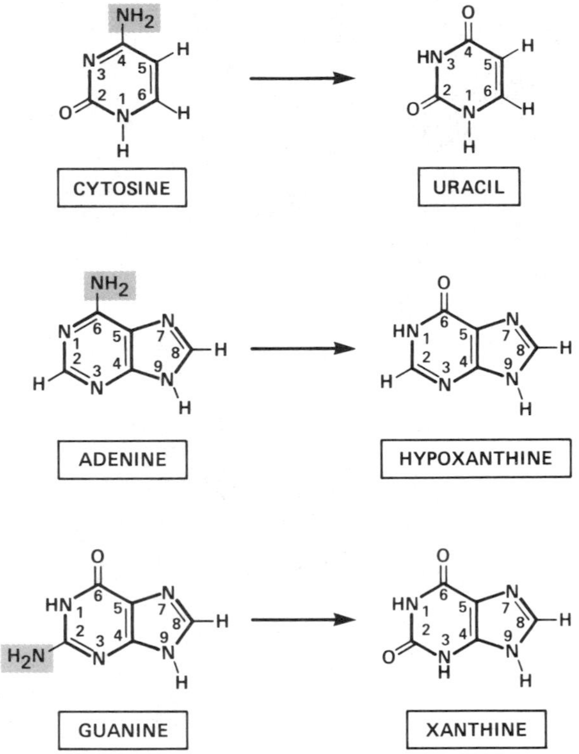

CYTOSINE → URACIL

ADENINE → HYPOXANTHINE

GUANINE → XANTHINE

FIGURE 1-6
The products formed from the deamination of bases in DNA.

involves the direct attack at the 4 position of the pyrimidine ring by a hydroxyl ion. The other postulated pathway involves an addition-elimination reaction with the formation of dihydrocytosine as an intermediate. The hydrolytic deamination of cytosine in nucleotides and polynucleotides occurs at a measurable rate when these compounds are incubated at elevated temperatures in buffers at physiological ionic strength and pH (48). The rate constants for these reactions and their dependence on temperature have been determined. By extrapolation, the rate of deamination of cytosine in single-strand DNA at 37°C was calculated to be $k = 2 \times 10^{-10}$/s (48) (see Table 1-3). What does this value tell us about the biological relevance of spontaneous deamination in living organisms? This question is difficult to answer quantitatively since the rate of spontaneous deamination of cytosine in *duplex* DNA in vitro is less than 1 percent of that in single-strand DNA (48). Even though it is not known how much of the genome of a living cell is in a single-strand configuration at any given time, it is clear that the processes of repli-

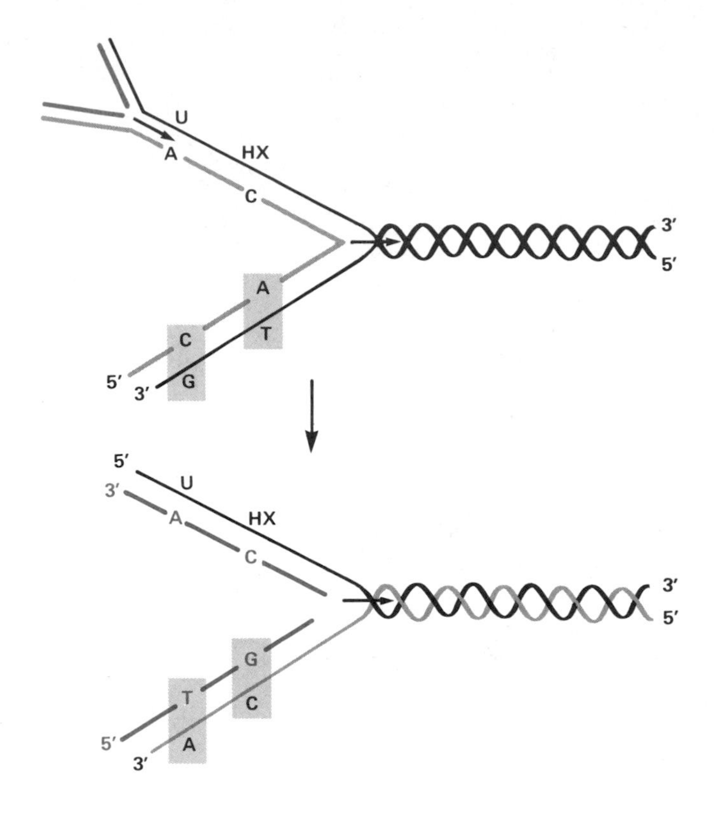

FIGURE 1-7
Deamination of cytosine into uracil (U) and of adenine into hypoxanthine (HX) can result in base pair transitions. The top part of the figure shows a replicating DNA molecule in which U and HX have already mispaired with A and C, respectively. A second round of DNA replication is just beginning. As this second replication fork proceeds (lower figure), replication of the template strand containing A and C results in *transition* of G·C and T·A base pairs to A·T and C·G base pairs respectively.

(I)
CYTIDINE
(II)
CYTIDINE HYDRATE
(III)
(IV)
URIDINE
(V)
URIDINE HYDRATE
H_2O
NH_4^+

FIGURE 1-8
Proposed mechanisms for the hydrolytic deamination of cytidine into uridine. The path I → III → IV is analogous to the hydrolysis of an amide. It is called the *direct route* and involves a direct attack at the 4 position of the pyrimidine ring by a hydroxyl ion. Loss of ammonia yields uridine. The path I → II → V → IV is called the *addition-elimination mechanism*. It involves addition of water to the 5,6 double bond of protonated cytidine to yield cytidine hydrate (dihydrocytidine) (II). Further attack by water is followed by the loss of ammonia, yielding uridine hydrate (dihydrowidine) (V), which is dehydrated to uridine (IV). (After R. Shapiro, ref. 12.)

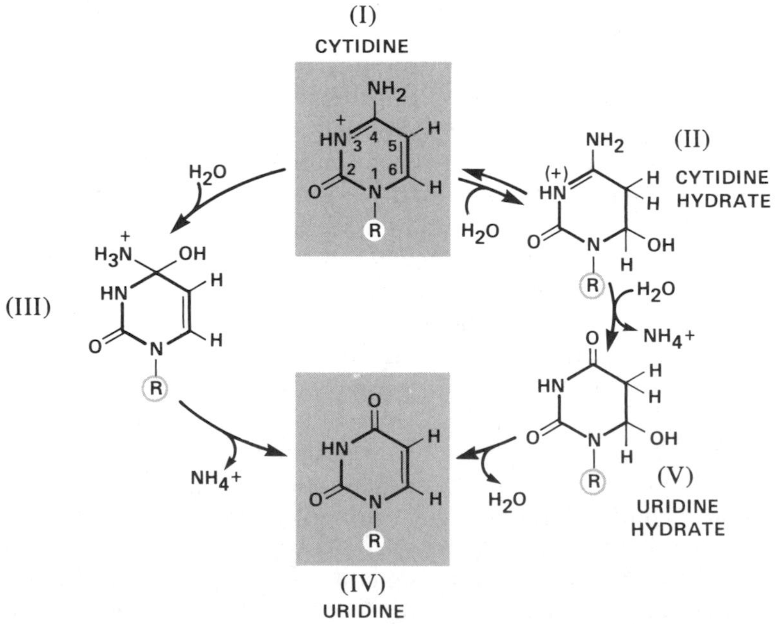

TABLE 1-3
The time needed for a single event at pH 7.4 at 37°C (hours)

Event	Single-Strand DNA (2×10^6 Bases)	Double-Strand DNA (10^6 Base Pairs)
Depurination	2.5	10
Depyrimidination	50.8	200
Deamination of C	2.8	700
Deamination of A	140.0	?

After R. Shapiro, ref. 12. Data computed from the rate constants for these events.

cation, recombination and transcription involve transient localized denaturation of DNA that could accelerate cytosine deaminations (8, 44). Furthermore, duplex DNA undergoes spontaneous localized denaturation, or "breathing," that could increase the rate of spontaneous deamination of cytosine (8, 44). If cytosine deamination in vitro occurs by the addition-elimination mechanism discussed above, this could involve the formation of dihydrocytosine and dihydrouracil as intermediates (Fig. 1-8), either of which could represent distinct (albeit transient) forms of DNA damage not amenable to the same mechanism of repair that deals with uracil in DNA. However, these compounds, if formed, may constitute substrates for other DNA repair processes (44).

The deamination of cytosine can also be promoted by a number of chemical agents, notably nitrous acid (49) and sodium bisulfite (50). The former reagent is relatively nonspecific since it also results in the deamination of adenine and guanine residues in DNA (44) and additionally promotes the cross-linking of DNA strands (51). Nitrous acid attacks cytosine residues in double-strand DNA almost as efficiently as in single-strand DNA (49). In contrast, sodium bisulfite effects the deamination of cytosine exclusively in single-strand regions of DNA and is specific for cytosine residues under defined experimental conditions (50). Purines are not attacked, and although bisulfite-thymine adducts can occur, these are transient and readily reversible (50). Bisulfite converts cytosine residues in DNA into uracil by an acid-catalyzed addition-elimination reaction with the intermediate formation of 5,6-dihydrocytosine-6-sulfonate and 5,6-dihydrouracil-6-sulfonate (50) (Fig. 1-9). At alkaline pH the latter compound is converted into uracil (Fig. 1-9). Yet another mechanism for the conversion of cytosine into uracil in DNA is by exposure to strong alkali (52). Finally, certain bases not normally present in DNA but that bear a strong structural resemblance to normal nitrogenous bases can be incorporated from the appropriate triphosphate precursor during DNA synthesis. These compounds are called base analogues, one of which is 5-bromouracil (5-BU), an analogue of thymine. When DNA

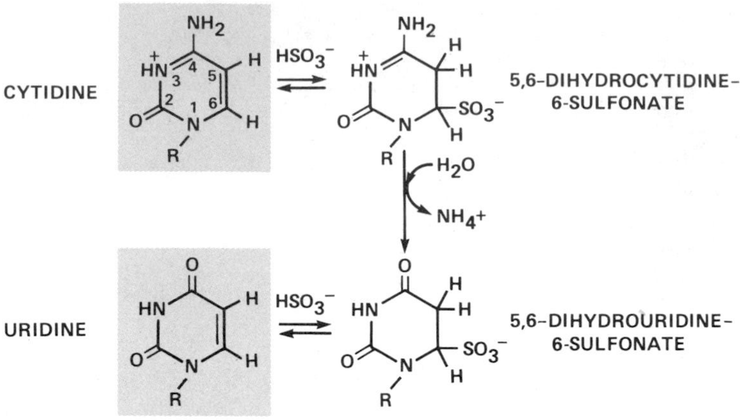

FIGURE 1-9
Mechanism of deamination of cytidine by bisulfite. Cytidine is converted into a sulfonated derivative (5,6-dihydrocytidine-6-sulfonate), which is then hydrolytically deaminated at acid pH to yield a sulfonated uridine derivative (5,6-dihydrouridine-6-sulfonate). At alkaline pH this derivative is converted into uridine. (After R. Shapiro, ref. 12.)

containing 5-BU is exposed to 313 nm radiation in the presence of cysteamine, it undergoes debromination to yield uracil in the DNA (53). Chapter 5 will deal with specific examples of the utility of 5-BU-substituted DNA in the study of the repair of DNA damage.

Incorporation of Uracil into DNA during Semiconservative Replication

Uracil can also arise in DNA during semiconservative synthesis. This phenomenon is observed in organisms that normally contain thymine as a principal pyrimidine and also in a rare class of viruses in which uracil replaces thymine completely. With respect to the former, the presence of U·A rather than T·A base pairs should not alter the replicational fidelity of DNA. However, subtle but biologically relevant parameters of the transcription of such DNA may be affected. Additionally, it is possible that significant replacement of T by U in DNA could affect the recognition of substrate nucleotide sequences by various enzymes and/or regulatory DNA binding proteins.

Small amounts of dUMP are incorporated into the DNA of *E. coli* (54–61) and of other prokaryotic cells such as *Bacillus subtilis* (62, 63) (Fig. 1-10). The extent of this incorporation is apparently directly related to the size of the intracellular dUTP pool, since the K_ms of *E. coli* DNA polymerase III for dUTP and TTP are not sig-

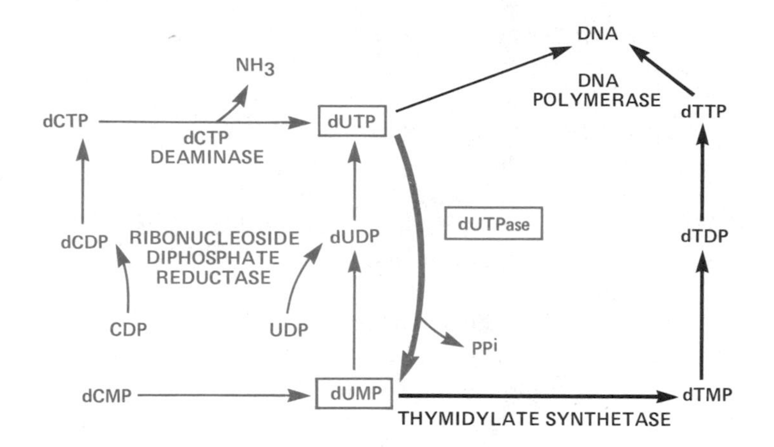

FIGURE 1-10
Uracil can be incorporated into DNA from dUTP during semiconservative DNA synthesis. The dUTP pool is generated from both dCTP and the phosphorylation of dUDP. In wild-type cells the pool size of dUTP is small relative to that of dTTP since most dUTP is degraded to dUMP by dUTPase. (From A. Kornberg, ref. 19.)

nificantly different (64). In wild-type cells, incorporated uracil is excised very rapidly by an enzyme called uracil DNA glycosylase (65) (see Section 3-3), and none of this base can normally be detected in DNA isolated from such cells (59). However, in mutants (*ung*) defective in the DNA glycosylase activity, dUMP is incorporated into DNA at a frequency of about one in 2000 to 3000 nucleotides (59). In double mutants (*ung dut*) defective in both uracil DNA glycosylase and dUTPase activities (the latter mutation prevents the degradation of dUTP into dUMP and hence increases the size of the dUTP pool) the frequency of uracil in DNA is as high as 0.5 percent of all bases (59). When such mutants are infected with bacteriophage T4, as much as 30 percent of the thymine in the phage DNA can be replaced by uracil (66).

The incorporation of uracil into DNA has been observed in a number of viruses, including polyoma virus (67) and adenovirus (68), and in human lymphocytes in culture (69). Presumably most, if not all, organisms that synthesize TMP from dUMP incorporate small amounts of uracil into their DNA. In bacteria dUMP is largely derived from dUTP; hence a pool of the latter normally exists in these cells (Fig. 1-10). In mammalian cells dUMP is generated mainly from dCMP and can be readily converted into the triphosphate form. In human lymphocytes inhibition of the biosynthesis of TMP from dUMP increases the pool size of dUTP relative to that of TTP and thus promotes significant incorporation of uracil relative to thymine during DNA replication. For example, during the synthesis of TMP from dUMP by the enzyme thymidylate synthetase, N^5,N^{10}-methylenetetrahydrofolate contributes the methyl group to dUMP and is converted into dihydrofolate (19). Regeneration of tetrahydrofolate is essential for continued TMP synthesis by thymidylate synthetase. This regeneration is catalyzed by the enzyme dihydrofolate reductase, an enzyme that is inhibited by the folate antagonist 4-amino-10-methylfolate (*amethopterin, methotrexate*) (19) (Fig. 1-11). Treatment of cells with amethopterin results in reduced utilization of dUMP by thymidylate synthetase, causing an increased pool size of this nucleotide, with an attendant drop in the concentration of TTP (70, 71). These relative pool-size changes result in deregulation of other aspects of pyrimidine nucleotide metabolism, which augment this effect (69). As a net result, intracellular dUMP in human lymphocytes in culture is increased about 1000-fold, and, despite the presence of dUTPase, dUTP also increases by at least three orders of

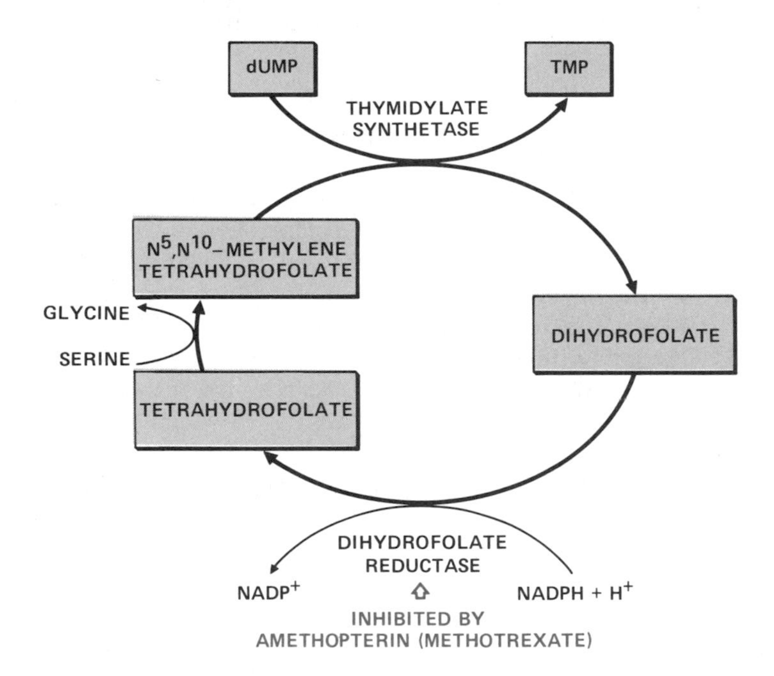

FIGURE 1-11
The formation of thymidylate from dUMP is catalyzed by the enzyme thymidylate synthetase. During this reaction N^5,N^{10}-methylenetetrahydrofolate is converted into dihydrofolate and regeneration of tetrahydrofolate is catalyzed by dihydrofolate reductase. Inhibition of dihydrofolate reductase by amethopterin (methotrexate) results in reduced levels of tetrahydrofolate and hence reduced conversion of dUMP into TMP. (After A. Kornberg, ref. 19.)

magnitude (69, 72). The TTP pool size drops about 50-fold in these cells and together with the attendant increase in dUTP, results in a significant incorporation of dUMP into DNA. In the presence of free uracil, which is an inhibitor of uracil DNA glycosylase (see Section 3-3), uracil accumulates as a stable component in DNA (69).

The related bacteriophages PBS1 and PBS2 normally contain uracil instead of thymine in their DNA (73). Following infection of their natural host *B. subtilis,* these phages induce a number of new enzyme activities, including dTMPase (74), dCTP deaminase (75) and dUMP kinase (74) activities. These activities significantly alter the normal host pathways of deoxynucleoside triphosphate biosynthesis, resulting in an increase in the dUTP pool size relative to that of TTP, thereby facilitating the incorporation of uracil instead of thymine during phage DNA synthesis (76).

Deamination of Cytosine Analogues

Analogues of cytosine are encountered naturally in the DNA of some living forms. For example, 5-methylcytosine occurs as a small fraction of the total cytosine content of a number of prokaryotes, including *E. coli,* as well as in all higher organisms (44). Consider that the rate of heat-induced deamination of 5-methyl-dCMP is about four times as fast as that of dCMP (48) and that the rate of spontaneous deamination of cytosine in dCMP is approximately the same as that in single-strand DNA (44). If it is assumed that the same parallelism holds for 5-methylcytosine, the deamination of this base could account for as much as 10 percent of the total spontaneous deamination events occurring in the DNA of the average mammalian cell (44). The deamination of 5-methylcytosine in DNA results in the formation of thymine and hence of T·G mispairs. Unlike the U·G mispair considered earlier arising from cytosine deamination, T·G mispairs in *nonreplicating* DNA are apparently nonrepairable lesions that result in mutational "hot spots" in *E. coli* (77). This is not too surprising, since unless a DNA repair mechanism for the removal of thymine could specifically distinguish between T·G and T·A base pairs, all thymine in DNA would be potentially subject to removal.

In the bacteriophages T2, T4 and T6, cytosine in DNA is completely replaced by 5-hydroxymethylcytosine, which is glucosylated to varying extents in the respective phages (19). The deamination of 5-hydroxymethylcytosine yields 5-hydroxymethyluracil. Interestingly, the enzyme in *E. coli* that catalyzes the removal of uracil from DNA (uracil DNA glycosylase; see Section 3-3) does not recognize the 5-hydroxymethylated derivative (78), nor is there evidence that phage T2, T4 or T6 codes for a different DNA glycosylase with the appropriate substrate specificity. This is consistent with the observation that at elevated temperatures, bacteriophage T4 accumulates mutations due to G·C→ A·T transitions at a significantly increased rate (79).

Deamination of Adenine and Guanine

Deamination of adenine and guanine also occurs under physiological conditions in vitro, but at rates much lower than that for cytosine (44). In single-strand DNA at raised temperatures and at pH 7.4, the conversion of adenine into hypoxanthine occurs at about 2 percent of the rate of conversion of cytosine into uracil (44). No detailed studies have been reported on the acidic reactions, but from the limited data available (80) it has been estimated that deamination of adenine and guanine occurs at rates of about 10^{-4} that of the rate of loss of these bases from DNA (depurination; see next section). If this ratio of relative reactivities holds at neutral pH, then the deamination reaction would not be very significant biologically (12). DNA adenine residues can be deaminated by nitrous acid at a rate similar to that of cytosine (49). Hypoxanthine in DNA is potentially mutagenic since it can base-pair with cytosine during DNA replication, thus giving rise to A·T → G·C transitions (44). Regarding the incorporation of hypoxanthine into DNA during *semiconservative* synthesis, it is not clear whether misincorporation of dIMP (the nucleotide form of hypoxanthine) instead of dGMP occurs, analogous to the situation discussed above with respect to dUMP incorporation from existing dUTP pools. Although IMP is a key metabolite in purine biosynthesis in *E. coli*, this organism (and probably others) does not have a detectable kinase activity that can convert IMP into IDP, from which dIDP and hence dITP could be generated (44). Xanthine in DNA, arising from the deamination of guanine, is unable to pair stably with either cytosine or thymine (42) and thus may result in the arrest of DNA synthesis on templates containing this base.

Loss of Bases: Depurination and Depyrimidination

The loss of purines and pyrimidines from DNA has been most extensively studied at acid pH; however, depurination and depyrimidination can also occur at appreciable rates at neutral or alkaline pH (6, 8, 12, 44, 81). The rate of depurination and depyrimidination of single- and double-strand DNA can be observed by incubating these substrates at various temperatures and pHs and measuring the rate of release of specifically labeled bases (82). Guanine is released from DNA about 1.5 times faster than adenine at both acid and neutral pH, but in alkali dAMP is hydrolyzed more rapidly than dGMP (82).

Figure 1-12 shows a pH-rate profile summarizing the behavior of the major deoxyribonucleosides of DNA in acid.

The chemical mechanism of hydrolytic DNA depurination at acid pH is believed to be the same as that established for acid hydrolysis of deoxynucleosides, that is, protonation of the base followed by direct cleavage of the glycosyl bond (82, 83). The mechanism of hydrolysis of nucleosides at neutral and alkaline pHs (52, 84) is less well characterized (44). By extrapolation of Arrhenius plots derived by direct measurements at high temperatures, the rate of depurination of duplex DNA at physiological pH and ionic strength is calculated to be about $k = 3 \times 10^{-11}$/s at 37°C (82). In vivo, this value corresponds to the loss of approximately one purine per *E. coli* genome per generation, given a DNA doubling time of about an hour (44) (Table 1-3). By the same argument, a thermophilic organism such as *Thermus thermophilus*, which grows optimally at a temperature of about 85°C, may lose as many as 300 purines per genome per generation (44). For mammalian cells, in which genomes are much larger and replication times longer, estimates of purine loss at the rate of 10,000 per cell generation have been made (44).

Pyrimidine nucleosides are considerably more stable than purine nucleosides with respect to the glycosylic linkage of the base to

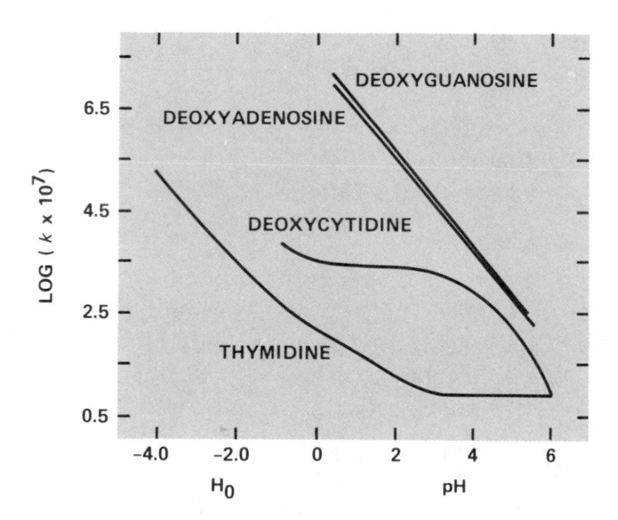

FIGURE 1-12
Dependence of the logarithms of the rate constants (s^{-1}) on pH and H_0 (a parameter used to indicate acidity of pH less than one) at 95°C for deoxyribonucleoside hydrolysis. At acid pHs depurination occurs more rapidly than depyrimidination. (From R. Shapiro, ref. 12.)

deoxyribose. The basic mechanism of depyrimidination is the same as for depurination, but cytosine and thymine are lost at rates only $\frac{1}{20}$ that for adenine or guanine (85). This still translates into the loss of hundreds of pyrimidines per mammalian cell generation. The influence, if any, of the packaging of DNA into nucleosomes and of the various levels of folding that chromatin assumes in the cell on the rate of spontaneous base loss has not been extensively explored.

The deoxyribose residues that are left at sites of base loss in DNA exist in equilibrium between the closed *furanose* form and the open *aldehyde* form (86). The 3′ phosphodiester bonds associated with the latter are labile and can be hydrolyzed by a β-elimination reaction in which the pentose carbon beta to the aldehyde is activated at alkaline pH and at elevated temperatures (86) (Fig. 1-13). The same reaction proceeds at a reduced rate at neutral pH. In buffered physiological saline a site of base loss in DNA has an average lifetime of approximately 400 hours at physiological pH and temperature (87). The presence of Mg^{2+} and of primary amines promotes phosphodiester bond cleavage by β-elimination (88), reducing the average lifetime of sites of base loss to about 100 hours (44). Polyamines further promote the rate of cleavage of the deoxyribose-phosphate backbone (87, 89); nonetheless, it is apparent that the integrity of the DNA backbone would probably be maintained for several generations in the absence of DNA repair in *E. coli* (44). In vitro this β-elimination reaction can be prevented by reducing the pentose aldehyde into an alcohol with a reducing agent such as sodium borohydride.

Clastogenic Factors as a Source of Spontaneous DNA Damge

Agents that cause chromosomal breakage are often referred to as *clastogens*. Most of these derive from well-characterized exogenous sources of DNA damage such as ionizing radiation (see Sec. 1.5). However, some clastogens may be generated in the course of cellular metabolism in certain mammalian cells. Thus, for example, co-cultivation of lymphocytes (with plasma) from patients suffering from the disease ataxia telangiectasia (see Section 9-3) with those of normal individuals results in a significant increase in chromosomal damage in the normal cells (90). Tissue culture medium from cultivated skin fibroblasts of ataxia telangiectasia patients also significantly increases chromosomal breakage in normal lymphocytes previously stimulated with phytohemagglutinin (PHA) (90). Clastogenic activity has also been detected in medium from cultures of fibroblasts of patients with Bloom's syndrome (91) (Table 1-4), another autosomal recessive disease characterized by chromosomal abnormalities, in which DNA repair defects have been implicated (see

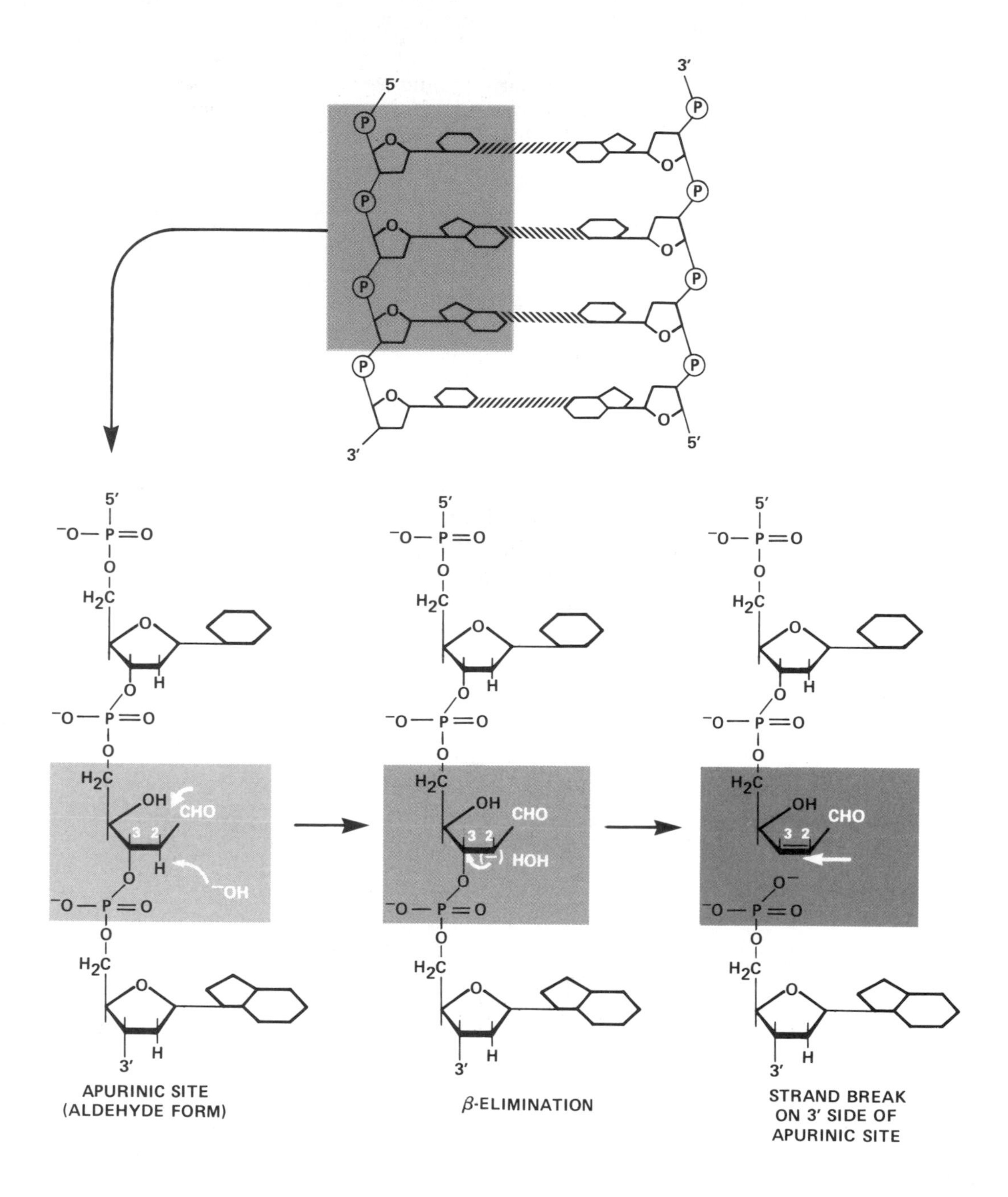

FIGURE 1-13
Mechanism of strand breakage in DNA by β-elimination. Deoxyribose residues at sites of base loss exist in equilibrium between the open (aldehyde) form (shown in the figure) and the closed (furanose) form (not shown). In the aldehyde form 3′ phosphodiester bonds are readily hydrolyzed by a β-elimination reaction in which the pentose carbon beta to the aldehyde is activated at alkaline pH, as shown.

TABLE 1-4
Chromosomal aberrations (per mitosis) induced in PHA-stimulated normal human lymphocytes by concentrated ultrafiltrates of Bloom's syndrome and of normal fibroblast culture media

Concentration of Ultrafiltrate	Bloom's Syndrome Culture Medium	Normal Fibroblast Culture Medium
10x	0.18	0.00
	0.22	0.02
	0.22	—
50x	0.38	0.06

From I. Emerit and P. Cerutti, ref. 91.

Section 9-3). The clastogenic factor from Bloom's syndrome cells has been identified as having a molecular weight between 1000 and 10,000, the same size range as that of clastogenic factors identified in the serum of patients with systemic lupus erythematosis (92) and in certain strains of NZB mice with high frequencies of spontaneous chromosomal aberrations (93). Addition of bovine superoxide dismutase to cultures of phytohemagglutinin-stimulated lymphocytes from normal individuals suppresses the clastogenic activity present in concentrated ultrafiltrates of media in which Bloom's syndrome

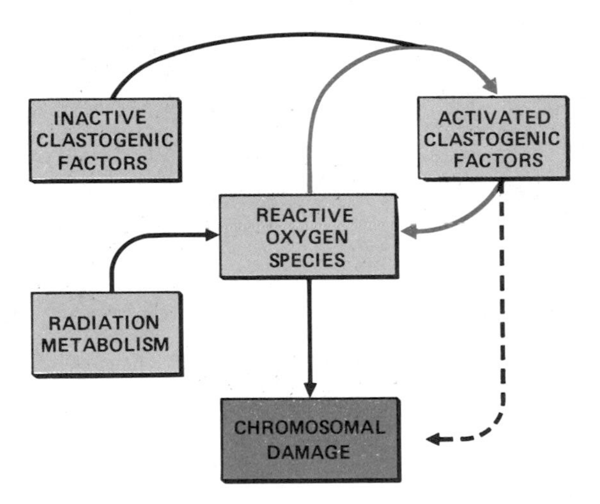

FIGURE 1-14
Clastogenic factors may be activated by reactive oxygen species generated by ionizing radiation and/or by normal cellular metabolism. It is postulated that both activated clastogenic factors and reactive oxygen species cause DNA damage. (After I. Emerit and P. Cerutti, ref. 91.)

cells are grown (91). Since superoxide dismutase converts superoxide into hydrogen peroxide and molecular oxygen, this observation suggests that clastogens may promote the formation of reactive oxygen species, (perhaps by oxidation of certain membrane lipids [94]), that result in the DNA damage (Fig. 1-14). In addition, superoxide dismutase inhibits the transformation to neoplasia of hamster embryo cells treated with X rays or bleomycin (an antibiotic that causes DNA damage). These observations suggest that the oncogenic action of these DNA-damaging agents may (at least in part) be mediated by free radicals (95).

1-3 Environmental Damage to DNA

Chemical Agents

Perhaps the earliest impetus for the study of the interactions of chemicals with DNA was the potential for using chemicals as lethal and injurious agents in warfare. A far more humane stimulus came from the field of cancer chemotherapy, which is based on the simple idea that damage to DNA can interfere with normal DNA synthesis, leading to the replicative arrest of rapidly dividing cell populations such as cancer cells. In more recent years increasing public awareness of environmental mutagens and carcinogens has led to an intense interest in the study of the mechanisms whereby genotoxic chemicals interact with and damage DNA. At the present time there is an enormous wealth of literature on chemical damage to DNA. The goal in this text is to provide illustrative examples of some classes of chemicals whose interaction with DNA is well characterized and that have been widely used as models for the study of the repair of DNA damage.

Alkylating Agents

Alkylating agents are electrophilic compounds with affinity for nucleophilic centers in organic macromolecules (11, 13, 96–99). These agents can be either monofunctional or bifunctional. The former have a single reactive group and thus covalently interact with single (although varied) nucleophilic centers in DNA. Some alkylating agents that cause DNA damage are bifunctional, i.e., they have two reactive groups and each molecule is potentially able to react with two sites in DNA. Numerous potential reaction sites for alkylation have been identified in all four bases, although not all of them have equal reactivity. The sites of reaction in DNA for many monofunctional alkylating agents include the following (Fig. 1-15): in

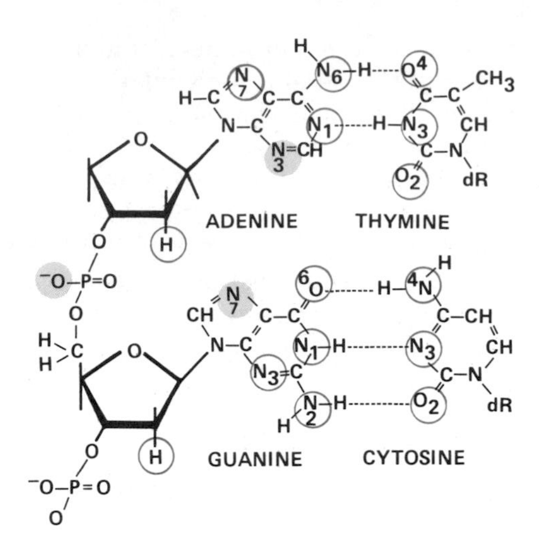

FIGURE 1-15
Nucleophilic centers in DNA that are the most reactive to alkylating agents. In general the ring nitrogens of the bases are more reactive than the ring oxygens. Alkylations at phosphodiester linkages (to yield phosphotriesters), at N^7 of guanine and at N^3 of adenine are the most frequently encountered.

adenine, N^1, N^3, N^6 and N^7; in guanine, N^1, N^2, N^3, N^7 and O^6; in cytosine, N^3, N^4 and O^2, and in thymine, N^3, O^2 and O^4 (11, 13). In general, the ring nitrogens of the bases are more nucleophilic than the oxygens, with the N^7 position of guanine and the N^3 position of adenine being the most reactive (11). Alkylation of oxygen in phosphodiester linkages results in the formation of phosphotriesters.

The reactivity of a given alkylating agent for particular chemical groups in DNA is roughly correlated with a constant, s, often referred to as the Swain-Scott constant (100). Reagents of low s value tend to react more extensively with less nucleophilic centers such as the O^6 position of guanine and the phosphodiester groups of the DNA backbone, whereas those compounds with higher Swain-Scott constants tend to react with the more nucleophilic nitrogen atoms such as the highly reactive N^7 nitrogen of guanine (11). Table 1-5 shows the relative extent of the reaction of the purine bases with two monofunctional alkylating agents (methyl- and ethylmethanesulfonate) frequently used experimentally as mutagens. Note that methylmethanesulfonate has a significantly lower affinity for the O^6 position of guanine than ethylmethanesulfonate does.

The reaction of alkylating agents with specific sites in DNA may also be governed by steric effects. For example, when DNA is in the

TABLE 1-5
Relative extents of reaction of the purine bases with monofunctional alkylmethanesulfonates

	Extent of Reaction*	
Reaction product	MMS	EMS
1-Alkyladenine	1.1	0.8
3-Alkyladenine	9.8	5.6
7-Alkyladenine	0.3	1.6
3-Alkylguanine	0.7	1.9
6-Alkylguanine	0.3	2.0
7-Alkylguanine	82.0	69.6

*The extents of reaction are the percentages of total alkylation products in the purine bases identified chromatographically. MMS, methylmethanesulfonate; EMS, ethylmethanesulfonate. (After J. J. Roberts, ref. 11.)

normal B (right-handed) helical configuration (Fig. 1-16), both the O^6 and N^7 atoms of guanine are in the more accessible major groove, whereas the N^3 position of adenine lies in the relatively less accessible minor groove (see Fig. 3-16). However, portions of DNA molecules may assume a *left-handed* helical conformation called the Z form (Fig. 1-16). This conformation occurs in the alternating copolymer poly(dG-dC) at high salt concentrations (101–104). In fact, this copolymer, when fully methylated at the 5 position of cytosine, undergoes a transition from B to Z at approximately physiological concentrations (105). A left-handed configuration also occurs in poly(dT-dG)·poly(dC-dA) (106). Interestingly, a sequence of 50 alternating dT and dG residues is present within one of the introns of a human cardiac muscle actin gene, and a probe specific for poly(dT-dG) sequences reveals that these potentially Z-DNA-forming sequences are highly repeated in the human genome (106). Under physiological ionic conditions, blocks of dC-dG in bacterial plasmid DNA are in a left-handed state when the negative superhelical density of the plasmid is greater than 0.072 (107). In addition, antibodies to Z DNA bind to polytene chromosomes of *Drosophila melanogaster* and *Chironomus thummi thummi*, and Z-DNA immunoreactivity has also been detected in the nuclei of rat cerebellum, liver, kidney and testis (108). Soon after the discovery of Z DNA it became apparent that certain atoms such as the O^6, N^7 and C^8 positions of guanine are much less sterically hindered in that structure than in the B form (Fig. 1-25). The relevance of this observation to the interaction of DNA with specific chemicals is discussed later in this section.

As pointed out above, bifunctional alkylating agents can react with

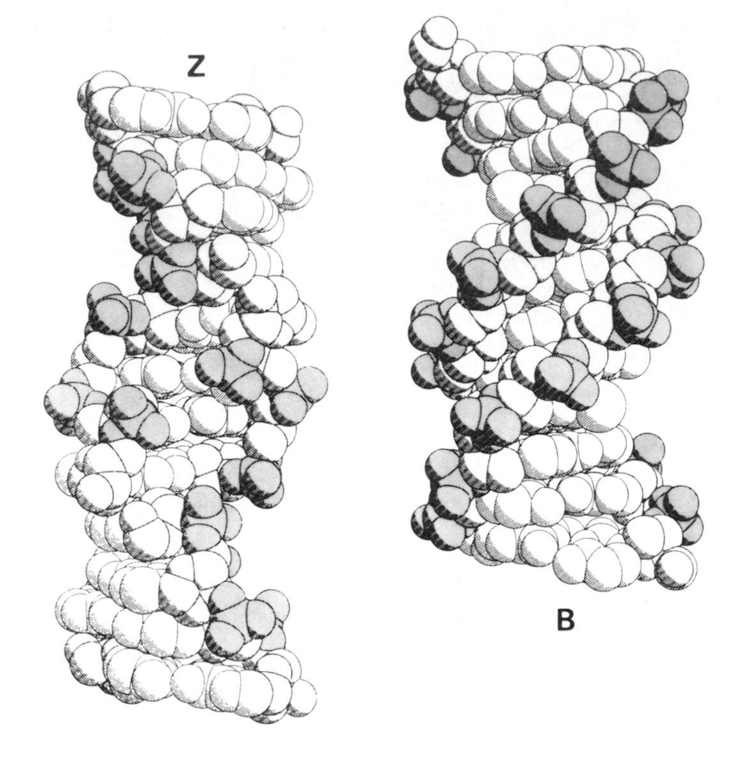

FIGURE 1-16
Space-filling drawings of Z DNA and B DNA. The irregularity of the Z DNA backbone (hence the designation Z DNA) is illustrated by the pattern of the shaded phosphate residues along the chain. In contrast in B DNA a smooth helical pattern of the shaded phosphate residues can be observed. In B DNA both O^6 guanine and N^7 guanine lie in the major groove, whereas N^3 adenine lies in the minor groove (see Fig. 3-16). In Z DNA O^6 guanine, N^7 guanine and C^8 guanine (see Fig. 1-25) are less sterically hindered than in the B form. (From A. Kornberg, ref. 38.)

two different nucleophilic centers in DNA (11, 13, 96–99). If these sites are situated on the same polynucleotide chain of a DNA duplex, the reaction product is referred to as an *intra*strand cross-link (Fig. 1-17). If the two sites are on opposite polynucleotide strands, *inter*strand cross-links result (Fig. 1-17).

DNA-DNA Cross-Links

Interstrand DNA cross-links represent an important class of chemical damage to DNA since they prevent DNA strand separation and hence can constitute complete blocks to DNA replication and transcription. It is precisely for this reason that a number of agents such as nitrous acid (109), mitomycin C (110), nitrogen and sulfur mustard (111, 112), various platinum derivatives such as *cis*-diamminedichloro-

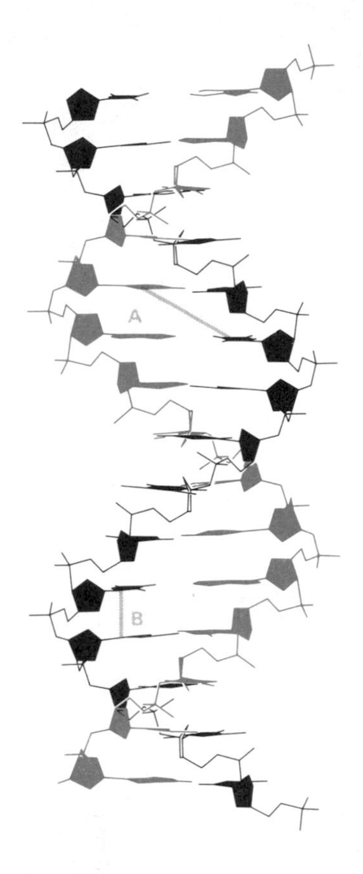

FIGURE 1-17
Skeletal model of double-strand DNA showing a diagrammatic representation of interstrand (A) and intrastrand (B) cross-links. (After L. Stryer, ref. 223.)

platinum (II) (113) and certain photoactivated psoralens (114) have been used extensively in cancer chemotherapy. The reactions of photoactivated psoralens with DNA are discussed in Section 1-5. In addition, UV radiation at about 254 nm (115) and ionizing radiation (116) can result in the formation of intermolecular DNA cross-links as minor products. In most cases the structures formed by these various agents have not been completely defined; however, nitrogen and sulfur mustard are well-studied exceptions. When bacterial DNA is reacted with concentrations of nitrogen mustard as low as 5×10^{-7} *M*, approximately 0.005 percent of the bases are alkylated (111). Studies with radiolabeled nitrogen mustard show that only a small fraction (approximately 4 percent) of the DNA-bound mustard molecules become effective interstrand cross-links between the N^7 positions of guanine on opposite DNA strands (111) (Fig. 1-18).

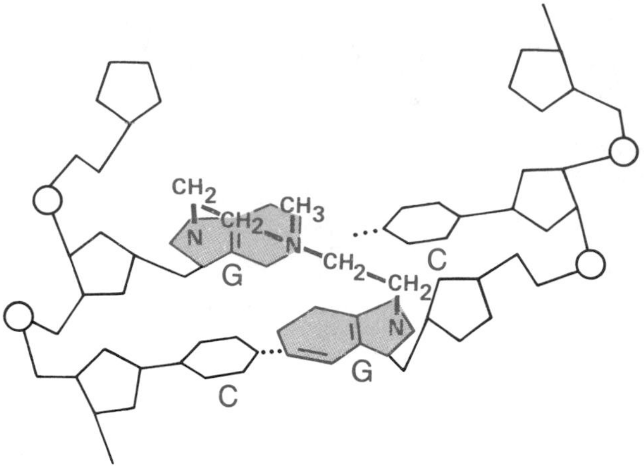

FIGURE 1-18
Schematic representation of the cross-linking of DNA by nitrogen mustard through the N^7 positions of two guanine moieties on opposite strands of the duplex.

The covalent interaction between the two strands constituted by cross-links in duplex DNA molecules facilitates their detection by a number of techniques, including alkaline elution of DNA and velocity and special density-labeling procedures (11). For example, DNA can be labeled in one of the two strands by allowing replication in the presence of the thymidine analogue [^{3}H]5-bromo-2′-deoxyuridine (BrdU). This compound is converted into the triphosphate form in living cells and is incorporated into DNA as bromodeoxyuridine monophosphate, during semiconservative DNA synthesis (40) (Fig. 1-19). The analogue imparts increased density to the regions (strands) of DNA in which it is present. Thus, if only *one* strand of a DNA duplex is density labeled, following denaturation of the DNA and sedimentation to equilibrium in a cesium chloride density gradient, all fragments representing the unlabeled strand will have a normal (light) density and all those containing BrdU will have a heavy density. However, if some pieces of duplex DNA contain covalent intermolecular cross-links, these will prevent strand separation during denaturation and the DNA will sediment as molecules of intermediate density (117) (Fig. 1-19).

Since cross-links prevent complete separation of the two strands of the DNA complex following exposure to denaturing agents, the DNA becomes rapidly renaturable. This property can be measured by the differential binding affinity of single- and double-strand DNA to hydroxylapatite. The reversibly denatured (cross-linked) DNA behaves as a duplex and can be separated from totally denatured (non-cross-linked) DNA by differential elution (118). Benzoylated-

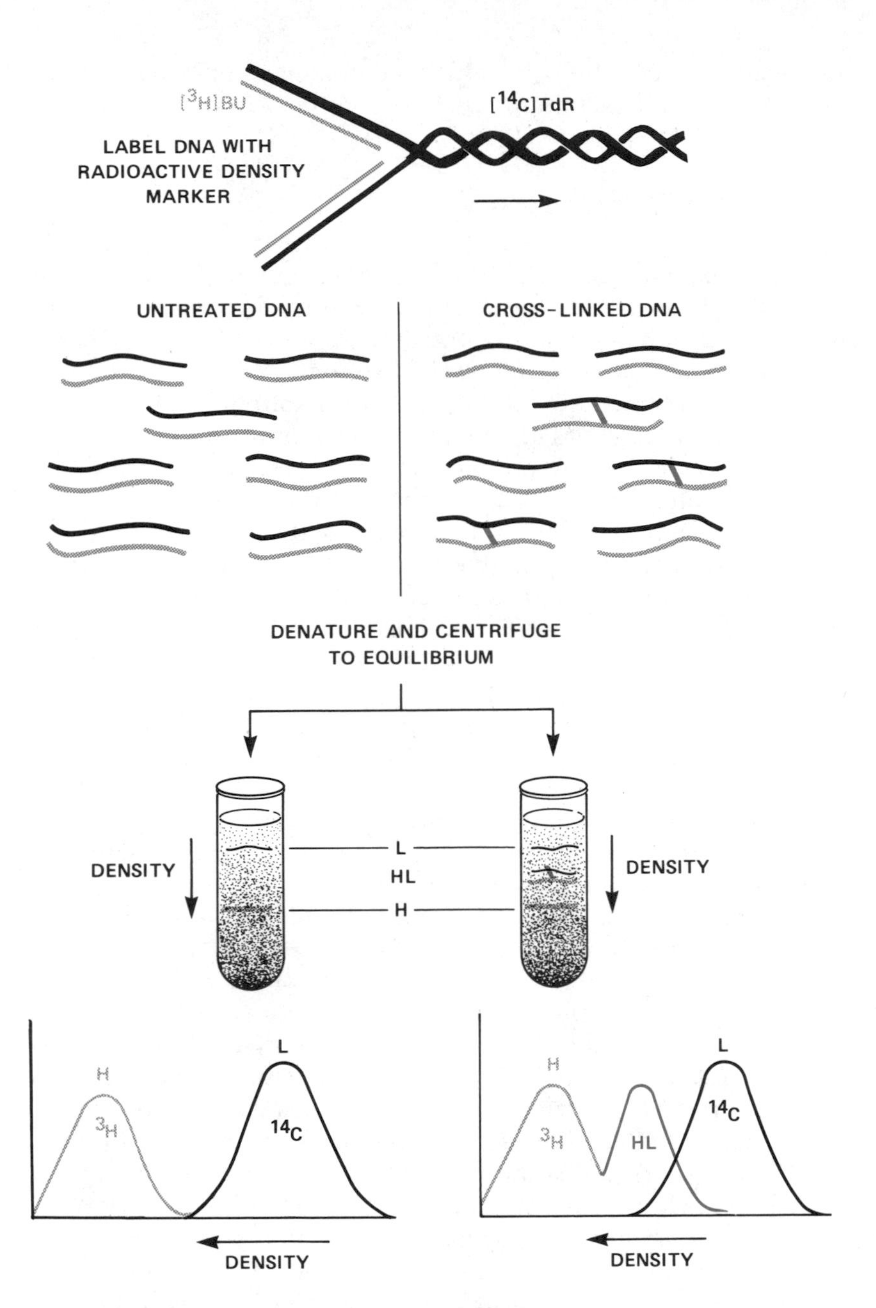

FIGURE 1-19
Detection of interstrand cross-links in DNA by isopycnic sedimentation. DNA uniformly labeled with [^{14}C]thymidine is replicated in the presence of [^{3}H]BrdU to generate DNA of intermediate density in which one strand is light and the other heavy. In the absence of cross-linking (left) denaturation of the DNA and sedimentation in alkaline cesium chloride yields ^{3}H (heavy) and ^{14}C (light) peaks of radioactivity. However, when the two DNA strands are cross-linked (right), DNA of intermediate density (HL) results.

napthylated DEAE-cellulose is another chromatographic matrix from which single- and double-strand DNA can be differentially eluted (119).

DNA-Protein Cross-Links

After living cells are exposed to a variety of DNA-damaging agents, including alkylating agents, UV radiation or ionizing radiation, the ease with which the DNA can be extracted by deproteinizing procedures such as phenol-salt treatment is diminished (120). In addition, the rate of elution of DNA extracted from UV-irradiated human cells through nitrocellulose filters in alkali (alkaline elution of DNA) is enhanced if the DNA is first incubated with proteases (121). These observations have been interpreted as evidence for the cross-linking of DNA to protein in vivo. This phenomenon has also been studied in vitro with purified DNA and proteins and with free bases and amino acids. For example, the monofunctional alkylating agent β-propiolactone reacts with DNA principally at the N^7 position of guanine (122). Incubation of β-propiolactone with DNA and purified proteins results in DNA-protein cross-linking (123). Similarly, if a mixture of uracil and cysteine is exposed to UV radiation, 5-S-cysteine-6-hydrouracil can be isolated (124). Cysteine also reacts photochemically with poly(rU), poly(rC), poly(dC) and poly(dT) and with RNA and DNA (120). Other amino acids such as serine, cystine, methionine, lysine, arginine, histidine, tryptophan, phenylalanine and tyrosine are also able to add photochemically to uracil (120), and presumably also to bases present in DNA.

A number of observations attest to the biological importance of DNA-protein cross-links. For example, the sensitivity of bacterial cells to killing by UV radiation as a function of growth in different media correlates with the degree of DNA-protein cross-linking, suggesting that these lesions are biologically important (120). In addition, the relative sensitivity of *E. coli* to UV radiation at temperatures ranging from +21°C down to −196°C correlates with this parameter (Fig. 1-20) (125). When *E. coli* is exposed to UV radiation at different wavelengths (see Sec. 1.5), the extent of killing of cells and the values of other biological end points such as inhibition of DNA synthesis or increase in mutation frequency vary as a function of the wavelength of the radiation (126, 127). A graphic representation of such measurements is called an *action spectrum*, and the wavelength at which the maximal effect occurs is typically at 260 nm (the absorption maximum of DNA), suggesting that for the biological parameters mentioned above DNA is the most photosensitive target (128) (Fig. 1-21). However, the action spectrum for the killing and for the inhibition of DNA synthesis in *Micrococcus radiodurans* shows a com-

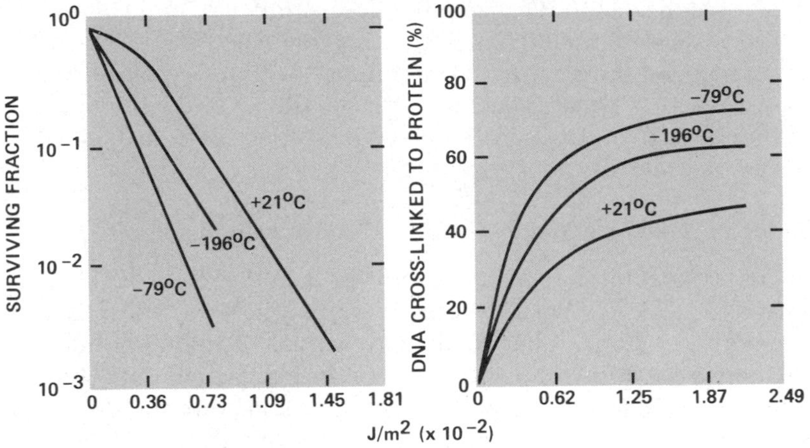

FIGURE 1-20
The extent of DNA-protein cross-linking during UV irradiation of *E. coli* at different temperatures (right) correlates with the sensitivity of *E. coli* to killing when irradiated at these temperatures (left). (From K. C. Smith and P. C. Hanawalt, ref. 120.)

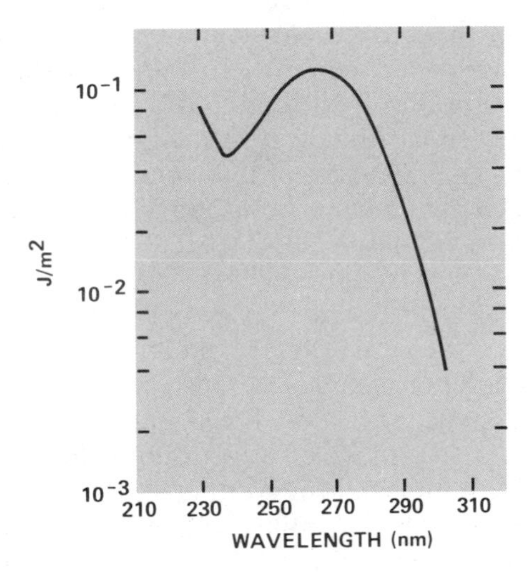

FIGURE 1-21
The action spectrum for the killing of *E. coli* by UV radiation. The curve relates the reciprocal of the incident energies required for 50% lethality, to the wavelength of the radiation. Note that the maximal lethality occurs at wavelengths close to the absorption maximum for DNA (approximately 260 nm). (After K. C. Smith and P. C. Hanawalt, ref. 120.)

ponent of high sensitivity at 280 nm (absorption maximum of typical proteins) as well as at 260 nm, suggesting that reactions of DNA with photosensitized proteins may be an important class of damage in this organism (120). The mechanism of various DNA-protein interactions and their effects on DNA metabolism are important aspects of DNA damage that merit further exploration.

Chemicals That Are Metabolized to Electrophilic Reactants

One of the most interesting recent advances in the study of chemical damage to DNA has been the discovery that a variety of relatively nonpolar (and hence chemically unreactive) compounds undergo metabolic activation to more reactive forms, which, like typical alkylating agents, interact with nucleophilic centers in DNA (129, 130). Many of these compounds are potent mutagens and carcinogens. Indeed, it was largely through the study of them as carcinogens that their metabolism by susceptible species was elucidated. For example, it was known for many years that some carcinogenic aromatic amines produce tumors at sites other than that of immediate administration (so-called remote carcinogenesis), such as the urinary bladder (131). The frequency of affliction of this particular organ led to the suggestion that the parent compounds were metabolized to water-soluble forms excreted in the urine (132). Another example was the observation that N,N-dimethyl-4-aminoazobenzene (butter yellow), a potent liver carcinogen in rats, did not itself bind to rat liver proteins, but a metabolite of this compound did (133).

It is now known that the metabolic activation of these compounds and of many other carcinogens is effected by the induction of specific metabolizing enzymes in the affected cells (134) (Fig. 1-22). The biological function of this enzyme system is to protect the cell against cytotoxicity effects by converting potentially toxic nonpolar chemicals into water-soluble excretable forms (134). Although most of the products of these reactions do indeed enjoy this innocuous fate, some of them become activated to electrophilic forms that are particularly reactive with nucleophilic centers in organic macromolecules such as DNA (Fig. 1-22) (129, 130, 135). Thus, although these agents are no longer directly cytotoxic, they have been converted into potent genotoxic forms. The activating enzymes are generally among a series of membrane-bound proteins containing numerous monooxygenase activities. Because of its strong absorbance at about 450 nm, this complex, in combination with one or more membrane-bound flavoprotein reductases, is frequently referred to as the cytochrome P-450 system (Fig. 1-22) (134). These multicomponent enzyme systems require NADPH and atmospheric oxygen. An initial reaction sequence usually metabolizes hydrophobic nonpolar sub-

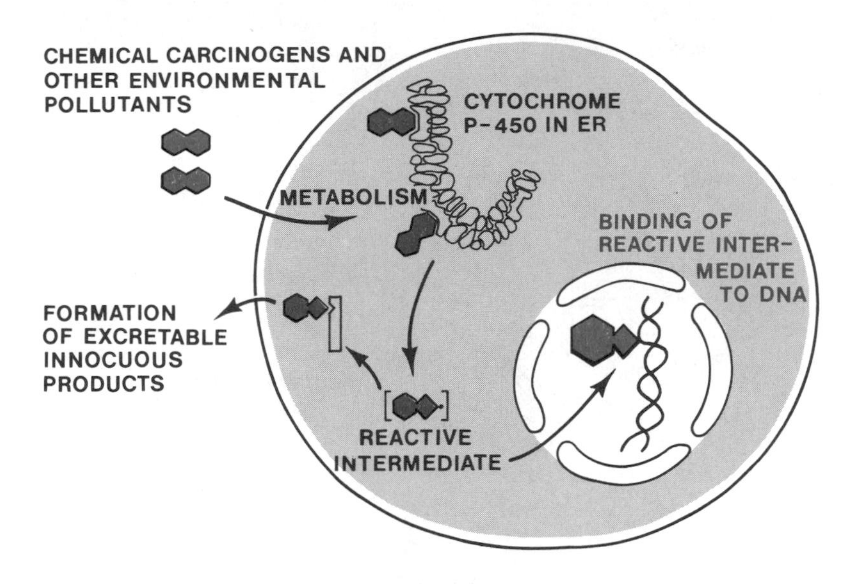

FIGURE 1-22
Scheme for the metabolic activation of nonpolar polycylic chemicals by the cytochrome P-450 system in a typical mammalian cell to form reactive intermediates that bind to nucleophilic centers in DNA (ER, endoplasmic reticulum). (After D. W. Nebert et al., ref. 252.)

strates to more polar oxygenated intermediates and products, which are substrates for secondary reactions with enzymes catalyzing the formation of ester conjugates that are rapidly excreted from the cell and from the body (Fig. 1-22). The following paragraphs detail specific examples of these activation reactions.

Metabolic Activation of N-2-Acetyl-2-aminofluorene (AAF). AAF belongs to a class of compounds known as the aromatic amines, exposure to many of which is associated with an increased incidence of cancer in humans. The first step in the metabolic activation of this compound (originally used as an insecticide) is the formation of an N-hydroxy derivative (129, 135) (Fig. 1-23). This intermediate (called a *proximate carcinogen*) is relatively unreactive with nucleic acids, but following the enzyme-catalyzed formation of a sulfate, phosphate or acetate ester, it becomes a highly reactive alkylating agent. Thus, for example, N-acetoxy-N-2-acetylaminofluorene reacts readily with guanine residues in nucleosides and nucleic acids at the C^8 and N^2 positions to yield N-(deoxyguanosin-8-yl)-N-acetyl-2-aminofluorene and 3-(deoxyguanosin-N^2-yl)-N-acetyl-2-aminofluorene as the major and minor components in DNA respectively (136, 137) (Fig. 1-23).

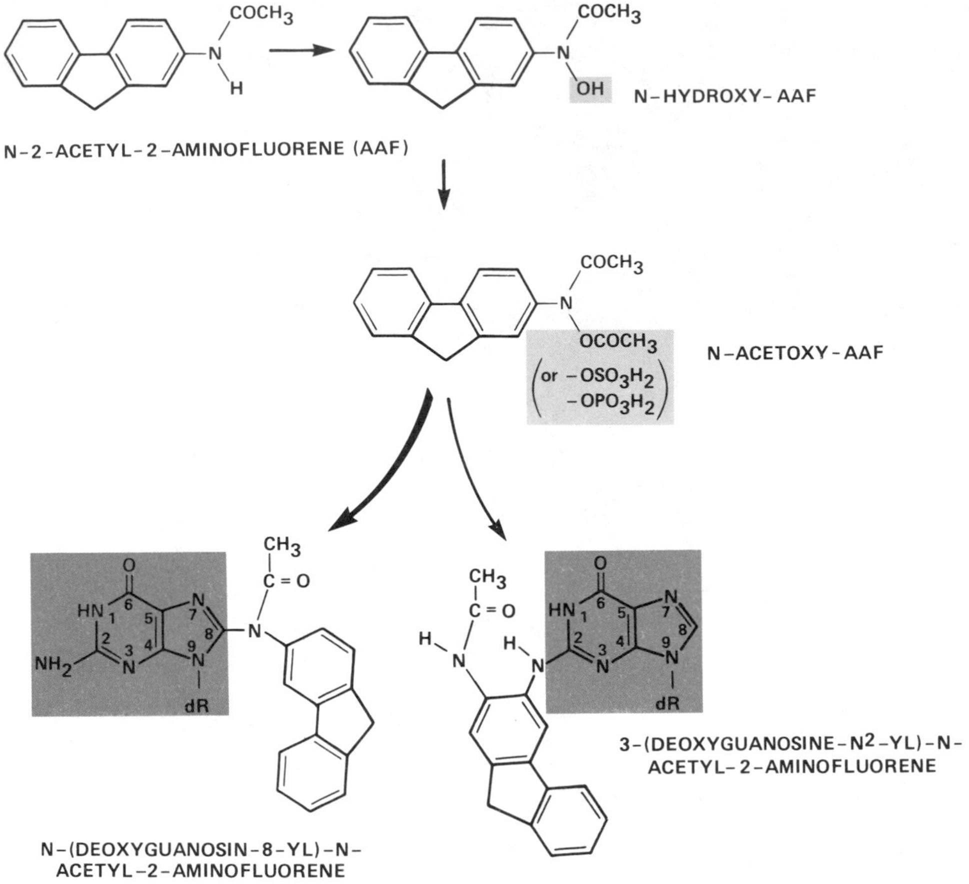

N-2-ACETYL-2-AMINOFLUORENE (AAF)

N-HYDROXY-AAF

N-ACETOXY-AAF

N-(DEOXYGUANOSIN-8-YL)-N-ACETYL-2-AMINOFLUORENE

3-(DEOXYGUANOSINE-N^2-YL)-N-ACETYL-2-AMINOFLUORENE

FIGURE 1-23
Metabolic activation of N-2-acetyl-2-aminofluorene (AAF) proceeds through the formation of an N-hydroxy intermediate before the formation of N-acetoxy-AAF (or other esterified forms). This compound is highly reactive with the C^8 position (lower left) and to a lesser extent with the N^2 position (lower right) of guanine in DNA.

Normally, the C^8 position of guanine in duplex DNA in the B conformation is relatively hindered sterically and is rather inaccessible to a bulky adduct such as activated AAF (138). However, it has been suggested that rotation of the base around the N-glycosylic bond from the *anti* to the *syn* conformation (see Fig. 4-32) allows for attack at the C^8 position of guanine in B DNA (138). During this conformational shift the modified guanine residue can be displaced from its normal coplanar relation to adjacent bases and its position subsumed by the fluorene (the so-called base displacement model) (138) (Fig. 1-24). When DNA is in the Z conformation, the C^8 position of guanine is highly accessible to reaction with activated AAF, because deoxy-

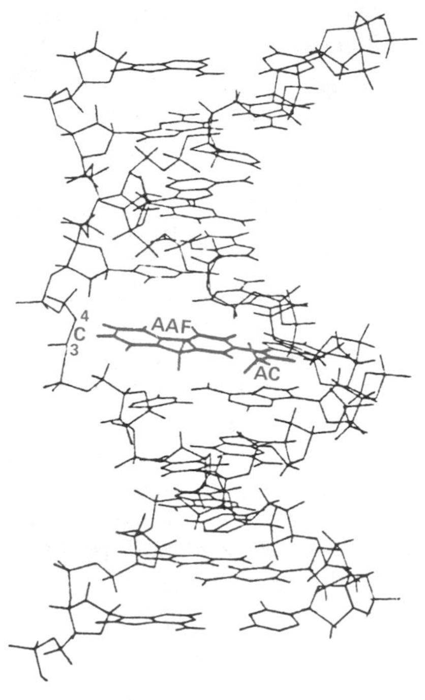

FIGURE 1-24
A stereoscopic view of the base displacement model of an AAF-DNA adduct. AC denotes the acetyl group of AAF. The guanine to which the adduct is covalently linked has been rotated out of the helix and the AAF moiety is inserted into the helix and stacked with the bases above and below. The complementary cytosine residue (C) on the opposite strand would overlap with the AAF residue and has been removed for convenience; the 3′ and 4′ carbons of the corresponding deoxyribose are indicated, however. (From D. Grunberger and R. M. Santella, ref. 138.)

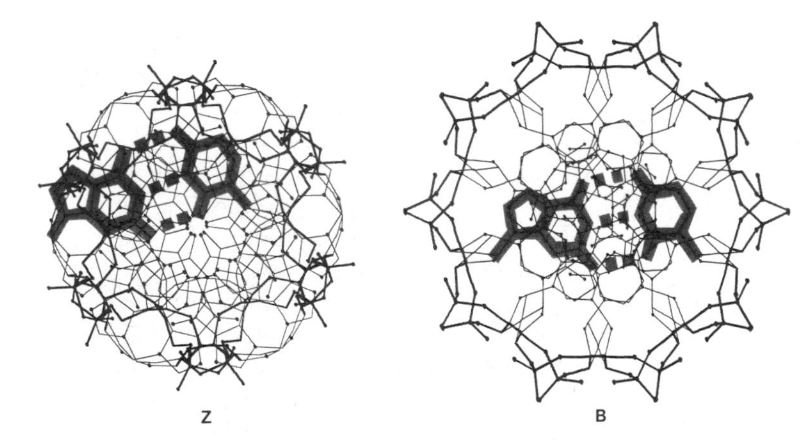

FIGURE 1-25
End views of the regular, idealized helical forms of Z and B DNA. Heavier lines indicate the deoxyribose-phosphate backbone, and the G-C base pairs are shown by shading. Note that the base pair is buried near the center of the B DNA but is at the periphery of Z DNA. (Courtesy of Dr. A. H-J. Wang.)

guanosine in Z DNA exists in the *syn* conformation while base paired to deoxycytidine (139) (Fig. 1-25). Since AAF modification of deoxyguanosine in B DNA is purportedly associated with the rotation of guanine from the *anti* to the *syn* conformation, a similar conformational change might be expected in poly(dG-dC)·poly(dG-dC) modified with AAF. This might force the conformation of the entire polymer into the Z form. Circular dichroism studies of AAF-modified poly(dG-dC)·poly(dG-dC) show this to be the case (138, 140, 141). These results suggest that if stretches of alternating G·C base pairs in DNA were modified by AAF, this might favor localized transitions to the Z form and perhaps alter gene function in that region (138).

Metabolic Activation of Benzo(a)pyrene. The association between cancer and hydrocarbon exposure goes back nearly 200 years, when Percival Pott, an English surgeon, observed the remarkable correlation between cancer of the skin of the scrotum and the occupation of chimney sweeping. Chimney sweeps in those times were exposed to very high levels of polycyclic aromatic hydrocarbons derived from the combustion of coal in the presence of air. Pott's observation provided the basis for the first preventive measures against environmental cancer since, shortly after its publication in 1775, the Danish chimney sweepers' guild urged its members to take daily baths and by 1892 a lower scrotal cancer incidence in northern European chimney sweeps relative to that in England was noted (142). It was not until some years later, however, that a very potent carcinogenic polycyclic aromatic hydrocarbon called benzo(a)-pyrene was identified in and isolated from crude coal tar (143, 144). Despite the subsequent identification of a large number of other carcinogens from industrial products, including coal tar, benzo-(a)pyrene remains to this day one of the most highly carcinogenic compounds known. Exposure to coal tar is no longer the public health hazard it represented at the time of the Industrial Revolution in England; however, benzo(a)pyrene exposure is still prevalent from sources such as cigarette smoke and automobile exhaust fumes (145).

Unmodified benzo(a)pyrene is an unreactive nonpolar compound with a planar configuration (Fig. 1-26). That configuration facilitates its intercalation between the H-bonded base pairs in duplex DNA, and this was initially thought to be a likely mode of DNA damage by the compound. It is now known that components of the P-450 system known as arylhydrocarbon hydroxylases can metabolize benzo(a)pyrene and other polycyclic aromatic hydrocarbons to phenols and dihydrodiols, which together with their corresponding ester conjugates are excretable (Fig. 1-26) (146, 147). However, some of the products of benzo(a)pyrene metabolism are electrophilic epoxides, and it is now reasonably well established that the ultimate carcinogenic form of this hydrocarbon is an *anti* diol epoxide called 4-7,*t*-8-dihydroxy-*t*-9-10-oxy-7,8,9,10-tetrahydrobenzo(a)pyrene (147) (Fig. 1-26). At least one major site of reaction of this product with DNA is the 2-amino group of guanine, which can covalently interact with the C^{10} position of the benzo(a)pyrene *anti* diol epoxide (147, 148) (Fig. 1-27).

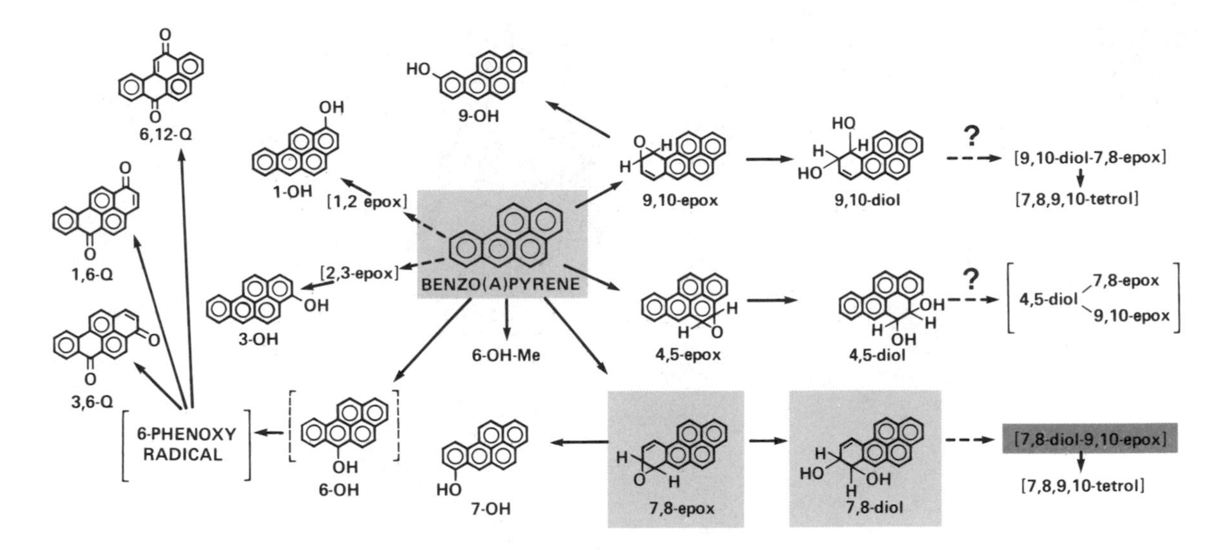

FIGURE 1-26
The metabolic products of the metabolism of benzo(a)pyrene by microsomal mixed-function oxygenases. Some of the products are electrophilic epoxides that have a high reactivity for nucleophilic centers in DNA. The 7,8-diol-9,10-epoxide is thought to be the ultimate carcinogenic form of benzo(a)pyrene. (From J. K. Selkirk et al., ref. 147.)

FIGURE 1-27
Covalent interaction of the *anti* benzo(a)-pyrene dihydrodiol epoxide (7,8-diol-9,10-epoxide; see Fig. 1-26) with the exocyclic amino group of deoxyguanosine. (From J. K. Selkirk et al., ref. 147.)

Metabolic Activation of Aflatoxins. Aflatoxins are among the most potent liver carcinogens known and are an interesting example of DNA-damaging agents that have their origin as products of natural metabolism (149). Aflatoxins are mycotoxins produced by the fungi *Aspergillus flavus* and *Aspergillus parasiticus* (149), and human and animal exposure is usually a result of contamination of food by these fungi, peanuts being a notorious source (149).

Chemically, the aflatoxins consist of a difurofuran ring system fused to a substituted coumarin moiety, with a methoxy group attached at the corresponding benzene ring (Fig. 1-28) (150). Among

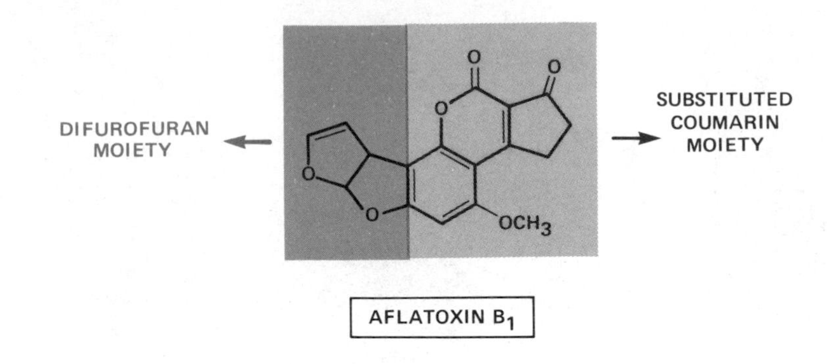

FIGURE 1-28
Chemical structure of aflatoxin B_1.

TABLE 1-6
Summary of reactions of chemical compounds or their metabolites with nucleic acids

Cytotoxic, Mutagenic or Carcinogenic Agents	Probable Metabolic Activation Process or Reactive Intermediate	Defined or Proposed Sites of Reaction in Nucleic Acids
ALKYLATING AGENTS		
Mustards		Inter-and intrastrand cross-links in DNA; DNA-protein cross-links
Difunctional	Direct acting	
Monofunctional	Direct acting	
β-Propiolactone	Direct acting	N^7, N^3, N^1 of adenine
Alkyl sulfonates	Direct acting	N^7, N^3, N^1, O^6 of guanine
Alkyl nitrosamides	Direct acting (alkali catalyzed)	N^1, N^3, O^2 of cytosine; N^3, O^4 of thymine
Alkyl nitrosamidines	Direct acting (alkali or thiol catalyzed)	Phosphate esterification
Dimethylnitrosamine	Oxidative demethylation; methyldiazonium hydroxide	Relatively more O^6 of guanine, O^2 of cytosine and phosphate alkylation with nitrosamides, amidines and dialkyl nitrosamines, and S_N1-type compounds as compared with S_N2-type agents
1-Aryl-3,3-dialkyl-triazenes	Oxidative demethylation	
POLYCYCLIC HYDROCARBONS		
Benzo(a)pyrene	Dihydrodiol epoxide	N^2 of guanine (to 10 position of hydrocarbon)
7,12-Dimethylbenzo(a)-anthracene	5,6-Oxide (K-region epoxide)	N^2 of guanine (to 5 and 6 positions of hydrocarbons)
7-Bromomethylbenzo(a)-anthracene	Direct acting	N^6 of adenine; N^2 of guanine, probably N^4 of cytosine

the aflatoxins of fungal origin, aflatoxin B_1 (Fig. 1-28) has been shown to be the most potent hepatocarcinogen when tested in the rat and rainbow trout (150). Aflatoxin B_1 is known to be oxidized by the mixed-function oxygenases of the P-450 system present in the microsomal fraction of liver extracts, giving rise to a number of hydroxylation, O-demethylation, hydration and epoxidation products (150). Aflatoxin B_1 is also a substrate for certain cytoplasmic reductases. Which of these (or possibly other metabolic products) is the form most reactive with DNA is not yet certain. Studies using DNA substrates of known nucleotide sequence have suggested that certain guanine residues in double-strand DNA are preferentially attacked by aflatoxin B_1, in a way predictable from a knowledge of vicinal nucleotides (151).

A summary of the major types of chemical damage to DNA discussed in this chapter and selected examples of other forms of chemically induced base damage are provided in Table 1-6.

TABLE 1-6 (Continued)

Cytotoxic, Mutagenic or Carcinogenic Agents	Probable Metabolic Activation Process or Reactive Intermediate	Defined or Proposed Sites of Reaction in Nucleic Acids
AROMATIC AMINES		
N-2-acetyl-2-amino-fluorene (AAF)	N-Hydroxylation followed by esterification	C^8 of guanine (to N atom of AAF), N^2 of guanine (to 3 position of AAF)
AZO DYES		
4-Methylaminoazo-benzene (MAB)	N-Hydroxylation followed by esterification	C^8 of guanine (to N atom of MAB)
MISCELLANEOUS COMPOUNDS		
4-Nitroquinoline-1-oxide	Reduction to 4-hydroxy-aminoquinoline-1-oxide (4-HAQO), followed by activation, possibly by seryl tRNA synthetase	Reduction to 4-HAQO followed by (?) (to 3 position of quinoline ring); two guanine adducts
Mitomycin C	Reduction to semiquinone radical	Cross-links in DNA; O^6 of guanine(?)
cis-diamminedichloro-platinum (II)	Direct acting	Cross-links from O^6 of guanine to N^7 of guanine
Safrole	Ester of 1′-hydroxysafrole and/or 1′-hydroxy-2′, 3′ oxide of safrole	O^6 of guanine
Urethane	Not known	5-Carboxyethylation of cytosine(?), phosphate ethylation (?)
Aflatoxin B_1	2,3-Oxide of aflatoxins	Possible sites: O^6 or N^7 of guanine; N^1, N^3 of adenine

After J. J. Roberts, ref. 11.

1-4 Base Analogues as a Source of DNA Damage

Certain analogues of the four naturally occurring bases in DNA can be incorporated from the appropriate triphosphate substrates during DNA replication. These are mentioned here for the sake of completeness, since they are used experimentally in studies of DNA damage and of the cellular responses to such damage. Among the most extensively studied base analogues are the halogenated uracil derivatives, 5-bromouracil, 5-fluorouracil and 5-iodouracil (Fig. 1-29), all of which are thymine analogues that can result in transition mutations when present in template DNA undergoing replication (152–154). The adenine analogue, 2-aminopurine (Fig. 1-29), is also mutagenic. This compound can produce transition mutations by at least two mechanisms during DNA replication: by a stable incorporation opposite thymine (A·T → 2AP·T) or by the stable incorporation of cytosine opposite 2-aminopurine (2AP·T → 2AP·C) (155, 156).

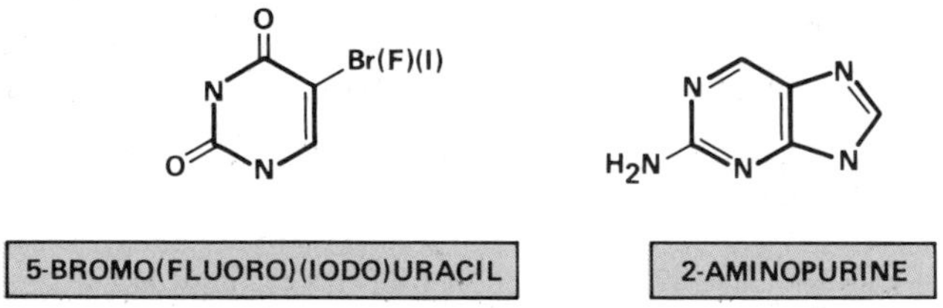

O
Br(F)(I)
N
O N

5-BROMO(FLUORO)(IODO)URACIL

N N
H_2N N N

2-AMINOPURINE

FIGURE 1-29
Some base analogues that can be incorporated into DNA. The halogenated uracil derivatives are thymine analogues, while 2-aminopurine can substitute for adenine.

1-5 Physical Agents That Damage DNA

Ultraviolet Radiation

Historically, the investigation of UV radiation damage to DNA marks the beginning of the study of the repair of DNA damage. The exposure of cells to UV radiation is probably the best-studied and most extensively used model system for investigating the biological consequences of DNA damage and of its repair and tolerance. Two of the many attributes of UV radiation as a means for producing DNA damage are that at 254 nm it is readily available from an ordinary germicidal lamp, and instrumentation for accurately measuring its intensity is commonplace. In addition, it is highly relevant biologically, because living organisms have had to contend with the genotoxic effects of solar UV radiation since the beginning of the evolution of life on this planet. Indeed, it has been suggested that unattenuated UV radiation during the pre-Phanarozoic period (from 3.5 billion until half a billion years ago) might have precluded the development of terrestrial life as we know it (157).

Cyclobutane-Type Dipyrimidines (Pyrimidine Dimers)

When DNA is exposed to radiation at wavelengths approaching its absorption maximum (about 260 nm), adjacent pyrimidines become covalently linked by the formation of a four-membered ring structure resulting from the saturation of their respective 5,6 double bonds (9, 120, 158–168). The structure formed by this photochemical cycloaddition is referred to as a cyclobutane-type dipyrimidine, or *pyrimidine dimer* (Fig. 1-30). These dimers can theoretically exist in 12 isomeric forms; however, only four of them, those with the configurations *cis-syn cis-anti*, *trans-syn* and *trans-anti*, are formed in significant amounts (169). In B DNA pyrimidine dimers are thought to exist exclusively in the *cis-syn* form (9).

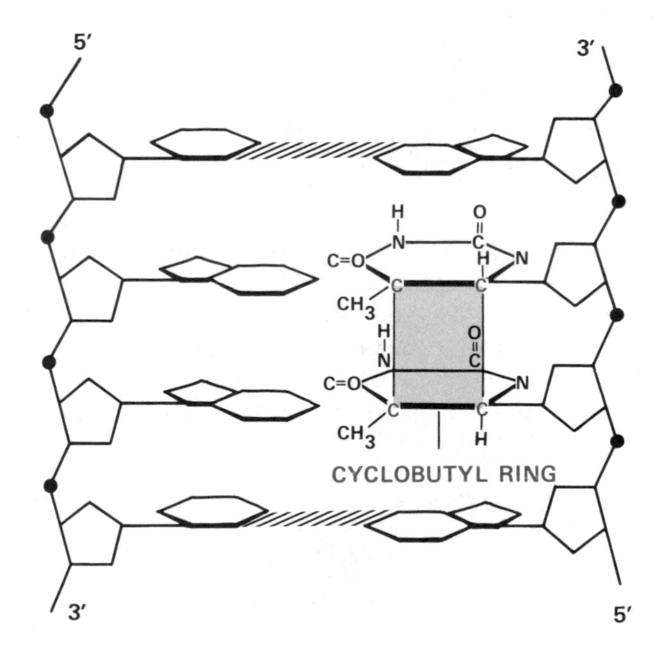

FIGURE 1-30
The cyclobutyl pyrimidine dimer is formed in DNA by the covalent interaction of two adjacent pyrimidines in the same polynucleotide chain. Saturation of their respectives 5,6 double bonds results in the formation of a 4-membered cyclobutyl ring (blue area) linking the two pyrimidines.

Dimerized pyrimidines are extraordinarily stable to extremes of pH and temperature (163). Thus these lesions survive total acid hydrolysis of DNA and can be separated from thymine in such hydrolysates by a variety of chromatographic techniques (170). Since thymine in DNA (unlike the other major bases) can be conveniently and specifically radiolabeled with ^{3}H or ^{14}C, one can directly measure the thymine-containing pyrimidine dimer content in a given sample of DNA by employing one or more of these techniques.

The formation of pyrimidine dimers during the irradiation of DNA is a reversible process that can be represented as

$$\mathrm{Py} + \mathrm{Py} \xrightleftharpoons{\mathrm{UV}} \mathrm{Py} <> \mathrm{Py}$$

Since under normal conditions the equilibrium is shifted far to the right, dimer formation is favored over dimer reversal (165). However, if *E. coli* DNA radiolabeled in thymine is continuously irradiated at 254 nm, the thymine-containing pyrimidine dimer content (thymine-thymine plus thymine-cytosine dimers) of the DNA does not increase beyond about 7 percent of the total thymine content (171). This steady state reflects a dynamic equilibrium in which the rates of dimer formation (which is pseudo zero order, to a good approximation) and reversal (which is first order in dimer content) are equal (171).

Formation of Pyrimidine Dimers in DNA Is Influenced by Nucleotide Composition

It was known for many years from measurements in bulk DNA that at moderately high doses of UV radiation the yield of C <> C is significantly less than that of C <> T or T <> T (165) (Table 1-7). However, aside from these quantitative observations, it was tacitly assumed that thymine-containing pyrimidine dimers are randomly distributed in DNA and that all sites in which thymine is adjacent to another pyrimidine are sites in which dimerization can occur with

TABLE 1-7
Distribution of pyrimidine dimers in UV irradiated DNA

Source of DNA	Wavelength (nm)	Dose (J/m²)	Distribution of Dimers (%)		
			C <> C	C <> T	T <> T
Haemophilus influenzae	265	2×10^{2}	5	24	71
(high AT)	268	4×10^{3}	3	19	78
E. coli	265	2×10^{2}	7	34	59
(GC≈AT)	280	4×10^{3}	6	26	68
M. luteus	265	2×10^{2}	26	55	19
(high GC)	280	4×10^{3}	23	50	27

After R. B. Setlow and W. L. Carrier, ref. 248.

equal probability. The advent of techniques for sequencing DNA has facilitated studies on the detailed distribution of a variety of forms of DNA damage. Pyrimidine dimers are an interesting and important case in point. The detailed distribution of pyrimidine dimers has been studied in a segment of DNA of known nucleotide sequence from the operator-promoter region of the *E. coli lacI* gene, present in a plasmid as a recombinant DNA insert (172).

A radiolabeled 117-base-pair (bp) segment of the *lacI* gene was isolated from plasmid DNA by digestion with the appropriate restriction enzymes. The DNA fragment was exposed to varying doses of UV radiation and then incubated with saturating amounts of an enzyme probe that specifically recognizes pyrimidine dimers in DNA and catalyzes the formation of a single-strand break (nick) in the DNA at each dimer site. (This enzyme is called the *M. luteus* pyrimidine dimer (PD) DNA glycosylase-AP endonuclease; its properties are discussed in detail in Section 3-4.) Thus, assuming that *all* sites of dimer formation are converted into nicks in DNA by the action of the enzyme probe, the precise location of dimers in the DNA sequence could be determined by denaturing the DNA into separate strands and comparing the electrophoretic distribution of each fragment with those generated from unirradiated DNA exposed to the Maxam-Gilbert sequencing technique (173) (Fig. 1-31). The frequency of a particular dimer could also be determined from the amount of radioactivity associated with each fragment (represented as a band) in the gel. The effect of UV radiation on single-strand DNA was examined by denaturing the 117-bp fragment and harvesting the labeled single strands prior to UV irradiation.

Such studies show that at high doses of UV light, the extent of dimer formation reaches a maximum, which is unaffected by further irradiation (172). This reflects the establishment of a steady state between the formation of dimers and their monomerization by photoreversal. The dose at which the maximum is reached varies for different types of dimers. The level of C <> C plateaus at doses of approximately 500 J/m^2, whereas for T <> T doses of greater than 2000 J/m^2 are required. This result is not surprising because it is known that the quantum yields for formation differ for different pyrimidine dimers (165). The absolute level of dimerization also varies at individual dimer sites. For example, different sites of potential T <> T attain different steady-state levels that vary between 4 and 16 percent (Fig. 1-32) (172). The steady-state level of dimer formation is also influenced by the nature of the nucleotides flanking potential dimer sites. In general, the equilibrium level of dimers is greater for TT sites flanked on both sides by A than for TT sites flanked on the 5′ side by A and on the 3′ side by G (172) (Fig. 1-33). However, this immediate nucleotide flanking effect is not sufficient

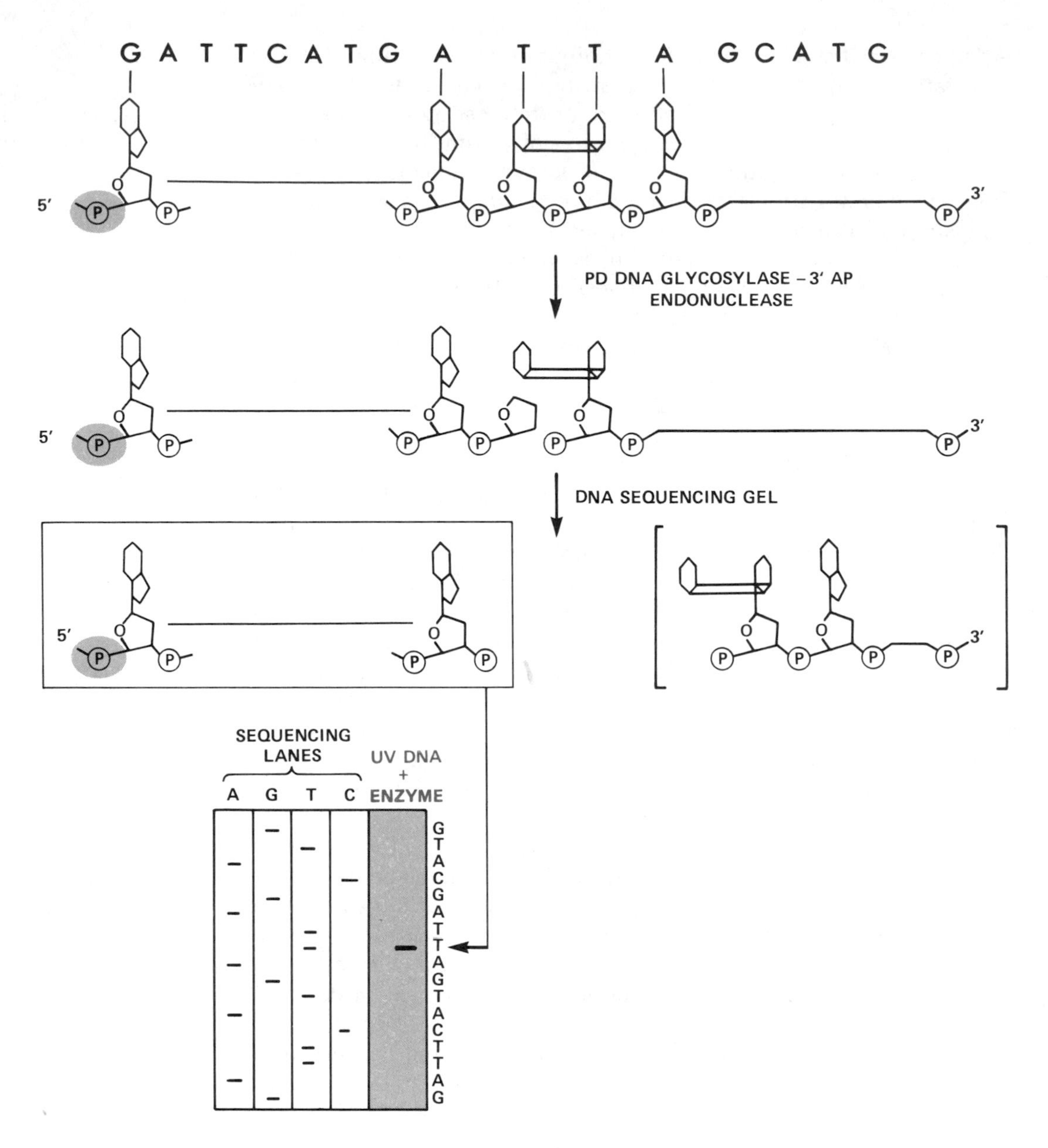

FIGURE 1-31
Determining the location of pyrimidine dimers in UV-irradiated DNA by using a dimer-specific enzyme probe. *E. coli* duplex DNA of known sequence (only one strand of the DNA duplex is shown) is radiolabeled at the 5′ ends and incubated with saturating amounts of an enzyme (PD DNA glycosylase-AP endonuclease) that specifically recognizes pyrimidine dimers in DNA. The enzyme cuts the 5′ glycosyl bond of the dimer and also the 3′ phosphodiester bond as shown. The DNA is then loaded onto a denaturing polyacrylamide sequencing gel alongside (in separate lanes of the gel) standard DNA sequencing lanes. This procedure hydrolyzes the 3′-terminal deoxyribose residues, leaving stable 5′-labeled DNA fragments, the lengths of which bear a precise relation to the sites of pyrimidine dimers. The representation of the sequencing gel has been simplified for clarity; normally G bands are also present in the A lane and C bands are also present in the T lane. (After L. K. Gordon and W. A. Haseltine, ref. 172.)

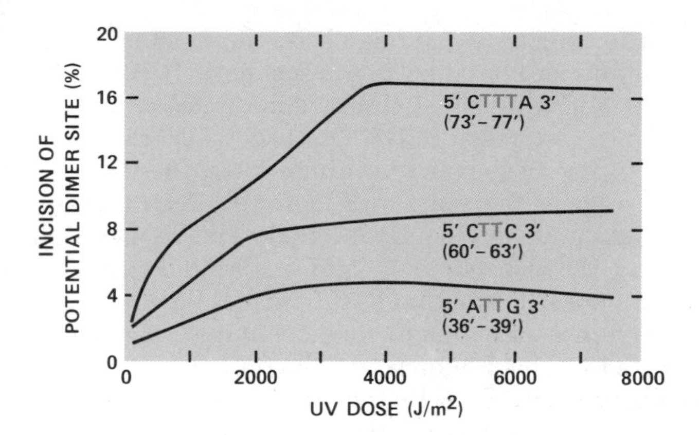

FIGURE 1-32
Dose response for thymine dimer formation in *E. coli* DNA of known nucleotide sequence. The different curves show the dose response for individual dimer sites, which were identified by their precise location in the sequenced DNA (numbers in parentheses). Dimers were quantitated from the amount of radioactivity present in electrophoretic bands using the technique shown in Fig. 1-31. (From L. K. Gordon and W. A. Haseltine, ref. 172.)

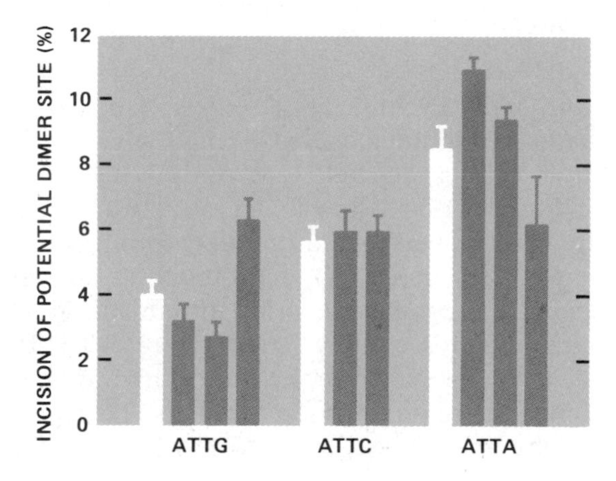

FIGURE 1-33
The influence of DNA sequence on the formation of pyrimidine dimers in UV-irradiated DNA. Each bar represents the effect of nucleotides flanking the dimer sites shown, in different regions of *E. coli lacI* DNA. The bar on the left of each set represents the mean of all determinations for the particular sequence, taken from all regions where that sequence occurred. The frequency of dimers was determined from the frequency of incision of UV-irradiated DNA by a dimer-specific enzyme probe (see Fig. 1-31). (From L. K. Gordon and W. A. Haseltine, ref. 172.)

to account for all the variations observed. Thus, for example, the frequency of dimer formation in the sequence TGAA*ATTG*TTAT is significantly higher than that observed in the sequence CTAT*ATTG*-ATTA or in the sequence TGGA*ATTG*TGAG (172).

In general, for most potential dimer sites, the dose response of dimer formation is the same in *E. coli lacI* DNA regardless of the secondary structure of the DNA (172). This is surprising, since single-strand DNA absorbs UV light more efficiently than double-strand DNA does. Furthermore, differences in the overall rate of pyrimidine dimer formation in single- and double-strand DNA have been detected in other studies (9, 158–168).

These experiments show that the formation of pyrimidine dimers in DNA is not a completely random phenomenon, as was believed for many years. Further, pyrimidine dimer contents in DNA for any given dose of UV radiation cannot be accurately extrapolated from measurements at any other dose and cannot be accurately extrapolated for any single dose from one DNA source to another. These limitations notwithstanding, the pyrimidine dimer has served well as the classic "test lesion" for analysis of DNA repair, and much of the discussion in the remainder of this book centers around the cellular responses to this particular lesion in DNA.

Other Photoproducts in DNA

Noncyclobutane-Type Pyrimidine Adducts. In irradiated solutions of free bases and of nucleosides, formation of noncyclobutane-type pyrimidine lesions referred to as pyrimidine adducts can be detected (9). Figure 1-34 shows the structure of 5-thyminyl-5,6-dihydrothymine, a major photoproduct produced in the UV-irradiated spores of *B. subtilis* (174). As much as 30 percent of the thymine in spore DNA can be converted into this product following exposure to very high doses of UV radiation (120). The formation of this lesion appears to be related to the state of hydration of the DNA, since the so-called spore photoproduct can also be generated by UV irradiation of dehydrated DNA, that is, DNA in the A form (120).

The Pyrimidine-Pyrimidine (6-4) Lesion. Alkali-labile lesions at positions of cytosine (and, much less frequently, thymine) 3′ to pyrimidine nucleosides are also present in UV-irradiated DNA (175). These photoproducts, referred to as pyrimidine-pyrimidine (6-4) lesions or simply as (6-4) lesions (176), are apparently identical or very similar to a noncyclobutane type of pyrimidine adducts (6,4′,5′-methylpyrimidin-2′,1′-thymine) that has been detected following the irradiation of thymine in frozen aqueous solution (177, 178) (Fig.

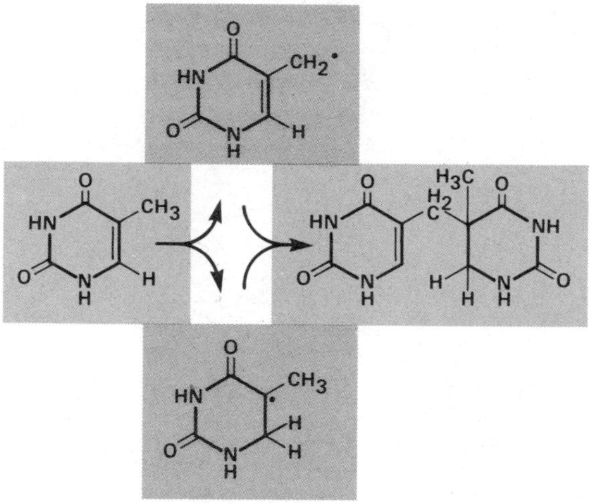

FIGURE 1-34
The formation of 5-thyminyl-5,6-dihydrothymine (spore photoproduct) by the addition of two different radicals of thymine generated by UV radiation. (From K. C. Smith, ref. 249.)

1-35). The (6-4) adducts of TC, CC and TT sequences are observed in UV-irradiated DNA, whereas that of CT is not. These lesions are insensitive to enzyme probes for pyrimidine dimers or for apurinic DNA but can be detected by their lability in hot alkali (175, 176).

As is true for pyrimidine dimers, the number of (6-4) lesions formed is proportional to the incident UV dose in the dose range of 100 to 500 J/m^2 (175, 179). When the frequencies of (6-4) photoproducts and pyrimidine dimers were compared at specific sites in the *E. coli lacI* gene (179), the frequency of the former varied between 10- and 15-fold at different sites. The incidence of (6-4) TC lesions was generally greater than that of CC lesions. TT lesions were only detected at very high UV doses (greater than 5 kJ/m^2). At most sites in the DNA the (6-4) lesions occurred at a frequency severalfold lower than that of the pyrimidine dimers. However, at some sites the lesion was detected at levels equal to or greater than that of the dimers.

UV-induced mutations that measured reversion of *amber, ochre* and TGA chain-terminating codons in the *E. coli lacI* gene included five "hot spots" in the gene, which accounted for over 30 percent of these nonsense mutations (77). A good correlation exists between the frequency of nonsense mutations and that of UV-induced damage at TC and CC sequences (176, 179). This correlation is signficantly better for (6-4) photoproducts than for pyrimidine dimers (176, 179). Thus, the (6-4) lesions may be biologically important photoproducts in UV-irradiated cells.

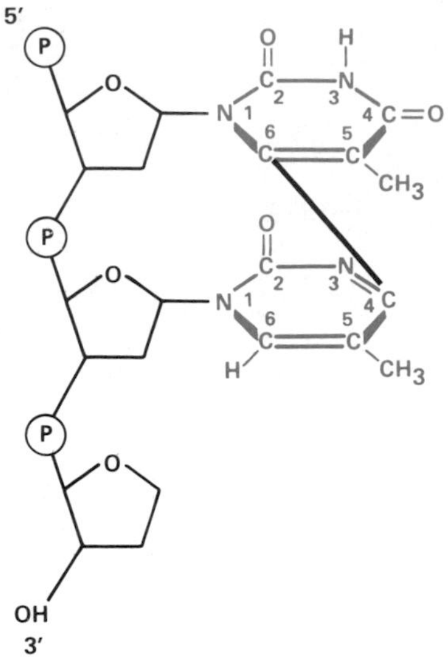

FIGURE 1-35
Schematic representation of a photoproduct produced by linkage between the C^6 position of one thymine and the C^4 position of the adjacent thymine. This covalent linkage is unstable to hot alkali. In DNA the 3′ pyrimidine in such so-called (6-4) lesions is typically cytosine rather than thymine. Hence, these photoproducts are sometimes referred to as Py-C lesions (where Py = pyrimidine).

Monomeric Base Damage Induced by UV Radiation

Pyrimidine Hydrates. Another type of photochemical reaction of a pyrimidine base is the addition of a molecule of water across the 5,6 double bond to form a 5,6-dihydro-6-hydroxy (hydrated) derivative (Fig. 1-36). The quantum yield for the formation of cytosine hydrates in UV-irradiated DNA and in UV-irradiated polymers is significantly greater in single-strand than in duplex DNA (9). Pyrimidine hydrates in deoxyribopolymers alter the template properties for transcription by RNA polymerase (9); however, there is no evidence that pyrimidine hydrates are biologically significant forms of DNA damage in vivo. This is consistent with the fact that these products have short half-lives and can readily dehydrate and revert to the parent form (9).

Thymine Glycols. A more stable lesion resulting from saturation of the 5,6 double bond of some pyrimidines is the so-called thymine glycol or 5,6-dihydroxydihydrothymine-type lesion (Fig. 1-36). This lesion is one of the major forms of base damage to DNA induced by *ionizing radiation* (see below), but it can also result from UV radiation (180, 181). *E. coli* contains an enzyme activity that catalyzes the excision of this type of base damage from DNA (182). The properties of this enzyme are discussed in Section 3-4.

Miscellaneous UV-Induced Base Damage

As indicated earlier, UV radiation can result in the cross-linking of DNA to proteins. Cross-links between different duplex DNA molecules (DNA-DNA cross-links) have been observed when DNA is irra-

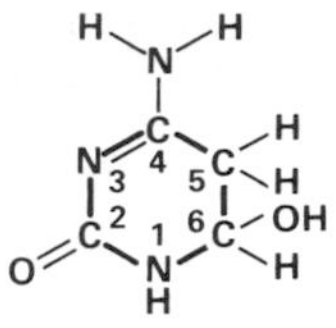

5,6-DIHYDRO-6-HYDROXY-CYTOSINE

CYTOSINE HYDRATE

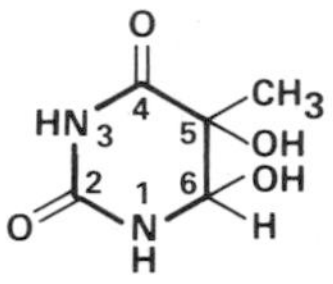

5,6-DIHYDROXY-5,6-DIHYDROTHYMINE

THYMINE GLYCOL

FIGURE 1-36
Examples of monomeric pyrimidine base damage caused by UV radiation at approximately 254 nm.

diated in the dry state or in an extremely densely packed state, such as in the heads of salmon sperm (183). Irradiation of DNA at 254 nm can also result in breakage of the polynucleotide chain. However, the amount of UV radiation required to reduce the molecular weight of *Diplococcus pneumoniae* DNA by 50 percent is about 100 times that required to reduce the transforming activity of the streptomycin resistance marker of that organism to the same extent (120). In addition, no chain breaks are detected in phage T7 DNA exposed to doses of UV that inactivate almost 100 percent of the phage population (120). Thus the formation of DNA strand breaks by UV is apparently of little (if any) biological consequence.

Sensitized Photoreactions of DNA

The UV radiation-induced damage discussed so far results principally from the direct absorption of photons by bases in DNA. DNA damage can also result from wavelengths in the electromagnetic spectrum that, although they are not absorbed directly by bases, are absorbed by other molecular species (sensitizer molecules), which then transfer energy to the bases in DNA. This phenomenon is referred to as *photosensitization* and can occur by a variety of possible reaction pathways (9, 184). Photosensitization reactions can be broadly classified into those in which the sensitizing molecule behaves as a true photocatalyst and those in which the sensitizing molecule is consumed (184).

The best example of a naturally occurring photosensitization process that is truly catalytic is called *enzymatic photoreactivation.* This is a ubiquitous biological reaction for the repair of *cis-syn* pyrimidine dimers in DNA; it is discussed in detail in Section 2-2. During this repair process pyrimidine dimers are monomerized in the presence of near-UV light (300 to 600 nm) by an enzyme called photoreactivating enzyme or DNA photolyase. The dimers are not affected by these wavelengths of light in the absence of the enzyme, but when a dimer site forms a complex with the enzyme, light is absorbed, dimers are split and the enzyme dissociates, free to bind at another dimer site.

The photosensitized reactions of greatest interest with respect to DNA *damage* are those that result in the *formation* of thymine-thymine dimers in DNA by triplet excitation transfer. Various ketones promote dimer formation in this way, a notable example being acetophenone (9, 184, 185). The lowest triplet energy state of acetophenone is slightly higher than that of thymine, but lower than the triplet states of the other DNA bases (Fig. 1-37). On irradiation of DNA in the presence of acetophenone at wavelengths of approximately 300 nm, the triplet energy of the photosensitizer is transferred to thymine, facilitating the formation of thymine-thymine dimers (9, 184).

A particular advantage of the use of photosensitizers in photobiological research is that they promote the formation of thymine

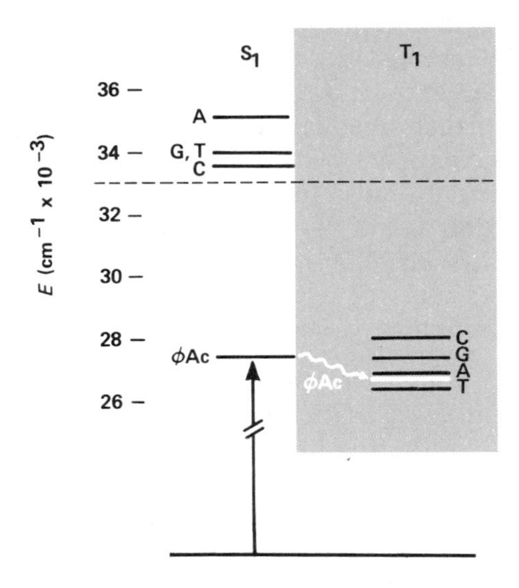

FIGURE 1-37
The energy levels (E) of the lowest excited singlet states (S_1) and lowest triplet states (T_1) of adenine (A), guanine (G), cytosine (C) and thymine (T) along with that of acetophenone (ϕAc). The lowest triplet energy state of ϕAc is slightly higher than that of thymine but lower than that of the other DNA bases. Thus, on irradiation of DNA at 300 nm the triplet energy of ϕAc is transferred to thymine, thereby facilitating the formation of dimers between adjacent thymines. (After A. Lamola, ref. 184.)

dimers at wavelengths of UV light that do not drive the reverse reaction (photoreversal). Thus, in contrast to the situation described earlier in which irradiation of DNA at 254 nm results in a steady state when approximately 7 percent of the thymine in DNA is dimerized, irradiation at 300 nm in the presence of a molecular photosensitizer such as acetophenone can yield thymine-thymine dimer contents close to the theoretical maximum (186). Hence photosensitizers such as acetophenone are extremely useful in the study of the repair of thymine dimers in DNA. In addition to the quantitative considerations just described, their use makes possible the study of the repair of a homogeneous class of base damage, since at the longer wavelengths of UV radiation employed for photosensitized reactions, other pyrimidine dimers as well as nondimer photoproducts produced at 254 nm are avoided (186).

Perhaps the best-studied reactions in which the photosensitizer is consumed involve the photoaddition of certain planar compounds to bases in DNA. For example, furocoumarins with planar tricyclic configurations (psoralens) (Fig. 1-38) can intercalate into DNA and, on subsequent photoactivation by long-wavelength UV radiation, may form covalent adducts to the pyrimidines, principally by addition across the 5,6 double bond of thymine (187). Psoralens with a totally planar organization of the three aromatic rings (such as 8-methoxypsoralen) are able to react with pyrimidines at one or both ends to form monofunctional or difunctional adducts respectively (Fig. 1-39). The latter join a pyrimidine in one DNA strand above the plane of the intercalated psoralen to an appropriately situated pyrimidine below in the other strand, thus cross-linking the two strands (187). This cross-linking reaction requires two independent UV absorption events. Some furocoumarins such as angelicin can form only monofunctional photoadducts since the unreacted end is not appropriately juxtaposed with a pyrimidine in the native DNA helix due to an angular arrangement of the three rings (187) (Fig. 1-40).

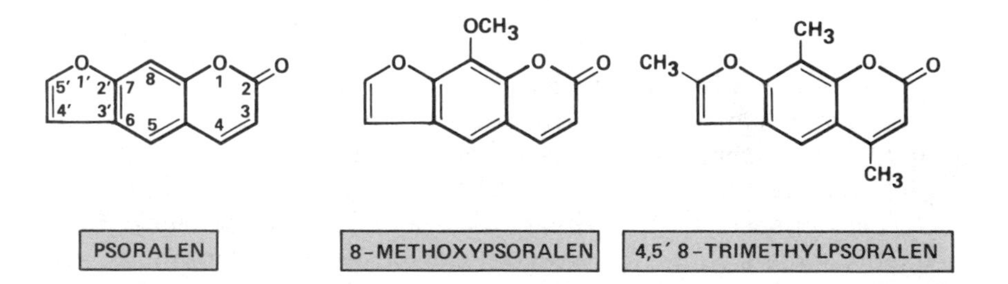

FIGURE 1-38
The structure of psoralen and some psoralen derivatives.

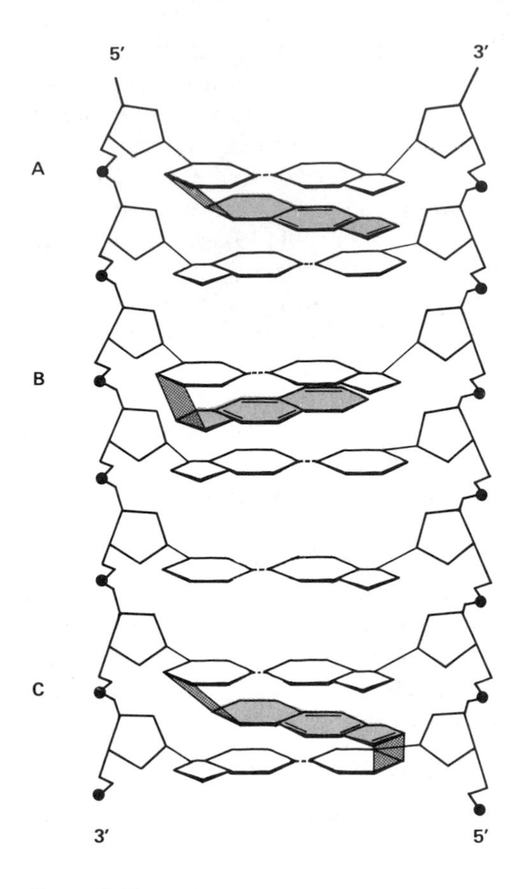

FIGURE 1-39
Interaction of psoralen with DNA to form two types of monoadducts (A and B) or a diadduct (interstrand cross-link) (C). Two types of monoadducts can result because the 5,6 double bond of thymine can photoreact with psoralen at either its 3,4 double bond or its 4′,5′ double bond (see Figs. 1-38 and 1-40). The formation of the cross-link requires independent UV absorption events at each reactive end. (After M. A. Pathak et al., ref. 187.)

The psoralen-plus-UV-light reaction is highly specific for native DNA, and the induced cross-links are stable to alkali. Psoralen monoadducts to pyrimidine bases are formed at about a three-fold higher yield than that for cross-links (187–190). However, the latter class of damage appears to be primarily responsible for biological inactivations induced by the treatment of bacteria, bacteriophage and eukaryotic cells with photoactivated psoralen (187–190). For this reason, psoralen-induced cross-links have become a favored model for the study of the repair of interstrand DNA cross-links in a variety

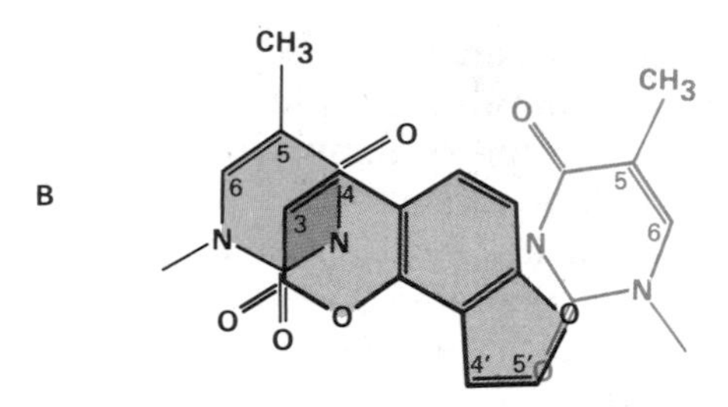

FIGURE 1-40
Projection of psoralen (A) and angelicin (B) molecules intercalated between two base pairs in DNA. In each case the thymines shown are on opposite strands of the DNA duplex. Note that angelicin cannot cross-link two DNA strands because one end of the molecule is not appropriately juxtaposed with one of the thymines due to its angular configuration. (From L. Musajo et al., ref. 250.)

of biological systems (191–196). The ability of photoactivated psoralens to arrest DNA replication of dividing cells has also facilitated the use of these compounds in the treatment of the human disease psoriasis, which is characterized by a marked proliferative disturbance of certain epithelial cells in the skin (187).

The Effect of Substitution by Halogenated Pyrimidines on the Sensitivity of DNA to UV Radiation

While the general topic of photosensitized reactions in DNA is being discussed, it is appropriate to consider some of the effects of the substitution of thymine by certain halogenated pyrimidines. As indicated earlier in this chapter, substitution of thymine in DNA by 5-BU is an effective means of perturbing the density of DNA, and this technique has contributed much to the current understanding of both semiconservative and nonsemiconservative modes of DNA replication. In addition to altering the density of DNA by its presence, 5-BU imparts a markedly increased sensitivity to irradiation at about 313 nm, resulting in polynucleotide strand breaks (BU photolysis) and the production of alkali-labile sites (197). The observed breaks are thought to arise from photochemical debromination followed by free-radical attack on the sugar or sugar-phosphate backbone (197). In Section 5-2 the use of BU-substituted DNA for the detection and measurement of DNA repair by both the density shift and the photolysis technique is discussed.

Ionizing Radiation Damage to DNA

Ionizing radiation constitutes another source of naturally occurring physical damage to DNA that living organisms on this planet have had to contend with since the beginning of biological evolution. Unlike UV radiation at approximately 260 nm, which is preferentially absorbed by nucleic acids, ionizing radiation can cause damage to all cellular components, including DNA, the probability of damage being largely a reflection of the target size of a particular cellular component (7, 198–204).

The deposition of energy from ionizing radiation results in the formation of excited molecules and ionized species, and traditionally, radiation damage to DNA has been ascribed to both "direct" and "indirect" effects of these molecules and species (7). So-called direct effects result from the direct interaction of the radiation energy with DNA. Indirect effects result from the interaction of reactive species formed by the radiation with DNA. Since in living cells DNA exists in an environment containing numerous molecular species, inorganic ions and water, a large variety of potential sources of reactive species can exist as excited molecules, ion radicals or free radicals, which are ultimately converted into chemically stable products by subsequent decay reactions. This multiplicity of reaction mechanisms and the even greater number of potential reactants create the possibility for a large spectrum of ionizing radiation products in DNA. However, the predominance of water in biological systems suggests that species formed by the radiolysis of water are major potential sources of damage (7). These species have been well characterized and consist of hydrogen peroxide, hydrogen atoms, hydrated electrons and hydroxyl radicals (7). Correlations between specific lesions in DNA produced by ionizing radiation and biological end points such as cell killing and mutagenesis are not as well established as for some of the other forms of DNA damage discussed in this chapter.

Radiation Products of Free Purine and Pyrimidine Bases

Thymine. The major radiolysis products of thymine in aerated aqueous solution are *trans*(*cis*)-5-hydroperoxy-6-hydroxy-5,6-dihydrothymine (Fig. 1-41, I, R = H) and *cis*-5-hydroxy-6-hydroperoxy-

5,6-dihydrothymine (Fig. 1-41, II, R = H) (205). These may in part react to 5,6-dihydroxy-5,6-dihydrothymine (Fig. 1-41, VII, R = H) and 5-methyl-5-hydroxyhydantoin, on standing or exposure to high radiation doses (205). Irradiation under anoxic conditions yields 5,6-dihydrothymine (Fig. 1-41, IV, R = H) and 6-hydroxy-5,6-dihydrothymine (Fig. 1-41, V, R = H) as major products (205). All radiolysis products of N^1-substituted thymine of the 6-(hydroxy or hydroperoxy)-5,6-dihydrothymine type, i.e., compounds I, II, V and VII in Fig. 1-41, and probably 5-methyl-5-hydroxyhydantoin (e.g., R =

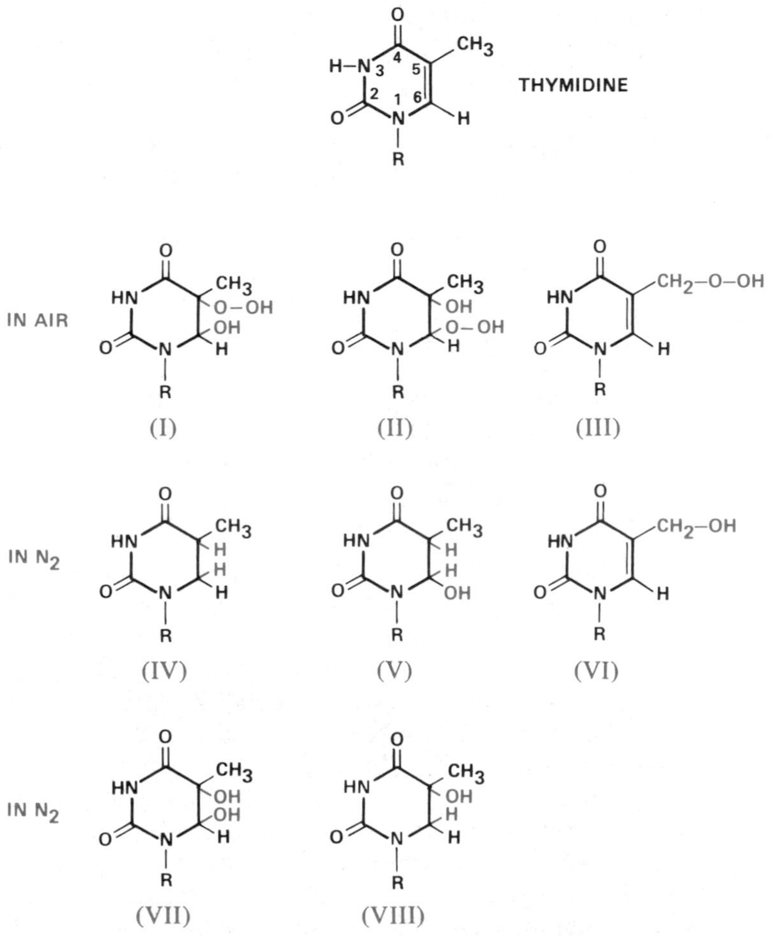

FIGURE 1-41
The major radiolysis products of thymidine when it is exposed to aerobic (top row) or anoxic (bottom two rows) conditions. (From P. V. Hariharan and P. A. Cerutti, ref. 205.)

deoxyribose and deoxyribose-phosphate), are expected to be reductively cleaved by sodium borohydride, giving rise to a urea derivative plus 2-methylglycerol (from I, II and VII), 1,3-dihydroxy-2-methylpropane from V and possible 1,2-dihydroxypropane from 5-hydroxy-5-methylhydantoin (Fig. 1-41) (205). The amount of 4- (and 3) carbon fragments so produced is a direct measure of the amount of the thymine radiolysis products of the 6-(hydroxy or hydroperoxy)-5,6-dihydrothymine type present in a radiolysis solution. On the basis of this reaction a sensitive procedure for the determination and characterization of thymine radiolysis has been developed, using thymine derivatives radioactively labeled in the methyl substituent (205). All radiolysis products of thymine that can be measured by this reductive assay are designated as *t*. Some (if not all) *t* products are excised from the genome of living cells exposed to ionizing radiation (206, 207). In addition, excision of *t* products from irradiated DNA can be demonstrated with cell-free extracts (208). Thymine glycol (5,6-dihydroxy-5,6-dihydrothymine) can also be very sensitively measured with a radioimmunoassay using antibody against single-strand DNA containing this lesion (209).

Other thymine radiation products have been detected in DNA exposed to ionizing radiation. These include formylpyruvylurea and hydroxybarbituric acid. Formylpyruvylurea can undergo further degradation to yield urea and N-substituted urea derivatives that largely remain stably attached to deoxyribose residues in DNA (205). These products are presumably of some biological import since an enzyme activity capable of excising free urea from DNA containing this thymine degradation product has been detected in extracts of *E. coli* (210) (see Section 3-3).

Cytosine and the Purines. There have been fewer studies on the effects of ionizing radiation damage to cytosine and to the purines, although it has been established with respect to cytosine that the major site for attack by hydroxyl radicals is the 5,6 double bond, as is the case with thymine (211). Although hydroxyl radicals react with purines as rapidly as with pyrimidines, the rate of destruction of the former is lower (212). Destruction of the imidazole ring of purines to yield 5-formamidopyrimidine compounds has been shown to occur after exposure to ionizing radiation in the absence of oxygen (213, 214). The scheme shown in Fig. 1-42 has been proposed to explain these reactions. Note that each reaction proceeds through a carbinolamine-type intermediate before the formation of the formamidopyrimidines. As will be seen in Section 3-3, an enzyme activity that can excise 2,6-diamino-4-hydroxy-5-formamidopyrimidine (FaPy) from DNA is present in extracts of *E. coli* (215).

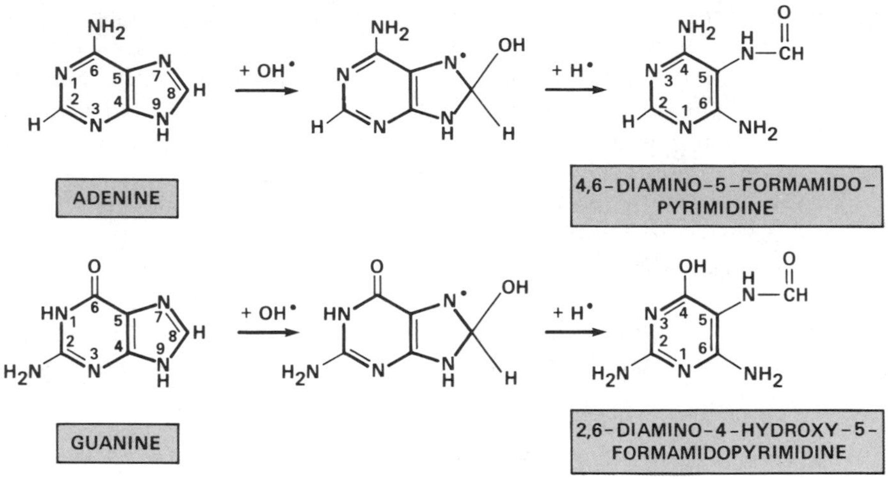

ADENINE

4,6-DIAMINO-5-FORMAMIDO-PYRIMIDINE

GUANINE

2,6-DIAMINO-4-HYDROXY-5-FORMAMIDOPYRIMIDINE

FIGURE 1-42
Imidazole ring opening of adenine and guanine following ionizing radiation of DNA can yield the products shown. (After C. J. Chetsanga et al., ref. 251.)

Base Damage in DNA Caused by Ionizing Radiation

Measurements of damage in DNA (as opposed to free bases or nucleosides) indicate that all four bases are subject to ionizing radiation damage (7, 216). However, the macromolecular organization of DNA apparently affects the quantitative yields of base damage. For example, the total yield of base destruction depends markedly on DNA concentration, being greater at higher DNA concentrations. This may result from the relatively low rate constant for the reaction of hydroxyl radicals with DNA, so that at low DNA concentrations scavenging of hydroxyl radicals occurs due to competing reactions such as

$$OH^* + OH^* \rightarrow H_2O_2$$

or

$$\text{Impurity} + OH^* \rightarrow \text{Impurity}^* + H_2O$$

A variety of hydroxyl-scavenging molecules such as cysteine and glutathione exist in living cells and can compete with DNA for the highly reactive free radicals (7). In addition, most hydroxyl radicals result from the radiolysis of water, and since these have a fairly small effective radius, hydrophobic regions of macromolecules are significantly protected from attack by radicals. Thus the stacked orientation of the bases in the interior of the DNA duplex protects bases from ionizing radiation damage to some extent. When polyadenylic acid is irradiated, the yield of adenine destruction by hydroxyl radical attack is the same as for free adenine. However, the yield of base destruction is much lower when poly(dA)·poly(dT) is subject to the same treatment (217).

Due largely to the extreme instability of base damage in DNA caused by ionizing radiation and hence the difficulty in its detection, it is probable that a major proportion of such base damage has not yet been characterized. In view of the fact that ionizing radiation constitutes a significant real and potential source of environmental damage to the genetic material of living cells, this area of investigation merits considerably more attention.

Strand Breaks in DNA Produced by Ionizing Radiation

An important biological consequence of ionizing irradiation of DNA is the formation of strand breaks (7). The majority of these are caused by breakage of phosphodiester linkages in one of the polynucleotide chains, but destruction of the deoxyribose ring can also result in interruptions of the deoxyribose-phosphate backbone (7). The presence of a radiation-induced single-strand break in DNA can result in very localized denaturation in the vicinity of the break, thereby increasing the probability of free radical-mediated attack at that site due to the loss of the protective effect of base stacking interactions.

Quantitative studies on DNA strand breaks have depended very largely on the evolution of numerous and progressively more refined techniques for measuring these events (170). Mechanistic studies have relied largely on the use of deoxynucleotides and alkylphosphates as model substrates in which the equivalent of a strand-break-producing reaction liberates inorganic phosphate, which can be readily quantitated (7). Studies with nucleotides suggest that deoxyribose-phosphate bond breakage occurs via a β-elimination reaction; however, the mechanism by which the alkali-labile deoxyribose moiety is derived is dependent on the presence or absence of oxygen (7). Thus, the ratio of the yield of immediate phosphate release to labile phosphate esters in thymidylate is lower in the absence of oxygen than in its presence; that is, alkali-labile sites are more frequent under anoxic radiation conditions (7). This phenomenon may account in part for the much discussed and extensively investigated "oxygen effect" of ionizing radiation on living cells: cells are significantly more sensitive to the lethal effects of ionizing radiation when irradiated in the presence of oxygen (218).

Like base damage, DNA strand breaks can result from direct or indirect effects of ionizing radiation. In the former case the deposition of approximately 9 eV of energy is sufficient to break the deoxyribose-phosphate backbone (219, 220). As regards indirect effects, hydroxyl radicals appear to be of primary importance in the hydrolysis of phosphodiester bonds (221). The precise chemical mechanisms of DNA strand breakage induced by ionizing radiation are varied and complex, as are most radiolytic processes. They may involve direct ionization of the phosphodiester bond, or fragmentation reactions involving the bases or sugars, coupled with electron rearrangements that ultimately result in hydrolysis of labilized phosphodiester bonds.

With the exception of viruses containing single-strand-DNA genomes, the extent of single-strand break formation correlates poorly with the lethality caused by ionizing radiation. A far better correlation of lethality is obtained with double-strand breaks in the DNA of living cells (221). These can arise from very closely juxtaposed single-strand scissions of independent origin on opposite strands of the DNA duplex, or by a single event resulting in hydrolysis of phosphodiester bonds in both DNA strands. Either phenomenon can result from the direct or indirect effects of ionizing radiation. The yield of both single- and double-strand breaks in DNA can be modified by the presence of free-radical scavengers and of oxygen (221).

1-6 DNA Damage and Chromatin Structure

This chapter has been very largely devoted to the consideration of damage to bases in DNA, without regard to the higher levels of organization of DNA molecules in the genomes of living cells. However, in eukaryotes DNA is associated with both histone and nonhistone chromosomal proteins to form chromatin. The association with histones results in the organization into repeating units called *nucleosomes*, which consist of a core of 140 bp of DNA wrapped around an octomer of histones, which in turn is made up of two each of the histones H2A, H2B, H3 and H4 (222). (See Figure 6-7.) Variable (in different organisms) amounts of DNA ranging from 20 to 60 bp are referred to as linker DNA, which is loosely associated with another histone, H1 (222). There is evidence that in living cells the nucleosomes are organized into further levels of folding to provide the structure of chromosomes. The highly compact organization of DNA in chromosomes accounts for the packaging of 5.3×10^9 bp of DNA, corresponding to a contour length of 180 cm, into 46 chromosomes whose total contour length is 200 μm in a human diploid cell (223).

Certain forms of damage are not random in their distribution in the nucleosome. For example, base damage produced by a variety of chemicals occurs selectively in linker regions (224, 225). On the other hand, thymine-containing pyrimidine dimers resulting from UV irradiation of living cells appear to be uniformly distributed in both linker and core nucleosomal DNA (226).

The influence of chromatin structure on the concentration and distribution of various forms of DNA damage is an important consideration that merits further detailed study. As will be seen in Section 4-3, this phenomenon has significant bearing on the responses of living cells to damage, since there is evidence that sites of DNA base damage in chromatin are not all equally accessible to DNA repair enzymes.

1-7 Detection and Measurement of Base Damage in DNA

The discussion of DNA damage in this chapter has included brief descriptions of selected forms of damage and their detection and quantification. This topic has not been extensively covered, partly because of space constraints, but primarily because the essential focus of this book is the molecular biology of DNA damage and the responses that it elicits in living cells. Technical and methodological descriptions are available from other sources in the literature (170, 227, 228). Nonetheless, the issues of quantitation of DNA damage and the sensitivity of its detection are important ones. Their importance relates partly and quite obviously to the furthering of information on DNA damage and its repair in general. However, in addition it would be extremely valuable to be able to establish accurate and meaningful quantitation of genotoxicity (i.e., the amount of DNA damage at any given time) in human populations in general, and in selected populations after specific genotoxic exposure in particular. To achieve the latter goals it would be necessary to isolate human cells that are sufficiently long-lived so that their DNA would reflect the accumulation of DNA damage over an extended period of time. Long-lived G_0 lymphocytes isolated from peripheral blood may be useful for such purposes, particularly since their isolation would involve a relatively noninvasive procedure. Alternatively (though perhaps logistically more complex) one might conceive of isolating small numbers of terminally differentiated nondividing cells such as myocytes for this purpose.

DNA Sequencing for Measuring DNA Damage

The detection and quantitation of some of the types of base damage discussed above has been facilitated by a number of technical approaches widely used in modern molecular biology. One of the most powerful modern tools for studying nucleic acid biochemistry is the various DNA-sequencing techniques that have been developed and refined in the past decade. In Section 1-2 the use of DNA sequencing was discussed as a technique for measuring the distribution of pyrimidine dimers and other photoproducts in DNA exposed to UV radiation. The Maxam-Gilbert technique (173) relies on specific chemical damage to DNA that results in the labilization of the N-glycosylic bonds linking the various bases to the deoxyribose-phosphate backbone of DNA. Thus, selective loss of T, C, A or G can be achieved, creating apyrimidinic or apurinic sites. Conversion of these sites into strand breaks by β-elimination allows for the precise positioning of a particular base because the resulting DNA strand breaks are reflected by the generation of fragments of precise length, which can be resolved by electrophoresis in polyacrylamide gels. The technique is extremely precise and allows for the sequencing of hundreds of nucleotides in a single gel. Hence rare mispairing of bases or other alterations in base sequence can be detected directly (provided they are represented in the entire population of DNA molecules being analyzed), and in practice the sensitivity of their detection is limited only by the number of nucleotides in a given genome that can conveniently be sequenced.

Specific Antibodies for Measuring DNA Damage

Conventional antisera are immunologically reactive against a variety of forms of damage (229), including pyrimidine dimers (230, 231), benzo(a)pyrene adducts (232), O^6-methylguanine (233), 7-methylguanine (234), N-acetoxy–N-2-acetylaminofluorene (235), 7,8-dihydro-8-oxyadenine(8-hydroxyadenine) (236), *trans*-7,8-dihydrobenzo(a)pyrene-7,8-diol-9,10-epoxide (237), guanine imidazole ring-opened adducts produced by aflatoxin B_1 (238) and *cis*-diamminedichloroplatinum (II)–modified DNA (239). The advent of hybridoma technology and the capacity for producing monoclonal antibodies will surely have a major impact on the field of quantitative genotoxicity. Monoclonal antibodies to UV-irradiated DNA (240) and to aflatoxin B_1-modified DNA (241) have already been described.

Enzyme Probes for Measuring DNA Damage

Uracil and hypoxanthine originating in DNA from the deamination of cytosine and adenine respectively, as well as a number of other forms of base damage to DNA discussed in this chapter, can be detected by highly specific enzyme probes belonging to the class of DNA glycosylases that recognize these bases in DNA and catalyze their excision by hydrolysis of the relevant N-glycosylic bonds (44, 242) (see Chapter 3). Once again we have a situation in which sites of base damage are converted into sites of base loss, which in turn can be translated into DNA strand breaks either by virtue of their alkali-lability or by a different class of enzymes that specifically recognize sites of base loss in DNA. The latter enzymes, called AP (apurinic/apyrimidinic) endonucleases are also discussed in detail in Chapter 3. Thus, the sensitivity for the detection of those forms of base damage that can be converted into DNA strand breaks is limited only by the sensitivity with which the DNA strand breaks can be detected and quantitated. By the judicious selection of appropriate DNA molecules for study, the detection of one strand break in some entire genomes is now technically feasible. For example, one strand break in a form I (covalently closed circular) DNA molecule results in the relaxation of the molecule to an open circular (form II) configuration. Form I and II DNA molecules can be readily separated by electrophoresis in agarose gels (243) or by sedimentation (244) and can also be distinguished by visualization in the electron microscope (245).

Other repair enzymes (discussed in Chapter 4) are potentially even more useful, since they recognize a large spectrum of DNA base damage, perhaps because the specific substrate for such enzymes is a conformational distortion of the secondary structure of the DNA duplex rather than a particular form of base damage. In addition, some nucleases not involved in DNA repair but more probably in DNA recombination are absolutely specific for single-strand DNA and may even be able to recognize "denatured" regions created by interrupted H bonding in a single base pair. Examples are the so-called S1 nuclease from *Aspergillus oryzae* (246) and BAL 31 nuclease from *Alteromonas* (247).

1-8 Summary

In this chapter a number of mechanisms of damage to DNA have been reviewed. In living cells that have sustained such damage a variety of cellular responses can ensue, many of which apparently do not operate in cells without damaged DNA. In general the most is known about those forms of DNA damage to which specific cellular responses have been identified, particularly responses that eliminate the damage. Indeed, a particularly fruitful approach to the discovery of new DNA repair mechanisms has been to first identify a particular form of base damage in DNA and then to ask whether living cells can remove that damage, and if so, how. Frequently the discovery of a new repair enzyme or biochemical pathway for repair provides the incentive to characterize and quantitate the substrate (damaged DNA) in even greater detail. A particularly good example of this relation between the study of DNA damage and of DNA repair is the pyrimidine dimer. Additionally, the discovery of enzymes that catalyze the removal of certain forms of base damage produced by ionizing radiation damage has focused interest in defining these forms of damage more precisely.

DNA damage is an inescapable aspect of life in the biosphere. Of particular import is the damage caused by agents that have been present in the environment long enough to have provided selective pressures for the evolution of mechanisms for the repair and tolerance of the damage by organisms. Indeed, it is likely that there are biological responses to all forms of such DNA damage, and the continued exploration of it may be expected to uncover further examples of DNA repair and damage tolerance mechanisms. Nonrepairable lesions in DNA are also very important, since these are the most likely to be mutagenic and/or lethal. In this regard the many synthetic products of modern technology merit special attention. Many of these may produce types of DNA damage that are recognized by evolutionarily long-established repair mechanisms. However, many may not, and thus a detailed understanding of their genotoxic potential is vital to the continued health of all living forms on this planet.

Suggestions for Further Reading

Major review articles on DNA damage include refs. 11 (chemical damage); 12 (spontaneous hydrolysis in DNA); 13 (chemical damage); 18 (replicational fidelity of DNA); 44 (spontaneous hydrolysis in DNA); 50 (bisulfite interactions with DNA); 99 (alkylation of DNA); 138 (interaction of AAF with DNA); 9, 159, 161, 163–165, 176 (photochemistry of DNA); and 198–204, (ionizing radiation damage).

References

1. Nevers, P., and Saedler, H. 1977. Transposable genetic elements as agents of gene instability and chromosomal rearrangements. *Nature* 268:109.
2. Cohen, S. N., and Shapiro, J. A. 1980. Transposable genetic elements. *Sci. Am.* 242:40.
3. Calos, M. P., and Miller, J. H. 1980. Transposable elements. *Cell* 20:579.

4. Kleckner, N. 1981. Transposable elements in prokaryotes. *Ann. Rev. Genet.* 15:341.

5. Cerutti, P. A. 1975. Repairable damage in DNA: overview. In *Molecular mechanisms for repair of DNA*. P. C. Hanawalt and R. B. Setlow, eds., p. 3. New York: Plenum.

6. Lindahl, T., and Ljungquist, S. 1975. Apurinic and apyrimidinic sites in DNA. In *Molecular mechanisms for repair of DNA*. P. C. Hanawalt and R. B. Setlow, eds., p. 31. New York: Plenum.

7. Ward, J. F. 1975. Molecular mechanisms of radiation-induced damage to nucleic acids. *Adv. Rad. Biol.* 5:181.

8. Lindahl, T. 1977. DNA repair enzymes acting on spontaneous lesions in DNA. In *Cellular senescence and somatic cell genetics. DNA repair processes*. W. W. Nichols and D. G. Murphy, eds., p. 225. Miami: Symposia Specialists Inc.

9. Kittler, L., and Löber, G. 1977. Photochemistry of the nucleic acids. *Photochem. Photobiol. Rev.* 2:39.

10. Cerutti, P. A. 1978. Repairable damage in DNA. In *DNA repair mechanisms*. P. C. Hanawalt, E. C. Friedberg and C. F. Fox, eds., p. 1. New York: Academic.

11. Roberts, J. J. 1978. The repair of DNA modified by cytotoxic, mutagenic, and carcinogenic chemicals. *Adv. Rad. Biol.* 7:211.

12. Shapiro, R. 1981. Damage to DNA caused by hydrolysis. In *Chromosome damage and repair*. E. Seeberg and K. Kleppe, eds., p. 3. New York: Plenum.

13. Singer, B., and Kuśmierek, J. T. 1982. Chemical mutagenesis. *Ann. Rev. Biochem.* 52:655.

14. Loeb, L. A., Weymouth, L. A., Kunkel, T. A., Gopinathan, K. P., Beckman, R. A., and Dube, D. K. 1978. On the fidelity of DNA replication. *Cold Spring Harbor Symp. Quant. Biol.* 43:921.

15. Radman, M., Villani, G., Boiteux, S., Kinsella, A. R., Glickman, B. W., and Spadari, S. 1978. Replicational fidelity: mechanisms of mutation avoidance and mutation fixation. *Cold Spring Harbor Symp. Quant. Biol.* 43:937.

16. Radman, M., Wagner, R. E., Glickman, B., and Meselson, M. 1980. DNA methylation mismatch correction and genetic stability. In *Progress in environmental mutagenesis*. M. Alačević, ed., p. 121. Amsterdam: Elsevier/North-Holland Biomedical Press.

17. Radman, M., Dohet, C., Bourgingnon, M-F., Doubleday, O. P., and Lecomte, P. 1981. High fidelity devices in the reproduction of DNA. In *Chromosome damage and repair*. E. Seeberg and K. Kleppe, eds., p. 431. New York: Plenum.

18. Loeb, L. A., and Kunkel, T. A. 1981. Fidelity of DNA synthesis. *Ann. Rev. Biochem.* 52:429.

19. Kornberg, A. 1980. *DNA replication*. San Francisco: Freeman.

20. Brutlag, D., and Kornberg, A. 1972. Enzymatic synthesis of deoxyribonucleic acid XXXVI. A proofreading function for the $3' \rightarrow 5'$ exonuclease activity in deoxyribonucleic acid polymerases. *J. Biol. Chem.* 247:241.

21. DeWaard, A., Paul, A. V., and Lehman, I. R. 1965. The structural gene for deoxyribonucleic acid polymerase in bacteriophages T4 and T5. *Proc. Natl. Acad. Sci.* (USA) 54:124.

22. Speyer, J. F. 1965. Mutagenic DNA polymerase. *Biochem. Biophys. Res. Comm.* 21:6.

23. Speyer, J. F., and Rosenberg, D. 1968. The function of T4 DNA polymerase. *Cold Spring Harbor Symp. Quant. Biol.* 33:345.

24. Drake, J. W., and Allen, E. F. 1968. Antimutagenic DNA polymerases of bacteriophage T4. *Cold Spring Harbor Symp. Quant. Biol.* 33:339.

25. Gillin, F. D., and Nossal, N. G. 1976. Control of mutation frequency by bacteriophage T4 DNA polymerase I. The CB120 antimutator DNA polymerase is defective in strand displacement. *J. Biol. Chem.* 251:5219.

26. Gillin, F. D., and Nossal, N. G. 1976. Control of mutation frequency by bacteriophage T4 DNA polymerase. I. Accuracy of nucleotide selection by the L88 mutator, CB120 antimutator, and wild type phage T4 DNA polymerases. *J. Biol. Chem.* 251:5225.

27. Goodman, M. F., Gore, W. C., Muzyczka, N., and Bessman, M. J. 1974. Studies on the biochemical basis of spontaneous mutations. III. Rate model for DNA polymerase-effected nucleotide misincorporation. *J. Mol. Biol.* 88:423.

28. Galas, D. J., and Branscomb, E. W. 1978. Enzymatic determinants of DNA polymerase accuracy. Theory of coliphage T4 polymerase mechanisms. *J. Mol. Biol.* 124:653.

29. Clayton, L. K., Goodman, M. F., Branscomb, E. W., and Galas, D. J. 1979. Error induction and correction by mutant and wild type T4 DNA polymerases. Kinetic error discrimination mechanisms. *J. Biol. Chem.* 254:1902.

30. Weymouth, L. A., and Loeb, L. A. 1978. Mutagenesis during *in vitro* DNA synthesis. *Proc. Natl. Acad. Sci.* (USA) 75:1924.

31. Kunkel, T. A., Eckstein, F., Mildvan, A. S., Koplitz, R. M., and Loeb, L. A. 1981. Deoxynucleoside [1-thio] triphosphates prevent proofreading during *in vitro* DNA synthesis. *Proc. Natl. Acad. Sci.* (USA) 78:6734.

32. Sirover, M. D., and Loeb, L. A. 1976. Infidelity of DNA synthesis *in vitro:* screening for potential metal mutagens or carcinogens. *Science* 194:1434.

33. Kunkel, T. A., Silber, J. R., and Loeb, L. A. 1982. The mutagenic effect of deoxynucleotide substrate imbalances during DNA synthesis with mammalian DNA polymerases. *Mutation Res.* 94:413.

34. Kunkel, T. A., and Loeb, L. A. 1980. On the fidelity of DNA replication. The accuracy of *Escherichia coli* DNA polymerase I in copying natural DNA *in vitro*. *J. Biol. Chem.* 255:9961.

35. Fersht, A., and Knill-Jones, J. W. 1981. DNA polymerase accuracy and spontaneous mutation rates: frequencies of purine·purine, purine·pyrimidine, and pyrimidine·pyrimidine mismatches during DNA replication. *Proc. Natl. Acad. Sci.* (USA) 78:4251.

36. Kunkel, T. A., Schaaper, R. M., Beckman, R. A., and Loeb, L. A. 1981. On the fidelity of DNA replication. Effect of the next nucleotide on proofreading. *J. Biol. Chem.* 256:9883.

37. Kunkel, T. A., Meyer, R. R., and Loeb, L. A. 1979. Single-strand binding protein enhances fidelity of DNA synthesis *in vitro*. *Proc. Natl. Acad. Sci.* (USA) 76:6331.

38. Kornberg, A. 1982. *1982 Supplement to DNA replication*. San Francisco: Freeman.

39. Drake, J. W. 1969. Comparative rates of spontaneous mutation. *Nature* 221:1132.

40. Pettijohn, D. and Hanawalt, P. C. 1964. Evidence for repair-replication of ultraviolet damaged DNA in bacteria. *J. Mol. Biol.* 9:395.

41. Cooper, P. K. 1982. Characterization of long patch excision repair of DNA in ultraviolet-irradiated *Escherichia coli:* an inducible function under rec-lex control. *Mol. Gen. Genet.* 185:189.

42. Drake, J. W. 1970. *The molecular basis of mutation*. San Francisco: Holden-Day Inc.

43. Watson, J. D. 1976. *Molecular biology of the gene*. 3rd ed. Menlo Park, Calif.: W. A. Benjamin, Inc.

44. Lindahl, T. 1979. DNA glycosylases, endonucleases for apurinic/apyrimidinic sites and base excision repair. *Prog. Nucleic Acids Res. Mol. Biol.* 22:135.

45. Duncan, B. K., and Weiss, B. 1978. Uracil-DNA glycosylase mutants are mutators. In *DNA repair mechanisms*. P. C. Hanawalt, E. C. Friedberg and C. F. Fox, eds., p. 183. New York: Academic.

46. Duncan, B. K., and Miller, J. 1980. Mutagenic deamination of cytosine residues in DNA. *Nature* 287:560.

47. Shapiro, R., and Klein, R. S. 1966. The deamination of cytidine and cytosine by acidic buffer solutions. Mutagenic implications. *Biochemistry* 5:2358.

48. Lindahl, T., and Nyberg, B. 1974. Heat-induced deamination of cytosine residues in DNA. *Biochemistry* 13:3405.

49. Schuster, H. 1960. The reaction of nitrous acid with deoxyribonucleic acid. *Biochem. Biophys. Res. Comm.* 2:320.

50. Hayatsu, H. 1976. Bisulfite modification of nucleic acids and their constituents. *Prog. Nucleic Acids Res. Mol. Biol.* 16:75.

51. Shapiro, R., Dubelman, S. Feinberg, A. M., Crain, P. F., and Closkey, J. A. M. 1977. Isolation and identification of cross-linked nucleosides from nitrous acid treated deoxyribonucleic acid. *J. Amer. Chem. Soc.* 99:302.

52. Ullman, J. S., and McCarthy, B. J. 1973. Alkali deamination of cytosine residues in DNA. *Biochim. Biophys. Acta.* 294:396.

53. Lion, M. B. 1968. Search for a mechanism for the increased sensitivity of 5-bromouracil-substituted DNA to ultraviolet light. *Biochim. Biophys. Acta.* 155:505.

54. Geider, K. 1972. DNA synthesis in nucleotide-permeable *Escherichia coli* cells. The effects of nucleotide analogues on DNA synthesis. *Eur. J. Biochem.* 27:554.

55. Konrad, E. B., and Lehman, I. R. 1975. Novel mutants of *Escherichia coli* that accumulate very small DNA replicative intermediates. *Proc. Natl. Acad. Sci.* (USA) 72:2150.

56. Konrad, E. B. 1977. Method for the isolation of *Escherichia coli* mutants with enhanced recombination between chromosomal duplications. *J. Bacteriol.* 130:167.

57. Tye, B-K., Nyman, P-O., Lehman, I. R., Hochhauser, S., and Weiss, B. 1977. Transient accumulation of Okazaki fragments as a result of uracil incorporation into nascent DNA. *Proc. Natl. Acad. Sci.* (USA) 74:154.

58. Tye, B-K., and Lehman, I. R. 1977. Excision repair of uracil incorporated in DNA as a result of a defect in dUTPase. *J. Mol. Biol.* 117:293.

59. Tye, B-K., Chien, J., Lehman, I. R., Duncan, B. K., and Warner, H. R. 1978. Uracil incorporation: a source of pulse-labeled DNA fragments in the replication of the *Escherichia coli* chromosome. *Proc. Natl. Acad. Sci.* (USA) 75:233.

60. Hochhauser, S. J., and Weiss, B. 1978. *Escherichia coli* mutants deficient in deoxyuridine triphosphatase. *J. Bacteriol.* 134:157.

61. Olivera, B. M. 1978. DNA intermediates at the *Escherichia coli* replication fork. Effect of dUTP. *Proc. Natl. Acad. Sci.* (USA) 75:238.

62. Makino, F., and Munakata, N. 1978. Deoxyuridine residues in DNA of thymine-requiring *Bacillus subtilis* strains with defective N-glycosidase activity for uracil-containing DNA. *J. Bacteriol.* 134:24.

63. Tamanoi, F., and Okazaki, T. 1978. Uracil incorporation into nascent DNA of thymine-requiring mutant of *B. subtilis* 168. *Proc. Natl. Acad. Sci.* (USA) 75:2195.

64. Shlomai, J., and Kornberg, A. Unpublished data quoted in Tye, B-K., et al. (59).

65. Lindahl, T. 1974. An N-glycosidase from *Escherichia coli* that releases free uracil from DNA containing deaminated cytosine residues. *Proc. Natl. Acad. Sci.* (USA) 71:3649.

66. Warner, H. R., and Duncan, B. K. 1978. *In vivo* synthesis and properties of uracil-containing DNA. *Nature* 272:32.

67. Brynolf, K., Eliasson, R., and Reichard, P. 1978. Formation of Okazaki fragments in polyoma DNA synthesis caused by misincorporation of uracil. *Cell* 13:573.

68. Ariga, H., and Shimojo, H. 1979. Incorporation of uracil into the growing strand of adenovirus 12 DNA. *Biochem Biophys Res. Comm.* 87:588.

69. Goulian, M., Bleile, B., and Tseng, B. Y. 1980. Methotrexate-induced misincorporation of uracil into DNA. *Proc. Natl. Acad. Sci.* (USA) 77:1956.

70. Myers, C. E., Young, R. C., and Chabner, B. A. 1975. Biochemical determinants of 5-fluorouracil response *in vivo*. The role of deoxyuridylate pool expansion. *J. Clin. Invest.* 56:1231.

71. Jackson, R. C. 1978. The regulation of thymidylate biosynthesis in Novikoff hepatoma cells and the effects of amethopterin, 5-fluorodeoxyuridine, and 3-deazauridine. *J. Biol. Chem.* 253:7440.

72. Goulian, M., Bleile, B., and Tseng, B. Y. 1980. The effect of methotrexate on levels of dUTP in animal cells. *J. Biol. Chem.* 255:10630.

73. Takahashi, I., and Marmur, J. 1963. Replacement of thymidylic acid by deoxyuridylic acid in the deoxyribonucleic acid of a transducing phage for *Bacillus subtilis*. *Nature* 197:794.

74. Kahan, F. M. 1963. Novel enzymes formed by *Bacillus subtilis* infected with bacteriophage. *Fed. Proc.* 22:406.

75. Tomita, F., and Takahashi, I. 1969. A novel enzyme dCTP deaminase, found in *Bacillus subtilis* infected with phage PBS1. *J. Virology* 15:1073.

76. Price, A. R. 1976. Bacteriophage-induced inhibitor of a host enzyme. In *Microbiology 1976*. D. Schlessinger, ed., p. 290. Washington, D.C.: American Soc. Microbiol. Publications.

77. Coulondre, C., Miller, J. H., Farabaugh, P. J., and Gilbert, W. 1978. Molecular basis of base substitution hotspots in *Escherichia coli*. *Nature* 274:775.

78. Friedberg, E. C., Ganesan, A. K., and Minton, K. 1975. N-glycosidase activity in extracts of *Bacillus subtilis* and its inhibition after infection with bacteriophage PBS2. *J. Virology* 16:315.

79. Baltz, R. H., Bingham, P. M., and Drake, J. W. 1976. Heat mutagenesis in bacteriophage T4: the transition pathway. *Proc. Natl. Acad. Sci.* (USA) 73:1269.

80. Jordon, D. O. 1960. *The chemistry of nucleic acids*. Washington, D.C.: Butterworth.

81. Greer, S., and Zamenhof, S. 1962. Studies on depurination of DNA by heat. *J. Mol. Biol.* 4:123.

82. Lindahl, T., and Nyberg, B. 1972. Rate of depurination of native deoxyribonucleic acid. *Biochemistry* 11:3610.

83. Zoltewicz, J. A., Clark, F. O., Sharpless, T. W., and Grahe, G. 1970. Kinetics and mechanism of the acid-catalyzed hydrolysis of some purine nucleosides. *J. Am. Chem. Soc.* 92:1741.

84. Garrett, E. R., and Mehta, P. J. 1972. Solvolysis of adenine nucleosides. II. Effects of sugars and adenine substituents on alkaline solvolysis. *J. Am. Chem. Soc.* 94:8542.

85. Lindahl, T., and Karlström, O. 1973. Heat-induced depyrimidination of DNA. *Biochemistry* 12:5151.

86. Jones, A. S., Mian, A. M., and Walter, R. T. 1968. The alkaline degradation of deoxyribonucleic acid derivatives. *J. Chem. Soc.* (C):2042.

87. Lindahl, T., and Andersson, A. 1972. Rate of chain breakage of apurinic sites in double-stranded DNA. *Biochemistry* 11:3618.

88. Tamm, C., Shapiro, H. S., Lipschitz, R., and Chargaff, E. 1953. Distribution density of nucleotides within a deoxyribonucleic acid chain. *J. Biol. Chem.* 203:673.

89. Male, R., Fosse, V. M., and Kleppe, K. 1982. Polyamine-induced hydrolysis of apurinic sites in DNA and nucleosomes. *Nucleic Acids Res.* 10:6305.

90. Shaham, M., Becker, Y., and Cohen, M. M. 1980. A diffusable clastogenic factor in ataxia telangiectasia. *Cytogenet. Cell Genet.* 27:155.

91. Emerit, I., and Cerutti, P. 1981. Clastogenic activity from Bloom syndrome fibroblast cultures. *Proc. Natl. Acad. Sci.* (USA) 78:1868.

92. Emerit, I., Michelson, A. M., Levy, A., Camus, J. P., and Emerit, J. 1980. Chromosome-breaking agent of low molecular weight in human systemic lupus erythematosus. Protector effect of superoxide dismutase. *Human Genetics* 55:341.

93. Emerit, I., Levy, A., and DeVaux Saint Cyr, C. 1980. Chromosome damaging agent of low molecular weight in the serum of New Zealand black mice. *Cytogenet. Cell Genet.* 26:41.

94. Emerit, I., and Cerutti, P. 1982. Tumor promoter phorbol 12-myrisate 13-acetate induces a clastogenic factor in human lymphocytes. *Proc. Natl. Acad. Sci.* (USA) 79:7509.

95. Borek, C., and Troll, W. 1983. Modifiers of free radicals inhibit *in vitro* the oncogenic actions of X-rays, bleomycin and the tumor promoter 12-O-tetradecanoylphorbol 13-acetate. *Proc. Natl. Acad. Sci.* (USA) 80:1304.

96. Ross, W. C. J. 1962. *Biological alkylating agents.* London: Butterworth.

97. Lawley, P. D. 1966. Effects of some chemical mutagens and carcinogens on nucleic acids. *Prog. Nucleic Acids Res. Mol. Biol.* 5:89.

98. Loveless, A. 1966. *Genetic and allied effects of alkylating agents.* London: Butterworth.

99. Singer, B. 1975. The chemical effects of nucleic acid alkylation and their relation to mutagenesis and carcinogenesis. *Prog. Nuc. Acid Res. Mol. Biol.* 15:219.

100. Swain, C. G., and Scott, C. B. 1953. Quantitative correlation of relative rates. Comparison of hydroxide ion with other nucleophilic reagents toward alkyl halides, esters, epoxides and acyl halides. *J. Am. Chem. Soc.* 75:141.

101. Wang, A. H-J., Quigley, G. J., Kolpak, F. J., Crawford, J. L., van Boom, J. H., van der Marel, G., and Rich, A. 1979. Molecular structure of a left-handed double helical DNA fragment at atomic resolution. *Nature* 282:680.

102. Arnott, S., Chandrasekaran, R., Birdsall, D. L., Leslie, A. G. W., and Ratliff, R. L. 1980. Left-handed DNA helices. *Nature* 283:743.

103. Drew, H., Takano, T., Tanaka, S., Itakura, K., and Dickerson, R. E. 1980. High salt d(CpGpCpG), a left-handed Z DNA double helix. *Nature* 286:567.

104. Crawford, J. L., Kolpak, F. J., Wang, A. H-J., Quigley, G. J., van Boom, J. H., van der Marel, G., and Rich, A. 1980. The tetramer d(CpGpCpG) crystallizes as a left-handed double helix. *Proc. Natl. Acad. Sci.* (USA) 77:4016.

105. Behe, M., and Felsenfeld, G. 1981. Effects of methylation on a synthetic polynucleotide: the B-Z transition in poly(dG-m^{5}dC)·poly(dG-m^{5}dC). *Proc. Natl. Acad. Sci.* (USA) 78:1619.

106. Hamada, H., and Kakunaga, T. 1982. Potential Z-DNA forming sequences are highly dispersed in the human genome. *Nature* 298:396.

107. Singleton, C. K., Klysik, J., Stirdivant, S. M., and Wells, R. D. 1982. Left-handed Z-DNA is induced by supercoiling in physiological ionic conditions. *Nature* 299:312.

108. Leng, M., Freund, A-M., Malfoy, B., Malinge, J-M., Pilet, J., and Rio, P. 1983. Z-DNA: implications for DNA damage and repair. In *Cellular responses to DNA damage*. E. C. Friedberg and B. A. Bridges, eds., p. 23. New York: Alan R. Liss.

109. Geiduschek, E. 1961. "Reversible" DNA. *Proc. Natl. Acad. Sci.* (USA) 47:951.

110. Iyer, V. N., and Szybalski, W. 1963. A molecular mechanism of mitomycin action: linking of complementary DNA strands. *Proc. Natl. Acad. Sci.* (USA) 50:355.

111. Kohn, K. W., Spears, C. L., and Doty, P. 1966. Inter-strand crosslinking of DNA by nitrogen mustard. *J. Mol. Biol.* 19:288.

112. Chun, E. H. L., Gonzales, L., Lewis, F. S., Jones, J., and Rutman, R. J. 1969. Differences in the *in vivo* alkylation and cross-linking of nitrogen mustard-sensitive and resistant lines of Lettré-Ehrlich ascites tumors. *Cancer Res.* 29:1184.

113. Roberts, J. J., and Pascoe, J. M. 1972. Cross-linking of complementary strands of DNA in mammalian cells by anti-tumor platinum compounds. *Nature* 235:282.

114. Cole, R. 1970. Light-induced cross-linking of DNA in the presence of a furocoumarin (psoralen). Studies with phage λ, *Escherichia coli*, and mouse leukemia cells. *Biochim. Biophys. Acta.* 217:30.

115. Marmur, J., and Grossman, L. 1961. Ultraviolet light induced linking of deoxyribonucleic acid strands and its reversal by photoreactivating enzyme. *Proc. Natl. Acad. Sci.* (USA) 47:778.

116. Lett, J. J., Stacey, K. A., and Alexander, P. 1961. Crosslinking of dry deoxyribonucleic acids by electrons. *Radiat. Res.* 14:349.

117. Ball, C. R., and Roberts, J. J. 1971. Estimation of interstrand DNA cross-linking resulting from mustard gas alkylation of HeLa cells. *Chem.-Biol. Interact.* 4:297.

118. Fujiwara, Y. 1983. Measurement of interstrand cross-links produced by mitomycin C. In *DNA repair: a laboratory manual of research procedures*. E. C. Friedberg and P. C. Hanawalt, eds., vol. 2, p. 143. New York: Dekker.

119. Strauss, B. S. 1981. Use of benzoylated naphthoylated DEAE cellulose. In *DNA repair: a laboratory manual of research procedures*. E. C. Friedberg and P. C. Hanawalt, eds., vol 1B, p. 319. New York: Dekker.

120. Smith, K. C., and Hanawalt, P. C. 1969. *Molecular photobiology*. New York: Academic.

121. Kohn, K. W., Ewig, R. A. G., Erickson, L. C., and Zwelling, L. A. 1981. Measurement of strand breaks and cross-links by alkaline elution. In *DNA repair:*

a laboratory manual of research procedures. E. C. Friedberg and P. C. Hanawalt, eds., vol. 1B, p. 379. New York: Dekker.

122. Boutwell, R. K., Colburn, N. H., and Muckerman, C. C. 1969. *In vitro* reactions of β propiolactone. *Ann. N.Y. Acad. Sci.* 163:751.

123. Nietert, W. C., Kellicutt, L. M., and Kubinski, H. 1974. DNA-protein complexes produced by a carcinogen, β-propiolactone. *Cancer Res.* 34:859.

124. Smith, K. C., and Aplin, R. T. 1966. A mixed photoproduct of uracil and cysteine (5,S-cysteine-6-hydrouracil). A possible model for the *in vivo* cross-linking of deoxyribonucleic acid and protein by ultraviolet light. *Biochemistry* 5:2125.

125. Smith, K. C., and O'Leary, M. E. 1967. Photoinduced DNA-protein cross-links and bacterial killing: a correlation at low temperatures. *Science* 155:1024.

126. Setlow, R. B., Swenson, P. A., and Carrier, W. L. 1963. Thymine dimers and inhibition of DNA synthesis by ultraviolet irradiation of cells. *Science* 142:1464.

127. Witkin, E. M. 1966. Radiation-induced mutations and their repair. *Science* 152:1345.

128. Gates, F. L. 1930. A study of the bacterial action of ultraviolet light. III. The absorption of ultraviolet light by bacteria. *J. Gen. Physiol.* 14:31.

129. Miller, E. C. 1978. Some current perspectives on chemical carcinogenesis in humans and experimental animals: presidential address. *Cancer Res.* 38:1479.

130. Hiatt, H. H., Watson, J. D., and Winston, J. A., eds. 1977. *Origins of human cancer,* book B. New York: Cold Spring Harbor Lab.

131. Rehn, L. 1895. Blasengeschwülste bei Fuchsin-Arbeitern. *Arch Klin. Chir.* 50:588.

132. Clayson, D. B. 1962. *Chemical carcinogenesis.* Boston: Little, Brown.

133. Miller, E. C., and Miller, J. A. 1947. The presence and significance of bound aminoazo dyes in the livers of rats fed *p*-dimethylaminoazobenzene. *Cancer Res.* 7:468.

134. Ullrich, V., Roots, I., Hildebrandt, A., Estabrook, R. W., and Conney, A. H., eds. 1977. *Microsomes and drug oxidations.* Oxford: Pergamon Press.

135. Miller, J. A., and Miller, E. C. 1966. In *Biochemical pathology.* E. Farber and P. N. Magee, eds., p. 217. Baltimore: Williams & Wilkins.

136. Kriek, E. 1972. Persistent binding of a new reaction product of the carcinogen N-hydroxy-2-acetylaminofluorene with guanine in rat liver DNA *in vivo. Cancer Res.* 32:2042.

137. Westra, J. G., Kriek, E., and Hittenhausen, H. Identification of the persistently bound form of the carcinogen N-acetyl-2-aminofluorene to rat liver DNA *in vivo. Chem.-Biol. Interact.* 15:149.

138. Grunberger, D., and Santella, R. M. 1982. Alternative conformations of DNA modified by N-2-acetylaminofluorene. In *Mechanisms of chemical carcinogenesis.* C. C. Harris and P. A. Cerutti, eds., p. 155. New York: Alan R. Liss.

139. Wang, A. H-J., Quigley, G. J., Kolpak, F. J., van der Marel, G., and van Boom, J. H. 1981. Left-handed double helical DNA: variations in the backbone conformation. *Science* 211:171.

140. Leng, M., Sage, E., and Rio, P. 1981. DNA chemically modified by N-acetoxy-N-2-acetylaminofluorene: nature of the adducts and conformation of the modified DNA. In *Chromosome damage and repair.* E. Seeberg and K. Kleppe, eds., p. 31. New York: Plenum.

141. Sage, E., and Leng, M. 1980. Conformation of poly(dG·dC)·poly(dG·dC) modified by the carcinogens N-acetoxy-N-2-acetylaminofluorene and N-hydroxy-N-2-aminofluorene. *Proc. Natl. Acad. Sci.* (USA) 77:4597.

142. Butlin, H. T. 1892. Cancer of the scrotum in chimney-sweeps and others. II. Why foreign sweeps do not suffer from scrotal cancer. *Brit. Med. J.* 2:1.

143. Kennaway, E. L. 1925. Experiments on cancer-producing substances. *Brit. Med. J.* 2:1.

144. Kennaway, E. L., and Huger, I. 1930. Carcinogenic substances and their fluorescence spectra. *Brit. Med. J.* 1:1044.

145. Albert, R. E., and Burns, F. J. 1977. Carcinogenic atmospheric pollutants and the nature of low-level risks. In *Origins of human cancer.* H. H. Hiatt, J. D. Watson and J. A. Winston, eds., p. 289. New York: Cold Spring Harbor Lab.

146. Levin, W., Lu, A. Y. H., Ryan, D., Wood, A. W., Kapitulnik, J., West, S., Huang, M-T., Conney, A. H., Thakker, D. R., Holder, G., Yogi, H., and Jerina, D. M. 1977. Properties of the liver microsomal monoxygenase system and epoxide hydrase: factors influencing the metabolism and mutagenicity of benzo(a)pyrene. In *Origins of human cancer.* H. H. Hiatt, J. D. Watson and J. A. Winston, eds., book B, p. 659. New York: Cold Spring Harbor Lab.

147. Selkirk, J. K., Macleod, M. C., Moore, C. J., Mansfield, B. K., Nikbakht, A., and Dearstone, K. 1982. Species variance in the metabolic activation of polycyclic hydrocarbons. In *Mechanisms of chemical carcinogenesis.* C. C. Harris and P. A. Cerutti, eds., p. 331. New York: Alan R. Liss.

148. Weinstein, I. B., Jeffrey, A. M., Jennette, K. W., Blobstein, S. H., Harvey, R. G., Harris, C., Autrup, H., Kasai, H., and Nakanishi, K. 1976. Benzo(a)pyrene diolepoxides as intermediates in nucleic acid binding *in vitro* and *in vivo*. *Science* 193:592.

149. Campbell, T. C., and Stoloff, L. 1974. Implications of mycotoxins for human health. *J. Agric, Food Chem.* 22:1006.

150. Hsieh, D. P. H., Wong, J. J., Wong, A., Michas, C., and Ruebner, B. H. 1977. Hepatic transformation of aflatoxin and its carcinogenicity. In *Origins of human cancer.* H. H. Hiatt, J. D. Watson and J. A. Winston, eds., Book B, p. 697. New York: Cold Spring Harbor Lab.

151. Muench, K. F., Misra, R. P., and Humayun, M. Z. 1983. Sequence specificity in aflatoxin B_1–DNA interactions. *Proc. Natl. Acad. Sci.* (USA) 80:6.

152. Freese, E. 1959. The specific mutagenic effect of base analogues on phage T4. *J. Mol. Biol.* 1:87.

153. Freese, E. 1959. The difference between spontaneous and base analogue induced mutation of phage T4. *Proc. Natl. Acad. Sci.* (USA) 45:622.

154. Freese, E. 1959. On the molecular explanation of spontaneous and induced mutations. *Brookhaven Symp. Biol.* 12:63.

155. Salts, Y., and Ronen, A. 1971. Neighbor effects in the mutation of *ochre* triplets in the T4rII gene. *Mutation Res.* 13:109.

156. Ripley, L. S. 1981. Influence of diverse gene 43 DNA polymerases on the incorporation and replication *in vivo* of 2-aminopurine at A·T base-pairs in bacteriophage T4. *J. Mol. Biol.* 150:197.

157. Berkner, L. V., and Marshall, L. C. 1964. The history of oxygenic concentrations in the earth's atmosphere. *Discuss. Faraday Soc.* 37:122.

158. McLaren, A. D., and Shugar, D. 1964. *Photochemistry of proteins and nucleic acids.* Oxford: Pergamon Press.

159. Smith, K. C. 1964. Photochemistry of nucleic acids. *Photophysiol.* 2:329.

160. Wang, S. Y. 1965. Photochemical reactions of nucleic acid components in frozen solutions. *Fed. Proc.* 24:71.

161. Setlow, J. K. 1966. The molecular basis of biological effects of ultraviolet radiation and photoreactivation. In *Current topics in radiation research.* M. Ebert and A. Howard, eds., vol. II, p. 195. Amsterdam: North-Holland.

162. Johns, H. E. 1967. Photoproducts produced in nucleic acids by ultraviolet light. In *Radiation research.* Proc. Third Int. Congress Radiat. Res., G. Silini, ed., p. 733. Amsterdam: North-Holland.

163. Setlow, R. B. 1966. Cyclobutane-type pyrimidine dimers in polynucleotides. *Science* 153:379.

164. Burr, J. G. 1968. Advances in the photochemistry of nucleic acid derivatives. *Adv. Photochem.* 6:193.

165. Setlow, R. B. 1968. The photochemistry, photobiology, and repair of polynucleotides. *Prog. Nuc. Acid. Res. Mol. Biol.* 8:257.

166. Varghese, A. J. 1972. Photochemistry of nucleic acids and their constituents. *Photophysiol.* 7:207.

167. Löber, G., and Kittler, L. 1977. Selected topics in photochemistry of nucleic acids. Recent results and perspectives. *Photochem. Photobiol.* 25:215.

168. Wacker, A., Dellweg, H., Träger, L., Kornhauser, A., Lodemann, E., Türck, G., Selzer, R., Chandra, P., and Ishimoto, M. 1964. Organic photochemistry of nucleic acids. *Photochem. Photobiol.* 3:369.

169. Khattak, M. N., and Wang, S. Y. 1972. The photochemical mechanism of pyrimidine cyclobutyl dimerization. *Tetrahedron* 28:945.

170. Friedberg, E. C., and Hanawalt, P. C., eds. 1981. *DNA repair—a laboratory manual of research procedures.* vol IA. New York: Dekker.

171. Radany, E. H., Love, J. D., and Friedberg, E. C. 1981. The use of direct photoreversal of UV-irradiated DNA for the demonstration of pyrimidine dimer–DNA glycosylase activity. In *Chromosome damage and repair.* E. Seeberg and K. Kleppe, eds., p. 91. New York: Plenum.

172. Gordon, L. K., and Haseltine, W. A. 1982. Quantitation of cyclobutane pyrimidine dimer formation in double- and single-stranded DNA fragments of defined sequence. *Radiat. Res.* 89:99.

173. Maxam, A. M., and Gilbert, W. 1980. Sequencing end-labeled DNA with base-specific chemical cleavages. In *Methods in enzymology.* L. Grossman and K. Moldave, eds., 65:499.

174. Varghese, A. J. 1970. 5-Thyminyl-5,6-dihydrothymine from DNA irradiated with ultraviolet light. *Biochem. Biophys. Res. Comm.* 38:484.

175. Lippke, J. A., Gordon, L. K., Brash, D. E., and Haseltine, W. A. 1981. Distribution of UV light–induced damage in a defined sequence of human DNA: detection of alkaline-sensitive lesions at pyrimidine nucleoside-cytidine sequences. *Proc. Natl. Acad. Sci.* (USA) 78:3388.

176. Haseltine, W. A. 1983. Site specificity of ultraviolet light induced mutagenesis. In *Cellular responses to DNA damage.* E. C. Friedberg and B. A. Bridges, eds., p. 3. New York: Alan R. Liss.

177. Varghese, A. J., and Wang, S. Y. 1967. Ultraviolet irradiation of DNA *in vitro* and *in vivo* produced a third thymine-derived product. *Science* 156:955.

178. Varghese, A. J., and Wang, S. Y. 1968. Thymine-thymine adduct as a photoproduct of thymine. *Science* 160:186.

179. Brash, D. E., and Haseltine, W. A. 1982. UV-induced mutation hotspots occur at DNA damage hotspots. *Nature* 298:189.

180. Yamane, T., Wyluda, B. J., and Shulman, R. G. 1967. Dihydrothymine from UV-irradiated DNA. *Proc. Natl. Acad. Sci.* (USA) 58:439.

181. Demple, B., and Linn, S. 1982. 5,6-Saturated thymine lesions in DNA: production by ultraviolet light or hydrogen peroxide. *Nucleic Acids. Res.* 10:3781.

182. Demple, B., and Linn, S. 1980. DNA N-glycosylases and UV repair. *Nature* 287:203.

183. Smith, K. C. 1968. The biological importance of UV-induced DNA-protein cross-linking *in vivo* and its probable chemical mechanism. *Photochem. Photobiol.* 7:651.

184. Lamola, A. 1974. Fundamental aspects of the spectroscopy and photochemistry of organic compounds; electronic energy transfer in biologic systems; and photosensitization. In *Sunlight and man.* M. A. Pathak, L. C. Harber, M. Seiji and A. Kukita, eds., p. 17. University of Tokyo Press.

185. Lamola, A. A., and Yamane, T. 1967. Sensitized photodimerization of thymine in DNA. *Proc. Natl. Acad. Sci.* (USA) 58:44.

186. Lamola, A. A. 1969. Specific formation of thymine dimers in DNA. *Photochem. Photobiol.* 9:291.

187. Pathak, M. A., Kramer, D. M., and Fitzpatrick, T. B. 1974. Photobiology and photochemistry of furocoumarins (psoralens). In *Sunlight and man.* M. A. Pathak, L. C. Harber, M. Seiji and A. Kukita, eds., p. 335. Tokyo University Press.

188. Ashwood-Smith, M. J., and Grant, E. 1977. Conversion of psoralen DNA monoadducts in *E. coli* to interstrand DNA crosslinks by near UV light (320–360 nm): inability of angelicin to form crosslinks *in vivo*. *Experientia* 33:384.

189. Cole, R. S. 1971. Psoralen monoadducts and interstrand cross-links in DNA. *Biochim. Biophys. Acta.* 254:30.

190. Sinden, R. R., and Cole, R. S. 1981. Measurement of cross-links formed by treatment with 4,5′,8-trimethylpsoralen and light. In *DNA repair—a laboratory manual of research procedures.* E. C. Friedberg and P. C. Hanawalt, eds., vol. IA, p. 69. New York: Dekker.

191. Cole, R. S. 1973. Repair of DNA containing interstrand cross-links in *Escherichia coli:* sequential excision and recombination. *Proc. Natl. Acad. Sci.* (USA) 70:1064.

192. Cole, R. S., Sinden, R. R., Yoakum, G. H., and Broyles, S. 1978. On the mechanism for repair of cross-linked DNA in *E. coli* treated with psoralen and light. In *DNA repair mechanisms.* P. C. Hanawalt, E. C. Friedberg and C. F. Fox, eds., p. 287. New York: Academic.

193. Magaña-Schwencke, N., Henriques, J-A. P., Chanet, R., and Moustacchi, E. 1982. The fate of 8-methoxypsoralen photo-induced crosslinks in nuclear and mitochondrial yeast DNA: comparison of wild-type and repair-deficient strains. *Proc. Natl. Acad. Sci.* (USA) 79:1722.

194. Miller, R. D., Prakash, L., and Prakash, S. 1982. Genetic control of excision of *Saccharomyces cerevisiae* interstrand DNA cross-links induced by psoralen plus near-UV light. *Mol. Cell Biol.* 2:939.

195. Kaye, J., Smith, C. A., and Hanawalt, P. C. 1980. DNA repair in human cells containing photoadducts of 8-methoxypsoralen or angelicin. *Cancer Res.* 40:696.

196. Hanawalt, P. C., Kaye, J., Smith, C. A., and Zolan, M. 1981. Cellular responses to psoralen adducts in DNA. In *Psoralens in cosmetics and dermatology*. J. Cahn, P. Forlot, C. Grupper, A. Meybeck and F. Urbach, eds., p. 133. France: Pergamon Press.

197. Hutchinson, F. 1973. The lesions produced by ultraviolet light in DNA containing 5-bromouracil. *Quart. Rev. Biophys.* 6:201.

198. Adams, G. E. 1972. Radiation chemical mechanisms in radiation biology. *Adv. Rad. Chem.* 3:125.

199. Cerutti, P. A. 1975. DNA base damage induced by ionizing radiation. In *Photochemistry and photobiology of nucleic acids*. S. Y. Wang, ed., p. 375. New York: Academic.

200. Emmerson, P. T. 1972. X-ray damage to DNA and loss of biological function: effect of sensitizing agents. *Adv. Rad. Chem.* 3:209.

201. Latarjet, R. 1972. Interaction of radiation energy with nucleic acids. *Curr. Top. Radiat. Res. Quart.* 8:1.

202. Myers, L. S., Jr. 1972. Radiation chemistry of nucleic acids, proteins, and polysaccharides. In *The radiation chemistry of macromolecules*. M. Dole, ed., vol. 2, p. 323. New York: Academic.

203. Scholes, G. 1975. The radiation chemistry of pyrimidines, purines and related substances. In *Photochemistry and photobiology of nucleic acids*. S. Y. Wang, ed., p. 521. New York: Academic.

204. Smith, K. C. 1975. The radiation-induced addition of proteins and other molecules to nucleic acids. In *Photochemistry and photobiology of nucleic acids*. S. Y. Wang, ed., p. 187. New York: Academic.

205. Hariharan, P. V., and Cerutti, P. A. 1972. Formation and repair of γ-ray-induced thymine damage in *Micrococcus radiodurans*. *J. Mol. Biol.* 66:65.

206. Cerutti, P. A., and Remsen, J. F. 1976. Gamma-ray excision repair in normal and diseased human cells. In *Biology of radiation carcinogenesis*. J. M. Yuhas, R. W. Tennant and J. D. Regan, eds., p. 93. New York: Raven Press.

207. Mattern, M. R., Hariharan, P. V., and Cerutti, P. A. 1975. Selective excision of gamma ray damaged thymine from the DNA of cultured mammalian cells. *Biochim. Biophys. Acta.* 395:48.

208. Cerutti, P. A., and Hariharan, P. V. 1974. Excision of damaged thymine residues from gamma-irradiated poly (dA-dT) by crude extracts of *Escherichia coli*. *Proc. Natl. Acad. Sci.* (USA) 71:3532.

209. West, G. J., West, I. W-L., and Ward, J. R. 1982. Radioimmunoassay of a thymine glycol. *Radiat. Res.* 90:595.

210. Breimer, L., and Lindahl, T. 1980. A DNA glycosylase from *Escherichia coli* that releases free urea from a polydeoxyribonucleotide containing fragments of base residues. *Nucleic Acids Res.* 8:6199.

211. Holian, J. Q., and Garrison, W. 1966. Radiation-induced oxidation of cytosine and uracil in aqueous solution of copper (II). *Nature* 212:394.

212. Scholer, G., Ward, J. F., and Weiss, J. J. 1960. Mechanism of the radiation-induced degradation of nucleic acids. *J. Mol. Biol.* 2:379.

213. Hems, G. 1958. Effect of ionizing radiation on aqueous solution of guanylic acid and guanosine. *Nature* 181:1721.

214. Van Hemmen, J. J., and Bleichradt, J. F. 1971. The decomposition of adenine by ionizing radiation. *Radiat. Res.* 46:444.

215. Chetsanga, C. J., and Lindahl, T. 1979. Release of 7-methylguanine residues whose imidazole rings have been opened, from damaged DNA by a DNA glycosylase from *Escherichia coli*. *Nucleic Acids Res.* 6:3673.

216. Hems, G. 1960. Effects of ionizing radiation on aqueous solutions of inosine and adenosine. *Radiat. Res.* 13:777.

217. Ward, J. F., and Urist, M. M. 1967. γ-Irradiation of aqueous solutions of polynucleotides. *Int. J. Radiat. Biol.* 12:209.

218. Koch, C. J. 1979. The effect of oxygen on the repair of radiation damage by cells and tissues. *Adv. Rad. Biol.* 8:273.

219. Bonura, T., Youngs, D. A., and Smith, K. C. 1975. R.b.e. of 50 KVp X-rays and 660 KeV γ-rays (^{137}Cs) with respect to the production of DNA damage, repair and cell-killing in *Escherichia coli* K-12. *Int. J. Radiat. Biol.* 28:539.

220. Youngs, D. A., and Smith, K. C. 1976. The yield and repair of X-ray-induced single-strand breaks in the DNA of *Escherichia coli* K-12 cells. *Radiat. Res.* 68:148.

221. Bonura, T., and Smith, K. C. 1976. The involvement of indirect effects in cell-killing and DNA double-strand breakage in γ-irradiated *Escherichia coli* K-12. *Int. J. Radiat. Biol.* 29:293.

222. Kornberg, R. D. 1977. Structure of chromatin. *Ann. Rev. Biochem.* 46:931.

223. Stryer, L. 1981. *Biochemistry*. 2nd ed. San Francisco: Freeman.

224. Lieberman, M. W., Smerdon, M. J., Tlsty, T. D., and Oleson, F. B. 1979. The role of chromatin structure in DNA repair in human cells damaged with chemical carcinogens and ultraviolet radiation. In *Environmental carcinogenesis*. P. Emmelot and E. Kriek, eds., p. 345. Amsterdam: Elsevier/North-Holland Biomedical Press.

225. Hanawalt, P. C., Cooper, P. K., Ganesan, A. K., and Smith, C. A. 1979. DNA repair in bacteria and mammalian cells. *Ann. Rev. Biochem.* 48:783.

226. Williams, J. I., and Friedberg, E. C. 1979. Deoxyribonucleic acid excision repair in chromatin after ultraviolet irradiation of human fibroblasts in culture. *Biochemistry* 18:3965.

227. Friedberg, E. C., and Hanawalt, P. C., eds. 1981. *DNA repair—a laboratory manual of research procedures*. Vol. 1B. New York: Dekker.

228. Friedberg, E. C., and Hanawalt, P. C. eds. 1983. *DNA repair—a laboratory manual of research procedures*. Vol. 2. New York: Dekker.

229. Muller, R., and Rajewsky, M. F. 1981. Antibodies specific for DNA components structurally modified by chemical carcinogens. *J. Cancer Res. Clin. Oncol.* 102:99.

230. Mitchell, D. L., and Clarkson, J. M. 1981. The development of a radioimmunoassay for the detection of photoproducts in mammalian cell DNA. *Biochim. Biophys. Acta.* 655:54.

231. Ley, R. D. 1983. Immunological detection of two types of cyclobutane pyrimidine dimers in DNA. *Cancer Res.* 43:41.

232. Hsu, I-C., Poirier, M. C., Yuspa, S. H., Grunberger, D., Weinstein, I. B., Yolken, R. H., and Harris, C. C. 1981. Measurement of benzo(a)pyrene-DNA adducts by enzyme immunoassays and radioimmunoassay. *Cancer Res.* 41:1091.

233. Briscoe, W. T., Spizizen, J., and Tan, E. M. 1978. Immunological detection of O^6-methylguanine in alkylated DNA. *Biochemistry.* 17:1896.

234. Sawicki, D. L., Beiser, S. M., Srinivasan, D., and Srinivasan, P. R. 1976. Immunochemical detection of 7-methylguanine residues in nucleic acids. *Arch. Biochem. Biophys.* 176:457.

235. Leng, M., Sage, E., Fuchs, R. P. P., and Duane, M. P. 1978. Antibodies to DNA modified by the carcinogen N-acetoxy-N-2-acetylaminofluorene. *FEBS Lett.* 92:207.

236. West, G. J., West, I. W-I., and Ward, J. F. 1982. Radioimmunoassay of 7,8-dihydro-8-oxyadenine (8-hydroxyadenine). *Int. J. Radiat. Biol.* 42:481.

237. Seidman, M., Mizusawa, H., Slor, H., and Bustin, M. 1983. Immunological detection of carcinogen-modified DNA fragments after *in vivo* modification of cellular and viral chromatin. *Cancer Res.* 43:743.

238. Hertzog, P. J., Lindsay Smith, J. R., and Garner, R. C. 1982. Production of monoclonal antibodies to guanine imidazole ring-opened aflatoxin B_1 DNA, the persistent DNA adduct *in vivo*. *Carcinogenesis* 3:825.

239. Poirier, M. D., Lippard, S. J., Zwelling, L. A., Ushay, H. M., Kerrigan, D., Thill, C. C., Santella, R. M., Grunberger, D., and Yuspa, S. H. 1982. Antibodies elicited against *cis*-diamminedichloroplatinum(II)-modified DNA are specific for *cis*-diamminedichloroplatinum(II)-DNA adducts formed *in vivo* and *in vitro*. *Proc. Natl. Acad. Sci.* (USA) 79:6443.

240. Boyle, J. M., and Strickland, P. T. 1981. Characterization of two monoclonal antibodies specific for dimerised and non-dimerised adjacent thymidines in single stranded DNA. *Photochem. Photobiol.* 34:595.

241. Haugen, A., Groopman, J. D., Hsu, I-H., Goodrich, G. R., Wogan, G. N., and Harris, C. C. 1981. Monoclonal antibody to aflatoxin B_1–modified DNA detected by enzyme immunoassay. *Proc. Natl. Acad. Sci.* (USA) 78:4124.

242. Duncan, B. K. 1981. DNA glycosylases. In *The enzymes*. 3rd ed. P. D. Boyer, ed., vol. XIV, p. 565. New York: Academic.

243. Seawell, P. C., and Ganesan, A. K. 1981. Measurement of strand breaks in supercoiled DNA by gel electrophoresis. In *DNA repair—a laboratory manual of research procedures*. E. C. Friedberg and P. C. Hanawalt, eds., vol. I, part B, p. 425. New York: Dekker.

244. Maniatis, T., Fritsch, E. F., and Sambrook, J. 1982. *Molecular cloning: a laboratory manual*. New York: Cold Spring Harbor Lab.

245. Clayton, D. A. 1981. Measurement of strand breaks in supercoiled DNA by electron microscopy. In *DNA repair—a laboratory manual of research procedures*. E. C. Friedberg and P. C. Hanawalt, eds., vol I, part B, p. 419. New York: Dekker.

246. McCormick, J., Heflich, R. H., and Maher, V. M. 1983. Use of S_1 endonuclease to quantify carcinogen-induced lesions in DNA. In *DNA repair—a laboratory manual of research procedures*. E. C. Friedberg and P. C. Hanawalt, eds., vol. 2, p. 3. New York: Dekker.

247. Wei, C-F., Legerski, R. J., Robberson, D. L., Alianell, G. A., and Gray, H. B., Jr. 1983. In *DNA repair—a laboratory manual of research procedures*. E. C. Friedberg and P. C. Hanawalt, eds., vol. 2, p. 13. New York: Dekker.

248. Setlow, R. B., and Carrier, W. L. 1966. Pyrimidine dimers in ultraviolet-irradiated DNA's. *J. Mol. Biol.* 17:237.

249. Smith, K. C. 1974. Molecular changes in the nucleic acids produced by ultraviolet and visible radiation. In *Sunlight and man*. M. A. Pathak, L. C. Harber, M. Seiji and A. Kukita, eds., p. 57. University of Tokyo Press.

250. Musajo, L., Rodighiero, G., Caporale, G., Dall'acqua, F., Marciani, S., Bordin, F., Baccichetti, F., and Bevilacqua, R. 1974. Photoreactions between skin-photosensitizing furocoumarins and nucleic acid. In *Sunlight and man*. M. A. Pathak, L. C. Harber, M. Seiji and A. Kukita, eds., p. 369. University of Tokyo Press.

251. Chetsanga, C. J., Lozon, M., Makaroff, C., and Savage, L. 1981. Purification and characterization of *Escherichia coli* formamidopyrimidine—DNA glycosylase that excises damaged 7-methylguanine from deoxyribonucleic acid. *Biochemistry* 20:5201.

252. Nebert, D. W., Negishi, M., Lang, M. A., Eisen, H. J., and Okey, A. B. 1980. Multiple forms of inducible drug-metabolizing enzymes: a reasonable mechanism by which any organism can cope with environmental adversity. *Abstracts: 1980 Intra-science symposium—new directions in cancer causation.* February 6–8, 1980, Santa Monica, Calif.

2

DNA Repair by Reversal of Damage

2-1 Introduction

Now that we have some idea of the spectrum of damages that the cellular genome may sustain, we can begin a consideration of the cellular responses to such damage. In this regard it must be noted that the term *DNA repair* is used throughout this text in a strict biochemical sense to specify only *those cellular responses associated with the restoration of the normal nucleotide sequence and chemistry of DNA following damage.* A variety of other cellular responses are described in the literature as examples of DNA repair because they are usually accompanied by enhanced survival of the affected cells relative to cells that are genetically or biochemically blocked in these responses. Some of these responses may be accompanied by the true repair of DNA strand breaks and/or gaps in the nucleotide sequence that *result* from biochemical events associated with the response(s) to DNA damage. However, these repair events do not usually include the physical removal of the *initial* damage present in the DNA, and for this reason they are considered as part of a series of DNA damage *tolerance* mechanisms rather than as DNA repair mechanisms. These tolerance mechanisms are discussed in Chapters 7 and 8.

TABLE 2-1
Cellular responses to DNA damage

1. Repair of DNA damage
A. Reversal of damage
Enzymatic photoreactivation
Repair of O^6-alkylguanine
Purine insertion
Ligation of DNA strand breaks
B. Excision of damage
Excision mediated by DNA glycosylases with or without AP endonucleases (base excision repair)
Excision mediated by direct-acting damage-specific endonucleases (nucleotide excision repair); mismatch excision repair
2. Tolerance of DNA damage
A. Replicative bypass of template damage with gap formation
B. Translesion DNA synthesis

Table 2-1 summarizes in general terms the spectrum of responses by living cells to DNA damage. Each of these is discussed in the ensuing chapters. Note that DNA repair as defined here can occur by one of two fundamental cellular responses that involve either the *reversal* of DNA damage or its *excision*. To use a crude analogy in order to clarify this distinction at the outset, consider damage to be represented by a knot in a length of twine (DNA). In some cases it may be possible to undo the knot (reversal of damage); in others it may be necessary to cut out a piece of string containing the knot and replace it with a new segment of twine (excision of damage and synthesis of new nucleotides).

In principle, the simplest biochemical mechanism by which damage to DNA might be repaired is one in which a single enzyme catalyzes a single reaction that restores the structure of the genome to its normal state. The requirement for only one biochemical event should provide a significant kinetic and energetic advantage over more complex, multistep reactions involving a number of different enzymes. In addition, the potential for introducing errors into the coding elements of the damaged gene during the course of repair would be expected to be reduced, and with some mechanisms, perhaps eliminated entirely. In the present chapter we shall consider four known examples of this phenomenon, all of which are characterized by the direct reversal of DNA damage. In this discussion, as well as in later discussions of excision repair, particular emphasis is placed on *base* damage in DNA. The repair of these moieties has been most extensively studied in view of their obvious biological importance as the coding elements of the genome. However, other forms of DNA damage, such as strand breaks, are also of significance inasmuch as they can interfere with the normal transaction of DNA functions such as replication and transcription.

2-2 Enzymatic Photoreactivation of Pyrimidine Dimers

As indicated in the previous chapter, cyclobutane-type pyrimidine dimers are believed to constitute a major source of biologically relevant genetic damage following the exposure of living cells to UV radiation at wavelengths near the absorption maximum of DNA. Pyrimidine dimers in template DNA strands block both the progression of the constitutive DNA polymerases of *E. coli* and DNA replication (1–6). A particularly sensitive and elegant technique for demonstrating this in vitro (shown in Fig. 2-1) once again exploits the resolving power of DNA-sequencing technology (6). The obvious physiological implication of these observations (1–6) is that unless pyrimidine dimers are removed from DNA prior to replication, or can be tolerated as nonblocking lesions during replication, affected cells are faced with reproductive death. In view of the central role that UV radiation has played as a source of genetic damage during biological evolution, it is not surprising that most (if not all) living organisms have evolved a highly specific mechanism for the repair of pyrimidine dimers. This mechanism is called *enzymatic photoreactivation* (EPR) of DNA (7–11) and is the first example to be discussed of DNA repair by the direct reversal of base damage.

EPR is a light-dependent process involving the enzyme-catalyzed monomerization of *cis-syn* cyclobutyl pyrimidine dimers (Fig. 2-2). This phenomenon should not be confused with a number of other light-dependent processes that do not involve enzymes by which pyrimidine dimers can be monomerized (Table 2-2). For example, irradiation of DNA containing pyrimidine dimers at wavelengths between 200 and 300 nm results in the direct reversal of some fraction of the dimers until a new equilibrium between monomerization and dimerization is attained (12). This phenomenon is referred to as *direct photoreversal*, but it is not believed to be a biologically important process. Another non-enzyme-dependent reaction that splits pyrimidine dimers in DNA is called *sensitized photoreversal*. This process has been observed with tryptophan and tryptophan-containing oligopeptides such as lys-trp-lys (13–17), as well as with a protein coded by gene *32* of phage T4, which is particularly rich in tryptophan (18). The mechanism of this reaction is thought to in-

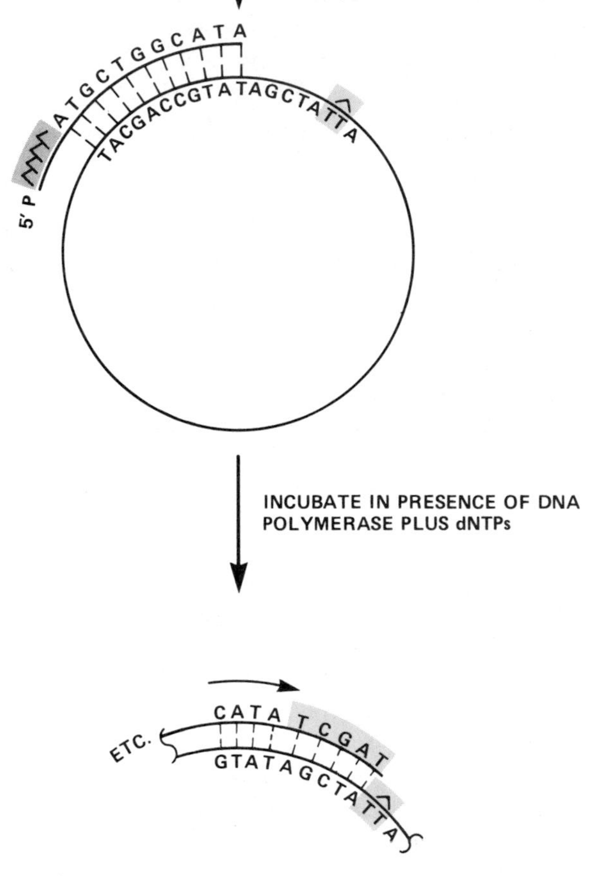

FIGURE 2-1
Demonstration of replicative arrest in vitro at sites of pyrimidine dimers in template DNA. A single-strand DNA template of known nucleotide sequence is exposed to UV radiation to produce pyrimidine dimers. A defined primer labeled at its 5′ end is annealed to the template and is extended in the 5′ → 3′ direction by DNA synthesis, using a well-characterized DNA polymerase such as *E. coli* DNA I polymerase or phage T4 DNA polymerase. Following completion of synthesis the extended primer is recovered and its sequence established to determine the sites of replicative arrest relative to the location of dimers in the template.

volve an electron transfer from the excited indole ring of tryptophan to the dimer, possibly mediated by base-stacking interactions of the tryptophanyl residue with the pyrimidine dimer in DNA. The photosensitized splitting of pyrimidine dimers is not observed at wavelengths greater than 300 nm. This wavelength criterion has been used to distinguish the phenomenon from "true" EPR, which always utilizes photons of wavelength greater than 300 nm (17, 19). The simple tripeptide-mediated reaction is interesting and from an evolutionary

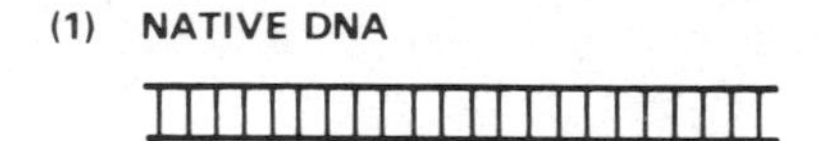

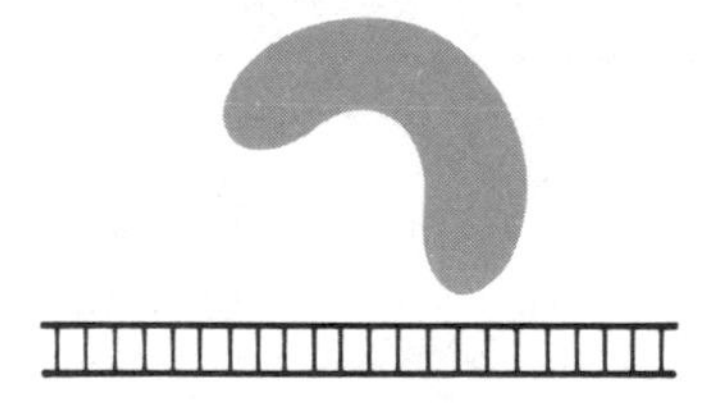

FIGURE 2-2
Schematic illustration of the enzyme-catalyzed monomerization of pyrimidine dimers, an example of DNA repair by the reversal of base damage. (Courtesy of Dr. P. C. Hanawalt.)

TABLE 2-2
Light-dependent monomerization of pyrimidine dimers in DNA

	Direct Photoreversal	Sensitized Photoreversal	Enzymatic Photoreactivation
Wavelength dependence	~240 nm	<200 nm	>300 nm
Enzyme requirement	No	No	Yes
Photosensitizer requirement	No	Yes	Yes*

*The photosensitizer is the enzyme.

standpoint may have served as the start of the development of EPR involving more complex proteins.

Historically, EPR was the first DNA repair mode to be discovered. In the late 1940s Albert Kelner, then at the Cold Spring Harbor Biological Laboratory, was studying the effects of UV radiation on certain strains of *Streptomyces griseus*, a fungus (20). While investigating the influence of postirradiation temperature on survival, Kelner was plagued by another experimental variable, the systematic examination of which ultimately led to the recognition of the role of light in the recovery of UV-irradiated cells. Kelner's own description of this fascinating discovery merits quotation (20):

> Careful consideration was made of variable factors which might have accounted for such tremendous variation. We were using a glass-fronted water bath placed on a table near a window, in which were suspended transparent bottles containing the irradiated spores. The fact that some of the bottles were more directly exposed to light than others suggested that light might be a factor. Moreover, the greatest and most consistent recovery in our preliminary experiments had taken place in suspensions stored in transparent bottles at room temperature on an open shelf exposed to diffuse light from a window. Experiment showed that exposure of ultra-violet irradiated suspensions to light resulted in an increase in survival rate or a recovery of 100,000- to 400,000-fold. Controls kept in the dark . . . showed no recovery at all.
>
> The magnitude of the light effect can hardly be overemphasized. The recovery was so much more complete than any previously observed, that we felt we were dealing here with a key factor in the mechanism causing inactivation and recovery from ultra-violet irradiation.

At about the same time, Renato Dulbecco, then in the Department of Bacteriology at Indiana University, reported the same phenomenon in the T group of coliphages and coined the term *photoreactivation* (21). He wrote:

> The occurrence of photo-reactivation of ultra-violet irradiated phage was noticed accidentally a few weeks after [I received] a personal communication from Dr. A. Kelner that he had discovered recovery of ultra-violet treated spores of *Actinomycetes* upon exposure to visible light. I am informed by Dr. Kelner that his results are in course of publication. My observation indicates the correctness of Dr. Kelner's suggestion that the phenomenon discovered by him may be of general occurrence for a number of biological objects.

It was not until the late 1950s, however, that studies with extracts of *E. coli* and of baker's yeast demonstrated that photoreactivation of pyrimidine dimers is an enzyme-mediated phenomenon (22, 23).

The enzyme activity that catalyzes photoreactivation of dimers is referred to as photoreactivating enzyme or *DNA photolyase* (7–11, 24). DNA photolyase activity has been detected in a large number of plant and animal cell extracts. These include algae, bacteria, yeast, fungi, protozoa, molluscs, arthropods, teleosts, amphibians, reptiles, birds, marsupials and placental mammals (9). Table 2-3 provides acomprehensive summary of the distribution in nature of both the biological phenomenon of EPR and the occurrence of DNA photolyase activity. In the phylogenetic survey shown in the table, the presence of DNA photolyase activity was in some instances demonstrated in cell-free extracts using one of a variety of enzyme assays,

TABLE 2-3
Occurrence of DNA Photoreactivating Enzyme Activity

Phylum or Class	Species	Tissues Possessing Enzyme Activity	Tissues Failing to Show Activity
Cyanophyta	*Plectonema boreanum* (blue-green alga)	Cells	—
	Anacystis nidulans (blue-green alga)	Cells	—
Schizomycophyta	*Escherichia coli* (colon bacterium)	Cells	—
	Streptomyces griseus (soil actinomycete)	Cells	—
	Serratia marcescens (bacterium)	Cells	—
	Haemophilus influenzae (respiratory tract bacterium)	—	Cells
	Diplococcus pneumoniae (respiratory tract bacterium)	—	Cells
	Bacillus subtilis (soil bacterium)	—	Vegetative cells
	Micrococcus luteus (bacterium)	—	Cells
	Micrococcus radiodurans (bacterium)	—	Cells
Eumycophyta	*Saccharomyces cerevisiae* (baker's yeast)	Cells	—
	Schizosaccharomyces pombe (fission yeast)	—	Cells
	Neurospora crassa (bread mold)	Condidia	—

TABLE 2-3 (Cont.)

Phylum or Class	Species	Tissues Possessing Enzyme Activity	Tissues Failing to Show Activity
Euglenophyta	*Euglena gracilis*	Cells	—
Angiospermae	*Phaseolus vulgaris* (pinto bean)	Plumule, hypocotyl, cotyledon of dark-grown sprout, young leaves	Radicles, older leaves
	Phaseolus lunatus (lima bean)	Whole sprout	—
	Zea mays (maize)	Harvested pollen	—
	Nicotiana tabacum (tobacco)	Cultured cells	—
	Phaseolus aureus (mung bean)	—	Whole sprout
	Haplopappus gracilis	—	Cultured cells
Gymnospermae	*Gingko biloba* (gingko tree)	Cultured cells	—
Protozoa	*Paramecium aurelia* (paramecium)	Cells	—
	Tetrahymena pyriformis	Cells	—
Mollusca	*Physa sp* (pond snail)	Muscle	—
Echinodermata	*Arbacia punctulata* (sea urchin)	Testis, eggs, ovary	Sperm
	Echinarachnius parma (sand dollar)	Egg, early embryo	—
Arthropoda	*Anagasta kiihniella* (flower moth)	Adult female abdomen	—
	Gecarcinus lateralis (land crab)	Testis, ovary, epithelium, somatic and heart muscle	Midgut gland
	Artemia salina (brine shrimp)	Whole nauplii	—
	Homarus americanus (lobster)	Ovary	—
Chordata			
Bony fishes (Teleost)	*Haemulon sciurus* (bluestriped grunt)	Established dorsal fin line	—
(Teleost)	*Poecilia formosa*	Whole fish	—
(Teleost)	*Pimephales promelas* (fathead minnow)	Established epithelial cell line	—
Amphibia	*Bufo marinus* (cowflop toad)	White blood cells	Blood serum, red blood cells
	Xenopus laevis (African clawed toad)	Established liver cell line	—
	Rana pipiens (frog)	Skeletal and cardiac muscle, sciatic nerve, brain, liver	—

Table 2-3 (Cont.)

Phylum or Class	Species	Tissues Possessing Enzyme Activity	Tissues Failing to Show Activity
Reptilia	*Terrapene carolina* (box turtle)	Established heart cell line	—
	Iguana iguana (lizard)	Established heart cell line	—
	Gekko gekko (lizard)	Established lung cell line	—
Aves (birds)	*Gallus gallus* (domestic chicken)	Primary fibroblasts, whole 4-day embryos, adult brain, primary embryo cultures	Kidney, liver, skeletal muscle, egg white, egg yolk
Mammalia			
(Marsupial)	*Didelphis marsupialis* (American opossum)	Brain, liver, kidney, testis, heart, lung	—
(Marsupial)	*Caluromys derbianus* (South American woolly opossum)	Established kidney line	—
(Marsupial)	*Potorous tridactylis* (Tasmanian rat, kangaroo, or potoroo)	Two established cell lines (male and female)	—
(Rodent)	*Mus musculus* (mouse)		
(Rodent)	*Rattus novegicus* (rat)	[Evidence has been provided for the presence of DNA photolyase activity in extracts of cells and tissues from a variety of placental mammals, including those listed here. The phenomenon of EPR in living cells and tissues from these species is still controversial however.]	
(Rodent)	*Cricetulus griseus* (Chinese hamster)		
(Lagomorph)	*Oryctolagus cuniculus* (rabbit)		
(Ungulate)	*Bos taurus* (domestic beef)		
(Primate)	*Homo sapiens* (human)		

After C. S. Rupert, ref. 9.

discussed below. In other cases the presence of the enzyme was inferred from studies on intact cells in which the light-dependent loss of dimers from DNA or the light-dependent biological recovery from the effects of UV radiation was observed. However, it must be remembered (as pointed out above) that following UV irradiation of cells, not all light-dependent recovery phenomena necessarily reflect EPR specifically. Despite these varied experimental approaches, there are a number of organisms at different levels of biological organization in which there is as yet no evidence for the presence of the

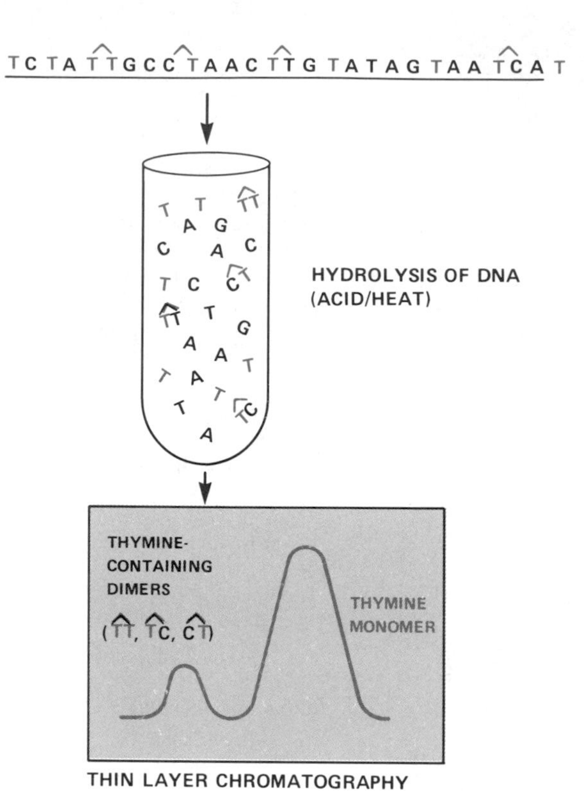

FIGURE 2-3
Schematic illustration of the measurement of thymine-containing pyrimidine dimers in DNA with thin layer chromatography. DNA radiolabeled in thymine is exposed to UV radiation to produce pyrimidine dimers, some of which contain radiolabeled thymine. The DNA is hydrolyzed in strong acid at high temperature, resulting in the preservation of structurally intact radiolabeled thymine monomer and of pyrimidine dimers. These species can be resolved and thymine-containing pyrimidine dimers quantitated with a variety of chromatographic techniques, including the use of thin layer plates. The thymine dimer content of the DNA is typically expressed as the fraction of total radioactivity present as thymine-containing dimers.

enzyme (Table 2-3). This does not necessarily imply its absence in these cases, but rather the failure to demonstrate its presence by the techniques used so far.

Assays for DNA Photolyase Activity

Restoration of Transforming Ability of DNA. In this assay a purified transforming DNA carrying a known genetic marker is UV-irradiated and incubated in vitro with DNA photolyase in the presence of photoreactivating light. The DNA is then used to transform an appropriate recipient cell defective in the same genetic marker (22, 23, 25, 26). The extent of EPR is expressed quantitatively as the enhanced acquisition of the transformed phenotype by the recipient cells as a function of increasing photolyase concentration in the incubation mixture.

Membrane-Binding Assay. Photoreactivating wavelengths of light are required for the splitting (monomerization) of pyrimidine dimers by all known DNA photolyases (7–11, 24). However, the binding of

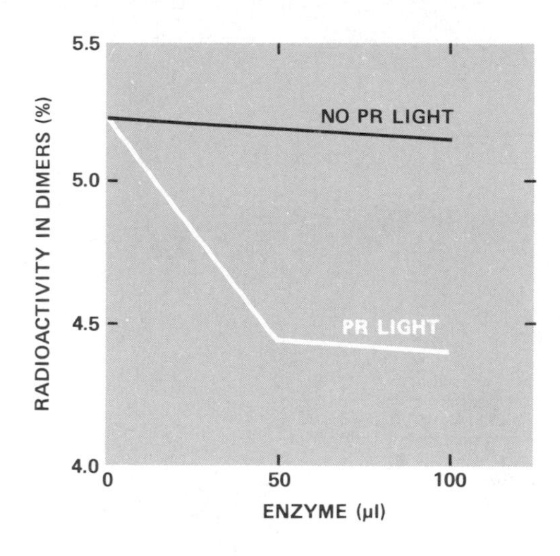

FIGURE 2-4
Loss of thymine-containing pyrimidine dimers from acid-precipitable DNA after incubation of UV-irradiated DNA with DNA photolyase from human cells in the presence of photoreactivating (PR) light. (From B. M. Sutherland, ref. 73.)

the enzyme to dimers in duplex DNA occurs in the absence of photoreactivating light (7–11, 24). Because of this property it is possible to assay photolyase activity by measuring the binding of complexes of enzyme with radioactively labeled DNA to filters, under conditions in which noncomplexed DNA is not retained on the filters. Purified yeast enzyme and [^{3}H]T7 DNA at a concentration of one photolyase-dipyrimidine complex per DNA molecule bind with an efficiency of 20 to 25 percent (27, 28). Although this assay is useful for measuring the binding and dissociation constants of purified enzyme, it has limited application for enzyme purification since other proteins that bind to DNA may cause significant interference.

Light-Dependent Loss of Thymine-Containing Pyrimidine Dimers from DNA. A number of procedures have been established for the chromatographic separation of radiolabeled thymine from thymine-containing pyrimidine dimers following the acid hydrolysis of UV-irradiated DNA at high temperature (Fig. 2-3) (29, 30). Thus, after incubation of UV-irradiated DNA with photolyase, the light-dependent loss of dimers from the acid-insoluble fraction can be quantitated (Fig. 2-4). Since the most convenient way of specifically labeling DNA is with ^{3}H or ^{14}C thymine, such procedures can measure the photoreactivation of T<>T and C<>T, but not of C<>C. A modified form of this assay utilizes DNA labeled in the deoxyribose-phosphate backbone with ^{32}P (Fig. 2-5) (31, 32). In UV-irradiated DNA the intradimer phosphodiester bond is uniquely resistant to hydrolysis by pancreatic DNase plus snake venom phosphodiesterase, so that these phosphates are left in covalent linkage to two

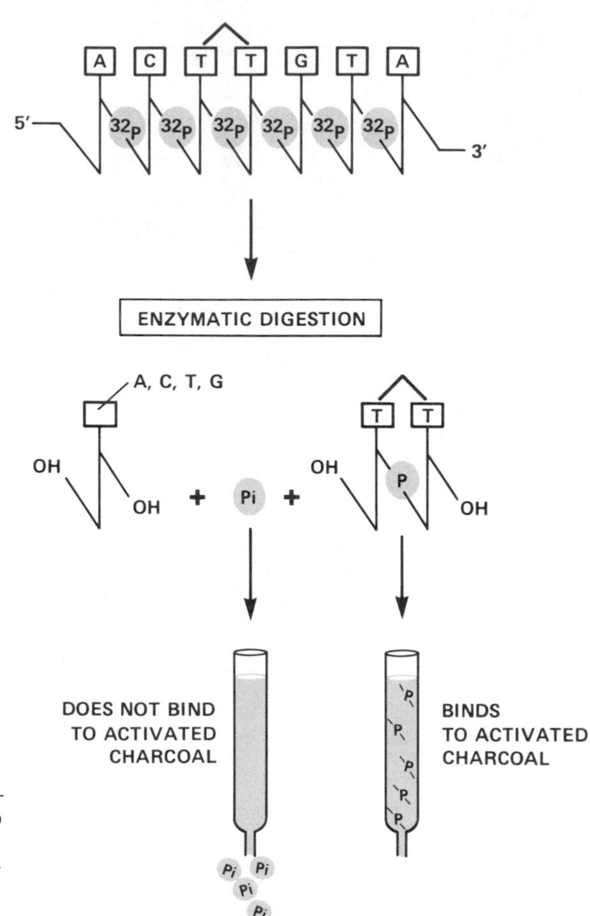

FIGURE 2-5
Measurement of pyrimidine dimers in DNA can be achieved by using DNA radiolabeled with ^{32}P. The DNA is enzymatically digested with nucleases and alkaline phosphatase to yield free inorganic phosphate and dinucleoside monophosphates containing radioactive phosphate associated with pyrimidine dimers (organic phosphate). The inorganic phosphate does not bind to activated charcoal, whereas the organic phosphate does. Thus the amount of bound ^{32}P is representative of the concentration of pyrimidine dimers in the DNA.

pyrimidine nucleoside residues after treatment with the enzymes (33). The remainder of the DNA can be digested to mononucleotides by these two enzymes and then to nucleosides by further incubation with alkaline phosphatase. Following the complete enzymatic digestion of such DNA, the only labeled phosphate that binds to activated charcoal is that associated with pyrimidine dimers, since inorganic phosphate does not bind to this substance. The charcoal-bound radioactivity can then be eluted and quantitated as a measure of *all* pyrimidine dimer species in DNA before and after photoreactivation (31, 32).

TABLE 2-4
Properties of purified DNA photolyase activities

	Yeast	*S. griseus*	*E. coli*	Human Cells
Molecular weight	Multiple forms with MW of 30 kDa, 130 kDa, 51 kDa reported.	~43 kDa	35.2 kDa	40 kDa
Structural components	30 kDa enzyme has uncharacterized associated "small" component. 130 kDa enzyme has subunits of 60 and 85 kDa and is possibly associated with low MW "activator." 51 kDa enzyme has an associated low MW flavin moiety.	No subunits; associated with low MW fluorescent cofactor	Associated with low MW RNA cofactor	
pH optimum		7.0	7.2	7.2
pI			5–5.1	5.4
Optimal ionic strength for enzyme activity (molar)		0.04	0.18	0.05
Absorption of light at photoreactivating wavelengths by purified enzyme	51 kDa enzyme absorbs light maximally at 380 nm.	Absorbs light maximally at 554 nm	None detected	None detected

Increased DNA Template Activity for RNA Polymerase. UV irradiation of DNA reduces the template activity for RNA transcription by *E. coli* RNA polymerase in vitro (34, 35). Enzymatic photoreactivation of the DNA results in the monomerization of pyrimidine dimers and hence allows for increased ribonucleotide incorporation into RNA. This provides a rapid and simple assay for DNA photolyase activity (36).

The Properties and Mechanism of Action of DNA Photolyases

DNA photolyase has been purified and extensively characterized from baker's yeast, *E. coli*, *S. griseus* and human cells. It is instructive to consider in some detail the properties of the activity from each of these sources (Table 2-4).

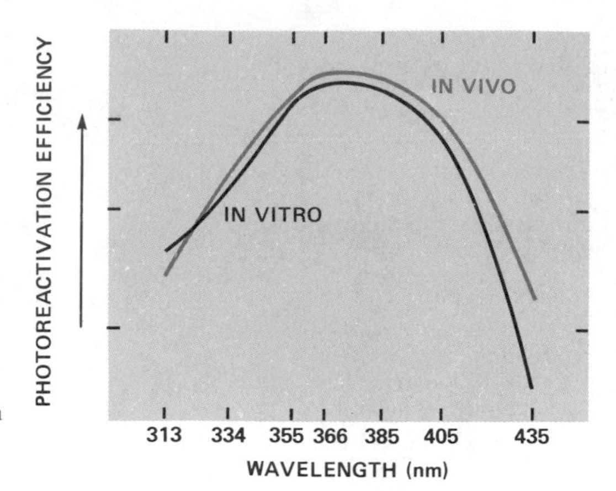

FIGURE 2-6
The action spectra for photoreactivation in the yeast S. *cerevisiae* in vivo and with yeast extracts. (After W. Harm, ref. 51.)

DNA Photolyase from Yeast. Yeast is one of the organisms in which DNA photolyase was first investigated in cell-free extracts (23), but its mechanism of action is still not understood. Different studies have yielded rather different results, and it is not clear whether these reflect the presence of multiple enzymes in yeast or a significant biochemical complexity associated with a single form of DNA photolyase.

Early studies demonstrated a requirement for light at wavelengths of about 350 nm (37), and this observation, coupled with the results of action spectrum analysis (Fig. 2-6), suggested that the purified enzyme contains a prosthetic group or *chromophore* that absorbs light at photoreactivating wavelengths (38). DNA photolyase activity was purified over 3000-fold from commercial baker's yeast and showed the presence of both a large component of molecular weight 30,000 and a low molecular weight component (39). However, the large component did not absorb at photoreactivating wavelengths, even under conditions in which the enzyme and UV-irradiated DNA were preincubated in the dark. The smaller component did absorb light, but at optimal wavelengths much longer than the optimum for enzymatic photoreactivation (39).

In a different study, extensive purification of DNA photolyase from yeast with affinity chromatography on UV-irradiated DNA yielded a single protein, as determined by gradient gel electrophoresis and by sedimentation in 5 to 20% sucrose gradients (40). The molecular weight of this apparently homogeneous enzyme is 130,000. Following polyacrylamide gel electrophoresis in sodium dodecylsulfate, two components with molecular weights of 60,000 and 85,000 were resolved (40). Subunits with these approximate molecular weights

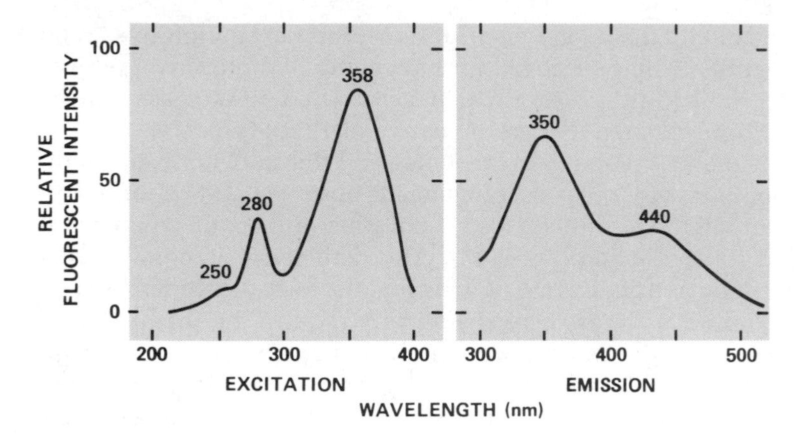

FIGURE 2-7
The fluorescence spectra of an aqueous solution (pH 7.0) of activator III of yeast DNA photolyase. The activator absorbs at 248–250 nm in neutral solution. However, there is only a small fluorescence excitation maximum in this region (left figure). The principal excitation peaks are at 280 and 358 nm. Following excitation of activator III at 290 nm, emission maxima are observed at 350 and 440 nm (right). (After H. Werbin and J. J. Madden, ref. 42.)

were also observed following sedimentation of the enzyme in sucrose containing a high concentration of salt (1.0 *M* KCl). Neither of the individual subunits alone was active; however, mixing of the two fractions resulted in a time-dependent restoration of DNA photolyase activity (40). This purified enzyme also did not show absorbance of light in the photoreactivating wavelength range, but it did show an excitation maximum at 358 nm for a fluorescence emission maximum at 440 nm. This observation led to the postulation that the fluorescent material associated with the purified enzyme might be the chromophore responsible for photoreactivation (41, 42).

Purification of this enzyme by affinity chromatography was consistently accompanied by a considerable loss of activity, suggesting that the putative chromophore may be noncovalently bound and hence susceptible to loss during protein fractionation. When enhancement of DNA photolyase activity was used as an assay, activators of the enzyme activity were isolated from acidified cell-free extracts of yeast (42–44). A component referred to as *activator III*, with a molecular weight of 450 daltons, emits fluorescence at wavelengths of 350 nm and 440 nm when excited at 290 nm (Fig. 2-7), and at 440 nm when excited at 358 nm. In neutral or acid solution the activator absorbs maximally at 248 to 250 nm. However, in alkaline solution or in relatively nonpolar media there is a large increase in the absorptivity of activator III in the near-UV region (about 360 nm), where photoreactivation is maximal. Whether or not this activator is a true fluorescent chromophore of yeast DNA photolyase is difficult to evaluate at present. It does enhance enzyme activity at concentrations that are roughly equimolar with those of the enzyme, and it activates only the highly purified and not the crude enzyme. However, although both the activator and the photolyase fluoresce, their excitation spectra do not reflect the action spectrum for enzymatic photoreactivation (see Fig. 2-6). In addition activators of yeast DNA photolyase can be isolated from mutants that are defective in EPR (44).

Despite these apparent inconsistencies, if the activator is the natural chromophore for the yeast enzyme, its observed absorption at 248 to 250 nm might shift to 260 to 265 nm (that of the enzyme) when it is bound to DNA photolyase (43). Alternatively, it is possible that absorption by active enzyme at wavelengths near the photoreactivation maximum for yeast may specifically require the interaction of the enzyme with its substrate. As indicated above, near-UV absorption can be induced in the activator by lowering the dielectric constant of the medium, and possibly a similar phenomenon occurs on binding of the enzyme to its substrate in vivo. Evidence that a chromophore is constituted following the binding of a particular DNA photolyase to pyrimidine dimers in DNA is presented in Section 2-2, "DNA Photolyase from *E. coli*." It is instructive to note at this juncture that the *E. coli* DNA photolyase also contains a dissociable low molecular weight cofactor that is absolutely required for enzymatic activity.

A single protein of molecular weight 51,000 with an absorption maximum at 380 nm has been isolated from baker's yeast by yet another purification protocol (45). On denaturation of this species by heat or 8 *M* urea, oxidized flavin adenine dinucleotide was identified by paper chromatography. This observation and the similarities in the UV and emission spectra of this form of yeast DNA photolyase with those of proteins in which the flavin has been identified as *4a,5-reduced flavin* suggest that the reduced form of this compound may also be present in DNA photolyase (45).

Our understanding of DNA photolyases from yeast is obviously very incomplete, and further study is required in this area. Conceivably yeast contains more than one such enzyme, or perhaps a single species is composed of subunits that can be individually purified, giving rise to the conflicting results described. In this respect it is of interest that two genetically distinct mutants of the yeast *Saccharomyces cerevisiae* defective in EPR exist (46, 47), and two distinct DNA photolyases have been isolated from yeast by chromatography on phosphocellulose (48). It is to be hoped that the detailed characterization of these two enzymes will clarify some of the apparent contradictions in the literature.

KINETICS OF DIMER MONOMERIZATION BY YEAST DNA PHOTOLYASE. Yeast enzyme has been used both to measure the rate constant for the formation of the intermediate constituted by the binding of DNA photolyase to pyrimidine dimers in DNA in the dark and to measure the rate constants for the formation of the reaction product, i.e., DNA

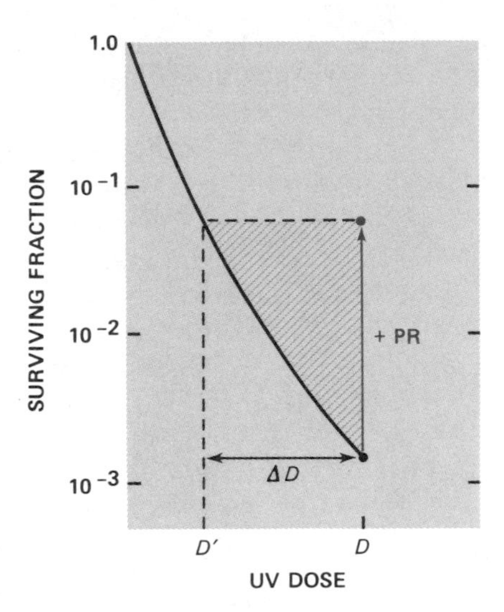

FIGURE 2-8
The parameter ΔD is derived as follows. A survival curve is generated over a range of UV doses. For a selected dose D, the survival of cells exposed to photoreactivating conditions (+PR) after UV irradiation is also measured to yield the data point shown in blue. In order to establish the dose of UV light that would have resulted in the same level of survival without PR, the extrapolation indicated by the dotted lines is made. This generates the UV dose D'. ΔD is the difference between D' and D. (After W. Harm, ref. 51.)

containing monomerized dimers (49–51). The reaction scheme for yeast DNA photolyase with DNA containing pyrimidine dimers can be described by the equation

$$E + S \underset{k_2}{\overset{k_1}{\rightleftharpoons}} ES \xrightarrow[k_3]{\text{light}} E + P$$

where S is the substrate (pyrimidine dimer), P is the product (monomerized pyrimidine) and E is the enzyme (51).

The fact that the light-dependent step is preceded by a light-independent one during which the enzyme-substrate (ES) complex forms has facilitated a detailed kinetic analysis of these steps by the use of intense flashes of photoreactivating light of millisecond duration (49–51). Continuous illumination cannot be used for such studies since enzyme molecules liberated by photolysis of existing ES complexes will enter the reaction again by forming new complexes. A single flash of light of adequate intensity permits photoreactivation by all ES complexes existing at that time. Therefore, the number of complexes can be determined by measuring the disappearance of pyrimidine dimers from DNA after a *single* light flash. In practice this parameter can be calculated from the extent of a selected biological effect, such as increased survival of UV-irradiated bacteria or phage, or increased activity of UV-irradiated transforming DNA. This biological effect can be expressed quantitatively by the parameter ΔD (the difference between the UV dose D actually applied and a smaller UV dose D' that would have led to the same effect without subsequent EPR) (Fig. 2-8). Thus ΔD (the photoreactivatable fraction) is an expression of the amount of UV radiation (or number of pyrimidine dimers) whose effect is mitigated by a single light flash. Under conditions where all enzyme is bound to substrate as ES complex (achieved by establishing a state of substrate excess), the maximal level of ΔD obtained by a single light flash is a measure of the number of DNA photolyase molecules.

When light flashes are given in relatively rapid sequence, the short time interval between flashes permits formation of only a few *ES* complexes relative to the number of enzyme molecules and substrate pyrimidine dimers (*S*) in the DNA present in a reaction. Hence the concentration of free enzyme (*E*) is roughly constant and approximately equal to the *total* concentration of enzyme. Under these conditions and with the assumption that the dissociation of *ES* complexes in the absence of photoreactivating light is negligible (see below), the kinetics of the loss of dimers with time is essentially first order and can be described by the equation

$$[S]_t/[S]_0 = e^{-k_1[E]t}$$

where $[S]_t/[S]_0$ is the experimentally obtained value of $1 - (\Delta D/\Delta D_{max})$, with $[S]_t$ = substrate concentration at time *t* and $[S]_0$ = starting substrate concentration. (Note that ΔD_{max} corresponds to the complete loss of dimers from DNA, or $[S]_t = 0$). The k_1 value so obtained for yeast enzyme in vitro is 6×10^7 L·mol^{-1}·s^{-1} at room temperature (51).

To measure the rate constant k_2, a great excess of competing substrate is introduced into the reaction after reaching the dark equilibrium of complex formation, so that enzyme that dissociates in the dark is likely to bind with the competing DNA, the repair of which is not measured by the assay. The resultant decrease in the number of *ES* complexes is measured as a decrease in EPR achieved after the light flash. The values obtained with yeast enzyme are on the order of 10^{-2} to 10^{-3} s^{-1} (51). The high k_1 and low k_2 values indicate an extremely high degree of stability of *ES* complexes under nonphotoreactivating conditions. Thus the equilibrium constant for complete complex formation in the dark is on the order of 10^{10} L·(L·mol^{-1}) mol^{-1} (51).

Finally, the rate constant k_3 for photolysis of pyrimidine dimers in the *ES* complex can be expressed by the product k_pI, where *I* equals light intensity and k_p equals the "photolytic constant." When all substrate is complexed with the enzyme prior to illumination (experimentally achieved by using *excess* enzyme), the application of light flashes results in the loss of *ES* complex as a function of light dose (*L*). Thus

$$[ES]_L/[ES]_0 = e^{-k_pL}$$

from which k_p can be calculated.

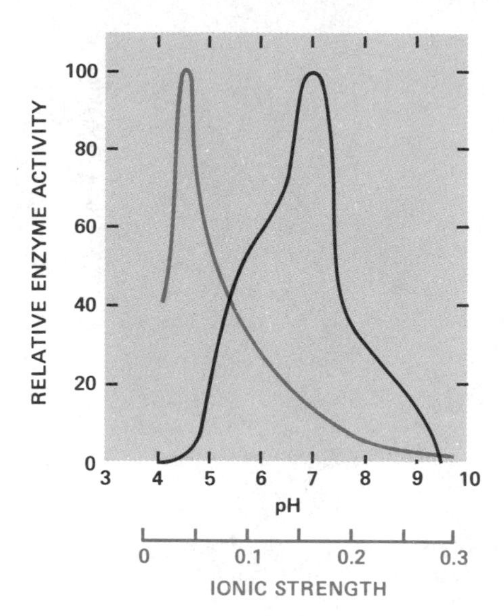

FIGURE 2-9
The effect of pH and ionic strength on the activity of DNA photolyase from S. *griseus*. Note that both parameters drastically affect enzyme activity over relatively narrow ranges. Thus the demonstration of DNA photolyase activity in cell-free extracts is critically dependent on the choice of correct ionic strength and pH conditions. (From A. P. M. Eker, ref. 53.)

The parameter k_p is a measure of the *efficiency* of the use of light in the reaction and reflects two characteristics (51): (1) the extent to which complexes absorb incident photons (expressed by their molar extinction coefficient ϵ) and (2) the probability with which an absorbed photon leads to monomerization of a dimer (expressed by the quantum yield ϕ). Experimentally, k_p measurements at the optimal wavelength for photoreactivation by yeast DNA photolyase yield values of about 10^4 L·mol^{-1}·cm^{-1}. Since the quantum yield (ϕ) is reasonably estimated at between 10^{-1} and 1 (51), such k_p values indicate that the absorption of light and its use in dimer monomerization are highly efficient.

DNA Photolyase from S. griseus. It will be recalled that the mold S. *griseus* was the first organism in which the phenomenon of EPR was recognized (20). In contrast to the yeast DNA photolyase, in which the moiety that absorbs light at photoreactivating wavelengths is still controversial, the enzyme purified and characterized from *S. griseus* presents a much clearer picture. The molecular weight of the purified enzyme is estimated at 43,000 by gel filtration and at 49,000 by polyacrylamide gel electrophoresis in sodium dodecylsulfate (52,

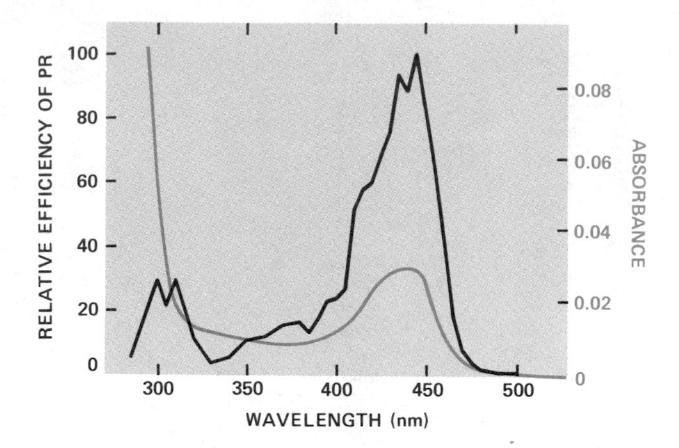

FIGURE 2-10
The absorption spectrum and the action spectrum of purified native DNA photolyase from *S. griseus* show a good correspondence, suggesting that the chromophoric moiety that absorbs maximally at 445 nm is involved in the reaction catalyzed by the enzyme.

53). Thus there is no evidence for subunit structure in this protein. The activity has a significant dependence on pH, with a sharp optimum at pH 7.0 (Fig. 2-9). The enzyme is also very sensitive to ionic strength. Maximal activity is observed at ionic strength (μ) = 0.04. At values of μ of about 0.08, 60 percent of the optimal activity is lost (Fig. 2-9) (53).

DNA photolyase from *S. griseus* exhibits an absorption spectrum in the visible region with a maximum of 445 nm and a shoulder at 425 nm, coinciding very well with its action spectrum (Fig. 2-10) (53, 54). Freshly prepared native enzyme has a fluorescence emission that is considerably enhanced by enzyme denaturation. Consistent with this observation, denaturation of the enzyme is accompanied by the release of a highly fluorescent low molecular weight compound (Fig. 2-11). This low molecular weight, slightly yellow-colored cofactor shows strong absorption in the visible region. The observed variations of absorption maxima of the cofactor with pH are very similar to those of synthetic 7,8-didemethyl-8-hydroxy-5-deazariboflavin Ib (Fig. 2-12) and to those of a substance called SF420 (Fig. 2-12), which can be isolated in relatively large amounts from *S. griseus* cells (55). The fluorescence emission spectra of these three compounds are identical, and the close structural relationship between the factor released from denatured DNA photolyase and SF420 has been further substantiated by the demonstration of very similar pK_a values. In fact, purified SF420 alone can photosensitize the monomerization of

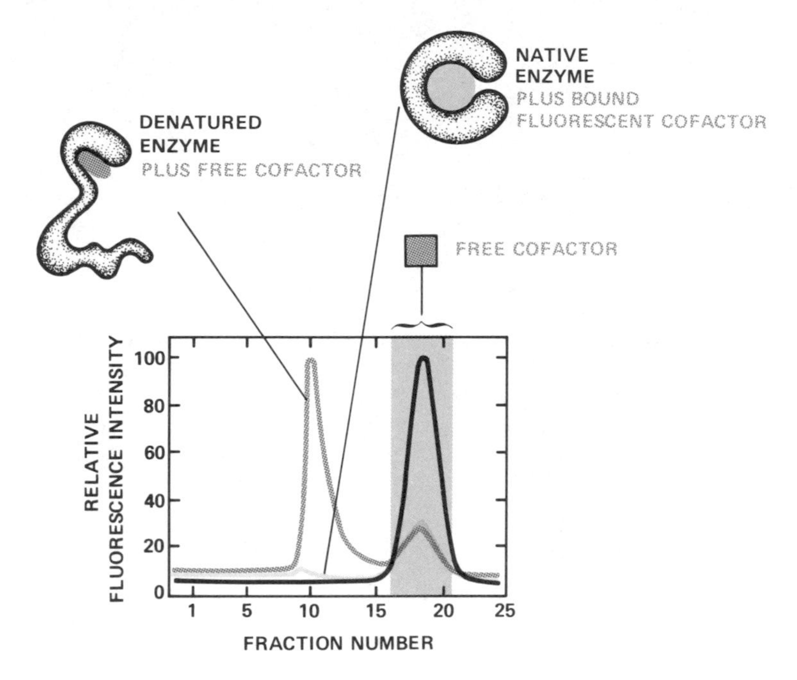

FIGURE 2-11
Release of a fluorescent compound (presumably the chromophore) following heat denaturation of native DNA photolyase from S. *griseus*. Enzyme was filtered through a Sephadex G-50 column and subjected to various postfiltration temperatures with a heating coil placed between the column and a fluorimeter. When the coil temperature is 10°C, (nondenaturing conditions) native enzyme retains bound chromophore and yields a very small peak of fluorescence at about fraction 10 gray curve. Some low molecular weight fluorescent material (presumably free chromophore that was resolved from the enzyme by gel filtration) is also identified in later fractions. When the coil temperature is 65°C, (denaturing conditions) the enzyme denatures and the free chromophore is identified by its intense fluorescence in the fractions corresponding to the molecular weight of native enzyme (fractions 10 through 12) (blue curve). If the enzyme is denatured *prior* to gel filtration, all fluorescent material that filters through the Sephadex column is in the free form and is observed at the expected position of the low molecular weight chromophore (black curve). (After A. P. M. Eker, ref. 53.)

thymine-thymine dimers (in the free-base form) in the presence of blue light (55).

This cofactor appears to be the chromophore required for EPR in S. *griseus*. It is apparently not a contaminant of the purified enzyme since cochromatography of EPR activity in vitro and of the enzyme-bound chromophore occurs in each of a number of different chromatographic systems (54). In addition, the cofactor must be bound to the protein in order for EPR activity to be demonstrable (54). Release of cofactor and loss of activity show a linear relation (Fig. 2-13), and

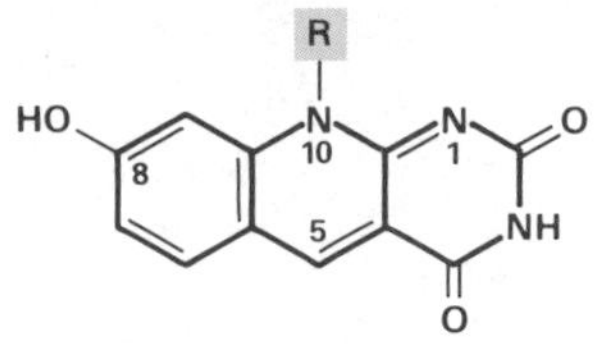

R		
=	UNIDENTIFIED	(DNA PHOTOLYASE COFACTOR)
=	RIBITYL	(7,8-DIDEMETHYL-8-HYDROXY-5-DEAZARIBOFLAVIN 1b)
=	RIBITYL-P-LACTYL (glu)$_n$ (where n = 1, 2 or 3)	(SF420)

FIGURE 2-12
The absorption spectra of the *S. griseus* cofactor at various pHs are very similar to the spectra of synthetic 7,8-didemethyl-8-hydroxy-5-deazariboflavin Ib and of a substance called SF420 normally present in *S. griseus*. Findings from absorption spectroscopy, proton NMR spectroscopy and the pK_a values derived from partial acid hydrolysis indicate that the proposed structures of these related compounds are those shown above. (From A. P. M. Eker, ref. 55.)

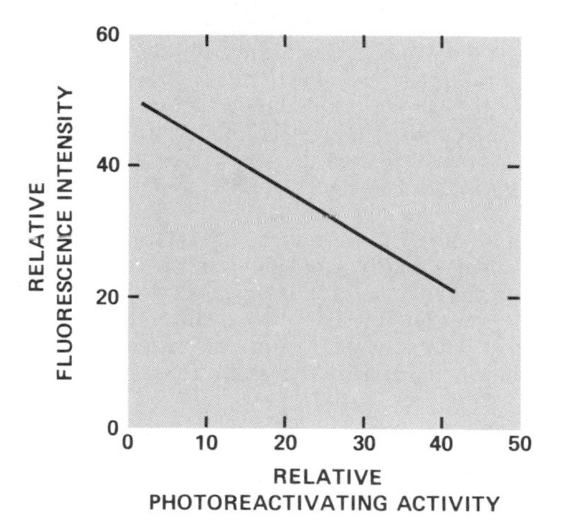

FIGURE 2-13
A direct correlation exists between the loss of DNA photolyase activity and the increase in fluorescence intensity of a solution of enzyme from *S. griseus*. The solution was kept at 25°C and at regular intervals samples were monitored for enzyme activity and fluorescence intensity. (From A. P. M. Eker, ref. 54.)

this release is almost completely inhibited by the addition of UV-irradiated DNA. Finally, as indicated above, there is a good correspondence between the action spectrum for EPR and the absorption spectrum of highly purified native DNA photolyase in the visible and near-UV region of the electromagnetic spectrum, despite the fact that the emission spectra of the native enzyme and the cofactor isolated from the denatured enzyme do not coincide. Attempts to achieve reconstitution between the isolated protein and chromophore have not yet been successful, nor is it clear exactly how this enzyme catalyzes monomerization of pyrimidine dimers in DNA (53–55).

DNA Photolyase from E. coli. Although *E. coli* was the first organism in which DNA photolyase activity was identified in vitro (22), it was not until relatively recently that this enzyme was extensively purified and characterized. The primary reason for this experimental difficulty is the presence of very low levels of DNA photolyase under normal growth conditions. This problem was overcome by constructing a λ bacteriophage carrying the *E. coli phr* gene (the gene for DNA photolyase) and introducing it into *E. coli*, where it becomes integrated into the genome as a prophage. After induction of the lysogen, about 2000 times as much enzyme activity is produced compared with that in the parental nonlysogenic strain (56). The construction of this lysogen led to the observation that the *phr* gene is located between *gal* and *att* at 17 minutes on the *E. coli* genetic map (56). However, the *phr* gene has also been mapped at about 15.7 minutes on the *E. coli* chromosome (57, 58). This apparent discrepancy has been investigated further. Strains carrying a deletion in the region of DNA including 17 but not 15.7 minutes on the *E. coli* genetic map contain approximately 20 percent of normal DNA photolyase levels and exhibit photoreactivation of UV-induced killing at approximately 20 percent of the normal rate (59). The enzyme activity from a strain carrying the deletion has an apparent K_m two- to threefold greater than that of enzyme preparations from wild-type strains and also has a markedly increased sensitivity to heat. These results suggest the presence of multiple loci affecting photoreactivation in *E. coli*. The gene that maps at 17 minutes is thus designated *phrA* and that mapping at 15.7 minutes *phrB* (Fig. 2-14) (59). The question as to whether or not multiple *phr* genes exist in *E. coli* certainly merits further exploration.

With the advent of recombinant DNA technology, the limitations to obtaining sufficient DNA photolyase from *E. coli* for biochemical and enzymologic studies have been potentially circumvented by the successful cloning of the *phrB* gene into the plasmid pMB9 (60–62). The recombinant plasmid (pCSR604) exists at a level of about 10 copies per cell, and the number of DNA photolyase molecules in transformed cells is reflected by this gene dosage effect. Further

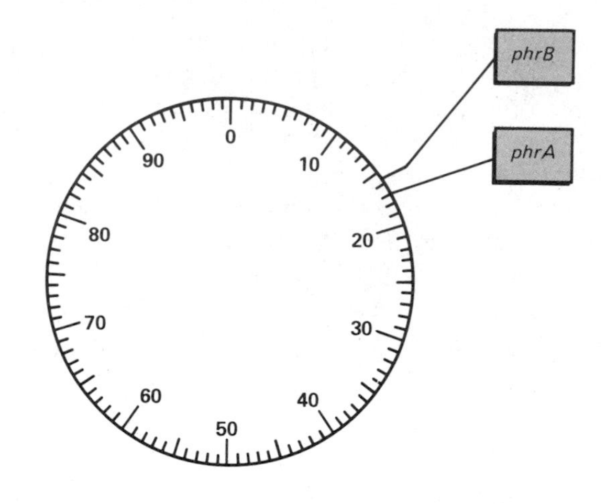

FIGURE 2-14
The linkage map of *E. coli* from zero to one hundred minutes showing the location of the *phrA* and *phrB* genes. (Adapted from A. Kornberg, ref. 185.) The location of other genes involved in DNA repair and in the tolerance of DNA damage is shown in a subsequent series of figures throughout the text.

amplification of expression of the cloned gene can be achieved by transformation of an *E. coli* host with a mutation in the *purA* gene. This increases *phr* gene expression by at least two orders of magnitude by a mechanism that is not understood (62).

DNA photolyase from *E. coli* carrying λ prophage with the putative *phrA* gene has been purified to apparent homogeneity (63–65). The purified enzyme consists of an apoprotein of molecular weight 35,200 and of a low molecular weight *cofactor* that can be readily dissociated from the apoprotein by dialysis or heating (63, 64). The native enzyme has a molecular weight of 36,800 and a sedimentation coefficient ($S^{\circ}_{20,w}$) of 3.75 (65). Amino acid analysis reveals an apparent absence of tryptophan and a low content of aromatic residues. The purified enzyme contains up to 13 percent carbohydrate by weight (65). Neither the apoprotein nor the cofactor exhibits appreciable absorption at wavelengths greater than 300 nm, whereas the optimal wavelength for EPR in *E. coli* is 360 nm (66, 67). Removal of the cofactor from the holoenzyme results in loss of enzyme activity, which is restored by readdition of the cofactor to the apoprotein. The cofactor has been identified as RNA with partial duplex character, which melts at 14°C in 0.01 *M* ionic strength at pH 7.0 (64, 65, 68). The difference spectrum at 4°C between the absorption spectra of native enzyme and heat-treated enzyme fits a superimposition of reference spectra for denaturation of A·U and G·C base pairs derived

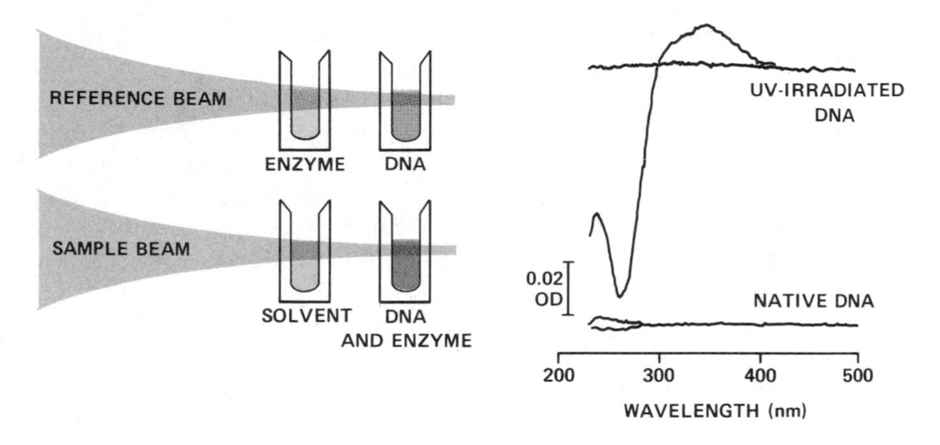

FIGURE 2-15
Absorption difference spectra of *E. coli* DNA photolyase mixed with either UV-irradiated or native DNA. The difference spectra were obtained by using a double-beam spectrophotometer, in which the reference beam measures the absorption of enzyme and DNA independently while the sample beam measures the absorption of DNA in the presence of enzyme. When the DNA is UV-irradiated, a new absorption band is detected at approximately 360 nm. This band is not present when the enzyme is incubated with native DNA. The baselines for each experiment were obtained by placing identical absorbing solutions in both the sample and the reference beam of the double-beam spectrophotometer. (After J. C. Sutherland, ref. 71.)

from model polynucleotides. Calculations from the reference spectra suggest that A·U and G·C base pairs are present in equal concentrations in *E. coli* DNA photolyase and that approximately 20 percent are in a double-strand (presumably RNA) conformation in the native enzyme (68).

There is a difference in the ratio of circular dichroism to absorption for the cofactor and for the holoenzyme, suggesting that when the cofactor is bound to the apoprotein, it is in a different conformation from that when it is free in solution (68). The *E. coli* DNA photolyase activity has a pH optimum at 7.2, an ionic strength optimum of 0.18 and an isoelectric point of 5 to 5.1. The enzyme has a strong tendency to concentration-dependent self-association, forming multiples of the monomeric molecular weight (63, 64).

A chromophore specifically associated with purified *E. coli* DNA photolyase has not been identified. How, then, does photoreactivating light interact with the enzyme? By absorption difference spectroscopy, absorption of light is detected in the spectral region required for enzyme catalysis (approximately 360 nm) when DNA photolyase is mixed with UV-irradiated DNA (69, 70) (Fig. 2-15). This spectral shift is accompanied by a concomitant decrease in absorption at wavelengths greater than 300 nm. In addition, a reduction in the magnitude of the difference spectrum accompanies the enzyme-catalyzed monomerization of pyrimidine dimers, a result expected if the absorption at 360 nm reflects the photoreactivation mechanism (69, 70). These observations are consistent with the idea that the *E. coli* DNA photolyase acts by forming a *charge-transfer complex* between a pyrimidine dimer in DNA and some portion of the enzyme when both components are in their ground electronic states (Fig. 2-16). In this state there could be partial transfer of an electron from one component to the other, resulting in the appearance of a new charge-transfer absorption band, rendering the dimer partially ionized, unstable and thus monomerizable by incident radiant energy (71) (Fig. 2-16).

Plasmid pCSR604, carrying the cloned *phr B* gene, complements *phr* point mutants as well as deletion mutants that map near 15.7 minutes on the *E. coli* chromosome (62). There is evidence that this plasmid, as well as derivative plasmids, carries the structural gene for the apoenzyme. Whether they also carry the gene(s) for a RNA cofactor has not been established (62).

The Chromophore Moiety in DNA Photolyases: Concluding Comments

The failure to detect a distinct chromophoric cofactor in yeast, *E. coli* and human (see below) DNA photolyases suggests that the presence of an identifiable dissociable chromophoric moiety in the *S. griseus* enzyme may be the exception rather than the rule for this class of enzymes. In this regard it may be useful to reevaluate the interpretation of the action spectra for these enzymes. The *absolute* action spectra for photoenzymatic splitting of pyrimidine dimers in natural DNA and in the synthetic polymers poly(dA)·poly(dT) and poly(dG)·poly(dC) differ profoundly with any single preparation of enzyme (72), suggesting that this parameter does not represent exclusively the absorption spectrum of a putative chromophore carried by a given DNA photolyase, nor is it solely determined by the nature of the substrate photoproducts. Rather it should be considered as a reflection of the overall polynucleotide structure (including its exact helical dimensions and the pattern of bases) determining a ground-state interaction between the enzyme and substrate in the enzyme-substrate complex (72).

Another unsolved problem is the mechanism of recognition of pyrimidine dimers in DNA by DNA photolyases. If it is assumed that the dimer itself is directly involved in the recognition step, it is pertinent to ask, What common structural features do all pyrimidine dimers that are monomerized by the enzyme share? In T<>T, T<>C, C<>T and C<>C and in dimers containing uracil, only positions 1, 2 and 6 in the pyrimidine ring are common to all (69)

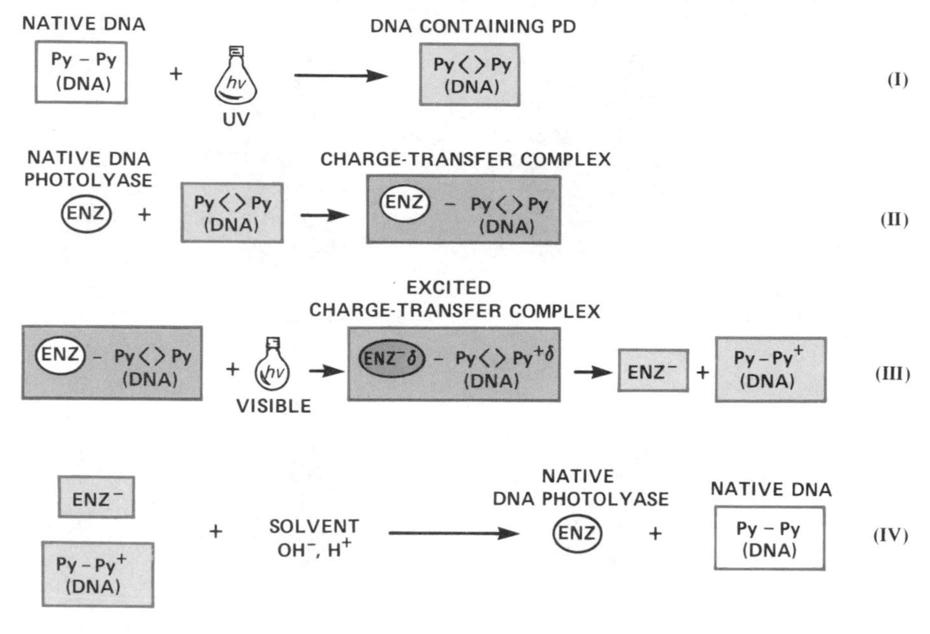

FIGURE 2-16
Hypothetical scheme for the monomerization of pyrimidine dimers by *E. coli* DNA photolyase. Native DNA with adjacent pyrimidines (Py) contains pyrimidine dimers (Py <> Py) after exposure to UV radiation (I). Native DNA photolyase binds to pyrimidine dimers in the dark to form a charge-transfer complex (II) that absorbs light at approximately 360 nm (III). The excited charge-transfer complex is unstable and undergoes partial ionization (III), which facilitates monomerization of the dimers and dissociation of the enzyme (III). Interaction with solvent ions restores the charge of the enzyme and the DNA to their respective native states (IV). (After J. C. Sutherland, ref. 71.)

(Fig. 2-17). Position 2 of the ring is theoretically the most reactive and would be accessible to the enzyme in all dimer species if DNA photolyase bound in the minor groove of the DNA duplex. Consistent with this model, the peptide antibiotic netropsin, which is known to bind in the minor groove of DNA, inhibits the binding of *E. coli* DNA photolyase to UV-irradiated DNA (69). Alternatively, the enzyme might recognize a unique modification of the secondary structure of the DNA helix introduced by any of the dimers.

DNA Photolyase from Mammalian Cells. For many years no photobiological or biochemical evidence for EPR existed in organisms evolutionarily more advanced than marsupials. However, an enzyme activity that catalyzes the monomerization of pyrimidine dimers in DNA in the presence of visible light at wavelengths between 300 and 600 nm has been isolated from human leukocytes and from human fibroblasts (73–76). The activity from leukocytes is a protein with an apparent molecular weight of 40,000 and a pI of 5.4. Like DNA

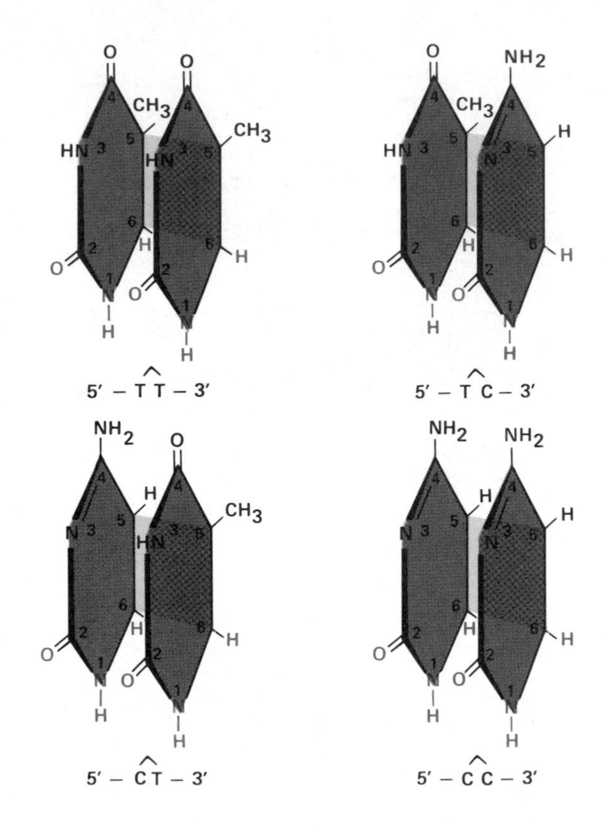

FIGURE 2-17
Positions N^1, C^2 and C^6 in the pyrimidine rings are the only ones common to all pyrimidine dimers in DNA.

photolyase from a number of other sources, the activity from human leukocytes has very stringent pH and ionic strength requirements. The optimum pH is 7.2, and at pH values of 8.1 and 6.9 only 29 percent and 49 percent respectively of the optimal activity remain (73). The *E. coli* and human enzymes demonstrate maximal activity at ionic strengths of 0.18 *M* and 0.05 *M* respectively. At the former concentration of ions the human enzyme has only 20 percent of its optimal activity (73). This ionic strength dependence may well account for the difficulty in demonstrating the presence of this enzyme in earlier studies. As yet no distinct chromophore has been identified in the enzyme.

Fractionation of leukocytes and of other cells from human peripheral blood shows DNA photolyase activity to be concentrated chiefly in monocytes and polymorphonuclear leukocytes (74). Lymphocytes have low levels of the enzyme, and serum and erythrocytes have no significant activity. Enzyme activity has also been detected in ex-

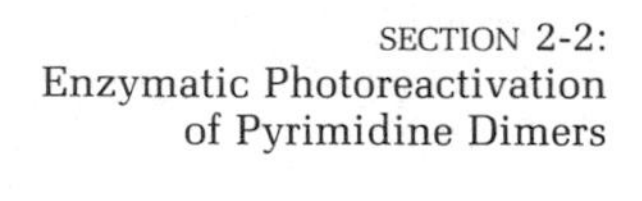

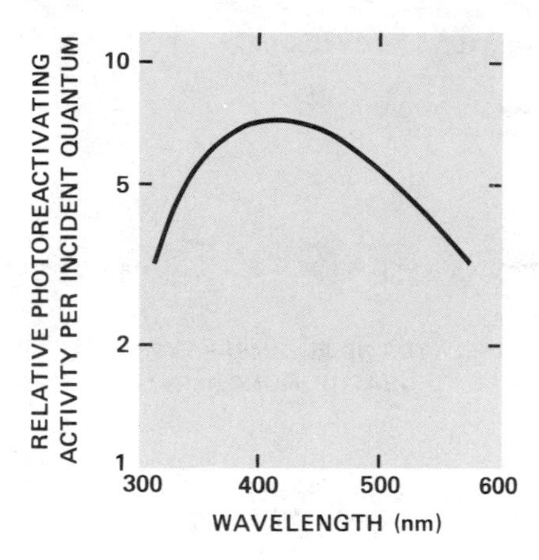

FIGURE 2-18
The action spectrum for DNA photolyase from human leukocytes is shifted into the red zone of the electromagnetic spectrum relative to that observed with the enzymes from yeast (see Figure 2-6), *E. coli* or *S. griseus* (see Figure 2-10). (From B. M. Sutherland et al., ref. 74.)

tracts of bovine bone marrow and murine and human fibroblasts in culture (74). The phenomenon of the light-dependent loss of pyrimidine dimers from DNA has also been demonstrated in living cells in culture and in intact human skin (77–80). In mouse skin, a light-dependent loss of dimers occurs in the dermis but not in the epidermis of neonatal animals and has not been demonstrated in either site in adult animals (81). Thus the expression of DNA photolyase in some animals may be developmentally regulated.

A detailed analysis of the action spectrum of the human leukocyte enzyme shows maximal activity at 400 nm (74, 76) (Fig. 2-18). The demonstration of activity at wavelengths as long as 600 nm is interesting and unique. [Since the human enzyme is active in the presence of yellow light it is necessary to resort to the use of red light for studies in which it is desired to eliminate EPR (76)].

The monomerization of pyrimidine dimers in DNA treated with human DNA photolyase has been directly demonstrated by treating UV-irradiated [^{3}H]cytosine-labeled DNA under conditions that promote the selective deamination of cytosine in pyrimidine dimers, thereby yielding a proportion of dimers containing uracil (74). Following enzymatic photoreactivation by the human enzyme, the formation of monomeric uracil quantitatively mirrors the disappearance of C <> C from the DNA (74) (Fig. 2-19).

A curious and interesting feature of human DNA photolyase is the apparent dependency of the level of enzyme in cultured fibroblasts

UV IRRADIATED [^{3}H] CYTOSINE-LABELED DNA

ON HEATING, CYTOSINE IN DIMERS IS PREFERENTIALLY DEAMINATED TO URACIL. MONOMERIC CYTOSINE IS MORE STABLE.

AFTER TREATMENT WITH DNA PHOTOLYASE SOME DIMERS ARE MONOMERIZED

(4)

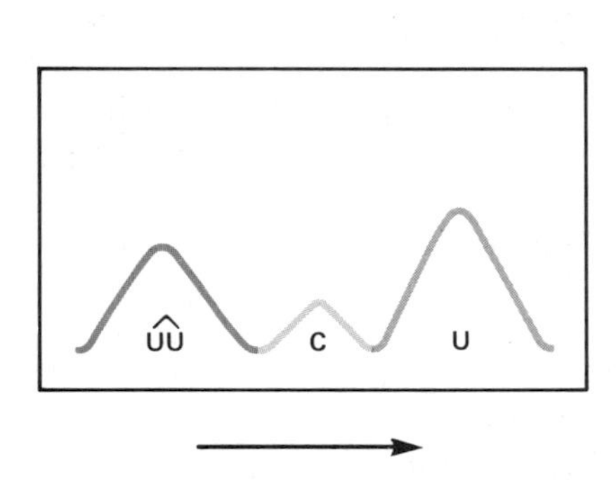

LABELED URACIL-CONTAINING DIMERS, URACIL AND CYTOSINE CAN BE SEPARATED CHROMATOGRAPHICALLY

(5)

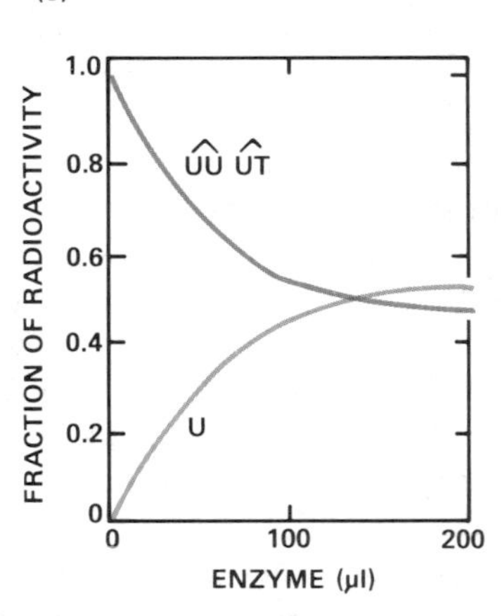

CONCOMITANT DISAPPEARANCE OF PYRIMIDINE DIMERS AND APPEARANCE OF CORRESPONDING MONOMERS

FIGURE 2-19
The formation of monomeric uracil in DNA derives directly from the enzymatic photoreactivation of cytosine-containing pyrimidine dimers. Cytosine-containing dimers in DNA (1) preferentially (relative to monomeric cytosine) deaminate into uracil-containing dimers on heating (2). After incubation with DNA photolyase and monomerization of dimers, uracil is present in the DNA (3). The uracil can be quantitated chromatographically (4) (see Figure 2-3) and shown to be related to the loss of cytosine-containing dimers from DNA (5). (From B. M. Sutherland et al., ref. 74.)

TABLE 2-5
DNA photolyase activity in extracts of cells from xeroderma pigmentosum patients

Cell Line	DNA Photolyase Activity (% of Normal)
XP12BE	36.3
XP1LO	<1.0
XP4BE	10.8
XP7BE	11.0
XP11BE	18.0
XP15BE	40.0

Data compiled from B. M. Sutherland and R. Oliver. ref. 199, and F. K. Wagner et al., ref. 77.

on the conditions of the cell culture (79, 82). Cells grown in Eagle's minimal essential medium supplemented with fetal bovine serum contain very low levels of DNA photolyase activity. However, parallel cultures, grown in Dulbecco's modified Eagle's minimal medium, contain higher levels of enzyme. The exact components in the latter medium that determine this difference have not been established, and some studies have failed to identify a photoreactivating activity in human cells in vivo (83–85). Thus the possibility should be considered that this enzyme is expressed under very specific conditions. In addition, as indicated in the introduction to this chapter, despite the fact that such reactions in general are not observed at wavelengths greater than 300 nm, some proteins and oligopeptides can effect the noncatalytic sensitized reversal of pyrimidine dimers in DNA, and could be erroneously identified as DNA photolyases in vivo (13–19).

Human patients suffering from the inherited disorder xeroderma pigmentosum (XP) have a defect in the repair of DNA damage caused by UV radiation (see Section 9-2). Cultured skin fibroblasts from such individuals contain decreased levels of enzyme activity in all cases, with values ranging from 0 to 40 percent of normal (78, 86, 87) (Table 2-5). The growth rate, age of cell donor and passage number of normal human cells does not influence enzyme activity (86, 87), and these variables apparently do not account for the deficiencies observed in extracts of XP cells compared with cells from normal humans.

Photoreactivation of RNA

The essential focus of this book is on the repair of *DNA* damage; however, it should be borne in mind that in some viruses genetic information is encoded in an RNA genome. Although the damage

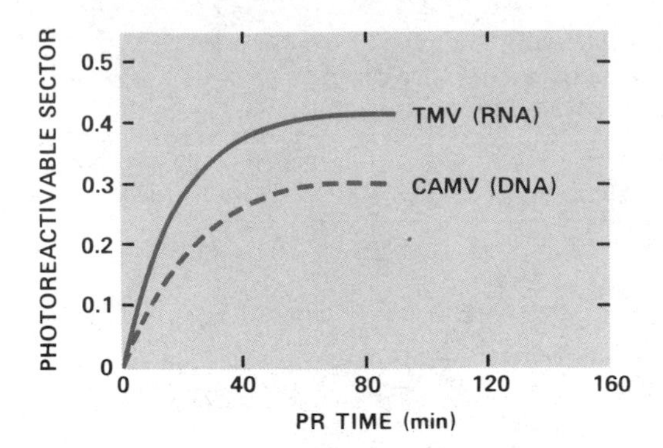

FIGURE 2-20
Dose response curves for photoreactivation (PR) of UV-irradiated tobacco mosaic virus (TMV), containing RNA, and for cauliflower mosaic virus (CAMV), containing DNA. The photoreactivatable sector is a measure of PR efficiency and expresses the fraction of the UV dose that is rendered ineffective in terms of a measurable endpoint such as lethality, by photoreactivation. (From L. Towill et al., ref. 90.)

and repair of RNA have not received anything like the attention that the damage and repair of DNA have, enzymatic photoreactivation mechanisms for the repair of pyrimidine dimers in RNA have been documented. For example, when a number of UV-irradiated RNA plant viruses and/or free viral RNAs are assayed on appropriate hosts, an increase in specific infectivity occurs when the assay plant is illuminated immediately after the application of the infectious material (Fig. 2-20) (88–90). This result is not observed when heat-inactivated infectious material is used; additionally, preillumination of the assay host has no effect (90). An activity in cell-free extracts of tobacco plants catalyzes the in vitro photoreactivation of tobacco mosaic virus (TMV) RNA, but it has not been purified or characterized because of its extreme lability (89). DNA photolyases from yeast or from pinto bean seedlings are inactive on UV-irradiated TMV RNA (89).

2-3 Repair of O^6-Guanine Alkylation

We now move from the repair of pyrimidine dimers in DNA to a specific form of chemical damage in order to consider a second well-documented example of DNA repair by direct reversal (91–93). As indicated in Section 1-3, certain strongly mutagenic monofunctional agents such as N-methyl-N′-nitro-N-nitrosoguanidine (MNNG), methylmethanesulfonate (MMS) and N-methyl-N-nitrosourea (MNU) react with DNA to produce both O-alkylated and N-alkylated products. One reaction product in the former class, O^6-methylguanine, is mutagenic since it can mispair with cytosine during semiconservative DNA synthesis (94, 95). The repair of this miscoding lesion in *E. coli* is understood in considerable detail and is the first example we shall consider of *induction* by DNA damage of gene products required for the processing of that damage. Other examples of inducible responses to DNA damage are considered in later chapters.

Adaptation to Alkylation Damage in *E. coli*

For many years it was known that a curious relation exists between the induction of mutations in *E. coli* and the time of exposure to mutagens such as MNNG. Specifically, the mutation frequency increases with time of exposure and then dramatically reaches a plateau as if some sort of *antimutagenic* mechanism were operative (96–99). Trivial explanations such as decay of the mutagen in the medium in which the *E. coli* cells were grown were readily ruled out. Additionally, this result could not be due to saturation of all possible mutable sites in DNA, since the use of small doses of mutagen that are unlikely to be saturating yields the same qualitative result with a smaller plateau level (98).

In the late 1970s John Cairns and his colleagues observed that when cultures of *E. coli* are exposed to very low levels of MNNG and subsequently challenged with a much higher dose of the alkylating agent, there is a marked resistance to both the lethal and the mutagenic effects of the chemical in the "adapted" cells relative to unadapted controls (100–102) (Fig. 2-21). This resistance is dependent on active protein synthesis by the cells prior to the challenge dose, indicating that it involves the induction of one or more genes in response to exposure to low levels of the alkylating agent (100). This phenomenon is called the *adaptive response* to alkylation damage, and as noted above, it is also an example of an inducible response to DNA damage in *E. coli*. The adaptation to killing and to mutagenesis have been shown to be distinct cellular responses, involving different genetic loci. Strains defective in the *polA* gene (which codes for DNA polymerase I of *E. coli*) and strains defective in a gene called *alkA* (which codes for an inducible DNA glycosylase; see Section 3-3) are deficient in the former response (103–106), whereas mutants designated as *ada* are deficient in *both* the antimutagenic adaptation and the adaptation to killing (104, 107). The *ada* gene thus appears to be regulatory in nature and maps at about 47 minutes on the *E. coli* K-12 genetic map (108) (Fig. 2-22). A gene called *alkB* that maps close to *nalA* has also been identified in *E. coli*. Mutations in the *alkB*

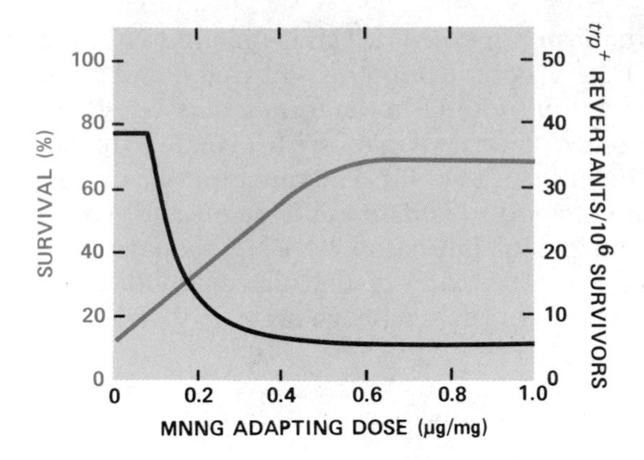

FIGURE 2-21
Adaptation to cell killing and to mutagenesis in *E. coli*. A culture of *trp E. coli* cells was grown for 90 minutes in the presence of various amounts of N-methyl-N′-nitro-N-nitrosoguanidine (MNNG) (adapting doses). At the end of this time, samples were exposed to 100 μg/ml of MNNG for five minutes (challenge dose) and the surviving fraction and *trp*$^+$ reversion frequency were determined. In adapted cells survival increased and mutation frequency (normalized to survival) decreased. (From P. F. Schendel, ref. 93.)

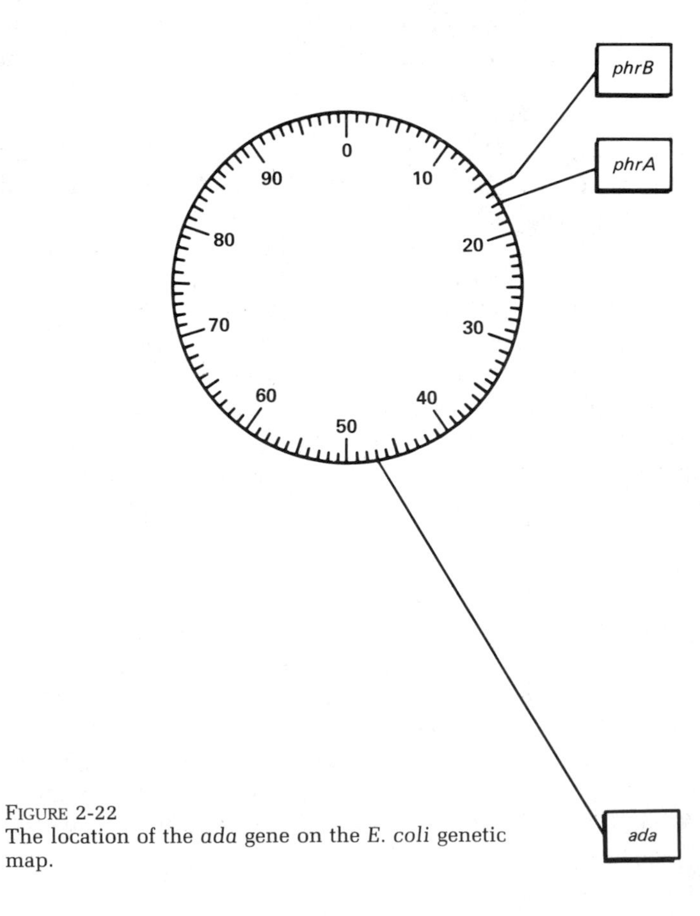

FIGURE 2-22
The location of the *ada* gene on the *E. coli* genetic map.

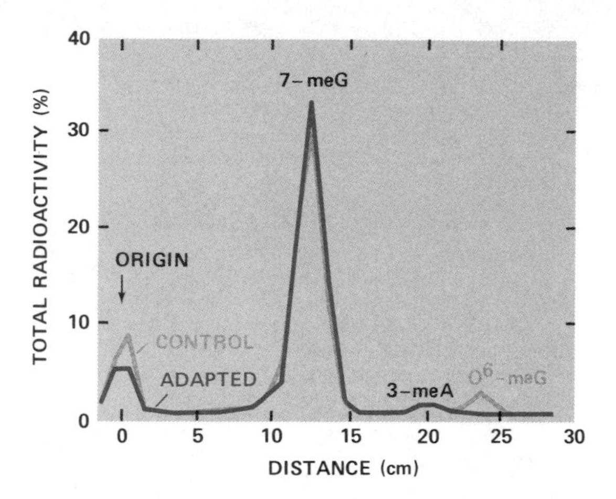

FIGURE 2-23
Preferential loss of O^6-methylguanine from the DNA of adapted cells. *E. coli* cells were exposed to adapting or nonadapting conditions and then challenged with a larger dose of radiolabeled alkylating agent. Following a period of incubation, DNA from both groups of cells was isolated and hydrolyzed. The alkylated bases 3-methyladenine (3-mA), 7-methylguanine (7-meG) and O^6-methylguanine (O^6meG) in the hydrolysate were identified by paper chromatography. (From P. F. Schendel and P. E. Robins, ref. 110.)

gene also confer enhanced sensitivity to killing by MMS, but not by MNNG (109). Such mutants have a normal mutational adaptive response to MNNG treatment, however (109).

The enzymatic mechanisms responsible for the adaptation to cell killing and to mutation are quite different (103–106). The discussion in this chapter focuses on the *mutational* adaptation since it represents an example of DNA repair by reversal of base damage. The adaptation to cell killing involves a DNA repair pathway by which sites of alkylation in DNA other than the O^6 position of guanine are *excised* from the DNA; it is considered in Section 3-3.

Adaptation to Mutagenesis in E. coli
Involves the Repair of O^6-Methylguanine in DNA

A quantitative comparison of the major alkylation products present in precipitated DNA of adapted and nonadapted cells subsequently challenged with radioactively labeled MNNG shows that the levels of 7-methylguanine and 3-methyladenine are indistinguishable. However, a significant reduction in the amount of O^6-methylguanine (labeled in the methyl group) in DNA is observed (110) (Fig. 2-23). Interestingly, the radiolabeled O^6-methylguanine lost from the DNA is not recovered in the soluble fractions from which the DNA was

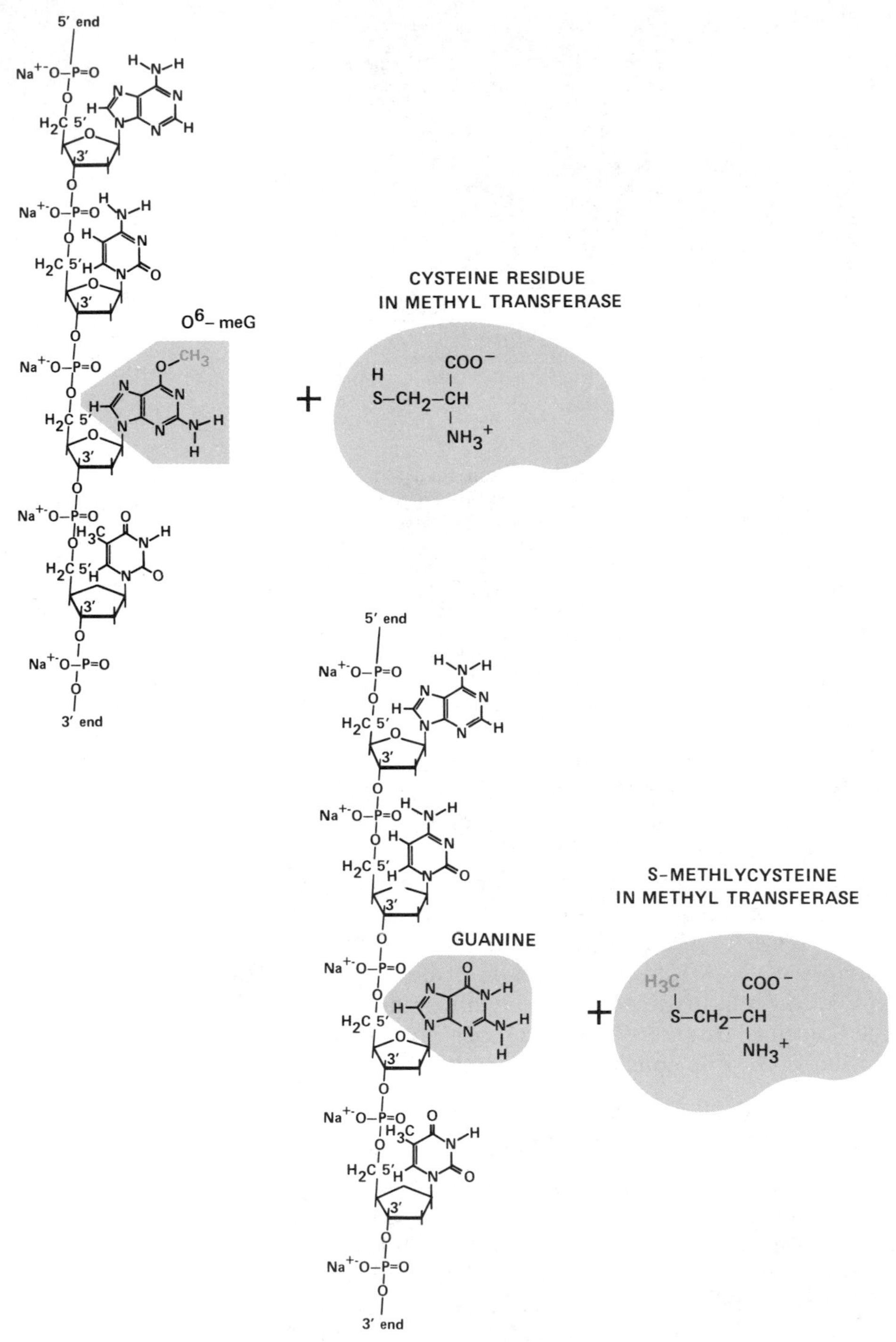

FIGURE 2-24
A specific transferase activity in *E. coli* transfers the methyl (or ethyl) group from the O^6 position of guanine in DNA to a cysteine residue in the protein, thereby repairing the base damage by direct reversal.

precipitated, as might be anticipated if the repair of this lesion involved the enzyme-catalyzed removal of the damaged base (110, 111). Furthermore, transfer of just the labeled methyl group to the acid-soluble fraction of adapted cells is not observed either (110).

Cell-free preparations of previously adapted strains of *E. coli* also promote the selective loss of O^6-methylguanine from DNA (112, 113). When such extracts [or extracts of a particular *ada* mutant of *E. coli* called *adc-1* (108), which is *constitutive* for the adaptation response (114)] are incubated with DNA containing radiolabeled O^6-methylguanine, the labeled methyl groups lost from the DNA are recovered bound to *protein* (112). This result explains the failure to find radioactivity in acid-soluble or ethanol-soluble fractions from intact cells. The protein has an unusual methyl transferase activity that both removes the methyl group from O^6-methylguanine in DNA and transfers it to one of its own cysteine residues, giving rise to S-methylcysteine (115) (Fig. 2-24). The S-methylcysteine has been detected by amino acid analysis of an enzymatic hydrolysate of the *E. coli* methyl acceptor protein after reaction with alkylated DNA (115). In addition, if the protein hydrolysate is oxidized prior to amino acid analysis, the radioactive material eluting at the position of intact S-methylcysteine disappears and a new major species appears at the position of S-methylcysteine sulfone (115) (Fig. 2-25).

Further evidence for the repair of O^6-methylguanine by enzymatic demethylation comes from studies with a synthetic DNA polymer, poly(dC·dG-8-[^{3}H]me^{6}dG), which contains O^6-methyl-8-[^{3}H]guanine; that is, the radiolabel is present in the purine ring rather than in the O^6 methyl group. Following incubation with extracts of *E. coli* containing methyl transferase activity this polymer substrate contains unsubstituted radiolabeled guanine, hence verifying the direct reversal of base damage by removal of the offending methyl group and reconstitution of intact guanine (113).

The O^6-Methylguanine (O^6-meG) Methyl Transferase of E. coli

The O^6-meG methyl transferase has been purified to 95 percent homogeneity (116, 117) from the *E. coli adc-1* mutant, which constitutively expresses the adaptive response to alkylation-induced mutation (108). Gel electrophoresis of the purified protein in the presence of sodium dodecylsulfate shows a single polypeptide with a molecular weight of approximately 18,000 (115–120). The active form of the enzyme is a monomer, since the same molecular weight is obtained with SDS gel electrophoresis or with velocity sedimentation analysis under nondenaturing conditions (116). The enzyme has no requirement for divalent cations or other known cofactors. There is no indication of peptide bond cleavage associated with the transfer of the methyl group, since the mass of the methylated protein is the same

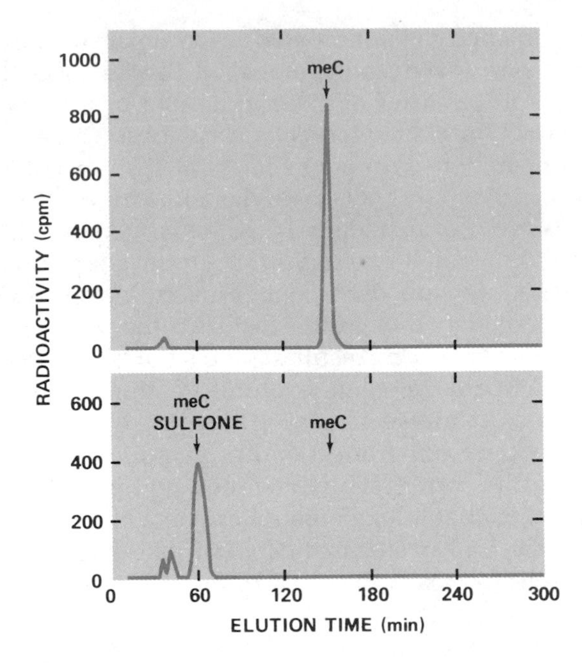

FIGURE 2-25
When S-methylcysteine is oxidized with H_2O_2, it is converted into a derivative with the chromatographic properties of S-methylcysteine sulfone. In the experiment shown here a protein hydrolysate containing bound radioactive methylated moieties was incubated with H_2O_2 prior to amino acid analysis. The radioactive material eluting at the position of intact S-methylcysteine (meC) (top) disappeared; a new major peak appeared at the position of S-methylcysteine sulfone (bottom). (From M. Olsson and T. Lindahl, ref. 115.)

as that of the unmethylated form. Amino acid analysis indicates the presence of four (or possibly five) cysteine residues per molecule, but only one of these is the methyl acceptor (116).

Adaptation in *E. coli* confers resistance to mutation against ethylating, propylating and butylating agents as well as against methylating agents (93). Consistent with these observations, the activity from *E. coli* also repairs O^6-ethylguanine in DNA in vitro, with the concomitant formation of an S-ethylcysteine residue (121). However, the rate of disappearance of this lesion from DNA in vivo is only about $\frac{1}{10}$ of that for methyl groups in guanine (121). The enzyme apparently can also catalyze the removal of O^6-chloroethyl groups from guanine in DNA, since its presence prevents the appearance of interstrand cross-links in DNA treated with the chloroethylating agents 1,3-bis(2-chloroethyl)-1-nitrosourea (BCNU) or 1-(2-chloroethyl)-3-cyclohexyl-1-nitrosourea (CCNU) (120). Presumably the inhibition of cross-linking results from repair of O^6-chloroethylguanine monoadducts before the second step of the cross-linking reaction can occur (120). Experiments with living cells suggest that alkyl groups as large as propyl can be removed by this enzyme, but butyl and amyl groups are not (122). Parenthetically it should be noted that alkyl adducts larger than methyl groups at the O^6 position of guanine in DNA can also be repaired in *E. coli* by a process called *excision repair* (see Chapters 3 through 6), in which the entire nucleotide is removed from the DNA (123).

The methyl transferase activity is consumed in the reaction it catalyzes (118, 119). Thus, when a limiting amount of enzyme is added to a reaction mixture, the reaction kinetics show that efficient transfer of some of the available methyl groups occurs very rapidly and then essentially ceases (118, 119). This is not due to nonspecific inactivation of the enzyme or to unusual refractoriness of the remaining substrate to attack by the enzyme. These results strongly suggest that the methylated protein acceptor cannot be regenerated and thus the enzyme is apparently expended in the reaction; that is, the methyl transferase is not catalytic in the usual sense of the term. Although the possibility that the methyl transferase activity donates the methyl group to a separate acceptor protein cannot be completely excluded, the available evidence indicates that the transferase and acceptor functions are resident in the same protein (118). An enzyme, strictly defined, is a catalyst and should therefore not be consumed in the reaction with its substrate. However, the methyl transferase reaction can be considered to be analogous to that between a suicide enzyme inactivator and its target enzyme, other examples of which are known and are discussed below (92, 120). The removal of the methyl group from O^6-methylguanine in DNA is also unusual in that it represents an example of enzymatic methyl transfer to a protein under conditions where S-adenosylmethionine is not the methyl donor (92). Finally, the methyl transferase is an enzyme that modifies itself in the course of the reaction it catalyzes.

It is not without precedent in biochemistry to find a protein-modifying enzyme that uses itself as the main target for modification (92). For example, a number of protein kinases that catalyze autophosphorylation are known, and the major acceptor for poly(ADP-ribose) in mammalian cell nuclei is the poly(ADP-ribose) synthetase itself (124, 125). Although these reactions do not involve inactivation of the mediating enzyme molecule, several enzymes are known to be irreversibly inactivated by the formation of dead-end complexes due to enzyme-catalyzed covalent binding of substrate analogues at the active site (120). For example, GMP reductase, which catalyzes the deamination of GMP to TMP, is irreversibly inhibited by the 6-chloro and 6-mercapto analogues of GMP by covalent binding with a sulfhydryl group at the nucleotide binding site of the enzyme (120, 126). In addition, 5-fluoro-dUMP inactivates thymidylate synthetase by the generation of a stable covalent bond to a specific cysteine residue at the dUMP binding site of the enzyme (120, 127). It has been pointed out that the unusual feature of the suicide inactivation exhibited by the O^6-methylguanine-DNA methyl transferase is that in this case the reaction occurs between the enzyme and its "natural" substrate, rather than with a substrate analogue (120). This leads to the interesting speculation that the enzyme might have some other "natural" substrate in the cell (120).

About 100 to 200-fold lower levels of the same enzyme activity are present in unadapted wild-type bacteria (128, 129). Whether or not this represents a constitutive level of enzyme or the response of the cells to persistent low-grade inducing stimuli is not clear. In fact, it might be very difficult to distinguish between these alternatives (129). The latter situation might result from spontaneous (i.e., nonenvironmental) methylation events at the O^6 position of guanine in DNA, caused by S-adenosylmethionine and/or other intracellular methyl donors. Interestingly, if radiolabeled highly purified S-adenosylmethionine is incubated with DNA in vitro, small amounts of alkylation products can be detected in the DNA; however, these levels are too low to determine whether or not O^6-methylguanine can be formed in such a reaction (130). It is unlikely that the presence of low levels of methyl transferase activity in unadapted cells is due to the presence of small amounts of alkylating agents in the growth medium, since these should be destroyed at the high temperatures associated with sterilization of media (129).

The *E. coli* O^6-meG transferase acts poorly on single-strand DNA containing O^6-alkylguanine moieties (119). This property of the enzyme provides a reasonable explanation for the well-known propensity of agents such as MNNG and MNU to produce a greater concentration of mutations near replication forks in *E. coli* than in nonreplicating regions of the genome (131). Thus alkylation of guanine present in parental DNA strands at replication forks may be relatively refractory to repair until replication restores the duplex structure by misincorporation in the daughter strand opposite O^6-alkylguanine (120).

Correlations between the Enzymology and Cellular Biology of Adaptation in E. coli

A priori the discovery that the O^6-meG methyl transferase of *E. coli* is a suicide protein may seem rather surprising. However, in retrospect any other result would be inconsistent with the information gleaned from the careful analysis of the biology of the adaptive phenomenon in living cells, which shows that adaptation to mutation in *E. coli* K-12 is a saturable phenomenon (111). Thus, for example,

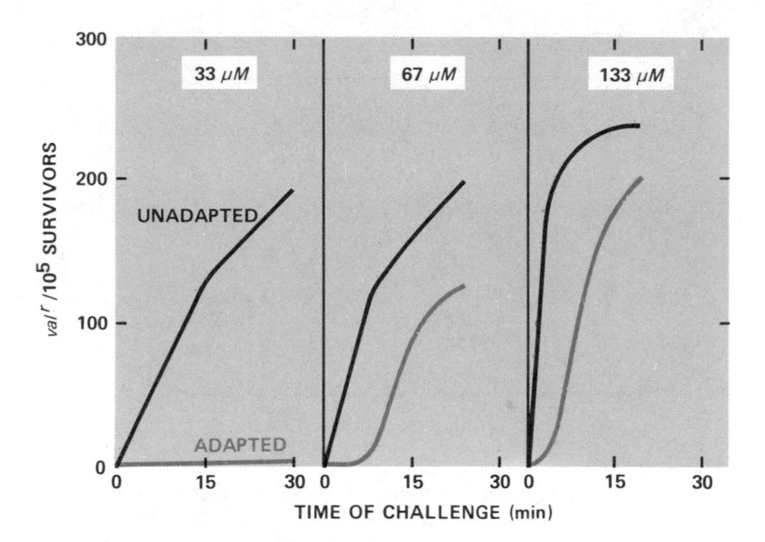

FIGURE 2-26
Frequency of valine-resistant mutants of *E. coli* in adapted and unadapted cells after exposure to increasing concentrations of MNNG. At the higher concentrations of alkylating agent the adaptation phenomenon saturates progressively more rapidly and the mutation frequency then increases at rates comparable to those observed in unadapted cells. (From P. F. Schendel et al., ref. 132.)

when adapted cells are exposed for a constant time period to MNNG at different concentrations, the yield of mutants is not linearly related to the concentration of the mutagen (132, 133). Similarly, when such cells are exposed to a moderate concentration of mutagen for varying periods of time, a nonlinear response is observed; initially the mutation frequency is low, but at later times it increases rapidly, as if an antimutagenic process had become saturated (132). In addition, detailed kinetic analyses using varying MNNG concentrations show a good correlation between the period of relative resistance to mutation and the amount of mutagen administered (132) (Fig. 2-26).

These and other biological experiments led to the model shown in Fig. 2-27 (111). It was assumed (based on the observation of saturation) that each adaptation molecule can dispose of only one O^6-methylguanine lesion. Given this assumption and knowing the level of O^6-methylguanine present in DNA as well as the kinetics of its loss, it was calculated that there are 100 repairing (methyl transferase) molecules synthesized per minute per cell during growth in adapting concentrations. Thus bacteria that divide every 30 minutes could accumulate no more than 3000 such molecules during adapta-

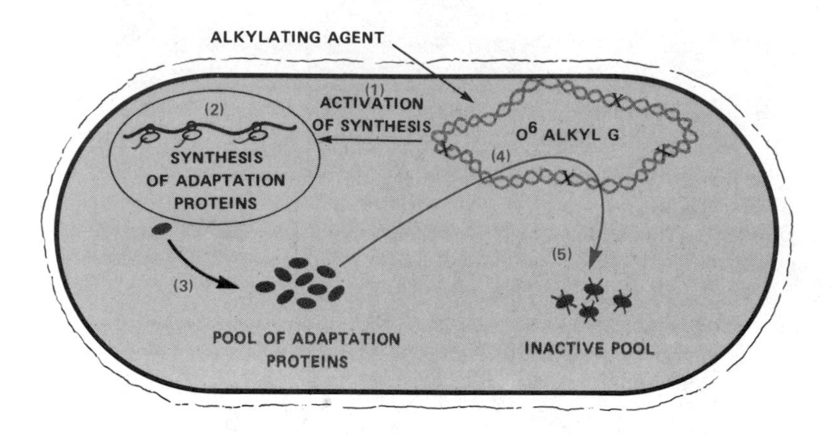

FIGURE 2-27
The process of adaptation to alkylation damage is illustrated schematically. Alkylation damage to DNA activates the synthesis of a special class of proteins (1) (adaptation proteins) whose function is to repair O^6-alkylguanine. These proteins accumulate within the cell until a pool of several thousand molecules has been formed (2, 3). During subsequent exposures to alkylating agent, the preformed adaptation proteins repair O^6-alkylguanine lesions (4) but are inactivated in the process (5). They must then be replaced by newly synthesized protein. If the flux of alkylating agent produces repairable lesions more slowly than protein synthesis generates adaptation protein, mutagenesis is allayed. If the alkylation flux produces lesions more rapidly than the rate at which adaptation protein can be synthesized, the pool will eventually be exhausted and the cell will be vulnerable to alkylation mutagenesis. (After P. F. Schendel, ref. 93.)

tion. This is enough material to deal immediately with a challenge of up to 3000 residues of O^6-methylguanine per cellular genome. Any lesions in excess of this amount formed in adapted bacteria can only be handled at a rate of about 100 per minute (111).

The Regulation of the Adaptive Response to Alkylation-Induced Mutation in E. coli

There is still much to be learned about the regulation of the adaptive response in *E. coli* (93). Little is known about the details of the induction event or of the regulation and maintenance of the response. It is of interest that a variety of alkylating agents are effective inducers (101, 102), yet there is no obvious relation between their effectiveness as inducers and the amount of O^6-alkylguanine they produce in DNA. Some chemicals that produce very little O^6-alkylguanine in DNA are effective inducers, whereas others that produce more of this product are ineffective (107, 134, 135). Thus, the appearance in DNA of O^6-alkylguanine is perhaps a necessary, but insufficient, stimulus for optimal induction (93).

Cells that are already adapted, but whose presynthesized adaptation protein pool has been depleted by challenge with alkylating agent, synthesize new adaptation proteins more rapidly than control cells (93). Thus the rate of synthesis of adaptation proteins and the size of the protein pool can vary. Little is presently known about how such variations are regulated. Finally, the structural gene for the methyl transferase has not been identified. The *ada* locus is required for its induction; however, both wild-type and *ada* mutants contain comparable levels of enzyme (13 to 60 molecules) in nonadapted cells, suggesting that the *ada* locus is regulatory in nature and does not code for the methyl transferase (128). The *ada* gene has been cloned on a plasmid, and its protein product(s) has been identified in cell-free preparations by gel electrophoresis (120). Two polypeptides of molecular weight 37,000 and 27,000 disappear when the cloned gene is activated by transposons inserted at a number of different locations. It is presently unclear if the smaller of the two proteins is derived from the larger one by proteolysis or if ada^+ represents an operon containing two genes (120).

Analysis of the activities associated with the adaptive response has led to the discovery of several other methyl transferase activities in extracts of induced *E. coli* (122). Alkylation sites other than the O^6 position of guanine appear to be substrates for these activities. These sites include the O^4 position of thymine and the *triesters* produced on the phosphates of the DNA backbone (see Section 1-3). Conceivably, distinct, inducible methyl transferases can remove alkyl groups from most if not all sites of O alkylation in DNA (122).

The Adaptive Response in Other Prokaryotic Cells

B. subtilis cells also become more resistant to both killing and mutagenesis by a challenge dose of MNNG after adaptation with lower doses of the alkylating agent (136). In addition, cell extracts of both nonadapted and adapted cells contain a methyl transferase activity that appears to be very similar to that of the enzyme from *E. coli*. Interestingly, the level of enzyme in nonadapted cells (about 250 molecules per cell) is significantly higher than that present in *E. coli*. In adapted *B. subtilis* cells enzyme activity increases by an order of magnitude (136). In contrast both *S. cerevisiae* and *H. influenzae* lack detectable methyl transferase and an adaptive response to mutagenesis by MNNG (136).

The Repair of O^6-Alkylguanine in Mammalian Cells

O^6-alkylguanine is a lesion of considerable biological importance in mammalian cells and tissues. In cultured mammalian cells O^6-methylguanine is implicated in mutagenesis by alkylating agents (137), and there is considerable evidence that O^6-alkylguanine may be involved in the production of tumors in experimental animals by alkylating carcinogens. In general, alkylating agents that produce little O^6-alkylguanine in DNA are weak carcinogens (138). In addition, different rates of repair of this lesion in target and nontarget tissues can have a profound effect on tumor production. For example, the production of brain tumors in young rats treated with N-ethyl-N-nitrosurea is correlated with the persistence of O^6-alkylguanine in the target organ (139). Similarly, chronic treatment with N-methyl-N-nitrosurea specifically results in neural tumors in experimental animals and is accompanied by a progressive accumulation of O^6-methylguanine in the brain without any concomitant accumulation in other tissues (140).

It has long been known that certain mammalian cells and tissues can effect the removal of O^6-methylguanine and O^6-ethylguanine from DNA (139–152). An enzyme activity that transfers O^6 alkyl groups to cysteine, forming S-alkylcysteine in a protein acceptor molecule, is present in a variety of mammalian, including human, tissues (153–165). In rodents and humans, liver has the highest transferase activity (166), with hamster liver containing about 22,000 molecules per cell, rat liver about 60,000 molecules per cell and human liver 600,000 to 900,000 molecules per cell. Kidney and nonparenchymal liver cells have less activity than parenchymal liver cells, and brain has even less activity. Human cells in culture have about 100,000 molecules per cell.

The human enzyme has been extensively purified but has not yet been made homogeneous (163). It is a small protein with a molecular weight of approximately 20,000. The transferase and acceptor activities are apparently produced by the same protein, as judged by the inability to resolve these activities by chromatography (163); however, this issue will only be resolved by purification of the human enzyme to physical homogeneity. Nonetheless, the biochemical properties of the bacterial and human activities are so similar that it appears very likely that the latter is also a suicide enzyme (120). In particular, low concentrations of the human enzyme display the same unusual very rapid but limited reaction with alkylated DNA that the *E. coli* enzyme does (120).

Although the biochemistry of the repair of O^6-alkylguanine in *E. coli* and in mammalian cells appears to be the same, an issue that is far more controversial is whether the enzyme activity present in mammalian cells is *inducible*, and if so, whether adaptation to mutation analogous to that in *E. coli* exists in these cells. The adaptive response in mammalian cells and tissues has been addressed in numerous studies in efforts to draw direct parallels with the results in *E. coli* (146–148, 151, 155, 157, 161, 167–174). Both in Chinese hamster ovary (CHO) cells and in human skin fibroblasts transformed

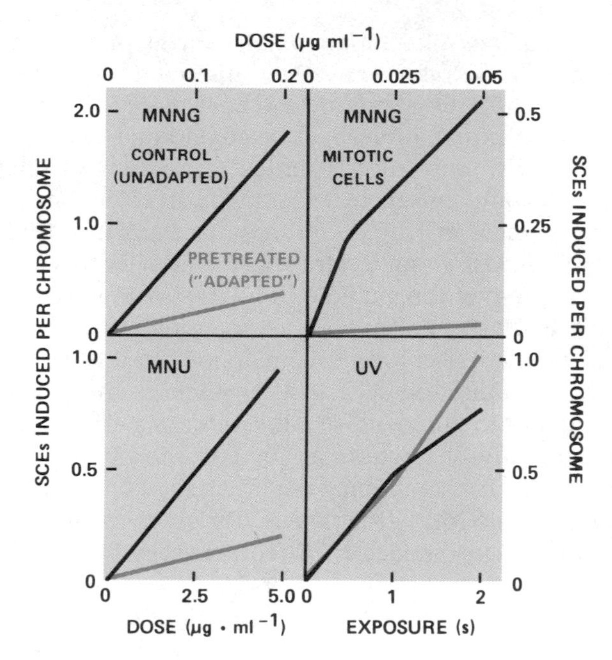

FIGURE 2-28
Pretreatment of CHO cells with N-methyl-N′-nitro-N-nitrosoguanidine (MNNG) (all panels) results in a reduced frequency of sister chromatid exchanges (SCEs) when mitotic or nonmitotic cells are subsequently treated with MNNG and when nonmitotic cells are treated with N-methyl-N-nitrosourea (MNU). This effect is not observed when the cells are treated with UV radiation (lower right panel). (From L. Samson and J. L. Schwartz, ref. 167.)

by SV40 virus exposure to very low doses of MNNG renders the cells resistant to the induction of sister chromatid exchanges (believed to be an indicator of persisting DNA damage; see Section 9-4) by further alkylation damage (167) (Fig. 2-28). CHO cells also apparently become more resistant to killing, but no adaptation to mutation has been observed (169). An enhanced resistance to cell killing also occurs in Chinese hamster V79 cells exposed to nontoxic doses of MNU before challenge with a toxic dose of this alkylating agent (168). However, neither the frequency of mutation into 6-thioguanine resistance nor the loss of O^6-methylguanine from DNA is affected in these cells (168). Rats chronically treated with unlabeled diethylnitrosamine and then challenged with a larger dose of radiolabeled compound selectively effect the loss of O^6-ethylguanine (but not 3-ethyladenine and 7-ethylguanine) from the DNA of their liver cells (146). In addition, the pretreatment is associated with increased levels of an enzyme activity from rat liver (presumably the alkyl transferase) that catalyzes the repair of O^6-alkylguanine in DNA (173, 174). However, in the rat liver system "inducers" are not confined to agents that produce O^6-alkylguanine in DNA (147, 151), and the important trivial explanation must be considered that the increased transferase activity may merely reflect a cellular proliferative response to cell death caused by the alkylation treatment (157, 175).

In summary, there is little doubt that mammalian cells contain an O^6-methylguanine DNA methyl transferase activity that appears to be very similar to that of the enzyme in *E. coli*. Whether or not expression of this activity is regulated in mammalian cells and, if so, whether its regulation is formally analogous to the inducible adaptive response to mutation in *E. coli* is unclear. In *E. coli* there are normally only 13 to 60 enzyme molecules per cell, but after induction of enzyme activity, the number increases to a maximum of about 3000. Animal cells have much higher constitutive levels of the enzyme, and the large induction of activity observed as a specific response to alkylation damage in *E. coli* does not seem to occur in mammalian cells or tissues.

Perhaps a major difficulty stems from the persistent attempts to extrapolate experimental approaches directly from *E. coli*. It is surely not unreasonable to expect that the regulation of mammalian genes is quite different from that in bacteria, and a definition of the regulation of O^6-methylguanine DNA methyl transferase in mammalian cells based on criteria established in *E. coli* is both arbitrary and unwarranted.

A potentially interesting model system exists for the study of the possible regulation of the repair of O^6-alkylguanine in mammalian cells. Human cell lines in culture differ in their capacity to repair O^6-alkylguanine in DNA. This phenomenon was first observed with respect to the competence of cells to reactivate adenovirus treated with MNNG (164, 176). Cells able to carry out such host-cell reactivation are designated as Mer^+. The majority of human fibroblast cell lines fall into this category, but many cell lines transformed with DNA viruses such as SV40 are Mer^- (164, 177). In addition, a number of human *tumor* cell lines are Mer^- (164, 176–178). The Mer phenotype appears to correlate with a phenotype designated Mex, which distinguishes cells (Mex^+) that effect efficient removal of O^6-methylguanine from their DNA from those (Mex^-) that do not (179–181).

Extracts of Mer^+ HeLa cells contain repair activity equivalent to approximately 100,000 molecules of methyl transferase per cell (162), if it is assumed that each molecule is catalytically active only once. Extracts of phytohemagglutinin(PHA)-stimulated Mer^+ lymphocytes apparently contain more O^6-methylguanine transferase activity than do extracts of unstimulated cells, and in general T cell lymphocytes have more activity than B cells do. However, there is great variability in the amount of enzyme activity in lymphocytes, ranging from 14,000 to more than 100,000 molecules per cell (154) and there is no solid evidence for the inducible expression of alkyl transferase activity in these or in Mer$^-$ cells.

2-4 Repair of Single-Strand Breaks by Direct Rejoining: A Further Example of DNA Repair by Reversal of Damage

Agents that promote the hydrolysis of phosphodiester bonds in duplex DNA are discussed in Chapter 1. Primary among these are ionizing radiation such as X rays. The repair of DNA strand breaks in *E. coli* requires DNA synthetic and/or recombinative events in the majority of cases, since such repair does not occur in mutants defective in these functions (182). However, in vitro, some fraction of single-strand breaks in DNA produced by ionizing radiation under anoxic conditions is repaired by simple rejoining of the ends (183), and such repair may be considered an example of the direct reversal of DNA damage. In these in vitro reactions the incubation of irradiated DNA with the enzyme *polynucleotide ligase* results in the loss of a fraction of the total DNA strand breaks as measured by sedimentation velocity in alkaline sucrose gradients (183). Polynucleotide ligase is a highly specific enzyme that is ubiquitous in its distribution and plays a role in most known biochemical pathways that require the rejoining of strand breaks in DNA (184, 185). A detailed discussion of this enzyme is presented in Section 5-5, where we shall consider the rejoining of newly synthesized DNA to extant DNA during the process of excision repair. The enzyme from *E. coli* has an absolute requirement for NAD (nicotinamide adenine dinucleotide) and for Mg^{2+} as cofactors (184, 185). All DNA ligases require free ends in duplex DNA, with no missing nucleotides at the site of the break, and the presence of adjacent 3′-OH and 5′-P termini (185) (Fig. 2-29). Thus only strand breaks with these particular characteristics produced by DNA damage are subject to repair by direct reversal.

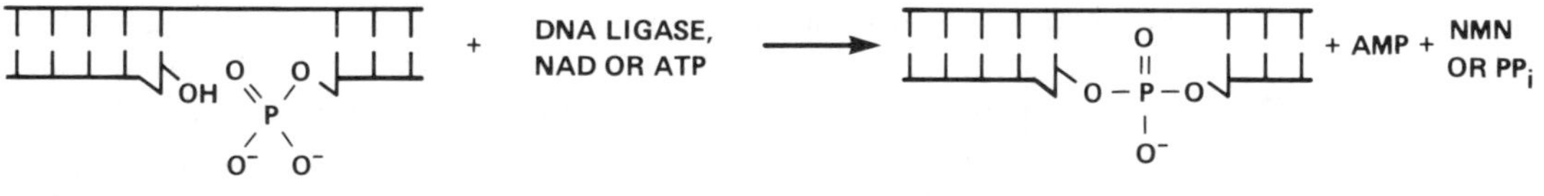

FIGURE 2-29
DNA ligase catalyzes the joining of strand breaks that contain juxtaposed 3′-OH and 5′-P termini in DNA. The enzyme from *E. coli* requires NAD as a cofactor and forms NMN plus AMP. The enzyme encoded by phage T4 requires ATP and forms PPi plus AMP. (NMN, nicotinamide mononucleotide). (From I. R. Lehman, ref. 184.)

2-5 Repair of Sites of Base Loss by Direct Insertion of Purines

In Section 1-2 depurination (the loss of purines from DNA) as a specific type of DNA damage was discussed. It was pointed out that such damage can occur spontaneously in living cells at 37°C (186) and that certain chemical modifications of purines that labilize the N-glycosylic bond linking the modified base to the deoxyribose-phosphate backbone can increase the rate of spontaneous depurination (186). Depurination can also result from enzyme-catalyzed hydrolysis of N-glycosylic bonds in DNA containing certain *damaged* purines (186). This class of enzymes, referred to as DNA glycosylases, is discussed in detail in the next chapter.

There appear to be multiple mechanisms for the repair of sites of base loss in duplex DNA. One of these involves the action of a class of enzymes called apurinic/apyrimidinic (AP) endonucleases, which specifically recognize sites of base loss in DNA (186, 187). The details of this repair pathway will also be discussed in conjunction with the discussion on DNA glycosylases in the next chapter. However, an alternative possible mechanism for the repair of apurinic sites is appropriately considered here, since it may represent yet another example of the direct reversal of DNA damage. The reaction in question involves the purported reinsertion of the appropriate purines into apurinic sites in duplex DNA, catalyzed by an enzyme activity called *DNA purine insertase* (188–193) (Fig. 2-30). Although there have been speculations about such a pathway ever since the discovery of DNA glycosylases, it is only in recent years that any direct, albeit controversial, biochemical evidence for such a repair mode has emerged.

A protein with a molecular weight greater than 120,000 daltons is reported to be present in human cell lines in tissue culture and binds specifically to duplex DNA containing apurinic sites (191, 192). In the presence of K^+, fractions containing this protein also catalyze the incorporation of purine bases in N-glycosylic covalent linkage in duplex apurinic DNA, utilizing either the corresponding free bases or deoxynucleosides as substrates. However, it is not yet certain that the purine insertase activity and the apurinic DNA-binding activity arise from the same protein (191, 192). The human insertase activity has no requirement for Mg^{2+} and is inhibited in the presence of either EDTA or caffeine and by dialysis or freezing. It demonstrates a remarkable specificity since it catalyzes the insertion of guanine but not adenine into depurinated poly(dG-dC) and of adenine but not guanine into depurinated poly(dA-dT). After incorporation into depurinated DNA, guanine has been reisolated as dGMP, suggesting its presence in glycosylic linkage with the deoxyribose-phosphate back-

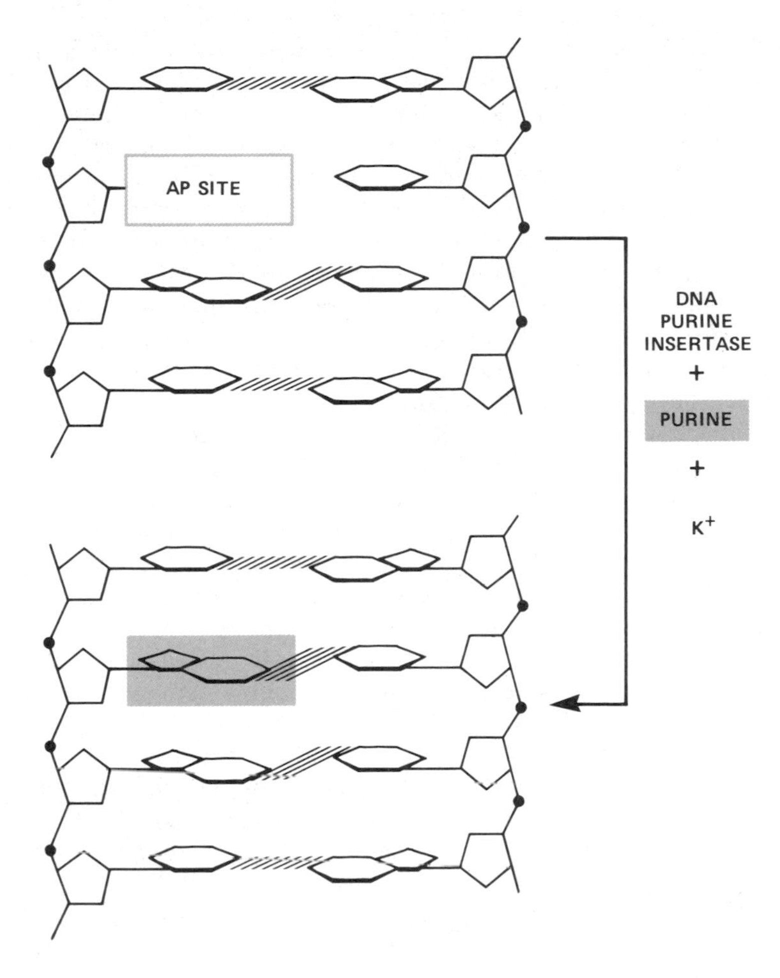

FIGURE 2-30
Schematic representation of the action of the putative human DNA purine insertase activity. In the presence of K^+ the enzyme catalyzes the direct insertion of the appropriate purine into an apurinic site in duplex DNA.

bone. Purine base insertion with the human enzyme obeys normal enzyme kinetics with regard to the concentration of substrate bases and yields an apparent K_m of 5 μM for both guanine and adenine (192).

A puzzling aspect of the purine insertase activity isolated from extracts of human cells is the absence of any obvious requirement for an energy source in the reaction. A possibility that is difficult to eliminate is that the extract contains trace amounts of an as yet unidentified energy donor. Alternatively, free energy may be provided by the reformation of normal base pairing and base stacking at the site of the reaction (191). A known reaction that bears some similarity to that catalyzed by the human DNA purine insertase is performed by the tRNA guanylation enzyme, which replaces the Q base of certain tRNAs with guanine (191). Like the putative DNA purine insertase, this enzyme has no apparent energy-donating cofactor. However, the tRNA reaction is a *replacement* one that involves both the cleavage and the reformation of a glycosylic bond, whereas the DNA purine insertion involves only the formation of such a bond. In addition, the guanylation reaction does not appear to require a duplex nucleic acid, whereas the purine insertion has an absolute requirement for duplex DNA containing apurinic sites (191, 192).

A more plausible reaction mechanism involves a purine insertase activity identified in crude extracts of *E. coli* (193). This activity, like that present in human cell extracts, is specific for purines and catalyzes the insertion of the correct missing purine into apurinic sites in duplex DNA. In contrast to the human cell enzyme, that from at least one strain of *E. coli* has an absolute requirement for the corresponding deoxynucleoside triphosphate as the purine donor, as well as a requirement for Mg^{2+} (193). This activity is inhibited in the presence of EDTA or caffeine and in the presence of ATP (188, 190, 193).

As yet no mutants defective in purine insertase activity have been isolated from either prokaryotes or eukaryotes, so that it is difficult to evaluate the biological significance of these biochemical findings. In fact, the existence of a DNA purine insertase in *E. coli* has been disputed by the demonstration that the preferential incorporation of adenine or guanine from dATP and dGTP respectively requires *E. coli* DNA polymerase I; i.e., it may result from the *excision repair* of apurinic sites in DNA (see Chapters 3 and 5) rather than by the direct insertion of purines (194).

Indirect support for a possible role of purine insertase in base excision repair comes from studies with xeroderma pigmentosum (XP) cells. As indicated earlier in this chapter (Section 2-2), XP is a hereditary condition in which affected individuals manifest a variety of clinical symptoms indicative of abnormal photosensitivity (195). Details of the biological and biochemical findings in this disease, particularly as they pertain to DNA repair, are discussed in Section 9-2. For the purposes of the present discussion, it should be noted that certain XP cell lines apparently lack a major AP endonuclease activity present in extracts of normal human fibroblasts (196, 197). However, these lines are more effective than normal cells in the reactivation of adenoviruses containing depurinated DNA, possibly reflecting the action of DNA purine insertase activity (198).

Thus far the published reports on DNA purine insertase activity have been limited, but this potentially fascinating mode of DNA repair by direct reversal of base damage merits much more extensive study. In the same vein, it is worth searching for enzyme activities that can catalyze the direct insertion of pyrimidines into apyrimidinic sites in DNA.

2-6 Summary and Conclusions

The reversal of damage in DNA is obviously the most direct mode of DNA repair and suggests a number of distinct possible advantages to a living cell.

1. In each of the four examples discussed above, only a single gene product is required—a highly economical use of genetic information. However, the alkyl transferase mode of repair is energetically expensive since an entire protein molecule is expended for each reaction.
2. As a corollary of the previous statement, the reversal mode of DNA repair is kinetically advantageous since it presumably occurs more rapidly than multistep biochemical pathways such as excision repair (see Chapters 3 through 6).
3. In general, these processes would tend to be relatively error-free due to their high degree of specificity, although in the case of purine insertion there is an obvious potential for insertion of the incorrect purine, or even of a pyrimidine instead of a purine. Whether or not these enzymes do have infidelity in vivo remains to be established. Furthermore, their actual contribution to biological end points such as cell survival relative to that facilitated by other repair pathways is yet to be established.

In the light of these considerations one might anticipate that reversal of DNA damage in single-step reactions would be a highly selected mode of DNA repair and that more examples are yet to be discovered in both prokaryotes and eukaryotes.

Suggestions for Further Reading

Photoreactivation, refs. 7–12, 19, 24, 84; repair of O^6-alkylguanine, 91–93, 120, 129; DNA purine insertase, 188.

References

1. Setlow, R. B., Swenson, P. A., and Carrier, W. L. 1963. Thymine dimers and inhibition of DNA synthesis by ultraviolet irradiation of cells. *Science* 142:1464.
2. Swenson, P. A., and Setlow, R. B. 1966. Effects of ultraviolet radiation on macromolecular synthesis in *E. coli*. *J. Mol. Biol.* 15:201.
3. Benbow, R. M., Zuccarelli, A. J., and Sinsheimer, R. L. 1974. A role for single-strand breaks in bacteriophage ϕX174 genetic recombination. *J. Mol. Biol.* 88:629.

4. Caillet-Fauquet, P., Defais, M., and Radman, M. 1977. Molecular mechanisms of induced mutagenesis. Replication *in vivo* of bacteriophage ϕX174 single-stranded, ultraviolet light–irradiated DNA in intact and irradiated host cells. *J. Mol. Biol.* 117:95.

5. Villani, G., Boiteux, S., and Radman, M. 1978. Mechanisms of ultraviolet-induced mutagenesis: extent and fidelity of *in vitro* DNA synthesis on irradiated templates. *Proc. Natl. Acad. Sci.* (USA) 75:3037.

6. Moore, P., and Strauss, B. S. 1979. Sites of inhibition of *in vitro* DNA synthesis in carcinogen- and UV-treated ϕX174 DNA. *Nature* 278:664.

7. Cook, J. S. 1970. Photoreactivation in animal cells. *Photophysiol.* 5:191.

8. Rupert, C. S., and Harm, W. 1966. Reactivation after photobiological damage. *Adv. Rad. Biol.* 2:1.

9. Rupert, C. S. 1975. Enzymatic photoreactivation: overview. In *Molecular mechanisms for repair of DNA.* P. C. Hanawalt and R. B. Setlow, eds., part A, p. 73. New York: Plenum.

10. Sutherland, B. M. 1978. Enzymatic photoreactivation of DNA. In *DNA repair mechanisms.* P. C. Hanawalt, E. C. Friedberg, C. F. Fox, eds., p. 113. New York: Academic.

11. Sutherland, B. M. 1977. Symposium on molecular mechanisms in photoreactivation. Introduction: fundamentals of photoreactivation. *Photochem. Photobiol.* 25:413.

12. Setlow, R. B. 1968. The photochemistry, photobiology, and repair of polynucleotides. *Prog. Nuc. Acids Res. Mol. Biol.* 8:257.

13. Chen, J., Huang, C. W., Hinman, L., Gordon, M. P., and Deranleau, D. A. 1976. Photomonomerization of pyrimidine dimers by indoles and proteins. *J. Theor. Biol.* 62:53.

14. Hélène, C., and Charlier, M. 1977. Photosensitized splitting of pyrimidine dimers by indole derivatives and by tryptophan-containing oligopeptides and proteins. *Photochem. Photobiol.* 25:429.

15. Toulmé, J-J., and Hélène, C. 1977. Specific recognition of single-stranded nucleic acids. Interaction of tryptophan-containing peptides with native, denatured, and ultraviolet-irradiated DNA. *J. Biol. Chem.* 252:244.

16. Hélène, C., Charlier, M., Toulmé, J-J., and Toulmé, F. 1978. Photosensitized splitting of thymine dimers in DNA by peptides and protein containing tryptophanyl residues. In *DNA repair mechanisms.* P. C. Hanawalt, E. C. Friedberg and C. F. Fox, eds., p. 141. New York: Academic.

17. Sutherland, J. C., and Griffin, K. P. 1980. Monomerization of pyrimidine dimers in DNA by tryptophan-containing peptides: wavelength dependence. *Radiat. Res.* 83:529.

18. Hélène, C., Toulmé, F., Charlier, M., and Yaniv, M. 1976. Photosensitized splitting of thymine dimers in DNA by gene 32 protein from phage T4. *Biochem. Biophys. Res. Comm.* 71:91.

19. Hélène, C. 1978. Workshop summary: mechanism and diversity of photoreactivation. In *DNA repair mechanisms.* P. C. Hanawalt, E. C. Friedberg and C. F. Fox, eds., p. 123. New York: Academic.

20. Kelner, A. 1949. Effect of visible light on the recovery of *Streptomyces griseus* conidia from ultraviolet irradiation injury. *Proc. Natl. Acad. Sci.* (USA) 35:73.

21. Dulbecco, R. 1949. Reactivation of ultraviolet inactivated bacteriophage by visible light. *Nature* 163:949.

22. Rupert, C. S., Goodgal, S. H., and Herriott, R. M. 1958. Photoreactivation *in vitro* of ultraviolet inactivated *Hemophilus influenzae* transforming factor. *J. Gen. Physiol.* 41:451.

23. Rupert, C. S. 1960. Photoreactivation of transforming DNA by an enzyme from baker's yeast. *J. Gen. Physiol.* 43:573.

24. Werbin, H. 1977. Yearly review. DNA photolyase. *Photochem. Photobiol.* 26:675.

25. Rupert, C. S. 1962. Photoenzymatic repair of ultraviolet damage in DNA. I. Kinetics of the reaction. *J. Gen. Physiol.* 45:703.

26. Rupert, C. S. 1962. Photoenzymatic repair of ultraviolet damage in DNA. II. Formation of an enzyme-substrate complex. *J. Gen. Physiol.* 45:725.

27. Madden, J. J., Werbin, H., and Denson, J. 1973. A rapid assay for DNA photolyase using a membrane-binding technique. *Photochem. Photobiol.* 18:441.

28. Madden, J. J., and Werbin, H. 1974. Use of membrane binding technique to study the kinetics of yeast deoxyribonucleic acid photolyase reactions. Formation of enzyme-substrate complexes in the dark and their photolysis. *Biochemistry* 13:2149.

29. Friedberg, E. C., and Hanawalt, P. C., eds. 1981. *DNA repair—a laboratory manual of research procedures.* Vol. I, part A. New York: Dekker.

30. Friedberg, E. C., and Hanawalt, P. C., eds. 1983. *DNA repair—a laboratory manual of research procedures.* Vol. 2. New York: Dekker.

31. Sutherland, B. M., and Chamberlin, M. J. 1973. A rapid and sensitive assay for pyrimidine dimers in DNA. *Anal. Biochem.* 53:168.

32. Farland, W. H., and Sutherland, B. M. 1981. Analysis of pyrimidine dimer content of isolated DNA by nuclease digestion. In *DNA repair—a laboratory manual of research procedures.* E. C. Friedberg and P. C. Hanawalt, eds., vol I, part A, p. 45. New York: Dekker.

33. Setlow, R. B., Carrier, W. L., and Bollum, F. J. 1964. Nuclease-resistant sequences in ultraviolet-irradiated deoxyribonucleic acid. *Biochim. Biophys. Acta.* 91:446.

34. Fox, C. F., and Weiss, S. B. 1964. Enzymatic synthesis of ribonucleic acid. II. Properties of the deoxyribonucleic acid-primed reaction with *Micrococcus lysodeikticus* ribonucleic acid polymerase. *J. Biol. Chem.* 239:175.

35. Hagen, U., Keck, K., Kröger, H., Zimmerman, F., and Lücking, T. 1965. Ultraviolet light inactivation of the priming ability of DNA in the RNA polymerase system. *Biochim. Biophys. Acta.* 95:418.

36. Piessens, J. P., and Eker, A. P. M. 1975. Photoreactivation of template activity of UV-irradiated DNA in an RNA-polymerase system. A rapid assay for photoreactivating enzyme. *FEBS Lett.* 50:125.

37. Wulff, D. L., and Rupert, C. S. 1962. Disappearance of thymine photodimer in ultraviolet irradiated DNA upon treatment with a photoreactivating enzyme from baker's yeast. *Biochim. Biophys. Res. Comm.* 7:237.

38. Setlow, J. K., and Boling, M. E. 1963. The action spectrum of an *in vitro* DNA photoreactivation system. *Photochem. Photobiol.* 2:471.

39. Muhammed, A. 1966. Studies on the yeast photoreactivating enzyme. I. A method for the large scale purification and some properties of the enzyme. *J. Biol. Chem.* 241:516.

40. Boatwright, D. T., Madden, J. J., Denson, J., and Werbin, H. 1975. Yeast DNA photolyase: molecular weight, subunit structure, and reconstitution of active enzyme from its subunits. *Biochemistry* 14:5418.

41. Werbin, H., and Madden, J. J. 1975. Purification of an inhibitor of DNA photolyase with fluorescent spectra similar to those of the enzyme. *Biochim. Biophys. Acta.* 383:160.

42. Werbin, H., and Madden, J. J. 1977. The subunit structure of yeast DNA photolyase and the purification of a fluorescent activator of the enzyme. *Photochem. Photobiol.* 25:421.

43. Madden, J. J., Denson, J., and Werbin, H. 1976. Purification from baker's yeast of an activator of DNA photolyase. *Biochim. Biophys. Acta.* 454:222.

44. Werbin, H., and Madden, J. 1978. Low molecular weight substances that enhance DNA photolyase activity. In *DNA repair mechanisms.* P. C. Hanawalt, E. C. Friedberg and C. F. Fox, eds., p. 133. New York: Plenum.

45. Iwatsuki, N., Joe, C. O., and Werbin, H. 1980. Evidence that deoxyribonucleic acid photolyase from baker's yeast is a flavoprotein. *Biochemistry* 19:1172.

46. Resnick, M. A. 1969. A photoreactivationless mutant of *Saccharomyces cerevisiae. Photochem. Photobiol.* 9:307.

47. MacQuillan, A. M., Herman, A., Coberly, J. S., and Green, G. 1981. A second photoreactivation-deficient mutation in *Saccharomyces cerevisiae. Photochem. Photobiol.* 34:673.

48. Madden, J. J., and Werbin, H. 1983. Personal communication.

49. Harm, W., Rupert, C. S., and Harm, H. 1972. Photoenzymatic repair of DNA. I. Investigation of the reaction by flash illumination. In *Molecular and cellular repair processes.* R. F. Beers, Jr., R. M. Herriott and R. C. Tilghman, eds., p. 53. Baltimore: Johns Hopkins University Press.

50. Rupert, C. S., Harm, W., and Harm, H. 1972. Photoenzymatic repair of DNA. II. Physical/chemical characterization of the process. In *Molecular and cellular repair processes.* R. F. Beers, Jr., R. M. Herriott and R. C. Tilghman, eds., p. 64. Baltimore: Johns Hopkins University Press.

51. Harm, W. 1975. Kinetics of photoreactivation. In *Molecular mechanisms for repair of DNA.* P. C. Hanawalt and R. B. Setlow, eds., p. 89. New York: Plenum.

52. Eker, A. P. M., and Fichtinger-Schepman, A. M. J. 1975. Studies on a DNA photoreactivating enzyme from *Streptomyces griseus.* II. Purification of the enzyme. *Biochim. Biophys. Acta.* 378:54.

53. Eker, A. P. M. 1978. Some properties of a DNA photoreactivating enzyme from *Streptomyces griseus.* In *DNA repair mechanisms.* P. C. Hanawalt, E. C. Friedberg and C. F. Fox, eds., p. 129. New York: Academic.

54. Eker, A. P. M. 1980. Photoreactivating enzyme from *Streptomyces griseus.* III. Evidence for the presence of an intrinsic chromophore. *Photochem. Photobiol.* 32:593.

55. Eker, A. P. M. 1981. Photoreactivating enzyme from *Streptomyces griseus.* IV. On the nature of the chromophoric cofactor in *Streptomyces griseus* photoreactivating enzyme. *Photochem. Photobiol.* 33:65.

56. Sutherland, B. M., Court, D., and Chamberlin, M. J. 1972. Studies on the DNA photoreactivating enzyme from *Escherichia coli.* 1. Transduction of the *phr* gene by bacteriophage lambda. *Virology* 48:87.

57. Youngs, D. A., and Smith, K. C. 1978. Genetic location of the *phr* gene of *Escherichia coli* K-12. *Mutation Res.* 51:133.

58. Sancar, A., and Rupert, C. S. 1978. Correction of the map location for the *phr* gene in *Escherichia coli* K-12. *Mutation Res.* 51:139.

59. Sutherland, B. M., and Hausrath, S. G. 1979. Multiple loci affecting photoreactivation in *Escherichia coli. J. Bacteriol.* 138:333.

60. Rupert, C. S., and Sancar, A. 1978. Cloning the *phr* gene of *Escherichia coli.* In *DNA repair mechanisms.* P. C. Hanawalt, E. C. Friedberg and C. F. Fox, eds., p. 159. New York: Academic.

61. Sancar, A., and Rupert, C. S. 1978. Cloning of the *phr* gene and amplification of photolyase in *Escherichia coli. Gene* 4:295.

62. Sancar, A., and Rupert, C. S. 1983. The *phr* gene of *Escherichia coli.* In *DNA repair—a laboratory manual of research procedures.* E. C. Friedberg and P. C. Hanawalt, eds., vol. 2, p. 241. New York: Dekker.

63. Sutherland, B. M., Chamberlin, M. J., and Sutherland, J. C. 1973. Deoxyribonucleic acid photoreactivating enzyme from *Escherichia coli. J. Biol. Chem.* 248:4200.

64. Snapka, R. M., and Fuselier, C. O. 1977. Photoreactivating enzyme from *Escherichia coli. Photochem. Photobiol.* 25:415.

65. Snapka, R. M., and Sutherland, B. M. 1980. *Escherichia coli* photoreactivating enzyme: purification and properties. *Biochemistry* 19:4201.

66. Jagger, J., and Latarjet, R. 1956. Spectres d'action de la photo-restauration chez *E. coli* B/r. *Ann. Inst. Pasteur* 91:858.

67. Jagger, J., Takebe, H., and Snow, J. M. 1970. Photoreactivation of killing in *Streptomyces:* action spectra and kinetic studies. *Photochem. Photobiol.* 12:185.

68. Cimino, G. D., and Sutherland, J. C. 1982. Photoreactivating enzyme from *Escherichia coli:* isolated enzyme lacks absorption in its actinic wavelength region and its ribonucleic acid cofactor is partially double stranded when associated with apoprotein. *Biochemistry* 21:3914.

69. Sutherland, J. C. 1978. Mechanism of action of the photoreactivating enzyme from *E. coli:* recent results. In *DNA repair mechanisms.* P. C. Hanawalt, E. C. Friedberg and C. F. Fox, eds., p. 137. New York: Academic.

70. Wun, K. L., Gih, A., and Sutherland, J. C. 1977. Photoreactivating enzyme from *Escherichia coli:* appearance of new absorption on binding to ultraviolet irradiated DNA. *Biochemistry* 16:921.

71. Sutherland, J. C. 1977. Photophysics and photochemistry of photoreactivation. *Photochem. Photobiol.* 25:435.

72. Rupert, C. S., and To, K. 1976. Substrate dependence of the action spectrum for photoenzymatic repair of DNA. *Photochem. Photobiol.* 24:229.

73. Sutherland, B. M. 1974. Photoreactivating enzyme from human leukocytes. *Nature* 248:109.

74. Sutherland, B. M., Runge, P., and Sutherland, J. C. 1974. DNA photoreactivating enzyme from placental mammals. Origin and characterization. *Biochemistry* 13:4710.

75. Sutherland, B. M. 1975. The human leukocyte photoreactivating enzyme. In *Molecular mechanisms for repair of DNA.* P. C. Hanawalt and R. B. Setlow, eds., p. 107. New York: Plenum.

76. Sutherland, J. C., and Sutherland, B. M. 1975. Human photoreactivating enzyme. Action spectrum and safelight conditions. *Biophys. J.* 15:435.

77. Wagner, E. K., Rice, M., and Sutherland, B. M. 1975. Photoreactivation of herpes simplex virus in human fibroblasts. *Nature* 254:627.

78. Sutherland, B. M., Oliver, R., Fuselier, C. O., and Sutherland, J. C. 1976. Photoreactivation of pyrimidine dimers in the DNA of normal and xeroderma pigmentosum cells. *Biochemistry* 15:402.

79. Sutherland, B. M., and Oliver, R. 1976. Culture conditions affect photoreactivating enzyme levels in human fibroblasts. *Biochim. Biophys. Acta.* 442:358.

80. Sutherland, B. M., Harber, L. C., and Kochevar, I. E. 1980. Pyrimidine dimer formation and repair in human skin. *Cancer Res.* 40:3181.

81. Ananthaswamy, H. M., and Fisher, M. S. 1981. Photoreactivation of ultraviolet radiation–induced pyrimidine dimers in neonatal BALB/c mouse skin. *Cancer Res.* 41:1829.

82. Mortelmans, K., Cleaver, J. E., Friedberg, E. C., Paterson, M. C., Smith, B. P., and Thomas, G. H. 1977. Photoreactivation of thymine dimers in UV-irradiated human cells: unique dependence on culture conditions. *Mutation Res.* 44:433.

83. Cleaver, J. E. 1966. Photoreactivation: a radiation repair mechanism absent from mammalian cells. *Biochem. Biophys. Res. Comm.* 24:569.

84. Cook, J. S. 1972. Photoenzymatic repair in animal cells. In *Molecular and cellular repair processes.* R. F. Beers, Jr., R. M. Herriott and R. C. Tilghman, eds., p. 79. Baltimore: Johns Hopkins University Press.

85. Fornace, A. J., Jr. 1982. Measurement of *M. luteus* endonuclease-sensitive lesions by alkaline elution. *Mutation Res.* 94:263.

86. Sutherland, B. M., and Oliver, R. 1975. Low levels of photoreactivating enzyme in xeroderma pigmentosum variants. *Nature* 257:132.

87. Sutherland, B. M., Rice, M., and Wagner, E. K. 1975. Xeroderma pigmentosum cells contain low levels of photoreactivating enzyme. *Proc. Natl. Acad. Sci. (USA)* 72:103.

88. Gordon, M. P. 1975. Photorepair of RNA. In *Molecular mechanisms for repair of DNA.* P. C. Hanawalt and R. B. Setlow, eds., p. 115. New York: Plenum.

89. Hurter, J., Gordon, M. P., Kirwan, J. P., and McLaren, A. D. 1974. *In vitro* photoreactivation of ultraviolet-inactivated ribonucleic acid from tobacco mosaic virus. *Photochem. Photobiol.* 19:185.

90. Towill, L., Huang, C. W., and Gordon, M. P. 1977. Photoreactivation of DNA-containing cauliflower mosaic virus and tobacco mosaic virus RNA on *Datura. Photochem. Photobiol.* 25:249.

91. Hanawalt, P. C., Cooper, P. K., Ganesan, A. K., and Smith, C. A. 1979. DNA repair in bacterial and mammalian cells. *Ann. Rev. Biochem.* 48:783.

92. Lindahl, T. 1982. DNA repair enzymes. *Ann. Rev. Biochem.* 51:61.

93. Schendel, P. 1981. Inducible repair systems and their implications for toxicology. *CRC Crit. Rev. Toxicol.* 8:311.

94. Loveless, A. 1969. Possible relevance of O-6 alkylation of deoxyguanosine to the mutagenicity and carcinogenicity of nitrosamines and nitrosamides. *Nature* 223:206.

95. Coulondre, C., and Miller, J. H. 1977. Genetic studies of the *lac* repressor. IV. Mutagenic specificity in the *lac*I gene of *Escherichia coli. J. Mol. Biol.* 117:577.

96. Cerdá-Olmedo, E., and Hanawalt, P. C. 1968. Diazomethane as the active agent in nitrosoguanidine mutagenesis and lethality. *Mol. Gen. Genet.* 101:191.

97. Neale, S. 1972. Effect of pH and temperature on nitrosamide-induced mutation in *Escherichia coli*. *Mutation Res.* 14:155.

98. Jiménez-Sánchez, A., and Cerdá-Olmedo, E. 1975. Mutation and DNA replication in *Escherichia coli* treated with low concentrations of N-methyl-N′-nitro-N-nitroso-guanidine. *Mutation Res.* 28:337.

99. Neale, S. 1976. Mutagenicity of nitrosamides and nitrosamidines in microorganisms and plants. *Mutation Res.* 32:229.

100. Samson, L., and Cairns, J. 1977. A new pathway for DNA repair in *Escherichia coli*. *Nature* 267:281.

101. Jeggo, P., Defais, M., Samson, L., and Schendel, P. 1977. An adaptive response of *E. coli*. to low levels of alkylating agent: comparison with previously characterized DNA repair pathways. *Mol. Gen. Genet.* 157:1.

102. Jeggo, P., Defais, M., Samson, L., and Schendel, P. 1978. An adaptive response of *E. coli* to low levels of alkylating agent. In *DNA synthesis: present and future*. I. Molineux and M. Kohiyama, eds., p. 1011. New York: Plenum.

103. Jeggo, P., Defais, M., Samson, L., and Schendel, P. 1978. An adaptive response of *E. coli* to low levels of alkylating agent: the role of *pol*A in killing adaptation. *Mol. Gen. Genet.* 162:299.

104. Jeggo, P. 1980. The adaptive response of *E. coli*: A comparison of its two components, killing and mutagenic adaptation. In *Progress in environmental mutagenesis*. M. Alečević, ed., p. 153. Amsterdam: Elsevier/North-Holland Biomedical Press.

105. Evensen, G., and Seeberg, E. 1982. Adaptation to alkylation resistance involves the induction of a DNA glycosylase. *Nature* 296:773.

106. Karran, P., Stevens, S., and Sedgwick, B. 1982. The adaptive response to alkylating agents; the removal of O^6-methylguanine from DNA is not dependent on DNA polymerase I. *Mutation Res.* 104:67.

107. Jeggo, P. 1979. Isolation and characterization of *Escherichia coli* K-12 mutants unable to induce the adaptive response to simple alkylating agents. *J. Bacteriol.* 139:783.

108. Sedgwick, B. 1982. Genetic mapping of *ada* and *adc* mutations affecting the adaptive response of *Escherichia coli* to alkylating agents. *J. Bacteriol.* 159:984.

109. Kataska, H., Yamamoto, Y., and Sekiguchi, M. 1983. A new gene (*alk*B) of *Escherichia coli* that controls sensitivity to methyl methane sulfonate. *J. Bacteriol.* 153:1301.

110. Schendel, P. F., and Robins, P. E. 1978. Repair of O^6-methylguanine in adapted *Escherichia coli*. *Proc. Natl. Acad. Sci.* (USA) 75:6017.

111. Robins, P., and Cairns, J. 1979. Quantitation of the adaptive response to alkylating agents. *Nature* 280:74.

112. Karran, P., Lindahl, T., and Griffin, B. 1979. Adaptive response to alkylating agents involves alteration *in situ* of O^6-methylguanine residues in DNA. *Nature* 280:76.

113. Foote, R. S., Mitra, S., and Pal, B. C. 1980. Demethylation of O^6-methylguanine in a synthetic DNA polymer by an inducible activity in *Escherichia coli*. *Biochem. Biophys. Res. Comm.* 97:654.

114. Sedgwick, B., and Robins, P. 1980. Isolation of mutants of *Escherichia coli* with increased resistance to alkylating agents: mutants deficient in thiols and mutants constitutive for the adaptive response. *Mol. Gen. Genet.* 180:85.

115. Olsson, M., and Lindahl, T. 1980. Repair of alkylated DNA in *Escherichia coli*. Methyl group transfer from O^6-methylguanine to a protein cysteine residue. *J. Biol. Chem.* 255:10569.

116. Demple, B., Jacobsson, A. Olsson, M., Robins, P., and Lindahl, T. 1982. Repair of alkylated DNA in *E. coli;* physical properties of O^6-methylguanine-DNA methyltransferase. *J. Biol. Chem.* 257:13776.

117. Demple, B., Jacobsson, A., Olsson, M., Karran, P., and Lindahl, T. 1983. Isolation of O^6-methylguanine-DNA methyltransferase from *E. coli*. In *DNA repair—a laboratory manual of research procedures*. E. C. Friedberg and P. C. Hanawalt, eds., vol. 2, p. 41. New York: Dekker.

118. Lindahl, T. 1981. DNA methyl transferase acting on O^6-methylguanine residues in adapted *E. coli*. In *Chromosome damage and repair*. E. Seeberg and K. Kleppe, eds., p. 207. New York: Plenum.

119. Lindahl, T., Demple, B., and Robins, P. 1982. Suicide inactivation of the *E. coli* O^6-methylguanine-DNA methyltransferase. *EMBO Journal* 1:1359.

120. Lindahl, T., Sedgwick, B., Demple, B., and Karran, P. 1983. Enzymology and regulation of the adaptive response to alkylating agents. In *Cellular responses to DNA damage*. E. C. Friedberg and B. A. Bridges, eds., p. 241. New York: Alan R. Liss.

121. Sedgwick, B., and Lindahl, T. 1982. A common mechanism for repair of O^6-methylguanine and O^6-ethylguanine in DNA. *J. Mol. Biol.* 154:169.

122. Schendel, P. F., Edington, B. V., McCarthy, J. G., and Todd, M. L. 1983. Repair of alkylation damage in *E. coli*. In *Cellular responses to DNA damage*. E. C. Friedberg and B. A. Bridges, eds., p. 227. New York: Alan R. Liss.

123. Warren, W., and Lawley, P. D. 1980. The removal of alkylation products from the DNA of *E. coli* cells treated with the carcinogens N-ethyl-N-nitrosourea and N-methyl-N-nitrosourea: influence of growth conditions and DNA repair defects. *Carcinogenesis* 1:67.

124. Jump, D. B., and Smulson, M. 1980. Purification and characterization of the major nonhistone protein acceptor for poly (adenosine diphosphate ribose) in HeLa cell nuclei. *Biochemistry* 19:1024.

125. Ogata, N., Ueda, K., Kawaichi, M., and Hayaishi, O. 1981. Poly (ADP-ribose) synthetase, a main acceptor of poly (ADP-ribose) in isolated nuclei. *J. Biol. Chem.* 256:4135.

126. Brox, L. W., and Hampton, A. 1968. Inactivation of GMP reductase by 6-chloro-, 6-mercapto-, and 2-amino-6-mercapto-9-β-D-ribofuranosylpurine 5′-phosphates. *Biochemistry* 7:398.

127. Bellisario, R. L., Maley, G. F., Galivan, J. H., and Maley, F. 1976. Amino acid sequence at the dUMP binding site of thymidylate synthetase. *Proc. Natl. Acad. Sci.* (USA) 73:1948.

128. Mitra, S., Pal, B. C., and Foote, R. S. 1982. O^6-methylguanine-DNA methyltransferase in wild-type and *ada* mutants of *Escherichia coli*. *J. Bacteriol.* 152:534.

129. Lindahl, T., Rydberg, B., Hjelmgren, T., Olsson, M., and Jacobsson, A. 1982. Cellular defense mechanisms against alkylation of DNA. In *Molecular and cellular mechanisms of mutagenesis*. J. F. Lemontt and W. M. Generoso, eds., p. 89. New York: Plenum.

130. Rydberg, B., and Lindahl, T. 1982. Nonenzymatic methylation of DNA by the intracellular methyl group donor S-adenosyl-L-methionine is a potentially mutagenic reaction. *EMBO Journal* 1:211.

131. Cerdá-Olmedo, E., Hanawalt, P. C., and Guerola, N. Mutagenesis of the replication point by nitrosoguanidine: map and pattern of replication of the *Escherichia coli* chromosome. *J. Mol. Biol.* 33:705.

132. Schendel, P. F., Defais, M., Jeggo, P., Samson, L., and Cairns, J. 1978. Pathways of mutagenesis and repair in *Escherichia coli* exposed to low levels of simple alkylating agents. *J. Bacteriol.* 135:466.

133. Schendel, P. F., Defais, M., Jeggo, P., Samson, L., and Cairns, J. 1978. Pathways involved in repair of alkylation damage in *E. coli*. In *DNA repair mechanisms*. P. C. Hanawalt, E. C. Friedberg and C. F. Fox, eds., p. 391. New York: Academic.

134. Lawley, P. D. 1976. Comparison of alkylating agent and radiation carcinogenesis: some aspects of the possible involvement of effects on DNA. In *Biology of radiation carcinogenesis*. J. M. Yuhas, R. W. Tennant and J. D. Regan, eds., p. 165. New York: Raven Press.

135. Lawley, P. D. 1974. Some chemical aspects of dose-response relationships in alkylation mutagenesis. *Mutation Res.* 23:283.

136. Hadden, C. T., Foote, R. S., and Mitra, S. 1983. Adaptive response of *Bacillus subtilis* to N-methyl-N′-nitro-N-nitrosoguanidine. *J. Bacteriol.* 153:756.

137. Newbold, R. F., Warren, W., Medcalf, A. S. C., and Amos, J. 1980. Mutagenicity of carcinogenic methylating agents is associated with a specific DNA modification. *Nature* 283:596.

138. Swan, P. F., and Magee, P. N. 1979. Induction of rat kidney tumours by ethyl methylsulfonate and nervous tissue tumours by methylmethane sulfonate. *Proc. Natl. Acad. Sci.* (USA) 71:639.

139. Goth, R., and Rajewsky, M. 1974. Persistence of O^6-ethylguanine in rat brain DNA: correlation with nervous system–specific carcinogenesis by ethylnitrosourea. *Proc. Natl. Acad. Sci.* (USA) 71:639.

140. Margison, G. P., and Kleihues, P. 1975. Chemical carcinogenesis in the nervous system: preferential accumulation of O^6-methylguanine in rat brain DNA during repetitive administration of methylnitrosourea. *Biochem. J.* 148:521.

141. Nicoll, J., Swann, P., and Pegg, A. 1975. Effect of dimethylnitrosamine on persistence of methylated guanines in rat liver and kidney DNA. *Nature* 254:261.

142. Kleihues, P., and Margison, G. P. 1976. Exhaustion and recovery of repair excision of O^6-methylguanine from rat liver DNA. *Nature* 259:153.

143. Buecheler, J., and Kleihues, P. 1977. Excision of O^6-methylguanine from DNA of various mouse tissues following a single injection of N-methyl-N-nitrosourea. *Chem.-Biol. Interact.* 16:327.

144. Scherer, E., Steward, A. P., and Emmelot, P. 1977. Kinetics of formation of O^6-ethylguanine in, and its removal from, liver DNA of rats receiving dimethylnitrosamine. *Chem.-Biol. Interact.* 19:1.

145. Pegg, A. E. 1978. Enzymatic removal of O^6-methylguanine from DNA by mammalian cells extracts. *Biochem. Biophys. Res. Comm.* 84:166.

146. Pegg, A. E., and Balog, B. 1979. Formation and subsequent excision of O^6-ethylguanine from DNA of rat liver following administration of diethylnitrosamine. *Cancer Res.* 39:5003.

147. Buckley, J. D., O'Conner, P. J., and Craig, A. W. 1979. Pretreatment with acetylaminofluorene enhances the repair of O^6-methylguanine in DNA. *Nature* 281:403.

148. Swan, P. F., and Mace, R. 1980. Changes in O^6-methylguanine disappearance from rat liver DNA during chronic dimethylnitrosamine administration. A possible similarity between the system removing O^6-methylguanine from DNA in rat liver and in *Escherichia coli* adapted to N-methyl-N′-nitro-N-nitrosoguanidine. *Chem.-Biol. Interact.* 31:239.

149. Smith, G. J., Kaufmann, D. G., and Grisham, J. W. 1980. Decreased excision of O^6-methylguanine and N^7-methylguanine during the S phase in $10T\frac{1}{2}$ cells. *Biochem. Biophys. Res. Comm.* 92:787.

150. Renard, A., and Verly, W. G. 1980. Kinetic analysis of O^6-ethylguanine disappearance from DNA catalyzed by the chromatin factor of rat liver. *FEBS Lett.* 122:271.

151. Chu, Y-H., Craig, A. W., and O'Conner, P. J. 1981. Repair of O^6-methylguanine in rat liver DNA is enhanced by pretreatment with single or multiple doses of aflatoxin B_1. *Br. J. Cancer* 43:850.

152. O'Conner, P. J., and Margison, G. P. 1981. The enhanced repair of O^6-alkylguanine in mammalian systems. In *Chromosome damage and repair*. E. Seeberg and K. Kleppe, eds., p. 233. New York: Plenum.

153. Teo, I. A., and Karran, P. 1982. Excision of O^6-methylguanine from DNA by human fibroblasts determined by a sensitive competition method. *Carcinogenesis* 3:923.

154. Waldstein, E. A., Cao, E-H., Bender, M. A., and Setlow, R. B. 1982. Abilities of extracts of human lymphocytes to remove O^6-methylguanine from DNA. *Mutation Res.* 95:405.

155. Montesano, R., Bresil, H., Drevon, C., and Piccoli, C. 1982. DNA repair in mammalian cells exposed to multiple doses of alkylating agents. *Biochimie* 64:591.

156. Metha, J. R., Ludlum, D. B., Renard, A., and Verly, W. G. 1981. Repair of O^6-ethylguanine in DNA by a chromatin fraction from rat liver: transfer of the ethyl group to an acceptor protein. *Proc. Natl. Acad. Sci.* (USA) 78:6766.

157. Pegg, A. E., and Perry, W. 1981. Stimulation of transfer of methyl groups from O^6-methylguanine in DNA to protein by rat liver extracts in response to hepatotoxins. *Carcinogenesis* 2:1195.

158. Bogden, J. M., Eastman, A., and Bresnick, E. 1981. A system in mouse liver for the repair of O^6-methylguanine lesions in methylated DNA. *Nucleic Acids Res.* 9:3089.

159. Lemâitre, M., Renard, A., and Verly, W. G. 1982. A common chromatin factor involved in the repair of O^6-methylguanine and O^6-ethylguanine lesions in DNA. *FEBS Lett.* 144:242.

160. Pegg, A. E., Roberfroid, M., von Bahr, C., Foote, R. S., Mitra, S., Bresil, H., Likhachev, A., and Montesano, R. 1982. Removal of O^6-methylguanine from DNA by human liver fractions. *Proc. Natl. Acad. Sci.* (USA) 79:5162.

161. Waldstein, E. A., Cao, E-H., and Setlow, R. B. 1982. Adaptive increase of O^6-methylguanine-acceptor protein in HeLa cells following N-methyl-N′-nitro-N-nitrosoguanidine treatment. *Nucleic Acids Res.* 10:4595.

162. Foote, R. S., Pal, B. C., and Mitra, S. 1983. Quantitations of O^6-methylguanine-DNA methyltransferase in HeLa cells. *Mutation Res.* 119:221.

163. Harris, A., Karran, P., and Lindahl, T. 1983. O^6-methylguanine-DNA methyltransferase of human lymphoid cells. *Cancer Res.* 43:3247.

164. Yarosh, D. B., Rice, M., Ziolkowski, C. H. J., Day, R. S. III, and Scudiero, D. A. 1983. O^6-methylguanine-DNA methyltransferase in human tumor cells. In *Cellular responses to DNA damage*. E. C. Friedberg and B. A. Bridges, eds., p. 261. New York: Alan R. Liss.

165. Renard, A., Lemâitre, M., and Verly, W. G. 1983. The O^6-alkylguanine transferase activity of rat liver chromatin. In *Cellular responses to DNA damage*. E. C. Friedberg and B. A. Bridges, eds., p. 255. New York: Alan R. Liss.

166. Cleaver, J. E. 1984. DNA repair and replication. In *Biochemical mechanisms of genetic disorders*. P. J. Benke, ed. New York: Dekker (in press).

167. Samson, L., and Schwartz, J. L. 1980. Evidence for an adaptive DNA repair pathway in CHO and human skin fibroblast cell lines. *Nature* 287:861.

168. Durrant, L. G., Margison, G. P., and Boyle, J. M. 1981. Pretreatment of Chinese hamster V79 cells with MNU increases survival without affecting DNA repair or mutagenicity. *Carcinogenesis* 2:55.

169. Jostes, R., Samson, L., and Schwartz, J. L. 1981. Kinetics of mutation and sister-chromatid exchange induction by ethyl methanesulfonate in Chinese hamster ovary cells. *Mutation Res.* 91:255.

170. Chang, M. J. W., Webb, T. E., and Koestner, A. 1979. Distribution of O^6-methylguanine in rat DNA following pretreatment *in vivo* with methylnitrosourea. *Cancer Lett.* 6:123.

171. Fox, M., Sultani-Makzoumi, C. M., and Boyle, J. M. 1982. A search for adaptive or inducible responses to DNA damage in V79 Chinese hamster cells. *Biochimie* 64:687.

172. Karran, P., Arlett, C. F., and Broughton, B. C. 1982. An adaptive response to the cytotoxic effects of N-methyl-N-nitrosourea is apparently absent in normal human fibroblasts. *Biochimie* 64:717.

173. Montesano, R., Bresil, H., Planche-Martel, G., Margison, C. P., and Pegg, A. E. 1980. Effect of chronic treatment of rats with dimethylnitrosamine on the removal of O^6-methylguanine from DNA. *Cancer Res.* 40:452.

174. Pegg, A. E. 1980. Formation and subsequent repair of alkylation lesions in tissues of rodents treated with nitrosamines. *Arch. Toxicol. Suppl.* 3:55.

175. Swenberg, J. A., Bedell, M. A., Billings, K. C., Umberhauer, D. R., and Pegg, A. E. 1982. Cell-specific differences in O^6-alkylguanine DNA repair activity during continuous exposure to carcinogen. *Proc. Natl. Acad. Sci.* (USA) 79:5499.

176. Day, R. III, and Ziolkowski, C. 1979. Human brain tumour cell strains with deficient host-cell reactivation of N-methyl-N′-nitro-N-nitrosoguanidine-damaged adenovirus 5. *Nature* 279:797.

177. Day, R. III, Ziolkowski, C., Scudiero, D., Meyer, S., Lubiniecki, A., Girardi, A., Galloway, S., and Bynum, G. 1980. Defective repair of alkylated DNA by human tumour and SV40-transformed human cell strains. *Nature* 288:724.

178. Day, R. III, and Ziolkowski, C. 1981. MNNG-pretreatment of a human kidney carcinoma cell strain decreases its ability to repair MNNG-treated adenovirus 5. *Carcinogenesis* 2:213.

179. Sklar, R., and Strauss, B. 1981. Removal of O^6-methylguanine from DNA of normal and xeroderma pigmentosum-derived lymphoblastoid lines. *Nature* 289:417.

180. Sklar, R., Brady, K., and Strauss, B. 1981. Limited capacity for the removal of O^6-methylguanine and its regeneration in a human lymphoma line. *Carcinogenesis* 2:1293.

181. Ayres, K., Sklar, R., Larson, K., Lindgren, V., and Strauss, B. 1982. Regulation of the capacity for O^6-methylguanine removal from DNA in human lymphoblastoid cells studied by cell hybridization. *Mol. Cell. Biol.* 2:904.

182. Town, C. D., Smith, K. C., and Kaplan, H. A. 1973. Repair of X-ray damage to bacterial DNA. *Curr. Top. Radiat. Res. Quart.* 8:351.

183. Jacobs, A., Bopp, A., and Hagen, U. 1972. *In vitro* repair of single-strand breaks in γ-irradiated DNA by polynucleotide ligase. *Int. J. Radiat. Biol.* 22:431.

184. Lehman, I. R. 1974. DNA ligase: structure, mechanism and function. *Science* 186:790.

185. Kornberg, A. 1980. *DNA synthesis.* San Francisco: Freeman.

186. Lindahl, T. 1979. DNA glycosylases, endonucleases for apurinic/apyrimidinic sites, and base-excision repair. *Prog. Nuc. Acids Res. Mol. Biol.* 22:135.

187. Friedberg, E. C., Bonura, T., Radany, E. H., and Love, J. D. 1981. Enzymes that incise damaged DNA. In *The enzymes.* 3rd ed. Nucleic acids, part A, P. D. Boyer, ed., p. 251. New York: Academic.

188. Livneh, Z., and Sperling, J. 1981. DNA base-insertion enzymes (insertases). In *The enzymes.* 3rd ed. Nucleic acids, part A, P. D. Boyer, ed., p. 549. New York: Academic.

189. Linn, S., Demple, B., Mosbaugh, D. W., Warner, H. R., and Deutsch, W. A. 1981. Enzymatic studies of base excision repair in cultured human fibroblasts and in *Escherichia coli.* In *Chromosome damage and repair.* E. Seeberg and K. Kleppe, eds., p. 97. New York: Plenum.

190. Livneh, Z., and Sperling, J. 1981. Base replacement mechanisms for the repair of unnatural and damaged bases in DNA. In *Chromosome damage and repair.* E. Seeberg and K. Kleppe, eds., p. 123. New York: Plenum.

191. Deutsch, W. A., and Linn, S. 1979. DNA binding activity from cultured human fibroblasts that is specific for partially depurinated DNA and that inserts purines into apurinic sites. *Proc. Natl. Acad. Sci.* (USA) 76:141.

192. Deutsch, W. A., and Linn, S. 1979. Further characterization of a depurinated DNA-purine base insertion activity from cultured human fibroblasts. *J. Biol. Chem.* 254:12099.

193. Livneh, Z., Elad, D., and Sperling J. 1979. Enzymatic insertion of purine bases into depurinated DNA *in vitro. Proc. Natl. Acad. Sci.* (USA) 76:1089.

194. Kataoka, H., and Sekiguchi, M. 1982. Are purine bases enzymatically inserted into depurinated DNA in *Escherichia coli*? *J. Biochem.* 92:971.

195. Friedberg, E. C., Ehmann, U. K., and Williams, J. I. 1979. Human diseases associated with defective DNA repair. *Adv Rad. Biol.* 8:86.

196. Kuhnlein, U., Penhoet, E. E., and Linn, S. 1976. An altered apurinic DNA endonuclease activity in group A and group D xeroderma pigmentosum fibroblasts. *Proc. Natl. Acad. Sci.* (USA) 73:1169.

197. Kuhnlein, U., Lee, B., Penhoet, E. E., and Linn, S. 1978. Xeroderma pigmentosum fibroblasts of the D group lack an apurinc DNA endonuclease species with a low apparent K_m. *Nucleic Acids Res.* 5:951.

198. Kudrna, R. D., Smith, J., Linn, S., and Penhoet, E. E. 1979. Survival of apurinic SV40 DNA in the D complementation group of xeroderma pigmentosum. *Mutation Res.* 62:173.

199. Sutherland, B. M., and Oliver, R. 1976. Inheritance of photoreactivating enzyme deficiencies in human cells. *Photochem. Photobiol.* 24:449.

3

Excision Repair. I. DNA Glycosylases and AP Endonucleases

3-1 Introduction

The forms of DNA damage that can be repaired by direct reversal are apparently limited in number. The most general DNA repair mode observed in nature is one in which damaged or inappropriate bases are excised from the genome and replaced by the normal nucleotide sequence and chemistry. This cellular response to DNA damage is appropriately referred to as *excision repair* and is the subject of this chapter and the succeeding three chapters. The most extensive information available on the topic stems from studies on prokaryote systems, particularly the enteric bacterium *E. coli* and the bacteriophage T4. However, comparisons will be made with other prokaryotes and with some lower and higher eukaryotes where relevant.

At present we know of a number of closely related but distinct enzymatic mechanisms by which the excision of damaged and/or inappropriate bases occurs in living cells (1–3). These are diagrammatically illustrated in Figs. 3-1 through 3-4. Some forms of base damage can be excised as the *free* base by the action of a distinct class of DNA repair enzymes called *DNA glycosylases* (1–6), which catalyze the hydrolysis of the N-glycosylic bonds linking bases to the deoxyribose-phosphate backbone (Fig. 3-1). This reaction still leaves apurinic or apyrimidinic sites in the DNA, and, whether these arise by the action of DNA glycosylases or by the spontaneous hydrolysis of N-glycosylic bonds, their removal by excision repair requires the action of one or more *nucleases*. Enzymes that specifically recognize sites of base loss in DNA are called *apurinic/apyrimidinic (AP) endonucleases* (1–3, 5, 7), and they effect the *incision* or *nicking* of DNA by catalyzing the hydrolysis of phosphodiester bonds at sites of base loss (Fig. 3-1).

In theory, the sequential action of a 5′-acting and a 3′-acting AP endonuclease (or vice versa) can result in the excision of apurinic or apyrimidinic deoxyribose-phosphate moieties, leaving a gap of just one nucleotide (Fig. 3-1). Alternatively, the action of just a single AP endonuclease (either 5′- or 3′-acting) will generate an incision (nick or strand break) that provides substrate sites for enzymes called *exonucleases*, which degrade DNA in the $5' \rightarrow 3'$ or $3' \rightarrow 5'$ direction at the free ends created by incisions (1–3, 8). These exonucleases are not uniquely involved in excision repair of DNA. They degrade DNA with free ends in a variety of metabolic processes, including replication and recombination (8). Hence the removal of damaged bases and/or deoxyribose-phosphate moieties can be viewed as a gratuitous result of exonucleolytic degradation.

The great majority of the known forms of base damage in DNA are not recognized by specific DNA glycosylases. How are these base damages excised? A third type of enzyme involved in excision repair of DNA belongs to a different class of endonucleases that specifically create incisions in the DNA backbone near sites of base damage (9, 10). Unlike the DNA glycosylases and AP endonucleases, which attack particular forms of base damage, this class of endonucleases apparently recognizes conformational distortions of the secondary structure of the DNA produced by a wide variety of base damages that in many instances bear no obvious structural relationship. This class of enzymes is generally referred to in this text as *damage-specific DNA incising activities*.

It is obvious from the above description that the incision of damaged DNA can occur by two rather distinct mechanisms. If the genome contains damaged bases (or inappropriate bases such as uracil) that are recognized by DNA glycosylases, incision occurs by a two-step reaction involving the *sequential* action of a DNA glycosylase and an AP endonuclease (Fig. 3-1). A second mechanism for the incision of damaged DNA is by the *direct* action of a damage-specific DNA incising activity; therefore in some contexts this type of enzyme

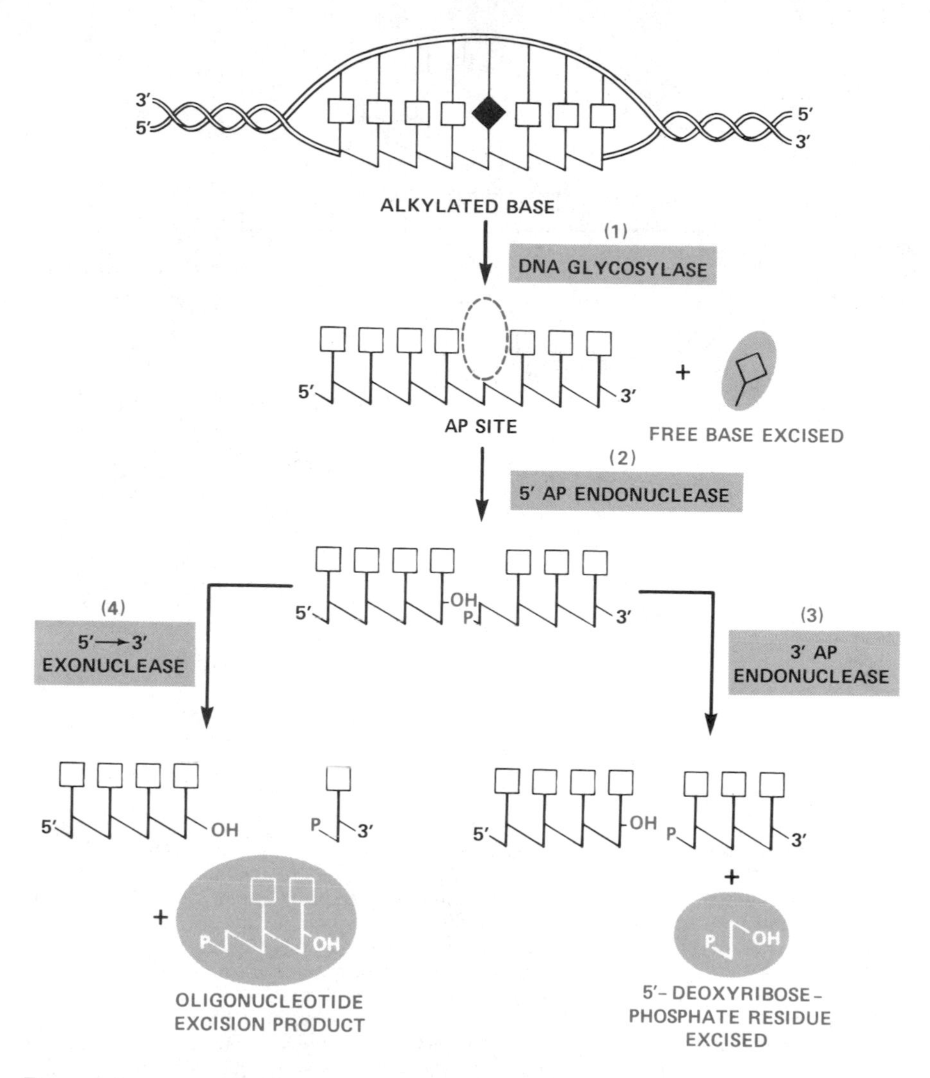

FIGURE 3-1
Schematic representations of excision repair of DNA are shown in this figure and in Figures 3-2, 3-3 and 3-4. In each of these figures the relevant region of only one of the two DNA strands is shown. Some forms of base damage in duplex DNA (e.g., alkylations at certain positions) are recognized by specific DNA glycosylases that catalyze excision of the base (1), leaving apurinic or apyrimidinic (AP) sites in the DNA. When such sites are attacked by 5′-acting AP endonucleases (2), the resulting 5′-terminal deoxyribose-phosphate moieties can be excised by the action of either a 3′ AP endonuclease (3) or a 5′ → 3′ exonuclease (4), leaving gaps in the affected strands.

will be referred to as a *direct-acting endonuclease* to distinguish its mode of action from the two-step mechanism described above.

For many years it was believed that damage-specific DNA incising activities catalyzed the formation of a single incision 5′ to the site of base damage and that the actual *excision* of the damaged nucleotides

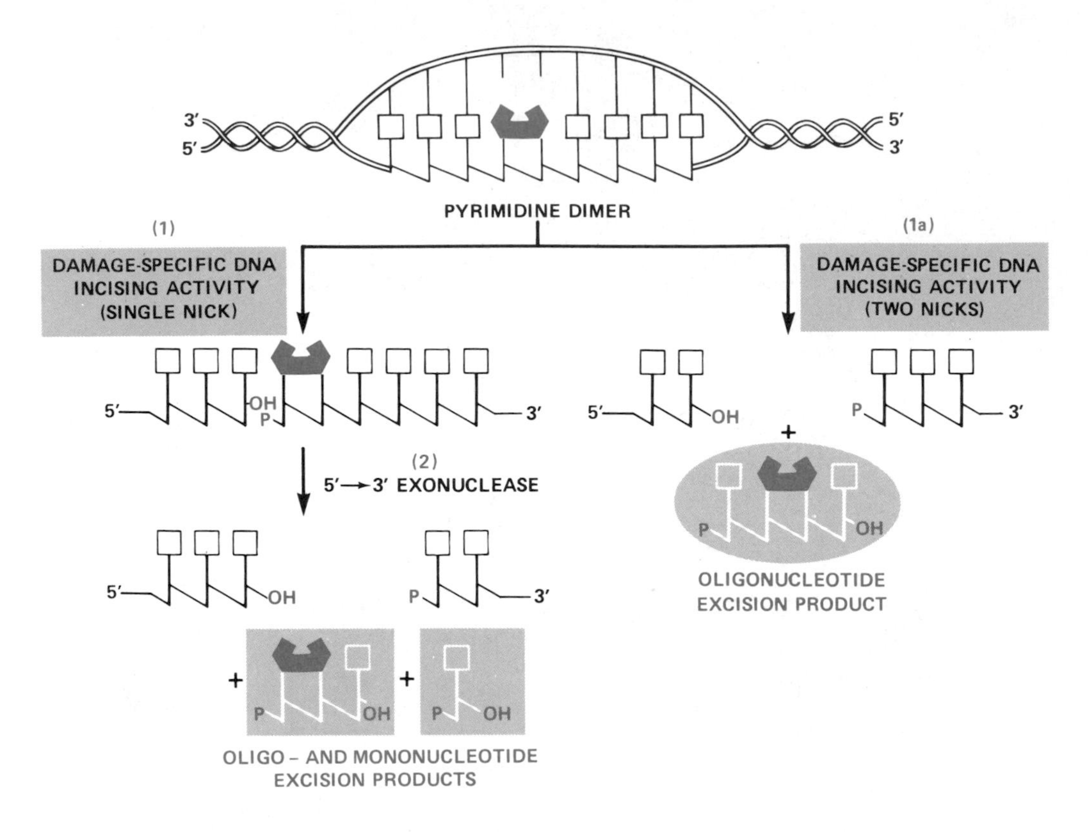

FIGURE 3-2
Duplex DNA containing bulky base damage such as pyrimidine dimers could in theory be specifically incised in a number of different ways. For many years it was believed that the damage-specific DNA incising activity of *E. coli* (uvrABC endonuclease; see Chapter 4) catalyzed the formation of *one* strand break immediately 5′ to the dimer (1), thus requiring subsequent 5′ → 3′ exonucleolytic degradation (2) to effect excision of the dimer as part of a small oligonucleotide. Depending on the extent of the degradation, varying amounts of mononucleotide may also be excised. It is now known that the *E. coli* uvrABC enzyme catalyzes the formation of strand breaks *on either side* of the dimer (1a), thus possibly facilitating concurrent incision of DNA and excision of the base damage. However, it remains to be established how general in nature the mechanism shown in (1a) is. Thus in other organisms endoculeases may attack DNA containing bulky base damage by the mechanism shown in (1) and (2).

was effected by subsequent exonucleolytic degradation of the DNA in the 5′ → 3′ direction (1–3) (Fig. 3-2). However, it has been recently demonstrated that in vitro a damage-specific DNA incising activity from *E. coli* creates incisions *on either side* of a damaged base. The nicks are separated by about 12 nucleotides and hence the damaged bases may be excised as part of an oligonucleotide structure without the obligatory requirement for exonucleolytic degradation (9, 10) (Fig. 3-2). This observation has led to the suggestion that this type of enzyme be designated an *excinuclease.* (10). However, it is not yet clearly established whether such a concerted incision-excision mechanism operates in vivo in *E. coli* or whether damage-specific DNA incising activities in other organisms operate in this way.

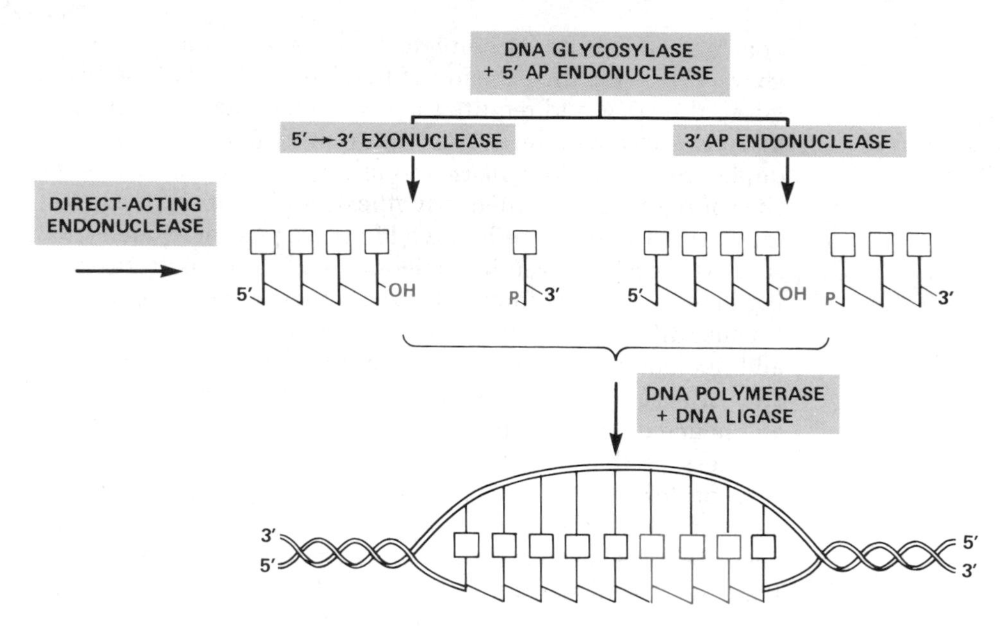

FIGURE 3-3
Gaps left in one strand of a DNA duplex by excision of nucleotides as shown in Figures 3-1 and 3-2, can be filled in by the action of a DNA polymerase provided that a suitable primer terminus (3′-OH nucleotide) is present. When the last nucleotide is inserted, covalent integrity of the DNA duplex is restored by the action of DNA ligase.

Irrespective of the particular mechanism of *excision* of damaged nucleotides or deoxyribose-phosphate moieties, the repair of DNA is not complete until the missing nucleotides are replaced by DNA synthesis and the covalent integrity of the genome is restored by the joining of the last newly synthesized nucleotide to the extant (parental) DNA. This is achieved by the action of *DNA polymerases* and *DNA ligases* respectively (Fig. 3-3) (1–3).

It would be very convenient for descriptive purposes if one could classify excision repair into discrete biochemical pathways. Attempts have been made in the literature to do just this. Thus, for example, the excision of damaged bases by the action of DNA glycosylases is often referred to as *base excision repair* (11), since the damage is usually excised as free bases (Fig. 3-1). Similarly, the excision of base damage following incision of DNA by damage-specific DNA incising activities (direct-acting endonucleases) is often referred to as *nucleotide excision repair* (11–13), since the damaged bases are excised as entire nucleotides (usually small oligonucleotides) (1–3) (Fig. 3-2). Though perhaps useful as a broad conceptual distinction, this classification is not totally satisfactory. For example, the use of the term base excision repair to describe the excision of free bases is inaccurate in the sense that the action of a DNA glycosylase alone does not constitute a complete DNA repair process. As indicated above, following the excision of free bases, sites of base loss are still left in the DNA and require the subsequent action of either specific insertases (see Section 2-5) or (more probably) incision of the DNA duplex by AP endonucleases to allow *postincision* excision of the sites of base loss and then resynthesis and ligation. Furthermore, in most cases the enzymes involved in these postincision events are the same as those involved in nucleotide excision repair, thereby clouding the absolute distinction between these two processes.

Other difficulties with the distinction between base excision repair and nucleotide excision repair can be cited. Consider the repair of sites of base loss that arise spontaneously in DNA (see Section 1-2) and hence obviously do not require the action of DNA glycosylases. The lesions that are eventually excised are neither free bases nor nucleotides; they are deoxyribose-phosphate moieties (Fig. 3-1). Finally, as will be seen later in this chapter, a particular DNA glycosylase that specifically recognizes pyrimidine dimers in DNA catalyzes the hydrolysis of the 5′ glycosyl bond in the dimer, leaving the 5′ pyrimidine covalently attached to its 3′ partner through the cyclobutane ring (3, 7) (Fig. 3-4). Even though incision of such DNA occurs by a two-step mechanism identical to that involving so-called base excision repair of other substrates recognized by DNA glycosylases and AP endonucleases (Fig. 3-1), in this particular case the pyrimidine dimer is excised *after* incision of the DNA, as part of a nucleotide structure, just as in typical nucleotide excision repair (Fig. 3-4). Should we consider the repair of pyrimidine dimers involving the activity of a pyrimidine dimer DNA glycosylase as an example of base excision repair or of nucleotide excision repair?

In view of these complexities, the terms base excision repair and nucleotide excision repair are not used in this book in any rigorous classificatory sense. Rather, the details of excision repair are approached through a consideration of the key enzymatic events that characterize this mode of DNA repair. Therefore, in the remainder of this chapter we shall consider excision repair involving the action of DNA glycosylases and also DNA incision at sites of base loss by AP endonucleases. DNA incision at sites of base damage by damage-specific DNA incising activities, as well as the postincision events in excision repair, are considered in the next three chapters. Chapters 5 and 6 also consider as discrete topics the excision repair of mismatched bases in DNA. These mismatches constitute substrates for yet another class of repair enzymes that are apparently utilized exclusively for the repair of this form of base damage.

Table 3-1 lists the genes of *E. coli* that are currently believed to be required for or involved in excision repair, including the repair of mismatched bases in DNA. The location of a number of these genes

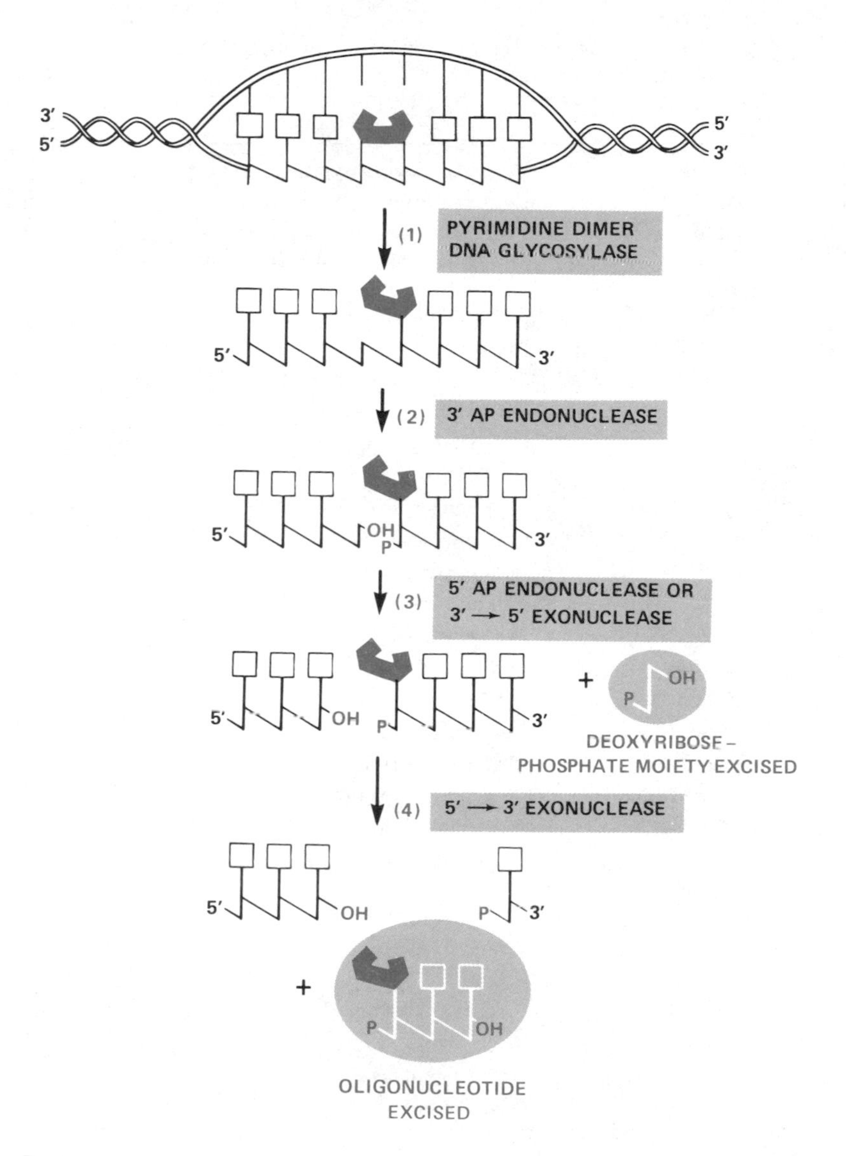

FIGURE 3-4
Pyrimidine dimer (PD) DNA glycosylases attack DNA by a mechanism that is quite distinct from that used by direct-acting damage-specific DNA incising activities such as the uvrABC endonuclease of *E. coli* (see Fig. 3-2). PD DNA glycosylases first hydrolyze the 5′ glycosyl bond in the dimer, thus generating an apyrimidinic site in the DNA (1). In vitro, this site is then attacked by an associated 3′ AP endonuclease, which incises the DNA (2) and leaves a 3′-terminal deoxyribose moiety (2). The subsequent action of either a 5′ AP endonuclease or a 3′ → 5′ exonuclease can effect excision of this moiety (3). The dimer can be independently excised as part of an oligonucleotide by a 5′ → 3′ exonuclease (4).

TABLE 3-1
E. coli genes involved in excision repair of DNA, including mismatch repair

Gene	Gene Product
ung	Uracil DNA glycosylase
?	Hypoxanthine DNA glycosylase
ada	Unknown; presumed regulator of adaptive response to alkylation damage
tag1	3-Methyladenine DNA glycosylase I
tag2	Unknown
alkA	3-Methyladenine DNA glycosylase II
?	FaPy DNA glycosylase
?	Urea DNA glycosylase
?	5-HT DNA glycosylase
xthA	Exonuclease III
?	Endonuclease IV
?	Endonuclease III
?	Endonuclease VII
uvrA	uvrA protein
uvrB	uvrB protein
uvrC	uvrC protein
polA	DNA polymerase I
polB	DNA polymerase II
polC (*dnaE*)	DNA polymerase III
xse	Exonuclease VII
uvrD (*uvrE*, *mutU*, *recL*)	Topoisomerase II
lig	DNA ligase
dam	DNA methylase
mutH	Unknown
mutL	Unknown
mutS	Unknown
recA	recA protein
lexA	lexA protein

on the *E. coli* genetic map is traced through figures presented in this and succeeding chapters. Even though this list is certain to be expanded since new repair genes continue to be discovered, the total number of genes tabulated already constitutes a significant fraction of the *E. coli* genome. This generous use of genetic information for the repair of DNA attests to the importance of this parameter of cellular metabolism in the maintenance and control of genetic fidelity in living cells. Despite this impressive armada of repair genes, the maintenance of genetic fidelity is clearly not perfect, since *E. coli* cells suffer on the average one mutation for every 10^9 to 10^{10} nucleotides replicated (see Section 1-2). Nor should one expect it to be, since on evolutionary grounds the utility of nucleotide sequence divergence arising from imperfect or absent DNA repair is intuitively obvious.

Much of our knowledge of the molecular mechanisms of excision repair has come from the identification, purification and characterization of enzymes that effect specific perturbations of damaged DNA in vitro. Such studies, particularly when accompanied by physiological and genetic investigations in organisms defective or deficient in these enzymes, have considerably advanced our understanding of the basic biochemical pathways illustrated in Figs. 3-1 through 3-4. However, it should be borne in mind that in the majority of such studies the substrate used is purified DNA, whereas in the living cell the substrate is the *genome*. This distinction is more than semantic, for even in simple prokaryotes the genome is a metabolically and structurally more complex entity than naked DNA in the test tube. The information gained from studies in cell-free systems is probably valid for excision repair in nonreplicating regions of the genome of living cells, i.e., those regions that are either well ahead of the replication fork or in which replication and postreplicational modifications of DNA (such as methylation) are completed. There are indications that excision repair of damaged bases may occur at or near replication forks (1, 14; see Section 7-1), but it is not known how the molecular mechanism(s) of such putative repair might differ from that in nonreplicating regions of the genome. In general, the cellular responses to DNA damage at or near replication forks are quite distinct from excision repair and are not usually associated with the removal or reversal of base damage but rather with the *tolerance* of such damage (1). These cellular responses, called *DNA damage tolerance mechanisms*, are considered in detail in Chapters 7 and 8.

In addition to replication, living cells undergo transcription and recombination, and we have little understanding as yet as to how these metabolic parameters perturb excision repair. Finally, the genome of living cells, particularly that of eukaryotes, is possessed of considerable structural complexity, including various levels of supercoiling of DNA and the association of DNA with histone and nonhistone proteins. Studies on the role of DNA topoisomerases, DNA helicases and other DNA binding proteins in excision repair are still in their infancy, although in recent years much attention has been devoted to the consideration of the role of chromatin conformation in excision repair in mammalian cells (1). This topic is explored in Chapters 4 and 6.

3-2 Recognition of Base Damage in DNA

In the cases that have been extensively studied, the recognition of base damage appears to be an integral component of the enzyme system involved in excision repair. This is true of the DNA glycosylases and probably also of the AP endonucleases. For example, following infection of *E. coli* under permissive conditions, bacteriophage T4 expresses an enzyme activity called the T4 pyrimidine dimer (PD) DNA glycosylase (see Section 3-4). As already indicated, this enzyme, as well as a very similar activity present in the bacterium *M. luteus*, attacks DNA containing pyrimidine dimers in a very specific way (3, 7) (Fig. 3-4). The details of this process and a discussion of DNA glycosylases in general are provided in Sections 3-3 and 3-4. For the present purposes, it should be noted that the specificity for recognition of pyrimidine dimers in DNA resides in the enzyme itself. Furthermore, it appears that PD DNA glycosylases recognize some component of the cis-syn cyclobutyl dimer rather than any conformational distortion of the DNA duplex imparted by the lesion. This conclusion stems from the observation that the enzyme activity recognizes pyrimidine dimers in single-strand DNA, where secondary structure is absent and the DNA exists as random coils (15, 16). In addition, these enzymes do not recognize any other known form of base damage in DNA (17–19).

In contrast, the DNA damage-specific incising activity from *E. coli* appears to have a recognition function that is distinct from a catalytic one, although the two are physically associated. This enzyme activity is coded by the *uvrA*, *uvrB* and *uvrC* genes of *E. coli;* it is discussed in detail in Section 4-2. These three genes are not closely linked on the *E. coli* chromosome, yet all three proteins are required for the incision of UV-irradiated DNA in cell-free systems (9, 10, 20). In the absence of the *uvrB* and *uvrC* gene products, however, the uvrA protein binds to UV-irradiated, but not to native, duplex DNA (21, 22). These observations suggest that the *uvrA* gene product may be a recognition protein for base damage that results in localized alteration of regions of the DNA duplex.

Examples of possible recognition proteins *without* associated catalytic activity have been observed in extracts of human cells. Preparations of human fibroblasts in culture contain a protein that binds very specifically to depurinated DNA (23). Fractions containing this protein have also been reported to catalyze the covalent insertion of purines into apurinic sites in duplex DNA (see Section 2-5); however, it is not clear whether the DNA binding and the DNA purine insertase activities reside in the same protein (23). Other DNA binding proteins selective for damaged DNA but without associated DNA repair activity have been isolated from eukaryote sources. A protein

TABLE 3-2
Recognition proteins for DNA containing base damage

Protein	Association with Catalytic Activities	Specificity
M. luteus and phage T4 PD DNA glycosylases	Are also PD DNA glycosylase–3′ AP endonucleases	Pyrimidine dimers in DNA
E. coli uvrA protein	Direct-acting endonuclease activity when uvrB and uvrC proteins are also present	Broad (e.g., pyrimidine dimers and bulky chemical base adducts)
E. coli uvrB protein in the presence of uvrA protein	Direct-acting endonuclease activity when uvrC protein is also present	Broad (e.g., pyrimidine dimers and bulky chemical base adducts)
Human fibroblast binding protein for AP sites in DNA	Possibly also has associated DNA purine insertase activity	DNA containing apurinic sites
Damage-specific DNA binding protein from human placenta	None detected	UV-treated DNA (not dimers), HSO_3-treated DNA, HNO_2-treated DNA, X-irradiated DNA, MNU-treated DNA
Damage-specific DNA binding protein from human placenta	None detected	DNA treated with N-acetoxy-AAF, MMS, MNU
Damage-specific DNA binding protein from HeLa cells	None detected	UV-treated DNA, DNA treated with N-acetoxy-AAF

with a molecular weight greater than 120,000 has been purified from extracts of human placenta and binds to UV-irradiated DNA at sites other than pyrimidine dimers, as well as to DNA treated with X irradiation, nitrous acid or sodium bisulfite (24–26). This protein also binds to UV-irradiated poly(dT) annealed to unirradiated poly(dA) and to irradiated poly(dA) annealed to unirradiated poly(dT). A protein with somewhat similar properties, also from human placenta, does not bind to UV-irradiated DNA but has a high affinity for DNA treated with the carcinogen N-acetoxy-N-2-acetylaminofluorene. This protein also binds to DNA treated with methylmethanesulfonate or with methylnitrosourea (27). Yet another DNA-binding protein with a molecular weight of approximately 20,000 has been partially purified from human HeLa cells (28). This protein binds preferentially to supercoiled phage PM2 DNA treated with UV light or with N-acetoxy-N-2-acetylaminofluorene, but not to native supercoiled PM2 DNA. Nicked or linear forms of PM2 DNA, either damaged or not, are not efficient substrates, suggesting a requirement for supercoiling for DNA binding. The dissociation equilibrium constant for the binding reaction to damaged DNA is estimated to be 4×10^{-11} *M* (28).

A summary of some recognition proteins for damaged DNA is provided in Table 3-2. This is an area of experimental study that has received relatively little attention. It is anticipated that the isolation and characterization of more of these proteins will provide insight into the mechanisms of the recognition of base damage in DNA and into the possible relation of recognition proteins to proteins that are catalytically active in excision repair.

3-3 DNA Glycosylases

The enzymatic event that characterizes the excision of free bases from DNA is the hydrolysis of the N-glycosylic bond linking damaged or inappropriate bases to the deoxyribose-phosphate backbone

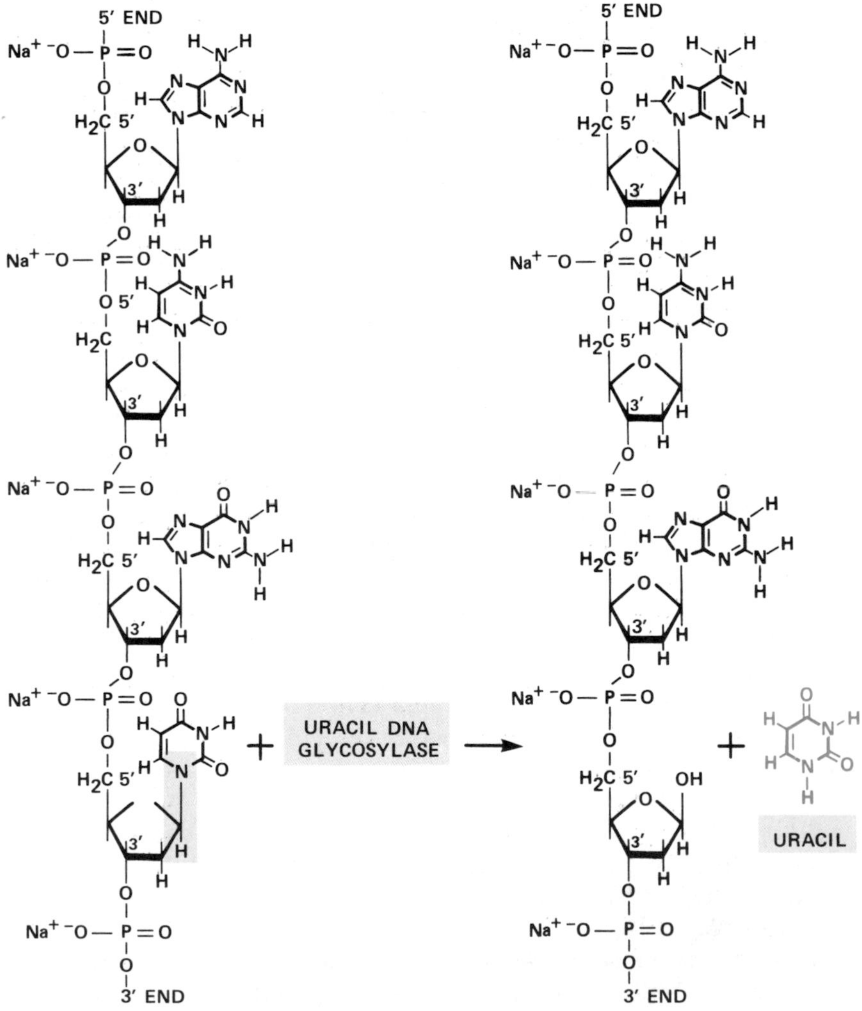

FIGURE 3-5
Structure of a single polynucleotide chain showing the action of a DNA glycosylase, in this case uracil DNA glycosylase. The resulting deoxyribose is shown in the closed (furanose) form, but it actually exists in equilibrium with the open (aldehyde) form (see Figure 1-13). (After A. Kornberg, ref. 8.)

of the DNA (2–7) (Fig. 3-5). The enzymes that catalyze this reaction are called *DNA glycosylases;* a list of known examples and their substrates is provided in Table 3-3. Most of these enzymes are highly specific for a particular form of monoadduct base damage in DNA, although the PD DNA glycosylase activities are clearly examples of enzymes that recognize diadduct damage. In addition, 3-methyladenine DNA glycosylase II recognizes a number of different alkylated bases as substrates (29).

E. coli is the organism in which most DNA glycosylases have been identified, although the majority of those enzymes are ubiquitously distributed in nature. In general, the *E. coli* enzymes are small proteins with a molecular weight of less than 30,000, with no subunit structure and no requirement for cofactors (2–7). The latter property has greatly facilitated the detection of these enzymes in crude extracts, since in the absence of added divalent cation and in the presence of EDTA, most nonspecific degradation of substrate DNA by enzymes that require metal cofactors is prevented. However, one should bear in mind the possibility that some DNA glycosylases may require metal cofactors and therefore would not be detected under such assay conditions.

If for the moment only monoadduct base damage to DNA is considered, the essential feature of the action of all known DNA glycosylases is the catalytic release of free modified or inappropriate bases as products of their reaction with DNA. The direct demonstration of

TABLE 3-3
DNA glycosylases

Enzyme	Substrate	Products
Ura DNA glycosylase	DNA containing uracil	Uracil + apyrimidinic sites
Hx DNA glycosylase	DNA containing hypoxanthine	Hypoxanthine + apurinic sites
3-mA DNA glycosylase I	DNA containing 3-methyladenine	3-Methyladenine + apurinic sites
3-mA DNA glycosylase II	DNA containing 3-methyladenine, 7-methylguanine or 3-methylguanine	3-Methyladenine, 7-methylguanine or 3-methyladenine + apurinic sites
FaPy DNA glycosylase	DNA containing formamidopyrimidine moieties	2,6-Diamino-4-hydroxy-5-N-methylformamidopyrimidine + apurinic sites
5,6-HT DNA glycosylase	DNA containing 5,6-hydrated thymine moieties	5,6-Dihydroxydihydrothymine or 5,6 dihydrothymine + apyrimidinic sites
Urea DNA glycosylase	DNA containing urea moieties	Urea + apyrimidinic sites
PD DNA glycosylase	DNA containing pyrimidine dimers	Pyrimidine dimers in DNA with hydrolyzed 5′ glycosyl bonds + apyrimidinic sites

the free base is thus the most definitive assay of DNA glycosylase activity. This is usually achieved by chromatographic analysis of either the entire incubation mixture or of the acid-soluble fraction of incubations containing DNA radiolabeled in the corresponding base (6). For many years the release of acid-soluble radioactivity from radiolabeled DNA during its incubation with extracts was an indication of DNA degradation exclusively by endonucleases and/or exonucleases. However, the discovery of DNA glycosylases led to the realization that the formation of acid-soluble radioactive products can also result from the action of these enzymes. Hence the simple measurement of total acid-soluble radioactivity does not distinguish between the release of free bases and the release of nucleotides from degraded DNA. A second potential source of confusion between DNA glycosylase and nuclease action on DNA stems from the historically well-established use of sedimentation of DNA in alkaline sucrose gradients for detection of the strand breaks produced by nucleases (30). However, the AP sites in DNA created by the action of DNA glycosylases are readily subject to β-elimination in the presence of strong alkali (31) (see Section 1-2). Thus, sites of base loss are converted into strand breaks during the sedimentation, resulting in the potential for mistaken evidence of endonuclease action on the DNA.

Uracil (Ura) DNA Glycosylase

The various mechanisms whereby uracil can arise in DNA are discussed in Section 1-2. Of these, the deamination of cytosine is particularly important since this event results in the formation of a U·G mispair in DNA. G·C → A·T transitions are increased in *E. coli* mutants defective in Ura DNA glycosylase (*ung*) (32). It is therefore not unreasonable to speculate that organisms, particularly those existing at high temperatures and low pH (conditions that favor hydrolytic deamination of cytosine), may have been under selective pressure to evolve mechanisms for the repair of such potentially mutagenic lesions. The evolution of a specific DNA glycosylase that removes uracil from DNA may have precluded the presence of this base in DNA in general and provokes speculation as to why most genomes contain thymine instead. However, as noted in Chapter 1, some organisms, such as phages PBS1 and PBS2, contain uracil instead of thymine (33). The natural host for these phages, *B. subtilis*, is one of the many organisms that contain Ura DNA glycosylase (34). How, then, is the DNA of these phages able to survive and replicate? We shall return to this question later in this section.

Ura DNA glycosylase was the first of many DNA glycosylases to be discovered in extracts of *E. coli* (35). The activity was independently discovered in extracts of *B. subtilis* (34, 36) and has since been shown to be present in other prokaryotes (37, 38). The activity is also present in extracts of the lower eukaryote *S. cerevisiae* (6) and of a variety of mammalian (including human) cells and tissues (6).

Properties of Ura DNA Glycosylases

Ura DNA glycosylase has been extensively purified and characterized from the prokaryotes *E. coli* (39), *B. subtilis* (34), *M. luteus* (37) and the thermophilic bacterium *B. stearothermophilus* (38); from the yeast *S. cerevisiae* (40); from calf thymus (41, 42), and from the cells of humans with acute myelocytic leukemia (43). Table 3-4 summarizes the principal properties of the enzyme from some of these sources. All are monomeric proteins with molecular weights that range between 19,400 (*M. luteus* enzyme) and 28,700 (calf thymus enzyme). None has a requirement for any known cofactor and all are fully active in the presence of EDTA.

All known Ura DNA glycosylases utilize as substrate either double- or single-strand DNA or deoxyribopolymers containing *deoxyuridine*. RNA containing uridine is not recognized as a substrate, nor is uracil excised from a ribopolymer annealed to a deoxyribopolymer. Neither 5-bromouracil nor 5-hydroxymethyluracil are recognized by the *B. subtilis* enzyme (34); however, the *E. coli* enzyme catalyzes the excision of 5-fluorouracil from DNA (44).

All Ura DNA glycosylases catalyze exclusively the hydrolysis of the N-glycosylic bond linking uracil to the deoxyribose-phosphate backbone; i.e., this hydrolysis is not accompanied by any detectable degradation of phosphodiester bonds. The enzyme from *E. coli* is encoded by the *ung* gene, which maps at 55 minutes on the *E. coli* circular map (6) (Fig. 3-6). It has a turnover number of more than 800 uracil residues released per minute, and it is estimated that about 300 enzyme molecules are present per cell (39). The enzyme has a K_m of 4×10^{-8} *M* for dUMP residues in PBS2 DNA (39); a similar value for dUMP residues in poly(dU) has been measured for the enzyme from *B. subtilis* (34). The *E. coli* enzyme is product inhibited by free uracil, which acts as a noncompetitive inhibitor with a K_i of 10^{-4} *M*. Deoxyuridine, dUMP, thymine, 5-bromouracil, 5-aminouracil, 2-thiouracil and orotic acid do not inhibit the *E. coli* enzyme (39). Sites of base loss in DNA lower the V_{max} of Ura DNA glycosylase for uracil in DNA (42, 45), suggesting that the enzyme has an affinity for such sites even though no associated AP endonuclease activity has been detected in vitro. We shall return to this observation in Section 3-4 when considering the biological relevance of some DNA glycosylases that *do* have associated AP endonuclease activity in vitro.

TABLE 3-4
Uracil DNA glycosylases

Source	Molecular Weight (kDa)	Cofactor Requirement	K_m (dUMP in DNA) (M)	K_i (uracil) (M)	pI	pH Optimum	Preference for SS or DS DNA as Substrate	Activators and Inhibitors (Other than Uracil)
E. coli	24.5	None	4×10^{-8}	1.2×10^{-4}		8.0	SS DNA	Inhibited by NaCl
B. subtilis	24.0	None	1.1×10^{-9}			7.3–7.8		Inhibited by heavy metals; stimulated by NaCl
M. luteus	19.4	None	7×10^{-8}	3.2×10^{-4}	7.0 ± 0.1	5.0–7.0	SS DNA	Activated by spermine and spermidine; inhibited by spermidine at high concentrations
B. stearothermophilus	28–30	None	4×10^{-7}					Inhibited by NaCl
Yeast	27.8	None				7.5–8.0		Inhibited by $CaCl_2$; stimulated by NaCl
Calf thymus	28.7	None	7×10^{-7}			7.2–8.6	DS DNA	Inhibited by NaCl, $MgCl_2$. $CaCl_2$, *p*-hydroxymercuribenzoate (PCMB), thymine, thymidine and TMP

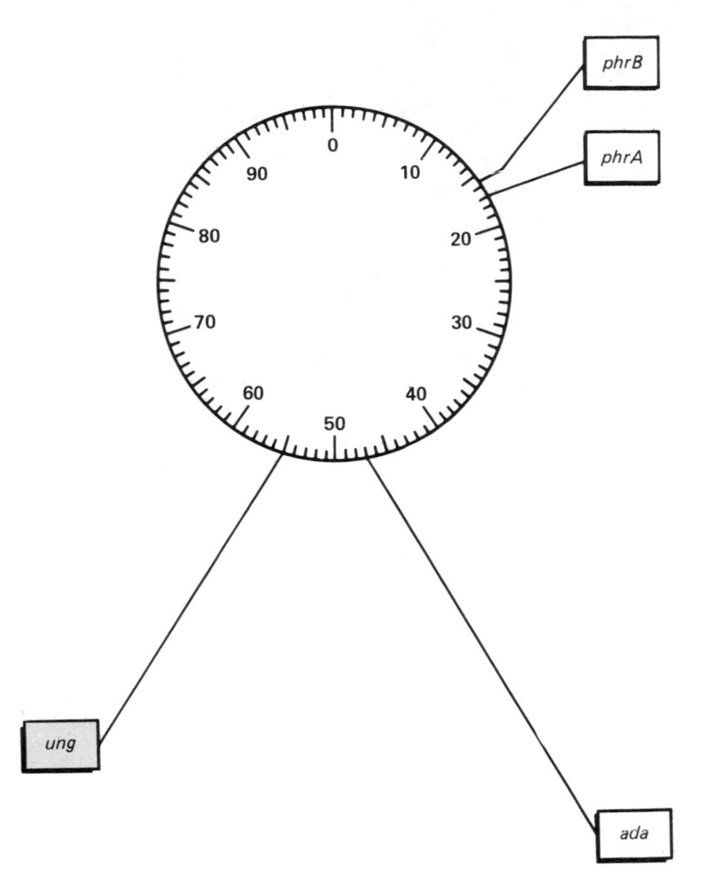

FIGURE 3-6
The location of the *ung* gene on the *E. coli* genetic map.

The enzyme from *M. luteus* is activated in the presence of spermine or spermidine (37). In addition drugs that intercalate in DNA such as ethidium bromide and ellipticine result in about a two-fold increase of the *M. luteus* enzyme activity (37).

Studies on human KB cells in culture indicate the presence of both nuclear and mitochondrial-associated activities (Fig. 3-7) (46), but it is not clear whether these are physically distinct proteins, nor whether the mitochondrial enzyme is encoded by a gene in mitochondrial DNA. In this respect it is of interest that in phytohemagglutinin (PHA)-stimulated human lymphocytes, two chromatographically distinguishable forms of the enzyme have been observed (Fig. 3-8) (47). In quiescent lymphocytes low levels of both forms of the enzyme are detected. However, following stimulation of cell division by PHA, a significant increase in the level of one of these forms occurs (Fig. 3-8). This observation is consistent with the notion that

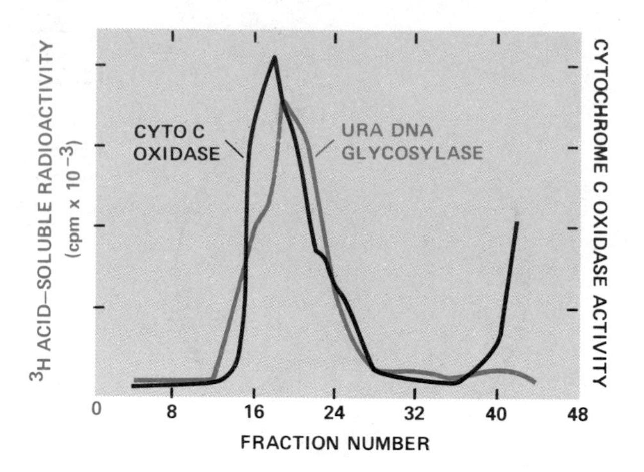

FIGURE 3-7
Cosedimentation of uracil DNA glycosylase and cytochrome C oxidase activities. A crude extract of human KB cells was prepared free of nuclei and sedimented in a neutral sucrose gradient. Fractions were assayed for both enzyme activities. The ^{3}H acid-soluble radioactivity measures the release of radiolabeled uracil from DNA catalyzed by uracil DNA glycosylase. Sedimentation was from right to left. (From C. T. M. Anderson and E. C. Friedberg, ref. 46.)

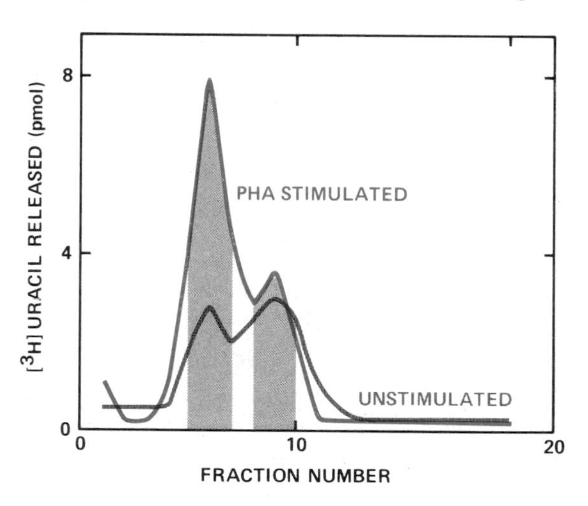

FIGURE 3-8
Extracts of unstimulated human lymphocytes contain two peaks of uracil DNA glycosylase activity identified by phosphocellulose chromatography. Only one of these is increased in cells stimulated to divide with phytohemagglutinin (PHA) and represents increased synthesis of the nuclear enzyme. The other peak may represent mitochondrial uracil DNA glycosylase, the activity of which is not increased in actively dividing cells. (After M. A. Sirover, ref. 47.)

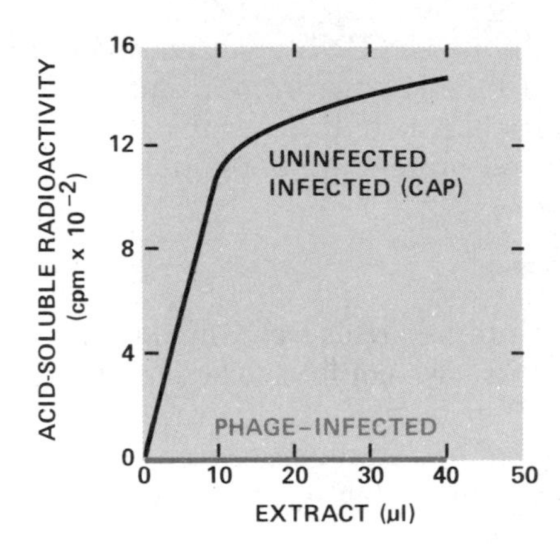

FIGURE 3-9
Infection of *B. subtilis* with phage PBS2 results in loss of host-cell uracil DNA glycosylase activity (bottom curve). This result requires protein synthesis after phage infection, since in the presence of chloramphenicol (CAP) (an inhibitor of protein synthesis) the level of uracil DNA glycosylase activity is the same as in uninfected cells (upper curve). (From E. C. Friedberg et al., ref. 36.)

one form, present in the nucleus, is regulated in the cell cycle and is induced during the S phase, whereas the other form, present in mitochondria, exists at a uniform level throughout the cell cycle. The two activities isolated from stimulated lymphocytes have molecular weights of approximately 42,500 and approximately 40,000 respectively (47). Ura DNA glycosylase activity partially purified from the cytoplasmic fraction of HeLa S_3 cells has a molecular weight of approximately 50,000 (48). On the other hand, highly purified enzyme from calf thymus has a molecular weight of only about 28,000 (41) (Table 3-4). When HeLa cells are infected with herpes simplex virus, there is a two- to six-fold increase in Ura DNA glycosylase activity (49). It is not known whether viral infection results in a modification of the host enzyme or if the virus induces a new enzyme.

Protein Inhibitors of Ura DNA Glycosylase

Since bacteriophages PBS1 and PBS2 normally contain uracil rather than thymine in their DNA, it is not surprising that very early after infection of their natural host *B. subtilis,* a potent inhibitor of the host Ura DNA glycosylase is made (36, 50) (Fig. 3-9). This protein also inhibits the *E. coli* enzyme, as well as Ura DNA glycosylase derived from a variety of other biological sources, including yeast and human cells (51). However, it does not inhibit other known DNA glycosylases from *E. coli* and is thus apparently specific for this particular enzyme. These results suggest that the mechanism of action of all Ura-DNA glycosylases is similar, if not identical, and indicates a significant evolutionary conservation of this enzyme. An inhibitor of Ura DNA glycosylase from *E. coli* is also present in extracts of bacteriophage T5-infected *E. coli* (52), even though this phage contains thymine in its DNA as a principal pyrimidine. Interestingly, the *E. coli* protein only inhibits Ura DNA glycosylase from this organism and has no effect on this enzyme from a number of other biological sources. The biological role of the T5 inhibitor and its distinction from the PBS2-induced protein are not known.

Hypoxanthine (Hx) DNA Glycosylase

As indicated in Section 1-2, hypoxanthine arises from the deamination of adenine in DNA. In theory this base could also arise in DNA following the incorporation of dIMP instead of dGMP during semiconservative DNA synthesis. However, such a phenomenon appears unlikely in *E. coli,* since in contrast to the potential for dUMP incorporation from the dUTP pool, this organism is devoid of detectable dITP and does not have a demonstrable deoxyinosine kinase activity (5). Nonetheless, dITP is readily incorporated into *E. coli* DNA in an in vitro replication system and is associated with significant fragmentation of the nascent DNA (53).

A hypoxanthine DNA glycosylase is present in extracts of *E. coli* (54), HeLa cells (55) and calf thymus (56). The activity in crude extracts of *E. coli* is much lower than that of Ura DNA glycosylase and the enzyme has not yet been extensively purified. In many respects, however, it resembles Ura DNA glycosylase. It is a small protein with an estimated molecular weight of about 30,000. The enzyme catalyzes the release of hypoxanthine, but not xanthine, adenine, guanine or uracil, from nitrous acid-treated DNA. Free hypoxanthine is released from the single-strand polymer poly(dA-[^{3}H]dI). DNA purine residues with other alterations in the 6 position are not substrates. Thus, no release of O^6-methyladenine from enzymatically methylated DNA or of O^6-methylguanine from DNA treated with methylating agents has been detected. In contrast to most Ura DNA glycosylases, hypoxanthine DNA glycosylase is not inhibited by the product of its reaction. Hypoxanthine analogues such as caffeine, xanthine and deoxyinosine are also without effect on enzyme activity (54).

3-Methyladenine (3-mA) DNA Glycosylases

These DNA glycosylases represent the second example of repair enzymes specific for alkylation damage in DNA. (See Section 2-3 for a discussion of O^6-methylguanine transferase activity.) It is not obvious what selective pressures have operated for the evolution in cells (particularly bacteria, which lack the activating enzymes necessary to convert many weakly reactive alkylating agents into strong electrophiles) of enzymes specific for alkylation damage. An interesting hypothesis is that methylating agents are continuously generated intracellularly from small molecules that serve functions in normal metabolism (57). For example, as indicated in Section 2-3, S-adenosylmethionine (the major intracellular methyl donor) is fairly unstable at neutral pH and hence could generate a short-lived alkylating agent derivative by spontaneous hydrolysis. Indeed, when purified DNA is incubated in neutral solution with highly radioactive [^{3}H]methyl-S-adenosylmethionine, small amounts of 3-methyladenine, 7-methylguanine (57) and O^6-methylguanine (58) have been detected in the DNA, although the origin of the O^6-methylguanine is controversial and may be due to a contaminant that is resistant to S-adenosylmethionine hydrolyase activity (59). At a level of S-adenosylmethionine similar to that present intracellularly (about 10^{-5} M), about one DNA base residue in 10^7 is methylated per six hours at 37°C (57).

The existence of enzymes in living cells capable of removing methylated purines from DNA was inferred from a number of early studies in which it was observed that following alkylation of cells in vivo, 3-methyladenine and 3-methylguanine are lost from DNA much more rapidly than could be accounted for from measurements of the spontaneous release of these bases (60–62) (Fig. 3-10). In fact,

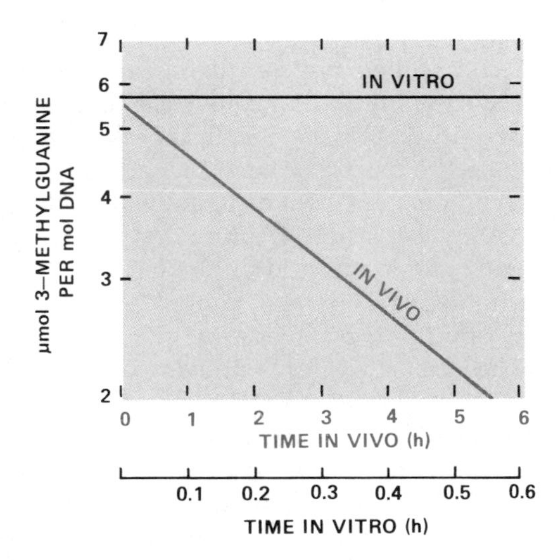

FIGURE 3-10
3-Methylguanine is lost more rapidly from the DNA of living cells than from DNA incubated in vitro. This suggests that loss of the alkylated base in vivo is effected by enzyme catalysis rather than by spontaneous hydrolysis. (From P. D. Lawley and W. Warren, ref. 60.)

the initial rate of elimination of 3-methyladenine from *E. coli* DNA in vivo is not precisely known, because after exposure of this organism to an alkylating agent such as dimethylsulfate for only a very brief period at 37°C, most of the 3-methyladenine residues initially formed during the period of alkylation are already released (5, 60).

E. coli *Has Two 3-Methyladenine DNA Glycosylases*

A distinct 3-mA DNA glycosylase has been purified from extracts of wild-type *E. coli* (63). This enzyme, like the two DNA glycosylases discussed above, is a small protein with an estimated molecular weight of 20,000. The enzyme is sensitive to inhibition by agents that inactivate SH groups and, like the hypoxanthine DNA glycosylase, it has a strong preference for double-strand DNA. The enzyme also resembles the other DNA glycosylases in the stringency of its substrate specificity. When alkylated DNA containing 7-methylguanine, 7-methyladenine, 3-methyladenine or O^6-methylguanine is incubated with the purified enzyme, only free 3-methyladenine is released (Fig. 3-11) (63). The enzyme does, however, also catalyze the excision of 3-*ethyl*adenine from alkylated DNA. No other normal or modified bases are known to be excised by this activity. The enzyme is product inhibited by free 3-methyladenine with an apparent K_i close to 1.0 m*M* (63).

Mutants of *E. coli* (called *tag*) are deficient, but not totally defective, in 3-methyladenine DNA glycosylase activity (64). The residual enzyme activity is not the result of leakiness of the mutant gene but represents the existence of a second form of 3-mA DNA glycosylase in *E. coli* encoded by a different gene. The latter enzyme (called 3-mA DNA glycosylase II) is present at similar levels in both wild-type cells and *tag* mutants and differs from the tag^+ gene product in being considerably more heat stable (29, 65). It is also insensitive to product inhibition by free 3-methyladenine (29, 65). This enzyme has also been purified from extracts of *E. coli* (65). It has the typical general characteristics of a DNA glycosylase, but unlike the enzymes considered above, it has a broader substrate specificity. Thus, in addition to 3-methyladenine, the enzyme catalyzes the excision of 3-methylguanine, 7-methylguanine (Fig. 3-11) and 7-methyladenine from alkylated DNA. In addition 3-mA DNA glycosylase II catalyzes the release of both N^1-carboxyethyladenine and N^7-carboxyethylguanine from DNA treated with β-propiolactone (65).

As far as the genes encoding these two DNA glycosylases are concerned, two mutants in the *tag* gene designated *tag1* and *tag2* have been identified; however, genetic analyses (64) indicate that these mutations are in two closely located but noncontinuous genes. It has been suggested that the *tag2* mutation (which maps at 47.2 minutes

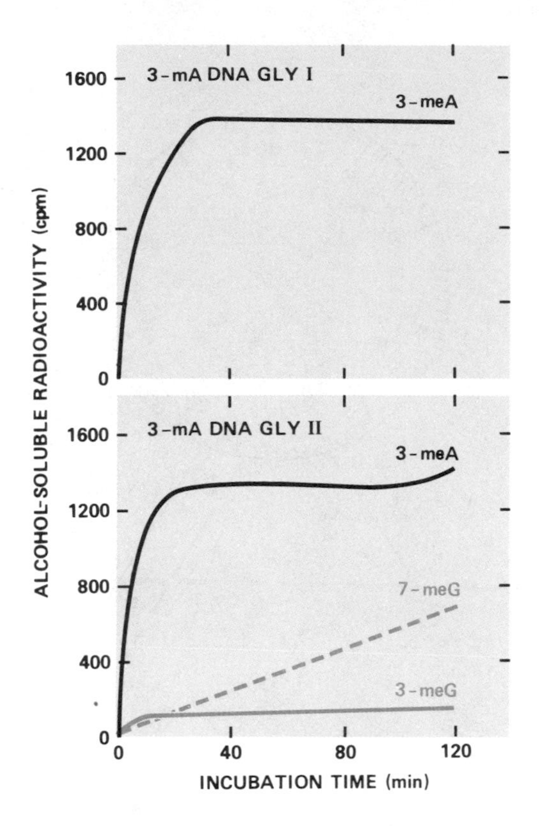

FIGURE 3-11
The enzyme 3-methyladenine DNA glycosylase I catalyzes the selective excision of 3-methyladenine from DNA (top panel). The enzyme 3-methyladenine DNA glycosylase II catalyzes the release of 3-methyladenine, 7-methylguanine and 3-methylguanine from DNA (lower panel). (From L. Thomas et al., ref. 65.)

on the *E. coli* chromosome) (Fig. 3-12) may reside in a *regulatory* gene that controls the expression of the *structural* gene (*tag1*) for 3-mA DNA glycosylase I, which maps between 43 and 46 minutes (64) (Fig. 3-12). As indicated later in this section, the structural gene for 3-mA DNA glycosylase II is called *alkA* and maps very close to the *tag1* gene.

3-mA DNA Glycosylase II of E. coli *Is Involved in the Adaptive Response to Alkylation Damage*

The adaptive response to alkylation damage in *E. coli* was discussed in Section 2-3, and it was pointed out that adaptation to survival and to mutation are governed by distinct biochemical pathways, since

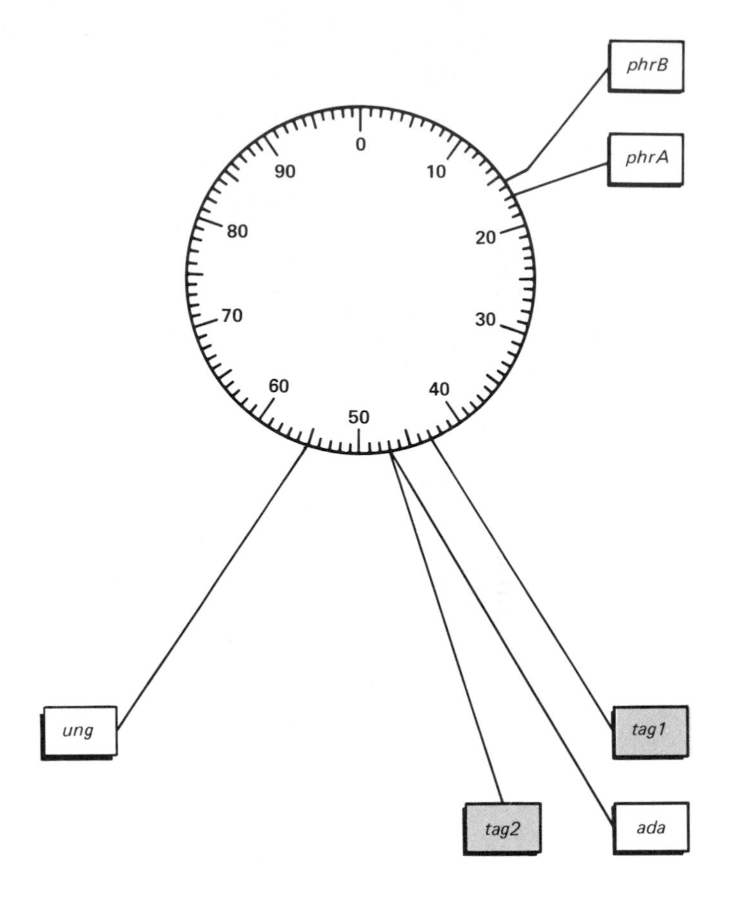

FIGURE 3-12
The approximate locations of the *tag1* and *tag2* genes on the *E. coli* genetic map.

only the former phenomenon requires an active *polA* gene (66, 67). As will be seen in Section 5-3, DNA polymerase I (the product of the *polA* gene) is believed to be the principal enzyme involved in the excision–resynthesis events that follow specific incision of damaged DNA by DNA glycosylases–AP endonucleases in *E. coli* (see Fig. 3-1). Thus the requirement for a functional *polA* gene was a clue that the enzymatic mechanism of adaptation to *survival* might involve *excision repair* of DNA, in contradistinction to the adaptation to *mutation* that occurs by the direct *reversal* of base damage.

Adaptation to survival is associated with the induction of increased synthesis of 3-mA DNA glycosylase II. This enzyme accounts for only 5 to 10 percent of the total enzyme activity in extracts of unadapted cells (29). However, in adapted cells the enzyme accounts for 50 to 70 percent of the total 3-mA DNA glycosylase activity; i.e., glycosylase II is present at about 20-fold higher levels in adapted

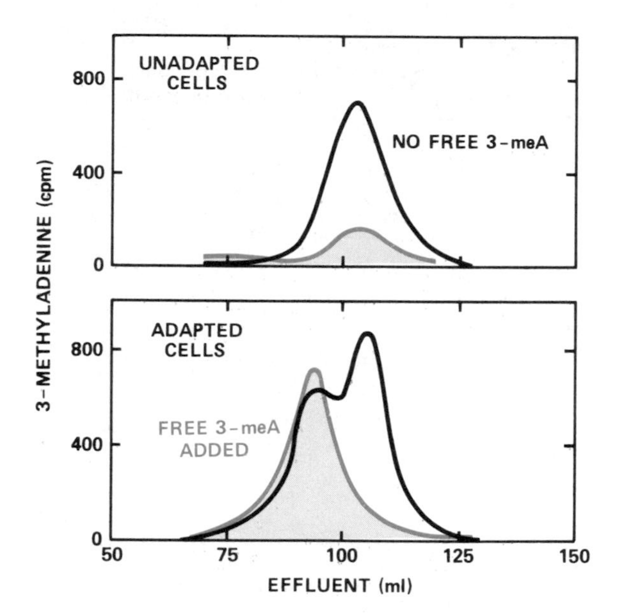

FIGURE 3-13
The activity of 3-mA DNA glycosylases I and II in extracts of unadapted (top panel) and adapted (bottom panel) *E. coli* following gel filtration through Sephadex G-75. Enzyme I is sensitive to product inhibition by free 3-methyladenine; thus it can be shown that most of the enzyme present in unadapted cells is glycosylase I. In adapted cells there is a significant increase in the amount of enzyme activity that is resistant to product inhibition (glycosylase II). (From P. Karran et al., ref. 29.)

relative to nonadapted cells (29) (Fig. 3-13). The identification of the induced enzyme as 3-mA DNA glycosylase II is confirmed by experiments with an *E. coli tag* mutant defective in glycosylase I. In the mutant, 3-mA DNA glycosylase activity was induced by exposure to adapting levels of MNNG (Table 3-5). The induced activity is resistant to inhibition by 3-methyladenine and is heat stable (29).

The biochemical demonstration of two 3-mA DNA glycosylases in *E. coli*, one of which is constitutive and one of which is inducible and under the regulatory control of the alkylation adaptation system, has been complemented by genetic evidence (68). A mutant of *E. coli* has been isolated that is extremely sensitive to the alkylating agent methylmethanesulfonate (MMS). The mutant is unable to repair phage λ previously treated with this alkylating agent and is deficient in *total* 3-mA DNA glycosylase activity (64). The mutant is defective in 3-mA DNA glycosylase I and in both the mutational and survival adaptive responses to alkylation damage, that is, it is *tag1 ada*. Elimination of the *ada*-phenotype by transduction rendered the strain

TABLE 3-5
DNA glycosylase activities in *E. coli* before and after adaptation by MNNG treatment

DNA Glycosylase	Specific Activity of Cell Extract (μU mg^{-1})	
	Unadapted	Adapted
Uracil	3800	3500
Hypoxanthine	2.1	2.0
3-Methyladenine I	3.6	3.6
3-Methyladenine II	0.22	4.1
3-Methylguanine*	~0.05	1.1
7-Methylguanine*	~0.02	0.2
Formamidopyrimidine	4.2	3.9
Thymine glycol	1.9	1.7
Urea	5.5	4.8

*3-Methylguanine and 7-methylguanine DNA glycosylases are the same as 3-methyladenine DNA glycosylase II.
From P. Karran et al., ref. 29.

resistant to the lethal effect of MMS, but only if the cells were grown under conditions in which protein synthesis (and hence expression of *induced* genes) could occur, for example, growth in complete medium (68). However, the strain was still unable to reactivate alkylated phage λ and showed considerable sensitivity to killing by MMS under conditions in which protein synthesis was inactive, for example, growth in buffer (68). These observations are consistent with the presence in *E. coli* of two pathways for the repair of 3-methyladenine: one that is constitutive and controlled by the *tag1* gene (DNA glycosylase I) and one that is inducible and is part of the adaptive response controlled by the *ada* locus (DNA glycosylase II).

The observation that the *ada* mutation prevents induction of two distinct enzymes (O^6-meG methyl transferase (Section 2-3), and 3-mA DNA glycosylase II) suggests that it has a regulatory function in the adaptive response and that other structural genes code for the enzymes themselves. As indicated previously, a gene called *alkA* (69, 70) is in fact the structural gene for 3-mA DNA glycosylase II and maps at approximately 43 minutes on the *E. coli* genetic map (69) (Fig. 3-14). Mutants defective in the *alkA* gene are very sensitive to MMS, despite the presence of normal levels of 3-mA DNA glycosylase I. This suggests that the *lethal* lesion in DNA caused by this alkylating agent is not just 3-methyladenine, but also one or more of the other alkylation products that are recognized as substrates by the DNA glycosylase II (29). If this interpretation is correct, one would expect that *alkA* mutants should not show adaptation to killing by methylating agents. As shown in Fig. 3-15, this is indeed observed (68).

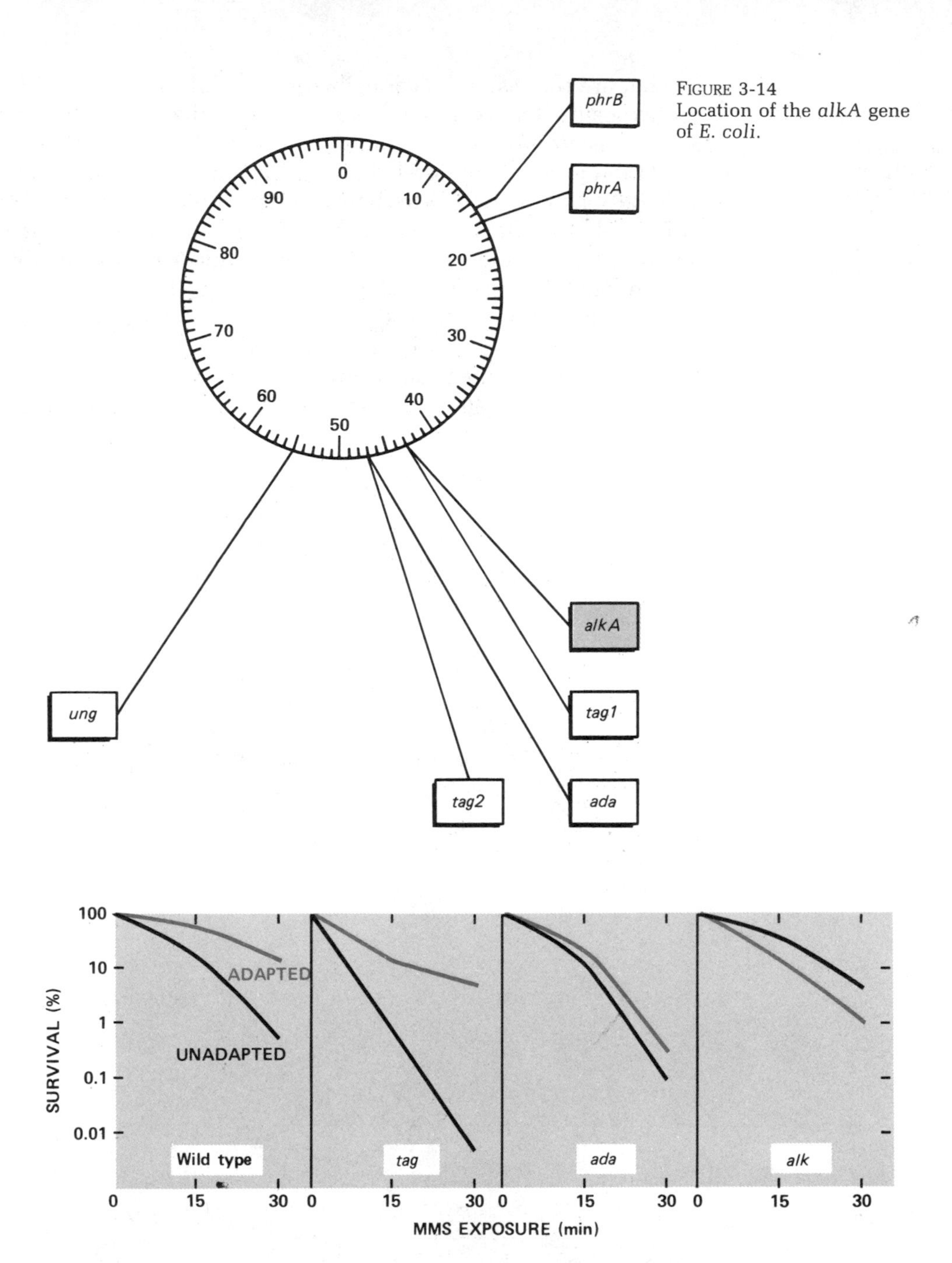

FIGURE 3-14
Location of the *alkA* gene of *E. coli*.

FIGURE 3-15
Wild-type cells and cells defective in 3-methyladenine DNA glycosylase I activity (*tag*) show retention of adaptation to survival. Thus, these cells show enhanced survival following treatment with methylmethane sulfonate (MMS) if they were previously adapted with *N*-methyl-*N′*-nitro-*N*-nitrosoguanidine (MNNG). However, cells defective in the *ada* gene or in 3-methyladenine DNA glycosylase II activity (*alk*) show little or no adaptation to survival. (From G. Evensen and E. Seeberg, ref. 68.)

What are the lesions responsible for the killing of *E. coli* following exposure to alkylating agents? Other than 3-methyladenine, 3-methylguanine and 7-methylguanine are the only other known alkylation products whose loss is significantly enhanced in adapted cells (29). Although 3-methylguanine is a minor alkylation product in DNA, it may have profound biological consequences if it is not removed from living cells. In DNA containing 3-methylguanine and 3-methyladenine the methyl groups occupy the minor groove of DNA, which, unlike the major groove, is normally free of methyl groups (Fig. 3-16). 3-Methylguanine accounts for only about 1 percent of the total alkylation products in DNA bases after treatment with methylating agents, whereas 3-methyladenine constitutes about

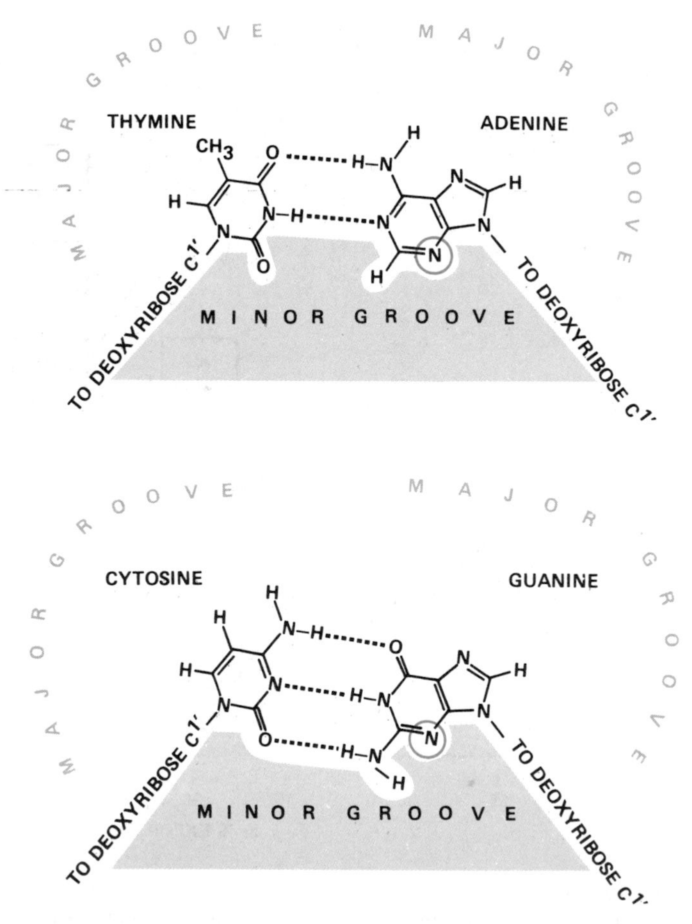

FIGURE 3-16
The N^3 positions of both adenine and guanine occupy the minor groove of the DNA helix. Other sites of alkylation in DNA such as O^6 guanine and N^7 guanine occupy the major groove. (After A. Kornberg, ref. 8.)

15 times as much product (29). Thus it seems reasonable to speculate that after the rapid removal of most of the 3-methyladenine by the constitutive DNA glycosylase I of *E. coli*, there would be approximately equal (though small) amounts of 3-methyladenine and 3-methylguanine present in the minor groove of the genome and that these lesions are lethal unless removed by DNA glycosylase II (29).

At first glance the broad substrate specificity of 3-methyladenine DNA glycosylase II is surprising, since all other known *E. coli* enzymes in this class are quite specific for very limited types of base damage in DNA. It has therefore been suggested that the enzyme might recognize a positively charged purine residue rather than alkylation at a particular site (3). Very recent studies have demonstrated that the *E. coli alkA* gene product (3-mA DNA glycosylase II) also removes O^2-methylthymine and O^2-methylcytosine, but not O^4-methylthymine, from alkylated DNA (71). Other DNA glycosylases specific for alkylated bases in DNA may yet be discovered.

3-mA DNA Glycosylases in Other Organisms

3-mA DNA glycosylase activities have been identified in other prokaryotes and in various eukaryote sources, but it is not clear whether two distinct forms of this enzyme exist in all of these. By analogy with *E. coli*, in which one enzyme form has a much broader substrate specificity, a number of observations suggest that a type II enzyme exists in other cells. For example, a DNA glycosylase from *M. luteus* that catalyzes the release of free 7-methylguanine from DNA is apparently distinct from a 3-mA DNA glycosylase in that organism (72). Like 3-mA DNA glycosylase II of *E. coli* (which, as indicated above, also excises 7-methylguanine), the 7-methylguanine activity from *M. luteus* is quite heat stable. A 3-mA DNA glycosylase activity has been purified about 100-fold from rat liver and releases 3-methyladenine and 7-methylguanine at the same rates (73). In addition, both catalytic functions are heat inactivated to the same extent. On the other hand, it has also been reported that a 3-mA DNA glycosylase from human lymphoblasts is essentially specific for 3-methyladenine in duplex DNA (74). A 3-mA DNA glycosylase activity is induced in rat liver during regeneration after partial hepatectomy (75); however, there is no evidence that this enzyme is directly comparable to 3-mA DNA glycosylase II of *E. coli*, and the increased enzyme activity may simply reflect the increased proliferation of cells associated with liver regeneration.

DNA Glycosylases in *E. coli* That Recognize Ionizing Radiation-Type Base Damage

Various forms of base damage in DNA can be produced either by direct exposure to ionizing radiation or by certain chemical treatments that mimic known or suspected ionizing radiation damage to bases (see Section 1-5). The following section deals with two different DNA glycosylases in *E. coli*, each of which recognizes distinct forms of ionizing radiation-type base damage. A third DNA glycosylase that recognizes specific types of ionizing radiation damage is discussed separately in Section 3-4, because unlike the DNA glycosylases discussed so far, which have no other known catalytic activities, this DNA glycosylase is one of several known examples of such enzymes that also demonstrate associated AP endonuclease activity in vitro.

DNA Glycosylase That Releases Purines with Opened Imidazole Rings from DNA: FaPy DNA Glycosylase

Treatment of DNA with dimethylsulfate followed by prolonged incubation at pH 11.4 results in the formation of 7-methylguanine with opened imidazole rings, generating 2,6-diamino-4-hydroxy-5-N-methylformamidopyrimidine (76, 77) (see Section 1-5). Similar substituted diaminopyrimidines can occur in DNA as a consequence of the exposure of cells to ionizing radiation under anoxic conditions (78, 79). A DNA glycosylase that catalyzes the excision of this formamidopyrimidine from DNA has been identified in extracts of *E. coli*, calf thymus and human fibroblasts and is designated form-

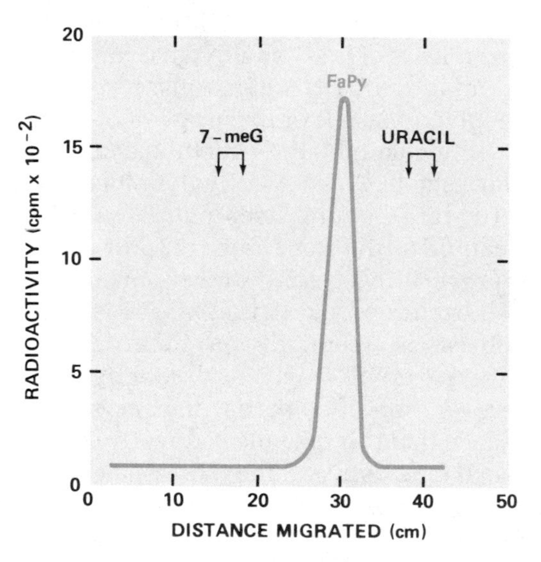

FIGURE 3-17
Release of free formamidopyrimidine (FaPy) from radiolabeled DNA by a specific DNA glycosylase in *E. coli* can be detected by chromatography of the ethanol-soluble fraction after incubation of DNA with the enzyme. The chromatographic positions of 7-methylguanine and of uracil are also shown. (From C. J. Chetsanga et al., ref. 77.)

amidopyrimidine (FaPy) DNA glycosylase (76, 77) (Fig. 3-17). Like other known DNA glycosylases, the enzyme from *E. coli* is a relatively small protein, with a molecular weight of approximately 30,000. The enzyme has a strong and possibly absolute specificity for duplex DNA as a substrate.

Urea DNA Glycosylase

As indicated in Section 1-5, ionizing radiation damage to DNA can result in attack at the 5,6 double bond of pyrimidines, resulting in the formation of unstable hydroperoxides. In the case of thymine, these can be converted into thymine glycols, which can degrade further to yield formylpyruvylurea, urea and N-substituted urea derivatives (80). The latter two forms of thymine base damage can be generated without ionizing radiation by oxidation of a model substrate such as poly(dA-[2^{14}C]dT) (Fig. 3-18). Incubation of this substrate with crude extracts of *E. coli* results in the release of free urea (80) (Fig. 3-19). A

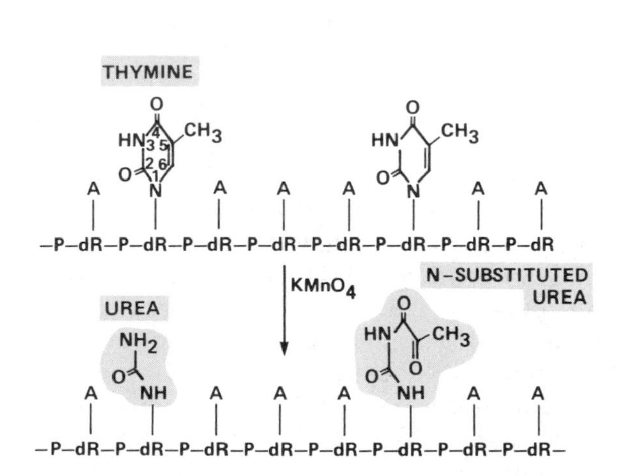

FIGURE 3-18
Schematic representation of a polynucleotide substrate for urea DNA glycosylase. Poly(dA-[2-^{14}C]dT) is treated with potassium permanganate, which oxidizes thymine to yield urea and N-substituted urea derivatives. (From L. Breimer and T. Lindahl, ref. 80.)

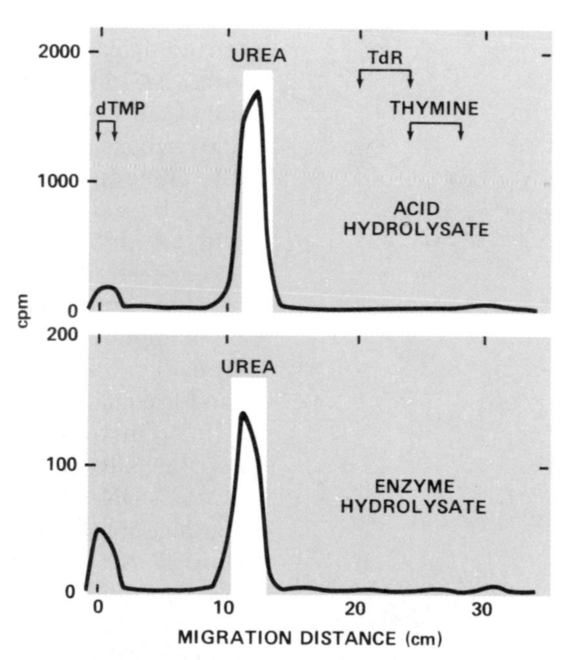

FIGURE 3-19
Release of free urea from poly(dA-[2-^{14}C]dT) containing ^{14}C-labeled fragmented thymine residues (see Fig. 3-18). Both acid treatment (top panel) and incubation with urea DNA glycosylase (lower panel) result in the release of a radiolabeled species that cochromatographs with authentic urea. The chromatographic positions of thymidylate, thymidine and thymine are also shown. (From L. Breimer and T. Lindahl, ref. 80.)

DNA glycosylase responsible for this activity has been partially purified and shows a strong preference for a duplex substrate. The enzyme exhibits no cofactor requirement and has a molecular weight of about 20,000 (80).

3-4 DNA Glycosylases with Associated AP Endonuclease Activity

The DNA glycosylases discussed thus far have no other known catalytic activities. Thus, following release of damaged bases, incision of the resulting apurinic or apyrimidinic (AP) sites must be completed by an unassociated AP endonuclease activity. The latter are ubiquitous in their distribution and are discussed in Section 3-5. However, some DNA glycosylases demonstrate an associated AP endonuclease activity in vitro and therefore (at least in theory) may be able to carry out both base excision *and* DNA incision in vivo.

At this juncture, some general comments are warranted concerning the biological relevance of AP endonuclease activity demonstrated in vitro. As indicated in Section 1-2, loss of bases from DNA creates increased lability of the phosphodiester bonds associated with the deoxyribose residues, particularly the 3′ bonds. Indeed, at alkaline pH, preferential hydrolysis of these bonds at sites of base loss can result by a β-elimination mechanism (31). In addition, polyamines (81), as well as high concentrations of some proteins such as pancreatic ribonuclease (82) and cytochrome C (5), can effect hydrolysis of phosphodiester bonds at sites of base loss in DNA. The same phenomenon has been demonstrated with oligopeptides containing aromatic amino acids, such as Lys-Trp-Lys (83–85). This tripeptide can form stacked complexes with native DNA; however, the stacking affinity increases by two orders of magnitude in DNA containing apurinic sites (83–85) and Lys-Trp-Lys can actually promote nicking of such DNA (Fig. 3-20) (83–85). The latter reaction results in a significant reduction in the binding affinity of the tripeptide for apurinic DNA. The reaction mediated by Lys-Trp-Lys may be a mechanistic model for AP endonucleases, i.e., such enzymes may make use of a particular aromatic amino acid residue to recognize apurinic sites in DNA and could facilitate their own dissociation by the enzymatic cleavage reaction (85). However, we must consider the possibility that these reactions with oligopeptides are *spurious*, i.e., that they are *not* mechanistic models for bona fide AP endonucleases and hence have no biological relevance. In this regard it would be instructive to know whether these reactions have selectivity for 3′ or 5′ phosphodiester bonds, whether they obey Michaelis-Menten kinet-

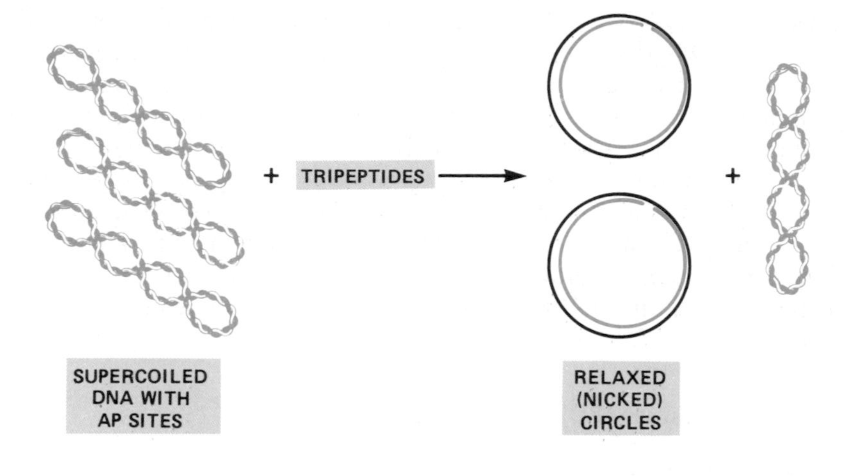

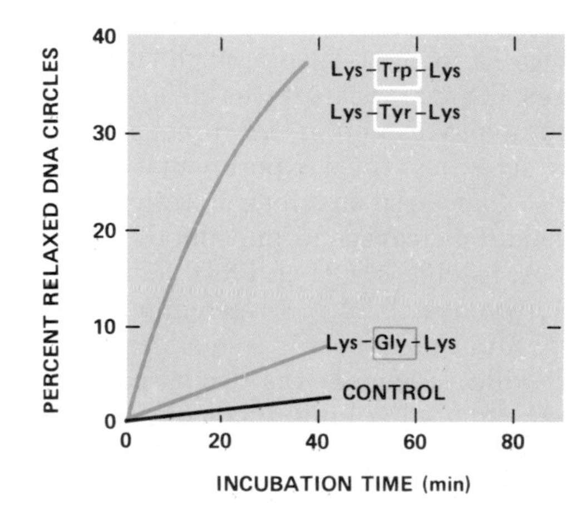

FIGURE 3-20
Nicking of supercoiled DNA containing apurinic sites can be detected by measuring the formation of relaxed circular DNA molecules. This nicking is effected by tripeptides containing aromatic amino acids (such as tryptophan or tyrosine) but to a much lesser extent by tripeptides without such amino acids (e.g., Lys-Gly-Lys). (After T. Behmoares et al., ref. 83.)

ics and how the V_{max} of these reactions compares with that of the reactions catalyzed by "true" AP endonucleases.

How, then, do we define "true" AP endonucleases, and is it possible to distinguish them from proteins (or oligopeptides) that effect hydrolysis of phosphodiester bonds at AP sites in DNA noncatalytically? We do not have satisfactory answers to these questions at this time. Clearly, AP endonuclease activities for which structural gene mutations confer some defective phenotype to the cell constitute examples of biologically relevant enzymes. One hopes that the substrate specificity, polarity for a particular phosphodiester bond (i.e., 5′ or 3′) and specific mechanism of action of such enzymes are faithfully reproduced in vitro. However, even if this is true, these parameters may not be distinct from spurious or biologically irrelevant reactions that are also demonstrable in vitro. Thus, for example, a true AP endonuclease may catalyze the hydrolysis of phosphodiester bonds 3′ to AP sites in vivo by a β-elimination mechanism, yet in vitro such a reaction may be effected spuriously by a protein that is not an AP endonuclease in the living cell. This dilemma will be reexamined in relation to some of the specific enzyme activities discussed below.

Pyrimidine Dimer (PD) DNA Glycosylases

Enzyme activities that catalyze the selective incision of DNA at sites of pyrimidine dimers in vitro were observed in extracts of *M. luteus* (86–88) and phage T4-infected *E. coli* (89, 90) before the discovery of DNA glycosylases. For some years it was believed that these enzymes represented prototypic examples of direct-acting DNA damage-specific incising activities, since experimental evidence was consistent with the idea that these enzymes *directly* (i.e., in a single-step reaction) catalyzed the cleavage of phosphodiester bonds 5′ to sites of dimers in DNA, leaving 3′-OH and 5′-P termini (15, 16, 91, 92).

In Section 1-5 the use of DNA sequencing gels for studying the distribution of pyrimidine dimers in a small fragment of *E. coli* DNA of known nucleotide sequence was discussed. In these studies a preparation of *M. luteus* UV endonuclease was used as a specific probe to detect pyrimidine dimers in DNA (93, 94). Because of the widely held assumption that the *M. luteus* enzyme attacks UV-irradiated DNA as shown in Fig. 3-21, a particular size distribution of DNA fragments after gel electrophoresis was expected. Instead, the fragments electrophoresed as if they were approximately one nucleotide larger. By this time, DNA glycosylases were a well-described enzymological entity, and it was suggested that the two-step DNA glycosylase-AP endonuclease reaction shown in Fig. 3-21 could explain this result (93, 94). This hypothesis appears to be correct, and this mode of action satisfactorily explains the mechanism of attack of UV-irradiated DNA by both the *M. luteus* and the phage T4 "UV endonucleases" (93–97).

Before discussing these two interesting enzymes, it should be noted that although they catalyze the hydrolysis of N-glycosyl bonds in DNA, thereby creating AP sites as products just as all other DNA glycosylases do, they effect the hydrolysis of only *one* of the two glycosyl bonds in pyrimidine dimers (98, 99) (Fig. 3-21). Thus the "liberated" bases are still covalently bound to the DNA through the cyclobutane ring that characterizes the structure of the dimers; i.e.,

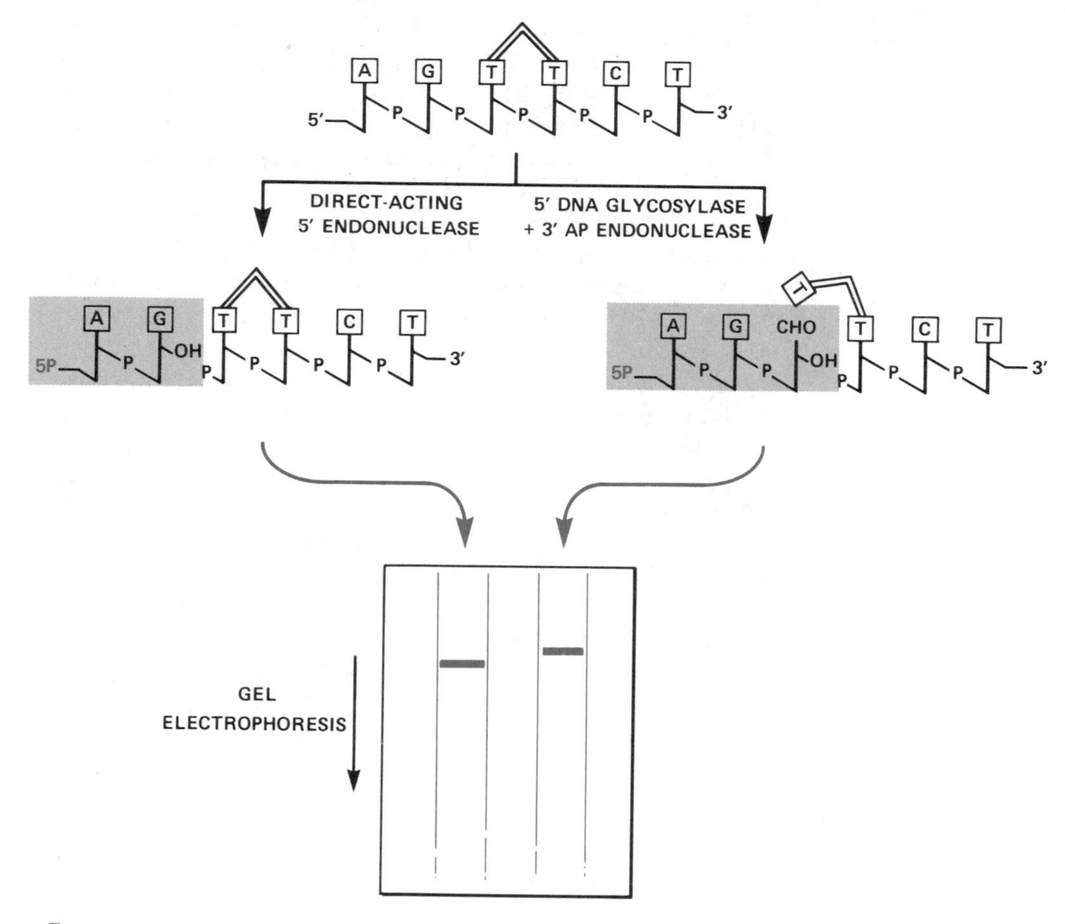

FIGURE 3-21
This figure schematically illustrates how the PD DNA glycosylase activity of the *M. luteus* "UV endonuclease" was discovered. If DNA radiolabeled at the 5′ ends is cleaved 5′ to pyrimidine dimers by a *direct-acting* endonuclease, DNA fragments of a particular size distribution are expected, as shown on the left. However, the fragments observed after gel electrophoresis were approximately one nucleotide bigger than expected, suggesting the cleavage mechanism shown on the right. Only one DNA chain is shown for simplicity. (Adapted from E. C. Friedberg et al., ref. 7.)

free bases are not products of this reaction. Excision of dimers that is initiated by a PD DNA glycosylase therefore occurs during *post-incision* exonucleolytic degradation of the DNA in the 5′ → 3′ direction, a process described in detail in Section 5-3.

The Phage T4 Pyrimidine Dimer DNA Glycosylase

Bacteriophage T4 is the only bacterial virus known to encode an enzyme specifically involved in the excision repair of DNA. All other phages rely on host enzymes for the repair of their genome, a phenomenon called *host-cell reactivation* (100). The discovery of the *denV* gene of phage T4 goes back almost 40 years, and its cultivated history is an interesting development in the field of DNA repair.

In 1947 Salvador Luria made the observation that phage T4 is about twice as resistant to ultraviolet light at 254 nm as phages T2 and T6 (101) (Fig. 3-22). At that time, he speculated that this might be due to an *absence* of a number of loci in T4, the presence of which in phages T2 and T6 was associated with increased UV sensitivity. Subsequent genetic crosses between T2 and T4 revealed that the UV sensitivity phenotype behaved genetically as a unit character; i.e., both the T2-like and the T4-like offspring fell into two distinct classes, whose levels of UV sensitivity corresponded to those of the parental types. This gene was mapped and termed *u* (presumably for ultraviolet), and in view of the earlier interpretation made by Luria, the T2 state (i.e., the more UV-sensitive state) was considered the wild-type or u^+ allele, and the T4 state the mutant or u^- allele (102). It was suggested that this gene could conceivably affect the UV sensitivity of the entire genome through the inhibition of some repair mechanism by the u^+ allele or through its stimulation by the u^- allele (102).

The survival of UV-irradiated u^+ phages in the presence of the u^- state was investigated by coinfecting *E. coli* with lightly UV-irradiated T2 phages (u^+) and *heavily* UV-irradiated T4 phages (u^-) (103). At the dose used for the heavy irradiation, the T4 phage sustained about 70 lethal hits per phage and its survival was negligible. However, the UV sensitivity of the u^- gene itself was sufficiently

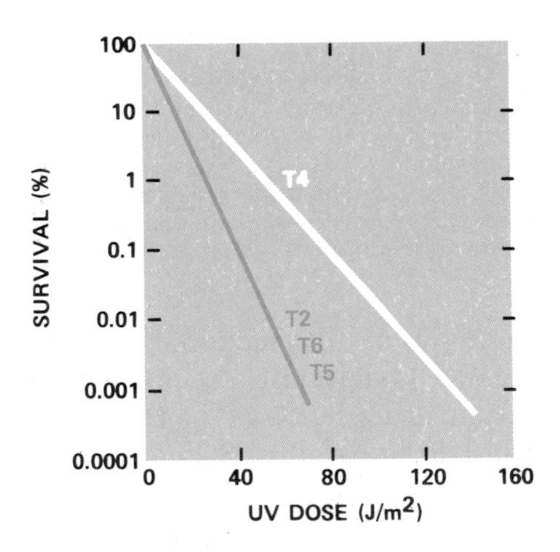

FIGURE 3-22
The relative sensitivity of phages T4, T2, T6 and T5 to killing by UV radiation. Phage T4 is significantly more UV radiation resistant than the other phages. (From S. E. Luria, ref. 101.)

small to ensure that many of the infected cells contained at least one undamaged copy. These experiments revealed a significant increase in the survival of the lightly irradiated phage T2, approximating the survival curve for T4 (Fig. 3-23). This effect (termed *u*-gene reactivation) is not observed during infection with UV-inactivated phage T2, nor is the survival of lightly irradiated phage T4 significantly affected by infection with inactivated phage T4. Single plaques originating from the surviving *u*-gene-reactivated T2 were isolated, and the phages were UV-irradiated again. The phage survival was now typical of the u^+ (T2) state, thereby precluding the possibility of marker rescue during the mixed infection as an explanation of the *u*-gene reactivation. Thus it was concluded that the u^- gene (T4 allelic state) actually codes for some function (possibly an enzyme) that influences the intracellular environment so as to *reduce* the lethality of UV-irradiated T2 phage. Since this interpretation was clearly inconsistent with the original allelic designation of u^-, the name of the *u* gene was changed to *v*, with the allelic state v^+ that of T4, and v^- the mutant or T2 state (103).

In the late 1960s, it was shown that the *v* gene is required for excision of thymine dimers from UV-irradiated T4 DNA in vivo

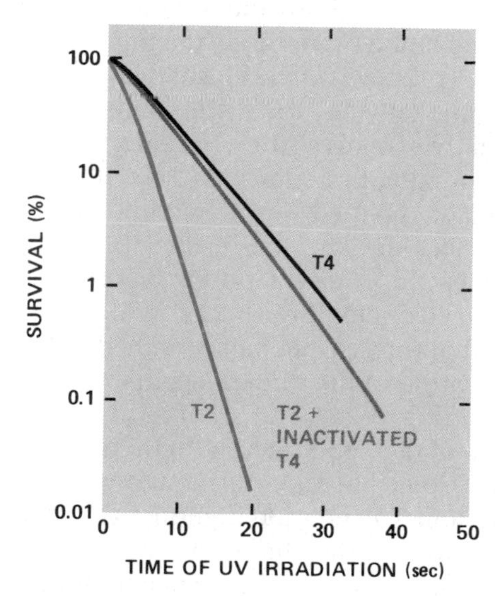

FIGURE 3-23
When phage T2 is irradiated with *modest* doses of UV light and infected into *E. coli* in the presence of *heavily* irradiated (inactivated) phage T4, the T2 phage shows enhanced survival relative to that observed in the absence of coinfection with phage T4. This result suggested that a product encoded by the phage T4 genome enhances the survival of the UV-irradiated phage T2. (After W. Harm, ref. 103.)

(104). Subsequently, a dependency on the *v* gene for thymine dimer excision was demonstrated in a cell-free system (105), and it was also shown that the *v* gene codes for an activity that incises UV-irradiated DNA but not unirradiated DNA in vitro (89, 106). This activity was referred to as T4 UV endonuclease (107) or endonuclease V of T4 (90). Studies using temperature-sensitive mutants defective in the *v* gene demonstrated unambiguously that this gene is a structural rather than a regulatory gene (108). When a revised nomenclature for a number of T4 genes was introduced (109), the designation *v* was changed to *denV* (for *D*NA *en*donuclease V). This nomenclature is still not strictly correct since the gene clearly codes for a DNA glycosylase and, as will be seen shortly, it is not firmly established that it also encodes a biologically active AP *endonuclease*.

The T4 PD DNA glycosylase has been purified to apparent physical homogeneity (110). Like other DNA glycosylases, the enzyme is a small monomeric protein (molecular weight, approximately 17,000), with no requirement for divalent cation or other cofactors (15, 110). It is apparently specific for cyclobutane dipyrimidines in DNA, since dimers are attacked when they are present in single-strand substrates, in which the secondary structure of the DNA presumably plays no role (15, 16, 111). Additionally, in duplex DNA containing dimers on only one of the two strands the dimer-containing strand is attacked uniquely (112). No other type of damaged DNA examined is recognized as a substrate (15–19). T4 mutants defective in the *denV* gene are abnormally sensitive to UV radiation (113–115) but not to chemicals or to ionizing radiation (17–19, 116).

A highly specific method for detecting PD DNA glycosylase activity is to subject the enzyme-reacted DNA to conditions that monomerize dimers (e.g., direct photoreversal or incubation with photoreactivating enzyme) (Fig. 3-24). When UV-irradiated DNA radiolabeled in thymine is incubated with the T4 enzyme, subsequent monomerization of the dimers results in the liberation of free thymine (95, 117) (Fig. 3-24). The amount of free thymine liberated is exactly one half of that associated with thymine-containing pyrimidine dimers lost from the DNA during the photoreversal (99) (Fig. 3-25). This stoichiometric relation demonstrates that only *one* of the two N-glycosylic bonds in the dimerized nucleotides is enzymatically hydrolyzed. Additionally, excision of thymine-containing pyrimidine dimers from DNA pretreated with T4 enzyme requires the action of a $5' \rightarrow 3'$ exonuclease activity (118), suggesting that the glycosyl bond cleaved is exclusively the $5'$ one.

The PD DNA glycosylase activity of the *denV* gene product has been observed in living cells by infecting UV-irradiated strains of *E. coli* defective in the *uvrA* or *uvrB* functions (and hence unable to catalyze incision of their DNA; see Section 4-2) with phage T4 (119). Under these conditions incision of *E. coli* DNA is mediated by the

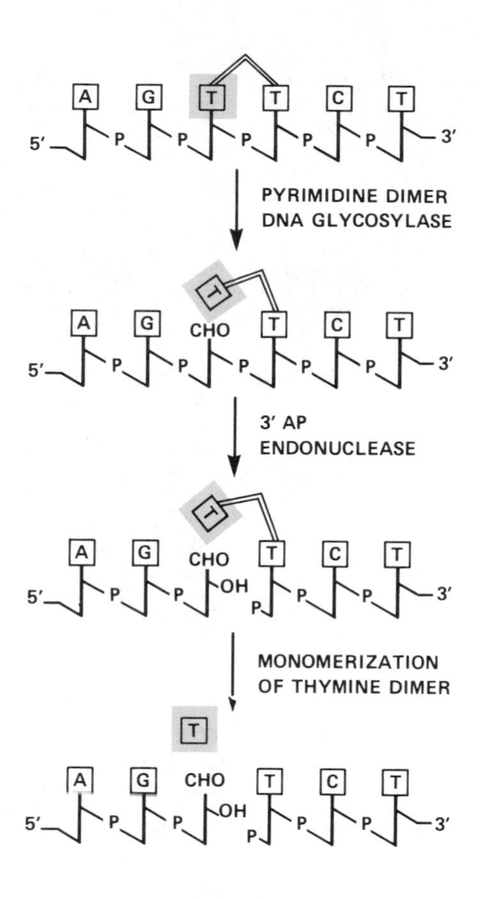

FIGURE 3-24
Cleavage of a single N-glycosyl bond in a pyrimidine dimer by a PD DNA glycosylase results in the formation of free thymine following monomerization of the dimer. The measurement of free thymine constitutes a convenient way of assaying PD DNA glycosylase activity. (From E. C. Friedberg et al., ref. 7.)

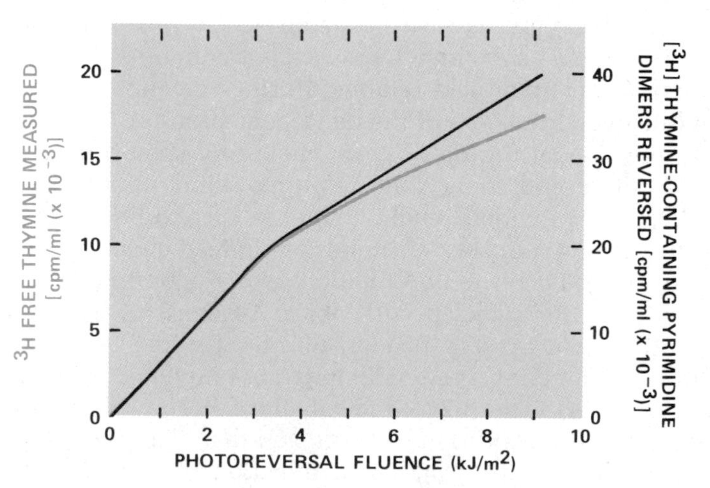

FIGURE 3-25
The amount of ^{3}H radioactivity released as free thymine from [^{3}H]thymine-labeled UV-irradiated DNA incubated with the phage T4 PD DNA glycosylase is one-half the radioactivity lost from the DNA as thymine-containing pyrimidine dimers (note the different scales). This result indicates that thymine-containing dimers are the probable source of the free thymine and that only *one* of the two glycosyl bonds in the dimer is cleaved by the enzyme. (From E. H. Radany et al., ref. 99.)

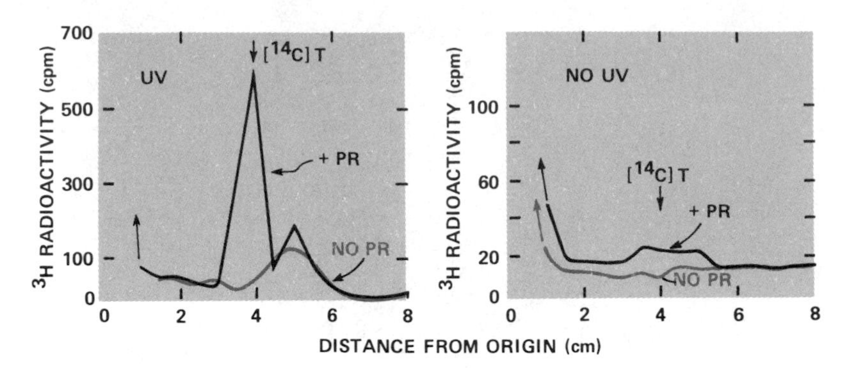

FIGURE 3-26
Chromatographic detection of the photoreversal (PR)-dependent formation of free thymine (see Sec. 2.2) from purified pyrimidine dimer-containing oligonucleotides (left panel). The oligonucleotides are excised during infection of UV-irradiated *E. coli* cells in which the DNA was prelabeled with [^{3}H]thymine. No significant amount of free thymine is observed without monomerization of dimers by photoreversal (left panel) or in cells not exposed to UV light (right panel). The position of free thymine in the chromatograms is determined with use of a [^{14}C]thymine standard. (From E. H. Radany and E. C. Friedberg, ref. 119.)

phage PD DNA glycosylase. The dimer-containing oligonucleotides subsequently excised from the *E. coli* chromosome contain thymine that is released following photoreversal (Fig. 3-26). The amount of free thymine so formed accounts quantitatively for all thymine-containing pyrimidine dimers excised (119).

The T4 PD DNA Glycosylase Has an Associated AP Endonuclease Activity In Vitro

Incision of UV-irradiated DNA catalyzed by the T4 enzyme occurs under conditions of neutral pH, in which apyrimidinic sites created by the activity of a DNA glycosylase are stable (15, 16, 90). Thus the glycosylase activity must somehow be associated with an AP endonuclease, either as a contaminant retained during enzyme purification or as an integral component of the *denV* gene product. Mutants of phage T4 defective in the *denV* gene, including some that are temperature sensitive and some that are suppressible (*amber* and *ochre* mutants), have provided genetic evidence that both PD DNA glycosylase and AP endonuclease activities are indeed coded by the *denV* gene (120–122). The level of AP endonuclease activity present in crude extracts of cells infected with phage T4 *denV*$^{+}$ is significantly greater than that present in extracts of cells infected with phages defective in the *denV* gene, which express no detectable PD DNA glycosylase (121) (Fig. 3-27). Indeed, the level of AP endonuclease under the latter conditions is no greater than that present in uninfected cells (Fig. 3-27). In addition, infection of *E. coli* with

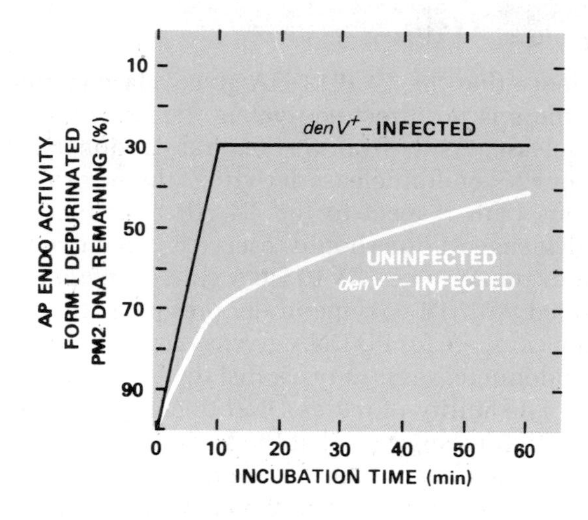

FIGURE 3-27
AP endonuclease activity, detected by measuring the loss of supercoiled (form I) DNA containing an average of one AP site per DNA molecule, is increased in crude extracts of T4 *denV*$^+$-infected cells relative to that present in uninfected or T4 *denV*$^-$-infected cells. (From T. Bonura et al., ref. 126.)

TABLE 3-6
Competition for T4 PD DNA glycosylase activity against UV-irradiated [^{3}H]poly(dT)·poly(dA) by apyrimidinic DNA

Expt.	Competing DNA	Specific Competing Site (AP or PD) (*M*)	PD DNA Glycosylase Activity*
1	None	0	51
	Native *E. coli* DNA	0	55
	Apyrimidinic PBS2 DNA	3.0×10^{-8}	23
	Apyrimidinic PBS2 DNA	10.0×10^{-8}	17
	Apyrimidinic PBS2 DNA	30.0×10^{-8}	4
2	None	0	97
	Native *E. coli* DNA	0	97
	UV-irradiated *E. coli* DNA	1.0×10^{-8}	57
	UV-irradiated *E. coli* DNA	3.0×10^{-8}	35
	UV-irradiated *E. coli* DNA	10.0×10^{-8}	6

*PD DNA glycosylase activity is expressed as femtomoles of [^{3}H]thymine-containing PD hydrolyzed per minute at 37°C.
From S. McMillan et al., ref. 121.

phage T4 containing an *amber* mutation in the *denV* gene results in the loss of both PD glycosylase and AP endonuclease activities (110). Aside from this genetic evidence, a physical association between PD DNA glycosylase and AP endonuclease activities has been inferred from the observation that *apyrimidinic* DNA competes almost as strongly for PD DNA glycosylase activity as does *UV-irradiated* DNA (121) (Table 3-6). In vitro this AP endonuclease function effects the cleavage of the phosphodiester bond situated 3′ with respect to the apyrimidinic site, leaving a 3′-terminal deoxyribose-phosphate moiety (97, 111) (Figs. 3-21, 3-24).

Is the T4 "AP Endonuclease" Utilized In Vivo?

In contrast to the evidence that the T4 PD DNA glycosylase is functional in living cells, there is no direct equivalent evidence for the associated AP endonuclease. Aside from the general cautionary remarks made earlier about AP endonuclease activities, there are some disquieting observations with respect to the T4 AP endonuclease activity in particular that suggest one should reserve judgment about its biological role. For example, when SV40 DNA containing pyrimidine dimers is incubated with T4 enzyme in the presence of *native* DNA, the latter does not compete for PD DNA glycosylase activity but does compete for AP endonuclease activity against the UV-irradiated DNA (121) (Table 3-7). The ability of native DNA to compete *differentially* for these two activities suggests that the hydrolysis of the 5′ glycosyl bond and of the 3′ phosphodiester bond at a pyrimidine dimer in DNA is not concerted in vitro; i.e., the enzyme dissociates from the DNA between these two events (Fig. 3-28). This is sur-

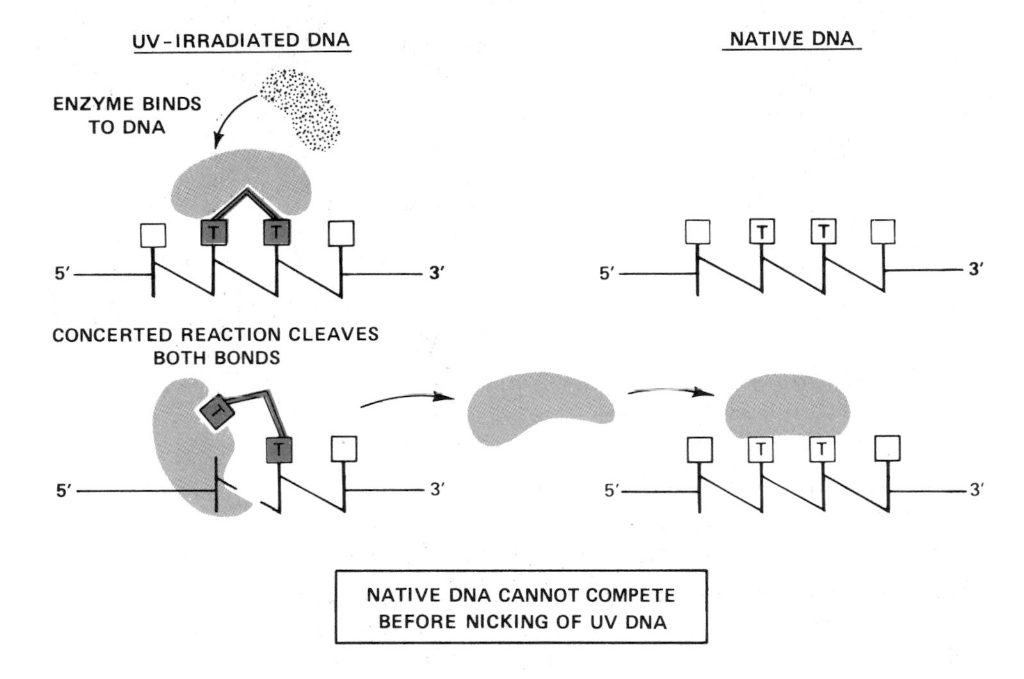

FIGURE 3-28
If the T4 PD DNA glycosylase–AP endonuclease attacks *both* the 5′ N-glycosyl and the 3′ phosphodiester bond at each enzyme-substrate encounter (concerted reaction mechanism), then native DNA cannot compete for AP endonuclease activity. However, if the enzyme dissociates from the DNA after hydrolysis of the N-glycosyl bond (nonconcerted reaction mechanism), then native DNA can compete for subsequent cleavage of the phosphodiester bond at the AP site, as shown on the opposite page.

TABLE 3-7
Competition for T4 AP endonuclease but not for PD DNA glycosylase by native DNA

Competing DNA	PD DNA Glycosylase Activity	AP Endonuclease Activity
None	100.0	100.0
Native	98.1	7.6

From T. Bonura et al., ref. 126.

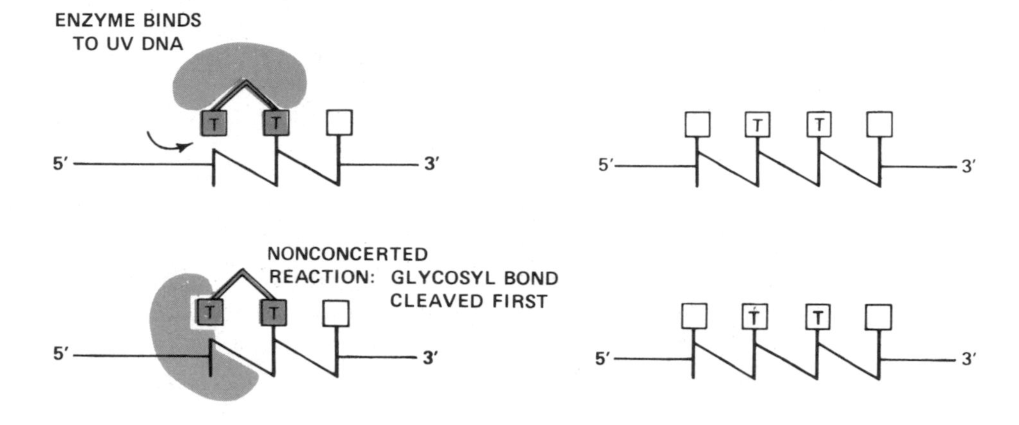

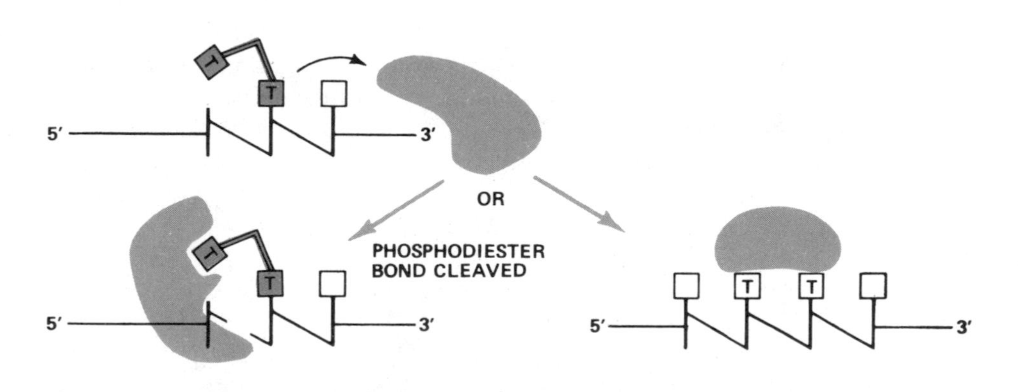

prising, since one would expect that if both enzymatic functions required for DNA incision are embodied in the same protein, the strand breakage mechanism would be a concerted one. If the non-concerted mechanism demonstrated in vitro operates in living cells, then conceivably other AP endonucleases known to be present in *E. coli* (see Section 3-5) can catalyze the strand breakage event in vivo.

Given this conclusion, one is prompted to ask the admittedly teleological question, Why does the T4 PD DNA glycosylase have an associated AP endonuclease activity? In this regard it is of interest that AP sites in DNA result in some inhibition of uracil DNA glycosylase activity (42, 45). As suggested earlier in this chapter, it is possible that all DNA glycosylases have some affinity for such sites (perhaps not unexpected, inasmuch as AP sites are the product of DNA glycosylase activity), and in some cases this may include the ability to effect hydrolysis of associated phosphodiester bonds. Since hydrolysis of phosphodiester bonds at AP sites in DNA (particularly those 3′ to AP sites) can result from β-elimination reactions produced by amines and amides (81, 82), one should be cautious in concluding that all proteins that produce this result are biologically relevant AP endonucleases.

PD DNA Glycosylases in Other Organisms

Like the enzyme from phage T4 (with which it has many properties in common), the PD DNA glycosylase from *M. luteus* also contains associated AP endonuclease activity in vitro (93, 94, 98, 111, 123) and catalyzes the cleavage of phosphodiester bonds 3′ to the apyrimidinic sites produced by the glycosylase, leaving 3′-OH and 5′-P termini (98, 123). Unfortunately, the lack of available mutants has precluded genetic evidence for a physical relationship between these two activities and it is not established whether the AP endonuclease activity is functional in vivo in this organism. However, the *M. luteus* PD DNA glycosylase has been purified to physical homogeneity, and during the final stages of purification the ratio of glycosylase to AP endonuclease remains constant. In addition, the two activities co-elute in a number of chromatographic systems (98).

The *M. luteus* PD DNA glycosylase is a small monomeric protein with a molecular weight of about 18,000 and has no known cofactor requirement. It prefers duplex DNA containing dimers to single-strand DNA with such lesions. This PD DNA glycosylase also catalyzes the hydrolysis exclusively of the 5′ glycosyl bond in dimerized pyrimidines (98) (Fig. 3-29) (Table 3-8). Interestingly, under conditions of substrate excess, the enzyme apparently prefers dimers with 5′ thymines to those containing 5′ cytosines (98).

Thus far, bacteriophage T4-infected *E. coli* and *M. luteus* are the only sources of cell-free extracts in which PD DNA glycosylase activ-

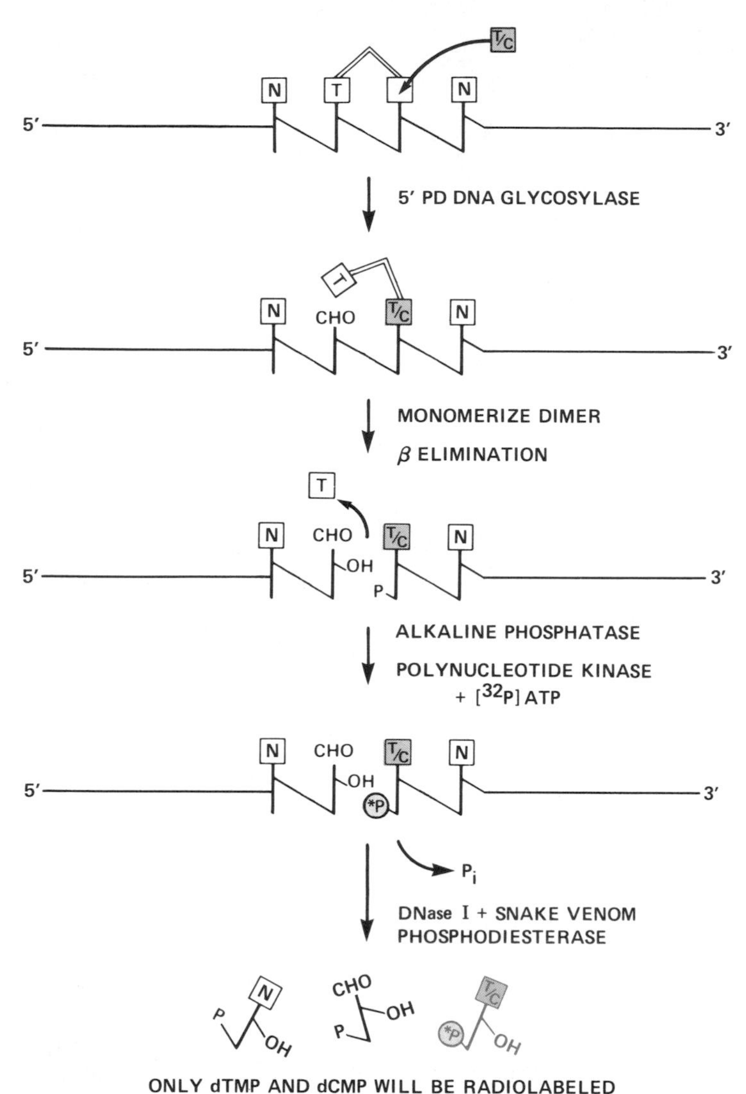

FIGURE 3-29
Diagram showing how the 5′ specificity of the *M. luteus* PD DNA glycosylase was established. If the 5′ glycosyl bond in a pyrimidine dimer is hydrolyzed by the enzyme, the dimer can be monomerized and strand breakage effected by a β-elimination reaction on the 3′ side of the AP site. *A 5′-terminal phosphate is now associated with the 3′ pyrimidine of the former dimer.* This phosphate can be replaced with ^{32}P by the use of alkaline phosphatase and polynucleotide kinase. Subsequent degradation of the DNA to mononucleotides shows that the radioactivity is associated exclusively with dCMP or dTMP (pyrimidine nucleotides). If the 3′ glycosyl bond had been hydrolyzed, β-elimination would have left a *terminal phosphate associated with the nucleotide 3′ to the dimer* and the ^{32}P would be randomly distributed among all the nucleotide products of DNA digestion. This result is not shown in the figure. (After R. H. Grafstrom et al., ref. 98.)

TABLE 3-8
Sites of cleavage by the
M. luteus PD DNA glycosylase

5′-Terminal Nucleotide	Total ^{32}P (%)
dTMP	52.6
dCMP	37.5
dGMP	2.0
dAMP	4.1

See Fig. 3-29 for an explanation of this experiment.
From R. H. Grafstrom et al., ref. 98.

ity has been observed in vitro. Both enzymes are also functional in vivo (119, 124). In theory it should be possible to demonstrate PD DNA glycosylase activity in other living cells. Such experiments indicate that neither yeast (125) nor human cells (124) contain PD DNA glycosylase activity. The photoreversal-dependent formation of free thymine in the acid-soluble fraction of post-UV-incubated *uninfected UV-irradiated E. coli* strains has been observed (126); however, a PD DNA glycosylase activity in *extracts* of uninfected *E. coli* has not yet been demonstrated.

Other DNA Glycosylases with Associated AP Endonuclease Activity In Vitro

Thymine Hydrate (TH) DNA Glycosylase of E. coli

Extracts of *E. coli* contain an activity that catalyzes the excision of free 5,6-dihydroxydihydrothymine and 5,6-dihydrothymine from DNA containing these bases (117, 127). Like the PD DNA glycosylases, this activity is apparently inseparable from an AP endonuclease activity in vitro called *endonuclease III* of *E. coli* (117). Endonuclease III is an activity that creates single-strand breaks in duplex UV-irradiated DNA at sites other than pyrimidine dimers (128, 129). The purified enzyme has a molecular weight of approximately 24,000, with no requirement for divalent cation, and it retains full activity in the presence of EDTA (128, 129). The enzyme also catalyzes the nicking of duplex DNA exposed to osmium tetroxide, hydrogen peroxide, ionizing radiation or acid. Acid-treated DNA contains alkali-labile (apurinic) sites, and there is a correlation between the number of alkali-labile and endonuclease III-sensitive sites in this substrate (129). However, the majority of endonuclease-sensitive sites in DNA

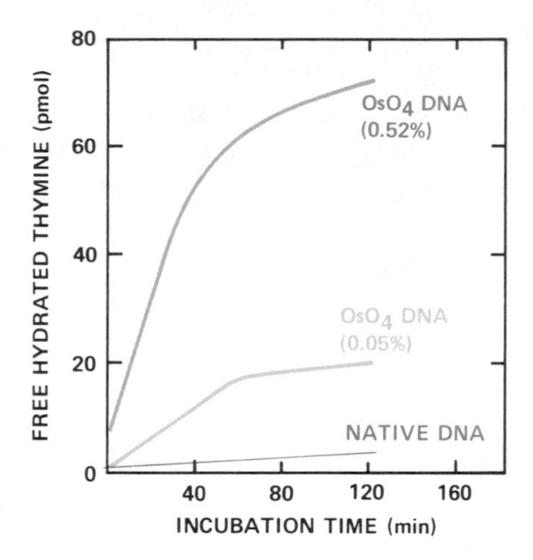

FIGURE 3-30
Release of free hydrated thymine (5,6-dihydroxydihydrothymine and 5,6-dihydrothymine) following incubation of osmium-tetroxide-treated DNA with thymine hydrate (TH) DNA glycosylase (or endonuclease III) of *E. coli*. (From B. Demple and S. Linn, ref. 117.)

exposed to UV radiation, hydrogen peroxide or osmium tetroxide are not alkali-labile (129). Purified preparations of endonuclease III also contain a DNA glycosylase activity that selectively catalyzes the excision of 5,6-dihydroxydihydrothymine and of 5,6-dihydrothymine (117, 127) (Fig. 3-30). Since both of these products are produced in DNA by treatment with osmium tetroxide, ionizing radiation or UV radiation (see Chapter 1), it is highly likely that they account for the alkali-stable lesions mentioned above. Whether or not this DNA glycosylase is an active site of the same protein that contains AP endonuclease activity is not yet certain. Although purification of the AP endonuclease has failed to resolve the two activities, the most definitive proof of their association in one protein would be the isolation of single-gene mutants of *E. coli* defective in both activities.

Concluding Comments on DNA Glycosylases

It is likely that additional DNA glycosylases remain to be discovered, the identification of which will depend significantly on the characterization of specific substrates in DNA. Although the total number of distinct DNA glycosylases in nature may amount to as much as a dozen or more, in the interests of genetic economy it is obviously unrealistic to expect a unique repair enzyme for each of the myriad forms of base damage to DNA known to be repairable in organisms such as *E. coli*. Evidence discussed in the next chapter suggests that most forms of base damage are probably dealt with by a different mode of DNA incision, involving fewer examples of a particular enzyme class, i.e., direct-acting damage-specific DNA incising activities.

3-5 AP Endonucleases Not Associated with DNA Glycosylases

Following the release of free damaged or inappropriate bases from DNA, apurinic or apyrimidinic sites are still left in the DNA as products of the reactions catalyzed by DNA glycosylases. In addition, such sites can arise spontaneously in native and particularly in alkylated DNA (5) (see Section 1-3). There is some suggestion that *apurinic* sites may be repaired by the direct reversal of such base damage, i.e., through the action of a DNA purine insertase activity (130) (see Section 2-5). However, it appears that a more general mode for the repair of sites of base loss in DNA is by their excision, as indicated in Fig. 3-1 (3, 5, 7, 11). DNA repair enzymes involved in the excision repair of AP sites are called AP endonucleases (3, 5, 7, 11). These enzymes selectively catalyze the incision of DNA at such sites (they do not degrade native DNA), thereby preparing the DNA for subsequent excision, repair synthesis and DNA ligation.

Proteins with AP endonuclease activity have been isolated from both prokaryotic and eukaryotic sources. Although some of these enzymes may be physically associated with DNA glycosylases described earlier in this chapter, the majority apparently are not. The most complete information available on this type of enzyme stems from studies with *E. coli*, and this organism will constitute the primary focus of the present discussion. In *E. coli* a number of apparently distinct AP endonuclease activities are detected in vitro (3, 7) (Table 3-9).

5′-Acting AP Endonucleases of *E. coli*

These enzymes catalyze incision of duplex DNA 5′ to sites of base loss (Fig. 3-31), leaving 3′-hydroxyl-nucleotide and 5′-deoxyribose-phosphate termini. Two such examples are known to be present in *E. coli* (Table 3-9).

TABLE 3-9
Apurinic/apyrimidinic endonucleases of *E. coli*

Enzyme	Principal Properties	Associated Catalytic Activities
Endonuclease III	No requirement for divalent cation; inhibited in presence of tRNA; 2.7 S; MW ~27,000; pH optimum ~7.0; requires duplex DNA with AP sites	DNA glycosylase activity that recognizes 5,6-saturated thymine photoproducts
Endonuclease IV	No requirement for divalent cation; not stimulated by Mg^{2+} or Ca^{2+}; no inhibition in presence of tRNA; 3.4 S; MW ~33,000; pH optimum ~8.0–8.5; inhibited in presence of PCMB; requires duplex DNA with AP sites	None detected
Endonuclease V	Requires Mg^{2+} for activity; inhibited in presence of tRNA; 2.3 S; MW ~20,000; pH optimum 9.25; requires duplex DNA with AP sites	Acts on a variety of damaged DNAs as well as certain native DNAs
AP endonuclease of exonuclease III	Requires Mg^{2+} for optimal activity; inhibited in presence of EDTA; 2.9 S; MW ~32,000; pH optimum ~8.5; requires duplex DNA with AP sites	$3' \rightarrow 5'$ exonuclease (exonuclease III); 3′ phosphatase; RNase H
Endonuclease VII	No requirement for divalent cation; stimulated by Mg^{2+} or Ca^{2+}; inhibited in presence of tRNA; 4.3 S; MW ~60,000; pH optimum ~7.0; insensitive to PCMB; attacks single-strand DNA and polydeoxypyrimidines with depyrimidinated sites; duplex DNA with apurinic sites is not a substrate	Not determined

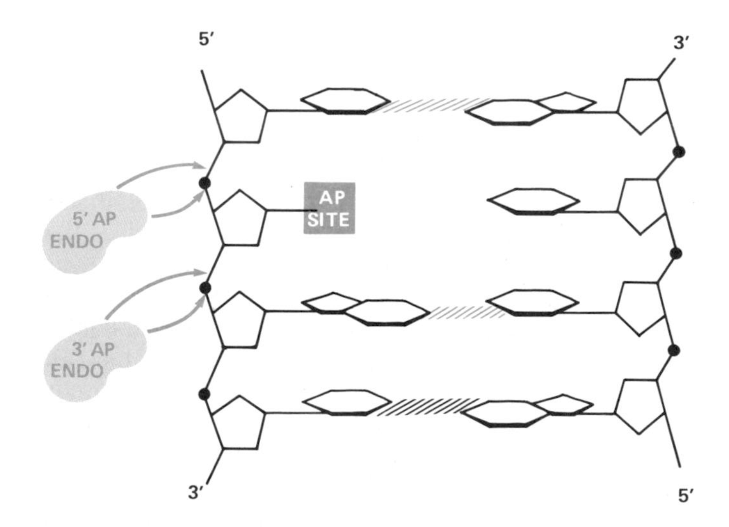

FIGURE 3-31
Schematic diagram showing the polarity of 5′- and 3′-acting AP endonucleases. In theory each can create either 3′-OH or 3′-P termini depending on which bond is cleared. All known 5′AP endonucleases of *E. coli* create 3′-OH termini.

The AP Endonuclease Function of Exonuclease III (131)

Exonuclease III of *E. coli* was originally characterized as an exonuclease with an associated phosphatase activity (132, 133). Exonuclease III specifically requires 3′-OH ends in duplex DNA, i.e., it is a 3′ → 5′ exonuclease. In light of the associated phosphatase activity, DNA containing either 3′-OH or 3′-P termini can be attacked exonucleolytically, since in the latter case the removal of the phosphase group creates a substrate for the exonuclease. In addition to these two catalytic functions, exonuclease III degrades a mixed copolymer of ribo- and deoxyribonucleotides, reflecting a ribonuclease H activity (an enzyme that degrades RNA in RNA-DNA hybrids) (134). A fourth catalytic function, and the one that primarily concerns us in the present discussion, is an AP endonuclease activity called the *AP endonuclease function of exonuclease III*. Before it was known that this AP endonuclease activity was a property of exonuclease III, it was variously referred to as endonuclease II (135, 136) or as endonuclease VI of *E. coli* (137). Mutants of *E. coli* defective in both the 3′-exonuclease and the associated 3′-phosphatase functions of exonuclease III have been isolated (138) and mapped to the *xth*A locus (139), which is situated between the *pnkB* and *pncA* loci of *E. coli* at approximately 38.5 minutes on the *E. coli* genetic map (Fig. 3-32). All of the known *xthA* mutants have been shown to be also defective in AP endonuclease activity (138). In addition, exonuclease III purified to greater than 98 percent homogeneity contains AP endonuclease activity that cannot be separated from exonuclease and phosphatase activities by electrophoresis, sedimentation or gel filtration (140).

The discovery of a new activity in a previously described enzyme is not without precedent. Both the 3′ → 5′ and 5′ → 3′ exonuclease activities of *E. coli* DNA polymerase I were discovered independently and originally designated as exonucleases II and VI of *E. coli* respectively (8). However, once it was clear that these were in fact catalytic functions associated with DNA polymerase I, these designations were dropped. In keeping with this nomenclatural precedent, the terms endonuclease II and VI are not used here and we shall refer instead to the AP endonuclease function of exonuclease III.

Purified exonuclease III is a protein with a molecular weight of approximately 28,000 (131, 141). The AP endonuclease activity has an absolute requirement for Mg^{2+} and is inhibited in the presence of EDTA (131, 141). The enzyme catalyzes the hydrolysis of duplex apurinic DNA on the 5′ side of sites of base loss, leaving 3′-OH and 5′-P termini (131, 141). It is generally assumed that the AP endonuclease function of exonuclease III attacks sites of pyrimidine or purine loss in duplex DNA with equal facility; however, quantitative

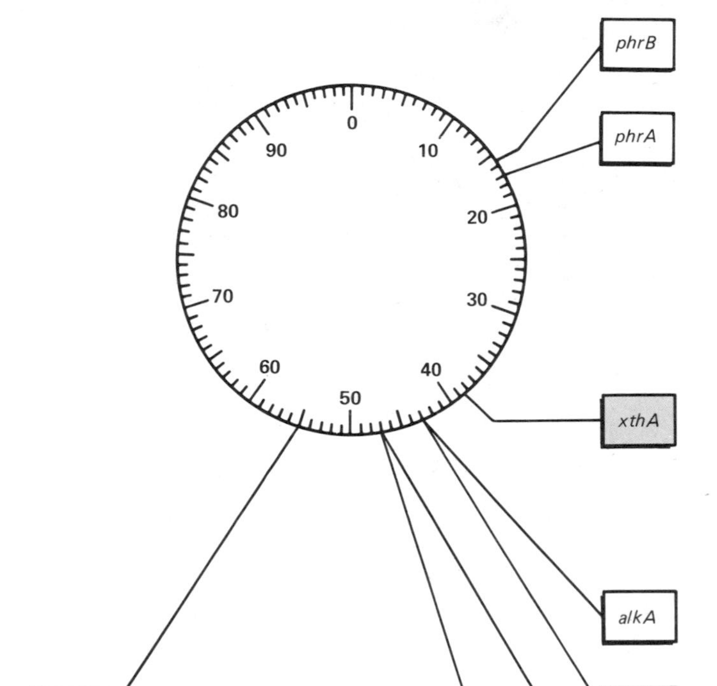

FIGURE 3-32
The location of the *xthA* gene of *E. coli*.

comparisons of this parameter have not been documented. The enzyme is at least as active on DNA that contains AP sites reduced with sodium borohydride as on the unreduced substrate (141), suggesting that the aldehyde group of C^1 in deoxyribose is not required for enzyme action. The mechanism of action of the enzyme is probably distinct from the β-elimination reaction catalyzed by alkali at sites of base loss in DNA; alkali-catalyzed β-elimination most frequently results in hydrolysis of phosphodiester bonds on the 3′ side of sites of base loss and has a requirement for the aldehyde group at C^1 of the deoxyribose residue, since reduction of C^1 to the alcohol prevents the reaction (31).

The presence of multiple catalytic functions associated with a single monomeric and relatively small protein suggests that a single active site catalyzes all the enzymatic reactions of exonuclease III. In this regard it has been proposed that the enzyme has three important

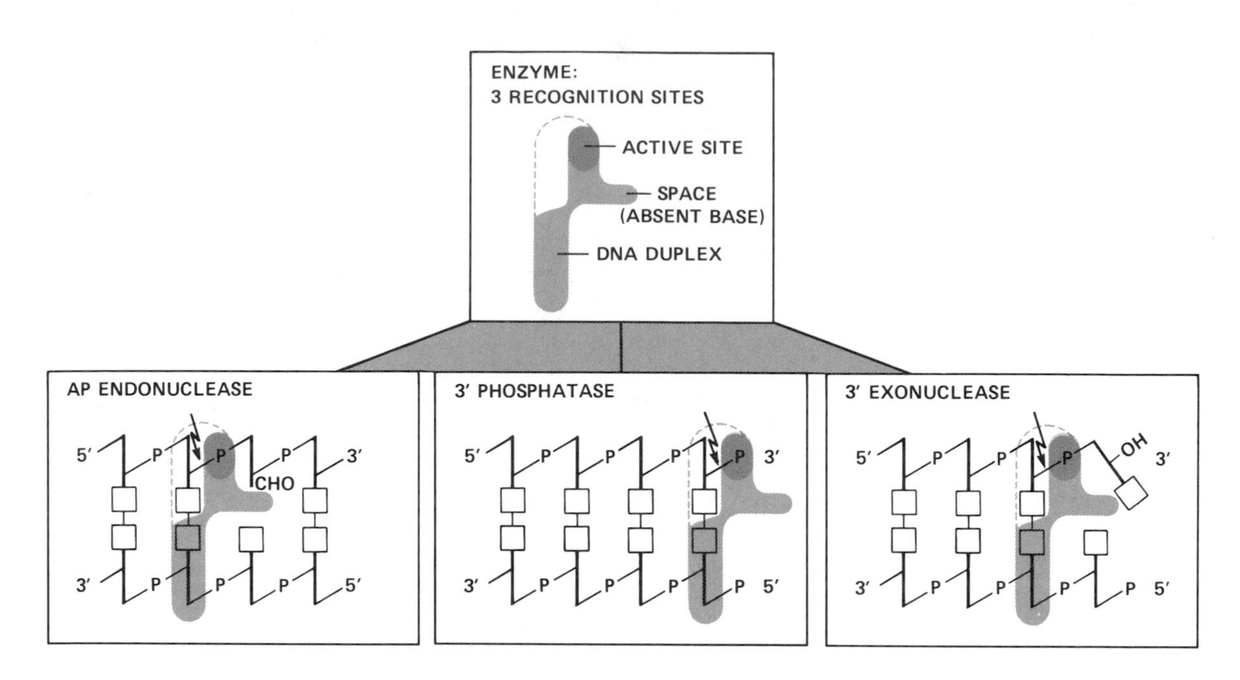

FIGURE 3-33
Schematic illustration of how exonuclease III of *E. coli* may function as a 3′ phosphatase, a 5′ AP endonuclease or a 3′ exonuclease. It is proposed that the enzyme has one domain that functions as the active site for nuclease and/or phosphatase activity. In addition, the enzyme has a second domain that recognizes a space in the DNA duplex that can be constituted by either a missing base (AP site) (left lower panel), an "absent" base at the end of a DNA molecule (middle lower panel) or an "absent" base due to fraying of the end of the DNA duplex (right lower panel). A third domain in the enzyme recognizes duplex DNA. (After B. Weiss, ref. 140.)

domains in its structure (140) (Fig. 3-33). One is the active site that catalyzes the cleavage of phosphodiester bonds in one strand of duplex nucleic acid. The second is a site for recognizing duplex structure in the nucleic acid and depends on the presence of deoxyribose in the strand opposite to that in which the active site works. (This idea is consistent with the absolute requirement for duplex DNA or for a DNA-RNA hybrid as substrate.) A third domain is postulated to recognize a "space" in the DNA duplex, which may be constituted by a missing base (thereby facilitating its AP endonuclease function), by the region beyond the end of a DNA strand or by the space created by the partial denaturation of the DNA duplex at the site of an internal nick (thereby providing substrate for the phosphatase-exonuclease function).

The exact biological role of the AP endonuclease activity of exonuclease III is not clear. Mutants in the *xthA* gene are exceptionally sensitive to killing by hydrogen peroxide (142). The nature of the lethal lesion(s) responsible for this phenotype is unknown at present, but a number of possibilities have been suggested (142). For example, the free radicals arising from H_2O_2 decomposition might generate DNA strand breaks with 3′-P termini. Such termini may be refractory to further processing unless the phosphate group is removed by the phosphatase function of exonuclease III. This enzyme accounts for more than 99 percent of the 3′ phosphatase activity present in *E. coli* (142). Surprisingly, mutants of *E. coli* (including deletion mutants) defective in exonuclease III (*xth*) are not particularly sensitive to treatment with alkylating agents such as methylmethanesulfonate (138, 143, 144). This may reflect the ability of other AP endonucleases to assume the essential function(s) of this enzyme during the repair of sites of base loss. Alternatively, repair of such sites may occur by the action of DNA purine insertase activity. In contrast, *E. coli* mutants defective in both the *xthA* and the *dut* gene (the latter codes for deoxyuridine triphosphatase) are only conditionally viable (145). In the absence of the AP endonuclease function of exonuclease III, the apyrimidinic sites created by incorporation into DNA of dUMP (due to increased levels of dUTP associated with the *dut* mutation; see Section 1-2), and the subsequent excision of uracil by Ura DNA glycosylase, may be lethal lesions. If so, the activity involved in the repair of *apurinic* sites in the absence of the AP endonuclease function of exonuclease III is apparently not able to deal as effectively with *apyrimidinic* sites in DNA. Such a conclusion is consistent with the apparent failure to discover DNA pyrimidine insertase activity in any biological source, but it is also consistent with the notion that some AP endonucleases may not attack apurinic and apyrimidinic sites in DNA with equal efficiency.

Endonuclease IV of E. coli

Endonuclease IV is a second example of a 5′-acting AP endonuclease in *E. coli* (Table 3-9). This activity is not readily detectable in wild-type *E. coli* because it constitutes only about 10 percent of the total AP endonuclease activity in crude extracts. Its identification was facilitated by the isolation of mutants defective in exonuclease III (*xth*). Endonuclease IV has been extensively purified and characterized (146). It catalyzes the formation of single-strand breaks at sites of base loss in duplex DNA containing either apurinic or apyrimidinic sites, although the relative K_m for these two substrates has not been determined. The enzyme has a sedimentation coefficient of 3.4 and a calculated molecular weight of 33,000. Full activity is retained in the presence of EDTA and the enzyme is not stimulated by $MgCl_2$ or $CaCl_2$. The enzyme activity is unusually resistant to high ionic strength (50 percent activity is retained in the presence of 0.56 *M* NaCl) and to heat (full activity is retained after heating to 60°C for five minutes). It is inhibited by SH-group inhibitors but is insensitive to inhibition by tRNA. The enzyme does not degrade native duplex DNA, single-strand depyrimidinated deoxyribopolymers or intact uracil-containing DNA. Although degradation of "aged" heavily UV-irradiated DNA as well as DNA exposed to ionizing radiation has been observed, these substrates probably contain sites of base loss, as evidenced by their alkali lability. The purified enzyme has no known associated exonuclease, phosphatase or DNA glycosylase activities and is present at wild-type levels in all mutants of *E. coli* thus far examined. These properties are of importance in terms of establishing its distinction from other enzymes with AP endonuclease activity.

Like the AP endonuclease function of exonuclease III, endonuclease IV attacks phosphodiester bonds 5′ to the sites of base loss in DNA. This has been inferred in the case of both enzymes by the type of experiment illustrated in Fig. 3-34. As indicated previously, the AP endonuclease activity in vitro associated with the *denV* gene product of phage T4 attacks the apyrimidinic sites created by the PD DNA glycosylase, leaving 3′ deoxyribose residues. When this substrate is incubated with *E. coli* DNA polymerase I in the presence of

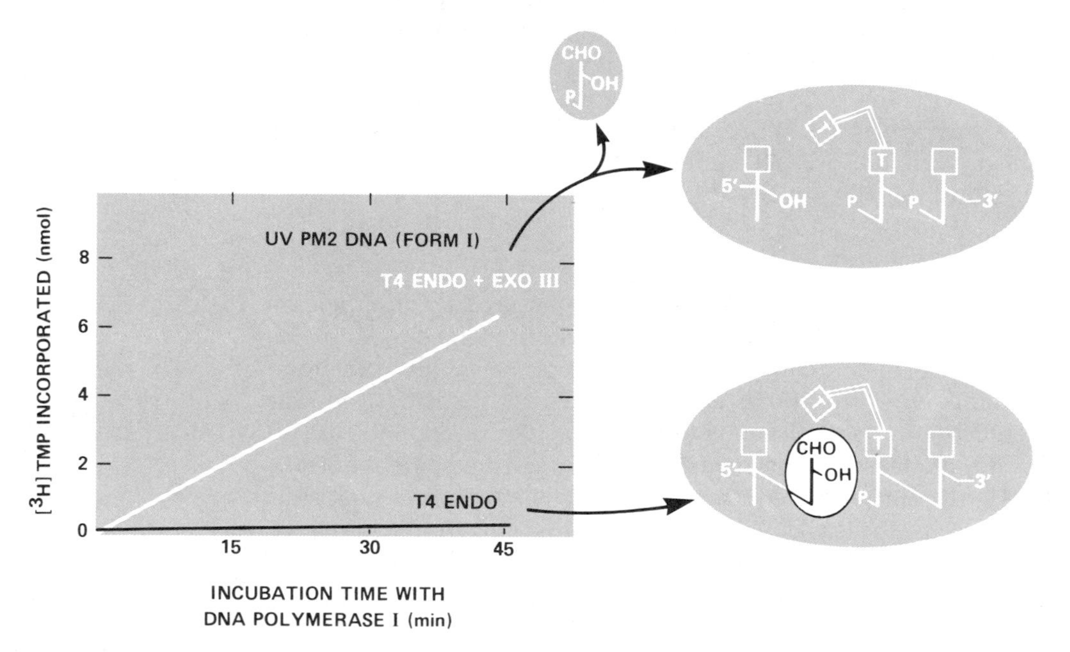

FIGURE 3-34
Incubation with exonuclease III (or endonuclease IV) of UV-irradiated DNA previously treated with T4 (or *M. luteus*) PD DNA glycosylase–AP endonuclease creates a normal primer terminus for DNA polymerase I by removal of the 3′-terminal deoxyribose moiety left by the 3′ AP endonuclease activity of the T4 (or *M. luteus*) enzyme (see Fig. 3-4). From this result it can be inferred that exonuclease III and endonuclease IV are 5′AP endonucleases. (After T. Bonura et al., ref. 126.)

the four usual deoxynucleoside triphosphates, a very low rate of DNA synthesis is observed despite the fact that this enzyme has an associated $3' \rightarrow 5'$ exonuclease activity (97). (The role of DNA polymerase I in the excision of deoxyribose-phosphate moieties is discussed fully in Section 5-4.) However, if the DNA is preincubated with either exonuclease III or endonuclease IV prior to incubation with DNA polymerase I, a much faster rate of DNA synthesis is observed (97). The interpretation given to these observations is that the 3′-terminal deoxyribose residue created by the T4 AP endonuclease activity is an inadequate primer for DNA polymerase I. However, the removal of this residue by a 3′ exonuclease or a 5′ AP endonuclease creates a normal 3′ nucleotide terminus, which can be utilized as an effective primer for DNA synthesis (97).

3′-Acting AP Endonucleases of *E. coli*

This class of enzymes produces 3′-deoxyribose and 5′-phosphomonoester-nucleotide termini (Fig. 3-31). Examples were considered earlier in this chapter (Section 3-4) in the discussion of AP endonucleases associated with the phage T4 and *M. luteus* PD DNA glycosylases. The only known enzyme in this category in uninfected *E. coli* is so-called endonuclease III, apparently associated with the thymine hydrate DNA glycosylase discussed earlier (Section 3-4) (127–129). The demonstration of 3′-acting AP endonuclease activity stems from the same experimental approach illustrated in Fig. 3-34. Thus, if the substrate created by the attack by an AP endonuclease at AP sites in duplex DNA is a poor primer-template for DNA synthesis by *E. coli* DNA polymerase I, and if this primer-template is activated by subsequent incubation with a known 5′ AP endonuclease, then the original AP endonuclease is presumed to be 3′-acting (97).

Endonuclease VII of *E. coli*

An interesting enzyme called endonuclease VII is apparently selective for single-strand DNA containing sites of base loss (147) (Table 3-9). The principal substrate used to identify and characterize the activity is the single-strand polymer poly(dU-[^{3}H]dT), in which varying degrees of controlled depyrimidination are produced by preincubation with purified Ura-DNA glycosylase (Fig. 3-35). Hydrolysis of phosphodiester bonds at depyrimidinated sites by endonuclease VII activity results in the formation of acid-soluble radioactively labeled fragments (Fig. 3-35). The enzyme degrades this substrate with nonlinear kinetics, and the distinct lag in the appearance of acid-soluble radioactivity suggests an endonucleolytic rather than an exonucleolytic mode of action (147). Such a mechanism has been confirmed by

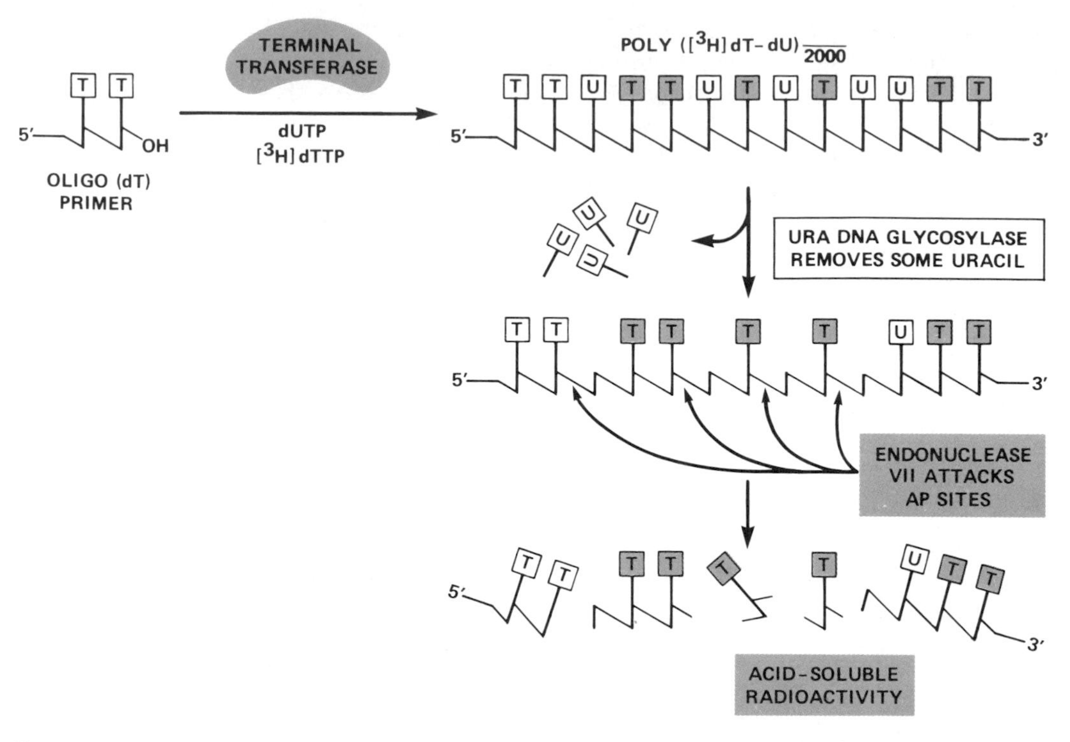

FIGURE 3-35
Assay of endonuclease VII of *E. coli*. Poly($[H^3]$dT-dU)$_{2000}$ is incubated with limiting amounts of uracil DNA glycosylase to remove some of the uracil and hence generate apyrimidinic sites in the polymer. When these sites are attacked by endonuclease VII, the polymer is degraded into radiolabeled acid-soluble products, which can be readily quantitated.

sedimentation of high molecular weight denatured PBS2 DNA containing an average of one to two depyrimidinated sites per molecule, as well as by the demonstration of selective degradation of depyrimidinated phage M13 single-strand circular DNA (147) (Fig. 3-36).

The specificity of endonuclease VII is indicated by the following observations. It does not degrade any of the intact single-strand substrates described above, nor does it produce any detectable nicks in heat or acid depurinated duplex PM2 DNA. Furthermore, whereas depyrimidinated PBS2 DNA is an effective substrate when denatured, it is not degraded in the duplex state (147). The enzyme has a sedimentation coefficient of 4.3 measured in the presence of 0.25 *M* NaCl, giving an estimated molecular weight of about 60,000 for a globular protein, a value significantly greater than that of any of the other AP endonucleases of *E. coli*. Endonuclease VII is fully active in the presence of EDTA and is stimulated about twofold by 5 m*M* $CaCl_2$. The activity is inhibited by tRNA and is much more sensitive to high ionic strength and to heat than is endonuclease IV. No asso-

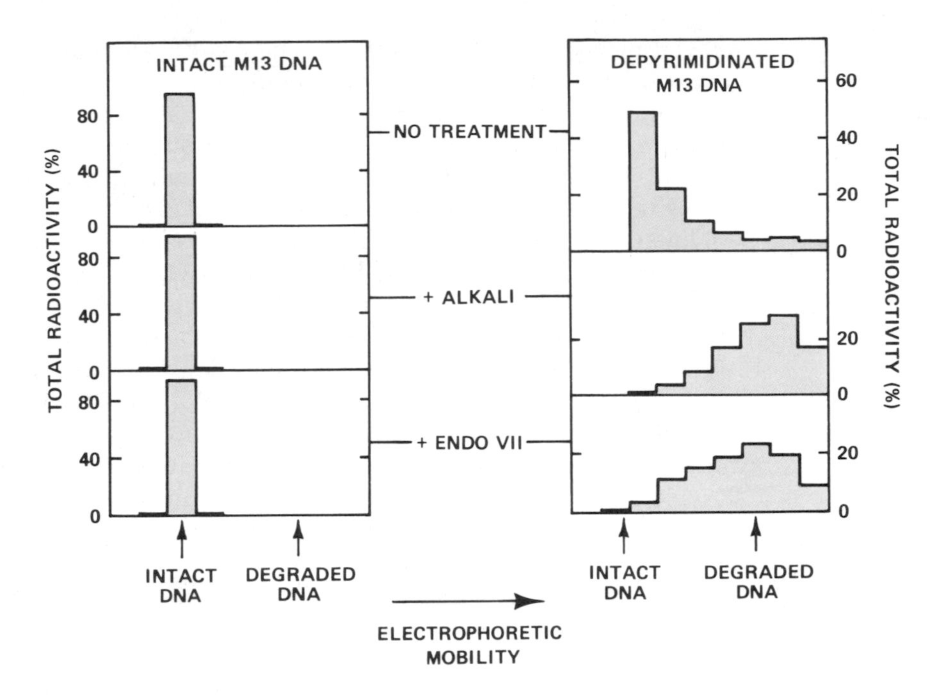

FIGURE 3-36
Endonuclease VII (or alkali) degrades depyrimidinated phage M13 DNA (right) but not DNA with intact bases (left). Degradation was detected by gel electrophoresis of [^{3}H]thymine-labeled M13 DNA. Following electrophoresis the gel was sliced into fractions, the DNA eluted and the radioactivity measured in each fraction. (From T. Bonura et al., ref. 147.)

ciated exonuclease or endonuclease activity on unirradiated DNA has been detected and no active substrate other than single-strand DNA containing sites of base loss has been identified thus far. The biological function of endonuclease VII is unknown; however, conceivably it could be involved in the repair of sites of base loss in partially denatured regions of the genome at which the affinity of other AP endonucleases specific for duplex DNA may be very low (147, 148).

Why So Many AP Endonucleases in *E. coli*?

It is not immediately obvious why so many AP endonucleases are present in *E. coli*. One general argument advanced earlier in this chapter relates to the current uncertainty as to whether or not all AP endonuclease activities demonstrated in vitro function as such in living cells. An obvious approach to this question would be the isolation of enzyme-defective mutants. The only mutants so far available are those defective in exonuclease III (*xth*) (138, 144). Not unexpectedly, many (but not all) of these mutants show sensitivity to alkylating agents such as methylmethanesulfonate and to hydrogen peroxide, but little else is known about them.

If the activities demonstrated in vitro are accurate reflections of biologically relevant enzymes, the observation that there exist both 5′- and 3′-acting AP endonucleases is intriguing. Especially interesting is the fact that both endonuclease III of *E. coli* and the T4 (and *M. luteus*) AP endonucleases (the only 3′ AP endonucleases thus far characterized) have associated DNA glycosylases, whereas the 5′ endonucleases apparently do not. If all the known DNA glycosylases of *E. coli* had associated 3′ AP endonuclease activity, it would obviously be tempting to speculate that their role was to create the initial (primary) phosphodiester bond break in AP DNA. The 5′ AP endonucleases might then be regarded as *secondary* enzymes needed specifically to remove the 3′-terminal deoxyribose residues in order to create functional primers for excision and resynthesis of DNA by DNA polymerase I. However, the majority of *E. coli* DNA glycosylases do *not* have associated 3′ AP endonuclease activity and therefore it is entirely likely that in living cells 5′ AP endonucleases function as primary endonucleases.

Aside from the isolation of mutants of *E. coli* defective in different AP endonucleases, it would be very informative to examine the substrate specificity of these enzymes in greater detail. It is intriguing that the DNA glycosylases that have associated 3′ AP endonuclease activity in vitro all create *apyrimidinic* sites. It is thus conceivable that AP endonucleases have varying specificity for apyrimidinic as opposed to apurinic sites in DNA. In addition, AP endonucleases may be differentially sensitive to the nucleotide sequence flanking sites of base loss in DNA.

AP Endonucleases from Organisms other Than *E. coli*

AP endonucleases have been isolated from a variety of other biological forms (Table 3-10). A survey of the properties of these enzymes suggests that there is no prototypic or "general" form of this enzyme in nature. For example, the AP endonucleases isolated (presumably as the quantitatively major form) from *B. subtilis* (149), *M. luteus* (150, 151), *B. stearothermophilus* (152) and the plant *Phaseolus multiflorus* (153) most closely resemble endonuclease IV of *E. coli* in terms of their general properties. All these enzymes are active in the presence of EDTA, none has associated exonuclease activity and all require duplex DNA as substrate. On the other hand, an AP endonuclease isolated from *Haemophilus influenzae* closely resembles exonuclease III of *E. coli* (154). The *Haemophilus* enzyme has a molecular weight of about 30,000, an absolute requirement for Mg^{2+} or Mn^{2+} and is completely inhibited in the presence of EDTA. In addition,

TABLE 3-10
AP endonucleases from a variety of biological sources

Source	Molecular Weight	Cofactor Requirement	Stimulation or Inhibition of Activity	General Comments
Phaseolus multiflorus	40,000	None	Mg^{2+} or Mn^{2+} stimulate about four-fold; inhibited in presence of NaCl	
B. stearothermophilus	28,000	None	Stimulated by monovalent cations; inhibited by divalent cations	Temperature optimum at 60° C
B. subtilis	56,000	None	Stimulated by 50 m*M* NaCl and by Mg^{2+}, Mn^{2+} or Ca^{2+}; inhibited in presence of 500 m*M* NaCl	
S. cerevisiae				
(Endonuclease D)		Mg^{2+}		Two activities identified in extracts
(Endonuclease E)		None		
S. cerevisiae		None	Inhibited in presence of 400 m*M* NaCl	Temperature optimum at 40°C
Calf thymus	35,000	Mg^{2+}	Inhibited by 1 m*M* EDTA	
Calf liver	28,000	Mg^{2+}	Inhibited by Ca^{2+}, EDTA or tRNA	Optimal pH at 9.5
Mouse epidermis	31,000	None	Stimulated by Mg^{2+} or KCl; inhibited by high ionic strength	
Human lymphocytes		Mg^{2+}	Inhibited by EDTA	
Human skin fibroblasts				The activity in XP group A and D cells has a five to 10-fold higher K_m than normal
Human placenta	27,000–31,000	None	Stimulated by Mg^{2+}, Mn^{2+}, Co^{2+}, Zn^{2+} or 40 m*M* KCl; inhibited by Ca^{2+} or EDTA	Six chromatographically separable forms identified

From E. C. Friedberg et al., ref. 7.

this enzyme has associated 3′ → 5′ exonuclease activity specific for duplex DNA, as well as associated 3′ phosphatase activity.

Mammalian AP Endonucleases

AP endonuclease activity is present in extracts of mammalian cells and tissues from a variety of sources. The most extensively purified and characterized of these are from calf thymus (155, 156), calf liver (157), human HeLa cells in culture (158), human fibroblasts in culture (159) and human placenta (160, 161). The enzymes from all these sources share in common a specificity for sites of base loss in duplex DNA and are all without associated exonuclease or phosphatase activity. All are active in the absence of added divalent cations but are inhibited in the presence of EDTA.

Some studies with human fibroblasts (159), human placenta (162) and rat liver (163) indicate the presence of multiple AP endonucleases from a single source; however, this has not been a general finding when different investigators have worked with the same tissue source. There are a number of possible explanations for the presence of enzyme heterogeneity. The most obvious are chromatographic artifacts and artifacts produced by proteolytic degradation during the course of enzyme extraction and purification. On the other hand, although more than 90 percent of the total 5′ AP endonuclease activity present in rat liver is found in the nuclei in intimate association with chromatin, activity is also detected in the nuclear sap and in association with nuclear membranes. Enzyme activity is also located in the cytosol, in association with mitochondria and in association with cell membranes (163). The properties of some of these forms vary according to their cellular location (163). Thus, the cell membrane-associated enzyme is activated by Triton, whereas the other forms are inhibited to varying degrees by this detergent. In addition, the enzyme present in the nuclear sap has a higher molecular weight and a greater resistance to heat inactivation than the chromatin-associated activity. These observations suggest the post-translational processing scheme represented in Fig. 3-37 (163), in which free ribosomes synthesize a protein that attaches to the inner side of the endoplasmic reticulum. This enzyme (which may be the form that is activated by Triton) may be transported to the nucleus as a membrane-bound form, where a hydrophobic moiety is split off by a protease and the soluble enzyme released into the nuclear sap. The latter could be degraded further by a second protease to yield a species that is bound to chromatin, the form of the enzyme most likely involved in the repair of AP sites in nuclear DNA.

Two Types of AP Endonuclease in Human Fibroblasts? Some studies with human fibroblasts in culture suggest the presence of two distinct AP endonucleases (159). One of them (designated human fibroblast AP endonuclease I) fails to bind to phosphocellulose and has a relatively high affinity for AP sites in DNA (K_m = 4.6 nM). This enzyme appears to be a 3′-acting AP endonuclease and is not detected in extracts of human fibroblasts from certain patients with the disease xeroderma pigmentosum (see Section 9-2). The other AP endonuclease (human fibroblast AP endonuclease II) is retained by phosphocellulose, has a relatively low affinity for AP DNA (K_m = 44 nM) and is a 5′-acting enzyme (159).

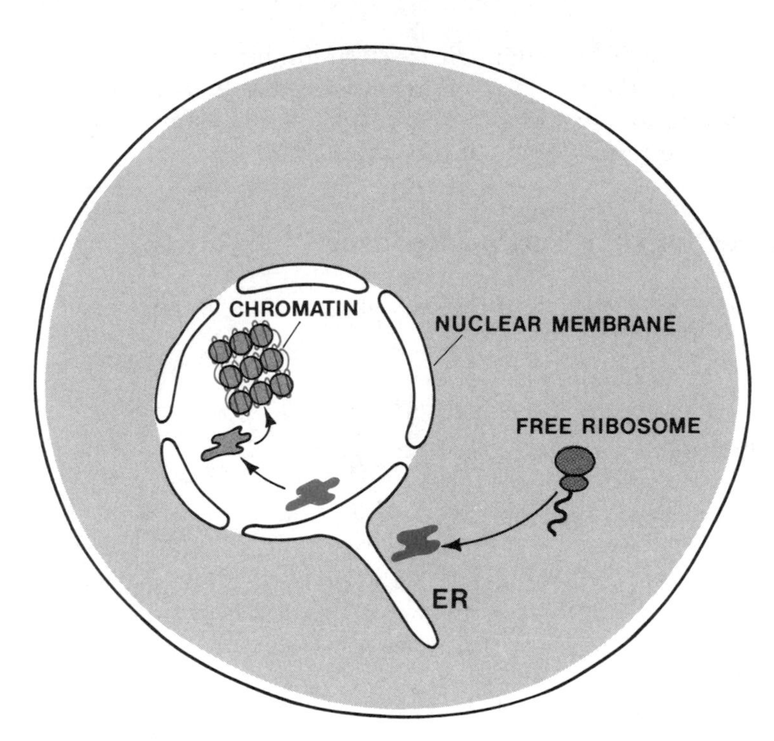

FIGURE 3-37
Hypothetical scheme for the processing and maturation of the major AP endonuclease of mammalian cells. The model suggests that during posttranslational modification the protein is associated with the endoplasmic reticulum as one active form of the enzyme. This form is transported into the nucleus as a membrane-bound protein, where a hydrophobic moiety is split off to yield a free intranuclear form that is also catalytically active. This state may be further modified before the mature (fully processed) enzyme associates with chromatin as a non-histone chromosomal protein. (After L. Thibodeau and W. G. Verly, ref. 163.)

AP Endonuclease from Human Placenta. A *single* species of AP endonuclease has been isolated and extensively purified from human placenta (160, 161). The endonuclease is a monomeric protein with an apparent molecular weight of about 37,000. The enzyme has a pI of 7.4 to 7.6, requires Mg^{2+} and is inhibited in the presence of EDTA. It has no detectable associated exonuclease or phosphomonoesterase activity. This enzyme can apparently hydrolyze phosphodiester bonds at *either* the 5′ or the 3′ side of sites of base loss in duplex DNA, but not at both (161). Given this possibility for other AP endonucleases, as well as that of posttranslational processing of proteins discussed above, it is easy to understand the relative confusion and apparent contradictions that currently pervade the literature on mammalian cell AP endonucleases. Careful future studies that avoid artifacts, that utilize the resolving power of well-characterized antibody probes and that include extensive peptide mapping of proteins should help resolve some of this confusion.

Suggestions for Further Reading

Excision repair, refs. 1–3; DNA glycosylases, refs. 5, 6, 11; AP endonucleases, refs. 5, 7, 131.

References

1. Hanawalt, P. C., Cooper, P. K., Ganesan, A. K., and Smith, C. A. 1979. DNA repair in bacteria and mammalian cells. *Ann. Rev. Biochem.* 48:783.
2. Lehmann, A. R., and Karran, P. 1981. DNA repair. *Int. Rev. Cytol.* 72:101.
3. Lindahl, T. 1982. DNA repair enzymes. *Ann. Rev. Biochem.* 51:61.
4. Lindahl, T. 1976. New class of enzymes acting on damaged DNA. *Nature* 259:64.
5. Lindahl, T. 1979. DNA glycosylases, endonucleases for apurinic/apyrimidinic sites, and base excision-repair. *Prog. Nuc. Acids Res. Mol. Biol.* 22:135.
6. Duncan, B. K. 1981. DNA glycosylases. In *The enzymes*. 3rd ed. P. D. Boyer, ed., vol. XIV, Nucleic acids, part A, p. 565. New York: Academic.
7. Friedberg, E. C., Bonura, T., Radany, E. H., and Love, J. D. 1981. Enzymes that incise damaged DNA. In *The enzymes*. 3rd ed. P. D. Boyer, ed., vol. XIV, Nucleic acids, part A, p. 251. New York: Academic.
8. Kornberg, A. 1981. *DNA Replication*. San Francisco: Freeman.
9. Rupp, W. D., Sancar, A., and Sancar, G. B. 1982. Properties and regulation of the UVR ABC endonuclease. *Biochimie* 64:595.
10. Sancar, A., and Rupp, W. D. 1983. A novel repair enzyme: UVR ABC excision nuclease of *Escherichia coli* cuts a DNA strand on both sides of the damaged region. *Cell* 33:249.
11. Friedberg, E. C., Bonura, T., Cone, R., Simmons, R., and Anderson, C. 1978. Base excision repair of DNA. In *DNA repair mechanisms*. P. C. Hanawalt, E. C. Friedberg and C. F. Fox, eds., p. 163. New York: Academic.
12. Hanawalt, P. C., Friedberg, E. C., and Fox, C. F., eds. 1978. *DNA repair mechanisms*. New York: Academic.
13. Friedberg, E. C., Cook, K. H., Duncan, J., and Mortelmans, K. 1977. DNA repair enzymes in mammalian cells. *Photochem. Photobiol. Rev.* 2:263.
14. Clark, A. J., and Volkert, M. R. 1978. A new classification of pathways repairing pyrimidine dimer damage in DNA. In *DNA repair mechanisms*. P. C. Hanawalt, E. C. Friedberg and C. F. Fox, eds., p. 57. New York: Academic.
15. Minton, K., Durphy, M., Taylor, R., and Friedberg, E. C. 1975. The ultraviolet endonuclease of bacteriophage T4. Further characterization. *J. Biol. Chem.* 250:2823.

16. Yasuda, S., and Sekiguchi, M. 1976. Further purification and characterization of T4 endonuclease V. *Biochim. Biophys. Acta* 442:197.

17. Friedberg, E. C. 1972. Studies on the substrate specificity of the T4 excision repair endonuclease. *Mutation Res.* 15:113.

18. Nishida, Y., Yasuda, S., and Sekiguchi, M. 1976. Repair of DNA damaged by methylmethanesulfonate in bacteriophage T4. *Biochim. Biophys. Acta* 442:208.

19. Ito, M., and Sekiguchi, M. 1976. Repair of DNA damaged by 4-nitroquinoline-1-oxide: a comparison of *Escherichia coli* and bacteriophage T4 repair systems. *Japan J. Genet.* 51:129.

20. Seeberg, E. 1981. Multiprotein interactions in strand cleavage of DNA damaged by UV and chemicals. *Prog. Nuc. Acid Res. Mol. Biol.* 26:217.

21. Seeberg, E. 1978. A DNA-binding activity associated with the uvrA$^+$ protein from *Escherichia coli*. In *DNA repair mechanisms*. P. C. Hanawalt, E. C. Friedberg and C. F. Fox, eds., p. 225. New York: Academic.

22. Seeberg, E., and Steinum, A-L. 1982. Purification and properties of the uvrA protein from *Escherichia coli*. *Proc. Natl. Acad. Sci.* (USA) 79:988.

23. Deutsch, W. A., and Linn, S. 1979. DNA binding activity from cultured human fibroblasts that is specific for partially depurinated DNA and that inserts purines into apurinic sites. *Proc. Natl. Acad. Sci.* (USA) 76:141.

24. Feldberg, R. S., and Grossman, L. 1976. A DNA binding protein from human placenta specific for ultraviolet damaged DNA. *Biochemistry* 15:2402.

25. Feldberg, R. S. 1980. On the substrate specificity of a damage-specific DNA binding protein from human cells. *Nucleic Acids Res.* 8:1133.

26. Feldberg, R. S., Lucas, J. L., and Dannenberg, A. 1982. A damage-specific DNA binding protein. Large scale purification from human placenta and characterization. *J. Biol. Chem.* 257:6394.

27. Moranelli, F., and Lieberman, M. W. 1980. Recognition of chemical carcinogen-modified DNA by a DNA-binding protein. *Proc. Natl. Acad. Sci.* (USA) 77:3201.

28. Tsang, S. S., and Kuhnlein, U. 1982. DNA-binding protein from HeLa cells that binds preferentially to supercoiled DNA damaged by ultraviolet light or N-acetoxy-N-acetyl-2-aminofluorene. *Biochim. Biophys. Acta* 697:202.

29. Karran, P., Hjelmgren, T., and Lindahl, T. 1982. Induction of a DNA glycosylase for N-methylated purines is part of the adaptive response to alkylating agents. *Nature* 296:770.

30. Lett, J. T. 1981. Measurement of single-strand breaks by sedimentation in alkaline sucrose gradients. In *DNA repair—a laboratory manual of research procedures*. E. C. Friedberg and P. C. Hanawalt, eds., vol. I, part B, p. 363. New York: Dekker.

31. Jones, A. S., Mian, A. M., and Walker, R. T. 1968. The alkaline degradation of deoxyribonucleic acid derivatives. *J. Chem. Soc.* (C):2042.

32. Duncan, B. K., and Weiss, B. 1982. Specific mutator effects of ung (uracil-DNA glycosylase) mutations in *Escherichia coli*. *J. Bacteriol.* 151:750.

33. Takahashi, I., and Marmur, J. 1963. Replacement of thymidylic acid by deoxyuridylic acid in the deoxyribonucleic acid of a transducing phage for *Bacillus subtilis*. *Nature* 197:794.

34. Cone, R., Duncan, J., Hamilton, L., and Friedberg, E. C. 1977. Partial purification and characterization of a uracil DNA-glycosidase from *B. subtilis*. *Biochemistry* 16:3194.

35. Lindahl, T. 1974. An N-glycosidase from *Escherichia coli* that releases free uracil from DNA containing deaminated cytosine residues. *Proc. Natl. Acad. Sci.* (USA) 71:3649.

36. Friedberg, E. C., Ganesan, A. K., and Minton, K. 1975. N-glycosidase activity in extracts of *Bacillus subtilis* and its inhibition after infection with bacteriophage PBS2. *J. Virology.* 16:315.

37. Leblanc, J-P., Martin, B., Cadet, J., and Laval, J. 1982. Uracil-DNA glycosylase. Purification and properties of uracil-DNA glycosylase from *Micrococcus luteus. J. Biol. Chem.* 257:3477.

38. Kaboev, O. K., Luchkina, L. A., Akhmedov, A. T., and Bekker, M. L. Uracil-DNA glycosylase from *Bacillus stearothermophilus. FEBS Lett.* 132:337.

39. Lindahl, T., Ljungquist, S., Siegert, W., Nyberg, B., and Sperens, B. 1977. DNA N-glycosidases: properties of uracil-DNA glycosidase from *Escherichia coli. J. Biol. Chem.* 252:3286.

40. Crosby, B., Prakash, L., Davis, H., and Hinkle, D. C. 1981. Purification and characterization of a uracil-DNA glycosylase from the yeast *Saccharomyces cerevisiae. Nucleic Acids Res.* 9:5797.

41. Borle, M-T., Clerici, L., and Campagnari, F. 1979. Isolation and characterization of a uracil-DNA glycosylase from calf thymus. *J. Biol. Chem.* 254:6387.

42. Borle, M-T., Campagnari, F., and Creissen, D. M. 1982. Properties of purified uracil-DNA glycosylase from calf thymus. An *in vitro* study using synthetic DNA-like substrates. *J. Biol. Chem.* 257:1208.

43. Caradonna, S. J., and Cheng, Y-C. 1980. Uracil-DNA glycosylase. Purification and properties of this enzyme isolated from blast cells of acute myelocytic leukemia patients. *J. Biol. Chem.* 255:2293.

44. Warner, H. R., and Rockstroh, P. A. 1980. Incorporation and excision of 5-fluorouracil from deoxyribonucleic acid in *Escherichia coli. J. Bacteriol.* 141:680.

45. Duker, N. J., Jensen, D. E., Hart, D. M., and Fishbein, D. E. 1982. Perturbations of enzymic uracil excision due to purine damage in DNA. *Proc. Natl. Acad. Sci.* (USA) 79:4878.

46. Anderson, C. T. M., and Friedberg, E. C. 1980. The presence of nuclear and mitochondrial uracil-DNA glycosylase in extracts of human KB cells. *Nucleic Acids Res.* 8:875.

47. Sirover, M. A. 1979. Induction of the DNA repair enzyme uracil-DNA glycosylase in stimulated human lymphocytes. *Cancer Res.* 39:2090.

48. Wist, E., Unhjem, O., and Krokan, H. 1978. Accumulation of small fragments of DNA in isolated HeLa cell nuclei due to transient incorporation of dUMP. *Biochim. Biophys. Acta* 520:253.

49. Caradonna, S. J., and Cheng, Y-C. 1981. Induction of uracil-DNA glycosylase and dUTP nucleotidohydrolase activity in herpes simplex virus–infected human cells. *J. Biol. Chem.* 256:9834.

50. Cone, R., Bonura, T., and Friedberg, E. C. 1980. Inhibitor of uracil-DNA glycosylase induced by bacteriophage PBS2. *J. Biol. Chem.* 225:10354.

51. Karran, P., Cone, R., and Friedberg, E. C. 1981. Specificity of the bacteriophage PBS2 induced inhibitor of uracil-DNA glycosylase. *Biochemistry* 21:6092.

52. Warner, H. R., Johnson, L. K., and Snusted, D. P. 1980. Early events after infection of *Escherichia coli* by bacteriophage T5. III. Inhibition of uracil-DNA glycosylase activity. *J. Virology.* 33:535.

53. Thomas, K. R., Manlapaz-Ramos, P., Lundquist, R., and Olivera, B. 1978. Formation of Okazaki pieces at the *Escherichia coli* replication fork *in vitro*. *Cold Spring Harbor Symp. Quant. Biol.* 43:231.

54. Karran, P., and Lindahl, T. 1978. Enzymatic excision of free hypoxanthine from polydeoxynucleotides and DNA containing deoxyinosine monophosphate residues. *J. Biol. Chem.* 253:5877.

55. Myrnes, B., Guddal, P-H., and Krokan, H. 1982. Metabolism of dITP in HeLa cell extracts, incorporation into DNA by isolated nuclei and release of hypoxanthine from DNA by a hypoxanthine-DNA glycosylase activity. *Nucleic Acids Res.* 10:3693.

56. Karran, P., and Lindahl, T. 1980. Hypoxanthine in DNA: generation of heat-induced hydrolysis of adenine residues and release in free form by a DNA glycosylase from calf thymus. *Biochemistry* 19:6005.

57. Rydberg, B., and Lindahl, T. 1982. Nonenzymatic methylation of DNA by the intracellular methyl group donor S-adenosyl-L-methionine is a potentially mutagenic reaction. *The EMBO J.* 1:211.

58. Barrows, L. R., and Magee, P. N. 1982. Nonenzymatic methylation of DNA by S-adenosylmethionine *in vitro*. *Carcinogenesis* 3:349.

59. Lindahl, T., and Karran, P. 1982. Enzymatic removal of mutagenic and lethal lesions from alkylated DNA. In *Progress in clinical and biological research*. E. A. Mirard, W. B. Hutchinson and E. Mihich, eds., vol. 132B, p. 241. New York: Alan R. Liss.

60. Lawley, P. D., and Warren, W. 1976. Removal of minor methylation products 7-methyladenine and 3-methylguanine from DNA of *Escherichia coli* treated with dimethylsulfate. *Chem.-Biol. Interact.* 12:211.

61. Lawley, P. D., and Orr, D. J. 1970. Specific excision of methylation products from DNA of *Escherichia coli* treated with N-methyl-N′-nitro-N-nitrosoguanidine. *Chem.-Biol. Interact.* 2:154.

62. Margison, G. P., and O'Conner, P. J. 1973. Biological implications of the instability of the N-glycosidic bond of 3-methyldeoxyadenosine in DNA. *Biochim. Biophys. Acta* 331:349.

63. Riazuddin, S., and Lindahl, T. 1978. Properties of 3-methyladenine-DNA glycosylase from *Escherichia coli*. *Biochemistry* 17:2110.

64. Karran, P., Lindahl, T., Øfsteng, I., Evensen, G. B., and Seeberg, E. 1980. *Escherichia coli* mutants deficient in 3-methyladenine-DNA glycosylase. *J. Mol. Biol.* 140:101.

65. Thomas, L., Yang, C-H., and Goldthwait, D. A. 1982. Two DNA glycosylases in *Escherichia coli* which release primarily 3-methyladenine. *Biochemistry* 21:1162.

66. Jeggo, P., Defais, M., Samson, L., and Schendel, P. 1978. An adaptive response of *E. coli* to low levels of alkylating agent: the role of *pol*A in killing adaptation. *Mol. Gen. Genet.* 162:299.

67. Karran, P., Stevens, S., and Sedgwick, B. 1982. The adaptive response to alkylating agents; the removal of O^6-methylguanine from DNA is not dependent on DNA polymerase I. *Mutation Res.* 104:67.

68. Evensen, G., and Seeberg, E. 1982. Adaptation to alkylation resistance involves the induction of a DNA glycosylase. *Nature* 296:773.

69. Yamamoto, Y., Katsuki, M., Sekiguchi, M., and Otsuji, N. 1978. *Escherichia coli* gene that controls sensitivity to alkylating agents. *J. Bacteriol.* 135:144.

70. Yamamoto, Y., and Sekiguchi, M. 1979. Pathways for repair of DNA damaged by alkylating agent in *Escherichia coli*. *Molec. Gen. Genet.* 171:251.

71. McCarthy, T. V., Karran, P., and Lindahl, T. 1984. Inducible repair of O-alkylated DNA pyrimidines in *Escherichia coli*. *The EMBO J.* 3:545.

72. Laval, J., Pierre, J., and Laval, F. 1981. Release of 7-methylguanine residues from alkylated DNA by extracts of *Micrococcus luteus* and *Escherichia coli*. *Proc. Natl. Acad. Sci.* (USA) 78:852.

73. Cathcart, R., and Goldthwait, D. A. 1981. Enzymatic excision of 3-methyladenine and 7-methylguanine by a rat liver nuclear fraction. *Biochemistry* 20:273.

74. Brent, T. P. 1979. Partial purification and characterization of a human 3-methyladenine-DNA glycosylase. *Biochemistry* 18:911.

75. Gombar, C. T., Katz, E. J., Magee, P. N., and Sirover, M. A. 1981. Induction of the DNA repair enzymes uracil-DNA glycosylase and 3-methyladenine-DNA glycosylase in regenerating rat liver. *Carcinogenesis* 2:595.

76. Chetsanga, C. J., and Lindahl, T. 1979. Release of 7-methylguanine residues whose imidazole rings have been opened from damaged DNA by a DNA glycosylase from *Escherichia coli*. *Nucleic Acids Res.* 6:3673.

77. Chetsanga, C. J., Lozon, M., Makaroff, C., and Savage, L. 1981. Purification and characterization of *Escherichia coli* formamidopyrimidine-DNA glycosylase that excises damaged 7-methylguanine from deoxyribonucleic acid. *Biochemistry* 20:5201.

78. Hems, G. 1960. Effects of ionizing radiation on aqueous solutions of inosine and adenosine. *Radiation Res.* 13:777.

79. Van Hemmen, J. J., and Bleichrodt, J. F. 1973. The decomposition of adenine by ionizing radiation. *Radiation Res.* 46:444.

80. Breimer, L., and Lindahl, T. 1980. A DNA glycosylase from *Escherichia coli* that releases free urea from a polydeoxyribonucleotide containing fragments of base residues. *Nucleic Acids Res.* 8:6199.

81. Lindahl, T., and Andersson, A. 1972. Rate of chain breakage at apurinic sites in double-stranded DNA. *Biochemistry* 11:3618.

82. McDonald, M. R., and Kaufmann, B. P. 1954. The degradation by ribonuclease of substrates other than ribonucleic acid. *J. Histochem. Cytochem.* 2:387.

83. Behmoaras, T., Toulmé, J-J., and Hélène, C. 1981. A tryptophan-containing peptide recognizes and cleaves DNA at apurinic sites. *Nature* 292:858.

84. Behmoaras, T., Toulmé, J-J., and Hélène, C. 1981. Specific recognition of apurinic sites in DNA by a tryptophan-containing peptide. *Proc. Natl. Acad. Sci.* (USA) 78:926.

85. Pierre, J., and Laval, J. 1981. Specific nicking of DNA at apurinic sites by peptides containing aromatic residues. *J. Biol. Chem.* 256:10217.

86. Strauss, B., Searashi, T., and Robbins, M. 1966. Repair of DNA studied with a nuclease specific for UV-induced lesions. *Proc. Natl. Acad. Sci.* (USA) 56:932.

87. Shimada, K., Nakayama, H., Okubo, S., Sekiguchi, M., and Takagi, Y. 1967. An endonucleolytic activity specific for ultraviolet-irradiated DNA in wild-type and mutant strains of *Micrococcus lysodeikticus*. *Biochem. Biophys. Res. Comm.* 27:539.

88. Kaplan, J. C., Kushner, S. R., and Grossman, L. 1969. Enzymatic repair of DNA I. Purification of two enzymes involved in the excision of thymine dimers from ultraviolet-irradiated DNA. *Proc. Natl. Acad. Sci.* (USA) 63:144.

89. Friedberg, E. C., and King, J. J. 1969. Endonucleolytic cleavage of UV-irradiated DNA controlled by the v^+ gene in phage T4. *Biochem. Biophys. Res. Comm.* 37:646.

90. Yasuda, S., and Sekiguchi, M. 1970. T4 endonuclease involved in repair of DNA. *Proc. Natl. Acad. Sci.* (USA) 67:1839.

91. Riazuddin, S., and Grossman, L. 1977. *Micrococcus luteus* correndonucleases. I. Resolution and purification of two endonucleases specific for DNA containing pyrimidine dimers. *J. Biol. Chem.* 252:6280.

92. Riazuddin, S., and Grossman, L. 1977. *Micrococcus luteus* correndonucleases. II. Mechanism of action of two endonucleases specific for DNA containing pyrimidine dimers. *J. Biol. Chem.* 252:6287.

93. Grossman, L., Riazuddin, S., Haseltine, W. A., and Lindan, C. 1979. Nucleotide excision repair of damaged DNA. *Cold Spring Harbor Symp. Quant. Biol.* 43:947.

94. Haseltine, W. A., Gordon, L. K., Lindan, C. P., Grafstrom, R. H., Shaper, N. L., and Grossman, L. 1980. Cleavage of pyrimidine dimers in specific DNA sequences by a pyrimidine dimer DNA-glycosylase of *M. luteus*. *Nature* 285:634.

95. Radany, E. H., and Friedberg, E. C. 1980. A pyrimidine dimer–DNA glycosylase activity associated with the *v* gene product of bacteriophage T4. *Nature* 286:182.

96. Seawell, P. C., Smith, C. A., and Ganesan, A. K. 1980 *den*V Gene of bacteriophage T4 determines a DNA glycosylase specific for pyrimidine dimers in DNA. *J. Virology*. 35:790.

97. Warner, H., Demple, B., Deutsch, W., Kane, C., and Linn, S. 1980. Apurinic/apyrimidinic endonucleases in repair of pyrimidine dimers and other lesions in DNA. *Proc. Natl. Acad. Sci.* (USA) 77:4602.

98. Grafstrom, R. H., Park, L., and Grossman, L. 1982. Enzymatic repair of pyrimidine dimer–containing DNA. A 5′ dimer DNA glycosylase: 3′ apyrimidinic endonuclease mechanism from *Micrococcus luteus*. *J. Biol. Chem.* 257:13465.

99. Radany, E. H., Love, J. D., and Friedberg, E. C. 1981. The use of direct photoreversal of UV-irradiated DNA for the demonstration of pyrimidine dimer–DNA glycosylase activity. In *Chromosome damage and repair*. E. Seeberg and K. Kleppe, eds., p. 95. New York: Plenum.

100. Rupert, C. S., and Harm, W. 1966. Reactivation after photobiological damage. *Adv. Radiat. Biol.* 2:1.

101. Luria, S. E. 1947. Reactivation of irradiated bacteriophage by transfer of self-reproducing units. *Proc. Natl. Acad. Sci.* (USA) 33:253.

102. Streisinger, G. 1956. The genetic control of ultraviolet sensitivity levels in bacteriophages T2 and T4. *Virology* 2:1.

103. Harm, W. 1961. Gene-controlled reactivation of ultraviolet-inactivated bacteriophage. *J. Cell. Comp. Physiol.* 58 (suppl. 1):69.

104. Setlow, R. B., and Carrier, W. L. 1968. The excision of pyrimidine dimers *in vivo* and *in vitro*. In *Replication and recombination of genetic material*. W. J. Peacock and R. D. Brock, eds., p. 134. Canberra: Australian Academy of Sciences.

105. Takagi, Y., Sekiguchi, M., Okubo, S., Nakayama, H., Shimada, K., Yasuda, S., Nishimoto, T., and Yoshihara, H. 1968. Nucleases specific for ultraviolet light-irradiated DNA and their possible role in dark repair. *Cold Spring Harbor Symp. Quant. Biol.* 33:219.

106. Sekiguchi, M., Yasuda, S., Okubo, H., Nakayama, K., Shimada, K., and Takagi, Y. 1970. Mechanism of repair of DNA in bacteriophage. I. Excision of pyrimidine dimers from ultraviolet-irradiated DNA by an extract of T4-infected cells. *J. Mol. Biol.* 47:231.

107. Friedberg, E. C., and King, J. J. 1971. Dark repair of ultraviolet-irradiated deoxyribonucleic acid by bacteriophage T4: purification and characterization of a dimer-specific phage-induced endonuclease. *J. Bacteriol.* 106:500.

108. Sato, K., and Sekiguchi, M. 1976. Studies on temperature-dependent ultraviolet light-sensitive mutants of bacteriophage T4: the structural gene for T4 endonuclease V. *J. Mol. Biol.* 102:15.

109. Wood, W. B., and Revel, H. R. 1976. The genome of bacteriophage T4. *Bacteriol. Rev.* 40:847.

110. Nakabeppu, Y., Yamashita, K., and Sekiguchi, M. 1982. Purification and characterization of normal and mutant forms of T4 endonuclease V. *J. Biol. Chem.* 257:2556.

111. Gordon, L. K., and Haseltine, W. A. 1980. Comparison of the cleavage of pyrimidine dimers by the bacteriophage T4 and *Micrococcus luteus* UV-specific endonucleases. *J. Biol. Chem.* 255:12047.

112. Simon, J. J., Smith, C. A., and Friedberg, E. C. 1975. Action of bacteriophage T4 ultraviolet endonuclease on duplex DNA containing one ultraviolet-irradiated strand. *J. Biol. Chem.* 250:8748.

113. Harm, W. 1963. Mutants of phage T4 with increased sensitivity to ultraviolet. *Virology* 19:66.

114. Ohshima, S., and Sekiguchi, M. 1975. Biochemical studies on radiation-sensitive mutations in bacteriophage T4. *J. Biochem.* 77:303.

115. Van Minderhout, L., Grimbergen, J., and de Groot, B. 1974. Nonsense mutants in the bacteriophage T4D *v* gene. *Mutation Res.* 29:333.

116. Mortelmans, K., and Friedberg, E. C. 1972. Deoxyribonucleic acid repair in bacteriophage T4: observations on the roles of the x and *v* genes and of host factors. *J. Virology* 10:730.

117. Demple, B., and Linn, S. 1980. DNA N-glycosylases and UV repair. *Nature* 287:203.

118. Friedberg, E. C., and Lehman, I. R. 1974. Excision of thymine dimers by proteolytic and amber fragments of *E. coli* DNA polymerase I. *Biochem. Biophys. Res. Comm.* 58:132.

119. Radany, E. H., and Friedberg, E. C. 1982. Demonstration of pyrimidine dimer–DNA glycosylase activity *in vivo:* bacteriophage T4–infected *Escherichia coli* as a model system. *J. Virology* 41:88.

120. Warner, H. R., Christensen, L. N., and Persson, M-L. 1981. Evidence that the UV endonuclease activity induced by bacteriophage T4 contains both pyrimidine dimer–DNA glycosylase and apyrimidinic/apurinic endonuclease activities in the enzyme molecule. *J. Virology* 40:204.

121. McMillan, S., Edenberg, H. J., Radany, E. H., Friedberg, R. C., and Friedberg, E. C. 1981. *den*V Gene of bacteriophage T4 codes for both pyrimidine dimer–DNA glycosylase and apyrimidinic endonuclease activities. *J. Virology* 40:211.

122. Nakabeppu, Y., and Sekiguchi, M. 1981. Physical association of pyrimidine dimer DNA glycosylase and apurinic/apyrimidinic DNA endonuclease essential for repair of ultraviolet-damaged DNA. *Proc. Natl. Acad. Sci.* (USA) 78:2742.

123. Gordon, L. K., and Haseltine, W. A. 1981. Early steps of excision repair of cyclobutane pyrimidine dimers by the *Micrococcus luteus* endonuclease. A three-step incision model. *J. Biol. Chem.* 256:6608.

124. LaBelle, M., and Linn, S. 1982. *In vivo* excision of pyrimidine dimers is mediated by a DNA N-glycosylase in *Micrococcus luteus* but not in human fibroblasts. *Photochem. Photobiol.* 36:319.

125. Friedberg, E. C., Bonura, T., Love, J. D., McMillan, S., Radany, E. H., and Schultz, R. A. 1981. The repair of DNA damage: recent developments and new insights. *J. Supramol. Struct. and Cell Biochem.* 16:91.

126. Bonura, T., Radany, E. H., McMillan, S., Love, J. D., Schultz, R. A., Edenberg, H. J., and Friedberg, E. C. 1982. Pyrimidine dimer–DNA glycosylases: studies on bacteriophage T4-infected and on uninfected *Escherichia coli*. *Biochimie* 64:643.

127. Demple, B., and Linn, S. 1982. 5,6-Saturated lesions in DNA: production by ultraviolet light or hydrogen peroxide. *Nucleic Acids Res.* 10:3781.

128. Radman, M. 1976. An endonuclease from *Escherichia coli* that introduces single polynucleotide chain scissions in ultraviolet-irradiated DNA. *J. Biol. Chem.* 251:1438.

129. Gates, F. T., III, and Linn, S. 1977. Endonuclease from *Escherichia coli* that acts specifically upon duplex DNA damaged by ultraviolet light, osmium tetroxide, acid, or X-rays. *J. Biol. Chem.* 252:2802.

130. Livneh, Z., and Sperling, J. 1981. DNA base-insertion enzymes (insertases). In *The enzymes*. 3rd ed. P. D. Boyer, ed., vol. XIV, Nucleic acids, part A, p. 549. New York: Academic.

131. Weiss, B. 1981. Exodeoxyribonucleases of *Escherichia coli*. In *The enzymes*. 3rd ed. P. D. Boyer, ed., vol. XIV, Nucleic acids, part A, p. 203. New York: Academic.

132. Richardson, C. C., and Kornberg, A. 1964. A deoxyribonucleic acid phosphatase-exonuclease from *Escherichia coli*. I. Purification of the enzyme and characterization of the phosphatase activity. *J. Biol. Chem.* 239:242.

133. Richardson, C. C., Lehman, I. R., and Kornberg, A. 1964. A deoxyribonucleic acid phosphatase-exonuclease from *Escherichia coli*. II. Characterization of the exonuclease activity. *J. Biol. Chem.* 239:251.

134. Keller, W., and Crouch, R. 1972. Degradation of DNA-RNA hybrids by ribonuclease H and DNA polymerases of cellular and viral origin. *Proc. Natl. Acad. Sci.* (USA) 69:3360.

135. Friedberg, E. C., and Goldthwait, D. A. 1968. Endonuclease II of *E. coli*. *Cold Spring Harbor Symp. Quant. Biol.* 33:271.

136. Friedberg, E. C., and Goldthwait, D. A. 1969. Endonuclease II of *E. coli*. I. Isolation and purification. *Proc. Natl. Acad. Sci.* (USA) 62:934.

137. Verly, W. G. 1978. Endonucleases specific for apurinic sites in DNA. In *DNA repair mechanisms*. P. C. Hanawalt, E. C. Friedberg and C. F. Fox, eds., p. 187. New York: Academic.

138. Yajko, D. M., and Weiss, B. 1975. Mutations simultaneously affecting endonuclease II and exonuclease III in *Escherichia coli*. *Proc. Natl. Acad. Sci.* (USA) 72:688.

139. White, B. J., Hochhauser, S. J., Citron, N. M., and Weiss, B. 1976. Genetic mapping of *xth*A, the structural gene for exonuclease III in *Escherichia coli* K-12. *J. Bacteriol.* 126:1082.

140. Weiss, B. 1976. Endonuclease II of *Escherichia coli* is exonuclease III. *J. Biol. Chem.* 251:1896.

141. Gossard, F., and Verly, W. G. 1978. Properties of the main endonuclease specific for apurinic sites of *Escherichia coli* (endonuclease VI). *Eur. J. Biochem.* 82:321.

142. Demple, B., Halbrook, J., and Linn, S. 1983. *Escherichia coli xth* mutants are hypersensitive to hydrogen peroxide. *J. Bacteriol.* 153:1079.

143. Kirtikar, D. M., Cathcart, G. R., White, J. G., Ukstins, I., and Goldthwait, D. A. 1977. Mutations in *Escherichia coli* altering an apurinic endonuclease, endonuclease II, and exonuclease III and their effect on *in vivo* sensitivity to methylmethanesulfonate. *Biochemistry* 16:5625.

144. Ljungquist, S., Lindahl, T., and Howard-Flanders, P. 1976. Methylmethanesulfonate-sensitive mutants of *Escherichia coli* deficient in an endonuclease specific for apurinic sites in deoxyribonucleic acid. *J. Bacteriol.* 126:646.

145. Weiss, B., Rogers, S. G., and Taylor, A. F. 1978. The endonuclease activity of exonuclease III and repair of uracil-containing DNA in *Escherichia coli*. In *DNA repair mechanisms*. P. C. Hanawalt, E. C. Friedberg and C. F. Fox, eds., p. 191. New York: Academic.

146. Ljungquist, S. 1977. A new endonuclease from *Escherichia coli* acting at apurinic sites in DNA. *J. Biol. Chem.* 252:2808.

147. Bonura, T., Schultz, R., and Friedberg, E. C. 1982. An enzyme activity from *Escherichia coli* that attacks single-stranded deoxyribopolymers and single-stranded deoxyribonucleic acid containing apyrimidinic sites. *Biochemistry* 21:2348.

148. Schultz, R. A., Friedberg, E. C., Moses, R. E., Rupp, W. D., Sancar, A., and Sharma, S. 1983. A mutant strain of *E. coli* allows measurement of single stranded AP endonuclease in crude extracts: studies with untransformed cells and cells transformed with plasmids containing the *uvr*C gene. *J. Bacteriol.* 154:1459.

149. Inoue, T., and Kada, T. 1978. Purification and properties of a *Bacillus subtilis* endonuclease specific for apurinic sites in DNA. *J. Biol. Chem.* 253:8559.

150. Pierre, J., and Laval, J. 1980. *Micrococcus luteus* endonucleases for apurinic/apyrimidinic sites in deoxyribonucleic acid. I. Purification and general properties. *Biochemistry* 19:5018.

151. Pierre, J., and Laval, J. 1980. *Micrococcus luteus* endonucleases for apurinic/apyrimidinic sites in deoxyribonucleic acid. 2. Further studies on the substrate specificity and mechanism of action. *Biochemistry* 19:5024.

152. Bibor, V., and Verly, W. G. 1978. Purification and properties of the endonuclease specific for apurinic sites of *Bacillus stearothermophilus*. *J. Biol. Chem.* 253:850.

153. Thibodeau, L., and Verly, W. G. 1977. Purification and properties of a plant endonuclease specific for apurinic sites. *J. Biol. Chem.* 252:3304.

154. Clements, J. E., Rogers, S. G., and Weiss, B. 1978. A DNase for apurinic/apyrimidinic sites associated with exonuclease III of *Hemophilus influenzae*. *J. Biol. Chem.* 253:2990.

155. Ljungquist, S., and Lindahl, T. 1974. A mammalian endonuclease specific for apurinic sites in double-stranded deoxyribonucleic acid. I. Purification and general properties. *J. Biol. Chem.* 249:1530.

156. Ljungquist, S., Andersson, A., and Lindahl, T. 1974. A mammalian endonuclease specific for apurinic sites in double-stranded deoxyribonucleic acid. II. Further studies on the substrate specificity. *J. Biol. Chem.* 249:1536.

157. Kuebler, J. P., and Goldthwait, D. A. 1977. An endonuclease from calf liver specific for apurinic sites in DNA. *Biochemistry* 16:1370.

158. Kane, C. M., and Linn, S. 1981. Purification and characterization of an apurinic/apyrimidinic endonuclease from HeLa cells. *J. Biol. Chem.* 256:3405.

159. Mosbaugh, D. W., and Linn, S. 1980. Further characterization of human fibroblast apurinic/apyrimidinic DNA endonucleases. *J. Biol. Chem.* 255:11743.

160. Shaper, N. L., Grafstrom, R. H., and Grossman, L. 1982. Human placental apurinic/apyrimidinic endonuclease. Its isolation and characterization. *J. Biol. Chem.* 257:13455.

161. Grafstrom, R. H., Shaper, N. L., and Grossman, L. 1982. Human placental apurinic/apyrimidinic endonuclease. Mechanism of action. *J. Biol. Chem.* 257:13459.

162. Linsley, W. S., Penhoet, E. E., and Linn, S. 1977. Human endonuclease specific for apurinic/apyrimidinic sites in DNA. Partial purification and characterization of multiple forms from placenta. *J. Biol. Chem.* 252:1235.

163. Thibodeau, L. and Verly, W. G. 1980. Cellular localization of the apurinic/apyrimidinic endonucleases in rat liver. *Eur. J. Biochem.* 107:555.

4

Excision Repair. II. Incision of DNA Containing Bulky Base Damage

4-1 Introduction

The majority of the known substrates for DNA glycosylases and AP endonucleases result from chemical and ionizing radiation damage to DNA (see Chapter 3). However, for many years before the discovery of these enzymes the most extensively used model lesion for the study of DNA repair was the pyrimidine dimer, a major photoproduct in DNA produced by UV radiation at approximately 254 nm (see Section 1-5). This form of DNA damage was (and continues to be) a highly convenient one to study, since it is easy to produce (requiring only a germicidal lamp), is chemically very stable in DNA, and reliable and sensitive methods exist for its quantitation (1, 2). In addition, it has long been known that strains of *E. coli* exposed to UV radiation undergo recovery processes in the course of the post-UV incubation that involve the repair of pyrimidine dimers (3–5). Recovery can be demonstrated as an increased survival of the bacteria themselves or as the ability to promote the survival of UV-irradiated bacteriophages. The latter phenomenon is called *host-cell reactivation* and is an extremely useful experimental system for the study of DNA repair, both in prokaryotes such as *E. coli* and in mammalian cells infected with viruses such as adenovirus and SV40 (3–5).

After the discovery of the enzymatic photoreactivation of pyrimidine dimers, DNA repair modes in *E. coli* were frequently distinguished as *light repair* (to indicate the dependence of photoreactivation on visible light) and *dark repair* (to indicate recovery phenomena other than photoreactivation). During the early 1960s a series of important biological experiments were independently reported by Richard Setlow and his colleagues (6) and by Paul Howard-Flanders and his coworkers (7) that led to major insights into the nature of dark repair. The former group irradiated a number of *E. coli* B strains and showed that in most strains thymine-containing pyrimidine dimers were lost from the acid-insoluble fraction of the DNA and appeared in the acid-soluble phase during post-UV incubation. The kinetics of this process correlated well with that of the resumption of DNA synthesis (8) and led to the suggestion that the loss of dimers from DNA may be directly related to the recovery of DNA synthetic capacity and hence to survival. The relevance of the loss of dimers from DNA to cellular recovery from the effects of UV radiation was directly indicated by the observation that in the UV-sensitive, host-cell reactivation-defective (*hcr*) strain $B_{S\text{-}1}$, thymine-containing dimers were *not* lost from DNA during post-UV incubation. Essentially identical results were reported by Howard-Flanders' group, using *E. coli* K-12 strains, including a UV-sensitive *hcr* strain defective at a genetic locus designated *uvrA* (Table 4-1).

TABLE 4-1
Distribution of radioactivity in thymine and thymine-containing pyrimidine dimers

	Acid-Insoluble Fraction			Acid-Soluble Fraction		
E. coli	Thymine dimers (cpm)	Thymine (cpm)	T<>T/T(%)	Thymine dimers (cpm)	Thymine (cpm)	T<>T/T(%)
uvr^+, no UV	100	247,000	0.04	40	6700	0.6
uvr^+, 96 J/m^2, no incubation	1250	304,000	0.40	40	12,000	0.3
uvr^+, 96 J/m^2, incubated	357	295,000	0.12	940	44,500	2.1
uvr, no UV	100	200,000	0.05	40	6250	0.6
uvr, 96 J/m^2, no incubation	805	260,000	0.31	40	6250	0.6
uvr, 96 J/m^2, incubated	890	267,000	0.33	40	4900	0.8

From R. P. Boyce and P. Howard-Flanders, ref. 7.

In addition to the *uvrA* locus, genetic loci called *uvrB* and *uvrC* (Fig. 4-1) are required for the excision of pyrimidine dimers from the DNA of *E. coli* K-12 (9–12). By the use of techniques (some of which are discussed in Section 1-7) that directly demonstrate the presence of single-strand breaks (nicks) in the DNA, it was established that the *uvrA*, *uvrB* and *uvrC* genes are all required for normal *incision* of DNA containing pyrimidine dimers (13, 14). Mutants defective in the *uvrC* gene do perform incision of UV-irradiated DNA in vivo, but strand breaks occur more slowly than in wild-type cells (15). Since strains carrying deletions in *uvrC* have not been isolated, it is likely that existing mutants in this gene are leaky. Cell-free systems have confirmed a requirement for all three genes for the incision of UV-irradiated DNA (16, 17).

These observations, together with the demonstration of repair synthesis of DNA by Philip Hanawalt and his coworkers (18, 19), were

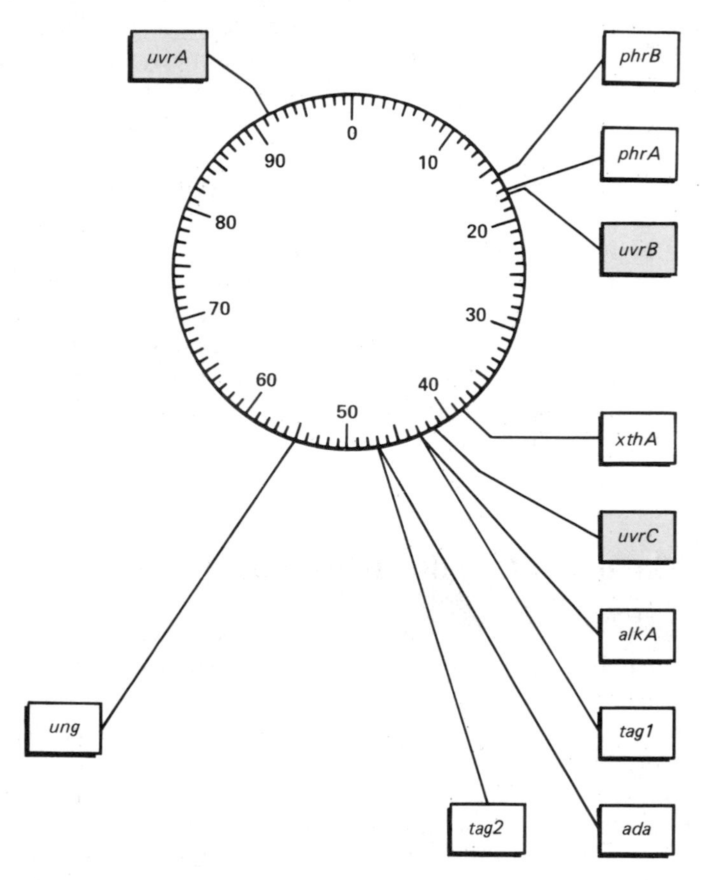

FIGURE 4-1
The location of the *uvrA*, *uvrB* and *uvrC* genes on the *E. coli* genetic map.

seminal in the evolution of a general molecular model for the removal of pyrimidine dimers from DNA, which was described in the previous chapter (see Fig. 3-1). An essential component of the model is that for dimers to be excised from a covalently intact DNA duplex, an early enzymatic event must involve breakage of the polynucleotide chain, i.e., the DNA must be *incised*. As indicated in Section 3-1, one mechanism for the enzymatic incision of DNA containing pyrimidine dimers is by the sequential action of a PD DNA glycosylase and an AP endonuclease. However, it was also noted in Section 3-4 that conclusive evidence for a mechanism mediated by DNA glycosylase–AP endonuclease for incision of DNA containing pyrimidine dimers is thus far confined to *M. luteus* and to *E. coli* infected with bacteriophage T4. The stringent substrate specificity of the T4 PD DNA glycosylase is totally consistent with the observation that phage mutants defective in the *denV* gene are not abnormally sensitive to other forms of physical or chemical damage to DNA (see Section 3-4). On the other hand, mutants of *E. coli* defective at the *uvrA*, *uvrB* or *uvrC* loci are not only abnormally sensitive to UV radiation but also to a variety of other agents, including mitomycin C, nitrogen mustard, photoactivated psoralen and 4-nitroquinoline-1-oxide (9, 11, 12, 20) (Fig. 4-2).

This chapter deals principally with an enzymatic mechanism for the incision of DNA in *E. coli*, in which a *damage-specific DNA incising activity* encoded by the *uvrA*, *uvrB* and *uvrC* genes recognizes DNA containing a wide variety of base damage and *directly* (i.e., in a single step reaction) catalyzes the hydrolysis of phosphodiester bonds at or near the sites of such damage. This mode of DNA incision is associated with the removal of damaged bases as nucleotides or oligonucleotides; hence the frequent use of the term *nucleotide excision repair* (see Section 3-1).

4-2 The uvrABC Endonuclease of *E. coli*

Although extensive biological studies on excision repair in *E. coli* have been going on for over 30 years, it is only relatively recently that some clues have been obtained as to the nature of the products of the *uvrA*, *uvrB* and *uvrC* genes. Some of these gene products have been extensively purified from extracts of *E. coli* (16, 17, 21); a prodigious task, since this organism normally contains extremely small amounts of these proteins (22–24). Indeed, it is largely this quantitative problem that for a long time limited progress on the biochemistry of the incision of damaged DNA in *E. coli*. Meanwhile, the pyrimidine-dimer-specific "UV endonucleases" from *M. luteus* and from phage-

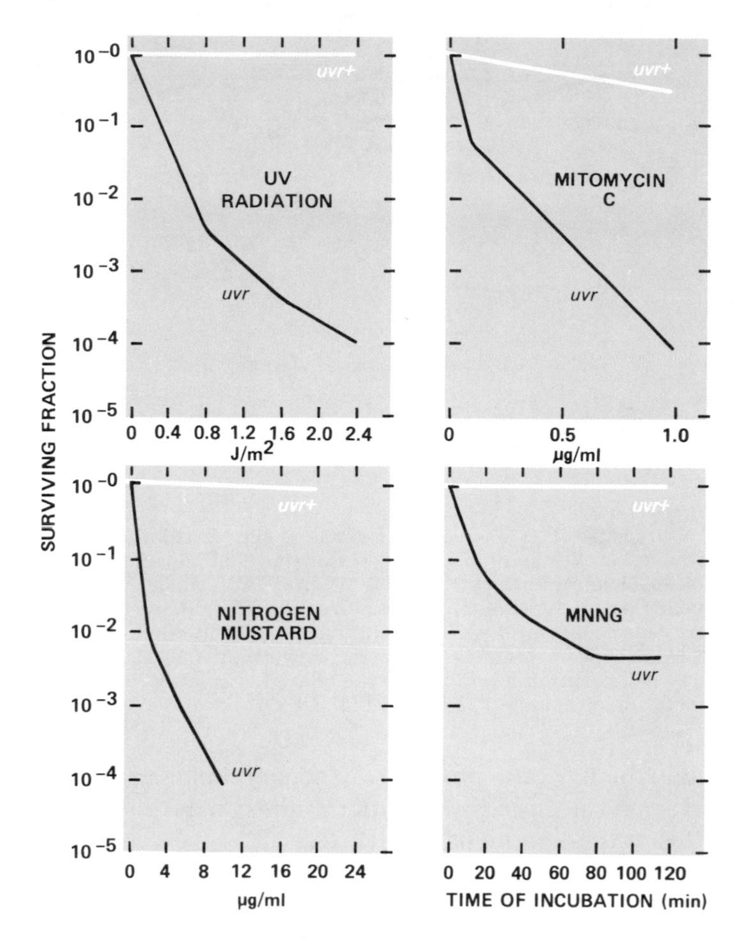

FIGURE 4-2
Excision repair-defective mutant (*uvr*) strains of *E. coli* are abnormally sensitive to killing by a wide variety of DNA-damaging agents, including UV radiation, mitomycin C, nitrogen mustard and N-methyl-N′-nitro-N-nitrosoguanidine (MNNG).

T4-infected *E. coli* (see Section 3-4) were readily identifiable in cell extracts and erroneously served as prototypic models of direct-acting DNA damage-specific incising activities.

When extracts of *E. coli* prepared by gentle lysis of cells are incubated with UV-irradiated or unirradiated radiolabeled covalently closed circular DNA, preferential nicking of irradiated compared to unirradiated DNA is observed in the presence of ATP and Mg^{2+}. This is not observed with extracts of mutant *uvrA*, *uvrB* or *uvrC* cells; however, extracts of these cells can be complemented for nicking activity by addition of extract from a different *uvr* mutant (16, 17, 21) (Fig. 4-3). This complementation provided the basis for an assay to

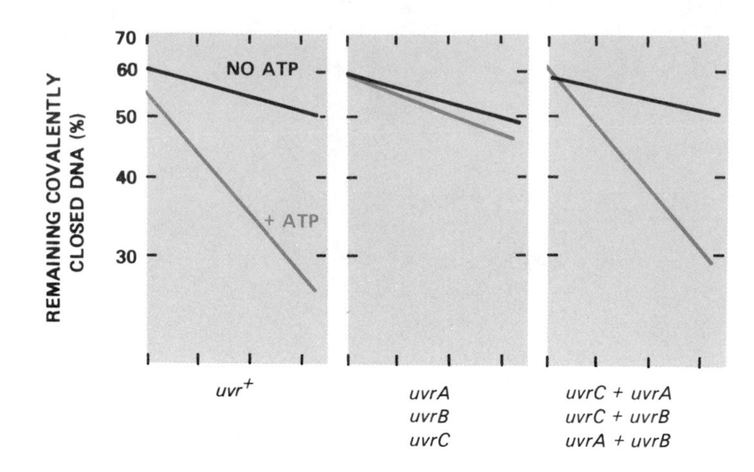

FIGURE 4-3
UV-irradiated DNA is nicked by extracts of *uvr*$^+$ *E. coli* cells in the presence of ATP and Mg^{2+} (left panel). Extracts of mutant (*uvr*) cells fail to demonstrate the preferential nicking of UV-irradiated DNA (center panel); however, mixtures of extracts from different mutant *uvr* cells complement each other for nicking activity (right panel). Nicking of DNA was measured by the conversion of covalently closed circular (form I) DNA into the relaxed circular or linear configurations. (From E. Seeberg et al., ref. 16.)

fractionate the *uvr* gene products by conventional enzyme purification and served as the basis for initial studies on the physicochemical properties of these proteins (16, 17, 21).

More recently, isolation of the *uvr* gene products has been greatly facilitated by the molecular cloning of all three genes. The use of multicopy plasmids as well as the tailoring of the cloned genes to place them under the control of strong promoters has allowed for the expression of vastly increased levels of individual gene products in *E. coli*. Before discussing the physicochemical and catalytic properties of the *uvr* gene products, however, it is instructive to consider what is known about the cloned genes themselves.

The Molecular Cloning and Characterization of the *uvrA*, *uvrB* and *uvrC* Genes

The *uvrA*, *uvrB* and *uvrC* genes have been isolated from *E. coli* gene pools cloned into various recombinant plasmids through their ability to restore UV resistance to the appropriate UV-sensitive *E. coli* mutants (22–40) after transformation with cloned DNA (Fig. 4-4). Most of these plasmids are multicopy in nature and therefore result in a significant increase in the level of individual gene products in *E. coli* without any further manipulations.

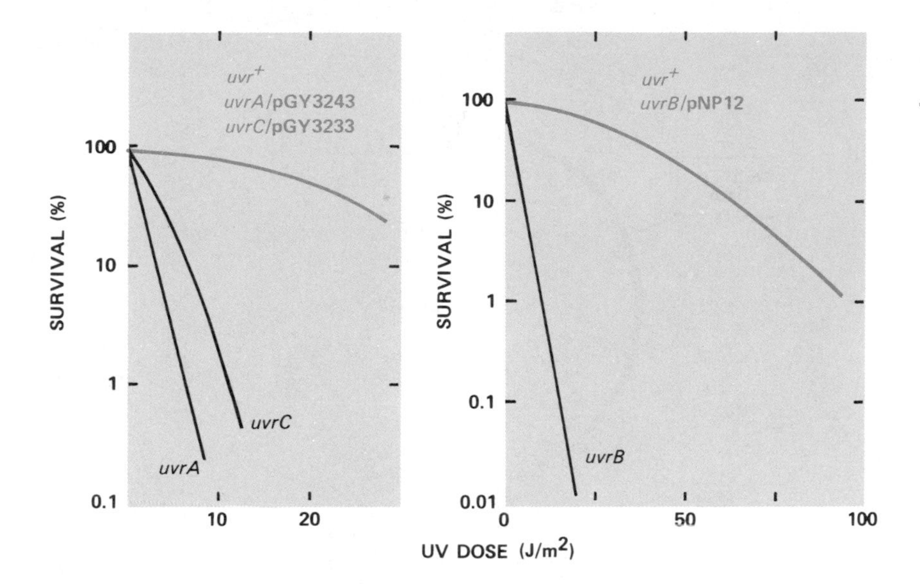

FIGURE 4-4
Plasmids pGY3243, pGY3233 and pNP12 complement the UV sensitivity of *E. coli uvrA*, *uvr C* (left) and *uvrB* (right) mutants, respectively. (From E. van den Berg et al., ref. 33, and G. H. Yoakum et al., ref. 30.)

The *uvrA* Gene

Transformation of *E. coli* strains defective in the *lexA* gene with plasmids containing the *uvrA* gene results in about a 100-fold increase in the level of uvrA protein (41) because the *uvrA* gene is inducible and under the control of the *recA-lexA* regulatory system (see Section 7-6). Expression of the *uvrA* gene can be further amplified by bypassing the endogenous *uvrA* regulatory elements through gene tailoring and linkage downstream from a regulatable promoter. An example is the λ phage promoter P_L present in a plasmid called pKC30 (42), transcription from which can be regulated by λ *c*I repressor, following transformation of λ lysogens. By transforming *E. coli* λ lysogens that express the thermoinducible repressor cI_{857}, expression of high levels of the *uvrA* gene in plasmid pKC30 can be controlled by shifting the temperature (42). Transformants can be grown to large quantities at 30°C, at which temperature stable *c*I repressor is synthesized and the *uvrA* gene is repressed. Thus potential cell killing resulting from massive overexpression of the gene is prevented. At 43°C, thermolabile inactive repressor is made, leading to derepression of the *uvrA* gene and markedly increased expression of uvrA protein (Fig. 4-5).

The regulatory region of the *uvrA* gene has been localized to a region of the cloned gene required for initiation of transcription (43).

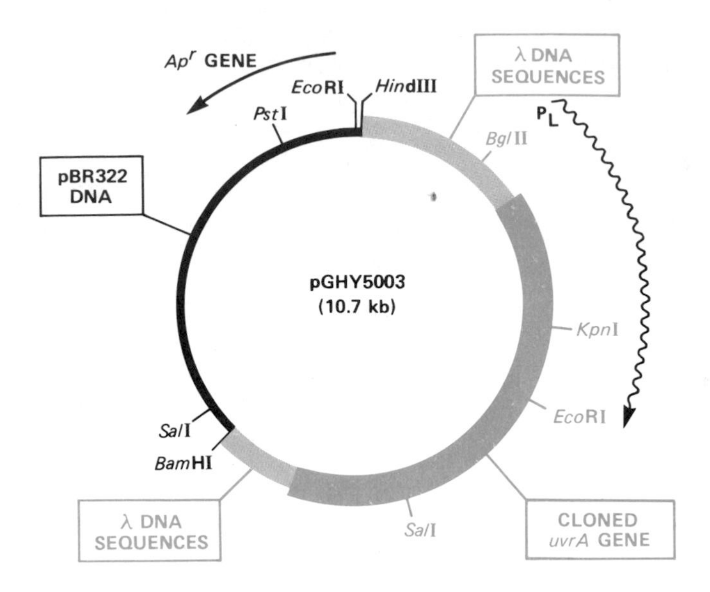

FIGURE 4-5
Plasmid pGHY5003 contains a 4.2 kb cloned insert that includes the *uvrA* gene of *E. coli*, in transcriptive phase with the λ cI_{857P_L} (thermosensitive) promoter. The squiggly arrow indicates the direction of transcription of *uvrA* and the straight arrow the direction of transcription of the ampicillin resistance gene. (After G. H. Yoakum et al., ref. 42.)

A 325-bp restriction fragment spanning this region yields the nucleotide sequence shown in Fig. 4-6 (43). The transcriptional start point in this sequence has been established from the sequencing of an RNA transcript made in vitro and is shown as nucleotide +1 in the figure. The two ubiquitous sequences defining a typical prokaryote promoter region (the so-called −35 sequence and the Pribnow or TATA box) are identified, as is the ATG translational initiation codon starting at nucleotide +63.

As indicated above, the *uvrA* gene is one of a series of genes in *E. coli* that are induced by agents that cause DNA damage, collectively referred to as "SOS" genes (44–46). The regulation of the SOS response is discussed in detail in Chapter 7. The expression of many, if not all, of the SOS genes is regulated by a repressor that is the product of a gene called *lexA*. A binding site for lexA protein at the operator-promoter region of the *uvrA* gene has been identified by DNase I protection or "footprinting" experiments (43), in which DNA containing the *uvrA* regulatory sequences to which lexA protein is bound in vitro is degraded with DNase I. The region of DNA to which the lexA protein is bound is protected from DNase I digestion and can thus be identified as a 25- to 30-bp segment that overlaps the *uvrA* promoter (Fig. 4-6). Not unexpectedly, similar sequences

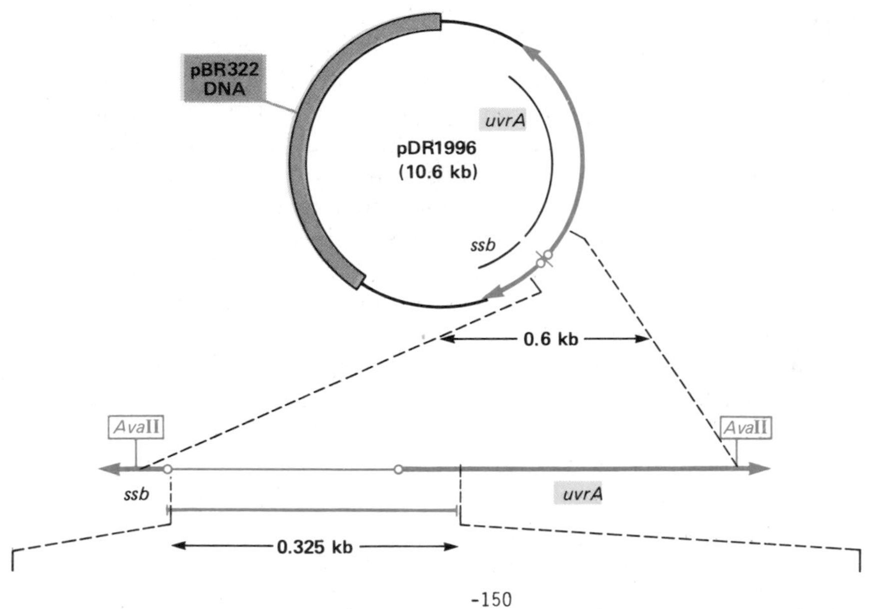

FIGURE 4-6
Plasmid pDR1996 contains the cloned *E. coli uvrA* and *ssb* genes (the latter encodes single-strand DNA binding protein), whose location and direction of transcription are shown in the top part of the figure. The 325-bp region, whose sequence is shown below, is from the 0.6kb *AvaII* fragment excised from the plasmid. Numbering of the nucleotides in the sequenced region is from the transcription start site (+1) of *uvrA*. The shaded region between about −16 and −45 is a region protected from DNase I digestion by binding of lexA protein. LexA protein is a repressor of a number of genes including *uvrA*. The details of this regulatory mechanism are described in Chapter 7. The Pribnow box and −35 sequences (putative promoter regions of the gene) are also identified. The *uvrA* translational start codon (ATG) begins at position +63 (dashed box). (After A. Sancar et al., ref. 43.)

have been identified in the regulatory regions of other genes that are involved in the SOS response, including the *uvrB* gene, which is also a member of the SOS regulon (see below and Section 7-7).

The *uvrB* Gene

Bacteriophage $\lambda b2c\mathrm{I}_{857}$*int am6* is a defective phage able to lysogenize susceptible *E. coli* strains at low frequency. This phage does not produce the active int^{+} gene product required for normal prophage excision (47). Thus, following heat induction an alternative mode of prophage excision occurs, dependent on host functions that mediate a specific recombination event between sites situated on either side of the normal λ attachment site on the *E. coli* chromosome (47). These sites, designated X_L and X_R, incorporate the *uvrB* gene (47). The resulting phage particles (called $\lambda b2att^{2}$) are thus specialized transducers of a segment of *E. coli* DNA that includes the *uvrB* gene (Fig. 4-7). This transducing phage was therefore very useful for cloning an *Eco*RI restricion fragment containing the *uvrB* gene into the plasmid pMB9 (27). Transformation of mutant *uvrB* bacteria with the recombinant plasmid results in complete restoration of the uvr^{+} phenotype (27).

The *uvrB* gene has also been cloned from the Clark-Carbon *E. coli* genome bank (23, 26). The regulatory region of the *uvrB* gene (which, as indicated above, is inducible by DNA damage) (48, 49) has been subcloned and sequenced (33, 50), and it reveals an unexpected complexity. The nucleotide sequence of a 540-bp restriction fragment derived from a plasmid containing the *uvrB* operator-promoter region is shown in Fig. 4-8. In vitro, the *uvrB* gene is transcribed from

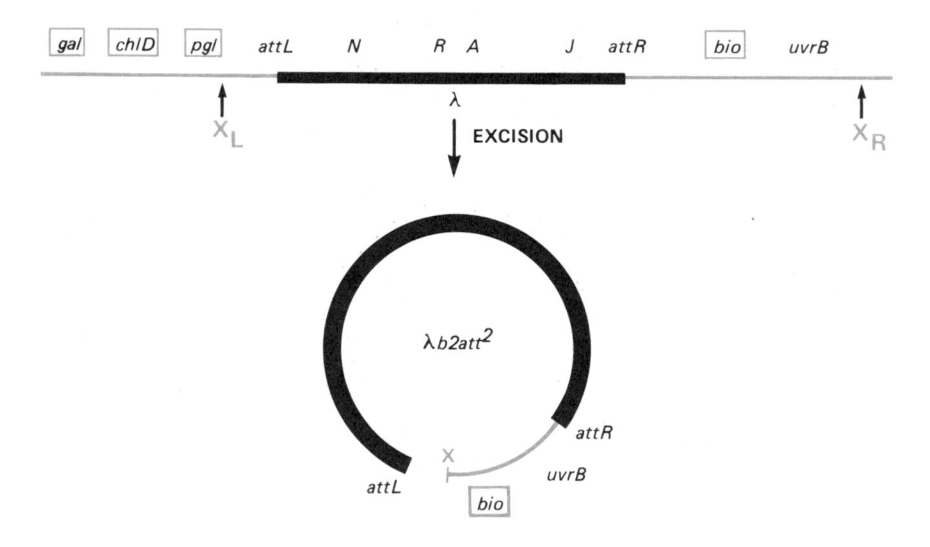

FIGURE 4-7
Excision of integrated phage λ at sites X_L and X_R generates the transducing phage $\lambda b2att^{2}$, which contains the the *E. coli uvrB* gene. (After M. Fiandt et al., ref. 47.)

two different promoters called P1 and P2. Pribnow boxes and −35 sequences for these promoters are shown in the figure. The transcriptional initiation sites for these are denoted S1 and S2 respectively and the nucleotides in the entire 540-bp sequence are numbered relative to S1. A lexA binding site (putative operator) extends from the region between −67 and −40 on the anticoding strand and between −63 and −36 on the coding strand (Fig. 4-8) (50). The lexA binding site apparently functions as the operator for P2, since in the presence of lexA protein transcription from P2 is inhibited in vitro, while that from P1 is unaffected, even though only several base pairs separate the lexA binding site from the −35 sequence of P1 (Fig. 4-8). Notice that P1 and P2 are actually overlapping promoters, since transcription from P2 is initiated at position −31 (S2 in Fig. 4-8), which is within the −35 regulatory sequence of P1. The sequences of the Pribnow boxes of P1 and P2 as well as the −35 sequence of P1 show close homology to the consensus sequences established from examination of a large number of other *E. coli* genes; however, the −35 sequence of P2 is a poor match to the consensus sequence (50).

The transcription of the *uvrB* gene in vitro suggests that only P2 transcription is regulated by lexA protein and that P1 is responsible for constitutive expression of the gene (50). However, studies have also been carried out in which the transcription from the *uvrB* promoters P1 and P2 was measured *in vivo* (51, 52). In cells harboring the *uvrB* gene on a multicopy plasmid, P1 transcripts were present in 10- to 20-fold excess over P2 transcripts. Both transcripts were induced strongly by UV irradiation, and after induction P1 remained the strongest promoter. Identical observations were made when transcripts from cells carrying only the *chromosomal uvrB* gene were analyzed. Analyses with plasmids carrying deletions in the part of the lexA binding sequence that overlaps with P2 indicate that in vivo P1 (the promoter closest to the coding region of *uvrB*) is responsible for both the constitutive and the induced levels of *uvrB* transcription and that P2 serves mainly as a target for regulation of P1. Thus, lexA protein bound to P2 may interfere with the binding of RNA polymerase to P1 or, alternatively, may interfere with the local denaturation of the DNA helix that precedes initiation of transcription (51, 52).

It is not clear why there is this discrepancy between the results of transcription of *uvrB* in vitro and in vivo. It has been suggested (52) that it may reflect differences in the conformation of the DNA templates used in the experiments. Linear DNA fragments were used as templates for in vitro transcription, whereas in vivo transcripts were measured from supercoiled plasmids or from the chromosome. It also remains to be established whether or not P1 is regulated by a repressor other than the lexA protein in vivo.

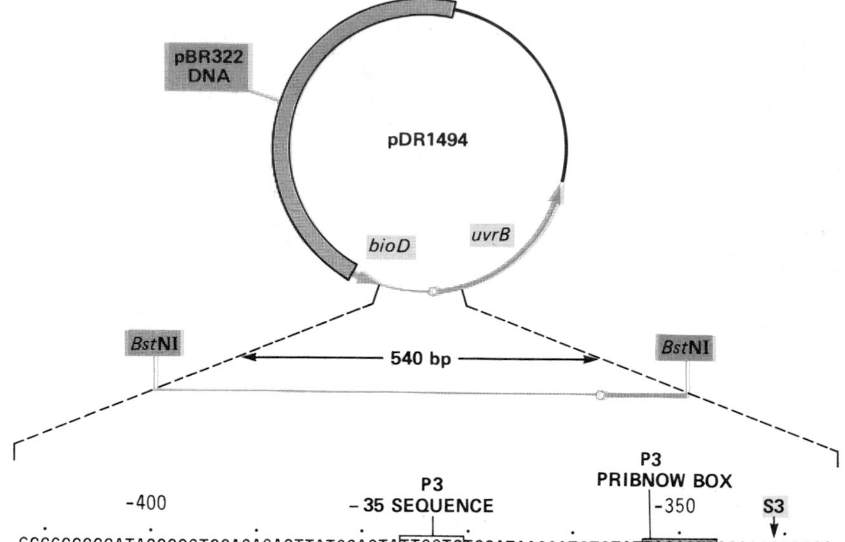

The 540-bp sequence that includes the *uvrB* gene contains a third promoter, P3, located 320-bp upstream from P2 (Fig. 4-8). In vitro, transcription from P3 is directed toward the *uvrB* structural gene but terminates in the region of the lexA protein binding site even in the absence of lexA repressor protein (50). The physiological role of this promoter and the nature of its transcript are unknown. There is pre-

FIGURE 4-8
Nucleotide sequence of the 540 bp *Bst*N1 fragment containing the *uvrB* regulatory regions. The transcription initiation sites for three mRNAs (RNA-1, RNA-2 and RNA-3) are denoted by S1, S2 and S3 respectively, with the numbering of nucleotides relative to S1 (+1). Pribnow boxes and −35 sequences are also identified, as is a region protected from DNase I attack by the binding of lexA protein.
It is not yet firmly established which promoters are used for transcription of the *uvrB* gene under constitutive and induced conditions. *In vitro*, the presence of lexA protein (a repressor which binds at the lexA binding site) inhibits transcription from promoter P2 but not from P1, suggesting the P1 is the promoter used for constitutive expression of the *uvrB* gene and P2 is the promoter used for inducible expression. However, *in vivo*, both promoters are used constitutively and following induction, and in both situations P1 is the stronger promoter. In addition to P1 and P2 a third promoter (P3) has been identified, the role of which is unknown.
The short double lines above the sequence at about +30 and at about +110 identify possible ribosome binding sites near ATG codons that are potential translational start sites. (Since this figure was drawn, personal communication from Dr. G. B. Sancar indicates that a C · G bp is present between positions +59 and +60.) (From G. B. Sancar et al., ref. 50.)

liminary evidence for even further complexity in the regulation of expression of the *uvrB* gene. In vitro, transcription of the *uvrB* gene is inhibited in the presence of uvrC protein (33). In addition uvrC protein reduces the biosynthesis of uvrB protein from a plasmid that lacks P2 (33). A survey of the sequence of the *uvrB* gene for inverted repeat sequences indicates several possible sites of modified secondary structure in and around the Pribnow box of P1 at which the uvrC protein might bind as a regulator of the *uvrB* gene (33).

The *uvrC* Gene

Plasmids carrying *E. coli* DNA inserts with genes that map very close to the *uvrC* gene have been exploited for cloning this gene (36). The *uvrC* gene has also been cloned from a bank of λ specialized transducing phage containing *E. coli* DNA sequences (35) and by transformation of *uvrC* mutants to UV resistance with recombinant pBR322. The *uvrC* gene has also been cloned from the Clark-Carbon *E. coli* genome bank of plasmids containing *E. coli* DNA inserts (24, 26, 30, 31, 34, 40). The structural gene has been localized to a 1.9-kilobase (kb) sequence (36–38); however there is still uncertainty about the regulation of the expression of this gene. Recent studies suggest that like the *uvrA* and *uvrB* genes, the *uvrC* gene of *E. coli* is inducible and under the control of the *lexA-recA* regulatory system (51). Plasmids carrying fusions between regulatory sequences upstream from the *uvrC* structural gene and the *E. coli galK* (galactokinase) gene have been constructed. The levels of galactokinase expressed from these plasmids under different conditions suggest that the putative *uvrC* promoter is inducible by mitomycin C and by UV radiation (Fig. 4-9) (51). This induction is *recA*-dependent and *lexA*-dependent, but in contrast to other damage-inducible genes under

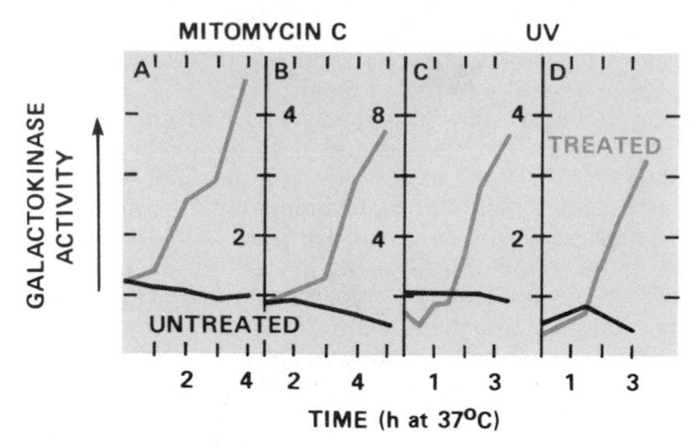

FIGURE 4-9
A fusion plasmid carrying the putative *uvrC* promoter and the *E. coli galK* structural gene was constructed and transformed into E. coli cells. The cells were either untreated or treated with mitomycin C (panels A and B) or UV radiation (panels C and D). Treatment with either of these DNA damaging agents was associated with enhanced expression of galactokinase activity, suggesting that the *uvrC* promoter is inducible by DNA damage. (From C. M. Backendorf, ref. 51.)

the control of this regulon, the induction of the *uvrC* promoter in the fusion plasmids is very slow (51). Independent studies have identified a promoter of *uvrC* that is approximately 1 kb upstream from the structural gene (37). Whether or not this is the same promoter as the inducible element just described remains to be established. It is also currently unclear whether other promoters regulate the *uvrC* gene and, if so, whether any of them are under inducible control.

The *uvrC* gene has been placed under the control of the *tac* promoter, a hybrid promoter containing the −35 region of the *E. coli trp* promoter and the Pribnow box of the *lac* promoter (25). The *tac* promoter retains regulation by the lac repressor, so that expression of the *uvrC* gene is regulated and inducible by agents such as lactose and isopropylthiogalactoside (IPTG) (a lactose analogue).

Properties of the *uvrA, uvrB* and *uvrC* Gene Products

The molecular cloning experiments discussed above have made possible the physical identification of the products of the *uvrA*, *uvrB* and *uvrC* genes, as well as the purification of these proteins, so that their roles in the incision of DNA catalyzed in vitro can be examined.

The rapid identification of individual plasmid-coded *uvr* gene products against the large background of the many host-coded proteins has been aided by the so-called maxicell procedure (39, 53)

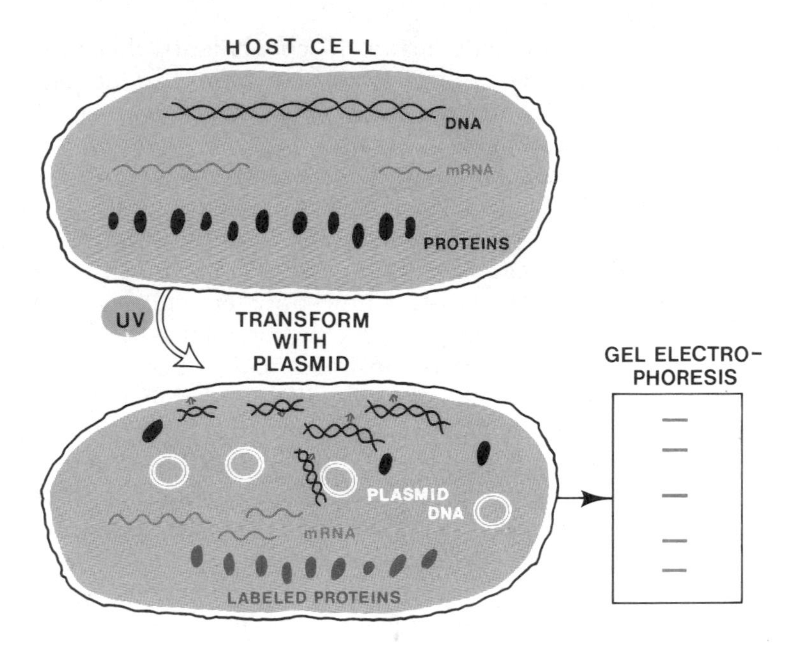

FIGURE 4-10
Schematic representation of the *E. coli* maxicell procedure for detecting the expression of plasmid-encoded proteins. When mutant *recA* cells are transformed with plasmid and then exposed to UV radiation, the host chromosome undergoes extensive degradation and few (if any) host proteins are synthesized. Plasmid DNA largely escapes irradiation damage because it is a much smaller target. Hence most newly synthesized proteins (detected by gel electrophoresis of radiolabeled protein) are of plasmid origin.

(Fig. 4-10). This technique is predicated on the long-established observation that when *E. coli* cells defective in the *recA* gene are exposed to UV radiation, they stop DNA synthesis and their chromosomal DNA is extensively degraded (12, 54). This effect can be produced at significantly lower doses of radiation by using a strain of *E. coli* that is also defective in the excision repair of pyrimidine dimers (e.g., a mutant *uvrA* strain). If such cells are transformed with a small ColE1-like multicopy plasmid (such as those used for most gene cloning experiments) and then UV-irradiated, those plasmid molecules that do not receive a "UV hit" (because of their small target size) continue to replicate, thereby increasing plasmid DNA levels about 10-fold by six hours after irradiation. Meanwhile, in these cells more than 80 percent of the chromosomal DNA is degraded (53). Thus, when *E. coli* chromosomal DNA degradation is *maximum*, the nondividing "*maxi*cells" contain mainly plasmid DNA and synthesize almost exclusively plasmid-coded proteins (53). These can be conveniently identified by radiolabeling with [^{35}S]methionine added to the growth medium. Parenthetically, it is amusing to note that the term "maxicell" conveniently distinguishes this technique from an alternative procedure for eliminating host proteins, in which cells that do not contain DNA ("minicells") are transformed with plasmids (55).

Even though the maxicell eliminates the expression of most, if not all, host proteins, [^{35}S]methionine-labeled preparations still contain a significant number of proteins, including all of those expressed by the recombinant plasmid. How, then, can one specifically identify the product of a cloned *uvr* gene? A procedure termed $\gamma\delta$ mapping is extremely useful for this purpose. F-episome-mediated transfer of the plasmid pBR322 results in the insertion of a sequence of F called $\gamma\delta$ at random sites in pBR322. If such sequences are inserted into a *uvr* gene, the gene is inactivated and the plasmid loses its ability to complement the appropriate UV-sensitive mutant (22, 23). In addition, in maxicell preparations the labeled protein normally expressed from the gene either disappears or is detected as a truncated polypeptide (Fig. 4-11). Selection of a sufficient number of plasmids with $\gamma\delta$ inserts allows for an unambiguous assignment of a given protein as the product of a given gene. In addition, restriction anal-

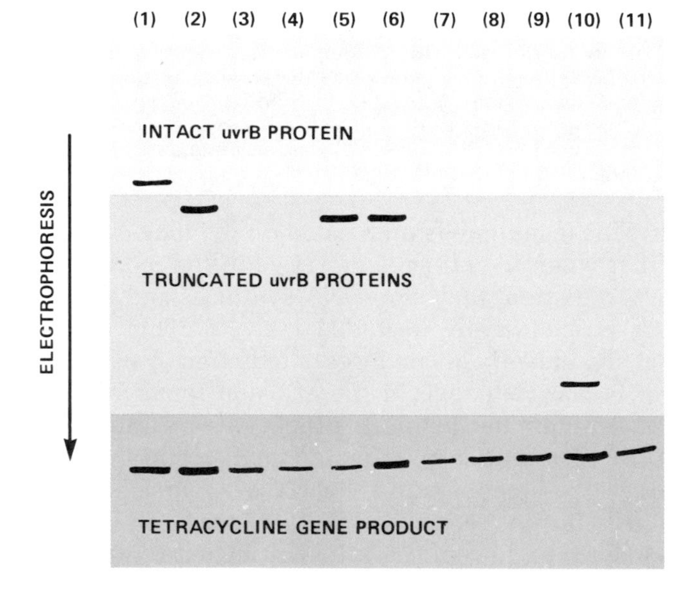

FIGURE 4-11
The maxicell procedure (see Fig. 4-10) allows positive identification of the uvrB protein following insertion of $\gamma\delta$ sequences into a plasmid containing the cloned *uvrB* gene and the *tet*^{r} gene. Sites of $\gamma\delta$ insertion result in the expression of truncated uvrB protein that can be detected by gel electrophoresis (lanes 2, 5, 6 and 10). In some cases (lanes 3, 4, 7–9 and 11) no uvrB protein can be detected. Lane 1 shows intact uvrB protein. (After A. Sancar et al., ref. 23.)

ysis of a sufficient number of plasmids with such inserts provides a localization of a given *uvr* gene in the cloned DNA insert and can also provide a reasonably accurate estimate of the size of the gene (Fig. 4-12). With the maxicell procedure, small amounts of different *uvr* gene products have been fractionated to radiochemical purity (22, 23, 31, 41).

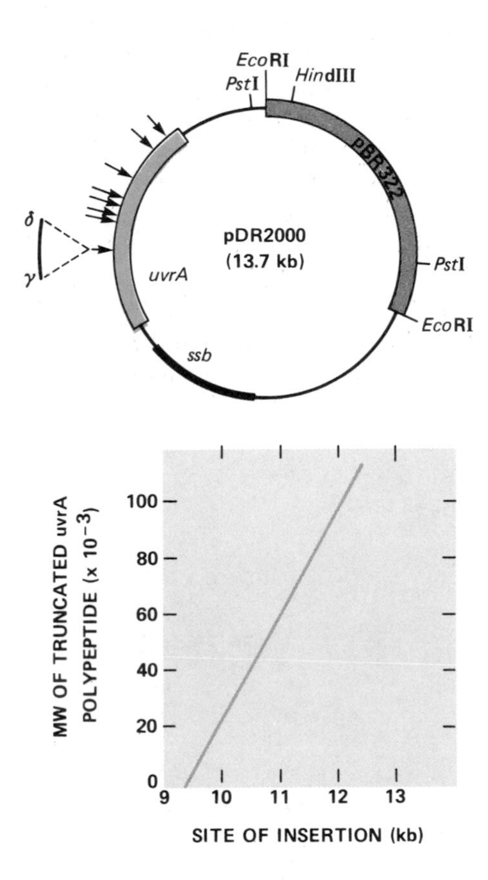

FIGURE 4-12
The size of the cloned *uvrA* gene can be estimated by inserting $\gamma\delta$ sequences at various sites in the gene (indicated by arrows in the upper figure) and correlating (lower figure) the sites of insertion, by restriction mapping, with the molecular weight of the truncated proteins produced in maxicells (see Fig. 4-11). As shown here, the molecular weight of uvrA protein is approximately 114,000. By extrapolation from the curve (the slope of which is drawn to correspond to one amino acid of molecular weight 118 per 3-bp) an approximate size of 2.9 kb for the *uvrA* gene is obtained. (After A. Sancar et al., ref. 22.)

As indicated previously, uvr proteins have been purified from untransformed cells (16, 17, 21) and from cells transformed with plasmids containing cloned genes (24–30). The following paragraphs describe the properties of the uvr proteins that have been elucidated by these studies.

The uvrA Protein

The uvrA protein has a molecular weight of 114,000 (22, 56, 57). It does not possess DNA incising activity when assayed in isolation, but the purified protein binds to single-strand DNA (either UV-irradiated or unirradiated) and to UV-irradiated duplex DNA with a higher affinity than to unirradiated duplex DNA (21, 56) (Fig. 4-13). The high affinity of uvrA protein for *UV-irradiated* duplex DNA possibly reflects the binding to sites of localized helix destabilization associated with the presence of pyrimidine dimers. The presence of Mg^{2+} is essential for the DNA binding activity of uvrA protein (21, 56). ATP stimulates binding to UV-irradiated DNA (Table 4-2), but ADP inhibits binding to both unirradiated and UV-irradiated DNA (21, 56) (Table 4-2). ATP[γS] (an analogue of ATP that is resistant to ATPase activity) also stimulates binding of uvrA protein to DNA;

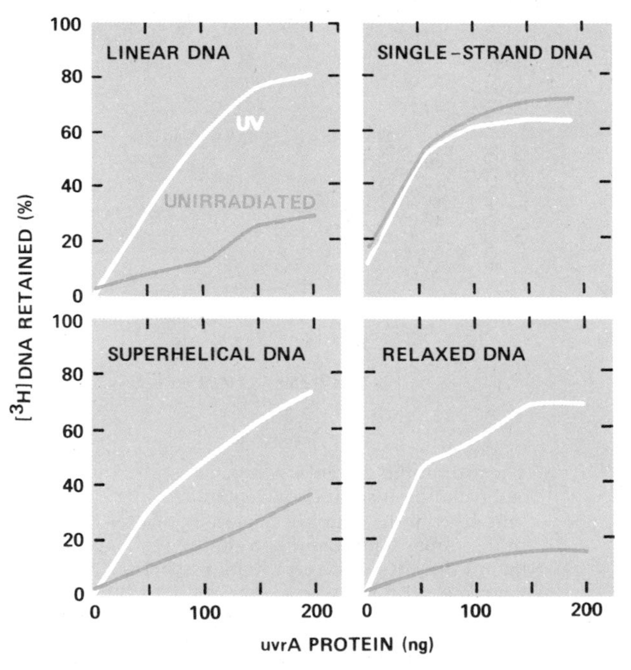

FIGURE 4-13
E. coli uvrA protein binds preferentially to *UV-irradiated duplex* linear, superhelical or relaxed circular DNA. There is no preferential binding to UV-irradiated single-strand DNA, however (upper right panel). (From E. Seeberg and A-L. Steinum, ref. 56.)

TABLE 4-2
Effect of ATP analogues on binding of uvrA protein to DNA

Addition	DNA Bound (%)	
	Unirradiated	UV-irradiated
No ATP	14	20
+ ATP	23	55
+ ADP	4	4
+ ATP[γS]	99	96
+ ATP[γS]*	44	41
+ ATP and ADP	17	30
+ ATP and ATP[γS]	97	100

*Reduced levels of uvrA protein were used to give less than normal binding.
From E. Seeberg and A-L. Steinum, ref. 56.

however, in the presence of this analogue no preferential binding to UV-irradiated DNA is observed (Table 4-2).

In the absence of DNA the *uvrA* gene product demonstrates ATPase activity (56) (Fig. 4-14). The K_m of this reaction corresponds very closely to the K_m for the ATP requirement for nicking of UV-irradiated DNA catalyzed by the combined uvrA, uvrB and uvrC proteins in vitro (21, 56). Both ADP and ATP[γS] are competitive inhibitors of the ATPase activity (21, 56). This observation is interesting since the incision of UV-irradiated DNA in vitro is also inhibited in the presence of ADP or ATP[γS] (21, 56).

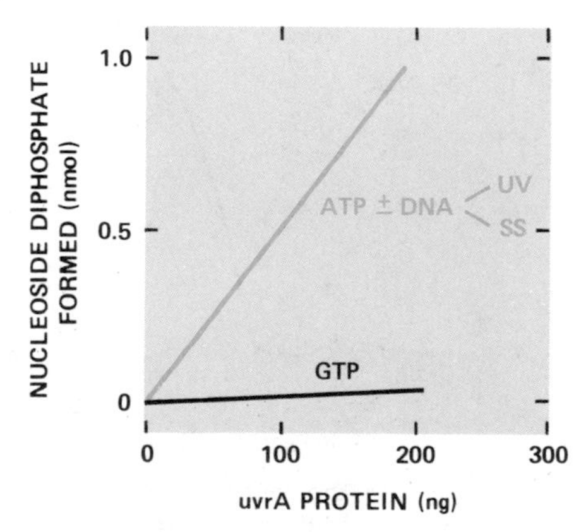

FIGURE 4-14
ATP is hydrolyzed to ADP by uvrA protein both in the presence and absence of UV-irradiated or single-strand DNA. GTP is not hydrolyzed by the protein. (From E. Seeberg and A-L. Steinum, ref. 56.)

The uvrB and uvrC Proteins

The *uvrB* gene product is a protein with a molecular weight of 84,000 (23). It has no known catalytic functions and by itself does not bind strongly to DNA. However, in the presence of uvrA protein tight binding to single-strand DNA is observed, thus demonstrating a direct cooperation between these two proteins in vitro (58) (Fig. 4-15). There are two major differences between the DNA binding activity of uvrA and uvrB protein when they are present together relative to that of uvrA protein alone (59). First, significantly more DNA–protein binding complex is obtained for a given amount of uvrA protein when uvrB protein is also present (Fig. 4-16). Hence apparently much less uvrA protein is needed to form the same amount of binding complex with UV-irradiated DNA when uvrB protein is also present. Second, the two proteins together bind to DNA that contains only one or two pyrimidine dimers per molecule, whereas uvrA

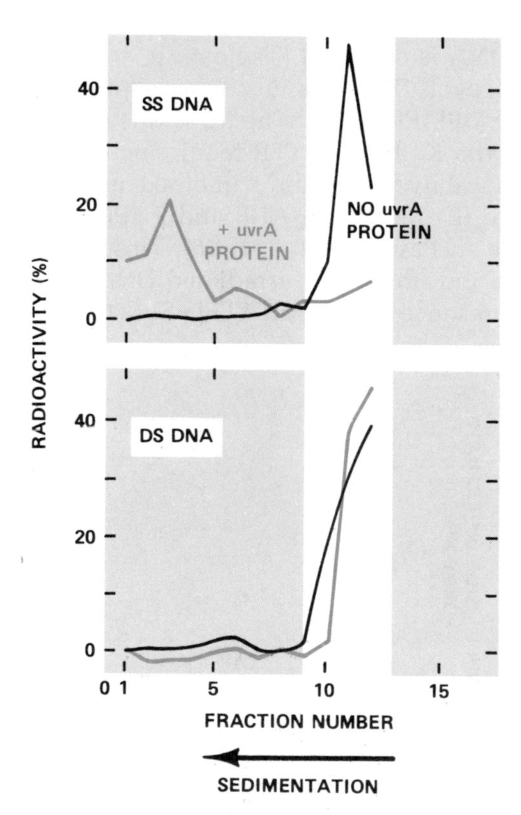

FIGURE 4-15
Radiolabeled *E. coli* uvrB protein binds to single-strand (top) but not to double-strand DNA (bottom) in the presence of uvrA protein. (After B. M. Kacinski and W. D. Rupp, ref. 58.)

protein alone binds less well to very lightly irradiated DNA. This suggests that the number of uvrAB protein complexes formed is stoichiometric with the number of substrate lesions (59).

The uvrC protein has a molecular weight of approximately 70,000 (24, 31, 36) and, like the uvrA protein, binds to single-strand DNA in the absence of the other uvr proteins (31). There are indications for two forms of protein that complement the DNA incising activity in extracts of *uvrC* mutants, one of which may be a modified form of uvrB protein (60). One of the complementing activities has a strong binding affinity for phosphocellulose, is apparently not associated with other uvr proteins and is thought to be the 70,000-dalton product of the *uvrC* gene (60). The other protein that complements extracts of *uvrC* mutants has chromatographic properties very similar to those of uvrB protein. The possibility that this is a fraction of uvrC protein that binds (perhaps nonspecifically) to the uvrB protein during purification of the latter is unlikely, since the molecular weight of this material determined by gel filtration is close to that of monomeric uvrB or uvrC protein (approximately 70,000) rather than that of a complex of these two. The protein cannot be isolated from extracts of mutant *uvrB* cells. It has thus been suggested that this complementing activity is a modified and activated form of the *uvrB* gene product (60). If so, the function responsible for its activation must be some form of the *uvrC* gene product, since no *uvrC*-complementing activity is associated with uvrB protein isolated from mutant *uvrC* cells (60).

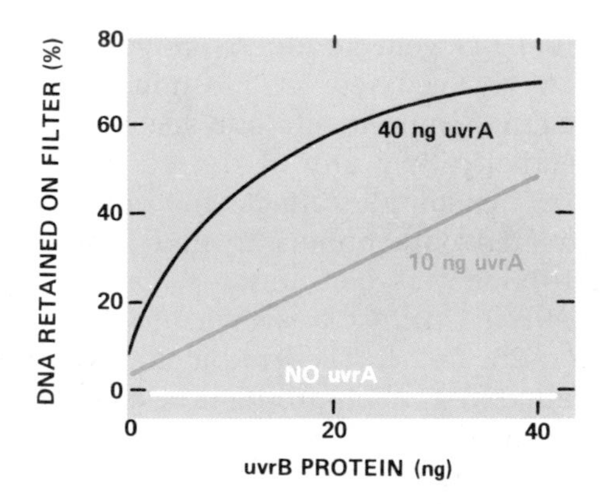

FIGURE 4-16
Cooperative binding of uvrA and uvrB proteins to UV-irradiated ColE1 DNA. In the presence of uvrB protein more DNA-protein complex is formed for a given concentration of uvrA protein than in the absence of uvrB protein. (From E. Seeberg and A-L. Steinum, ref. 59.)

Catalytic Activity of the uvrABC Complex

The mechanism of incision of damaged DNA by the uvrABC endonuclease is just beginning to be understood. A priori it might be anticipated that this complex may exist in different forms that may have different catalytic functions. As mentioned earlier, sequence analysis of the regulatory region of the *uvrB* gene indicates the potential for the expression of different amounts of uvrB protein under different conditions of induction. It is also possible that other specific mechanisms exist for the regulation of this process. For example, the sequence *AATTTGT*GTC*ATAATTAA* found in the *uvrA* promoter (Fig. 4-6) closely resembles the sequence *AATTTGT*TGGG*ATAATTAA* that extends from P2 into P1 in *uvrB* (22) (Fig. 4-8). The extensive conservation of this sequence strongly suggests that its presence in the control regions of both genes is not fortuitous. Thus, it might be expected that a protein other than lexA that binds specifically to this sequence could be a repressor of *uvrA* and of both promoters P1 and P2 in *uvrB*. Additionally, as mentioned earlier, more than one form of uvrC protein may exist, one of which may modify uvrB protein (60). Hence, depending on the exact nature and amount of each *uvr* gene product expressed under particular physiological conditions and the stoichiometry with which each associates with the others, multiple forms of a uvr protein complex may exist.

Such speculation notwithstanding, the available evidence indicates that under constitutive conditions the complex functions as a direct-acting endonuclease rather than by a DNA glycosylase–AP endonuclease mechanism (21, 25, 59, 61–63). First, both the complex and the individual *uvr* gene products are significantly larger than any known DNA glycosylases. In addition, the protein complex has specific cofactor requirements and has a much broader range of substrate specificity than any of the known DNA glycosylases (21). Thus in vitro the complex attacks not only UV-irradiated DNA but also DNA treated with photoactivated psoralen (Fig. 4-17), 4-nitroquinoline-1-oxide, *cis*-platinum, N-2-acetyl-2-aminofluorene, 1,3-bis(2-chloroethyl)-1-nitrosourea, benzo(a)pyrene-diolepoxide or mitomycin C (21, 25, 64). In UV-irradiated DNA, both pyrimidine dimers and so-called (6-4) photoproducts (see Section 1-5) have been identified as substrates (57, 62). Finally, when DNA containing psoralen adducts is incubated with the complex, monoadduct damage is excised as nucleotides, not as free bases, as would be expected from a DNA glycosylase mode of action (21).

Single proteins or mixtures of any two of the three uvr proteins do not result in incision of DNA containing the base damage described above. All three proteins are required, and the evidence from in vitro

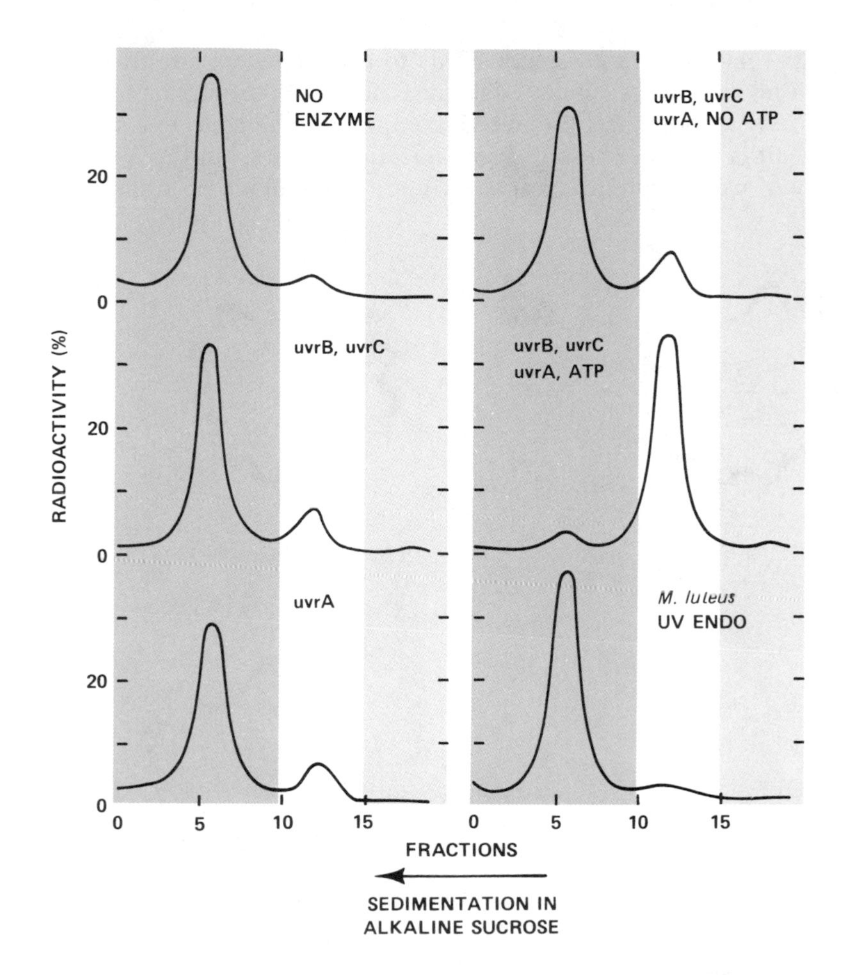

FIGURE 4-17
Nicking of radiolabeled covalently closed circular ColE1 DNA treated with photoactivated psoralen by the uvrABC enzyme of *E. coli*. The peaks on the left of each panel are at the sedimentation position of covalently closed circular DNA. Nicked (relaxed circular) DNA sediments more slowly. Enzyme-catalyzed nicking of the damaged DNA is only observed when all three uvr proteins and ATP are present (middle right panel). (From E. Seeberg, ref. 21.)

studies is that they function as a complex that is assembled at sites of base damage in DNA. Thus, in the presence of Mg^{2+} and ATP, uvrB protein binds very tightly to uvrA protein already associated with duplex DNA containing pyrimidine dimers (59, 61–63). In fact, such complexes are resistant to 1 *M* KCl at 0°C (61). The uvrC protein binds to the preformed uvrAB-dimer complex with a high affinity (59, 61–63) and is neither diluted nor inhibited by a 100-fold excess of native double-strand DNA (61). Incision proceeds very rapidly after the addition of uvrC protein to UV-irradiated DNA complexed with uvrA and uvrB protein (59, 61, 62), and once complete the uvr protein-DNA complex does not dissociate in vitro (61, 62). In vitro, both enzyme complex formation and enzyme activity are optimal when the three proteins are present in equal amounts (61, 62).

Based on these observations, a model for the mechanism of action of the uvrABC proteins has been proposed (59) (Fig. 4-18). The model suggests that uvrA protein first binds to a nonspecific site in DNA containing bulky base damage. The uvrB protein then associates with bound uvrA protein and the uvrAB complex translocates to a site of damage. It is not clear how this translocation occurs, but the ATPase activity of uvrA protein is apparently essential, since no transloca-

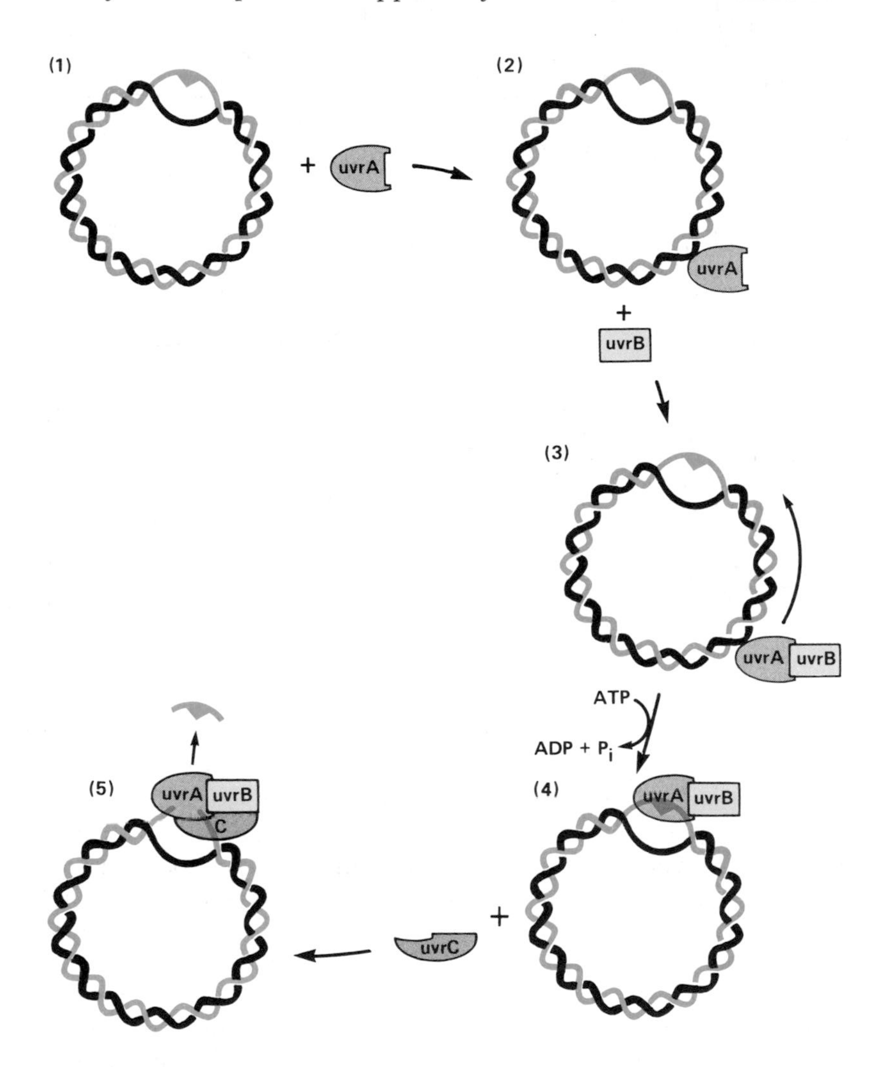

FIGURE 4-18
A model for the incision of DNA containing bulky base damage (1) by the uvrABC enzyme of *E. coli*. UvrA protein binds weakly to duplex DNA (2); however, in the presence of uvrB protein a tighter binding complex is formed (3). This complex translocates along the DNA by a mechanism that involves the hydrolysis of ATP (3). When a site of bulky base damage is encountered, a stable complex of DNA-uvrA-uvrB protein is formed (4). This complex is not catalytically active. However, in the presence of bound uvrC protein DNA incision occurs on either site of the damage (5), resulting in excision of a small oligonucleotide (see Fig. 4-19). (After E. Seeberg and A-L. Steinum, ref. 59.)

tion occurs when ATP[γS] replaces ATP. After translocation to a site of bulky base damage, the uvrAB complex forms a stable interaction with DNA and in the presence of uvrC protein will incise DNA even in the absence of ATP or in the presence of ATP[γS].

Sequencing gels have been used to define the precise location of the enzyme-catalyzed strand breaks relative to pyrimidine dimers in DNA (25, 57, 61, 62). Most unexpectedly, nicks are made on each side of the dimers. These nicks are 12 to 13 nucleotides apart; one seven nucleotides upstream from the 5′ member of the dimer and the other three or four nucleotides downstream from the 3′ pyrimidine (Fig. 4-19). The two cuts are apparently catalyzed by a concerted

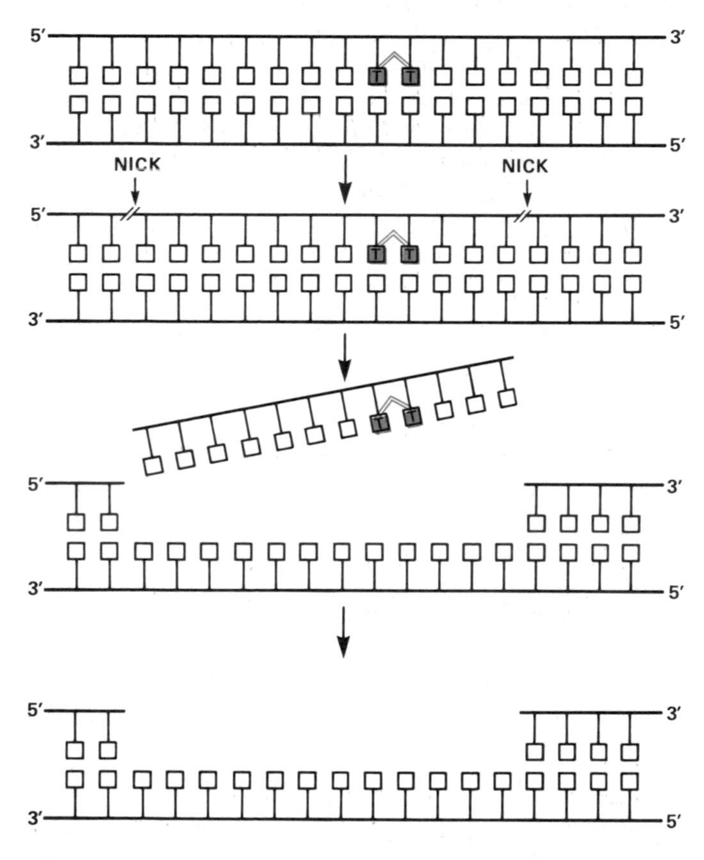

FIGURE 4-19
Postulated mode of coordinated incision and excision of base damage, such as pyrimidine dimers, by the *E. coli* uvrABC enzyme. The enzyme catalyzes the hydrolysis of phosphodiester bonds on either side of a dimer. The nick 5′ to the dimer is seven nucleotides upstream from the dimer, whereas that 3′ to the dimer is three or four nucleotides downstream. Thus an oligonucleotide fragment about 12 nucleotides in size is released from the DNA, leaving a gap in the DNA duplex. (From A. Sancar and W. D. Rupp, ref. 57.)

mechanism in vitro since, if under certain circumstances or at certain sequences only a single cut was made at a lesion, then sequencing gels would demonstrate some incisions at only one or the other side of the lesion. Instead, cuts on each side of all lesions are observed (57, 62). The uvrABC endonuclease always hydrolyzes the eighth phosphodiester bond 5′ to pyrimidine dimers or to (6-4) photoproducts in UV-irradiated DNA (57, 62). The site of incision 3′ to these lesions is more variable. Thus, incision occurs at the fourth phosphodiester bond 3′ to potential TC, CC and CT dimers, but at either the fourth or the fifth phosphodiester bond 3′ to potential TT dimers (57, 62). DNA containing psoralen monoadducts is also incised on each side of the adducts, although the precise distance separating these nicks is not yet clearly defined (57). In vitro, the oligonucleotide generated by the incisions flanking sites of bulky base damage is released from the DNA substrate (57). If the same is true in vivo, then it is likely that the uvrABC complex of *E. coli* catalyzes not only the *incision* of DNA containing bulky base damages but also the *excision* of such damage (Fig. 4-19)—a radical departure from earlier notions about the molecular mechanism of nucleotide excision repair in *E. coli*. This issue is discussed further in Section 5-1.

4-3 Damage-Specific DNA Incising Activities in Eukaryotic Cells

In Section 1-6 it was stressed that a consideration of the distribution of DNA damage in eukaryotes must take into account the fact that in such organisms the genome consists of DNA in intimate association with a number of proteins, forming chromatin, and that the structure of chomosomes reflects various levels of folding and coiling of the basic chromatin structural unit: the nucleosome. The same complexities apply to a consideration of the cellular responses to DNA damage. Our current understanding of how chromosomal structure and nucleosome conformation affect the enzymology of DNA incision and of postincision events in excision repair is still scanty (3). Nonetheless, there are indications that these structural elements may limit the access of repair enzymes to sites of damage and that alterations of chromatin structure may be necessary to effect excision repair in eukaryotic cells. For example, when UV-irradiated mammalian cells are permeabilized to the *M. luteus* PD DNA glycosylase–AP endonuclease activity, fewer sites are incised than in purified UV-irradiated DNA (65). However, additional enzyme-catalyzed nicks are detected by exposing the cells to concentrations of salt known to

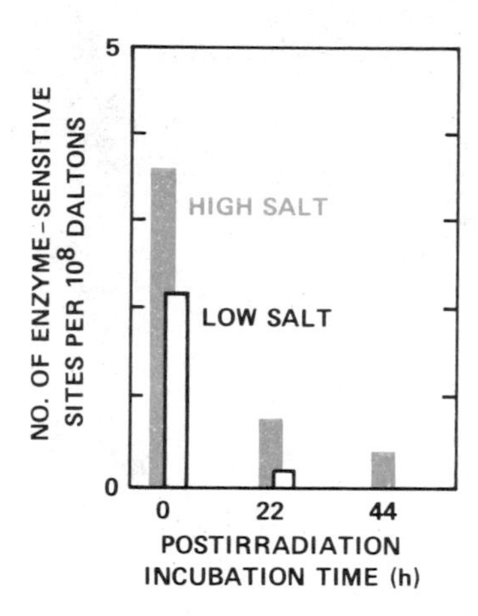

FIGURE 4-20
Immediately after UV irradiation of mammalian cells there are more pyrimidine dimer sites sensitive to the *M. luteus* PD DNA glycosylase–AP endonuclease in salt-denatured chromatin than in native chromatin of UV-irradiated cells. This suggests that in native chromatin some of the pyrimidine dimers are relatively inaccessible to the enzyme. After UV irradiation, enzyme-sensitive sites disappear with time due to excision repair in the cells. (From R. J. Wilkins and R. W. Hart, ref. 65.)

promote nucleosome unfolding (Fig. 4-20) (65). Similar results obtain with the phage T4 enzyme in UV-irradiated mammalian cells permeabilized by freezing and thawing (66). Specific examples of the possible complexity of the incision of damaged DNA due to the structural organization of chromatin stem from a consideration of the genetic complexity of nucleotide excision repair in a variety of eukaryotes, including humans.

The XP Complementation Groups

The human disease xeroderma pigmentosum (XP) has already been mentioned in several different contexts in this book and is discussed more fully in Section 9-2. Fusions between cells of different patients with this disease (Fig. 4-21) in various pairwise combinations have established the existence of at least eight genetic complementation groups, members of each of which are defective or deficient in the excision repair of pyrimidine dimers as well as various types of chemical damage to DNA (67, 68). If it is assumed that the cell fusion assay does not reflect *intra*genic complementation for any pair of cells, these results suggest the involvement of at least eight gene products in the excision repair of bulky base damage in normal human cells.

It appears that much, if not all, of the genetic complexity in XP is concentrated at very early steps in excision repair, specifically those involved in, or required for, the *incision* of DNA (68). The presence of strand breaks in the DNA of UV-irradiated normal human cells generated (presumably enzymatically) during post-UV incubation can be demonstrated by the technique of alkaline elution, in which cells in alkali are placed on a filter that is subject to a controlled

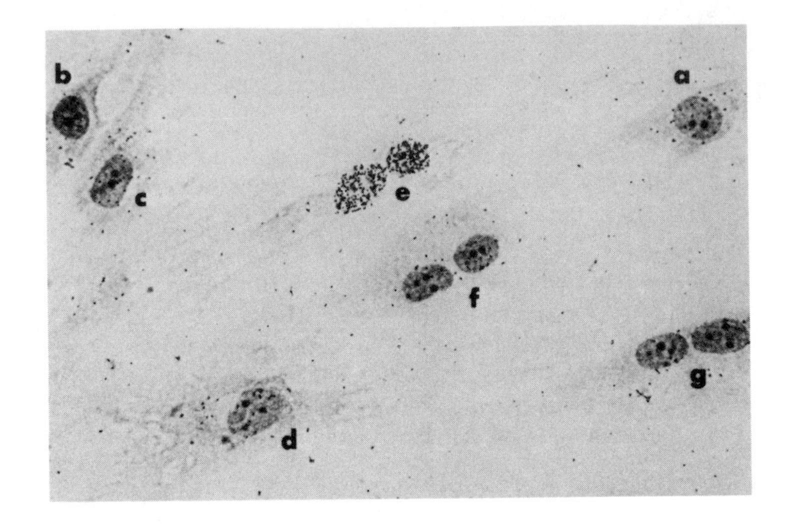

FIGURE 4-21
Complementation of unscheduled DNA synthesis (UDS) (nonsemiconservative DNA synthesis during excision repair) by fusion of XP cells from different genetic backgrounds: (a–d) monokaryons (unfused cells) showing no UDS; (e) a heterodikaryon showing normal levels of UDS; (f, g) homodikaryons showing no UDS. UDS is detected by incubating UV-irradiated cells in the presence of [^{3}H] thymidine and preparing autoradiograms (see Sec. 5.2). (Courtesy of Dr. Jay H. Robbins.)

gentle suction force (69). The alkali lyses the cells and also denatures the DNA. Thus the rate at which the DNA is eluted from the filters is a direct function of its single-strand molecular weight; i.e., the smaller the DNA (the more nicks), the faster it elutes (69). Studies with normal human cells show a very rapid accumulation of strand breaks, followed by a gradual restoration to normal molecular weight, indicative of the completion of excision repair (70) (Fig. 4-22). Cells from XP complementation groups A, B, C and D fail to accumulate significant numbers of breaks in their DNA during post-irradiation incubation, suggesting a defect in the incision of DNA in these cells (70) (Fig. 4-22). Similarly, when purified T4 PD DNA glycosylase–AP endonuclease (see Section 3-4) is introduced into permeabilized UV-irradiated cells, normal levels of repair synthesis of DNA are restored to cells from complementation groups A through E (71) (Fig. 4-23). Presumably the T4 enzyme catalyzes incision in DNA at pyrimidine dimers, effectively circumventing the endogenous incision defect. Introduction of the T4 enzyme into group F cells only partially restores unscheduled DNA synthesis (UDS); XP group G and H cells have not yet been examined by this procedure.

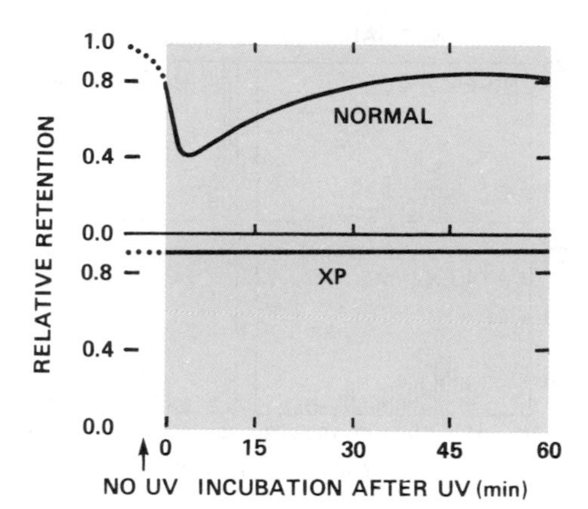

FIGURE 4-22
The incision of DNA in UV-irradiated normal and XP fibroblasts. The figure shows the relative retention of DNA on filters as measured by the alkaline elution technique. Normal human fibroblasts create strand breaks in their DNA by excision repair during post-irradiation incubation. This results in an increased rate of elution of the DNA through filters following denaturation of the DNA in alkali, due to the smaller size of the DNA fragments. Some DNA strand breaks that are independent of UV irradiation also occur (dotted lines). The gradual return to higher molecular weight DNA associated with the completion of excision repair is reflected by increased filter retention of the DNA. DNA from XP cells does not show any evidence of DNA incision by this technique. (From A. J. Fornace, Jr., et al., ref. 70.)

Another technique used for the detection of incisions in DNA at pyrimidine dimers is referred to as the endonuclease-sensitive-site (ESS) assay (72) (see Section 5-2). If DNA extracted from XP cells is not incised in vivo, it constitutes a substrate for subsequent incubation with the T4 or *M. luteus* enzymes that specifically recognize pyrimidine dimers (see Section 3-4). The extent to which the DNA retains sensitivity to these enzyme probes, i.e., the extent of loss of ESS, can be determined by procedures that measure the single-strand molecular weight of the XP DNA after treatment with enzyme. Such analyses have shown a profound defect in incision in XP cells of complementation groups A, B, C, D and G (73), whereas cells from XP complementation groups E and F are distinctly leaky by this parameter (73) (Fig. 4-24). Analysis of XP group H cells has not been reported.

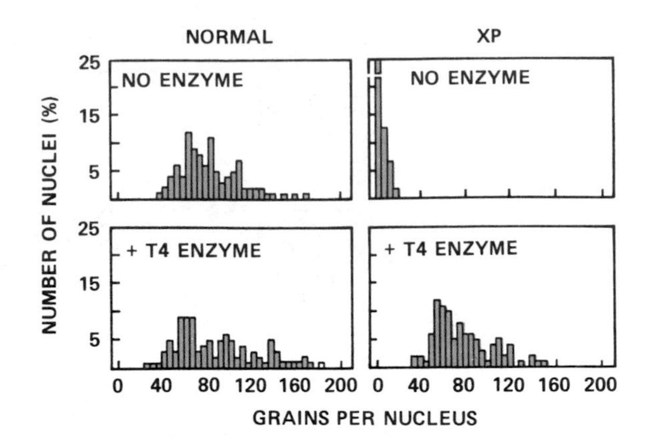

FIGURE 4-23
The introduction of T4 PD DNA glycosylase–AP endonuclease into human fibroblasts permeabilized by treatment with inactivated Sendai virus results in the restoration of apparently normal levels of repair synthesis in UV-irradiated XP cells. Repair synthesis (UDS) was measured by counting autoradiographic grains in the nuclei of cells labeled with [^{3}H]thymine (see Fig. 4-21). (After K. Tanaka et al., ref. 71.)

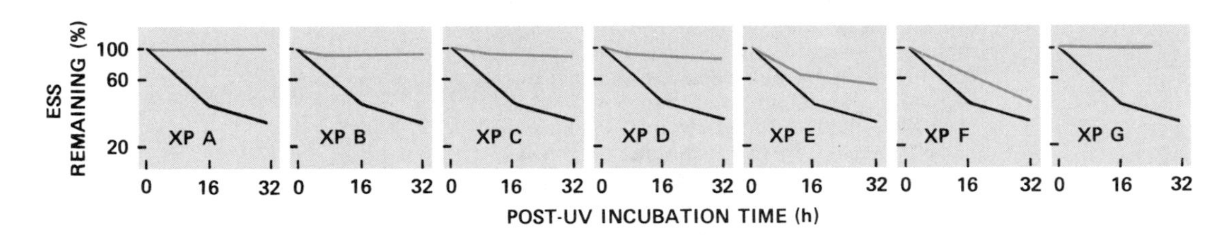

FIGURE 4-24
Kinetics of the loss of sites in DNA sensitive to the *M. luteus* PD DNA glycosylase–AP endonuclease activity in vitro in normal and XP cells. Normal cells show a significant loss of endonuclease-sensitive sites (ESS); the curve describing these results is reproduced in each panel. Note that cells from XP groups A, B, C, D and G are severely defective in the loss of sites sensitive to the *M. luteus* enzyme probe. Cells from complementation groups E and F, however, still effect significant loss of ESS from UV-irradiated DNA. (After B. Zelle and P. H. M. Lohman, ref. 73.)

Based on these results it would appear that some genes in human cells are absolutely required for the excision repair of bulky base damage in DNA. Since mutation in these genes (as in the disease XP) is reflected in a profound (possibly total) defect in DNA incision, it is probable that these genes encode proteins that are *required* for the incision of damaged DNA. Other genes may not be required for nucleotide excision repair but are apparently *involved* in this process in vivo. Perhaps these genes encode products functioning as accessory proteins that improve the efficiency and/or fidelity of excision repair in living cells. Why are so many genes (at least five) apparently *required* for the incision of DNA in human cells? At least two hypothetical possibilities merit consideration. First, one might imagine that the enzymology of DNA incision during excision repair in human cells is strictly analogous to that in *E. coli;* i.e., the five genes in question code for a direct-acting damage-specific DNA incising activity that catalyzes nicking of damaged DNA. Alternatively, some or all of these gene products may function specifically to create *accessibility* of sites of base damage in chromosomes to a catalytically active endonuclease.

Support for the latter model comes from studies in which the capacity for selective excision of thymine-containing pyrimidine dimers from chromatin and from deproteinized DNA was examined in cell-free extracts of human cells. Extracts of normal human fibroblasts prepared by the sonication of cells catalyze the excision of thymine-containing pyrimidine dimers from their endogenous chromatin or from exogenously added purified DNA (74–78) (Fig. 4-25). However, extracts of XP cells from complementation groups A, C and G selectively fail to carry out the *former* reaction (74–77) (Fig. 4-25). This defect is apparently not at the level of the chromatin in these cells, since the addition of extract from normal cells to those containing XP chromatin results in the loss of dimers (i.e., DNA incision) from the XP substrate. Thus it is possible that the products encoded by some of the human genes required for the incision of DNA are involved in the localized processing of the chromosome at sites of DNA damage, in preparation for an endonuclease encoded by a different subset of genes.

One approach to the study of the human genes required for the excision of bulky DNA adducts is to attempt the molecular cloning of these genes, with a view to expressing the proteins they encode in amounts suitable for direct biochemical study. Complementation of the phenotype of UV sensitivity of XP group A cells by transfection with DNA from normal (but not from XP group A) cells has been reported (79). However, in other very similar studies a high frequency of reversion of the phenotype of XP cells has been observed (80, 81).

In a different experimental approach, cell extracts from repair-proficient human cells were microinjected into XP cells from complementation groups A, D and G. This resulted in a temporary increase in the levels of unscheduled DNA synthesis in these cells (82, 83). The phenomenon is apparently specific, since extracts from XP group A cells did not complement the defect in repair synthesis in XP group A cells. The cell fractions that corrected the XP group A and G cells are stable to storage but sensitive to repeated freezing and

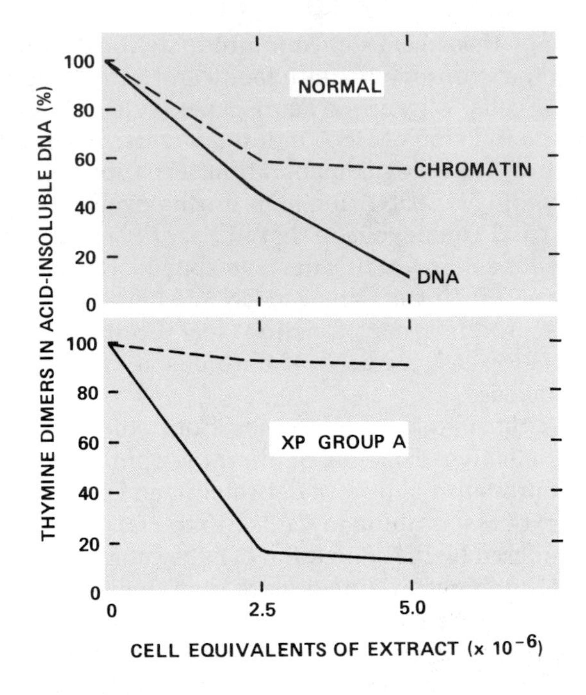

FIGURE 4-25
Extracts of normal human fibroblasts (top) catalyze the excision of thymine-containing pyrimidine dimers from both purified DNA and sonicated chromatin. Extracts of XP group A cells (bottom) are defective in the excision of dimers from chromatin but still catalyze the loss of dimers from purified DNA. (From K. Mortelmans et al., ref. 74.)

thawing. Both active principles are apparently proteins since they are sensitive to proteolysis.

DNA Incision in Other Eukaryotic Cells

Further evidence for genetic complexity in the incision of damaged DNA in eukaryotes comes from studies on other eukaryotic systems, including the yeast *S. cerevisiae*, the fruit fly *D. melanogaster* and rodent cells in culture.

The Yeast RAD3 *Epistasis Group*

S. cerevisiae is a eukaryotic organism that is genetically very well characterized, especially with respect to its response to agents that cause DNA damage (84). Over 30 *RAD* (*rad*iation sensitivity) loci have been identified in this organism, which generally fall into three

TABLE 4-3
Yeast epistasis groups for cellular responses to DNA damage*

RAD3	*RAD6*	*RAD52*
rad1	*rad5*	*rad50*
rad2	*rad6*	*rad51*
rad3	*rad8*	*rad52*
rad4	*rad9*	*rad53*
rad7	*rad15*	*rad54*
rad10	*rad18*	*rad55*
rad14	*rev1*	*rad56*
rad16	*rev3*	*rad57*
r_1^s	*umr1*	r_1^s
mms19	*umr2*	*cdc9*
cdc8	*umr3*	
	mms3	
	cdc8	
	pso1	

*Mutants in any single epistasis group do not show additive sensitivity to ionizing and/or UV radiation when a second mutation in the same epistasis group is present. When double mutants do demonstrate additive sensitivity, the two loci in question are classified in different epistasis groups. In addition to the loci shown here, about 60 other loci have been identified in yeast, mutation at which causes abnormal sensitivity to killing by chemical and/or physical agents. Most of these have not yet been organized into epistasis groups.

From R. H. Haynes and B. A. Kunz, ref. 84.

largely nonoverlapping or *epistasis* groups, thought to reflect three distinct cellular responses to DNA damage in this eukaryote (84) (Table 4-3). One of these (referred to as the *RAD3* epistasis group) (Table 4-3) constitutes a number of genetic loci, mutations at which are expressed phenotypically as abnormal sensitivity to UV radiation but not to ionizing radiation (84) (Fig. 4-26). In addition, such mutants are defective in the removal of thymine-containing pyrimidine dimers from their DNA during post-UV incubation (84).

In wild-type yeast, DNA incision at pyrimidine dimers is immediately followed by excision-resynthesis and DNA ligation; thus the number of incisions that can be detected at any given time is very small relative to the total number of dimers present in the genome (85). This makes it difficult to be certain whether or not a given *rad* mutant is totally defective in DNA incision. In order to circumvent this problem, a temperature-sensitive mutation in the gene for DNA ligase (*CDC9*) was introduced into a number of *rad* mutant strains (86). Because of the defect in DNA ligase activity, DNA incisions associated with excision repair of dimers cannot be sealed and many

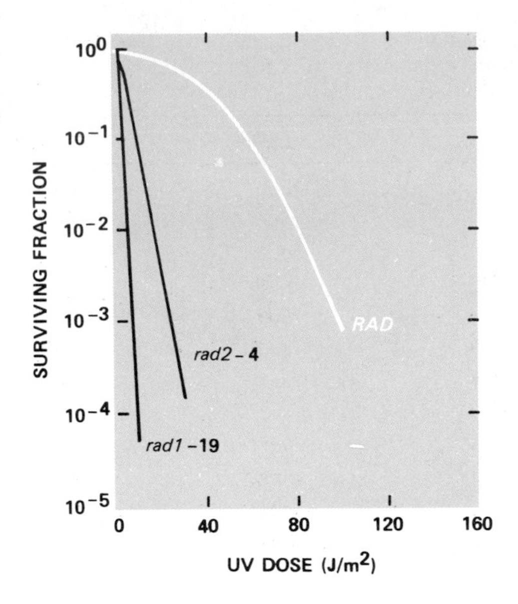

FIGURE 4-26
UV survival curves for wild type (*RAD*) and various excision-defective (*rad*) mutants of *S. cerevisiae*. (From R. J. Reynolds and E. C. Friedberg, ref. 85.)

more strand breaks are detected than with wild-type strains (Fig. 4-27). It is therefore possible to demonstrate that mutants in the *RAD1*, *RAD2*, *RAD3*, *RAD4* and *RAD10* loci are totally defective in the incision of UV-irradiated DNA in vivo (85, 86) (Table 4-4). On the other hand, mutants in the *RAD7*, *RAD14*, *RAD16* and *RAD23* loci are deficient, but not defective, in DNA incision in vivo (85, 86). Hence as in human cells, there appear to be two categories of genes involved in excision repair in yeast. Some genes (*RAD1*, *RAD2*, *RAD3*, *RAD4* and *RAD10*) are apparently *required* for DNA incision (87) whereas others (*RAD7*, *RAD14*, *RAD16*, *RAD23*, *MMS19*) appear to be involved in but not essential for this process. Mutants in the latter group are less UV-sensitive and show significant residual capacity for both DNA incision and pyrimidine dimer excision (87). The observation that both in human and in yeast cells *five* genes have been identified as essential for DNA incision may be a pure numerical coincidence and should not be taken literally at this time. Nonetheless, the general analogy is striking. It is also of interest that in one reported study, extracts of wild-type yeast cells and of mutants defective in the *RAD1*, *RAD2*, *RAD3*, *RAD4* or *RAD10* genes were *all* able to catalyze the incision of UV-irradiated purified DNA (88) (Table 4-5). Whether or not these extracts are specifically defective in the

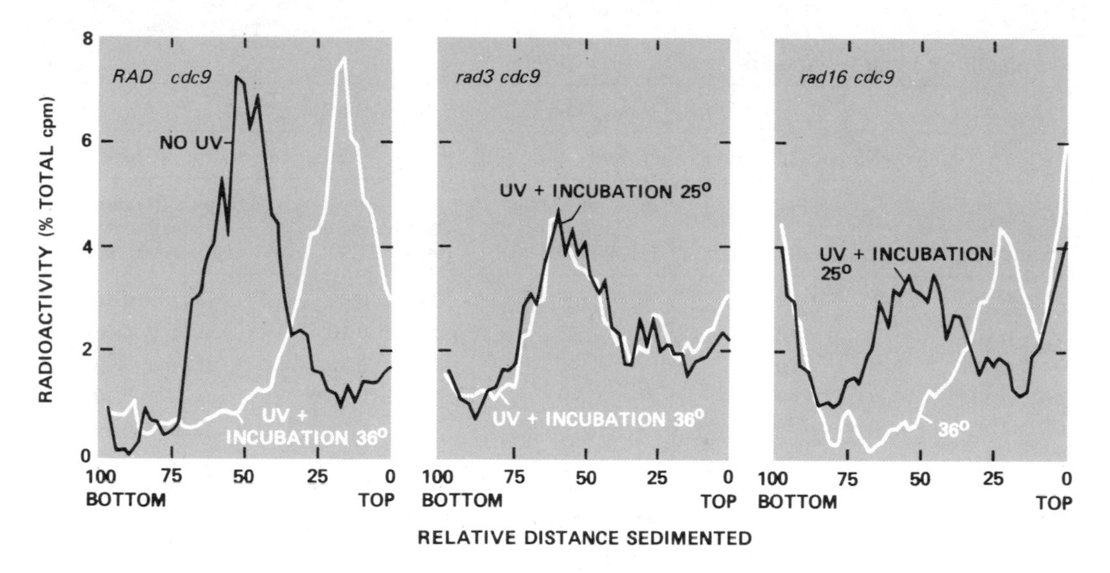

FIGURE 4-27
Defective incision of UV-irradiated DNA in a *rad3* mutant in vivo. The *RAD* strain (left) and a strain mutant in the *RAD16* gene (right) generate DNA incisions that shift the sedimentation of the DNA toward the top of the sucrose gradients (white curves). These incisions persist at 36°C because of the presence of the temperature-sensitive *cdc9* mutation that prevents DNA ligation and hence completion of excision repair. At 25°C, however, DNA ligase is active and excision repair can be completed. Thus the higher molecular weight DNA of wild type (not shown) and *rad16* cells sediments faster in the alkaline gradients. No DNA strand breaks are detected in the DNA of the *rad3* mutant (center), even at 36°C, indicating that this mutant is completely defective in the incision of UV-irradiated DNA. (After D. R. Wilcox and L. Prakash, ref. 86.)

TABLE 4-4
Strand breaks in DNA of UV-irradiated cells defective in *RAD* and *CDC9* genes*

Genotype	No. of Breaks Observed (%)
RAD CDC	0
RAD cdc9	100
rad1 cdc9	0
rad2 cdc9	0
rad3 cdc9	0
rad4 cdc9	0
rad10 cdc9	0
rad14 cdc9	29
rad16 cdc9	63

*Cells were irradiated and held at 36°C for one hour before measuring strand breaks by sedimentation velocity.
After D. R. Wilcox and L. Prakash, ref. 86.

TABLE 4-5
Endonuclease activity of cell-free extracts of *S. cerevisiae*

	No. of Single-Strand Breaks/ColE1 DNA Molecule				
	After Action of Cell-Free Extracts on:			Produced by Extract After:	
Strain	Native DNA	UV-treated DNA	AcP* plus 340-nm light-treated DNA	UV treatment (mean ± SD)	AcP plus 340-nm light treatment (mean ± SD)
RAD	0.34	0.60	0.61	0.26 ± 0.09	0.27 ± 0.10
rad1	0.46	0.68	0.67	0.22 ± 0.10	0.21 ± 0.08
rad2	0.42	0.70	0.66	0.28 ± 0.10	0.24 ± 0.03
rad3	0.40	0.68	0.68	0.28 ± 0.10	0.28 ± 0.10
rad4	0.36	0.57	0.61	0.21 ± 0.09	0.25 ± 0.04
rad10	0.42	0.67	0.74	0.25 ± 0.12	0.32 ± 0.10
rad16	0.38	0.63	0.55	0.25 ± 0.09	0.17 ± 0.05

*AcP = acetophenone. The date on the left half of the table show the number of strand breaks in both irradiated and unirradiated DNAs. The data on the right half of the table show the *net* effect on the irradiated DNAs.
From M. L. Bekker et al., ref. 88.

incision of UV-irradiated *chromatin* remains to be determined. Perhaps in yeast, too, one of the critical functions of the "repairosome" is not only to locate sites of base damage in DNA but also to promote the dissociation of DNA from structural proteins and thereby facilitate the physical access of catalytic functions to phosphodiester bonds that must be cleaved during excision repair.

An obvious approach to understanding the extraordinary complexity of the incision of damaged DNA in eukaryotes is to isolate the XP and/or *RAD* genes of interest and study their biochemistry. *S. cerevisiae* is particularly well suited for such studies because plasmid vectors for the cloning of yeast genes have been established (89). A number of yeast *RAD* genes have been isolated from a total yeast gene pool inserted into a yeast cloning vector that functions as a plasmid in yeast and replicates autonomously at a relatively high copy number. The vector (Fig. 4-28) contains the yeast chromosomal *URA3* gene; hence *ura3* transformants can be conveniently selected. Individual transformants containing specific *rad* mutations have been screened for enhancement of UV resistance, and plasmids containing the *RAD1*, *RAD2*, *RAD3* and *RAD10* genes have been individually isolated (87, 90–93). At present, little is known about the products of these genes. However, DNA sequencing of their coding regions as well as measurement of the size of their transcripts indicates that the *RAD1*, *RAD2* and *RAD3* genes could encode products of at least 120, 130 and 90 kDa respectively, assuming that none of these genes contains introns.

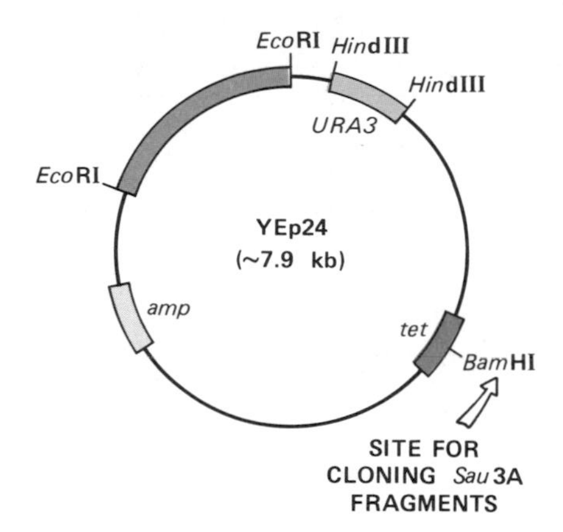

FIGURE 4-28
The *E. coli*–yeast shuttle vector YEp24 is a multicopy plasmid that replicates both in yeast and in *E. coli*. The *Eco*RI fragment contains a yeast autonomously replicating sequence (*ars*) and the *Hin*dIII fragment contains the yeast *URA3* gene for selection of the plasmid in yeast *ura3* mutants. The amp^R and tet^R genes allow for selection of the plasmid in *E. coli*. A yeast genomic library has been constructed by cloning *Sau*3A partial digests of yeast chromosomal DNA into the *Bam*HI site in the tet^R gene.

An interesting observation is that the *RAD3* gene of *S. cerevisiae* is essential for the viability of haploid cells in the absence of DNA damage (91, 92). The nature of this essential function is still unknown. However, it is apparently distinct from the function of *RAD3* during DNA incision, since some mutants in the *RAD3* gene are totally defective in DNA incision yet retain viability unconditionally. The demonstration that a gene required for the incision of damaged DNA is essential raises the intriguing possibility that the yeast and human cell mutants suggested above as having a *secondary* role in excision repair may in fact be leaky because they are mutated in essential genes.

The availability of these cloned *RAD* genes should facilitate the study of the biochemistry of DNA incision in this eukaryote and may help to define molecular events that are common to excision repair in all eukaryotic cells, including those of humans.

D. Melanogaster

Drosophila is another eukaryotic organism that has been extensively characterized genetically. About 30 genes have been identified in this organism, mutations at which effect increased sensitivity to killing by mutagens (94). Eleven loci have been mapped on the X and the third chromosomes and seven genes have been localized to the second chromosome (94).

Several mutants have been shown to be abnormally sensitive to UV radiation and demonstrate a reduced capacity for excision of pyrimidine dimers and for the formation of DNA strand breaks after exposure to UV radiation (94). Of these, mutants at two distinct loci designated *mei-9* and *mus(2)201* are completely defective in DNA incision (Table 4-6), and a number of other mutants show partial

TABLE 4-6
Repair capacities of excision-deficient mutants of *Drosophila*

Mutant	Dimer Excision Capacity (% of Control)	Formation of Strand Breaks (DNA Incision) (% of Control)
mei-9	~0	~0
mus(2)201	~0	~0
mus(2)205	47	98
mus(3)302	72	113
mus(3)304	43	90
mus(3)306	47	74
mus(3)308	24	51

From J. B. Boyd, et al., ref. 94.

deficiency in this parameter of excision repair (Table 4-6) (94–97). Mutant alleles of the *mei-9* and *mus(2)201* loci have been successfully introduced into permanent cell lines (94). These lines should facilitate more detailed biochemical studies on the role of these genes in the incision of damaged DNA in *Drosophila*.

Excision Repair Mutants from Rodent Cells

DNA repair-deficient mutants of Chinese hamster ovary (CHO) and of mouse L cells have also been isolated (98–102). For the isolation of CHO mutants a semiautomated procedure designed for very large-scale screening of mutagenized cells in culture has been utilized (98). Extremely large agar trays were inoculated with large numbers of mutagenized CHO cells, which grew to form visible colonies. The colonies were irradiated with low levels of UV light controlled by an integrating radiation meter. The dishes were photographed at various times after the UV challenge, and by simultaneous observation of the colony images from sequential photographies on the screen of a microfilm reader, colonies exhibiting little or no visible growth between photographies were identified as putative UV mutants (98). These colonies were isolated and further screened to determine UV sensitivity of colony-forming ability (101). Those mutant lines that are hypersensitive to killing and to mutagenesis by UV light have been analyzed by genetic complementation by measuring the frequency of UV-resistant cells after cell fusion with polyethylene glycol (101). Five classes (or complementation groups) have been identified thus far (101, 102).

Representative mutant clones from each of the five complementation groups have been examined for their ability to incise DNA during post-UV incubation (102). Under conditions in which repair synthesis of DNA is inhibited by the presence of 1-β-D-cytosine arabinoside (ara-C) and hydroxyurea, DNA strand breaks accumulate in wild-type cells and can be readily quantitated (102). Each of the mutants examined shows very low levels of DNA incision (less than 10 percent of that observed in normal CHO cells). In addition, none of the mutants is able to effect the excision of [^{3}H]-7-bromomethylbenz(a)anthracene adducts from DNA (102).

A question of obvious interest is whether or not the defects in human XP cells and in the CHO cell mutants are complementary. Two hybrid clones have been isolated by fusion of a CHO cell mutant designated *UV20* (from CHO complementation group 2) with XP group A and group C cells respectively (102). In both cases the hybrids show enhanced resistance to killing by UV radiation (Fig. 4-29) (102), suggesting that the defect in CHO complementation group 2 is distinct from that in XP groups A and C. More extensive studies of this type will be necessary before it can be established to what extent the human and rodent cell complementation groups overlap. Parenthetically, it should be noted that there may be significant qualitative

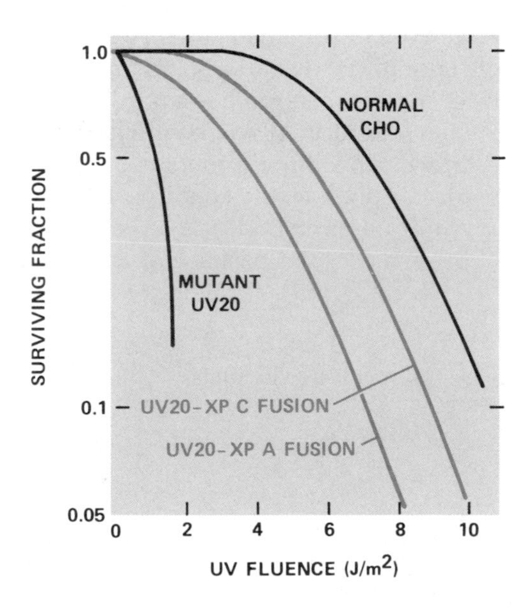

FIGURE 4-29
When a mutant Chinese hamster ovary cell line (UV20) is fused to XP group A or to XP group C cells, there is a significant complementation of the sensitivity to UV killing. This complementation suggests that the defect in UV20 is distinct from that in XP group A and C cells. (From L. Thompson and A. V. Carrano, ref. 102.)

differences in the excision repair of bulky base damage in rodent and human cells. The former cells carry out significantly less excision repair than do human cells, yet they survive equally as well when exposed to comparable levels of UV radiation (3).

Further aspects of the role of chromatin structure in the excision repair of DNA in mammalian cells are considered in the next chapter, which deals with postincision events. However, it should be noted that the differences in the distribution of these events may simply reflect differences in the distribution of incision due to the conformational considerations discussed above.

Damage-Specific DNA Incising Activities in Extracts of Mammalian Cells

Enzyme activities that preferentially degrade UV-irradiated DNA in vitro have been observed in extracts of mammalian cells or tissues (including those from humans) (103). However, with limited exceptions, none of these studies has adequately distinguished between pyrimidine dimers and other photoproducts (including AP sites) as the substrate recognized in DNA. As indicated above, extracts of wild-type human cells in culture can catalyze the selective loss (excision) of thymine-containing pyrimidine dimers from UV-irradiated DNA (103) (Table 4-7). This reaction has a requirement for Mg^{2+} and is stimulated by the addition of deoxynucleoside triphosphates (Table 4-7) (104). Since *excision* of dimers is based on *specific incision* of DNA, the extracts presumably contain a DNA incising activity that recognizes pyrimidine dimers; however, attempts to purify the activity have not been successful because of its extraordinary lability (87).

TABLE 4-7
Thymine dimer excision in cell-free extracts

			Thymine Dimers in Acid-Precipitable DNA (%)	
Experiment	Cell type	Additions*	0 min	60 min
1	HeLa	None	100.0	71.0
2	HeLa	None	100.0	50.0
	HeLa	4 dNTPs	100.0	30.0
3	WIL II	None	100.0	63.4
	WIL II	4 dNTPs	100.0	40.0

*Refers to additions other than crude extract.
From J. Duncan et al., ref. 104.

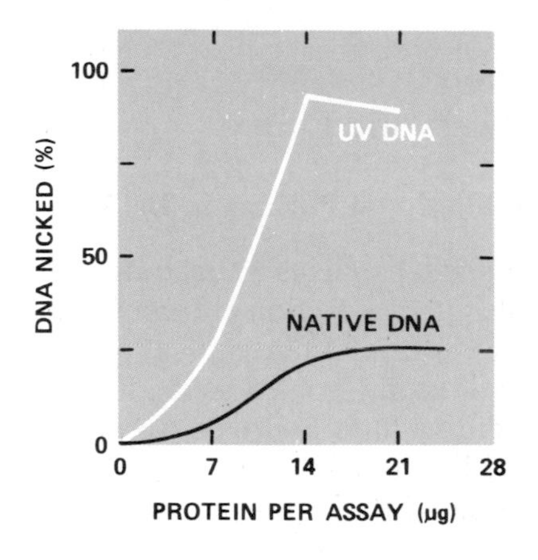

FIGURE 4-30
The preferential nicking of UV-irradiated covalently closed circular PM2 DNA by an extract from calf thymus (From E. A. Waldstein et al., ref. 105.)

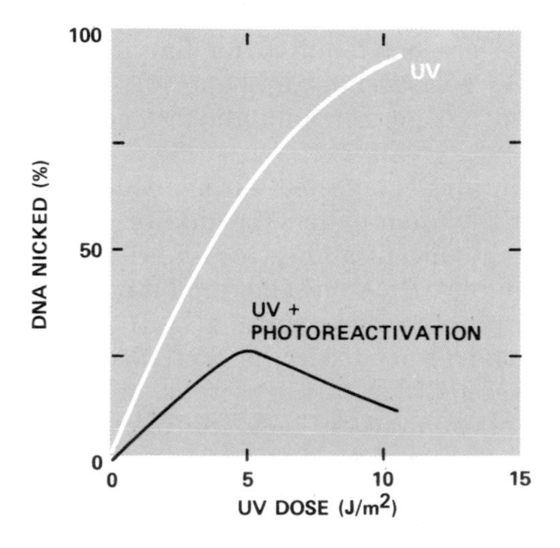

FIGURE 4-31
Prior photoreactivation of UV-irradiated covalently closed circular DNA reduces nicking by the calf thymus activity (see Fig. 4-30), suggesting that at least some fraction of the nicks are specifically at pyrimidine dimers. (From E. A. Waldstein et al., ref. 105.)

An enzyme activity that nicks phage PM2 DNA irradiated with UV doses producing only a few pyrimidine dimers per molecule (thereby largely if not completely eliminating minor photoproducts as possible substrate sites) has also been identified in extracts of calf thymus (105) (Fig. 4-30). This activity is less active on substrate DNA subjected to prior enzymatic photoreactivation, suggesting a specificity for pyrimidine dimers in UV-irradiated DNA (Fig. 4-31). The activity has a requirement for Mg^{2+} and is associated with a protein(s) of high molecular weight, although no numerical estimate of this parameter has been made. The activity is very labile, especially to freezing and thawing, and its purification and detailed characterization have not yet been achieved.

4-4 Other Damage-Specific DNA Incising Activities in Prokaryotic Cells

Endonuclease Activity against Photoalkylated Purines in DNA

An endonuclease activity from *M. luteus* recognizes a substrate produced by photoalkylation of DNA (106, 107). Reaction of form I PM2 DNA with isopropyl alcohol in the presence of a free-radical photoinitiator plus UV light at 300 nm leads to the specific substitution of purine moieties in the DNA, yielding 8-(2-hydroxy-2-propyl)-adenine and 8-(2-hydroxy-2-propyl)-guanine (106, 107) (Fig. 4-32). The endonuclease activity is not competed for by the simultaneous presence of single-strand DNA, single-strand UV- or single-strand γ-irradiated DNA or depurinated duplex DNA. However, the inclusion of *duplex* UV- or γ-irradiated DNA does result in inhibition of activity (Fig. 4-33). Divalent cations are not required for the activity, which is fully active in EDTA. ATP results in about 40 percent inhibition and caffeine in about 30 percent inhibition of activity (106, 107).

No release of free modified purines is detected during incubation of the photoalkylated DNA with *M. luteus* extract. In addition, when substrate is present at saturating concentrations, the addition of a 10-fold excess of depurinated duplex DNA does not inhibit enzyme activity. Both of these observations argue against a DNA glycosylase–AP endonuclease mechanism of incision of the DNA (106, 107). It has therefore been suggested that this activity is an endonuclease that directly attacks phosphodiester bonds in photoalkylated DNA (106, 107). What is the specific substrate recognized by

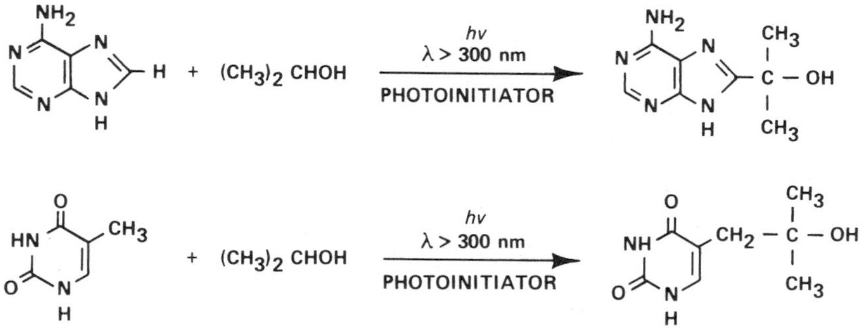

FIGURE 4-32
UV light-induced alkylation of adenine (top) and thymine (bottom) by isopropyl alcohol occurs in the presence of a free-radical photoinitiator. (From Z. Livneh et al., ref. 106.)

this endonuclease? Substitution in the C^8 position of purines by a group as small as bromine results in destabilization of the normal *anti* conformation of the affected nucleotide (108). Physical studies on such substitutions have shown that a stable configuration requires rotation around the N-glycosylic bond to assume the *syn* conformation (108) (Fig. 4-34). It is possible that the *syn* conformation assumed by any 8-substituted purine in DNA provides the molecular basis for the specific substrate recognized by this class of endonucleases. If so, one might expect that the purified enzyme would also attack DNA containing N-acetoxy-N-2-acetyl-2-aminofluorene adducts, since this compound has also been shown to bind at the C^8 position of guanine in DNA (see Section 1-3).

Endonuclease V of *E. Coli*

A small protein that degrades a variety of substrates that share no obvious structural similarity has been purified from *E. coli* and is termed endonuclease V (109, 110). This enzyme has an absolute requirement for $MgCl_2$ for activity. Among the substrates for the purified enzyme are duplex UV-irradiated DNA (at sites *other* than pyrimidine dimers), DNA treated with osmium tetroxide, heat or acid depurinated DNA and DNA containing adducts of 7-bromomethylbenz(a)anthracene (109, 110). In addition, native DNA from phage PBS2 (a DNA that contains uracil instead of thymine; see Section 1-2) and phage T5 DNA containing thymine substituted with uracil are degraded by the enzyme (109, 110). The enzyme also at-

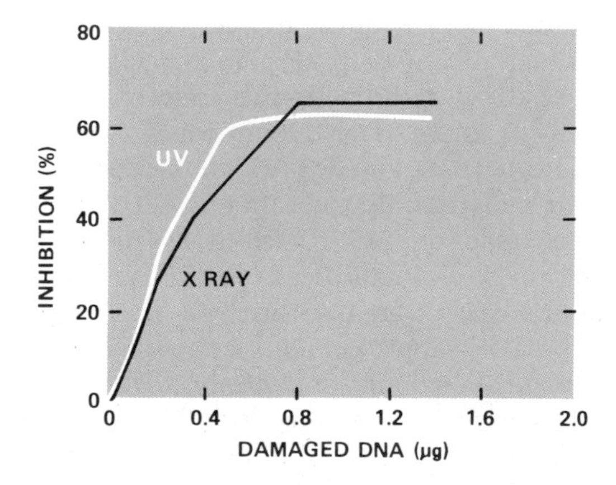

FIGURE 4-33
UV-irradiated (254 nm) or γ-irradiated DNA competes for endonuclease activity against photoalkylated DNA (From Z. Livneh et al., ref. 106.)

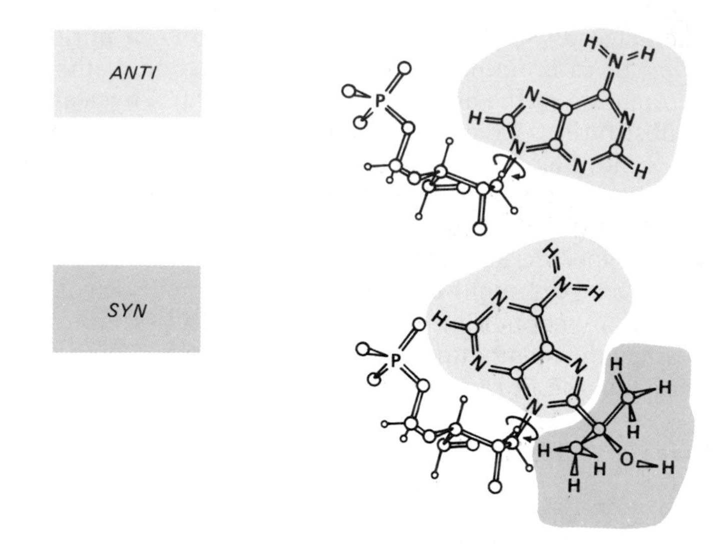

FIGURE 4-34
AMP *anti* conformation (top) and 8-(2-hydroxy-2-propyl)-AMP *syn* conformation (bottom). (From Z. Livneh et al., ref. 106.)

tacks the single-strand DNA from phage fd and duplex fd RFI DNA (109, 110).

It is difficult to identify a common substrate in all of these DNAs and the basis for the apparent broad substrate specificity of endonuclease V remains unclear. An interesting possibility is that the enzyme is single-strand specific and therefore recognizes single-strand areas in duplex DNA created by relatively nonspecific base damage. Endonuclease V is optimally active at pH 9.5, with little if any detectable activity at neutral pH. The alkaline pH optimum might facilitate helix destabilization in areas of localized conformational distortion in duplex DNA. This could explain the observed degradation of PBS2 DNA, since at pH 9.5 the disruption of A·U base pairs may occur more readily than that of A·T base pairs.

When enzyme reactions with various DNAs are sedimented in alkali to monitor single-strand breaks (nicks) and in neutral sucrose to measure double-strand breaks, the ratio of single-strand nicks to double-strand breaks is approximately 8:1 over a wide range of nicks and a sixfold range of enzyme concentration (Table 4-8) (110). Thus, the relatively low number of double-strand breaks do not arise from the random juxtaposition of two nicks, nor from an enzyme concentration-dependent alteration in the mechanism of endonucleolytic cleavage of DNA (110).

Endonuclease V has been detected at normal levels in all *E. coli* mutants examined, including a number of strains defective in func-

TABLE 4-8
Nicking of DNA by endonuclease V

Amount of Endonuclease (μL)	Total Double-Strand Breaks (fmol)	Total Single-Strand Breaks (fmol)	Ratio of Single-Strand to Double-Strand Breaks
5	107	920	8.6
10	201	1250	6.2
20	434	3720	8.6
30	497	3840	7.7

From B. Demple and S. Linn, ref. 110.

tions required for replication, recombination and repair. Although the multiplicity of substrates recognized by this activity might be interpreted in terms of its role in DNA repair, the determination of the true cellular function of endonuclease V must await the isolation of mutants defective in the activity (110).

Summary and Conclusion

Direct-acting endonucleases such as the uvrABC complex of *E. coli*, and probably the products of certain *RAD* genes in yeast and of the genes in human cells that are defective in XP patients, are singularly important enzymes in excision repair. Cellular-biological studies with mutants defective in the genes coding for these products suggest that these endonucleases are required for the incision of DNA containing a large spectrum of base damage. Thus it is probable that the nucleotide excision repair pathway initiated by these damage-specific enzymes is a much more general mode for excision repair of base damage than the base excision repair mechanism in which limited types of base damage are excised by specific DNA glycosylases.

Despite the fact that the excision of pyrimidine dimers in *E. coli* has been a principal model for the study of excision repair for many years, the promise of understanding the biochemical complexity of the "repairosome" in *E. coli* and in higher organisms is just beginning to emerge. The availability of highly purified gene products encoded by genes cloned into recombinant DNA molecules has unquestionably overcome the quantitative problems that hampered progress in this area of nucleic acid enzymology for so long. However, it is possible that the stoichiometry with which the *uvrA*, *uvrB* and *uvrC* gene products associate is critical in determining the functional state of the complex, and it remains to be seen whether reconstitution of these gene products in vitro will lead to catalytic activity that truly reflects the mechanism of incision of damaged DNA in vivo. Thus it is possible that in living *E. coli* cells different stoichiometric relations between the *uvrA*, *uvrB* and *uvrC* gene products determine different catalytic activities. The formation of distinctive putative protein complexes of these gene products is likely to be under strict regulatory control. Clearly the molecular mechanism(s) of the incision of damaged DNA by direct-acting specific endonucleases is a crucial area of DNA repair, about which more should be learned.

Suggestions for Further Reading

Excision repair in general, refs. 3, 4; the *E. coli* uvrABC endonuclease, refs. 21, 25, 57, 59, 61, 63; excision repair in yeast, refs. 84, 87; excision repair in *Drosophila*, ref. 94.

References

1. Friedberg, E. C., and Hanawalt, P. C., eds. 1981. *DNA repair—a laboratory manual of research procedures*. Vol. 1, part A. New York: Dekker.
2. Friedberg, E. C., and Hanawalt, P. C. eds. 1983. *DNA repair–a laboratory manual of research procedures*. Vol. 2. New York: Dekker.
3. Hanawalt, P. C., Cooper, P. K., Ganesan, A. K., and Smith, C. A. 1979. DNA repair in bacteria and mammalian cells. *Ann. Rev. Biochem.* 48:783.
4. Lehmann, A. R., and Karran, P. 1981. DNA repair. *Int. Rev. Cytol.* 72:101.
5. Friedberg, E. C., and Bridges, B. A., eds. 1983. *Cellular responses to DNA damage*. New York: Alan R. Liss.
6. Setlow, R. B., and Carrier, W. L. 1964. The disappearance of thymine dimers from DNA: an error-correcting mechanism. *Proc. Natl. Acad. Sci.* (USA) 51:226.
7. Boyce, R. P., and Howard-Flanders, P. 1964. Release of ultraviolet light–induced thymine dimers from DNA in *E. coli*. *Proc. Natl. Acad. Sci.* (USA) 51:293.
8. Setlow, R. B., Swenson, P. A., and Carrier, W. L. 1963. Thymine dimers and inhibition of DNA synthesis by ultraviolet irradiation of cells. *Science* 142:1464.
9. Howard-Flanders, P., Boyce, R. P., and Theriot, R. 1966. Three loci in *Escherichia coli* K-12 that control the excision of thymine dimers and certain other mutagen products from host or phage DNA. Genetics 53:1119.
10. Van de Putte, P., van Sluis, C. A., van Dillewijn, J., and Rörsch, A. 1965. The location of genes controlling radiation sensitivity in *E. coli*. *Mutation Res.* 2:97.
11. Mattern, I. E., van Winden, M. P., and Rörsch, A. 1965. The range of action of genes controlling radiation sensitivity in *E. coli*. *Mutation Res.* 2:111.
12. Howard-Flanders, P., and Boyce, R. P. 1966. DNA repair and genetic recombination: studies on mutants of *Escherichia coli* defective in these processes. *Radiat. Res. Supp.* 6:156.

13. Shimada, K., Ogawa, H., and Tomizawa, J. 1968. Studies on radiation-sensitive mutants of *E. coli*. I. Breakage and repair of ultraviolet irradiated intracellular DNA of phage lambda. *Mol. Gen. Genet.* 101:245.

14. Rupp, W. D., and Howard-Flanders, P. 1968. Discontinuities in the DNA synthesized in an excision-defective strain of *Escherichia coli* following ultraviolet irradiation. *J. Mol. Biol.* 31:291.

15. Seeberg, E., Rupp, W. D., and Strike, P. 1980. Impaired incision of ultraviolet-irradiated deoxyribonucleic acid in *uvrC* mutants of *Escherichia coli*. *J. Bacteriol.* 144:97.

16. Seeberg, E., Nissen-Meyer, J., and Strike, P. 1976. Incision of ultraviolet-irradiated DNA by extracts of *E. coli* requires three different gene products. *Nature* 263:524.

17. Seeberg, E. 1978. Reconstitution of an *Escherichia coli* repair endonuclease activity from the separated *uvrA*$^+$ and *uvrB*$^+$/*uvrC*$^+$ gene products. *Proc. Natl. Acad. Sci.* (USA) 75:2569.

18. Pettijohn, D., and Hanawalt, P. 1963. Deoxyribonucleic acid replication in bacteria following ultraviolet irradiation. *Biochim. Biophys. Acta* 72:127.

19. Pettijohn, D., and Hanawalt, P. 1964. Evidence for repair-replication of ultraviolet damaged DNA in bacteria. *J. Mol. Biol.* 9:395.

20. Murray, M. L. 1979. Substrate-specificity of *uvr* excision repair. *Environ. Mutagen.* 1:347.

21. Seeberg, E. 1981. Multiprotein interactions in strand cleavage of DNA damaged by UV and chemicals. *Prog. Nuc. Acid Res. Mol. Biol.* 26:217.

22. Sancar, A., Wharton, R. P., Seltzer, S., Kacinski, B. M., Clarke, N. D., and Rupp, W. D. 1981. Identification of the *uvrA* gene product. *J. Mol. Biol.* 148:45.

23. Sancar, A., Clarke, N. D., Griswold, J., Kennedy, W. J., and Rupp, W. D. 1981. Identification of the *uvrB* gene product. *J. Mol. Biol.* 148:63.

24. Yoakum, G. H., and Grossman, L. 1981. Identification of the *E. coli uvrC* protein. *Nature* 292:171.

25. Rupp, W. D., Sancar, A., and Sancar, G. B. 1982. Properties and regulation of the UVR ABC endonuclease. *Biochimie* 64:595.

26. Rupp, W. D., Sancar, A., Kennedy, W., Ayers, J., and Griswold, J. 1978. Cloning of *E. coli* DNA repair genes. In *DNA repair mechanisms*. P. C. Hanawalt, E. C. Friedberg and C. F. Fox, eds. p. 229. New York: Academic.

27. Pannekoek, H., Noordermeer, I. A., van Sluis, C. A., and van de Putte, P. 1978. Expression of the *uvrB* gene of *Escherichia coli: in vitro* construction of a pMB9 *uvrB* plasmid. *J. Bacteriol.* 133:884.

28. Sancar, A., and Rupp, W. D. 1979. Cloning of *uvrA*, *lexC* and *ssb* genes of *Escherichia coli*. *Biochem. Biophys. Res. Comm.* 90:123.

29. Auerbach J., and Howard-Flanders, P. 1979. The isolation and genetic characterization of lambda transducing phages of the *uvrA*$^+$ and *uvrC*$^+$ genes of *E. coli* K-12. *Mol. Gen. Genet.* 168:341.

30. Yoakum, G. H., Kushner, S. R., and Grossman, L. 1980. Isolation of plasmids carrying either the *uvrC*, or *uvrC uvrA* and *ssb* genes of *Escherichia coli* K-12. *Gene* 12:243.

31. Sancar, A., Kacinski, B. M., Mott, D. L., and Rupp, W. D. 1981. Identification of the *uvrC* gene product. *Proc. Natl. Acad. Sci.* (USA) 78:5450.

32. Brandsma, J. A., van Sluis, C. A., and van de Putte, P. 1981. Use of transposons in cloning poorly selectable genes of *Escherichia coli:* cloning of *uvr*A and adjacent genes. *J. Bacteriol.* 147:682.

33. Van den Berg, E., Zwetsloot, J., Noordermeer, I., Pannekoek, H., Dekker, B., Dijkema, R., and van Ormondt, H. 1981. The structure and function of the regulatory elements of the *Escherichia coli uvr*B gene. *Nucleic Acids Res.* 9:5623.

34. Van Sluis, C. A., and Brandsma, J. A. 1981. Plasmids carrying the *uvr*A and *uvr*C genes of *Escherichia coli* K-12: construction and properties. In *Chromosomal damage and repair.* E. Seeberg and K. Kleppe, eds., p. 293. New York: Plenum.

35. Blingsmo, O. R., Steinum, A. L., Rivedal, E. and Seeberg, E. 1981. Cloning of the *uvr*C^+ gene from *Escherichia coli* onto a plaque-forming phage vector. In *Chromosomal damage and repair.* E. Seeberg and K. Kleppe, eds., p. 303. New York: Plenum.

36. Sharma, S., Ohta, A., Dowhan, W., and Moses, R. E. 1981. Cloning of the *uvr*C gene of *Escherichia coli:* expression of a DNA repair gene. *Proc. Natl. Acad. Sci.* (USA) 78:6033.

37. Sharma, S., and Moses, R. E. 1983. Identification of the regulatory regions of the *uvr*C gene of *Escherichia coli.* In *Cellular responses to DNA damage.* E. C. Friedberg and B. A. Bridges, eds., p. 145. New York: Alan R. Liss.

38. Sharma, S., Dowhan, W., and Moses, R. E. 1982. Molecular structure of the *uvr*C gene of *Escherichia coli:* identification of DNA sequences required for transcription of the *uvr*C gene. *Nucleic Acids Res.* 10:5209.

39. Sancar, A., and Rupp, W. D. 1983. The *uvr*A and *uvr*B genes of *E. coli:* use of the maxicell procedure for detecting gene expression. In *DNA repair—a laboratory manual of research procedures.* E. C. Friedberg and P. C. Hanawalt, eds., vol. 2, p. 253. New York: Dekker.

40. Van Sluis, C. A., and Dubbeld, D. 1983. Cloning the *uvr*C gene of *Escherichia coli.* In *DNA repair—a laboratory manual of research procedures.* E. C. Friedberg and P. C. Hanawalt, eds., vol. 2, p. 267. New York: Dekker.

41. Kacinski, B. M., Sancar, A., and Rupp, W. D. 1981. A general approach for purifying proteins encoded by cloned genes without using a functional assay: isolation of the *uvr*A gene product from radiolabeled maxicells. *Nucleic Acids Res.* 9:4495.

42. Yoakum, G. H., Yeung, A. T., Mattes, W. B., and Grossman, L. 1982. Amplification of the *uvr*A gene product of *Escherichia coli* to 7% of cellular protein by linkage of the P_L promoter of pKC30. *Proc. Natl. Acad. Sci.* (USA) 79:1766.

43. Sancar, A., Sancar, G. B., Rupp, W. D., Little, J. W., and Mount, D. W. 1982. LexA protein inhibits transcription of the *E. coli uvr*A gene *in vitro. Nature* 298:96.

44. Radman, M. 1975. SOS repair hypothesis: phenomenology of an inducible DNA repair which is accompanied by mutagenesis. In *Molecular mechanisms for repair of DNA.* P. C. Hanawalt and R. B. Setlow, eds., p. 355. New York: Plenum.

45. Witkin, E. 1976. Ultraviolet mutagenesis and inducible DNA repair in *Escherichia coli. Bacteriol. Rev.* 40:869.

46. Walker, G. C., Elledge, S. J., Kenyon, C., Krueger, J. H., and Perry, K. L. 1982. Mutagenesis and other responses induced by DNA damage in *Escherichia coli. Biochimie* 64:607.

47. Fiandt, M., Gottesman, M. E., Shulman, M. J., Szybalski, E. H., Szybalski, W. and Weisberg, R. A. 1976. Physical mapping of coliphage λatt^2. *Virology* 72:6.

48. Fogliano, M., and Schendel, P. F. 1981. Evidence for the inducibility of the *uvr*B operon. *Nature* 289:196.

49. Schendel, P. F., Fogliano, M., and Strausbaugh, L. D. 1982. Regulation of the *Escherichia coli* K-12 *uvr*B operon. *J. Bacteriol.* 150:676.

50. Sancar, G. B., Sancar, A., Little, J. W., and Rupp, W. D. 1982. The *uvr*B gene of *Escherichia coli* has both *lex*A-repressed and *lex*A-independent promoters. *Cell* 28:523.

51. Backendorf, C. M., van den Berg, E. A., Brandsma, J. A., Kartasova, T., van Sluis, C., and van de Putte, P. 1983. *In vivo* regulation of the *UVR* and *SSB* genes in *Escherichia coli*. In *Cellular responses to DNA damage*. E. C. Friedberg and B. A. Bridges, eds., p. 161. New York: Alan R. Liss.

52. Van den Berg, E. A., Geerse, R. H., Pannekoek, H., and van de Putte, P. 1983. *In vivo* transcription of the *E. coli uvrB* gene: both promoters are inducible by UV. *Nucleic Acids Res.* 11:4355.

53. Sancar, A., Hack, A. M., and Rupp, W. D. 1979. Simple method for identification of plasmid-coded proteins. *J. Bacteriol.* 137:692.

54. Howard-Flanders, P. 1968. Genes that control DNA repair and genetic recombination in *Escherichia coli*. *Adv. Biol. Med. Phys.* 12:299.

55. Meagher, R. B., Tait, R. C., Betlach, M., and Boyer, H. W. 1977. Protein expression in *E. coli* minicells by recombinant plasmids. *Cell* 10:521.

56. Seeberg, E., and Steinum, A-L. 1982. Purification and properties of the uvrA protein from *Escherichia coli*. *Proc. Natl. Acad. Sci.* (USA). 79:988.

57. Sancar, A., and Rupp, W. D. 1983. A novel repair enzyme: UVRABC excision nuclease of *Escherichia coli* cuts a DNA strand on both sides of the damaged region. *Cell* 33:249.

58. Kacinski, B. M., and Rupp, W. D. 1981. *E. coli* uvrB protein binds to DNA in the presence of uvrA protein. *Nature* 294:480.

59. Seeberg, E., and Steinum, A-L. 1983. Properties of the uvrABC endonuclease from *E. coli*. In *Cellular responses to DNA damage*. E. C. Friedberg and B. A. Bridges, eds., p. 39. New York: Alan R. Liss.

60. Seeberg, E., Steinum, A-L., and Blingsmo, O. R. 1982. Two separable protein species which both restore uvrABC endonuclease activity in extracts from *uvr*C mutated cells. *Biochimie* 64:825.

61. Yeung, A. T., Mattes, W. B., Oh, E. Y., and Grossman, L. 1983. Enzymatic properties of the purified *Escherichia coli* uvrABC complex. In *Cellular responses to DNA damage*. E. C. Friedberg and B. A. Bridges, eds., p. 77. New York: Alan R. Liss.

62. Yeung, A. T., Mattes, W. B., Oh, E. Y., and Grossman, L. 1983. Enzymatic properties of purified *Escherichia coli* uvrABC proteins. *Proc. Natl. Acad. Sci.* (USA) 80:6157.

63. Grossman, L. 1983. DNA repair enzymes: workshop summary. In *Cellular responses to DNA damage*. E. C. Friedberg and B. A. Bridges, eds., p. 331. New York: Alan R. Liss.

64. Seeberg, E., Steinum, A-L, Nordenskjöld, M., Soderhall, S., and Jernstrom, B. 1983. Strand break formation in benz(a)pyrene diolepoxide modified DNA: quantitative cleavage by the *Escherichia coli* uvrABC endonuclease. *Mutation Res.* 112:139.

65. Wilkins, R. J., and Hart, R. W. (1974). Preferential DNA repair in human cells. *Nature* 247:35.

66. Van Zeeland, A. A., Smith, C. A., and Hanawalt, P. C. 1981. Sensitive determination of pyrimidine dimers in DNA of UV-irradiated mammalian cells. Introduction of T4 endonuclease V into frozen and thawed cells. *Mutation Res.* 82:173.

67. Setlow, R. B. 1978. Repair deficient human disorders and cancer. *Nature* 271:713.

68. Friedberg, E. C., Ehmann, U. K., and Williams, J. I. 1979. Human diseases associated with defective DNA repair. *Adv. Rad. Biol.* 8:86.

69. Kohn, K., Ewig, R. A. G., Erickson, L. C., and Zwelling, L. A. 1981. Measurement of DNA strand breaks and cross-links by alkaline elution. In *DNA repair—a laboratory manual of research procedures.* E. C. Friedberg and P. C. Hanawalt, eds., vol. 1, part B, p. 379. New York: Dekker.

70. Fornace, A. J., Jr., Kohn, K. W., and Kann, H. E., Jr., 1976. DNA single-strand breaks during repair of UV damage in human fibroblasts and abnormalities in xeroderma pigmentosum. *Proc. Natl. Acad. Sci.* (USA) 73:39.

71. Tanaka, K., Sekiguchi, M., and Okada, Y. 1975. Restoration of ultraviolet-induced unscheduled DNA synthesis of xeroderma pigmentosum cells by the concomitant treatment with bacteriophage T4 endonuclease V and HVJ (Sendai virus). *Proc. Natl. Acad. Sci.* (USA) 72:4071.

72. Paterson, M. C., Smith, B. P., and Smith, P. J. 1981. Measurement of enzyme-sensitive sites in UV- or γ- irradiated human cells using *Micrococcus luteus* extracts. In *DNA repair—a laboratory manual of research procedures.* E. C. Friedberg and P. C. Hanawalt, eds., vol. 1, part A, p. 99. New York: Dekker.

73. Zelle, B., and Lohman, P. H. M. 1979. Repair of UV-endonuclease-susceptible sites in the 7 complementation groups of xeroderma pigmentosum A through G. *Mutation Res.* 62:363.

74. Mortelmans, K., Friedberg, E. C., Slor, H., Thomas, G., and Cleaver, J. E. 1976. Evidence for a defect in thymine dimer excision in extracts of xeroderma pigmentosum cells. *Proc. Natl. Acad. Sci.* (USA) 73:2757.

75. Friedberg, E. C., Rudé, J. M., Cook, K. H., Ehmann, U. K., Mortelmans, K., Cleaver, J. E., and Slor, H. 1977. Excision repair in mammalian cells and the current status of xeroderma pigmentosum. In *DNA repair processes.* W. W. Nichols and D. G. Murphy, eds., p. 21. Miami: Symposia Spec., Inc.

76. Kano, Y., and Fujiwara, Y. 1983. Defective thymine dimer excision from xeroderma pigmentosum chromatin and its characteristic catalysis by cell-free extracts. *Carcinogenesis* 4:1419.

77. Fujiwara, Y., and Kano, Y. 1983. Characteristics of thymine dimer excision from xeroderma pigmentosum chromatin. In *Cellular responses to DNA damage.* E. C. Friedberg and B. A. Bridges, eds., p. 215. New York: Alan R. Liss.

78. Mansbridge, J. N., and Hanawalt, P. C. 1983. Domain-limited repair of DNA in ultraviolet irradiated fibroblasts from xeroderma pigmentosum complementation group C. In *Cellular responses to DNA damage.* E. C. Friedberg and B. A. Bridges, eds., p. 195. New York: Alan R. Liss.

79. Takano, T., Noda, M., and Tamura, T-A. 1982. Transfection of cells from a xeroderma pigmentosum patient with normal human DNA confers UV resistance. *Nature* 296:269.

80. Protic-Sabljic, M., Whyte, D. B., Fagan, J., and Kraemer, K. 1983. Transfection of xeroderma pigmentosum cells with cloned DNA. In *Cellular responses to DNA damage*. E. C. Friedberg and B. A. Bridges, eds. p. 647. New York: Alan R. Liss.

81. Ganesan, A., Spivak, G., and Hanawalt, P. 1983. Expression of DNA repair genes in mammalian cells. In *Manipulation and expression of genes in eukaryotics*. P. Nagley, A. W. Linnar, W. J. Peacock and J. A. Pateman, eds., p. 45. Australia: Academic.

82. Hoeijmakers, J. H. J., Zwetsloot, J. C. M., Vermeulen, W., de Jonge, A. J. R., Backendorf, C., Klein, B., and Bootsma, D. 1983. Phenotypic correction of xeroderma pigmentosum cells by microinjection of crude extracts and purified proteins. In *Cellular responses to DNA damage*. E. C. Friedberg and B. A. Bridges, eds., p. 173. New York: Alan R. Liss.

83. De Jonge, A. J. R., Vermeulen, W., Klein, B., and Hoeijmakers, J. H. J. 1983. Microinjection of human cell extracts corrects xeroderma pigmentosum defect. *The EMBO J.* 2:637.

84. Haynes, R. H., and Kunz, B. A. 1981. DNA repair and mutagenesis in yeast. In *The molecular biology of the yeast saccharomyces. Life cycle and inheritance*. J. Strathern, E. Jones and J. Broach, eds., p. 371. New York: Cold Spring Harbor Lab.

85. Reynolds, R. J., and Friedberg, E. C. 1981. Molecular mechanisms of pyrimidine dimer excision in *Saccharomyces cerevisiae:* incision of ultraviolet irradiated deoxyribonucleic acid *in vivo*. *J. Bacteriol.* 146:692.

86. Wilcox, D. R., and Prakash, L. 1981. Incision and post-incision steps of pyrimidine dimer removal in excision-defective mutants of *Saccharomyces cerevisiae*. *J. Bacteriol.* 148:618.

87. Friedberg, E. C., Naumovski, L., Yang, E., Pure, G., Schultz, R. A., Weiss, W., and Love, J. D. 1983. Approaching the biochemistry of excision repair in eukaryotic cells: the use of cloned genes from *Saccharomyces cerevisiae*. In *Cellular responses to DNA damage*. E. C. Friedberg and B. A. Bridges, eds., p. 63. New York: Alan R. Liss.

88. Bekker, M. L., Kaboev, O. K., Akhmedov, A. T., and Luchinka, L. A. 1980. Ultraviolet-endonuclease activity in cell extracts of *Saccharomyces cerevisiae* mutants defective in excision of pyrimidine dimers. *J. Bacteriol.* 142:322.

89. Botstein, D., and Davis, R. W. 1982. Principles and practice of recombinant DNA research with yeast. In *The molecular biology of the yeast saccharomyces. Metabolism and gene expression*. J. N. Strathern, E. W. Jones and J. R. Broach, eds., p. 607. New York: Cold Spring Harbor Lab.

90. Naumovski, L., and Friedberg, E. C. 1982. Molecular cloning of eukaryotic genes required for excision repair of UV-irradiated DNA: isolation and partial characterization of the *RAD*3 gene of *Saccharomyces cerevisiae*. *J. Bacteriol.* 152:323.

91. Naumovski, L., and Friedberg, E. C. 1983. A DNA repair gene required for the incision of damaged DNA is essential for viability in *S. cerevisiae*. *Proc. Natl. Acad. Sci.* (USA) 80:4818

92. Higgins, D. R., Prakash, S., Reynolds, P., Polakowska, R., Weber, S., and Prakash, L. 1983. Isolation and characterization of the *RAD3* gene of *Saccharomyces cerevisiae* and inviability of *rad3* deletion mutants. *Proc. Natl. Acad. Sci.* (USA) 80:5680.

93. Naumovski, L., and Friedberg, E. C. 1984. The *RAD2* gene of *Saccharomyces cerevisiae:* isolation, subcloning and partial characterization. *Mol. Cell. Biol.* 4:290.

94. Boyd, J. B., Harris, P. V., Presley, J. M., and Narachi, M. 1983. *Drosophila melanogaster:* a model eukaryote for the study of DNA repair. In *Cellular responses to DNA damage.* E. C. Friedberg and B. A. Bridges, eds., p. 107. New York: Alan R. Liss.

95. Nguyen, T. D., and Boyd, J. B. 1977. The meiotic-9 (*mei-9*) mutants of *Drosophila melanogaster* are deficient in repair replication of DNA. *Molec. Gen. Genet.* 158:141.

96. Boyd, J. B., Snyder, R. D., Harris, P. V., Presley, J. M., Boyd, S. F., and Smith, P. D. 1982. Identification of a second locus in *Drosophila melanogaster* required for excision repair. *Genetics* 100:239.

97. Harris, P. V., and Boyd, J. B. 1980. Excision repair in *Drosophila*, analysis of strand breaks appearing in DNA of *mei-9* mutants following mutagen treatment. *Biochem. Biophys. Acta* 610:116.

98. Busch, D. B., Cleaver, J. E., and Glaser, D. A. 1980. Large-scale isolation of UV-sensitive clones of CHO cells. *Somat. Cell. Genet.* 6:407.

99. Thompson, L. H., Rubin, J. S., Cleaver, J. E., Whitmore, G. F., and Brookman, K. 1980. A screening method for isolating DNA repair-deficient mutants of CHO cells. *Somat. Cell. Genet.* 6:391.

100. Shiomi, T., Hieda-Shiomi, N., and Sato, K. 1982. Isolation of UV-sensitive mutants of mouse LS178Y cells by a cell suspension spotting method. *Somat. Cell Genet.* 8:329.

101. Thompson, L. H., Busch, D. B., Brookman, K., Mooney, C. L., and Glaser, D. A. 1981. Genetic diversity of UV-sensitive DNA repair mutants of Chinese hamster ovary cells. *Proc. Natl. Acad. Sci.* (USA) 78:3734.

102. Thompson, L., and Carrano, A. V. 1983. Analysis of mammalian cell mutagenesis and DNA repair using *in vitro* selected CHO cell mutants. In *Cellular responses to DNA damage.* E. C. Friedberg and B. A. Bridges, eds., p. 125. New York: Alan R. Liss.

103. Friedberg, E. C., Cook, K. H., Duncan, J., and Mortelmans, K. 1977. DNA repair enzymes in mammalian cells. *Photochem. Photobiol. Rev.* 2:263.

104. Duncan, J., Slor, H., Cook, K., and Friedberg, E. C. 1975. Thymine dimer excision by extracts of human cells. In *Molecular mechanisms for repair of DNA.* P. C. Hanawalt and R. B. Setlow, eds. p. 643. New York: Plenum Press.

105. Waldstein, E. A., Peller, S., and Setlow, R. B. UV-endonuclease from calf thymus with specificity toward pyrimidine dimers in DNA. *Proc. Natl. Acad. Sci.* (USA) 76:3746.

106. Livneh, Z., Elad, D., and Sperling, J. 1979. Endonucleolytic activity directed towards 8-(2-hydroxy-2-propyl)-purines in double-stranded DNA. *Proc. Natl. Acad. Sci.* (USA) 76:5500.

107. Livneh, Z., and Sperling, J. 1981. An endonucleolytic activity from *M. luteus* directed towards 8-(2-hydroxy-2-propyl)-purines in DNA. In *Chromosome damage and repair.* E. Seeberg and K. Kleppe, eds., p. 159. New York: Plenum.

108. Howard, F. B., Frazier, J., and Miles, H. T. 1975. Poly (8-bromoadenylic acid): synthesis and characterization of an all-*syn* polynucleotide. *J. Biol. Chem.* 250:3951.

109. Gates, F. T., III, and Linn, S. 1977. Endonuclease V of *Escherichia coli. J. Biol. Chem.* 252:1647.

110. Demple, B., and Linn, S. 1982. On the recognition and cleavage mechanism of *Escherichia coli* endodeoxyribonuclease V, a possible DNA repair enzyme. *J. Biol. Chem.* 257:2848.

5

Excision Repair. III. Postincision Events and Mismatch Repair in Prokaryotic Cells

5-1 Introduction

The previous two chapters have been devoted to a consideration of enzymes that are specific for the excision repair of DNA damage. The sequential action of DNA glycosylases and AP endonucleases, or the action of direct-acting damage-specific DNA incising activities, leads to the incision of damaged DNA and the formation of strand breaks or nicks. It will be recalled from Section 4-2 that recent studies with the uvrABC endonuclease of *E. coli* in vitro (1–5) show the presence of *two* breaks on the same DNA strand; one on either side of each lesion. An obvious implication of this observation is that in vivo the products of the *uvrA*, *uvrB* and *uvrC* genes of *E. coli* may effect *excision* of base damage without the intervention of any further degradative enzymes. In vitro this direct excisional mode results in the release of DNA fragments that are only about 12 nucleotides long (1–5). Yet, as will be noted in this chapter, some of the "patches" created by repair synthesis in UV-irradiated cells are significantly longer than 12 nucleotides. Thus, apparently exonucleolytic degradation during or preceding repair synthesis of DNA is sometimes an integral component of excision repair, even if such degradation does not directly effect the physical excision of damaged bases. On the other hand, the incision of DNA by the sequential action of DNA glycosylases and AP endonucleases (see Chapter 3) creates only a *single* incision in duplex DNA in relation to each substrate lesion. Under these circumstances excision of the residual base damage (either AP sites or, in the case of pyrimidine dimers, apyrimidinic sites with dimerized pyrimidines still covalently linked to the DNA) apparently *always* requires postincision degradation of the DNA, usually by *exonuclease* activities that are physically distinct from the DNA incising activity.

This chapter focuses on *postincision* events during excision repair of DNA in prokaryotic cells. For the purposes of this discussion such events are defined as those that occur subsequent to the incision of DNA and that require the action of gene products *distinct* from those utilized for incision. By these criteria the excision of oligonucleotides containing bulky base damage that is coupled to incision catalyzed by the uvrABC endonuclease of *E. coli* is not considered a postincision event. Should future studies demonstrate that following incision on each side of lesions in DNA, other gene products are utilized in vivo to effect the release of the oligonucleotide fragments spanned by these nicks, this definition of incision and postincision events will require revision. This chapter therefore includes a consideration of the exonuclease-catalyzed excision of damaged nucleotides (e.g., pyrimidine dimers) and of apurinic and apyrimidinic sites that follows the incision of DNA by DNA glycosylase–AP endonucleases or by AP endonuclease action, as well as a consideration of the occasional apparent exonucleolytic degradation of DNA at gaps created by the incision of DNA by the uvrABC endonuclease of *E. coli* (see Figs. 3-1, 3-2 and 3-4). In addition, this chapter will consider the processes of *repair synthesis* (whereby the gaps in the DNA duplex created by the excision of nucleotides are filled in by the action of DNA polymerases) and of *DNA ligation* (in which the final covalent integrity of the genome is restored and excision repair is completed) (Fig. 3-3). Finally, this chapter addresses as a separate topic the repair of mismatched bases in prokaryotic cells. The biochemistry of postincision events and of mismatch repair in higher organisms is discussed in Chapter 6.

5-2 Measurement of Postincision Events in Excision Repair

Before discussing the cell biology and biochemistry of postincision events during excision repair, it is useful to know something about the more frequently used experimental techniques for detecting and measuring these events, since an understanding of at least the principles of these techniques provides the essential basis for a critical evaluation of some of the mechanistic models of excision repair that have been postulated.

Postincision Excision of Damaged Bases

Loss of Radiolabeled Pyrimidine Dimers from DNA

As indicated in Chapter 4, in bacteriophage T4-infected *E. coli* and in *M. luteus* the incision of DNA containing pyrimidine dimers is effected by the sequential action of a PD DNA glycosylase and one or more AP endonucleases (6–9). The excision of dimers from DNA incised in this way thus requires exonuclease-catalyzed degradation of the DNA at the sites of incision. In DNA radiolabeled in thymine (usually with ^{14}C or ^{3}H) the excision event can be readily monitored by demonstrating the transfer of labeled thymine-containing pyrimidine dimers to an ethanol- or acid-soluble fraction, since high molecular weight DNA is precipitated by ethanol or acid. It is of course necessary to resolve the radiolabel specifically associated with dimers from that associated with nondimer nucleotides that are also transferred to the soluble phase after precipitation of DNA. A number of techniques for achieving this resolution (10–13) are mentioned in Section 2-2.

These techniques do, however, have a number of limitations that must be kept in mind. For one thing, it is very difficult to reliably detect the excision of thymine-containing pyrimidine dimers from DNA when these lesions constitute less than approximately 0.05 percent of the total radioactivity in thymine. Yet many experiments require irradiation of cells at doses that produce less than this level. Secondly, since acid- or ethanol-soluble oligonucleotides containing pyrimidine dimers may be excised concomitantly with the incision of DNA in uninfected *E. coli* (and perhaps in other prokaryotic organisms that do not contain a PD DNA glycosylase for excision repair of dimers), the loss of dimers from high molecular weight DNA in these cases may *not* reflect a postincision event as defined here. Indeed, in general the demonstration of the transfer of *any* radiolabeled damaged bases from the insoluble to the soluble fraction of DNA does not distinguish excision that is coupled to incision from excision that is effected by postincision enzymatic events. Furthermore, unless specific identification is made of the nature of the soluble damaged bases, such techniques also do not distinguish the excision of bases as free bases from their excision as nucleotides.

Loss of Sites in DNA Sensitive to Specific Enzyme Probes

Another technique for monitoring postincision excision of pyrimidine dimers was mentioned in Section 4-3. The *M. luteus* and phage T4 "UV endonucleases" are specific for dimers in DNA (see Section 3-4) (6–9). If these lesions are removed during excision repair and the covalent integrity of the DNA is restored by repair synthesis and DNA ligation in living cells, isolated DNA will obviously no longer be sensitive to incision by these probes. Thus, when such DNA is incubated with one of these enzymes and then denatured and sedimented in alkaline sucrose gradients, it will have a higher single-strand molecular weight than DNA in which dimers are still present (Fig. 5-1) (14, 15). This technique is quantitatively more sensitive than the direct measurement of the loss of thymine-containing pyrimidine dimers from DNA and is particularly well suited to studies in which cells are exposed to very low levels of UV radiation. Furthermore, the loss of enzyme-sensitive sites measures the excision of *all* pyrimidine dimers and hence is not limited to those containing thymine. In principle, this technique is applicable to a consideration of any lesions in DNA for which specific enzyme probes exist. However, like the measurement of the transfer of damaged bases to the soluble fraction of DNA, it does not yield information about the precise mechanism of the excision repair.

Appearance and Disappearance of Strand Breaks in DNA

Damage-specific incision of DNA can be monitored with a variety of techniques that directly detect the strand breaks (or nicks) enzymatically produced in vivo (see Section 4-3) (16–22). In cells undergoing excision repair, postincision degradation, repair synthesis and DNA ligation restore the covalent integrity of the DNA and the strand breaks disappear. The disappearance of damage-specific DNA strand breaks is therefore a more specific measure of *postincision* biochemical steps than the techniques described above. However, since this methodology does not provide identification of the particular lesion being repaired, strand breaks that are spuriously introduced into the

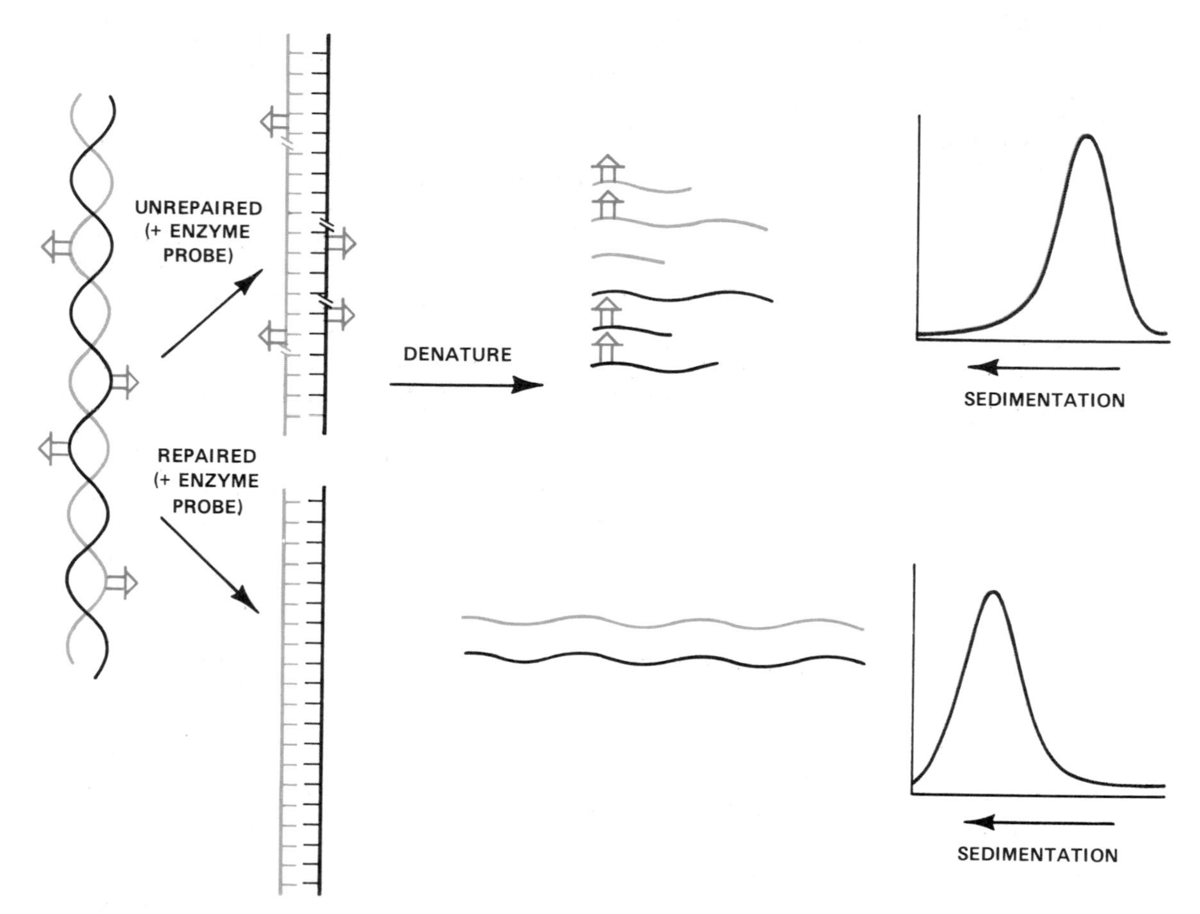

FIGURE 5-1
The presence of unrepaired pyrimidine dimers in the DNA of UV-irradiated cells can be detected with the use of dimer-specific enzyme probes such as the *M. luteus* or phage T4 PD DNA glycosylase–AP endonucleases. Radiolabeled DNA is extracted from the cells and incubated with enzyme. The enzyme catalyzes the formation of strand breaks at pyrimidine dimer (enzyme-sensitive) sites. This DNA will sediment more slowly in alkaline sucrose gradients relative to DNA containing no (or fewer) dimers.

DNA and that are thus unrelated to the excision repair of base damage may not be readily distinguished from those that are (6). The extensively studied model of DNA containing pyrimidine dimers serves as an example of how this problem can be overcome in specific instances. Pyrimidine dimers are subject to repair by enzymatic photoreactivation (see Section 2-2). One can therefore ask whether the strand breaks that appear in a given cell during excision repair fail to appear if the cells are subjected to photoreactivation prior to excision repair. If so, the breaks are presumably incisions at pyrimidine dimers. Alternatively, the specificity of DNA strand breaks observed in wild-type cells can be evaluated by direct comparison with mutants defective in incision of DNA at the lesions in question.

Measurement of Repair Synthesis in DNA

The excision of nucleotides at sites of base damage in DNA is normally accompanied by the restoration of bases by a nonsemiconservative mode of DNA synthesis referred to as repair synthesis (23, 24). The demonstration of repair synthesis is an extremely useful indicator of postincision events since its specific relation to excision repair does not require a prior knowledge of the lesion being repaired nor of the particular mechanism of incision or excision of DNA (6). The most general method used for measuring repair synthesis is density labeling of DNA (25) (Fig. 5-2). In this procedure, DNA synthesis following damage is carried out in the presence of 5-bromouracil or 5-bromodeoxyuridine. The incorporation in sufficient amounts of 5-bromouracil (5-BU) instead of thymine imparts an increased buoyant density to the DNA. Such increases are generally too small to be detectable in regions of nonsemiconservative (repair) synthesis because the incorporated 5-BU constitutes only a tiny fraction of the mass of DNA fragments isolated by the usual procedures. However, a significant increase in density can usually be detected in regions of semiconservative replication. Thus, with the use of radiolabeled 5-BU, DNA undergoing semiconservative synthesis can be separated from DNA undergoing exclusively repair synthesis by sedimentation in isopycnic gradients, because in the former case the BU is typically incorporated into an entire strand of a given DNA fragment and hence constitutes an appreciable fraction of its mass (Fig. 5-2). The radioactivity incorporated into *unreplicated* (*parental density*) DNA is then a measure of the total amount of repair synthesis (Fig. 5-2).

The accuracy of this measurement is influenced chiefly by the amount of background semiconservative DNA synthesis. Therefore, wherever feasible, selective inhibition of this mode is attempted. In prokaryotes such as *E. coli*, such selective inhibition is difficult to achieve, although semiconservative replication is reduced considerably by DNA damage itself (26). In mammalian cells the agent hydroxyurea is widely used for this purpose (27). This compound is an inhibitor of ribonucleotide diphosphate reductase (ribonucleotide reductase), an enzyme that facilitates the biosynthesis of deoxynucleoside triphosphates from the corresponding ribonucleoside diphosphates (28). Hydroxyurea has little or no effect on repair synthesis of DNA (27), probably because in mammalian cells there is a relatively small requirement in excision repair for DNA precursors. Thus the preexisting precursor pool is sufficient to allow repair synthesis without the need for further precursor production (27).

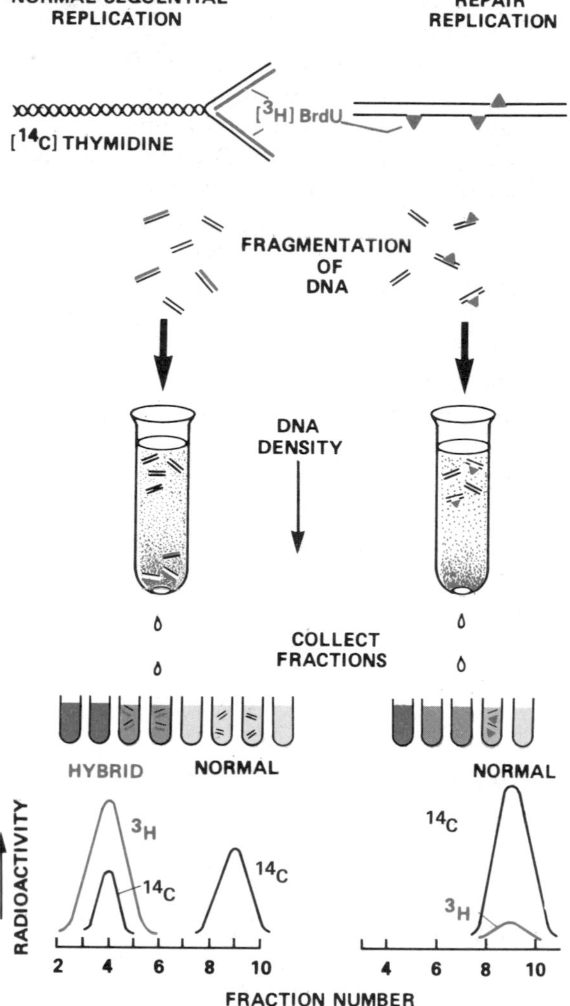

FIGURE 5-2
Schematic illustration of the detection of repair synthesis by buoyant density centrifugation of DNA containing 5-bromouracil(5-BU). DNA is prelabeled with [^{14}C]thymidine to provide a uniform label. Following exposure to UV radiation (or some other form of DNA damage) repair synthesis during excision repair takes place in the presence of [^{3}H]BrdU. DNA synthesized both by semiconservative and by nonsemiconservative modes will thus be density labeled. In order to distinguish these, the DNA is fragmented (by shearing) and sedimented to equilibrium density. Fragments of DNA containing strands synthesized semiconservatively will have a hybrid density detectable from the position of ^{3}H radioactivity (left). Repair synthesis patches do not alter the density of the DNA and appear at the position of normal density DNA (right). (From P. C. Hanawalt, ref. 154.)

Refinements of the density labeling technique allow for estimates of the *size* of the regions (*patches*) of repair synthesis in DNA. Most simply, the amount of radiolabel incorporated during repair, together with an independent determination of the number of repair events, can be used to calculate the *average repair patch size* (29). An alternative procedure involves shearing the DNA to a known small size by sonication of the isolated parental density fraction containing repair patches. The repair patches then constitute an appreciable fraction of the length of the DNA fragments. The DNA is then analyzed in *alkaline* isopycnic gradients so that the density shift of only the affected strands is measured. The observed increase in the density of the DNA fragments, together with the measured average size of the fragments, yields an estimate of the average size of the repair patches (30).

Another procedure for measuring the size of repair synthesis patches in DNA exploits the *photolytic sensitivity* of DNA containing 5-BU (31). As indicated in Section 1-5, when DNA containing this thymine analogue is exposed to radiation at 313 nm, debromination followed by free-radical attack of the deoxyribose or deoxyribose-phosphate backbone occurs, resulting in DNA strand breaks and/or alkali-labile sites (31). Alkaline sucrose gradient sedimentation can then be used to measure the extent of DNA fragmentation (Fig. 5-3).

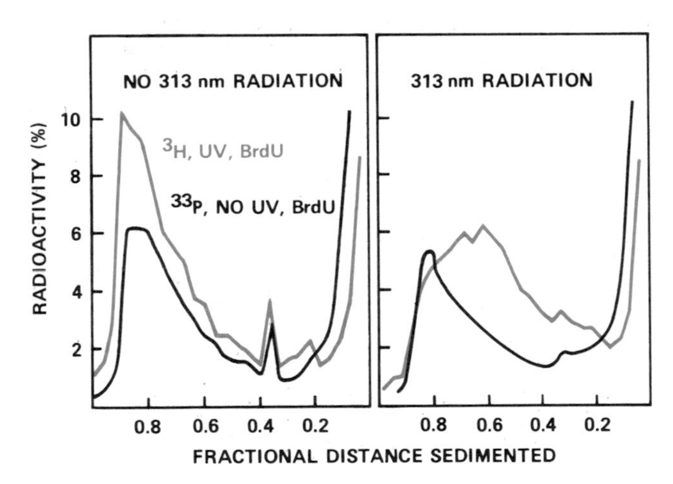

FIGURE 5-3
Sedimentation patterns in alkaline sucrose of labeled DNA from normal human fibroblasts treated with UV or not and allowed to repair in BrdU before exposure to 313 nm photolysis. As shown in the right panel, photolysis of incorporated 5-BU results in strand breakage of the DNA when the DNA is sedimented in alkali. No degradation of DNA is observed in the absence of photolytic irradiation (left panel). The amount of 313 nm irradiation required to cause strand breakage at *all* sites of 5-BU incorporation provides a means for estimating the size of the DNA synthesis (repair) patches. (From R. B. Setlow and J. D. Regan, ref. 31.)

If enough 313 nm light can be delivered to achieve a plateau level of fragmentation (i.e., to produce at least one break or alkali-labile site per repair patch), then the average patch size can be derived directly from the known efficiency of the 5-BU photolysis (31).

Unscheduled DNA Synthesis (UDS) in Mammalian Cells

At about the time the technique for measuring DNA repair synthesis in bacteria was developed, an autoradiographic procedure provided the first demonstration of repair synthesis (and, incidentally, of excision repair) in mammalian cells in culture (32). In this procedure advantage is taken of the fact that eukaryotic cells carry out semiconservative DNA synthesis only during a limited period of the cell cycle, a period designated as the S phase (Fig. 5-4). If mammalian cells in culture are allowed to grow for a short period in the presence of [^{3}H]thymidine and are then examined autoradiographically, DNA synthetic, or S, phase cells show intense labeling of their nuclei with silver grains, whereas cells not in the S phase show essentially no detectable grains (33). However, if the cells in culture undergo repair synthesis of DNA, non-S-phase cells also manifest the presence of silver grains (Fig. 5-5). By appropriate adjustment of the labeling and autoradiographic development times, the number of silver grains in cells carrying out repair (*unscheduled*) DNA synthesis can be quantitated (33). This value reflects the amount of repair synthesis in the cells.

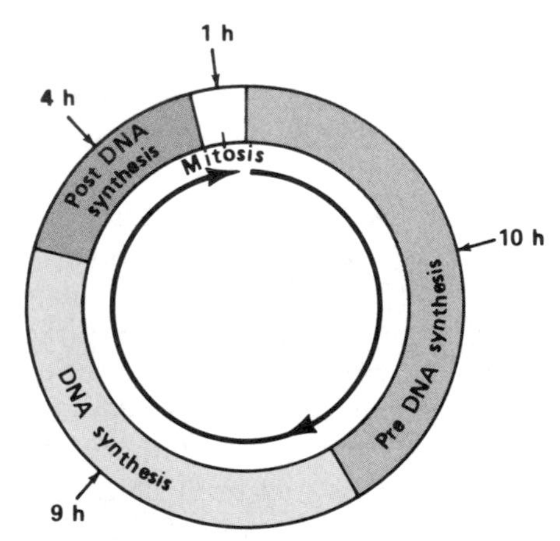

FIGURE 5-4
Phases in the life cycle of a typical mammalian cell growing in tissue culture and dividing once every 24 hours. DNA synthesis occurring during the DNA synthesis (S) phase of the cycle is called scheduled synthesis. That occurring outside of the S phase is called *unscheduled DNA synthesis* (UDS).

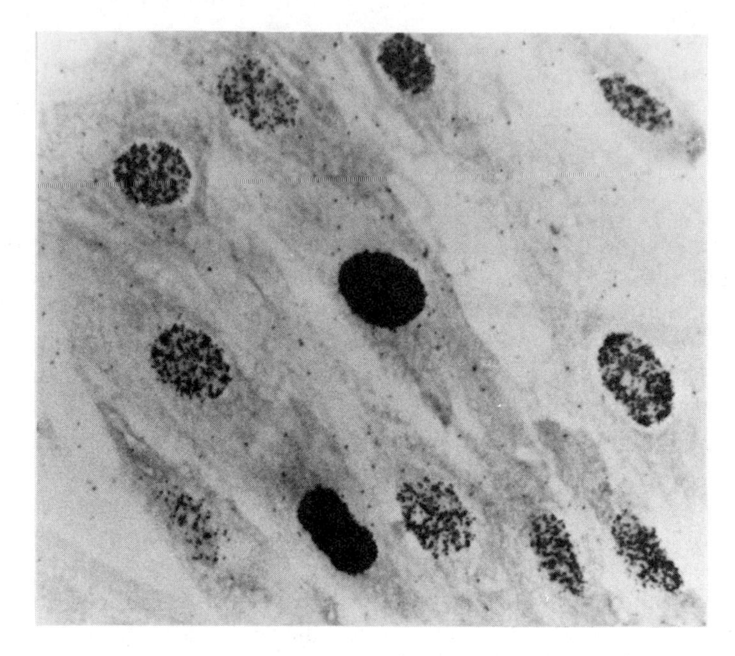

FIGURE 5-5
Human fibroblasts in culture showing unscheduled DNA synthesis (stippled nuclei) and also semiconservative (scheduled) DNA synthesis (intensely labeled nuclei). (Courtesy of Dr. J. E. Cleaver, University of California at San Francisco.)

5-3 Postincision Events during Excision Repair Mediated by the uvrABC Endonuclease of *E. coli*

As indicated in Section 4-1, UV-irradiated *E. coli* has long been an organism of choice for the study of excision repair of DNA, and most of the information on the biology and biochemistry of postincision events during excision repair mediated by direct-acting damage-specific DNA incising activities comes from studies with this organism. For many years such studies were based on the assumption that physical *excision* of dimers and repair synthesis of DNA were closely related biochemical events distinct from DNA incision (6). This notion was fostered by the use of some of the techniques described earlier for monitoring excision repair, which, as indicated previously, do not discriminate between excision that is coupled to incision and that which follows incision as an independent enzyme-catalyzed event. The postulated association between excision and repair synthesis was further reinforced by the observation that DNA polymerase I of *E. coli* catalyzes coupled exonucleolytic degradation

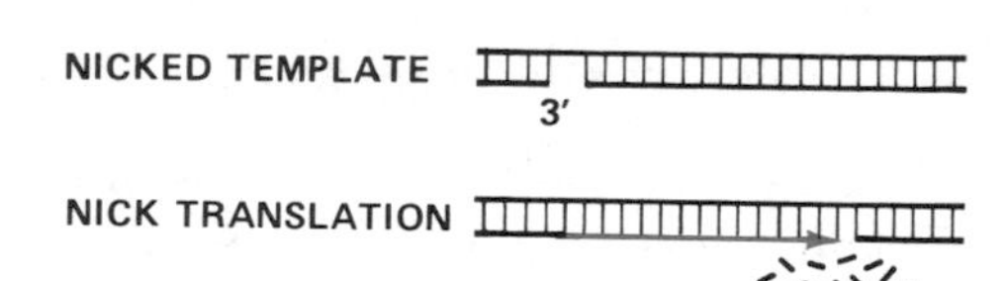

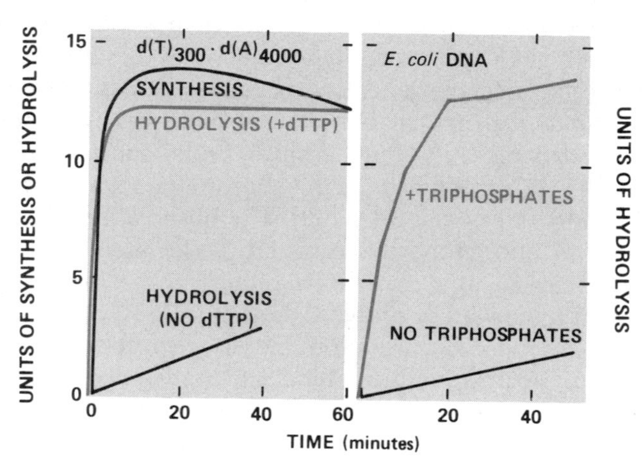

FIGURE 5-6
Nick translation by DNA polymerase I. When concurrent polymerization and $5' \rightarrow 3'$ exonuclease actions balance each other, a nick in the DNA is linearly advanced (translated) along the chain (top figure). Degradation is stimulated by concurrent synthesis (i.e., in the presence of the appropriate triphosphates) with polymer (lower left) or DNA (lower right) template-primers. (From A. Kornberg, ref. 46.)

and resynthesis of DNA in the $5' \rightarrow 3'$ direction from nicks with 3′-OH termini (a process called *nick translation* (34, 35) (Fig. 5-6), and by the results of studies in cell-free systems and in permeabilized cells (36–41). Such studies frequently utilized the phage T4 or *M. luteus* enzymes as a means of incising UV-irradiated DNA in order to study postincision events in excision repair, before it was known that these enzymes are DNA glycosylases–AP endonucleases that attack DNA containing pyrimidine dimers very differently from the way the uvrABC enzyme does. The recent recognition that the uvrABC damage-specific DNA incising activity of *E. coli* may catalyze both the incision and the excision of base damage suggests that repair synthesis is a discrete postincision event that is *not* biochemically linked to excision of bulky base damage in *E. coli* (see Fig. 3-1).

Repair Synthesis of DNA in UV-Irradiated *E. coli*

Measurements of the size of the repair patches in vivo in wild-type UV-irradiated *E. coli* are consistent with the revised model of excision repair mediated by the uvrABC enzyme. Thus, for example, a number of studies (42, 43) show that the vast majority (approximately 99 percent) of the excision repair events result in repair patches 20 to 30 nucleotides long, a value acceptably close to that of the oligonucleotides excised in vitro (approximately 12 nucleotides) (1–5). However, under normal conditions a very small fraction of repair synthesis patches are much longer than 20 to 30 nucleotides and can in fact be longer than 1500 nucleotides (42, 43). The process whereby these tracts are generated is referred to as *long patch excision repair* to distinguish it from the more general *short patch excision repair* mode (29). Furthermore, under some circumstances the size of the repair patches during short patch excision repair can extend to 80 to 100 nucleotides (43). The observation of long patch excision repair and of the heterogeneity associated with the short patch mode suggests that sometimes repair synthesis is preceded or accompanied by postexcision degradation of DNA; i.e., the excision gaps created by the uvrABC endonuclease are apparently expanded by exonucleolytic degradation. There is evidence that this patch size heterogeneity at least in part reflects the action of distinct DNA polymerases during repair synthesis (43). Hence before discussing the biology and biochemistry of repair synthesis of DNA it is relevant to consider briefly the enzymes that may be involved in this process.

The DNA Polymerases of E. coli

E. coli contains three distinct enzymes that can synthesize DNA, referred to as DNA polymerases I, II and III, encoded by the *polA*, *polB* and *polC* genes respectively. The principal properties of these three enzymes and distinctions between them are summarized in Table 5-1.

DNA Polymerase I. DNA polymerase I (so designated because it was the first DNA polymerase identified in extracts of *E. coli*) was discovered by Arthur Kornberg and his associates (44). The enzyme is the product of the *polA* gene of *E. coli*, which maps at 85 minutes on the *E. coli* genetic map (Fig. 5-7) (45). The enzyme has been purified to physical homogeneity and has been extensively characterized (46, 47). It is a monomeric protein with a molecular weight of 109,000 and has discrete catalytic functions for DNA polymerization, pyrophosphorolysis, pyrophosphate exchange, $3' \rightarrow 5'$ exonucleolytic degradation and $5' \rightarrow 3'$ exonucleolytic degradation (46, 47). A detailed discussion of this enzyme is beyond the province of

TABLE 5-1
Properties of DNA polymerases I, II and III of *E. coli*

	pol I	pol II	pol III
Functions			
Polymerization: 5′ → 3′	+	+	+
Exonuclease: 3′ → 5′	+	+	+
Exonuclease: 5′ → 3′	+	−	+
Pyrophosphorolysis and PP_i exchange	+		+
Template-primer			
Intact duplex	−	−	−
Primed single strands*	+	−	−
Nicked duplex (poly dAT)	+	−	−
Duplex with gaps or protruding single-strand 5′ ends of: <100 nucleotides	+	+	+
>100 nucleotides	+	−	−
Polymer synthesis de novo	+	−	−
Activity			
Effect of KCl (percent of optimal) 20 m*M*	60	60	100
50 m*M*	80	100	50
100 m*M*	100	70	10
150 m*M*	80	50	0
K_m for triphosphate	Low	Low	High
Inhibition by 2′-deoxyanalogues	−	+	+
Inhibition by arabinosyl CTP	−	+	−
Inhibition by sulfhydryl (SH) blocking agents	−	+	+
Inhibition by pol I antiserum	+	−	−
General			
Size (kDa)	109	120	140
Affinity for phosphocellulose: molarity of phosphate required for elution	0.15	0.25	0.10
Molecules per cell, estimated	~400	~40	10–20
Turnover number,† estimated	(1)	0.05	15
Structural genes	*polA*	*polB*	*polC*
Conditional lethal mutant	Yes	No	Yes

*A primed single strand is a long single strand with a short length of complementary strand annealed to it.
†Nucleotides polymerized at 37°C/min/molecule of enzyme, relative to pol I, which is near 600.
From A. Kornberg, ref. 46.

this book, and the interested reader should consult Kornberg's texts on DNA replication (46, 47). Nonetheless, a number of important features of DNA polymerase I merit emphasis here.

Polymerization, pyrophosphorolysis and pyrophosphate exchange are all features of the polymerization activity, the latter two simply being the reverse of the polymerization reaction (46). Like all known

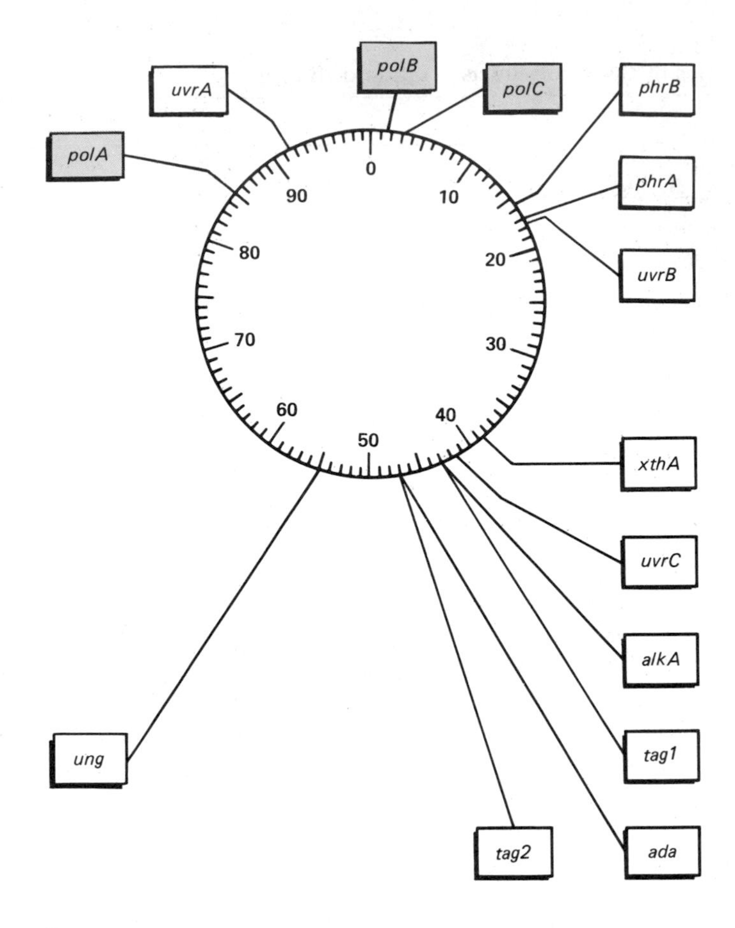

FIGURE 5-7
The location of the *polA*, *polB* and *polC* genes on the *E. coli* genetic map.

DNA polymerases, DNA polymerase I of *E. coli* catalyzes the growth of DNA chains only in the 5′ → 3′ direction. This polarity is dictated by its specificity for 3′-OH primer termini and for deoxynucleoside 5′ triphosphates as substrates (46).

A striking feature of DNA polymerase I, observed in no other *E. coli* polymerase, is its capacity to promote replication of DNA at a nick, unaided by other proteins (46). This requires unwinding of the duplex beyond the nick and progressive strand displacement of the 5′ chain. The 5′ → 3′ exonuclease function of *E. coli* DNA polymerase I can catalyze exonucleolytic degradation of DNA in the *absence* of any base damage, from nicks containing 3′-OH termini. Strand displacement coupled with degradation by the 5′ → 3′ exonuclease leads to transfer of the nick along the template (*nick translation*). In so doing the enzyme cleaves a phosphodiester bond only at base-

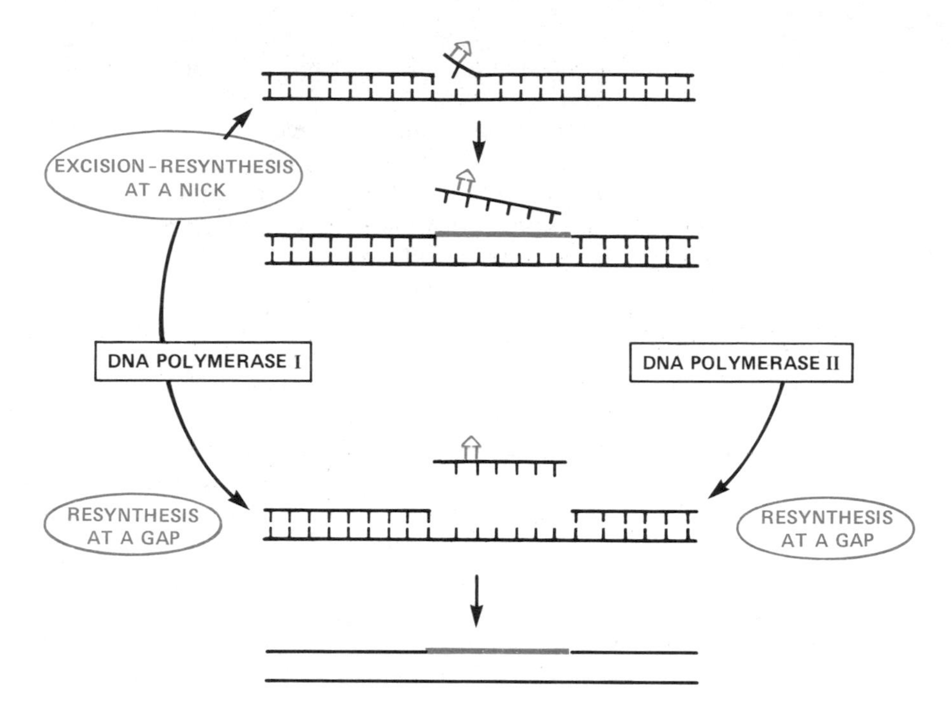

FIGURE 5-8
E. coli DNA polymerase I can catalyze repair synthesis at both nicked (top) and gapped (bottom) template-primers. DNA polymerase II, however, can only catalyze repair synthesis in gapped molecules.

paired regions (i.e., it requires duplex DNA for activity) (46). In addition the 5′ → 3′ exonuclease can excise terminal DNA fragments containing as many as 10 nucleotides. It is precisely these properties of the exonuclease that endow DNA polymerase I with the ability to remove oligonucleotides containing pyrimidine dimers or other forms of bulky base damage from DNA with nicks 5′ to the sites of damage (35, 48). This enzyme or its homologue in other prokaryotic cells is therefore a primary candidate for coupled excision-resynthesis of DNA from nicks exclusively 5′ to sites of base damage or base loss (Fig. 5-8).

The 3′ → 5′ exonuclease of DNA polymerase I is a component of the enzyme necessary for the recognition of a correctly base-paired primer terminus (46). It supplements the capacity of DNA polymerase I to match the nucleotide substrate to the template by base pairing and hence plays a critical role in the fidelity of repair synthesis during excision repair. In addition, this exonuclease function can potentially serve as an excision exonuclease for the removal of 3′-terminal damaged nucleotides. The roles of both the 3′ and the 5′ exonucleases of DNA polymerase I in the excision of nucleotides and of nucleotide residues (apurinic and apyrimidinic sites) are discussed further in Section 5-4.

In vitro, *E. coli* DNA polymerase I can be cleaved by gentle proteolysis into two fragments of 75,000 and 35,000 daltons (46) (Table 5-2). The former (*large fragment*) retains the polymerizing and 3′→ 5′ exonuclease activities, whereas the latter (*small fragment*) retains the 5′→ 3′ exonuclease activity (46) (Table 5-2).

E. coli DNA polymerase I is roughly spherical in shape, with a diameter of about 65 Å, and it can contact a DNA helix across its width (about 20 Å) for a length of nearly two helical turns (about 20 bp) (46). Thus, if the enzyme is *processive* in its action, i.e., if it does not dissociate from the DNA after each nucleotide incorporation, a single productive binding event could facilitate complete repair synthesis of excision tracts less than 20 nucleotides long (Fig. 5-8). Various experimental approaches have been utilized to establish whether *E. coli* DNA polymerase I is processive or distributive in its action (46). The general conclusion that has emerged from in vitro studies is that the enzyme binds very rapidly to suitable template-primers. The binding is followed by a relatively long delay before initiation of polymerization, during which time there is a shift in equilibrium from the inactive to the active enzyme. A cycle of about 20 polymerization steps takes place before the enzyme slowly dissociates (46). These observations suggest that during normal short patch excision repair the enzyme is essentially processive.

Processivity is influenced by many factors that affect the secondary structure of the DNA or the conformation of the enzyme. These include temperature, ionic conditions and interactions with other proteins (46). For example, DNA polymerase I from a mutant (*polA5*) incorporates only about one-fifth the number of nucleotides added by the wild-type enzyme before dissociating from the template (49). These factors are of obvious relevance to DNA repair, since the processivity of the enzyme during repair synthesis of DNA may contribute to the relative heterogeneity of the patches synthesized and could affect the kinetics and possibly even the fidelity of repair synthesis in vivo.

TABLE 5-2
Some properties of intact *E. coli* DNA polymerase I and its small and large proteolytic fragments

	Mol. Wt.	3′→ 5′ Exonuclease	5′→ 3′ Exonuclease
Intact enzyme	109,000	+	+
Small fragment	35,000	−	+
Large fragment	75,000	+	−

From A. Kornberg, ref. 46.

DNA Polymerase II. The isolation by John Cairns and his coworkers of a viable mutant of *E. coli* defective in DNA polymerase I activity (50) established that although this enzyme may be *involved* in DNA replication, it is clearly not the enzyme that catalyzes the majority of semiconservative DNA synthesis in vivo. A search for other DNA polymerases in *E. coli* uncovered two new enzymes called DNA polymerases II and III (46). DNA polymerase II, which is encoded by the *polB* gene (Fig. 5-7), is distinguished from DNA polymerase I by a number of criteria (Table 5-1). Most significantly from the point of view of its potential role in DNA repair, it is devoid of $5' \rightarrow 3'$ exonuclease activity and it cannot use a template-primer that is simply nicked or is extensively single strand (46). The optimal substrate for DNA polymerase II is a duplex with short gaps of less than 100 nucleotides. This substrate requirement suggests that DNA polymerase II would be ideally suited to a primary role in repair synthesis of the excision gaps created by the *E. coli* uvrABC enzyme (Fig. 5-8). However, in excision repair mediated by endonucleases that create single incisions, DNA polymerase II could only catalyze repair synthesis following the conversion of the nicks into gaps by exonucleolytic degradation (Fig. 5-8).

DNA Polymerase III. DNA polymerase III of *E. coli* is a complex, multicomponent enzyme that resembles RNA polymerase in its complexity (46). The *holoenzyme* core has been purified to virtual homogeneity (47) and contains α, ϵ and θ subunits of 140, 25 and 10 kDa respectively (Fig. 5-9). The α subunit is the product of the *polC* (*dnaE*) gene (Fig. 5-7). Holoenzyme preparations judged to be about 50 percent pure contain as many as 13 discrete polypeptides (47). In addition to those of the core they include β, γ, δ and τ subunits (Fig. 5-9). Alternative forms of DNA polymerase III include DNA polymerase III′, a subassembly of four subunits of the holoenzyme (Table 5-3), and DNA polymerase III*, a subassembly of six subunits (Table 5-3) (47).

Like DNA polymerase I, polymerase III of *E. coli* has both $3' \rightarrow 5'$ and $5' \rightarrow 3'$ exonuclease activities in vitro; however, some of their

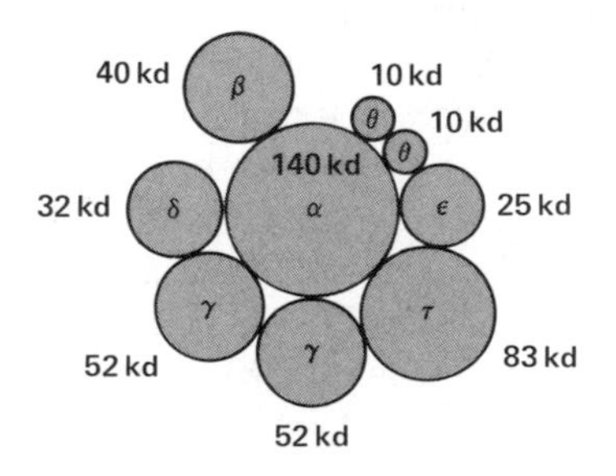

FIGURE 5-9
Diagrammatic representation of the various polypeptide components of *E. coli* DNA polymerase III holoenzyme. (From A. Kornberg, ref. 46.)

TABLE 5-3
Components of DNA polymerase III holoenzyme

Subunit	Mass (kDa)	Alternative Designations			
α	140	dnaE protein, polC protein	pol III (core)	pol III′	pol III*
ϵ	25	—			
θ	10	—			
τ	83	—			
γ	52	dnaZ protein			
δ	32	Factor III, dnaZ protein			
β	37	Factor I, copol III*, dnaN protein			

From A. Kornberg, ref. 47.

properties differ from those of DNA polymerase I. For example, the $3' \rightarrow 5'$ exonuclease of DNA polymerase III fails to degrade dinucleotides. Additionally, although the $5' \rightarrow 3'$ exonuclease activity can excise oligonucleotides from DNA, it differs from DNA polymerase I in its requirement for a single-strand substrate (46). However, once hydrolysis is initiated, the exonuclease proceeds into a duplex region. Hence the $5' \rightarrow 3'$ exonuclease function can catalyze the selective excision of thymine-containing pyrimidine dimers from UV-irradiated DNA previously incised with the *M. luteus* (and presumably the phage T4) PD DNA glycosylase–AP endonuclease (51) (Table 5-4).

The Role of E. coli *DNA Polymerases in Repair Synthesis.* The absence of detectable repair synthesis of DNA in UV-irradiated *E. coli* is only observed in mutants defective in all three DNA polymerases, suggesting that all of these enzymes are potentially able to perform this function in vivo (52). This is consistent with the observation that in vitro all three enzymes can utilize gapped DNA substrates of the type created by the uvrABC endonuclease. It is thus difficult to assess the relative contribution of each to repair synthesis in wild-type cells. Nonetheless, several observations suggest that *E. coli* DNA polymerase I occupies a *primary* role in repair synthesis under normal conditions. First, *polA* mutants are abnormally sensitive to UV radiation (45) (Fig. 5-10), although, not surprisingly in view of the redundancy just described, these mutants are not as sensitive as are cells defective in the *uvrA, uvrB* or *uvrC* genes (Fig. 5-10). As further evidence for the role of DNA polymerase I in repair synthesis, it is perhaps significant that in *E. coli* there is considerably more DNA polymerase I (about 400 molecules per cell) than there is DNA polymerase II (about 40 molecules per cell) or DNA polymerase III (about 10 molecules per cell) (53). Mutants at the *polB* gene are not

TABLE 5-4
Degradation of UV-irradiated DNA by *E. coli* DNA polymerase III $5' \rightarrow 3'$ exonuclease

Enzyme	[^{3}H]T7 DNA Unirradiated (pmol acid-soluble radioactivity)	[^{3}H]T7 DNA Ultraviolet irradiated (pmol acid-soluble radioactivity)
M. luteus PD DNA glycosylase–AP endonuclease	0.39	2.76
DNA polymerase III	2.50	0.27
M. luteus PD DNA glycosylase–AP endonuclease + DNA polymerase III	1.18	35.6

E. coli DNA polymerase III degrades UV-irradiated DNA in the $5' \rightarrow 3'$ direction (and thus excises pyrimidine dimers) if the DNA was preincised at dimer sites. (From D. M. Livingston and C. C. Richardson, ref. 51.)

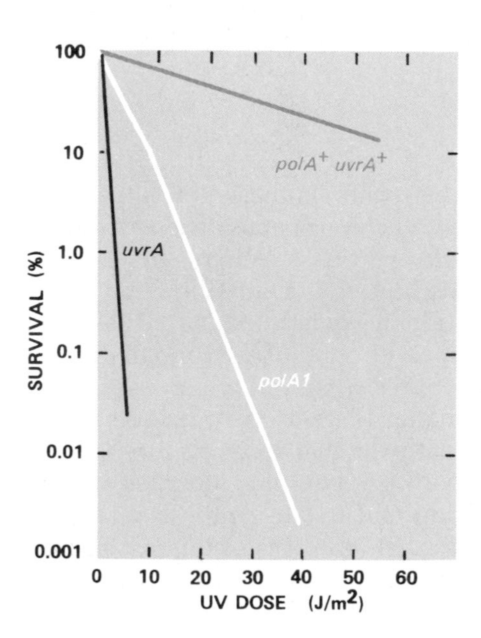

FIGURE 5-10
PolA mutants of *E. coli* are abnormally sensitive to killing by UV radiation. However, they are not as sensitive as *uvr* mutants.

abnormally UV sensitive (54). Since *polC* mutants exist only as conditional lethals it is not possible to assess their UV sensitivity under conditions in which the enzyme is not functional (6).

Mutants (*polAex*) defective in the $5' \rightarrow 3'$ exonuclease function of DNA polymerase I are also abnormally sensitive to UV radiation (55).

If it is assumed that the 5′ → 3′ exonuclease is required for excision of base damage during nick translation at nicks 5′ to such damage, its role in excision repair is difficult to reconcile with the observation that excision of dimers can be effected in vitro by the uvrABC enzyme alone (4). However, recent studies (56) implicate a role of the 5′ → 3′ exonuclease function of *E. coli* DNA polymerase I in efficient gap filling of DNA. In vitro, the filling of gaps approximately 20 nucleotides long and their subsequent ligation to extant DNA was found to be highly efficient when the gapped DNA was incubated with DNA polymerase I from wild-type cells and with DNA ligase. But when such DNA was incubated with DNA ligase and DNA polymerase I from a *polAex1* mutant, gap filling and ligation were much less efficient. Repair synthesis of most of the gapped molecules was incomplete, leaving a residual gap that could be filled in by T4 DNA polymerase and then ligated. This limited synthesis was masked, however, in the total population by a small fraction of DNA molecules that underwent repair synthesis associated with strand displacement, resulting in a *total* extent of incorporation per gap much greater than the actual gap size. These molecules, too, could not be ligated (56).

Repair Patch Heterogeneity

As indicated earlier, the size of the repair synthesis tracts (patches) in UV-irradiated *E. coli* can be quite heterogeneous. The vast majority of the excision repair events in *wild-type* cells result in repair patches 20 to 30 nucleotides long; however, a small fraction of the patches are at least 1500 nucleotides in length (42, 43). These two different-size classes apparently result from the operation of distinct enzymatic pathways for repair synthesis, both of which are dependent on the incision of UV-irradiated DNA by the products of the *uvrA*, *uvrB* and *uvrC* genes. The pathway that accounts for the short patches appears to be constitutive in *E. coli* and includes at least uvrABC-mediated incision-excision and repair synthesis catalyzed in all likelihood by DNA polymerase I (43). It is of interest historically that the long patch repair mode was discovered by the observation of *increased* total repair synthesis of DNA in UV-irradiated cells defective in DNA polymerase I, despite the repair of fewer pyrimidine dimers in these mutants (57). This led to the speculation that the primary biochemical distinction between short and long patch repair is that the former is mediated by DNA polymerase I and the latter by other DNA polymerases (57). The increased repair synthesis observed in polymerase-I-defective mutants is indeed accounted for by longer patch sizes, but these are about 100 nucleotides in length and are still considered examples of *short* patch repair. No truly "long" patches (longer than 1000 nucleotides) are observed under these conditions (43). In addition, strains defective in DNA polymerases II or III carry out essentially normal amounts of true long patch repair under conditions where this repair synthesis mode is optimally oberved (see below) (43). It appears, then, that DNA polymerase I is required for *both* short and long patch repair. In the absence of this enzyme the constitutive short patch repair mode is probably effected by DNA polymerases II and/or III and the resulting repair patches are larger (about 100 nucleotides).

Long patch repair also differs from short patch repair in that it has an absolute requirement for inducible functions in *E. coli* (58, 59). In Section 4-2 mention was made of the inducibility of the *uvrA*, *uvrB* and *uvrC* genes of *E. coli*. These genes are part of an inducible system that governs a number of the cellular responses to DNA damage and is controlled by two regulatory genes called *recA* and *lexA* (6). Details of the *recA-lexA*-dependent inducible responses are presented in Chapter 7. For the purposes of the present discussion it should be noted that long patch repair synthesis of DNA is one of the many responses associated with this induction phenomenon (58, 59). Aside from the demonstration of long repair patches, independent evidence for a *recA-lexA*-dependent pathway of excision repair comes from the observation that the completion of at least some fraction of the excision repair events initiated by the *uvrA*, *uvrB* and *uvrC* genes requires the *recA* and *lexA* genes, as well as the capacity for protein synthesis (60).

The frequency of long patches during excision repair is increased in mutants defective in the *recL* gene (61) (a recombinational function) and *polAex* gene (55) (the $5' \rightarrow 3'$ exonuclease function of DNA polymerase I). The reason for the effect of these mutations on the mode of repair synthesis is not understood.

The molecular mechanism and physiological significance of long patch excision repair are also not known. With respect to the former issue, one may ask, What determines repair patch size under any circumstances in UV-irradiated *E. coli*? A number of models have been proposed. For example, since the completion of repair synthesis is followed by DNA ligation (see Section 5-5) it is possible that the particular DNA polymerase catalyzing repair synthesis is displaced by DNA ligase, although this model does not provide an obvious explanation as to why the kinetics of displacement should vary under different conditions. An alternative model relates repair patch size to the intrinsic processivity of DNA polymerase I (62). In fact, the size of the repair patches generated by extracts of *E. coli* in specifically preincised UV-irradiated DNA correlates well with the intrinsic processivity of various forms of DNA polymerase I. The wild-type enzyme has a processivity of 15 to 20 nucleotides during replication of a nicked DNA template and inserts a repair patch of approximately 16 to 21 nucleotides in length in vitro (62). In contrast, the processivity of DNA polymerase I derived from the mutant *polA5*, which is defective in its ability to translocate along the DNA template during synthesis (49), is three to five nucleotides on a nicked DNA template-primer. Correspondingly, this enzyme inserts a repair patch of approximately five nucleotides in vitro (62).

The processivity of DNA polymerase I during repair synthesis may in turn be determined by the extent of the localized helix destabilization imparted by the precise mechanism of incision of DNA at pyrimidine dimers. In this regard it is perhaps relevant that in uninfected *E. coli* (in which DNA incision is mediated by the uvrABC enzyme) the patch size in vivo is 20 to 30 nucleotides, whereas in phage-T4-infected mutant *uvr* cells (in which incision of the UV-irradiated *E. coli* genome is mediated by the phage-coded PD DNA glycosylase–AP endonuclease), the repair patches average four to six nucleotides in length (63, 64). Another possibility worth considering is that in induced cells in which expression of the *recA* gene is activated, the binding of *recA* protein to DNA at sites of DNA incision may somehow determine the extraordinarily long length of the patches observed.

As regards the physiological significance of long patch excision repair in *E. coli* cells in which the *recA-lexA* system is induced, it is interesting that this excision repair mode apparently correlates with a more rapid recovery of semiconservative DNA synthesis and with more cell survival after UV irradiation than in uninduced cells (65). Thus, this mode of repair synthesis may reflect the operation of a distinct form of excision repair associated with the ability of *E. coli* to bypass pyrimidine dimers at or near replication forks. This topic is discussed in greater detail in Section 7-9.

Further Roles of DNA Polymerases and the Role of Other Proteins in Excision Repair in E. coli

DNA Polymerases. The model of excision repair in which incision of DNA and excision of base damage are both mediated by a damage-specific DNA incising activity such as the uvrABC enzyme and the resultant gaps are filled in by DNA polymerase I is consistent with the observation that at low doses of UV radiation (less than 50 J/m^2), strains of *E. coli* defective in the *polymerizing* function of DNA polymerase I show no detectable defect in dimer excision (55) (Fig. 5-11). However, a number of observations suggest that the rate and extent of pyrimidine dimer excision (defined by direct measurement of the loss of thymine-containing pyrimidine dimers from acid-insoluble fractions) *are* affected by mutations in the genes that en-

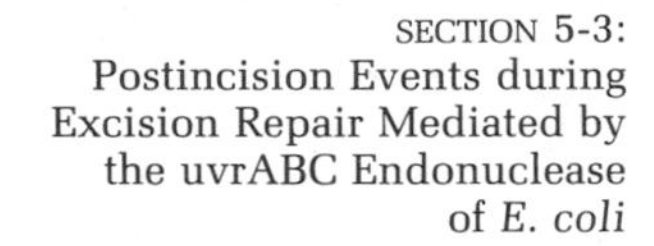

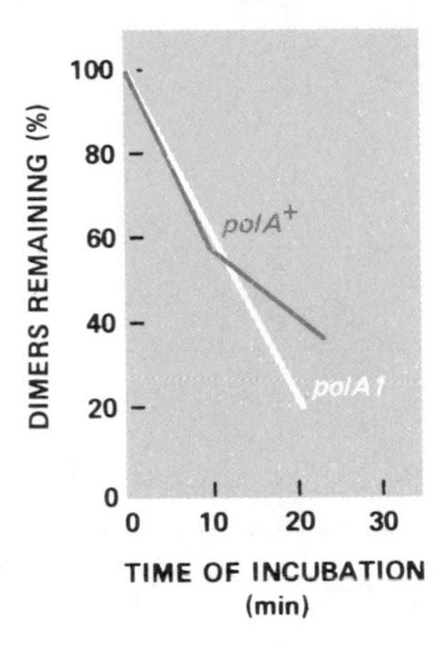

FIGURE 5-11
At relatively *low* doses of UV radiation (10 J/m^2 in the experiment shown here) the kinetics of the loss of thymine-containing pyrimidine dimers from the acid-insoluble DNA of *polA*$^+$ and *polA1* cells are not significantly different. (From P. Cooper, ref. 55.)

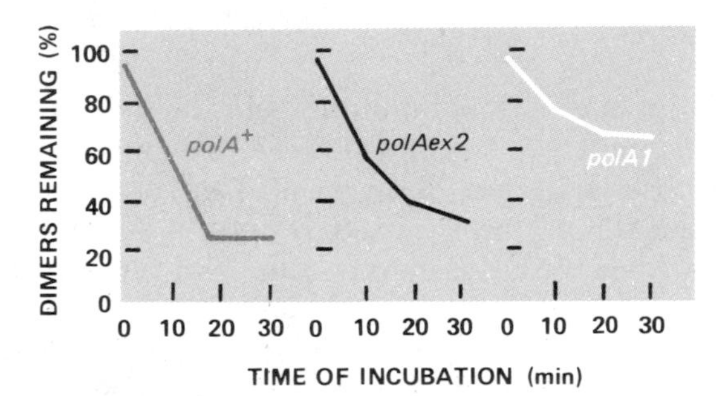

FIGURE 5-12
At high doses of UV radiation (50 J/m^2 in the experiments shown here) the kinetics and the extent of loss of thymine-containing pyrimidine dimers from DNA are slower in *polAex2* and particularly *polA1* mutants relative to that of wild type (*polA*$^+$) cells. (From P. Cooper, ref. 55.)

code some of the DNA polymerases, as well as by mutations in genes coding for other functions (6). Thus, for example, mutants defective in the polymerizing function of DNA polymerase I have a reduced rate and extent of dimer excision relative to that of wild-type strains after irradiation at *relatively high doses* of UV radiation (50 J/m^2 or greater) (Fig. 5-12) (55). The interpretation of this observation is complicated by the fact that in cells exposed to such high doses of radiation extensive degradation of their DNA occurs, mediated largely by exonuclease V, the product of the *recB* and *recC* genes (55). Since mutants defective in exonuclease V (*recBC*) show no detectable defect in dimer excision, it is possible to examine this parameter in *polA* mutants in which post-UV DNA degradation is reduced by

TABLE 5-5
Excision of thymine dimers in *E. coli*

UV Dose (J/m²)	Thymine Dimers Remaining in DNA After Incubation for 45 Min (%)				
	polA⁺ *recB*⁺ *recC*⁺	*polA*	*polA recB recC*	*polA polC*	*polAex recB recC*
25	15	25	25	65	
45		35	45	70	20
60	20	45	45	75	30
80	15	60	65	80	
100	15	70		90	50

From P. K. Cooper and J. G. Hunt, ref. 58.

the introduction of additional mutations in the *recB* and *recC* genes (55). However, under these conditions, a deficiency in dimer excision is still observed; after doses of 60 J/m^2 or greater, only half (or less) of the dimers are excised by *polA1* mutants, compared with essentially quantitative removal in wild-type *E. coli* (55, 58) (Table 5-5).

A defect in the 5′ → 3′ exonuclease function of DNA polymerase I (*polAex* mutants) (66, 67) also results in decreased excision capacity, which is the most pronounced at higher doses of UV radiation (Fig. 5-12). However, at any given dose, *polA1* mutants are less proficient in dimer excision than are *polAex* mutants (55). The participation of DNA polymerase III in dimer excision in *E. coli* is suggested by the observation that a mutant (*polA polC*) deficient in both polymerizing activities has less excision capability than the *polA* strain alone, even at low UV doses (58) (Table 5-5).

Exonuclease VII of E. coli. Exonuclease VII is a protein with a molecular weight of approximately 88,000 that is the product of the *E. coli xseA* gene (Fig. 5-13) and is specific for single-strand DNA or single-strand regions of duplex DNA (68, 69). Thus it degrades denatured DNA, single-strand regions extending from the ends of duplex DNA or displaced single-strand regions. The purified enzyme is active in the presence of EDTA but is slightly stimulated by $MgCl_2$ (69). The limit size of products of exonuclease VII action are oligonucleotides bearing 5′-P and 3′-OH termini, predominantly in the range of tetramers to dodecamers (68, 69). The enzyme acts processively, initially releasing large acid-insoluble oligonucleotides that can be degraded further to produce a limit-size product of acid-

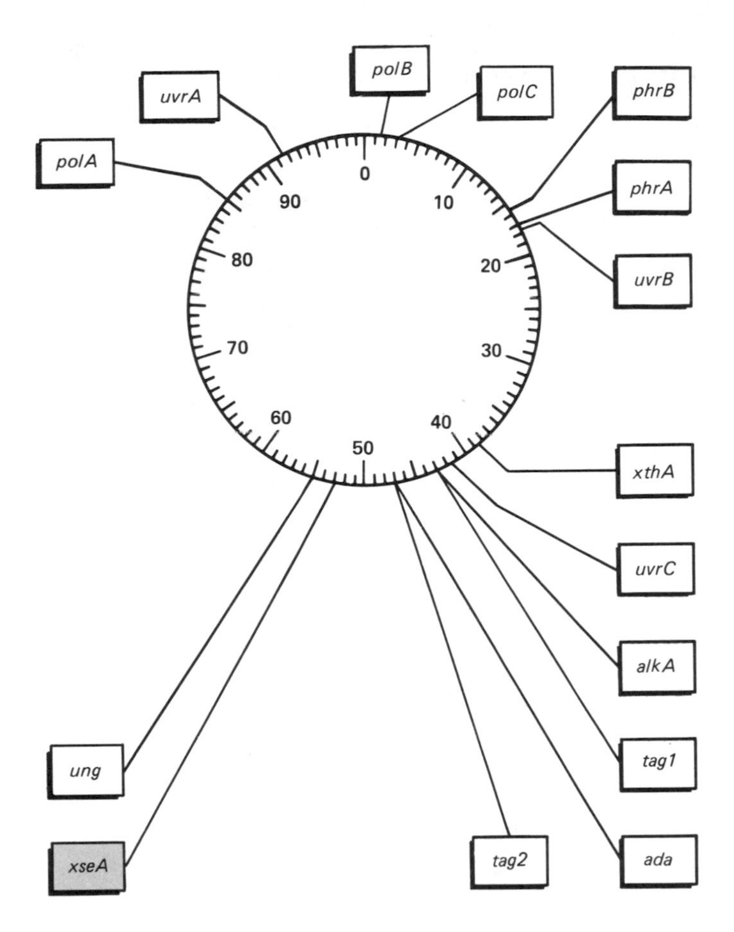

FIGURE 5-13
The location of the *xseA* gene on the *E. coli* genetic map.

soluble oligonucleotides. Hydrolysis of DNA by exonuclease VII can be initiated at either 3′ or 5′ termini; that is, the enzyme can act as both a 3′ → 5′ and a 5′ → 3′ exonuclease (68, 69).

Like *E. coli* DNA polymerases I and II, exonuclease VII catalyzes the excision of thymine-containing pyrimidine dimers in vitro from UV-irradiated DNA preincised with the T4 or *M. luteus* dimer-specific DNA glycosylase–AP endonucleases, i.e., DNA containing nicks 5′ to the dimers (69) (Table 5-6). Furthermore, a strain of *E. coli* defective in DNA polymerase I, exonuclease VII and exonuclease V (*polAex xseA ts recBC*) has a more pronounced excision deficiency than does a *polAex ts recBC* control strain (70) (Fig. 5-14), suggesting that exonuclease VII may play a role in postincision DNA degradation during excision repair.

TABLE 5-6
Degradation of preincised UV-irradiated DNA by *E. coli* or *M. luteus* exonuclease VII

Treatment	Unirradiated [^{3}H]ϕX174 RF DNA	Ultraviolet-irradiated [^{3}H]ϕX174 RF DNA
	pmol acid-soluble radioactivity	
M. luteus PD DNA glycosylase–AP endonuclease	0.89	0.93
M. luteus exonuclease	3.2	3.8
M. luteus PD DNA glycosylase–AP endonuclease + *M. luteus* exonuclease VII	4.6	38
Exonuclease VII of *E. coli*	1.2	1.8
M. luteus PD DNA glycosylase–AP endonuclease + *E. coli* exonuclease VII	1.9	12.5

From J. W. Chase and C. C. Richardson, ref. 69.

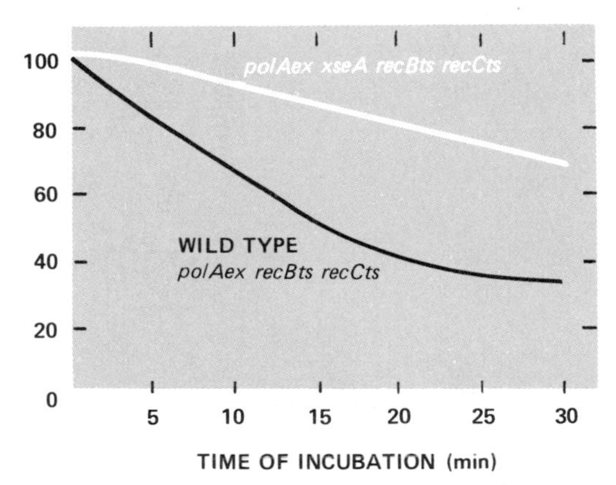

FIGURE 5-14
A mutant of *E. coli* defective in the 5′ → 3′ exonuclease function of DNA polymerase I, in the recBC enzyme (exonuclease V) and in *exonuclease VII* shows a reduced ability to remove thymine-containing pyrimidine dimers from acid-insoluble DNA relative to that of a wild-type strain or a strain that is isogenic except for the *xseA* mutation. (From W. E. Masker and J. W. Chase, ref. 70.)

The uvrD *Gene of* E. coli

Another genetic locus that apparently affects excision repair of pyrimidine dimers in *E. coli* is a function that has been independently identified in a number of different studies and hence has been given a number of redundant genetic designations. A UV-sensitive mutant

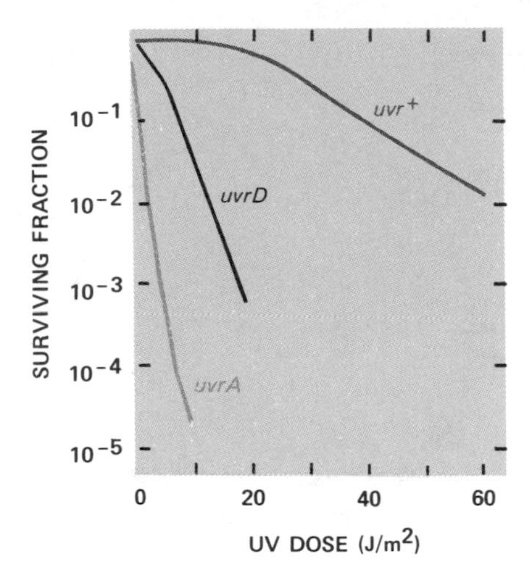

FIGURE 5-15
Mutants defective at the *uvrD* locus (*uvrE*, *mutU*, *recL*) are abnormally UV sensitive, but not as sensitive as mutants defective at *uvrA*. (From G. B. Smirnov et al., ref. 155.)

of *E. coli* was shown to be distinct from mutants defective at the *uvrA*, *uvrB* and *uvrC* loci, and the relevant gene was designated *uvrD* (Fig. 5-15) (71). Subsequently a mutant strain was isolated that showed increased UV sensitivity and an increased spontaneous mutation frequency; it was called *mutU* (72). At about the same time, mutants with very similar properties, called *uvrE*, were identified (73), and yet another locus that affects genetic recombination was called *recL* (74). This genetic complexity was resolved when it was shown that *uvrE*, *recL*, *mutU* and *uvrD* are alleles of the same gene, which is located at about 84 minutes on the *E. coli* genome (Fig. 5-16) (75).

UvrD mutants of *E. coli* are not defective in DNA incision. However, these mutants have abnormally slow rates of incision in semi-in-vitro experimental systems, for example, in cells permeabilized with the detergent Triton X-100 (76). In addition, intact UV-irradiated mutant *uvrD* cells have a defect in the *excision* of thymine-containing pyrimidine dimers (77–80) (Fig. 5-17). Both of these observations suggest that the *uvrD* gene product might directly interact with the uvrABC enzyme during concerted incision of DNA and excision of base damage.

There is also evidence that postincision events are abnormal in *uvrD* strains. Thus, a decrease in the rate and extent of the rejoining of repaired regions to parental DNA has been observed (81). In addition, some *uvrD* mutants show *enhanced* repair replication relative

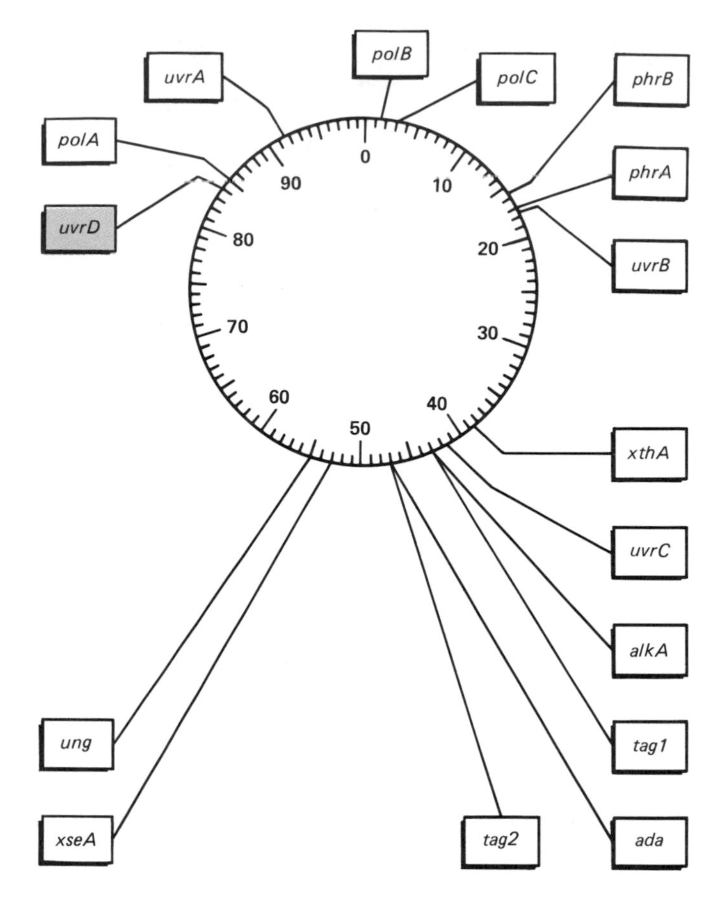

FIGURE 5-16
The location of the *uvrD* gene on the *E. coli* genetic map.

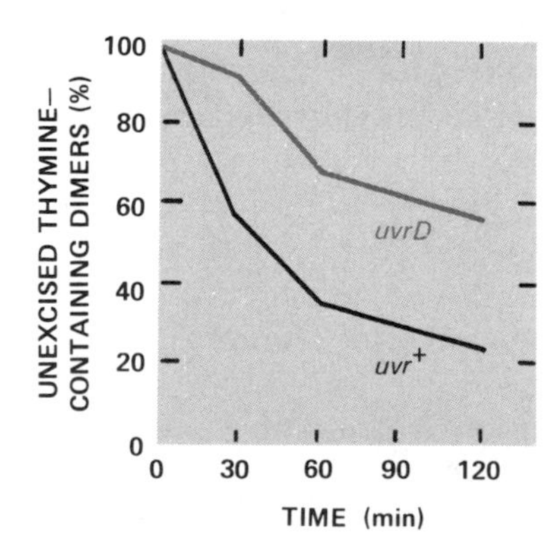

FIGURE 5-17
Mutants defective in the *uvrD* gene show a reduced rate and extent of loss of thymine-containing pyrimidine dimers from DNA. (From C. A. van Sluis et al., ref. 80.)

to wild-type cells (82). Repair patch size measurements have shown an effect primarily on the long patch component of excision repair synthesis; that is, the *uvrD101* mutant contains long patches of repair synthesis at a larger number of sites than are repaired by this mode in wild-type cells. In addition, these patches are longer than the average long patch in normal cells (82).

The *uvrD* gene has been cloned (83–85), and plasmid or phage vectors carrying the gene complement the UV sensitivity of recessive *uvrD, uvrE* and *recL* mutations (83–85). Expression of these vectors has made possible identification of the product of the *uvrD* gene as a single polypeptide of molecular weight approximately 75,000 with a DNA-dependent ATPase activity (83–88). A known DNA-dependent ATPase of *E. coli* of the same molecular weight is referred to as DNA helicase II (89), and since antiserum directed against DNA helicase II inhibits the *uvrD* gene product (87), it is probable that they are the same protein. Consistent with this conclusion, purified *uvrD* protein catalyzes the unwinding of duplex ϕX174 DNA in vitro (88). If this protein unwinds DNA during the excision repair of bulky base damage mediated by the uvrABC endonuclease of *E. coli,* it is certainly easy to understand how incisions separated by 12 nucleotides (see Section 4-2) would result in the loss (excision) of a dodecanucleotide fragment.

Indirect evidence in support of a role of the *uvrD* gene in the excision of base damage mediated by the uvrABC endonuclease is that like the *uvrA, uvrB* and *uvrC* genes, the *uvrD* gene is inducible by DNA damage. Thus, when growing *E. coli* cells are treated with mitomycin C or nalidixic acid (well-established inducers of *recA-lexA*-regulated genes; see Section 7-6), the level of DNA-dependent ATPase activity increases four- to sixfold (88). This increase is not observed in mutant *recA* cells, however (84). Furthermore, the 5′ end of the cloned *uvrD* gene has been sequenced and shown to contain a *lexA* binding site (87).

The mfd *and* dnaB *genes*

A poorly understood mutation at a locus in *E. coli* designated *mfd* (*m*utation *f*requency *d*ecline) results in slower kinetics of dimer excision relative to that of wild-type cells, even though this mutant is not abnormally sensitive to UV radiation (90). A specific *dnaB* mutant also has the phenotype of normal UV sensitivity but is hypermutable by UV radiation and shows a decreased rate of dimer excision in vivo (91).

The observation of only a moderate increase in UV sensitivity in some of the mutants discussed above (e.g., *polA1, uvrD* and *dnaB*) suggests that these and possibly other genes are *involved* in but not *essential* for excision repair in *E. coli* in vivo. In considering possible roles of these gene products a number of possibilities (not necessarily mutually exclusive) come to mind.

1. As suggested above with respect to the *uvrD* protein, the products of some of these genes may interact with damaged DNA during DNA incision-excision mediated by the uvrABC endonuclease and thereby improve the kinetics and efficiency of this process in living cells.
2. The reported differences in the size of the fragments excised by the uvrABC endonuclease in vitro (about 12 nucleotides) and the size of the repair patches (about 20 to 30 nucleotides) may simply reflect the limits of the experimental technique used for these determinations, as suggested previously. On the other hand, this size difference may be real and reflect the expansion of some or all of the excision gaps prior to or during repair synthesis; that is, postexcisional gap expansion mediated by the exonuclease functions discussed earlier in this section may be an inherent component of excision repair.
3. If the oligonucleotide fragments excised by the uvrABC endonuclease are 12 nucleotides or longer in vivo, they may not be acid soluble until further degraded intracellularly. Thus the demonstrated involvement of exonucleases in the loss of thymine-containing pyrimidine dimers from the acid-insoluble fraction of cells may reflect the degradation of the excised fragment into acid-soluble products rather than the excision repair process itself. For this reason it is important to make a clear distinction between the *excision of bases* and the *loss of bases* from the acid-insoluble fraction of DNA; that is, the appearance of damaged bases in a soluble fraction of DNA *always* implies their excision, but their excision from DNA need not always be accompanied by their immediate presence in the soluble phase.

5-4 Postincision Events during Excision Repair Mediated by AP Endonucleases

AP endonucleases with or without associated DNA glycosylases catalyze incisions in DNA only on one side of sites of base loss in DNA (8, 9). Thus the excision of pyrimidine dimers in which incision of DNA is mediated by the phage T4 or *M. luteus* PD DNA glycosylase–AP endonuclease requires the action of a $5' \rightarrow 3'$ exonuclease. In addition, irrespective of whether a given AP endonuclease hydrolyzes a phosphodiester bond $3'$ or $5'$ with respect to *any* AP site, the removal of the deoxyribose-phosphate moiety obviously requires further degradation of the DNA. This could be achieved by the action of a second AP endonuclease of opposite

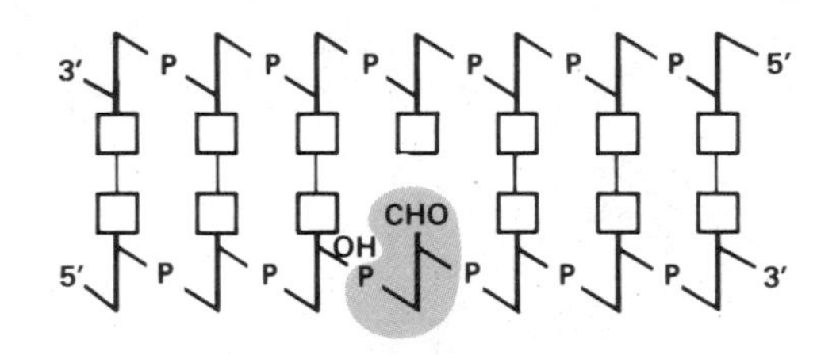

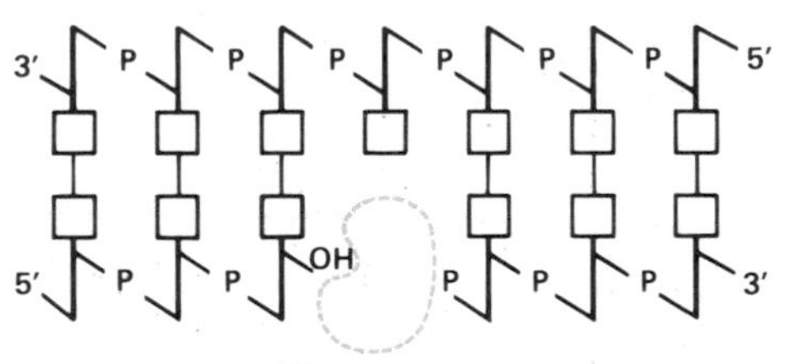

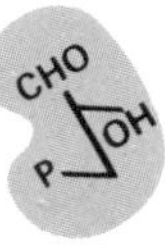

FIGURE 5-18
Incision of DNA containing sites of base loss by a 5′ AP endonuclease leaves a 5′-terminal deoxyribose-phosphate moiety in the DNA. This moiety can be excised by the action of either a 3′ AP endonuclease or a 5′ → 3′ exonuclease, leaving a single nucleotide gap in the DNA duplex.

polarity, and such an excision mechanism has been demonstrated in vitro (92) (Fig. 5-18). However, since the biological significance of AP endonucleases that hydrolyze 3′ phosphodiester bonds in vitro is not firmly established (see Section 3-4), it is necessary to consider that in living cells DNA incision at sites of base loss occurs exclusively 5′ to such lesions and hence that excision of nucleotide residues requires the action of 5′ → 3′ exonucleases (Fig. 5-18). Finally, it should be kept in mind that in *E. coli* and in other prokaryotic cells, damage-specific endonucleases that are distinct from the uvrABC enzyme (e.g., endonuclease V; see Section 4-4) may also catalyze the formation of only one strand break adjacent to each lesion. Hence, completion of excision repair initiated by these endonucleases would also require independent exonucleases or other endonucleases.

Excision of Pyrimidine Dimers in Phage T4-Infected Cells

Support for a primary role of DNA polymerase I in pyrimidine dimer excision in vivo stems from studies on *E. coli* infected with UV-irradiated phage T4. Not surprisingly, loss of thymine-containing pyrimidine dimers from acid-insoluble phage DNA has an absolute

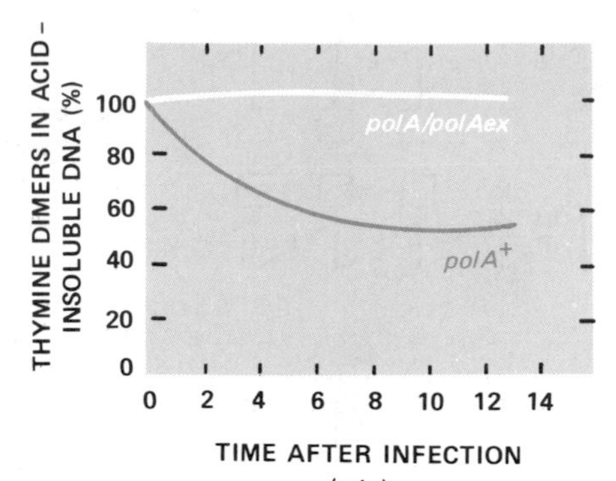

FIGURE 5-19
Mutants of *E. coli* defective in either the polymerizing (*polA*) or the 5′ → 3′ exonuclease (*polAex*) functions of DNA polymerase I do not support excision of thymine-containing dimers from phage T4 DNA in vivo. (From G. Pawl et al., ref. 93.)

dependence on a normal *denV* gene (93). However, no detectable dimer excision is observed in either *polA1* or *polAex* mutants of *E. coli* (Fig. 5-19) (93), despite the fact that phage T4 expresses an independent 5′ → 3′ exonuclease activity that catalyzes the excision of dimers from specifically preincised UV-irradiated DNA in vitro (94–96). Apparently this phage T4-coded exonuclease is not involved in pyrimidine dimer excision in vivo; rather, the repair of the phage genome is dependent on functional DNA polymerase I. Consistent with this interpretation, UV-irradiated phage T4 has a lower survival in *E. coli polA1* mutants than in cells that are wild type for this gene (97, 98) (Fig. 5-20). However, the sensitivity of T4 phage to killing by UV radiation in mutant *polA* cells is not as great as that of irradiated T4 *denV*$^-$ phage in wild-type cells (Fig. 5-20). Thus presumably other pathways for excision of dimers can operate in the absence of functional DNA polymerase I, and/or phage viability is partially restored by other mechanisms, e.g., recombination (99–102) or replicational tolerance of lesions in template DNA (103, 104).

If incision of UV-irradiated phage DNA occurs by a two-step mechanism that leaves 3′-terminal apyrimidinic sites, as suggested from studies in vitro (see Section 3-4), the observed dependence on DNA polymerase I for dimer excision from T4 DNA could derive from a requirement for the 3′ → 5′ exonuclease function of this enzyme to remove the 3′-terminal lesions. However, the effect of mutations in the *xthA* gene (which encodes the major 5′ AP endonuclease function of *E. coli;* see Section 3-5) on thymine dimer excision from UV-irradiated T4 DNA has not been determined; thus removal of putative 3′-terminal deoxyribose-phosphate moieties in vivo may equally well be catalyzed by this or by any other 5′ AP endonuclease in *E. coli* (see Fig. 3-4).

Repair synthesis during excision repair of UV-irradiated phage T4

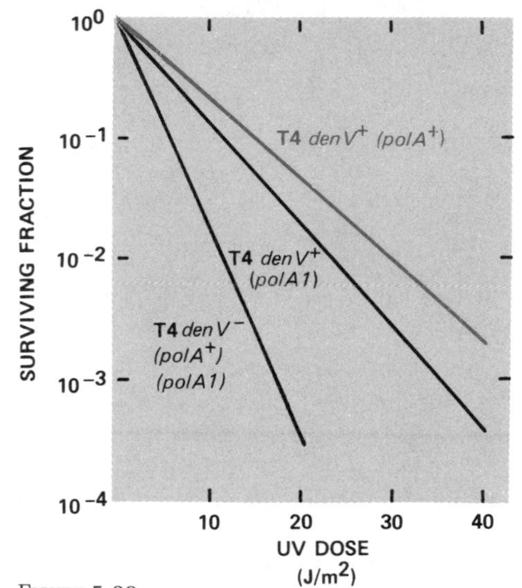

FIGURE 5-20
Phage T4 *denV*$^+$ is more sensitive to killing by UV radiation in an *E. coli polA1* than in a *polA*$^+$ strain, but it is not as sensitive as a T4 *denV*$^-$ phage in either *polA* strain.If survival of UV-irradiated phage T4 depended exclusively on excision repair involving *E. coli* DNA polymerase I, one would expect the sensitivity of T4 *denV*$^+$ strains in *polA1* cells to be the same as that of *denV*$^-$ phage strains. The observation that this is not the case implies that either other enzymes can subsume the role of DNA polymerase I in excision repair of T4 DNA, or phage survival is effected by mechanisms other that DNA polymerase I-dependent excision repair. (After K. Mortelmans and E. C. Friedberg, ref. 98.)

(63, 105) has not been extensively studied. The evidence presented above suggesting a primary role of *E. coli* DNA polymerase I in the excision of pyrimidine dimers from the phage genome leads to the obvious speculation that this enzyme is also a likely candidate for associated repair synthesis of DNA by nick translation. This is supported by observations on the excision of thymine-containing pyrimidine dimers in vitro when fragments of *E. coli* DNA polymerase I are used (48). The small fragment of the enzyme catalyzes the excision of dimers from UV-irradiated DNA previously incised with the T4 PD DNA glycosylase–AP endonuclease; however, the rate of this reaction is significantly slower than that catalyzed by the intact enzyme (48) (Fig. 5-21). Addition of the large fragment to the excision reaction stimulates the $5' \rightarrow 3'$ exonuclease activity (Fig. 5-21) and the rate of excision by the intact enzyme is stimulated even further by the addition of the four deoxynucleoside triphosphates, i.e., by conditions that allow for DNA synthesis (48) (Fig. 5-21). Collectively, these observations support a role of DNA polymerase I in excision

FIGURE 5-21
The small fragment of *E. coli* DNA polymerase I (containing just the 5′ → 3′ exonuclease function) excises thymine-containing pyrimidine dimers at a slow rate from preincised UV-irradiated DNA. The rate is increased by addition of the large fragment (containing the DNA polymerizing function) and is increased even further by addition of the large fragment and the four usual deoxynucleoside triphosphates; that is, under conditions permissive for DNA synthesis. These data suggest that excision of dimers from preincised DNA catalyzed by *E. coli* DNA polymerase I is optimal during concerted DNA synthesis and nick translation in the 5′ → 3′ direction. (From E. C. Friedberg and I. R. Lehman, ref. 48.)

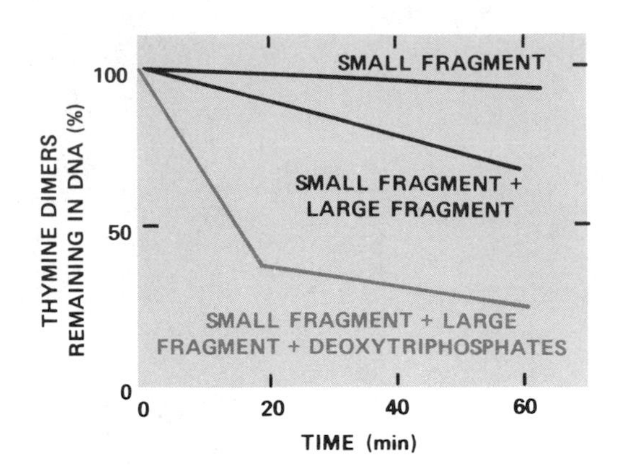

repair that is intuitively obvious based on its known properties; that is, excision and resynthesis of DNA are probably coordinated events that occur by nick translation of DNA and at no time leave a real gap in the DNA duplex. Despite these arguments, it is not possible to exclude roles for DNA polymerases II and/or III during repair synthesis of T4 DNA. In addition, gene *43* of T4 encodes a phage DNA polymerase (106). Mutants with a temperature-sensitive defect in gene *43* have been reported to show a reduced ability to repair UV-induced lethal lesions (105), but the possibility has been raised that this may be an experimental artifact (107).

Mutants defective in gene *32* of phage T4 (which encodes a major single-strand-binding protein) (108) are abnormally sensitive to UV radiation, and a temperature-sensitive mutant in this gene has been used to study its role in excision repair (109). At low temperatures at which semiconservative DNA synthesis is permissive, no repair synthesis of UV-irradiated T4 DNA is observed, suggesting that some component of excision repair requires the gene *32* product (109).

Excision of 3′-Terminal AP Sites

It has been mentioned on several occasions that the incision of DNA at pyrimidine dimers in vitro by both the phage T4 and the *M. luteus* PD DNA glycosylase–AP endonuclease activities creates 3′-terminal deoxyribose-phosphate moieties (9). It was also indicated in Section 3-5 that such moieties are poor template-primers for DNA synthesis by *E. coli* DNA polymerase I in vitro (92). Thus, if the same polarity of incision at the apyrimidinic sites created by these glycosylases operates in vivo, living cells presumably need to remove these 3′-terminal moieties by some type of excision event(s) in order for efficient repair synthesis to take place. The same considerations apply to the incision of DNA 3′ with respect to any other sites of base loss in the DNA in living cells, irrespective of their origin.

A number of *E. coli* enzymes can remove 3′-terminal moieties in vitro. These include the two 5′ AP endonucleases IV and the AP endonuclease function of exonuclease III (see Section 3-5) (9). In addition, despite the observation that 3′-terminal deoxyribose-phosphate moieties are poor template-primers for DNA polymerase I relative to normal 3′ termini, if the amount of polymerase is increased, or if the nicked DNA is preincubated for extended periods with polymerase, essentially normal rates of DNA synthesis are observed (92, 110) (Fig. 3-34). Thus it would appear that the 3′ → 5′ exonuclease function of this enzyme can also excise these 3′-terminal structures in vitro, albeit slowly. The 3′ → 5′ exonucleases of DNA polymerases II and III have not been characterized in this regard. There is also no information on the possible role of other 3′ exonucleases in *E. coli* (Table 5-7) in the repair of 3′-terminal lesions in duplex DNA. The 3′ exonuclease function of exonuclease III might also be able to remove 3′-terminal deoxyribose-phosphate moieties from DNA; however, it is not possible to distinguish this catalytic activity from that of the associated 5′ AP endonuclease (see Section 3-5), unless one or the other can be selectively inactivated.

TABLE 5-7
E. coli 3′ → 5′ exonucleases

Enzyme	Molecular Weight (kDa)	Single- or Double-Strand Specific	Comments
DNA polymerase I (3′ → 5′ exonuclease function)	109	Single	Acts on frayed or mismatched termini of DNA duplex
DNA polymerase II (3′ → 5′ exonuclease function)	120	Uncertain	
DNA polymerase III (3′ → 5′ exonuclease function)	Uncertain	Uncertain	
Exonuclease VII	~88	Single	Has both 3′ → 5′ and 5′ → 3′ polarity
Exonuclease V	268	Single and double	Also has 5′ → 3′ exonuclease and endonuclease activities
Exonuclease IV, A and B	Unknown	Single	Degrades oligonucleotides more rapidly than high molecular weight DNA
Exonuclease III	27.4	Double	Phosphomonoesterase on a 3′-phosphate terminus; exonuclease on duplex DNA at end or nick. Also has 5′ AP endonuclease activity
Exonuclease I	~70	Single	Single-strand specific; processive

From A. Kornberg, ref. 46.

5-5 DNA Ligation

The final postincisional biochemical event in all forms of excision repair is thought to be the joining of the last newly incorporated nucleotide to the polynucleotide chain; i.e., the sealing of the nick left following the completion of repair synthesis. In *E. coli* this event is catalyzed by an extensively characterized enzyme called *DNA ligase*. As is true of nucleases and the DNA polymerases involved in the earlier postincisional events in excision repair, DNA ligase is an enzyme that plays a role in aspects of DNA metabolism other than DNA repair, including genetic recombination and DNA replication (46).

The DNA ligase of *E. coli* is a product of the *lig* gene, which maps at about 51 minutes on the *E. coli* chromosome (111) (Fig. 5-22). A

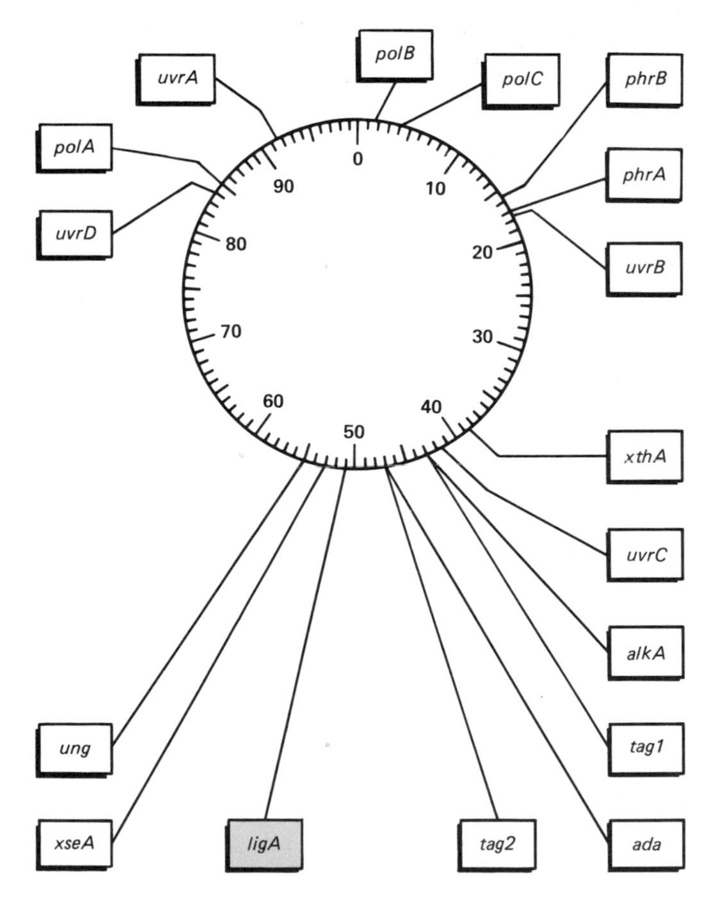

FIGURE 5-22
The location of the *ligA* gene on the *E. coli* genetic map.

conditional lethal mutant designated *ligts7* is abnormally sensitive to killing by alkylating agents such as methylmethanesulfonate or UV radiation, indicating its role in DNA repair (112). The enzyme is a monomeric protein with a molecular weight of 77,000. Wild-type *E. coli* has about 200 to 400 molecules per cell under optimal growth conditions and a turnover number of about 25 per minute has been calculated (113).

E. coli DNA ligase catalyzes the joining of nicks in duplex DNA with nucleotide termini containing 3′-OH and 5′-P groups, in a reaction that has a requirement for Mg^{2+} and for nicotinamide adenine dinucleotide (NAD) (113). During the reaction the NAD is hydrolyzed to yield nicotinamide mononucleotide (NMN) and AMP. The cleavage of a pyrophosphate bond leads to the synthesis of a phosphodiester bond in DNA by a sequence of three steps involving two covalently linked intermediates (113). The first step (Fig. 5-23) consists of a reaction of the enzyme with NAD to form a DNA ligase adenylate and NMN. Subsequently (Fig. 5-23), the adenyl group is transferred to DNA to generate a new pyrophosphate linkage between adenosine monophosphate and the 5′-phosphoryl terminus at

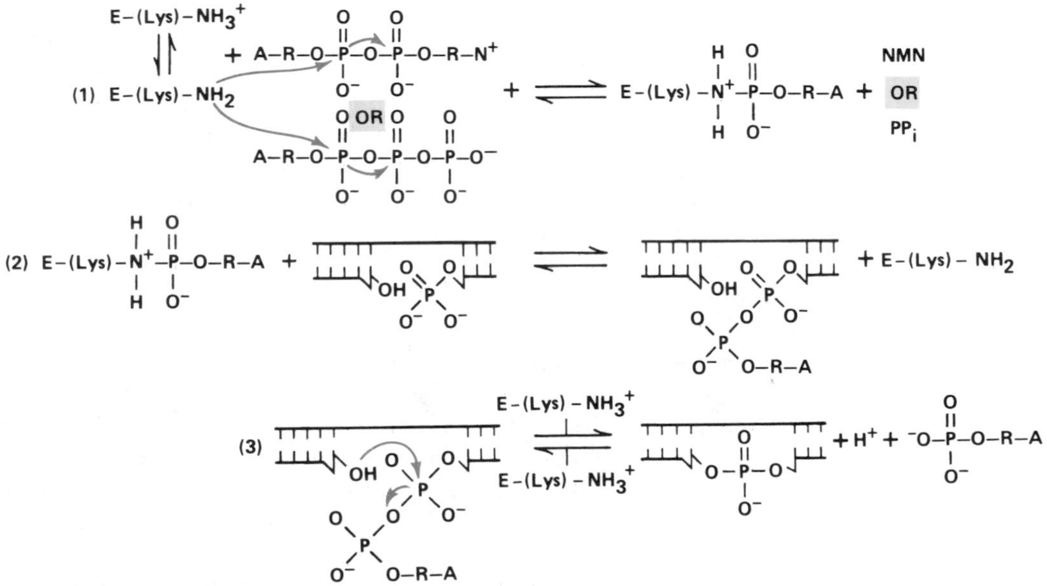

FIGURE 5-23
Mechanism of the DNA ligase reaction catalyzed by the *E. coli* (or phage T4) enzymes. These enzymes react with NAD (or ATP) to form a DNA-ligase–adenylate intermediate (through the ϵ-amino group of a single lysine residue in the protein) and NMN (or PP_i) (1). The adenyl group of this intermediate is transferred to DNA to form a pyrophosphate linkage between adenosine monophosphate and the 5′-P terminus at the DNA strand break (2). The 5′ P is then attacked by the opposing 3′-OH group to generate a phosphodiester bond, and free AMP is liberated (3). (E=enzyme; A=adenine; R=ribose.) (From I. R. Lehman, ref. 113.)

the nick. The 5′ phosphate is then attacked by the opposing 3′-OH group to generate a phosphodiester bond, and AMP is liberated (Fig. 5-23).

DNA ligase is a ubiquitous enzyme; those present in extracts of mammalian cells are discussed in the next chapter. Another prokaryote form of the enzyme that has been studied in detail is encoded by bacteriophage T4. It too is a single polypeptide chain, with a molecular weight of 63,000 (113). In contrast to the *E. coli* enzyme, the DNA ligase from phage T4 uses ATP rather than NAD as an energy source (113) and yields AMP and PP_i as reaction products (Fig. 5-23). The T4 enzyme can catalyze the joining of oligodeoxyribonucleotides or of oligoribonucleotides in RNA-DNA hybrid duplex molecules and can also promote the joining of DNA duplexes with flush ends. The rate of the latter reaction is lower relative to that of the joining of nicks in duplex DNA; however, it is an extremely useful reaction that is widely used in recombinant DNA technology.

5-6 Miscellaneous Functions Associated with Excision Repair

The involvement of other functions in excision repair (either prior to, concomitant with or following DNA incision) is suggested by a number of interesting preliminary studies. For example, mutants of *E. coli* in a gene designated *top*, which encodes DNA topoisomerase I, are abnormally sensitive to killing by UV radiation (114). In addition, the presence of pyrimidine dimers in DNA is associated with a reduction in the rate of relaxation of supercoiled plasmid DNA by *E. coli* DNA topoisomerase I in vitro (115). This has led to the suggestion that altered relaxation of damaged superhelical DNA in vivo resulting from reduced topoisomerase I activity might facilitate the recognition and binding of the uvrABC endonuclease to damaged DNA if this enzyme is sensitive to the superhelical density of substrate DNA (115). Other DNA topoisomerases such as DNA gyrase could also be involved in the excision repair of DNA. In fact, little is known about the role of DNA topology in DNA repair, and it is to be hoped that future years will provide enlightening information in this area.

The same basic postincision events of damage excision, repair synthesis and DNA ligation described in *E. coli* in this chapter are believed to occur in mammalian cells. However, there are important differences, which presumably reflect important distinctions between prokaryotes and eukaryotes. These differences as well as other aspects of postincision events in mammalian cells are discussed in Chapter 6.

5-7 Repair of Mismatched Bases in DNA

A consideration of the excision repair of DNA base damage would not be complete without a discussion of the repair of mismatched bases. It is very likely that at least some aspects of the biochemistry of *mismatch excision repair* include enzymes already discussed in this chapter. However, present information on the enzymology of the repair of this type of DNA damage is rudimentary, and for this reason the topic is discussed as a distinct form of excision repair.

Historically, an interest in the fate of mismatched sequences in DNA arose from studies on genetic recombination. Thus, for example, it had been noted since the mid-1950s that apparent genetic exchanges within very short intervals of certain fungal and bacteriophage genomes take place very often, suggesting the occurrence of multiple crossovers at much greater than random frequency (116–118). This phenomenon is termed localized or high-negative interference because in classical genetics studies it frequently resulted in an underestimate of the linkage distance between two closely linked markers. During the early 1960s the concept emerged that this as well as other genetic "peculiarities," such as the phenomenon of gene conversion (nonreciprocal recombination) observed in certain fungi, might be accounted for by the correction of mismatched sequences present in the heteroduplex regions of DNA generated during exchanges between two genomes (119). The notion that such correction may take place by excision repair stemmed directly from the basic model of the excision repair of pyrimidine dimers that evolved in the early 1960s. An instructional example of this extrapolation is contained in a paper on the topic of gene conversion in fungi published by Robin Holliday in 1964. In exploring various mechanisms by which gene conversion might occur, Holliday referred to the formation of "hybrid" DNA molecules by homologous pairing between portions of single strands from two different DNA duplexes. He stated (120) that if

> this part of the genetic material is homozygous then normal base pairing will occur in the hybrid region, but if the annealed region spans a point of heterozygosity—a mutant site—then mispairing of bases will occur at this site. It is further postulated that this condition of mispaired bases is unstable. . . . One or both of such bases may get involved in exchange reactions. . . [and] it is most reasonable to suppose that such exchange reactions would be enzyme mediated. There is a rather obvious connection between this suggestion and the growing evidence for mechanisms in the cell which can repair DNA damaged by mutagens. . . . If there are enzymes which can repair points of damage in DNA, it would seem possible that the same enzymes could recognize the abnormality of base pairing, and by exchange reactions rectify this.

The Use of Heteroduplex DNA for Studying Mismatch Repair

A powerful model system for studying the repair of mismatches involves the use of heteroduplex DNA molecules, usually of phage origin. Such molecules are called heteroduplexes (or hets) because they contain regions of noncomplementarity in their nucleotide sequence. They are generally constructed using techniques by which the two strands of a phage DNA duplex are separated on the basis of differential density and then reannealed with the complementary strands from a mutant phage (121). Following transfection of appropriate hosts, each DNA strand will of course serve as a template for the synthesis of two genetically distinct homoduplex molecules. A bacterial cell transfected with a het will thus yield two genotypically distinct populations of phage particles and ideally produce a completely mixed phenotype in the progeny, i.e., a 50-50 mixture. However, if mismatched (heteroduplex) regions are corrected *prior* to DNA replication, affected molecules will yield only one type of homoduplex and the phenotype will reflect a corresponding bias in its representation in progeny phage (Fig. 5-24), the extent of which will depend on the efficiency of the correction mechanism(s). Thus, following transfection of *E. coli* with heteroduplex DNA molecules, a conversion of the heteroduplex into homoduplex phenotype indicates the operation of a correctional mechanism (121). This approach has been used with hets constructed from phage λ (122, 123), phage ϕX174 (124) and the *B. subtilis* phage SPP1 (125), as well as with heteroduplex transfecting DNA from *Diplococcus pneumoniae* (126). In each case the results have been interpreted in terms of mismatch repair, although other explanations such as recombination following replication have been considered.

The possible contribution of recombination events to mismatch correction can be reduced by transfecting strains of *E. coli* defective in the *recA* gene (which is required for most generalized recombination in *E. coli;* see Section 7-3) with hets constructed from λ phages defective in the *int* and *red* genes (122). The latter are required for site-specific recombination (integration) and for phage-mediated generalized recombination respectively (46). A significant role of any residual generalized recombination in the conversion of heteroduplex transfections into the homoduplex state is unlikely, because the frequency of conversion of selected *amber* mutations is unrelated to the distance separating the mutation in the genome (122). The heteroduplex conversions observed under these restrictive conditions are therefore most likely the result of the repair of mismatches by excision-resynthesis of tracts of DNA greater than several hundred nucleotides (122). Well-separated pairs of mis-

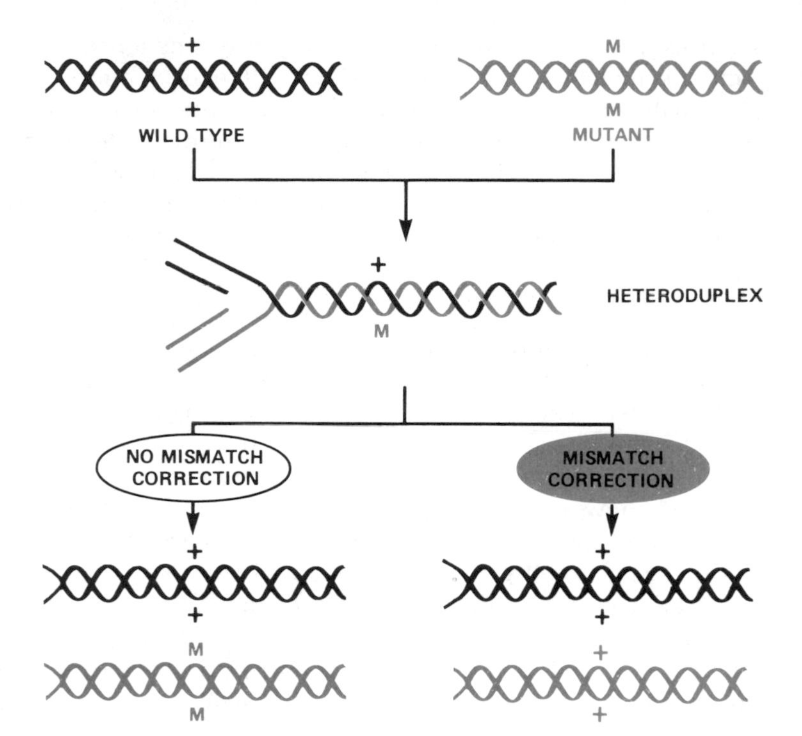

FIGURE 5-24
The use of heteroduplex molecules for measuring mismatch corrections. If the mismatch is corrected, all molecules generated by semiconservative DNA synthesis will be wild-type homoduplexes (right). However, in the absence of mismatch correction 50 percent of the progeny molecules will be mutant homoduplexes (left).

matches are repaired independently; however, sites less than about 3000 nucleotides apart can be repaired in a single event. Selection of appropriate genetic markers with respect to the polarity of the physical genome of phage λ has suggested that the repair tracts initiated at mismatches are propagated in the $5' \rightarrow 3'$ direction (127).

The Molecular Mechanism of Mismatch Excision Repair

The heteroduplex systems considered above are presumably valid models for mismatches arising in heteroduplex structures generated during genetic recombination in vivo. An additional source of mispaired bases is from replicative infidelity during semiconservative (or nonsemiconservative) DNA synthesis. In view of the fundamental distinction between these two possible mechanisms for the origin of mismatches, it is conceivable that the molecular mechanism(s) for their repair are partially or completely distinct. The phenomenon of mismatch excision repair has only recently been demonstrated in vitro and its molecular mechanism(s) is still poorly understood. However, with respect to the repair of mismatches arising during DNA synthesis in *E. coli*, a number of specific required genetic components have been identified, some of which have known biochemical functions that have provided important insights into the molecular biology of mismatch repair in this organism.

The Role of DNA Methylation

The excision of nucleotides at mismatched regions generated by replicative errors is determined in part by the extent of methylation of the DNA, in a way very reminiscent of the modification-restriction system that determines host specificity (128). An obvious question in the consideration of any mechanism for the repair of replicative mismatches is how a repair system could discriminate between the newly-replicated strand containing the incorrect base and the template strand. Since methylation of DNA as an antirestriction modification is known to be a postreplicative phenomenon, it was suggested that one way in which an excision repair system might be "instructed" as to which DNA strand to attack is by the specific recognition of the undermethylated strand during the interval between replication and DNA methylation (127) (Fig. 5-25).

This model is consistent with observations on mutants of *E. coli* containing reduced amounts of 6-methyladenine in their DNA. The undermethylation in these strains is the result of mutations in a gene called *dam*, located at 65 minutes on the *E. coli* genetic map (Fig. 5-26) and well separated from the *hsp* genes responsible for host specificity (129–131). The *dam* gene encodes a DNA methylase that catalyzes methylation at the 6 position of adenine moieties in GATC sequences of DNA (132, 133). DNA isolated from *dam* mutants contains single-strand breaks and/or alkali-labile lesions that are amplified in *dampolA* and *damlig* double mutants (129), suggesting that a function of *dam*-specified 6-methyladenine moieties in DNA is

FIGURE 5-25
A model for postreplicative mismatch correction of DNA. GATC sequences in DNA are normally methylated (Me) at the 6 position of adenine. During semiconservative DNA synthesis a G · T mismatch arises in one of the sister DNA duplexes. The enzymatic mechanism for repairing this lesion depends on discrimination between the newly synthesized (T) and the parental (G) strands. This is achieved by recognition of the transient lack of methylation of the newly synthesized strand before postreplicative DNA methylation takes place. The nonmethylated daughter strand containing the incorrect base is enzymatically attacked by mismatch correction enzymes and the misincorporated base is excised. Repair synthesis and daughter-strand methylation at GATC sequences restore the sister DNA duplexes to their native state.

the protection against an enzyme(s) that degrades undermethylated DNA. Another feature observed in the mutant *dam* phenotype consistent with a role of DNA methylation in postreplicative mismatch repair is that mutants defective in this gene are mutators (131) (Table 5-8). In addition, *dam* mutants are abnormally sensitive to agents

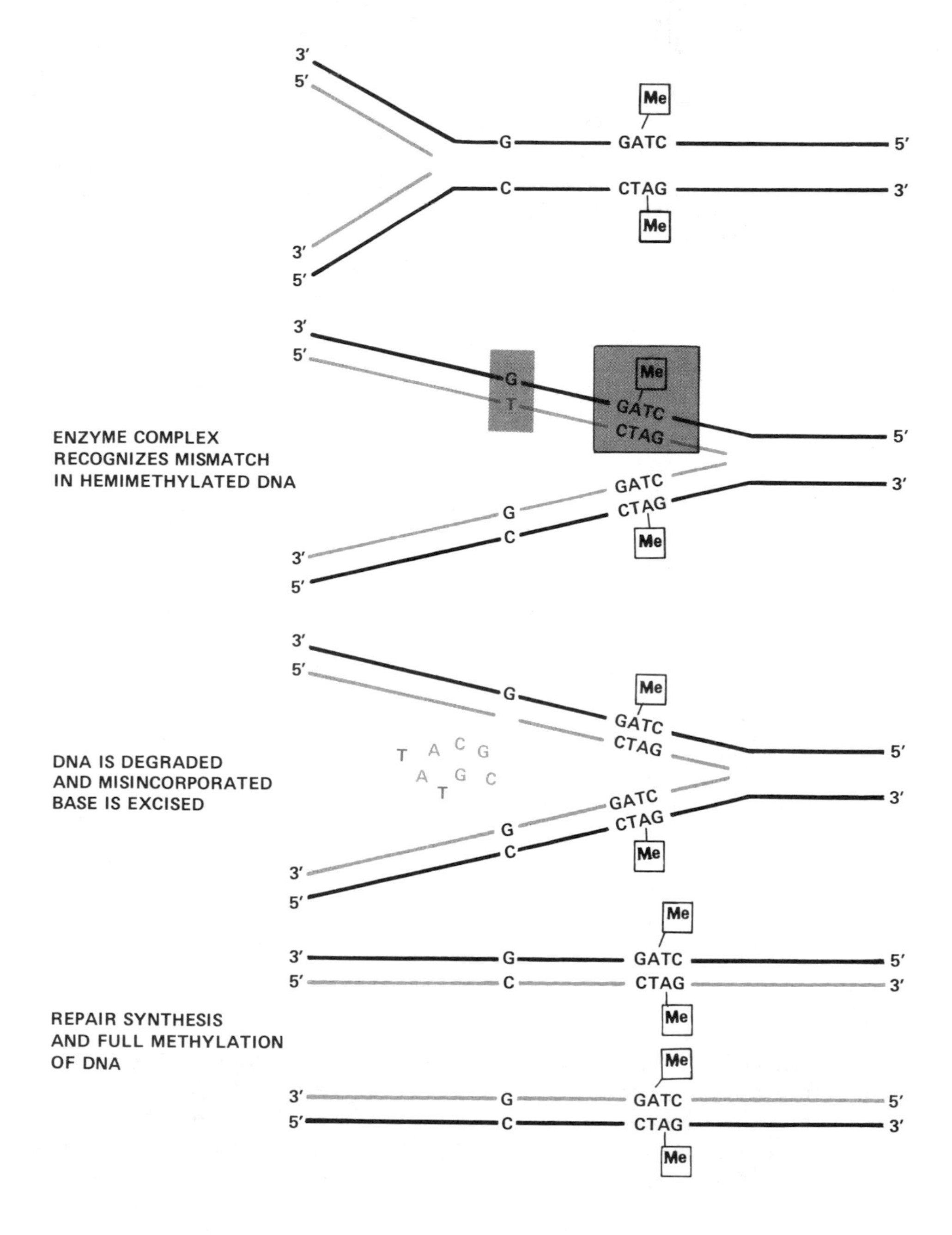

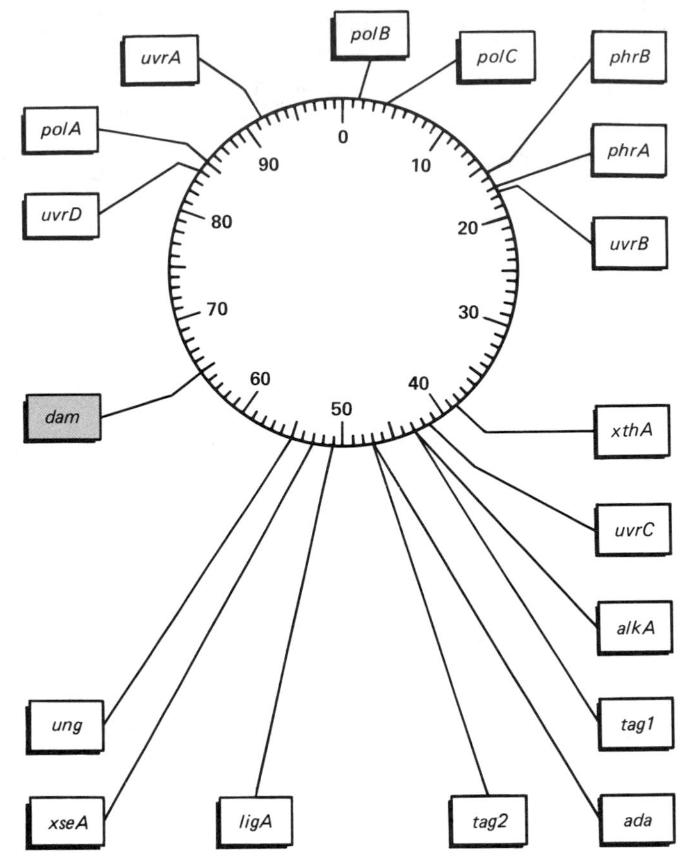

FIGURE 5-26
The location of the *dam* gene on the *E. coli* genetic map.

TABLE 5-8
Reversion frequencies ($\times 10^{-9}$) of identical markers in *dam-3* and wild-type strains of *E. coli*

	leu-6	*proA2*	*lacYl*	*strA*	*rif*
GM44 (wild)	3.67	0.16	4.5	0.16	4.0
GM45 (*dam-3*)	26.3	2.67	161.0	5.0	182.4

From M. G. Marinus and N. R. Morris, ref. 129.

that cause DNA strand breaks directly, or that result in strand breakage during the repair of the base damage they cause. These mutants also show increased recombination, are inviable in association with mutations in other genes normally required for genetic recombination (e.g., *recA*, *recB* or *recC*) and show increased induction of prophage λ (129–131). All these phenotypes are consistent with the model of mismatch excision repair shown in Fig. 5-25; that is, undermethylation of DNA in *both* strands would rob these strains of a mechanism for strand discrimination during mismatch excision repair. Thus, the excision of relatively long tracts of DNA as a consequence of attempted correction on opposite strands would lead to an increased probability of double-strand breaks, which could promote recombination, prophage induction and enhanced lethality. The mutant *dam* strains that have been isolated all contain residual 6-methyladenine and presumably owe their viability to leakiness of the *dam* mutation.

The *dam* gene encodes a DNA methylase that recognizes the symmetric tetranucleotide d(GATC) and introduces two methyl groups per duplex site, with the product of methylation being 6-methylaminopurine (132, 133). The *dam* gene has been cloned, and strains transformed with a plasmid (pGG503) carrying the cloned gene overproduce *dam* methylase (133). The availability of such plasmids has facilitated studies on cells containing elevated levels of the enzyme, which have shown that these strains are also hypermutable (134), presumably due to a reduction of the "time window" available for mismatch correction, i.e., between the events of daughter-strand DNA synthesis and its methylation at GATC sequences. On the other hand, a different multicopy plasmid (pMQ3) that also overexpresses the methylase reverses all the phenotypes of *dam* mutants, including hypermutability (135). This discrepancy may be a function of the absolute level of methylase expressed (135), since pGG503-containing strains of *E. coli* produce more than four times the amount of enzyme produced in pMQ3-containing strains (135).

A direct test of the hypothesis of methylation-instructed mismatch repair was carried out by transfecting *E. coli* with heteroduplex phage λ DNA molecules containing differentially methylated DNA strands (136–138). In order to simulate the hemimethylated state characteristic of newly replicated DNA in wild-type *E. coli* (in which the parental DNA strand is methylated and the daughter strand is not), phage λ carrying selected genetic markers *Pam3* and *cI* (Table 5-9) was grown in wild-type and mutant *dam* hosts and heteroduplexes were constructed with mismatched base pairs at the sites of the markers in all four possible combinations of strand methylation: both strands methylated; Watson strand methylated, Crick strand unmethylated; Watson strand unmethylated, Crick strand methylated; and both strands unmethylated. Table 5-9 shows the results of transfection of wild-type *E. coli* with doubly heterozygous DNA hets carrying well-separated mismatches. When the expected deviation due to slight contamination of the λ *l* and r DNA strand pools during strand separation is taken into account, the recovery of parental markers in all cases shows a profound bias in favor of the genotype

TABLE 5-9
Transfections of wild-type *E. coli* with doubly heterozygous DNA heteroduplexes

hets* NR 3752 (su^+)	su^+ cI^+	su^+ cI	su^- cI^+	su^- cI	Genotypes cI^+ *Pam3*	Genotypes cIP^+	Total Plaques Analyzed
1 + *Pam3* / r —— / *cI* +	254	258	16	248	238 (46.5%)	248 (48.4%)	512
1 + *Pam3* / r —— / *cI* +	584	21	20	13	564 (96.5%)	13 (2%)	605
1 + *Pam3* / r —— / *cI* +	40	760	21	758	19 (2.4%)	758 (94.75%)	800
1 + *Pam3* / r —— / *cI* +	339	289	58	271	281 (44.7%)	271 (43%)	628

*—— indicates methylated strands; —— represents unmethylated strands.
From M. Radman et al., ref. 136.

represented on the methylated strand; that is, it is consistent with excision of the mismatch on the *unmethylated* strand. The possible alternative interpretation that there is a preferential *loss* of the non-methylated strand was ruled out by control experiments. In addition, mutants other than the *dam* one that are defective in mismatch repair (see below) show no methylation-dependent bias in marker representation.

Other Genes Involved in Mismatch Repair in E. coli

Beside the *dam* methylase, what other proteins (presumably enzymes) are required for postreplicative strand-directed mismatch repair in *E. coli*? Mutations in the *uvrA*, *uvrB* and *uvrC* genes have no effect on the conversion of heteroduplexes into homoduplexes (121). However, when mutants defective in other genes known to be involved in cellular responses to DNA damage were systematically examined, it was shown that mutants defective in the *uvrD* gene are defective in mismatch repair of λ hets (139). This is consistent with the observation that such mutants are mutators. Indeed, as indicated earlier in this chapter (Section 5-3), one of the numerous designations for the *uvrD* locus before it was realized that these were redundant was *mut*[ator]*U* (72).

One might therefore anticipate that other *E. coli* mutator mutants are defective in mismatch repair, and experimental protocols for selecting mutator mutants have been established to examine this relation (140, 141). For example, *dam* mutants are sensitive to killing as the result of mismatches in DNA arising from misincorporation of 2-aminopurine (142). This provides a strategy for the isolation of mutants defective in adenine-methylation-instructed mismatch correction (141) (Fig. 5-27). Relatively closely spaced, newly incorporated 2-aminopurine residues result in overlapping excision repair tracts when excision occurs in *both* the parental and the newly synthesized strands, i.e., in a *dam* mutant. Mutants defective in the enzymes required for DNA degradation during mismatch excision repair would be expected to be resistant to killing by 2-aminopurine in a mutant *dam* background. Such mutants were isolated, and deriv-

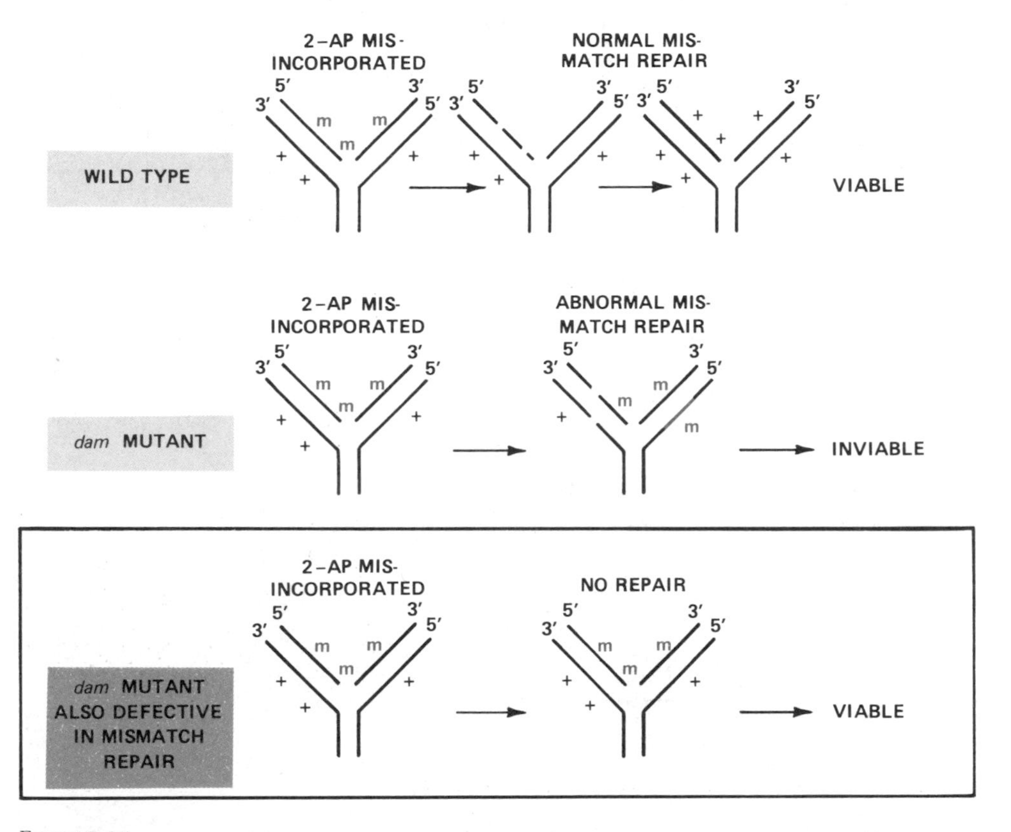

FIGURE 5-27
Scheme for the isolation of mutants that are defective in mismatch repair. In wild-type (dam^+) cells (top row) misincorporation of 2-aminopurine (m) results in normal mismatch repair as postulated in Fig. 5-25. In *dam* mutants (middle row) the strand discrimination for DNA degradation specified by the DNA methylation pattern is disrupted. Hence *both* strands are attacked by mismatch repair enzymes and there is a high probability of lethal double-strand breaks. However, if in addition to the *dam* mutation strains are also defective in genes required for excision repair of the mismatches, the DNA is not degraded and the mutants survive (bottom row). (From B. W. Glickman and M. Radman, ref. 141.)

ative strains constructed to eliminate the *dam* mutation are in fact mutators (141) and map genetically at one of three previously established mutator loci called *mutL*, *mutS* and *mutH* (Fig. 5-28) (143). The mutator potencies of double and triple *mut* mutants are no greater than that of individual mutants. This epistatic interaction suggests that these genes are involved in the same general mismatch repair pathway (141).

There are indications that the products of these mutator genes are specific for the particular biochemistry of the mismatch and of the specificity of the relation of this biochemistry to the strand bearing the appropriate methylation (137, 138). However, an explanation for these specificities must await the isolation and characterization of the products of the mutator genes. Information concerning these gene products is likely to be forthcoming, since recent studies have dem-

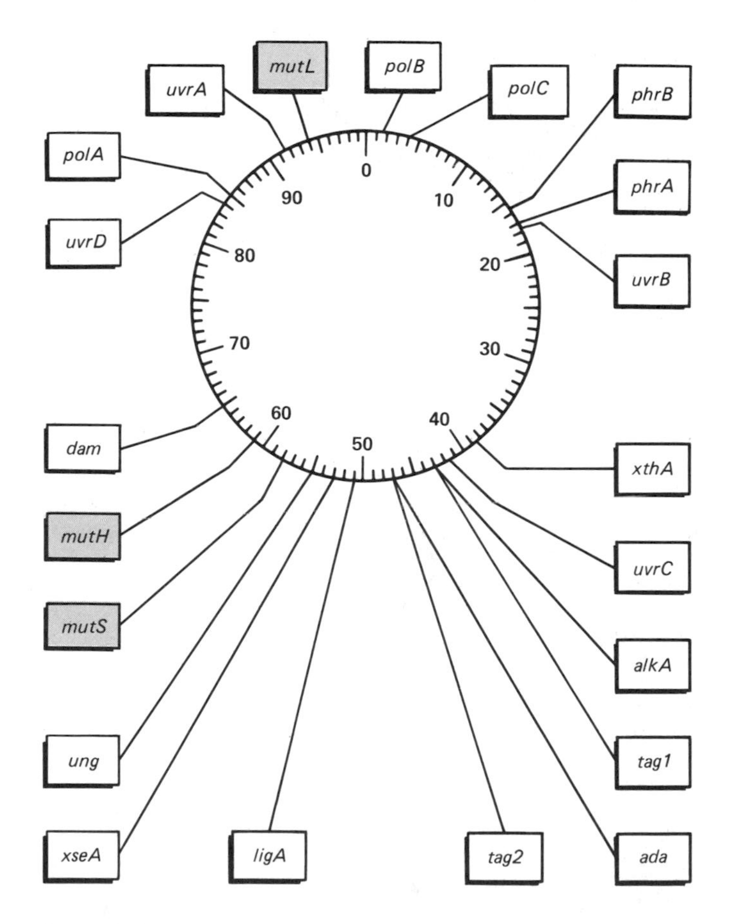

FIGURE 5-28
The location of the *mutL*, *mutS* and *mutH* genes on the *E. coli* genetic map.

onstrated the correction of mismatches in vitro (144, 145). Thus, the incubation of a hemimethylated DNA duplex containing a mismatch at a single restriction site for the enzyme *EcoRI* with crude extracts of *E. coli* results in restoration of the *EcoRI* site (144). This mismatch repair in vitro is defective in extracts of *mutL*, *mutS*, *mutH* and *uvrE* strains, but complementation of a given mutant extract with that from a different mutant restores the repair activity (144, 145). Such complementation could provide the basis for an assay to facilitate the purification and characterization of the proteins encoded by the *mut* genes mentioned above.

Progress in elucidating the biochemistry of mismatch repair can also be expected to come from studies with cloned genes. The *mutH* gene has been isolated and the product of its expression identified as a protein with a molecular weight of 25,000 (146). After the gene is placed under the control of the strong λ P_L promoter, thermal induction results in overexpression of *mutH* and has facilitated the purification of the gene product to better than 90 percent homogeneity. The purified protein does not bind to single- or double-strand DNA and has no intrinsic endonuclease, exonuclease or DNA-independent ATPase activity (146).

Mismatch Repair in Other Organisms

A discriminating strain of *Streptococcus pneumoniae* (*hex*) is transformed for various genetic markers with widely varying efficiencies. Two efficiency classes for a series of aminopurine-resistance mutants have been observed, referred to as high- and low-efficiency strains (147). It has been proposed that mismatch excision repair of donor-recipient complexes accounts for the low efficiency of transformation by some markers, and nondiscriminating mutant *hex* strains, which are transformed for most of these with equal efficiency, have provided support for the hypothesis (148–151). Little is known about the biochemistry and genetics of this system; however, there are indications that the *hex* system discriminates at the level of the particular chemistry of the mismatches. Thus, sequence analysis of DNA was carried out after transformation of *S. pneumoniae* with low- and high-efficiency markers (152), and it was observed that among the mismatches recognized by the *hex* system, all contained at least one pyrimidine. Thus, A·C, G·T and C·T were recognized, whereas A·G mismatches escaped correction by the *hex* system.

It has recently been shown that in *S. typhimurium*, *mutL*, *mutS*, *mutH* and *uvrD* mutants are particularly sensitive to mutagenesis produced by alkylating agents but not to mutagenesis resulting from treatment of cells with agents that produce bulky base damage, such as 4-nitroquinoline-1-oxide or UV radiation (153). A definitive explanation for this observation has not been established. However, it is likely that the *mut* genes referred to above are required for the repair of mismatches that originate from the replication of alkylated bases, with subsequent repair of the alkylation damage (153).

There is evidence suggesting the presence of mismatch repair of DNA in mammalian cells. This topic is discussed in the next chapter.

Suggestions for Further Reading

Postincision events during excision repair, refs. 6–8, 36; mismatch repair, refs. 136–138, 143, 144.

References

1. Rupp, W. D., Sancar, A., and Sancar, G. B. 1982. Properties and regulation of the UVRABC endonuclease. *Biochimie* 64:595.
2. Yeung, A. T., Mattes, W. B., Oh, E. Y., and Grossman, L. 1983. Enzymatic properties of the purified *Escherichia coli* uvrABC complex. In *Cellular responses to DNA damage*. E. C. Friedberg and B. A. Bridges, eds., p. 77. New York: Alan R. Liss.
3. Seeberg, E., and Steinum, A-L. 1983. Properties of the uvrABC endonuclease from *E. coli*. In *Cellular responses to DNA damage*. E. C. Friedberg and B. A. Bridges, eds., p. 39. New York: Alan R. Liss.
4. Sancar, A., and Rupp, W. D. 1983. A novel repair enzyme: UVRABC excision nuclease of *E. coli* cuts a DNA strand on both sides of the damaged region. *Cell* 33:249.
5. Yeung, A. T., Mattes, W. B., Oh, E. Y., and Grossman, L. 1983. Enzymatic properties of purified *Escherichia coli* uvrABC proteins. *Proc. Natl. Acad. Sci.* (USA) 80:6157.
6. Hanawalt, P. C., Cooper, P. K., Ganesan, A. K., and Smith, C. A. 1979. DNA repair in bacteria and mammalian cells. *Ann. Rev. Biochem.* 48:783.
7. Lehmann, A. R., and Karran, P. 1981. DNA repair. *Int. Rev. Cytol.* 72:101.
8. Lindahl, T. 1982. DNA repair enzymes. *Ann. Rev. Biochem.* 51:61.
9. Friedberg, E. C., Bonura,T., Radany, E. H., and Love, J. D. 1981. Enzymes that incise damaged DNA. In *The enzymes*. 3rd ed. P. D. Boyer, ed., vol. XIV, p. 251. New York: Academic.
10. Carrier, W. L. 1981. Measurement of pyrimidine dimers by paper chromatography. In *DNA repair—a laboratory manual of research procedures*. E. C. Friedberg and P. C. Hanawalt, eds., vol. I, part A, p. 3. New York: Dekker.
11. Reynolds, R. J., Cook, K. H., and Friedberg, E. C. 1981. Measurement of thymine-containing pyrimidine dimers by one-dimensional thin-layer chromatography. In *DNA repair—a laboratory manual of research procedures*. E. C. Friedberg and P. C. Hanawalt, eds., vol. I, part A, p. 11. New York: Dekker.
12. Sekiguchi, M., and Shimizu, K. 1981. Measurement of pyrimidine dimers by ion-exchange chromatography. In *DNA repair—a laboratory manual of re-*

search procedures. E. C. Friedberg and P. C. Hanawalt, eds., vol. I, part A, p. 23. New York: Dekker.

13. Love, J. D., and Friedberg, E. C. 1983. Measurement of thymine-containing pyrimidine dimers in DNA by high performance liquid chromatography (HPLC). In *DNA repair—a laboratory manual of research procedures*. E. C. Friedberg and P. C. Hanawalt, eds., vol. II, p. 87. New York: Dekker.
14. Ganesan, A. K., Smith, C. A., and Van Zeeland, A. A. 1981. Measurement of the pyrimidine dimer content of DNA in permeabilized bacterial or mammalian cells with endonuclease V of bacteriophage T4. In *DNA repair—a laboratory manual of research procedures*. E. C. Friedberg and P. C. Hanawalt, eds., vol. I, part A, p. 89. New York: Dekker.
15. Paterson, M. C., Smith, B. P., and Smith, P. J. 1981. Measurement of enzyme-sensitive sites in UV- or γ-irradiated human cells using *Micrococcus luteus* extracts. In *DNA repair—a laboratory manual of research procedures*. E. C. Friedberg and P. C. Hanawalt, eds., vol. I, part A, p. 99. New York: Dekker.
16. Lett, J. T. 1981. Measurement of single-strand breaks by sedimentation in alkaline sucrose gradients. In *DNA repair—a laboratory manual of research procedures*. E. C. Friedberg and P. C. Hanawalt, eds., vol. I, part A, p. 363. New York: Dekker.
17. Kohn, K. W., Ewig, R. A. G., Erickson, L. C., and Zwelling, L. A. 1981. Measurement of strand breaks and cross-links by alkaline elution. In *DNA repair—a laboratory manual of research procedures*. E. C. Friedberg and P. C. Hanawalt, eds., vol. I, part A, p. 379. New York: Dekker.
18. Ahnström, G., and Erixon, K. 1981. Measurement of strand breaks by alkaline denaturation and hydroxyapatite chromatography. In *DNA repair—a laboratory manual of research procedures*. E. C. Friedberg and P. C. Hanawalt, eds., vol. I, part A, p. 379. New York: Dekker.
19. Clayton, D. A. 1981. Measurement of strand breaks in supercoiled DNA by electron microscopy. In *DNA repair—a laboratory manual of research procedures*. E. C. Friedberg and P. C. Hanawalt, eds., vol. I, part A, p. 419. New York: Dekker.
20. Seawell, P. C., and Ganesan, A. K. 1981. Measurement of strand breaks in supercoiled DNA by gel electrophoresis. In *DNA repair—a laboratory manual of research procedures*. E. C. Friedberg and P. C. Hanawalt, eds., vol. I, part A, p. 425. New York: Dekker.
21. Hagen, U. F. W. 1981. Measurement of strand breaks by end labeling. In *DNA repair—a laboratory manual of research procedures*. E. C. Friedberg and P. C. Hanawalt, eds., vol. I, part A, p. 431. New York: Dekker.
22. Braun, A. G. 1981. Measurement of strand breaks by nitrocellulose membrane filtration. In *DNA repair—a laboratory manual of research procedures*. E. C. Friedberg and P. C. Hanawalt, eds., vol. I, part A, p. 447. New York: Dekker.
23. Pettijohn, D., and Hanawalt, P. 1963. Deoxyribonucleic acid replication in bacteria following ultraviolet irradiation. *Biochim. Biophys. Acta* 72:127.
24. Pettijohn, D., and Hanawalt, P. 1964. Evidence for repair-replication of ultraviolet damaged DNA in bacteria. *J. Mol. Biol.* 9:395.
25. Smith, C. A., Cooper, P. K., and Hanawalt, P. C. 1981. Measurement of repair replication by equilibrium sedimentation. In *DNA repair—a laboratory manual of research procedures*. E. C. Friedberg and P. C. Hanawalt, eds., vol. I, part B, p. 289. New York: Dekker.
26. Swenson, P. A., and Setlow, R. B. 1966. Effects of ultraviolet radiation on macromolecular synthesis in *Escherichia coli*. *J. Mol. Biol.* 15:201.

27. Collins, A. R. S., and Johnson, R. T. 1981. Use of metabolic inhibitors in repair studies. In *DNA repair—a laboratory manual of research procedures*. E. C. Friedberg and P. C. Hanawalt, eds., vol. I, part B, p. 341. New York: Dekker.

28. Thelander, L. 1973. Physicochemical characterization of ribonucleoside reductase from *Escherichia coli*. *J. Biol. Chem.* 248:4591.

29. Cooper, P. K., and Hanawalt, P. C. 1972. Heterogeneity of patch size in repair replicated DNA in *Escherichia coli*. *J. Mol. Biol.* 67:1.

30. Smith, C. A., and Hanawalt, P. C. 1978. Phage T4 endonuclease V stimulates DNA repair replication in isolated nuclei from ultraviolet-irradiated human cells, including xeroderma pigmentosum fibroblasts. *Proc. Natl. Acad. Sci. (USA)* 75:2598.

31. Setlow, R. B., and Regan, J. D. 1981. Measurement of repair synthesis by photolysis of bromouracil. In *DNA repair—a laboratory manual of research procedures*. E. C. Friedberg and P. C. Hanawalt, eds., vol. I, part B, p. 307. New York: Dekker.

32. Rasmussen, R. E., and Painter, R. B. 1964. Evidence for repair of ultra-violet damaged deoxyribonucleic acid in cultured mammalian cells. *Nature* 203:1360.

33. Cleaver, J. E., and Thomas, G. H. 1981. Measurement of unscheduled synthesis by autoradiography. In *DNA repair—a laboratory manual of research procedures*. E. C. Friedberg and P. C. Hanawalt, eds., vol. I, part B, p. 277. New York: Dekker.

34. Cozzarelli, N. R., Kelly, R. B., and Kornberg, A. 1969. Enzymic synthesis of DNA XXXIII. Hydrolysis of a 5′ triphosphate-terminated polynucleotide in the active center of DNA polymerase. *J. Mol. Biol.* 45:513.

35. Kelly, R. B., Atkinson, M. R., Huberman, J. A., and Kornberg, A. 1969. Excision of thymine dimers and other mismatched sequences by DNA polymerase of *Escherichia coli*. *Nature* 224:495.

36. Hanawalt, P. C., Cooper, P. K., and Smith, C. A. 1981. Repair replication schemes in bacteria and human cells. *Prog. Nuc. Acid Res. Mol. Biol.* 26:181.

37. Masker, W. E., Kuemmerle, N. B., and Dodson, L. A. 1981. *In vitro* packaging of damaged bacteriophage T7 DNA. *Prog. Nuc. Acid Res. Mol. Biol.* 26:227.

38. Masker, W. E., Simon, T. J., and Hanawalt, P. C. 1975. Repair replication in permeabilized *Escherichia coli*. In *Molecular mechanisms for repair of DNA*. P. C. Hanawalt and R. B. Setlow, eds., p. 245. New York: Plenum.

39. Ben-Ishai, R., Pugravitsky, E., and Sharon, R. 1978. Conditions for constitutive and inducible gap-filling of excision and post-replication repair in toluene treated *E. coli*. In *DNA repair mechanisms*. P. C. Hanawalt, E. C. Friedberg and C. F. Fox, eds., p. 267. New York: Academic.

40. Shimizu, K., Yamashita, K., and Sekiguchi, M. 1981. Restoration of defective cellular functions by supply of DNA polymerase I to permeable cells of *Escherichia coli*. *Biochem. Biophys. Res. Comm.* 101:15.

41. Deutsch, W. A., Dorson, J. W., and Moses, R. E. 1976. Excision of pyrimidine dimers in toluene-treated *Escherichia coli*. *J. Bacteriol.* 125:220.

42. Kuemmerle, N., Ley, R., and Masker, W. 1981. Analysis of resynthesis tracts in repaired *Escherichia coli* deoxyribonucleic acid. *J. Bacteriol.* 147:333.

43. Cooper, P. K. 1982. Characterization of long patch excision repair of DNA in ultraviolet-irradiated *Escherichia coli*: an inducible function under *rec-lex* control. *Mol. Gen. Genet.* 185:189.

44. Lehman, I. R., Bessman, M. J., Simms, E. S., and Kornberg, A. 1958. Enzymatic synthesis of deoxyribonucleic acid. I. Preparation of substrates and partial purification of an enzyme from *Escherichia coli*. *J. Biol. Chem.* 233:163.

45. Gross, J., and Gross, M. 1969. Genetic analysis of an *E. coli* strain with a mutation affecting DNA polymerase. *Nature* 224:1166.

46. Kornberg, A. 1981. *DNA replication*. San Francisco: Freeman.

47. Kornberg, A. 1982. *1982 Supplement to DNA replication*. San Francisco: Freeman.

48. Friedberg, E. C., and Lehman, I. R. 1974. The excision of thymine dimers by proteolytic and amber fragments of *E. coli* DNA polymerase I. *Biochem. Biophys. Res. Comm.* 58:132.

49. Matson, S. W., Capaldo-Kimball, F. N., and Bambara, R. A. 1978. On the processive mechanism of *Escherichia coli* DNA polymerase I. The *pol*A5 mutation. *J. Biol. Chem.* 253:7851.

50. De Lucia, P., and Cairns, J. 1969. Isolation of an *E. coli* strain with a mutation affecting DNA polymerase. *Nature* 224:1164.

51. Livingston, D. M., and Richardson, C. C. 1975. Deoxyribonucleic acid polymerase III of *Escherichia coli*. Characterization of associated exonuclease activities. *J. Biol. Chem.* 250:470.

52. Masker, W., Hanawalt, P., and Shizuya, H. 1973. Role of DNA polymerase II in repair replication in *Escherichia coli*. *Nature New Biol.* 244:242.

53. Gefter, M. 1975. DNA replication. *Ann. Rev. Biochem.* 44:45.

54. Campbell, J. L., Soll, L., and Richardson, C. C. 1972. Isolation and partial characterization of a mutant of *Escherichia coli* deficient in DNA polymerase II. *Proc. Natl. Acad. Sci.* (USA) 69:2090.

55. Cooper, P. 1977. Excision-repair in mutants of *Escherichia coli* deficient in DNA polymerase I and/or its associated $5' \rightarrow 3'$ exonuclease. *Mol. Gen. Genet.* 150:1.

56. Wahl, A. F., Hockensmith, J. W., Kowalski, S., and Bambara, R. A. 1983. Alternative explanation for excision repair deficiency caused by the *polAex*1 mutation. *J. Bacteriol.* 155:922.

57. Cooper, P. K., and Hanawalt, P. C. 1972. Role of DNA polymerase I and the *rec* system in excision repair in *Escherichia coli*. *Proc. Natl. Acad. Sci.* (USA) 69:1156.

58. Cooper, P. K., and Hunt, J. G. 1978. Alternative pathways for excision and resynthesis in *Escherichia coli*: DNA polymerase III role? In *DNA repair mechanisms*. P. C. Hanawalt, E. C. Friedberg and C. F. Fox, eds., p. 255. New York: Academic.

59. Cooper, P. K. 1981. Inducible excision repair in *Escherichia coli*. In *Chromosome damage and repair*. E. Seeberg and K. Kleppe, eds., p. 139. New York: Plenum.

60. Youngs, D. A., van der Schueren, E., and Smith, K. C. 1974. Separate branches of the *uvr* gene-dependent excision repair process in ultraviolet-irradiated *Escherichia coli* K-12 cells; their dependence upon growth medium and the *pol*A, *rec*A, *rec*B and *exr*A genes. *J. Bacteriol.* 117:717.

61. Rothman, R. H. 1978. Dimer excision and repair replication patch size in a *rec*L152 mutant of *Escherichia coli* K-12. *J. Bacteriol.* 136:444.

62. Matson, S. W., and Bambara, R. A. (1981). Short deoxyribonucleic acid repair patch length in *Escherichia coli* is determined by the processive mechanism of deoxyribonucleic acid polymerase I. *J. Bacteriol.* 146:275.
63. Yarosh, D. B., Rosenstein, B. S., and Setlow, R. B. 1981. Excision repair and patch size in UV-irradiated bacteriophage T4. *J. Virology* 40:465.
64. Radany, E. H., and Friedberg, E. C. 1983. Measurement of repair patch size by quantitation of nucleotides excised during DNA repair *in vivo*. *J. Virology*. 47:367.
65. Hanawalt, P. C. 1982. Perspectives on DNA repair and inducible recovery phenomena. *Biochimie* 64:847.
66. Heijneker, H. L., Ellens, D. J., Tjeerde, R. H., Glickman, B. W., van Dorp, B., and Pouwels, P. H. 1973. A mutant of *Escherichia coli*, K-12 deficient in the $5' \rightarrow 3'$ exonucleolytic activity of DNA polymerase I. *Molec. Gen. Genet.* 124:83.
67. Konrad, E. B., and Lehman, I. R. 1974. A conditional lethal mutant of *Escherichia coli* K-12 defective in the $5' \rightarrow 3'$ exonuclease associated with DNA polymerase I. *Proc. Natl. Acad. Sci.* (USA) 71:2048.
68. Chase, J. W., and Richardson, C. C. 1974. Exonuclease VII of *Escherichia coli*. Purification and properties. *J. Biol. Chem.* 249:4545.
69. Chase, J. W., and Richardson, C. C. 1974. Exonuclease VII of *Escherichia coli*. Mechanism of action. *J. Biol. Chem.* 249:4553.
70. Masker, W. E., and Chase, J. W. 1978. Pyrimidine dimer excision in exonuclease deficient mutants of *Escherichia coli*. In *DNA repair mechanisms*. P. C. Hanawalt, E. C. Friedberg and C. F. Fox, eds., p. 261. New York: Academic.
71. Ogawa, H., Shimada, K., and Tomizawa, J. 1968. Studies on radiation-sensitive mutants of *E. coli*. I. Mutants defective in the repair synthesis. *Mol. Gen. Genet.* 101:227.
72. Siegel, E. C. 1973. An ultraviolet-sensitive mutator strain of *Escherichia coli* K-12. *J. Bacteriol.* 113:145.
73. Smirnov, G. B., and Skavronskaya, A. G. 1971. Location of *uvr*502 mutation on the chromosome of *Escherichia coli*. *Mol. Gen. Genet.* 113:217.
74. Horii, Z. I., and Clark, A. J. 1973. Genetic analysis of *rec*F pathway of genetic recombination in *Escherichia coli*: isolation and characterization of mutants. *J. Mol. Biol.* 80:327.
75. Kushner, S. R., Sheperd, J., Edwards, G., and Maples, V. F. 1978. *UVR*D, *UVR*E and *REC*L represent a single gene. In *DNA repair mechanisms*. P. C. Hanawalt, E. C. Friedberg and C. F. Fox, eds., p. 251. New York: Academic.
76. Ben-Ishai, R., and Sharon, R. 1981. On the nature of the repair deficiency in *E. coli uvrE*. In *Chromosome damage and repair*. E. Seeberg and K. Kleppe, eds., p. 147. New York: Plenum.
77. Rothman, R. H., and Clark, A. J. 1977. Defective excision and postreplication repair of UV-damaged DNA in a *rec*L mutant strain of *E. coli* K-12. *Mol. Gen. Genet.* 155:267.
78. Rothman, R. H. 1978. Dimer excision and repair replication patch size in a *rec*L152 mutant of *Escherichia coli* K-12. *J. Bacteriol.* 136:444.
79. Kuemmerle, N. B., and Masker, W. E. 1980. Effect of the *uvr*D mutation on excision repair. *J. Bacteriol.* 142:535.
80. van Sluis, C. A., Mattern, I. E., and Paterson, M. C. 1974. Properties of *uvr*E mutants of *Escherichia coli* K-12. I. Effects of UV irradiation on DNA metabolism. *Mutation Res.* 25:273.

81. Sinzinis, B. I., Smirnov, G. B., and Saenko, A. A. 1973. Repair deficiency in *Escherichia coli* UV-sensitive mutator strain *uvr*502. *Biochem. Biophys. Res. Comm.* 53:309.

82. Kuemmerle, N. B., Ley, R. D., and Masker, W. E. 1982. The effect of mutations in the *uvr*D cistron of *Escherichia coli* on repair synthesis. *Mutation Res.* 94:285.

83. Oeda, K., Horiuchi, T., and Sekiguchi, M. 1981. Molecular cloning of the *uvr*D gene of *Escherichia coli* that controls ultraviolet sensitivity and spontaneous mutation frequency. *Mol. Gen. Genet.* 184:191.

84. Maples, V. F., and Kushner, S. R. 1982. DNA repair in *Escherichia coli:* identification of the *uvr*D gene product. *Proc. Natl. Acad. Sci.* (USA) 79:5616.

85. Arther, H. M., Bramhill, D., Eastlake, P. B., and Emmerson, P. T. 1982. Cloning of the *uvr*D gene of *E. coli* and identification of the product. *Gene* 19:285.

86. Oeda, K., Horiuchi, T., and Sekiguchi, M. 1982. The *uvr*D gene of *E. coli* encodes a DNA-dependent ATPase. *Nature* 298:98.

87. Kushner, S. R., Maples, V. F., Easton, A., Farrance, I., and Peramachi, P. 1983. Physical, biochemical, and genetic characterization of the *uvr*D gene product. In *Cellular responses to DNA damage*. E. C. Friedberg and B. A. Bridges, eds., p. 153. New York: Alan R. Liss.

88. Kumura, K., Oeda, K., Akiyama, M., Horiuchi, T., and Sekiguchi, M. 1983. The *uvr*D gene of *E. coli:* molecular cloning and expression. In *Cellular responses to DNA damage*. E. C. Friedberg and B. A. Bridges, eds., p. 51. New York: Alan R. Liss.

89. Abdel-Monem, M., Chanel, M. C., and Hoffman-Berling, H. 1977. DNA unwinding enzyme II of *Escherichia coli:* 1. Purification and characterization of the ATPase activity. *Eur. J. Biochem.* 79:33.

90. George, D. L., and Witkin, E. M. 1975. Ultraviolet light-induced responses of an *mfd* mutant of *Escherichia coli* B/r having a slow rate of dimer excision. *Mutation Res.* 28:347.

91. Bridges, B. A., Mottershead, R. P., and Lehmann, A. R. 1976. Error-prone DNA repair in *Escherichia coli* IV. Excision repair and radiation-induced mutation in a *dna*B strain. *Biol. Zentralbl.* 95:393.

92. Mosbaugh, D. W., and Linn, S. 1982. Characterization of the action of *Escherichia coli* DNA polymerase I at incisions produced by repair endodeoxyribonucleases. *J. Biol. Chem.* 257:575.

93. Pawl, G., Taylor, R., Minton, K., and Friedberg, E. C. 1976. The enzymes involved in thymine dimer excision in bacteriophage T4-infected *Escherichia coli*. *J. Mol. Biol.* 108:99.

94. Ohshima, S., and Sekiguchi, M. 1972. Induction of a new enzyme activity to excise pyrimidine dimers in *Escherichia coli* infected with bacteriophage T4. *Biochem. Biophys. Res. Comm.* 47:1126.

95. Friedberg, E. C., Minton, K., Pawl, G., and Verzola, P. 1974. The excision of thymine dimers by extracts of bacteriophage-infected *E. coli*. *J. Virology.* 13:953.

96. Shimizu, K., and Sekiguchi, M. 1976. 5′ → 3′ Exonucleases of bacteriophage T4. *J. Biol. Chem.* 251:2613.

97. Smith, S. M., Symonds, N., and White, P. 1970. The Kornberg polymerase and the repair of irradiated T4 bacteriophage. *J. Mol. Biol.* 54:391.

98. Mortelmans, K., and Friedberg, E. C. 1972. Deoxyribonucleic acid repair in bacteriophage T4: observations on the role of the *x* and *v* genes and of host factors. *J. Virology.* 10:730.

99. Harm, W. 1958. Multiplicity reactivation, marker rescue and genetic recombination in phage T4 following X-ray inactivation. *Virology* 5:337.

100. Symonds, N., and Ritchie, D. A. 1963. Multiplicity reactivation after decay of incorporated radioactive phosphorus in phage T4. *J. Mol. Biol.* 3:61.

101. Nonn, E., and Bernstein, C. 1977. Multiplicity reactivation and repair of nitrous acid–induced lesions in bacteriophage T4. *J. Mol. Biol.* 116:31.

102. Schneider, S., Bernstein, C., and Bernstein, H. 1978. Recombinational repair of alkylation lesions in phage T4. I. *N*-methyl-*N*′-nitro-*N*-nitrosoguanidine. *Mol. Gen. Genet.* 167:185.

103. Van Minderhout, L., Grimbergen, J., and de Groot, B. 1978. Non-essential UV-sensitive bacteriophage T4 mutants affecting early DNA synthesis: a third pathway of DNA repair. *Mutation Res.* 52:313.

104. Cupido, M. 1983. Bypass of pyrimidine dimers in DNA of bacteriophage T4 via induction of primer RNA. *Mutation Res.* 109:1.

105. Maynard-Smith, S., and Symonds, N. 1973. Involvement of bacteriophage T4 genes in radiation repair. *J. Mol. Biol.* 74:33.

106. Goulian, M., Lucas, Z. J., and Kornberg, A. 1968. Enzymatic synthesis of deoxyribonucleic acid XXV. Purification and properties of deoxyribonucleic acid polymerase induced by infection with phage T4. *J. Biol. Chem.* 243:627.

107. Schnitzlein, C. F., Albrecht, I., and Drake, J. W. 1974. Is bacteriophage T4 DNA polymerase involved in the repair of ultraviolet damage? *Virology* 59:580.

108. Alberts, B. M., and Trey, L. 1970. T4 bacteriophage gene 32: a structural protein in the replication and recombination of DNA. *Nature* 227:1313.

109. Wu, J-R., and Yeh, Y-C. 1973. Requirement of a functional gene 32 product of bacteriophage T4 in UV repair. *J. Virology.* 12:758.

110. Bonura, T., Radany, E. H., McMillan, S., Love, J. D., Schultz, R. A., Edenberg, H. J., and Friedberg, E. C. 1982. Pyrimidine dimer DNA glycosylases: studies on bacteriophage T4–infected and on uninfected *Escherichia coli*. *Biochimie* 64:643.

111. Gottesman, M. M., Hicks, M. L., and Gellert, M. 1973. Genetics and function of DNA ligase in *Escherichia coli*. *J. Mol. Biol.* 77:531.

112. Konrad, E. B., Modrich, P., and Lehman, I. R. 1973. Genetic and enzymatic characterization of a conditional lethal mutant of *Escherichia coli* K-12 with a temperature-sensitive DNA ligase. *J. Mol. Biol.* 77:519.

113. Lehman, I. R. 1974. DNA ligase: structure, mechanism, and function. *Science* 186:790.

114. Sternglanz, R., DiNardo, S., Voelkel, K. A., Nishimura, Y., Hirota, Y., Becherer, K., Zumstein, L., and Wang, J. C. 1981. Mutations in the gene coding for *Escherichia coli* DNA topoisomerase I affect transcription and transposition. *Proc. Natl. Acad. Sci.* (USA) 78:2747.

115. Pedrini, A. M., and Ciarrochi, G. 1983. Inhibition of *Micrococcus luteus* DNA topoisomerase I by UV photoproducts. *Proc. Natl. Acad. Sci.* (USA) 80:1787.

116. Pritchard, R. H. 1955. The linear arrangement of a series of alleles in *Aspergillus nidulans*. *Heredity* 9:343.

117. Chase, M., and Doermann, A. H. 1958. High negative interference over short segments of the genetic structure of bacteriophage T4. *Genetics* 43:332.

118. Amati, P., and Meselson, M. 1965. Localized negative interference in bacteriophage λ. *Genetics* 51:369.

119. Whitehouse, H. L. K. 1963. A theory of crossing-over by means of hybrid deoxyribonucleic acid. *Nature* 199:1034.

120. Holliday, R. 1964. A mechanism for gene conversion in fungi. *Genet. Res.* 5:282.

121. Doerfler, W., and Hogness, D. S. 1968. Gene orientation in bacteriophage lambda as determined from the genetic activities of heteroduplex DNA formed *in vitro*. *J. Mol. Biol.* 33:661.

122. Wildenberg, J., and Meselson, M. 1975. Mismatch repair in heteroduplex DNA. *Proc. Natl. Acad. Sci.* (USA) 72:2202.

123. White, R. L., and Fox, M. S. 1975. Genetic consequences of transfection with heteroduplex bacteriophage λ DNA. *Molec. Gen. Genet.* 141:163.

124. Baas, P. D., and Jansz, H. S. 1972. Asymmetric information transfer during ϕX174 DNA replication. *J. Mol. Biol.* 63:557.

125. Spatz, H. C., and Trautner, T. A. 1970. One way to do experiments on gene conversion? Transfection with heteroduplex SPP1 DNA. *Molec. Gen. Genet.* 109:84.

126. Roger, M. 1977. Mismatch excision and possible polarity effects result in preferred deoxyribonucleic acid strand of integration in pneumococcal transformation. *J. Bacteriol.* 129:298.

127. Wagner, R., Jr., and Meselson, M. 1976. Repair tracts in mismatched DNA heteroduplexes. *Proc. Natl. Acad. Sci.* (USA) 73:4135.

128. Arber, W. 1974. DNA modification and restriction. *Prog. Nuc. Acid Res. Mol. Biol.* 14:1.

129. Marinus, M. G., and Morris, N. R. 1974. Biological function for 6-methyladenine residues in the DNA of *Escherichia coli* K-12. *J. Mol. Biol.* 85:309.

130. Marinus, M. G., and Morris, N. R. 1975. Pleiotropic effects of a DNA adenine methylation mutation (*dam*-3) in *Escherichia coli* K-12. *Mutation Res.* 28:15.

131. Bale, A., d'Alarcao, M., and Marinus, M. G. 1979. Characterization of DNA adenine methylation mutants of *Escherichia coli* K-12. *Mutation Res.* 59:157.

132. Geier, G. E., and Modrich, P. 1979. Recognition sequence of the *dam* methylase of *Escherichia coli* K-12 and mode of cleavage of *Dpn* I endonuclease. *J. Biol. Chem.* 254:1408.

133. Herman, G. E., and Modrich, P. 1982. *Escherichia coli dam* methylase. Physical and catalytic properties of the homogeneous enzyme. *J. Biol. Chem.* 257:2605.

134. Herman, G. E., and Modrich, P. 1981. *Escherichia coli* K-12 clones that overproduce *dam* methylase are hypermutable. *J. Bacteriol.* 145:644.

135. Arraj, J. A., and Marinus, M. G. 1983. Phenotypic reversal in *dam* mutants of *Escherichia coli* K-12 by a recombinant plasmid containing the dam^+ gene. *J. Bacteriol.* 153:562.

136. Radman, M., Wagner, R. E., Jr., Glickman, B. W., and Meselson, M. 1980. DNA methylation, mismatch correction, and genetic stability. In *Progress in environmental mutagenesis.* M. Alacevic, ed., p. 121. Amsterdam: Elsevier/North-Holland Biomedical Press.

137. Radman, M., Dohet, C., Bourguignon, M-F., Doubleday, O. P., and Lecomte, P. 1981. High fidelity devices in the reproduction of DNA. In *Chromosomal damage and repair.* E. Seeberg and K. Kleppe, eds., p. 431. New York: Plenum.

138. Bourguignon-Van Horen, F., Brotcorn, A., Caillet-Fauquet, P., Diver, W. P., Dohet, C., Doubleday, O. P., Lecomte, P., Maenhaut-Michel, G., and Radman, M. 1982. Conservation and diversification of genes by mismatch correction and SOS induction. *Biochimie* 64:559.

139. Nevers, P., and Spatz, H-C. 1975. *Escherichia coli* mutants *uvr*D and *uvr*E deficient in gene conversion of λ heteroduplexes. *Molec. Gen. Genet.* 139:233.

140. Rydberg, B. 1978. Bromouracil mutagenesis and mismatch repair in mutator strains of *Escherichia coli. Mutation Res.* 52:11.

141. Glickman, B. W., and Radman, M. 1980. *Escherichia coli* mutator mutants deficient in methylation-instructed DNA mismatch correction. *Proc. Natl. Acad. Sci.* (USA) 77:1063.

142. Glickman, B., van den Elsen, P., and Radman, M. 1978. Induced mutagenesis in *dam* mutants of *Escherichia coli:* a role for 6-methyladenine residues in mutation avoidance. *Molec. Gen. Genet.* 163:307.

143. Cox, E. C. 1976. Bacterial mutator genes and the control of spontaneous mutation. *Ann. Rev. Genet.* 10:135.

144. Meselson, M., and Fox, M. S. 1983. Mismatch repair. Workshop report. In *Cellular responses to DNA damage.* E. C. Friedberg and B. A. Bridges, eds., p. 333. New York: Alan R. Liss.

145. Lu, A-L., Clark, S., and Modrich, P. 1983. Methyl-directed repair of DNA basepair mismatches *in vitro. Proc. Natl. Acad. Sci.* (USA) 80:4639.

146. Grafstrom, R. H., Bear, J., and Hoess, R. H. 1983. Cloning, identification and partial purification of *mut*H gene product. In *Cellular responses to DNA damage.* E. C. Friedberg and B. A. Bridges, eds., p. 299. New York: Alan R. Liss.

147. Ephrussi-Taylor, H., and Gray, T. C. 1966. Genetic studies of recombining DNA in pneumococcal transformation. *J. Gen. Physiol.* 49:211.

148. Lacks, S. 1970. Mutants of *Diplococcus pneumoniae* that lack deoxyribonucleases and activities possibly pertinent to genetic transformation. *J. Bacteriol.* 101:371.

149. Tiraby, G., and Fox, M. S. 1973. Marker discrimination in transformation and mutation of pneumococcus. *Proc. Natl. Acad. Sci.* (USA) 70:3541.

150. Tiraby, G., and Sicard, A. M. 1973. Integration efficiencies of spontaneous mutant alleles of *ami*A locus in pneumococcal transformation. *J. Bacteriol.* 116:1130.

151. Claverys, J-P., Roger, M., and Sicard, A. M. 1980. Excision and repair of mismatched base pairs in transformation of *Streptococcus pneumoniae. Mol. Gen. Genet.* 178:191.

152. Claverys, J-P., Mejean, V., Gasc, A-M., Galibert, F., and Sicard, A-M. 1981. Base specificity of mismatch repair in *Streptococcus pneumoniae. Nucleic Acids. Res.* 9:2267.

153. Shanabruch, W. G., Rein, R. P., Behlau, I., and Walker, G. C. 1983. Mutagenesis, by methylating and ethylating agents, in *mut*H, *mut*L, *mut*S and *uvr*D mutants of *Salmonella typhimurium J. Bacteriol.* 153:33.

154. Hanawalt, P. C. 1970. Repair replication of damaged DNA *in vivo.* In *Genetic concepts and neoplasia,* p. 528. Baltimore: Williams & Wilkins.

155. Smirnov, G. B., Filkova, E. V., and Skavronskaya, A. G. 1973. Ultraviolet sensitivity, spontaneous mutability and DNA degradation in *Escherichia coli* strains carrying mutations in *uvr* and *rec* genes. *J. Gen. Microbiol.* 76:407.

6

Excision Repair. IV. Postincision Events and Mismatch Repair in Mammalian Cells

6-1 What Are Postincision Events in Mammalian Cells?

Unlike the situation in some prokaryotic cells, there is no evidence for the existence of pyrimidine dimer (PD) DNA glycosylases in mammalian cells (1). Hence it is likely that all excision of pyrimidine dimers and of other types of bulky base damage in higher organisms is mediated by the action of general damage-specific DNA incising activities such as the uvrABC endonuclease of *E. coli*. However, as indicated in Section 4-3, such putative activities have not yet been characterized in any eukaryotic system, and it therefore remains an open question as to whether excision of damaged nucleotides is directly coupled to the incision of DNA or represents an independent postincision event. We are thus currently in the uncomfortable position of not being able to precisely define postincision events during excision repair of bulky base damage in eukaryotic cells, and for the purposes of this discussion, the experimental evidence that bears directly on excision of such damage will be evaluated in terms of both models of excision repair.

In general, the measurement of postincision events during excision repair in higher organisms suffers from the same interpretive limitations described in the previous chapter with respect to the prokaryote *E. coli*. The techniques most commonly employed include measurement of the loss of sites sensitive to the dimer-specific *M. luteus* or phage T4 enzyme probes (2–7), the loss of binding sites for dimer-specific antibodies (7–10), the loss of thymine-containing pyrimidine dimers from acid-insoluble DNA (11–20) and repair synthesis of DNA (20–32). As indicated previously (see Section 5-2), the retention of sites sensitive to pyrimidine dimer-specific enzyme probes implies that DNA incision does not occur at those sites in vivo. On the other hand, the loss of such sensitivity does not distinguish between DNA *incision* and *postincision* events since once incision occurs at a given site in vivo, that site will no longer be a substrate for the probe. In general, however, the loss of enzyme-sensitive sites from DNA is an indicator that excision repair of those sites is complete.

Although all these techniques provide results consistent with the existence of a nucleotide excision repair mode(s) in higher organisms, there is controversy concerning the relative kinetics of DNA incision, the loss of thymine-containing dimers from high molecular weight DNA and repair synthesis. Varying interpretations of different experiments have further complicated an understanding of the molecular mechanism of the excision repair of bulky base damage in general and of the details of postincision events in particular. There is little question that these controversies have arisen in part from the use of distinct experimental procedures, which, although designed to measure the same end points, have optimal reliabilities at different extents of DNA damage and hence may not be strictly comparable. However, it is important to bear in mind that some apparent experimental inconsistencies may reflect genuine biological distinctions in excision repair in different cell types.

After low levels of UV radiation, i.e., less than 20 J/m^2, it is difficult to accurately monitor excision repair by measuring the loss of thymine-containing pyrimidine dimers from the acid-precipitable fraction of cells, because the absolute amount of radioactivity associated with the dimers at these doses becomes limiting (33). Therefore

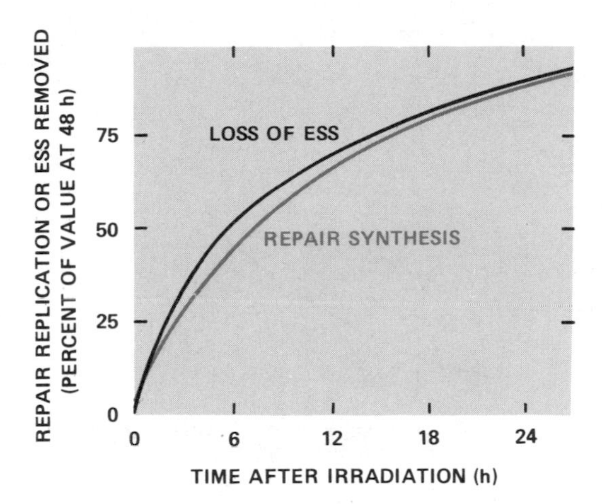

FIGURE 6-1
Comparison of the kinetics of repair synthesis after irradiation at 10 J/m^2 and of the loss of sites in DNA sensitive to the T4 pyrimidine dimer-specific enzyme (enzyme-sensitive sites or ESS) after irradiation at 5 J/m^2 in human diploid fibroblasts. (After C. A. Smith, ref. 169.)

excision repair at low doses of UV radiation is generally monitored by measuring the loss of sites sensitive to dimer-specific enzymes, or by measuring repair synthesis. In general such measurements are in reasonable agreement; a significant fraction of the total sites present are lost early after irradiation, and after eight to 10 hours at least half of the enzyme-sensitive sites are no longer detectable (2–7) (Fig. 6-1). However, the time required for the *complete* removal of these sites may be much longer and can be as long as 10 to 12 days in cells irradiated at 1 J/m^2 and 15 to 20 days in cells irradiated at 10 J/m^2 (4). In the few studies in which direct comparisons have been made under identical experimental conditions, the kinetics of the loss of enzyme-sensitive sites correlates well with that of repair synthesis (21) (Fig. 6-1), suggesting that both observations measure related biochemical events during excision repair.

Direct measurement of the loss of thymine-containing pyrimidine dimers from acid-insoluble DNA can be reliably carried out at levels of dimers produced by irradiation of cells at greater than 20 J/m^2. Many such studies suggest that the kinetics of the loss of dimers from DNA is significantly slower than that of repair synthesis (12, 14, 17, 19) (Fig. 6-2). There are at least two possible interpretations of this observation. One is based on the historically established model of excision repair in *E. coli* (see Fig. 3-2), in which it was postulated that DNA incision results in strand breakage exclusively on the 5′

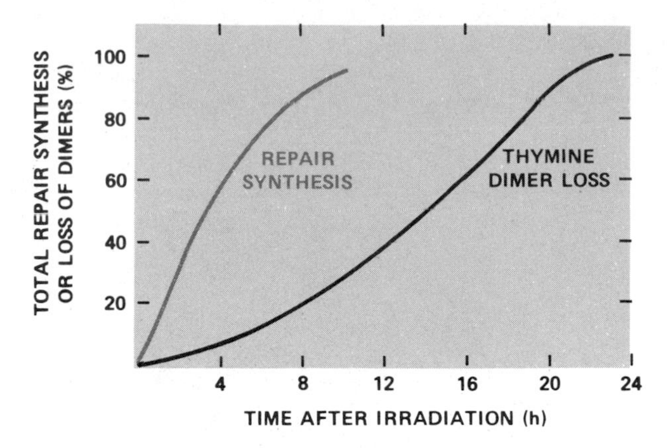

FIGURE 6-2
The rate of repair synthesis in human fibroblasts is often observed to be much faster than the rate of the loss of thymine-containing pyrimidine dimers from the acid-precipitable fraction of cells. (After U. K. Ehmann et al., ref. 12.)

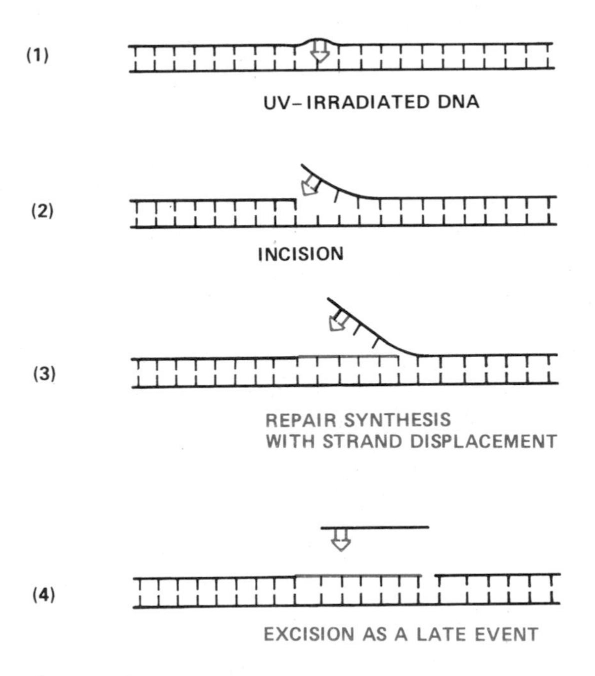

FIGURE 6-3
A model of excision repair of pyrimidine dimers in which the physical excision of base damage *follows* repair synthesis with strand displacement. According to this model actual loss of dimers from high molecular weight DNA would be a relatively late event and could explain the delayed kinetics of dimer loss relative to that of repair synthesis (see Fig. 6-2).

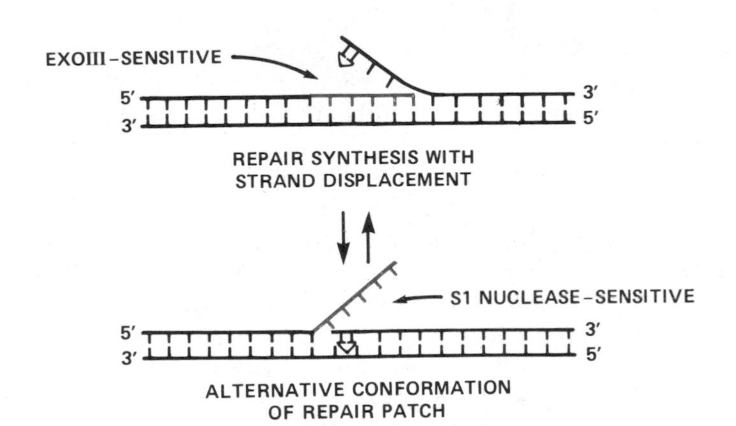

FIGURE 6-4
Postulated alternative conformations of DNA during excision repair. When the DNA is in the conformation shown in the upper illustration the repair label will be sensitive to exonuclease III but not to S1 nuclease. When the labeled DNA is in the conformation shown in the lower illustration, it will be degraded by S1 nuclease but insensitive to degradation by exonuclease III. (After J. E. Cleaver, ref. 34.)

side of dimers. If repair synthesis follows DNA incision by a strand displacement rather than by a nick translation mode (2, 12), then actual excision of DNA sequences containing pyrimidine dimers could be a relatively late event that occurs *after* most repair synthesis is completed at a given site in the genome (Fig. 6-3). According to this model, excision of bulky base damage would not be directly coupled to DNA incision, as is apparently true in *E. coli* (see Fig. 3-2).

Experimentally there is no direct support for such a mechanism, although indirect evidence of repair synthesis with strand displacement in the $5' \rightarrow 3'$ direction has been found in UV-irradiated cells undergoing excision repair in the presence of cytosine arabinoside (ara-C) (34). In these experiments repair synthesis in human diploid fibroblasts was inhibited by ara-C, and incomplete repair patches resulted. If a strand displacement mode of excision were operative, one would expect all of the radiolabeled nucleotides incorporated during repair synthesis to be sensitive to degradation by exonuclease III, since this enzyme attacks 3′-terminal regions in duplex DNA. None of the radiolabel should be sensitive to the single-strand-specific nuclease S1, on the other hand (Fig. 6-4). What is observed is that 50 percent of the label is sensitive to attack by *either* enzyme. A possible explanation of these results is that repair synthesis with strand displacement in the $5' \rightarrow 3'$ direction does indeed occur; however, the two conformations shown in Fig. 6-4 are equally probable.

Thus, about half the time the repair synthesis patch is protected from one or the other of the enzymes used for digestion (34). However, other interpretations of these results are possible and, in fact, it has been observed that cytosine arabinoside does *not* block repair synthesis at *all* sites in UV-irradiated cells (35).

An alternative explanation for the differences mentioned above in the kinetics of dimer loss from DNA and of repair synthesis (Fig. 6-2) is that pyrimidine dimers are excised rapidly but are initially present in large, acid-insoluble oligonucleotides that are slowly degraded into an acid-soluble state. In that case, measurement of the loss of thymine-containing dimers from acid-insoluble DNA really measures the kinetics of *degradation* of relatively large DNA fragments into small ones, rather than the kinetics of true dimer excision (2, 12). If in human cells excision of bulky base damage is effected by incision on either side of the lesions in DNA, as occurs in *E. coli* (see Chapters 3 through 5), rather than by independent and sequential incision and excision events, then it is distinctly possible that pyrimidine dimers may be released from DNA as part of relatively large acid-precipitable oligonucleotides. This phenomenon has indeed been observed in primary mouse cell cultures (36, 37), but attempts to resolve putative acid-precipitable oligonucleotides from bulk DNA in established fibroblast cultures undergoing excision repair have not been successful (12). A final possible explanation for the delayed appearance of thymine-containing pyrimidine dimers in the acid-soluble fraction of cells is that the oligonucleotides are rapidly excised (perhaps during DNA incision) but are initially tightly bound to the proteins that constitute the putative damage-specific DNA incising activity and hence are acid-precipitable (35).

The apparent complexity of excision repair in mammalian organisms is well illustrated by a detailed study of human diploid fibroblasts in which the loss of sites sensitive to the *M. luteus* enzyme probe was measured after exposure of the cells to UV radiation (4) (Fig. 6-5). The overall rate curves generated could be interpreted as the sum of two first-order reactions with rate constants that differed by almost an order of magnitude. This interpretation suggests that about 60 to 70 percent of the pyrimidine dimers originally present in the DNA are repaired relatively rapidly, but the residual population are removed significantly more slowly. At present, it is not known what parameters might account for this apparent complexity.

Another example of the complexity that emerges from kinetic measurements of excision repair in mammalian cells concerns the use of pyrimidine dimer-specific antibodies as a probe for these lesions in DNA. It has been shown that the sites of binding for a dimer-specific antibody disappear more rapidly than do sites sensitive to the T4 enzyme probe (7). Thus 50 percent of the antibody sites are lost from

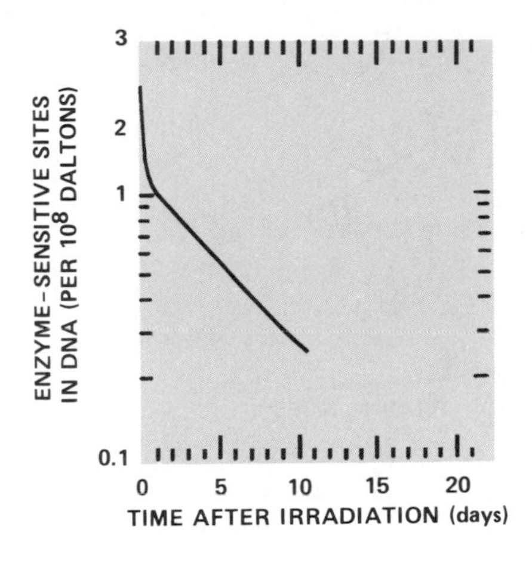

FIGURE 6-5
The kinetics of the loss of enzyme-sensitive sites from the DNA of normal human fibroblasts irradiated at 1 J/m^2 show two components to the slope. (From G. J. Kantor and R. B. Setlow, ref. 4.)

the DNA of UV-irradiated Chinese hamster ovary or HeLa cells within an hour of irradiation, and within four hours 80 percent of such sites are lost (7) (Fig. 6-6). This contrasts with the loss after eight hours of only about 40 percent of the sites sensitive to the T4 enzyme probe (7) (Fig. 6-6). The loss of antibody binding is not due to prior cleavage of the 5′ glycosyl bonds of the dimers, nor to hydrolysis of the phosphodiester bonds 3′ to the apyrimidinic sites created by the T4 PD DNA glycosylase, since when dimer-containing DNA is treated with T4 enzyme to effect cleavages at these bonds, dimers are still recognized by the antibody probe (7). Conceivably, pyrimidine dimers and/or the regions of DNA in which they reside in mammalian cell chromatin must undergo some sort of structural modification before they become accessible to the enzyme probe (see Section 4-3), but they are recognized by the antibody in the absence of such modification (7).

Repair Synthesis

Repair synthesis of DNA is unequivocally a postincision event. In mammalian cells repair synthesis is frequently described in terms of "short" and "long" patch repair modes (38), but these are not the mechanistic equivalent of the short and long patches in UV-irradiated *E. coli* (see Section 5-3). In mammalian cells short patches are estimated to be about four nucleotides long. They are thought to be associated with the repair of damage caused by ionizing radiation and by certain alkylating agents known to produce small alkyl adducts that can be repaired by the action of specific DNA glycosylases (39–41). However, the techniques used for the calculation of these short patch sizes are indirect, and their accuracy has been challenged

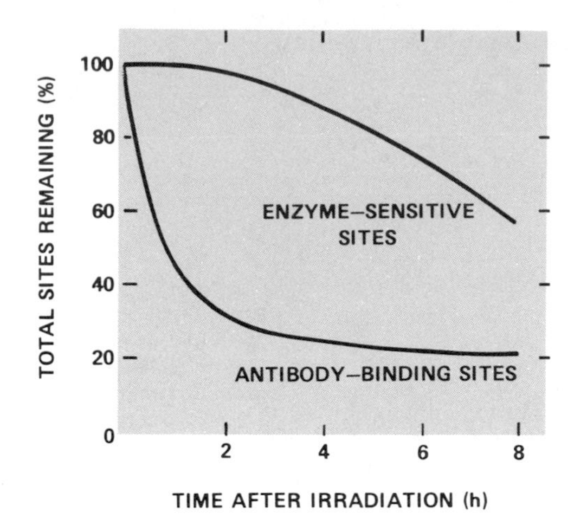

FIGURE 6-6
The rate of loss of sites sensitive to the T4 pyrimidine dimer-specific enzyme (enzyme-sensitive sites) and to a pyrimidine dimer-specific antibody probe in human HeLa cells exposed to 3.5 J/m^2 of UV radiation. Sites sensitive to the antibody disappear much faster than do those sensitive to the enzyme. (After D. L. Mitchell et al., ref. 7.)

(38). Furthermore, the notion that the repair of damage induced by certain agents in mammalian cells results in *very* short patches of three to four nucleotides has not been confirmed by the density shift method (see Section 5-2). In fact, measurements made using this technique have revealed repair patches of 30 to 40 nucleotides in human cells treated with either UV radiation or the alkylating agent methylmethanesulfonate (42, 43), suggesting that excision repair patch size is independent of the DNA damaging agent.

So-called *long* patch repair in mammalian cells is quantitatively equivalent to short patch excision repair in *E. coli* (38). These patches (on the order of 30 to 100 nucleotides) reflect the nucleotide excision repair pathway that presumably operates following the exposure of cells to UV radiation and to "UV-like" agents such as N-2-acetyl-2-aminofluorene, aflatoxin, and psoralen plus UV light (39, 44–49), i.e., agents that produce bulky adducts and significant localized helix distortion in DNA. No repair patches corresponding to the very long tracts (more than 1000 nucleotides) associated with the inducible excision repair system in *E. coli* (see Section 5-3) have been observed over a wide range of UV doses and postirradiation incubation times in mammalian cells (50).

6-2 The Role of Chromatin Structure: Distribution of Postincision Events in Nucleosomes

The current understanding of the organization of histones and DNA into nucleosomes stems in part from studies on the susceptibility of DNA in chromatin to digestion with specific nucleases (51–54). The regions of DNA most accessible to micrococcal nuclease (MN) are operationally defined as *linker* DNA; the core *nucleosomal* DNA is more resistant to digestion with this enzyme (Fig. 6-7). The sensitivity to digestion of DNA in chromatin by MN has been examined in a variety of mammalian cells exposed to DNA damage in efforts to establish whether the distribution of *repair synthesis* is the same or different in nuclease-sensitive and nuclease-resistant regions of the genome (55–66). One way of carrying out such experiments is to first *uniformly* label DNA with [^{14}C]thymidine so that the radiolabel is present in both core and linker DNA, expose cells to a DNA damaging agent and then incubate them in the presence of [^{3}H]thymidine (to label regions of repair synthesis) under conditions restrictive for semiconservative DNA synthesis. Isolated nuclei are then incubated with MN and the relative sensitivities of the ^{3}H and ^{14}C radiolabels to enzymatic digestion are compared.

When mouse mammary cells are treated with methylmethanesulfonate and exposed to this type of analysis, the DNA labeled during repair synthesis is decidedly more sensitive to the nuclease probe

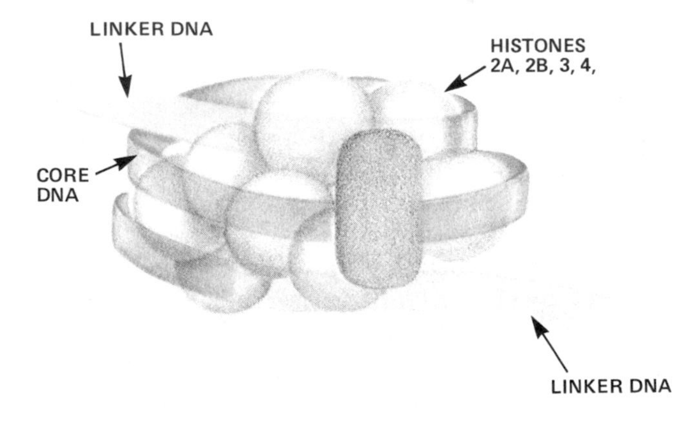

FIGURE 6-7
Diagrammatic representation of a typical nucleosome. The core DNA is in intimate association with an octomer of histones (two molecules of each of the histones H2A, H2B, H3 and H4) and is less accessible to micrococcal nuclease (MN) than is the linker DNA. Histone H1 is thought to be bound on the exterior of the nucleosome. (From A. Kornberg, ref. 110.)

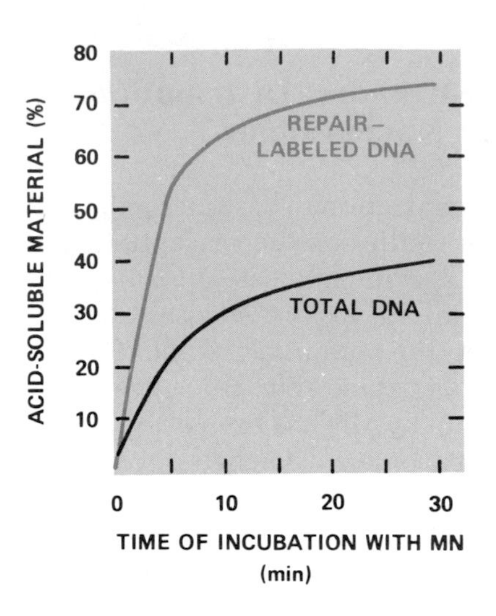

FIGURE 6-8
Kinetics of degradation by micrococcal nuclease (MN) of differentially radiolabeled bulk and repaired DNA in mouse mammary cells treated with methylmethanesulfonate (MMS). The bulk (total) DNA was prelabeled with [^{14}C]thymine. Repair synthesis following treatment of cells with MMS was carried out in the presence of [^{3}H]thymidine. Note that repair-labeled DNA is more sensitive to MN than is bulk DNA. (From W. J. Bodell, ref. 55.)

than is bulk DNA (55) (Fig. 6-8). In addition, almost all of the DNA labeled during DNA repair is nuclease sensitive, whereas bulk DNA loses its sensitivity to degradation by MN after approximately 35 percent is rendered acid-soluble, coincident with the time when most linker regions are digested (55) (Fig. 6-8). These observations suggest that DNA damage by methylmethanesulfonate and hence repair synthesis are localized largely, if not exclusively, to linker regions of the nucleosome (55). The small fraction of ^{3}H labeled DNA that is resistant to MN may represent semiconservative DNA synthesis that was initiated in core DNA during the [^{3}H]thymidine pulse or may result from rearrangement of nucleosomes such that a small fraction of repaired linker DNA becomes associated with core regions (55). This interpretation is consistent with independent observations suggesting that at least some forms of alkylation damage in the DNA of mammalian cells are indeed largely confined to nuclease-sensitive (i.e., linker) regions of nucleosomes (67).

In contradistinction to the apparent concentration of alkylation damage in linker regions of nucleosomes, pyrimidine dimers are thought to be uniformly distributed in chromatin (60). However, the interpretation of the results of MN digestion studies is less obvious. Once again the radioactivity incorporated immediately after irradiation is significantly more sensitive to MN digestion than that associated with bulk DNA. However, with increasing time after irradiation this difference is less apparent, until after about 24 hours almost as much repair synthesis is present in nuclease-resistant as in nuclease-sensitive regions of DNA (57, 60, 61) (Fig. 6-9). A number of interpretations of these results are tenable. For example, DNA repair synthesis may occur in both linker and core nucleosomal DNA, but initially preferentially in the former due to specific struc-

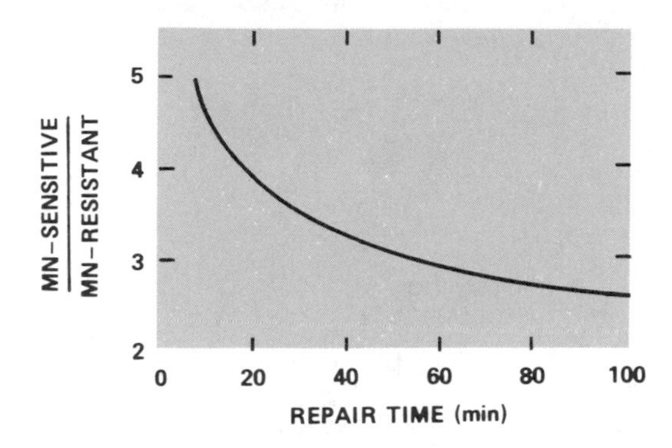

FIGURE 6-9
Ratio of the amount of repair synthesis per unit DNA in micrococcal nuclease (MN) sensitive and resistant regions as a function of repair time after UV irradiation of cells. Note that very soon after UV irradiation repair synthesis label is apparently distributed largely in MN-sensitive regions of the genome. However, at later times the label is more uniformly distributed between MN-sensitive and MN-resistant regions. (After M. J. Smerdon and M. W. Lieberman, ref. 170.)

tural and/or regulatory influences. Alternatively, at early times after UV radiation damage repair patches may be significantly smaller in core than in linker regions. The latter hypothesis is difficult to test experimentally because of the technical problems associated with measuring repair patches in DNA only 200 bp long. A third possible explanation is that the position of nucleosomes may shift during DNA excision repair. Thus, for instance, repair label could be inserted preferentially into core particles and then shift rapidly into nuclease-sensitive linker regions.

Perhaps one approach to reconciling some of these apparent contradictions is to recognize at the outset that the assumption that MN sensitivity of nucleosomes exclusively reflects the degradation of *linker* DNA may not be valid during DNA repair (63). It is possible that lesions and/or repair events produce local perturbations in nucleosome structure that result in increased sensitivity of *core* nucleosomal DNA to the nuclease probe (63) (Fig. 6-10). This model of transient alterations in nucleosome conformation associated with excision repair events, resulting in an obligatory local sensitivity to micrococcal nuclease, is supported by some experiments. For example, it has been shown that following UV irradiation of monkey kidney cells, even after a very short repair synthesis labeling time approximately 30 percent of the DNA repair label is *resistant* to the nuclease (64). If repair occurred exclusively in linker regions, after such short pulse times the label should be *completely* nuclease *sensitive*. Additionally, the size of the repair patches in nucleosomes has been estimated to be about 100 bases, which is closer to that of the entire nucleosome than to that of the linker region only (64).

There are other indications that in cells that have sustained DNA damage and are undergoing active DNA repair, MN-sensitive regions of chromatin do not necessarily define linker regions of DNA. Thus, when the repair of angelicin adducts, which are thought to be distributed almost entirely in linker DNA (68, 69), is compared with the repair of pyrimidine dimers, which are believed to be randomly

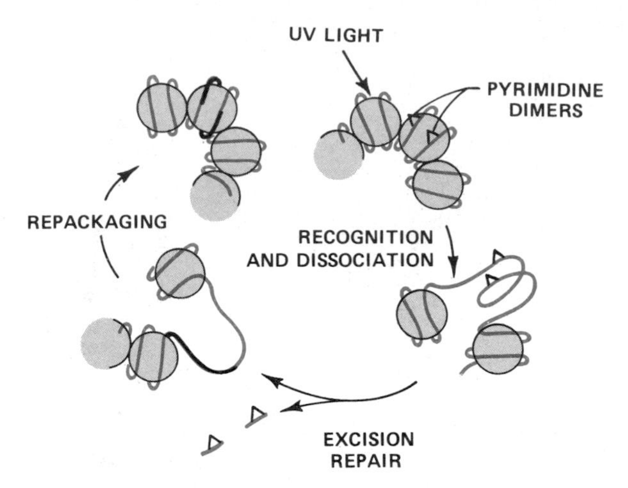

FIGURE 6-10
Schematic representation of how excision repair in nucleosomal core DNA could result in sensitivity of this DNA to micrococcal nuclease (MN). During excision repair of DNA a specific recognition signal may result in the dissociation of DNA from nucleosomal proteins to allow repair to occur. Before repackaging of the nucleosomes is complete, repaired sites of the DNA would be sensitive to MN relative to DNA in packaged nucleosomes. Hence DNA that is MN-sensitive during excision repair is not necessarily linker DNA. (After J. E. Cleaver, ref. 64.)

distributed in chromatin (60), the kinetics with which repair synthesis label shifts from MN-sensitive to MN-resistant regions of the genome are indistinguishable (60) (Fig. 6-11). Furthermore, the incubation of human fibroblasts for up to four hours in the presence of cytosine arabinoside results in an increase in the MN sensitivity of DNA synthesized during excision repair. As indicated earlier in this chapter, cytosine arabinoside causes an inhibition of repair synthesis of DNA (35, 70–74) and an accumulation of incomplete repair patches that remain sensitive to degradation by exonuclease III (75). The persistence of MN-sensitivity in cells with some incomplete repair patches therefore suggests that reassembly of the nucleosomal structure in human fibroblasts cannot occur properly until repair synthesis and DNA ligation are completed (75).

There are also hints that the mechanism of reassembly of nucleosomes following excision repair of DNA is different from that following semiconservative DNA synthesis. Thus, inhibition of protein synthesis by cyclohexamide has no effect on the kinetics of the appearance or disappearance of MN sensitivity of repair-labeled DNA in human fibroblasts (73). However, protein synthesis *is* required for the synthesis of new histones for assembly of nucleosomes following replicative synthesis (76).

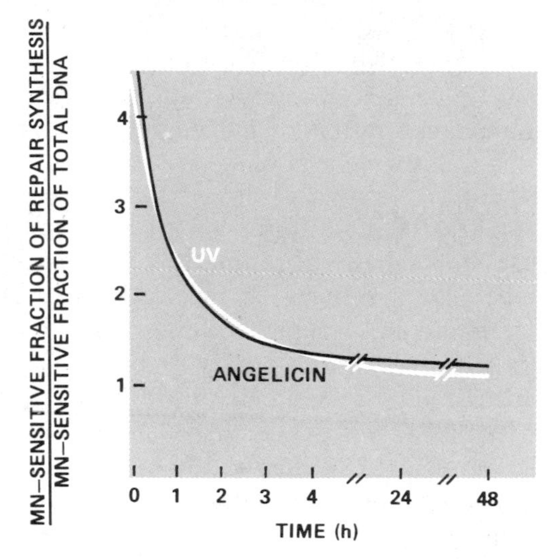

FIGURE 6-11
The sensitivity to digestion with micrococcal nuclease (MN) of regions of repair synthesis after treatment of cells with UV radiation or angelicin. Angelicin damage is concentrated in linker regions of the nucleosome, whereas UV radiation (pyrimidine dimer) damage is randomly distributed. Nonetheless, the kinetics with which radiolabel incorporated during repair synthesis shifts from an MN-sensitive to an MN-resistant state are identical for both forms of DNA damage. (From M. E. Zolan et al., ref. 66.)

6-3 Other Aspects of Genomic Organization during Excision Repair

Higher levels of structural organization of the genome do not appear to partition the DNA into accessible and inaccessible regions at the chromosomal level as apparently histones do at the nucleosomal level (38). This is demonstrated by the unimodal molecular weight distributions observed on alkaline sucrose gradients of DNA from UV-irradiated permeabilized cells probed with pyrimidine dimer-specific enzymes (38). On the other hand, some studies show that the distribution of repair synthesis is not uniform in chromosomes. Thus, for example, mouse satellite DNA (which is associated with condensed heterochromatin) is apparently repaired less efficiently than is euchromatic DNA in cells treated with alkylating agents (77). In addition, autoradiographs of nuclear sections of human lymphocytes exposed to UV or ionizing radiation show that most repair synthesis is distributed over the nuclear area adjacent to the nuclear membrane (78), and in confluent non-S-phase human diplod fibroblasts exposed to either N-methyl-N-nitrosourea, 7-bromo-methylbenz(a)anthracene, N-acetoxy-2-acetylaminofluorene or UV light, there is a greater concentration of autoradiographic grains in the central nuclear area relative to that of the area adjacent to the nuclear envelope (79).

Other studies suggest these differences are apparently not due to differences in excision repair in unique relative to repetitive DNA sequences, since repair synthesis is uniformly distributed within highly repeated, moderately repeated and single-copy DNA sequences (80–82). However, in the highly repeated α DNA sequence of cultured African green monkey cells, differences in the extent of repair synthesis *are* observed in α DNA relative to bulk DNA after treatment with chemicals (83). Whereas the rate and extent of pyrimidine dimer removal as judged by measurement of repair synthesis are similar for the two forms of DNA in cells treated with furocoumarins (e.g., psoralen) plus long-wavelength UV light, repair synthesis in α DNA is only 30 percent of that in bulk DNA, and in cells treated with N-acetoxy-2-acetylaminofluorene repair synthesis in α DNA is 60 percent of that in bulk DNA. Furthermore, excision repair of aflatoxin adducts is relatively reduced in α DNA (84), although the repair of damage resulting from dimethylsulfate or methylmethanesulfonate is not different from that in bulk DNA (85). The reduced repair observed with photoactivated psoralen is not the result of different initial amounts of base damage or of different sizes of repair patches; direct quantitation shows that fewer furocoumarin adducts are actually removed from α DNA than from bulk DNA (83). Thus the repair of different kinds of DNA damage apparently can be differentially affected by some as yet unidentified property or properties of heterochromatic DNA.

The descriptions just provided indicate quite clearly that the influence on DNA excision repair of nucleosome conformation and of higher-order structural organization in the mammalian cell nucleus are poorly understood. Many studies are contradictory, and techniques for exploring this important parameter of DNA repair are still indirect and subject to varying interpretations.

Methylation of DNA during Excision Repair

The level and/or distribution of 5-methylcytosine in DNA appears to be important for some aspects of the regulation of gene expression in higher organisms (86–91). Normally, methylation patterns in parental DNA are preserved during DNA replication by rapid methylation of newly synthesized daughter-strand DNA (92). This mechanism stabilizes the methylation pattern from one cell generation to the next. A question of considerable interest is whether in the course of repair synthesis, newly incorporated cytosine becomes normally methylated and, if so, with what kinetics. If it is assumed that modified (by methylation) gene-controlling sequences exist in DNA, then DNA damage in or adjacent to such sequences could have profound effects on gene expression. It has been proposed (93) that enzymes exist in higher organisms that recognize DNA with one methylated strand and add a methyl group to the other strand. However, *non*methylated DNA is not a substrate for the modification enzyme. Thus, under normal conditions replication of DNA results in half-methylated DNA and the daughter strands become methylated soon after DNA synthesis by the action of a maintenance methylase. If a DNA sequence has no methyl groups, then it remains unmethylated in the presence of this enzyme. In one state an affected gene is transcribed and in the other it is repressed (93). This difference is stably inherited, since a change in gene expression would only occur if the methylating enzyme were lost by mutation.

Such a regulatory mechanism could be upset during excision repair. For example, base damage located just ahead of the replication fork may be removed by excision repair (Fig. 6-12). The new patch of DNA generated during excision repair may not be immediately methylated and, if replication occurs before this event, alterations in gene expression may result (Fig. 6-12). Damage to a parental strand immediately after replication can produce the same effect (Fig. 6-12) (94).

In confluent human diploid fibroblasts exposed to UV radiation, N-methyl-N-nitrosourea or N-acetoxy-2-acetyl-2-aminofluorene, methylation of deoxycytidine incorporated by repair synthesis is slow and incomplete (95) (Fig. 6-13). In cells from cultures in logarithmic growth, 5-methylcytosine formation in nucleotide patches associated with the repair of pyrimidine dimers occurs faster and to a greater extent but still does not attain the level observed in replicating nondamaged DNA (Fig. 6-13). The hypomethylated repair patches in confluent cells are further methylated when the cells are stimulated to divide, but such regions may still not be fully methylated before cell division occurs (95). Hence DNA damage and repair apparently do lead to changes in methylation patterns in daughter cells.

There are also indications that methylation of cytosine in the DNA of mammalian cells is directly affected by the presence of certain forms of DNA damage. Thus, a diverse range of chemical carcinogens inhibit the transfer of methyl groups from S-adenosylmethionine to hemimethylated DNA, in a reaction catalyzed by mouse spleen methyl transferase in vitro (96), and some carcinogens have been shown to directly modify and inactivate the methyl transferase enzyme. In addition, DNA containing sites of base loss has reduced

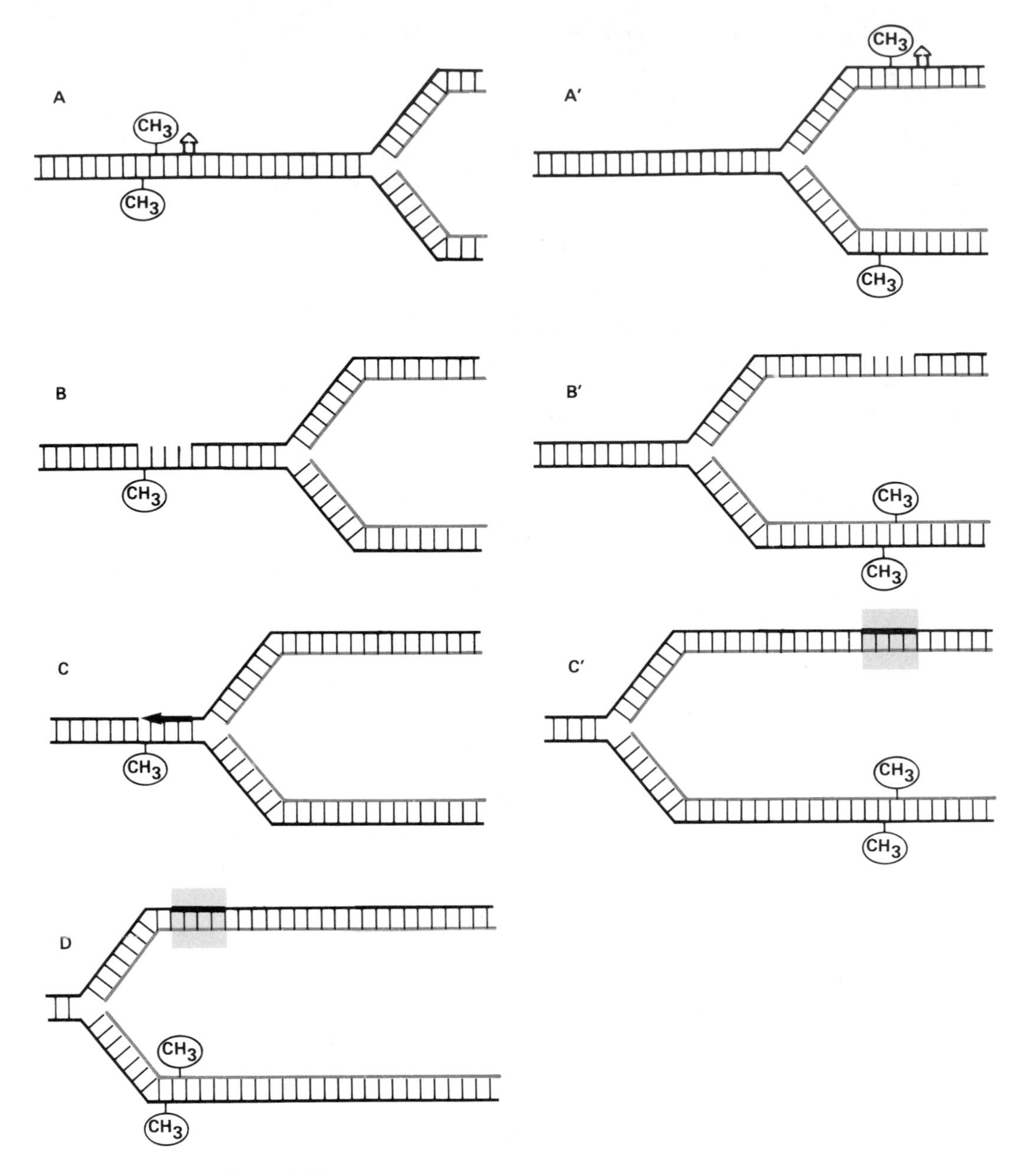

FIGURE 6-12
Excision repair of damage to DNA immediately in front of a replication fork (left). If the excision is initiated close to a methylated controlling sequence (A), a methylated base is removed (B) and repair synthesis will insert a nonmethylated base (C). Semiconservative replication of this region *before* methylation results in a nonmethylated DNA duplex, which is not a substrate for the maintenance methylase. The other DNA duplex will be normally methylated (D).

Repair of damage to DNA close to a controlling sequence immediately after replication (right). Excision repair before the daughter strand has been methylated (A′, B′), followed by repair synthesis, results in a nonmethylated duplex which is not a substrate for the maintenance methylase (C′). (From R. Holliday, ref. 94.)

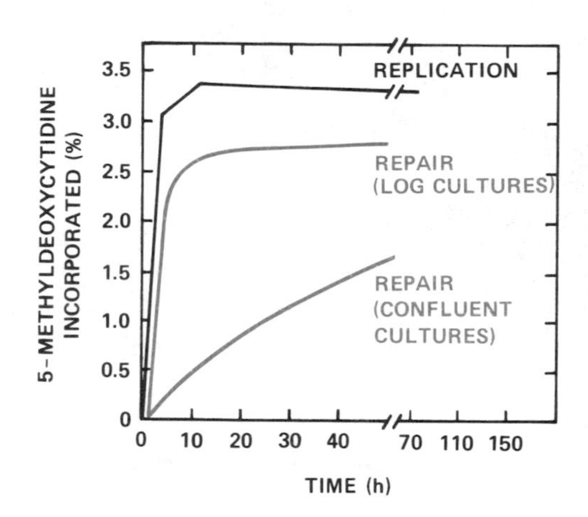

FIGURE 6-13
The methylation of DNA following excision repair in confluent and in logarithmic growth phase human fibroblasts is slower and less complete than in nondamaged cells undergoing normal semiconservative DNA replication. (From M. B. Kastan et al., ref. 95.)

ability to accept methyl groups. Carcinogenic agents may therefore cause heritable changes in 5-methylcytosine patterns by a variety of mechanisms (96).

6-4 Poly(ADP-Ribose) and DNA Repair

Nicotinamide adenine dinucleotide (NAD), also known as diphosphopyridine nucleotide (DPN), is the most abundant of the respiratory coenzymes (97). The classic biochemical studies of Harden, Young, Warburg and others established that the major biochemical function of NAD is as an electron carrier in various biological oxidation-reduction systems (97). Its structure was determined in 1936 by von Euler and his collaborators and was shown to consist of two mononucleotides (5′ AMP and nicotinamide mononucleotide) linked together by a pyrophosphate bond. However, the structure of NAD can also be viewed as an adenosine diphosphate–ribosyl moiety (ADP-ribose) attached covalently to the vitamin nicotinamide through an N-glycosylic linkage (Fig. 6-14). This linkage constitutes a high-energy bond, since its free energy of hydrolysis is approximately −8.2 kcal/mol at pH 7 and 25°C (97).

The energy of this bond is used to enzymatically transfer the ADP-ribosyl moiety of NAD to macromolecules. Thus, ADP-ribosyl transferase catalyzes the transfer of the ADP-ribosyl moiety of NAD to chromosomal proteins, with the concomitant release of nicotinamide and a proton. This reaction is termed *mono ADP-ribosylation* (97). Additionally, a polymer of ADP-ribose can be formed as a moiety attached to a macromolecular acceptor such as a chromosomal protein, in a reaction termed *poly ADP-ribosylation* (97). The catalytic

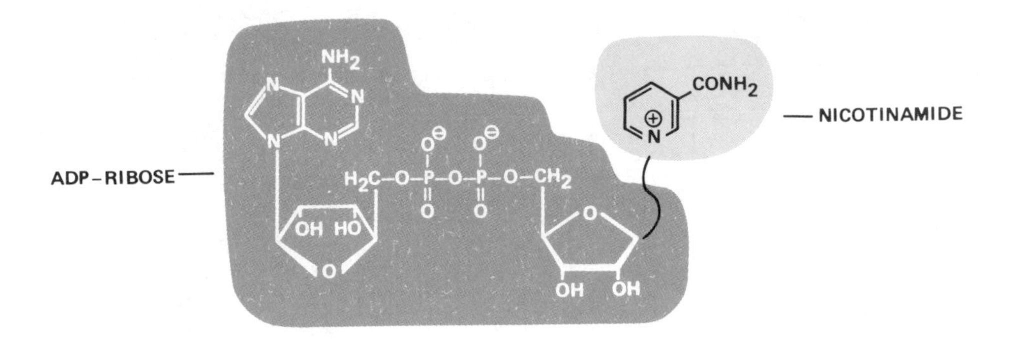

FIGURE 6-14
The structure of nicotinamide adenine dinucleotide (NAD) can also be viewed as ADP-ribosyl nicotinamide.

activity responsible for these reactions is termed poly(ADP-ribose) synthetase or polymerase. In normal (undamaged) cells about 90 percent of the ADP-ribose is present in the monomeric form; the small percentage present as polymers may contain up to 70 ADP-ribose units in a single chain. However, the *total* amount of constitutive ADP-ribose in the chromatin of undamaged cells is relatively small (one ADP-ribose residue per 5000 nucleotides) (97).

In eukaryotic cells all NAD synthesis occurs in the nucleus, and the vast majority of NAD formed is used for the biosynthesis of ADP-ribose and poly(ADP-ribose) (98). Poly(ADP-ribose) synthetase is tightly associated with chromatin and catalyzes the synthesis of acceptor-bound poly(ADP-ribose) in a reaction that has an absolute requirement for DNA (98). Furthermore, this DNA must have free ends. Covalently closed circular DNA does not activate the enzyme, but fragmentation of DNA increases its ability to activate the enzyme and the activation is proportional to the number of nicks in the DNA (98–100).

A variety of compounds inhibit poly(ADP-ribose) biosynthesis. These include 5-methylnicotinamide, methylxanthines, thymine, thymidine and benzamides such as 3-aminobenzamide (98). With the use of these inhibitors it can be shown that the biosynthesis of poly(ADP-ribose) is not needed for cell growth or for DNA, RNA or protein synthesis in cultures of exponentially growing cells not subject to DNA damage (98). However, a wide variety of DNA damaging agents cause a marked lowering of cellular NAD levels (101, 102). This response can be eliminated by inhibitors of poly(ADP-ribose) synthesis (102) (Fig. 6-15). In addition, the specific activity of the synthetase increases in damaged cells in a dose-dependent way. These observations indicate that DNA damaging agents somehow promote the utilization of cellular NAD for the biosynthesis of poly(ADP-ribose), and they have led to the hypothesis that this bio-

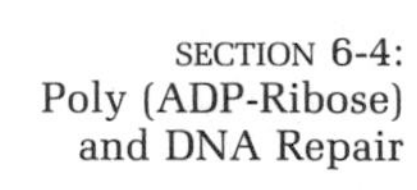

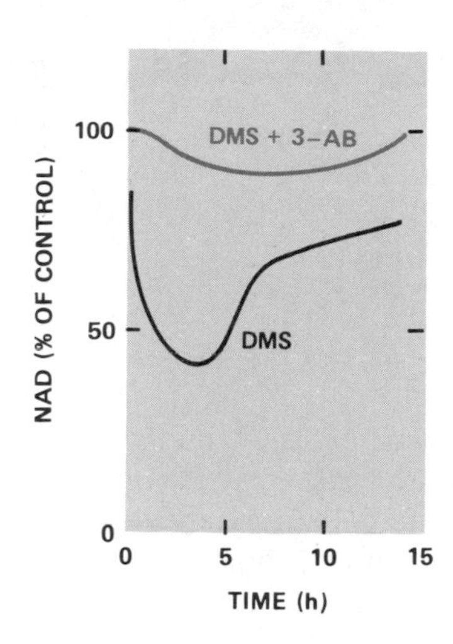

FIGURE 6-15
Dimethylsulfate (DMS) causes a dramatic decrease in the NAD content of mouse L cells in culture. This effect can be largely mitigated by inhibitors of poly(ADP-ribose) synthetase (polymerase) such as 3-aminobenzamide (3-AB). (From B. W. Durkacz et al., ref. 102.)

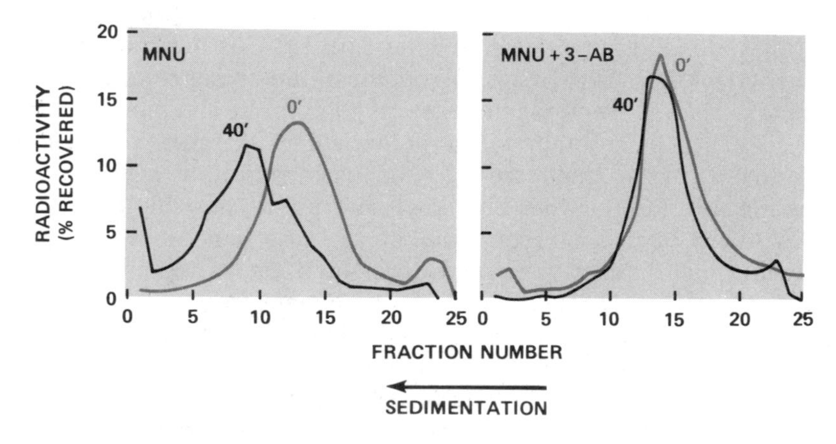

FIGURE 6-16
When cells treated with N-methyl-N-nitrosourea (MNU) are incubated for about 40 minutes and then lysed and sedimented in alkaline sucrose gradients, the DNA sediments near the bottom of the gradient relative to the DNA from unincubated cells (zero minutes of incubation). This indicates that repair of the strand breaks introduced by the MNU treatment took place (left panel). However, in the presence of 3-aminobenzamide (3-AB), an inhibitor of poly(ADP-ribose) polymerase, repair of DNA strand breaks is inhibited and the DNA continues to sediment nearer the top of the gradient (right panel). (From S. Shall, ref. 98.)

synthesis is a cellular response to DNA damage required for efficient DNA repair (98, 103).

This hypothesis has been tested in a number of ways. When DNA incision and rejoining are monitored as parameters of excision repair, there is a marked inhibition of the reconstitution of high molecular weight DNA if cells exposed to MNU are incubated in the presence of inhibitors of poly(ADP-ribose) synthetase (Fig. 6-16) (102, 103). Additionally, the ability of cells to carry out excision repair is

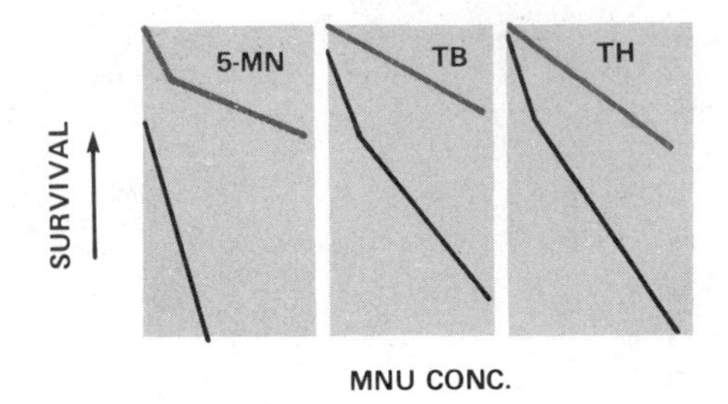

FIGURE 6-17
A synergistic potentiation of killing results from treatment of cells with MNU (upper curves) in the presence of 5-methylnicotinamide (5-MN), theobromine (TB) or theophylline (TH) (lower curves). (From S. Shall, ref. 98.)

depressed when cellular NAD levels are severely reduced by nutritional deprivation (98). When the cellular NAD level is reduced below K_m levels for the synthetase and the cells are exposed to γ irradiation or to alkylation damage, they still incorporate precursors for DNA, RNA and protein synthesis at normal rates but have a marked defect in the reconstitution of DNA strand breaks (98). Finally, in the presence of the inhibitor 5-methylnicotinamide, cell killing induced by treatment of cells with MNU is markedly potentiated (102–105). Similar results obtain in the presence of other inhibitors; such as methylxanthines, theobromine and theophylline (Fig. 6-17). These inhibitors also potentiate a synergistic increase in cell toxicity after treatment of cells with radiation or with neocarzinostatin (103). Other biological end points possibly related to DNA repair have been examined. In Chinese hamster ovary cells grown in nicotinamide-free medium (which results in a marked decrease in cellular NAD levels), the spontaneous frequency of certain chromosomal abnormalities termed sister chromatid exchanges (SCEs; see Section 9-4) is enhanced (106). Furthermore, treatment of cells with inhibitors of poly(ADP-ribose) synthetase in the *absence* of any exogenous DNA damaging agents results in a marked increase in SCEs and in other chromosomal aberrations (106).

Poly(ADP-ribose) is apparently not required for the incision of damaged DNA, nor for repair synthesis. However, it may activate a DNA ligase required for DNA repair in mammalian cells (107). Mammalian cells contain two forms of this enzyme, called DNA ligase I and II (see Section 6-5), and when cells are exposed to the alkylating agent dimethylsulfate, the total DNA ligase activity increases about twofold, most if not all of which reflects an increase in DNA ligase II (107). This increase is prevented by all known inhibitors of poly(ADP-ribose) synthetase. Exactly how poly(ADP-ribose) influences DNA ligase activity is not known. Conceivably, under normal conditions DNA ligase is an acceptor protein for ADP-ribose, in which state it is activated to carry out more efficient rejoining of nicks in DNA (107). However, this model fails to explain why the excision of pyrimidine dimers, which results in approximately the same frequency of DNA strand breaks over the same time intervals as do low doses of alkylating agents, does not apparently involve a ligation step that is sensitive to perturbations of the level of poly(ADP-ribose) by 3-aminobenzamide (108, 109). Either the precise chemical nature of the incisions produced by the excision repair of pyrimidine dimers makes them considerably less effective as a stimulus for synthesis of poly(ADP-ribose), or single-strand breaks by themselves are not the effective trigger for such synthesis (108). Thus, some other change associated with DNA damage by alkylating agents and/or its repair may be important. Finally, it is necessary to consider the possibility that inhibition of NAD biosynthesis by agents such as 3-aminobenzamide may have effects on cellular metabolism other than inhibition of poly(ADP-ribose) synthetase (108). For example, 3-aminobenzamide inhibits the incorporation of radiolabel from the C^1 pool into purines during de novo biosynthesis (109) and also inhibits DNA methylation by S-adenosylmethionine (108).

6-5 Enzymes Possibly Involved in Postincision Events during Excision Repair

The same general considerations described in the introduction to Chapter 3 with respect to the mechanism of the excision of bulky base damage in *E. coli* are appropriate here. Thus, irrespective of whether or not excision of such damage is directly coupled to the incision of DNA in eukaryotic cells, exonuclease-catalyzed excision of apurinic and apyrimidinic sites in DNA is likely, since DNA glycosylases and AP endonucleases are ubiquitous in such cells (see Chapter 3). Since most of the known mammalian cell DNA polymerases do not have associated exonucleases (110), such excision is presumably effected by independent exonucleases. Nonetheless, it is possible that in vivo one or more such exonucleases exist in intimate functional association with DNA polymerases involved in repair synthesis. Thus DNA degradation and repair synthesis may be tightly coordinated, as is true during some forms of excision repair in *E. coli* (see Section 5-4). A number of $5' \rightarrow 3'$ and $3' \rightarrow 5'$ exonucleases have been described in mammalian cells, and their principal properties are discussed in the ensuing paragraphs.

$5' \rightarrow 3'$ Exonucleases

Deoxyribonuclease (DNase) IV

DNase IV has been partially purified from rabbit bone marrow and from rabbit lung (111–114). The properties of the enzyme from both sources are identical. That from bone marrow has optimal activity at

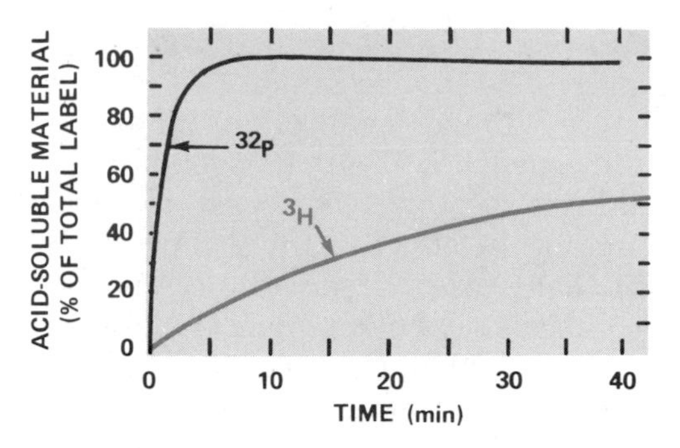

FIGURE 6-18
When ^{3}H uniformly labeled and ^{32}P terminally labeled poly(dA)·poly(dT) is degraded by DNase IV, the terminal label is released much faster than the bulk label. This indicates that the ends of the molecules are being attacked first; i.e., DNase IV acts *exonucleolytically*. (From T. Lindahl et al., ref. 111.)

pH 8.5 and requires a divalent cation for activity. At an optimal concentration of 2 m*M*, Mn^{2+} is twice as effective as Mg^{2+}. The enzyme is completely inactivated by 0.3-m*M* *p*-hydroxymercuribenzoate, and inhibition is observed in the presence of NaCl or potassium phosphate. After five minutes at 52°C, less than 1 percent of the activity is detectable. The molecular weight calculated from the sedimentation coefficient and the Stokes radius is about 42,000, a value that corresponds well with that measured with gel filtration (111).

After 35 percent hydrolysis of poly(dA)·poly([^{3}H]dT) by DNase IV, the major radioactive acid-soluble product is 5′ dTMP. Acid-soluble radioactivity as dinucleotide constitutes less than 10 percent of that in mononucleotide, and in oligonucleotide it is less than 3 percent of that in mononucleotide. Thus the enzyme acts exonucleolytically, releasing mononucleotides as the principal products (112). Additional support for an exonucleolytic mode of action comes from the observation that when 5 percent of poly(dA)·poly([^{3}H]dT) is degraded into an acid-soluble form, only a very small reduction in the sedimentation coefficient of the polymer is detected (112). Furthermore, the kinetics of degradation of terminally labeled polydeoxyribonucleotides indicate an exonucleolytic mechanism operating from the 5′ end exclusively (Fig. 6-18) (111).

DNase IV is not significantly inhibited by photoproducts in DNA or in poly(dA)·poly(dT) (Fig. 6-19). About 40 percent of the products of the degradation of the latter UV-irradiated substrate are oligonucleotides containing two to eight nucleotides. Thymine dimers are found in oligonucleotides five to eight residues in size (113). After incubation of DNase IV with a preparation of UV-irradiated DNA preincubated with the *M. luteus* PD DNA glycosylase–AP endonuclease, there is an enrichment of the ratio of thymine dimers to

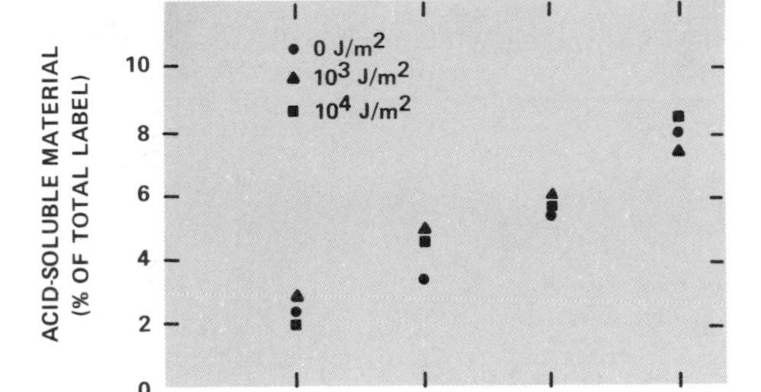

FIGURE 6-19
Enzymatic degradation of unirradiated and UV-irradiated *E. coli* DNA by DNase IV. The extent of DNA degradation into acid-soluble product is not inhibited by the presence of photoproducts in the DNA. Thus, like the 5′ → 3′ exonuclease of *E. coli* DNA polymerase I, DNase IV removes oligonucleotides (including those containing base damage) as well as mononucleotides during exonucleolytic degradation of DNA. (From T. Lindahl, ref. 114.)

thymine monomers in the acid-soluble fraction, indicating that dimers are removed from the DNA by degradation in the 5′ → 3′ direction (Table 6-1).

DNase IV bears a striking resemblance to the 5′ → 3′ exonuclease moiety of *E. coli* DNA polymerase I (see Section 5-3). Similar properties between the two activities (113) include

1. A preference for double-strand substrates
2. The sites of attack (5′ ends with either phosphate or hydroxyl end groups)
3. The ability to excise pyrimidine dimers from incised DNA
4. The more rapid degradation in vitro of double-strand synthetic polydeoxyribonucleotides of simple repeating base composition compared with DNA from natural sources

This comparison is of more than passing interest since, as indicated previously (see Section 5-3), the 5′ exonuclease activity in *E. coli* can be readily separated from the polymerizing moiety by limited proteolysis. The question therefore arises as to whether a similar association exists in living mammalian cells that is particularly susceptible to proteolysis during cell extraction. Indirect evidence (including the absence of detectable DNA polymerase activity in purified preparations of DNase IV) suggests this possibility is unlikely, but definitive proof that such an association does not exist in vivo is, of course, difficult to obtain.

TABLE 6-1
Excision of thymine-containing pyrimidine dimers by DNase IV from UV-irradiated DNA preincised with *M. luteus* "UV endonuclease"

	[^{14}C]Thymine (pmol Released)	
	Dimer Fraction	Monomer Fraction
1. DNase IV treatment, no previous *M. luteus* enzyme	12	220
2. *M. luteus* enzyme treatment, not followed by DNase IV treatment	28	58
3. *M. luteus* enzyme treatment, followed by DNase IV treatment	74	260
Material released from nicked DNA by DNase IV [3. minus 2.]	46	202

The DNA in each reaction contained 8000 pmol of thymine as monomer and 530 pmol as thymine-containing dimers.
From T. Lindahl, ref. 114.

5′ → 3′ Thymine Dimer Excising Activities from Human KB Cells

In order to directly detect enzymes in crude extracts of human cells that catalyze the *selective* excision of thymine-containing pyrimidine dimers in vitro, substrates have been constructed that consist of UV-irradiated duplex DNA preincised at pyrimidine sites with the phage T4 PD DNA glycosylase–AP endonuclease activity (Fig. 6-20). This substrate offers the advantage of *including* putative dimer excising activities that may fail to degrade UV-irradiated single-strand DNA and *excluding* from consideration those that fail to selectively excise dimers from duplex DNA, irrespective of any other catalytic activities. Thus the dimer excising activity is selected for as a primary activity of the enzyme (115, 116). Three activities have been detected in extracts of human KB cells that excise thymine-containing pyrimidine dimers from UV-irradiated preincised DNA. These are designated as activities A, B and C and have $S^{\circ}_{20,w}$ values of 3.2, 2.6 and 3.0 respectively (115, 116). Activity B appears to be distinct from activities A and C. It can be distinguished by its higher isoelectric point (9.0 compared with 6.0 for A and C) and its smaller sedimentation coefficient, and it does not significantly degrade single-strand DNA into acid-soluble products (Table 6-2). The distinction between activities A and C is less clear-cut. The physical parameters of sedimentation velocity and isoelectric point do not distinguish them, but their binding affinities for DEAE-cellulose and phosphocellulose are different. Furthermore, the inhibition of activity A by 75 m*M* sodium chloride is twice as great as that of activity C, and activity A is less sensitive to inhibition by agents that bind SH groups (Table 6-2). Both activities degrade UV-irradiated or unirra-

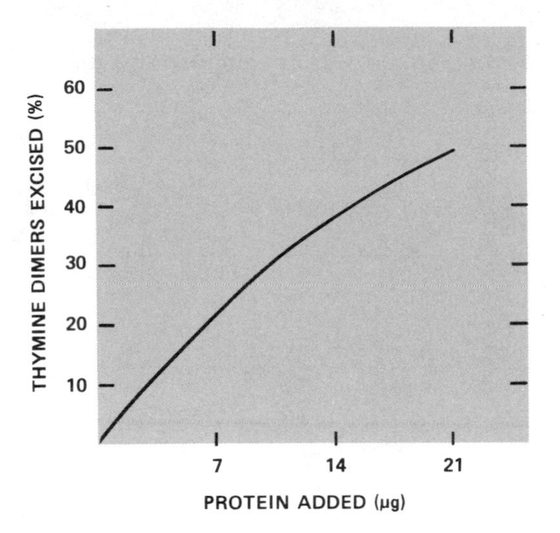

FIGURE 6-20
Excision of thymine-containing pyrimidine dimers from UV-irradiated *E. coli* DNA preincised with T4 PD DNA glycosylase–AP endonuclease is effected by extracts of human HeLa cells. (From K. H. Cook et al., ref. 171.)

TABLE 6-2
Activities from human KB cells that excise thymine-containing pyrimidine dimers from preincised UV-irradiated DNA

	Activity A	Activity B	Activity C
$S^{\circ}_{20,w}$	3.2	2.6	3.0
pI	6.0	9.0	6.0
Degradation of:			
Native DS DNA	–	–	–
Unincised UV'd DS DNA	–	–	–
Preincised UV'd DS DNA	+ + + (Preferential removal of dimers)	+ + + (Preferential removal of dimers)	+ + + (Preferential removal of dimers)
Unirradiated SS DNA	+ + +	±	+ + +
UV'd SS DNA	+ + +	±	+ + +
Poly d(A-T)	–	+ +	+ +
[NaCl] required to elute from phosphocellulose (*M*)	0.0	0.15	0.25
[NaCl] required to elute from DEAE-cellulose (*M*)	0.11	0.06	0.08
Inhibition by NaCl (75 m*M*)	50%	50%	25%
Inhibition by *p*-chloromercuriphenylsulfonic acid (pCMPSA) (10μM)	25%	80%	55%

TABLE 6-3
Selective excision of thymine-containing pyrimidine dimers by human KB cell activities

	Degradation of Preincised UV-Irradiated DNA	
Activity	Total nucleotide rendered acid soluble (%)	Thymine-containing dimers rendered acid soluble (%)
A	0.26	4.3
B	0.10	20.0
C	0.29	10.5

From K. H. Cook and E. C. Friedberg, ref. 115.

diated single-strand but not duplex DNA, and both excise thymine dimers from preincised duplex DNA (Table 6-3). It is possible that activities A and C represent heterogeneous forms of the same gene product, perhaps due to posttranslational modification of one form. Alternatively, the modification may be artifactual due to nonspecific binding of some subcellular component or due to proteolysis.

$5' \rightarrow 3'$ *Enzyme Activities from Human Placenta and from Mammalian Cells in Culture Other than Human KB Cells*

Two $5' \rightarrow 3'$ exonucleases have been purified from human placenta that resemble in a number of respects activities A and C from human KB cells described above. These are referred to as the human placental correxonuclease (*corr*ectional exonuclease) (117) and DNase VIII (118). The correxonuclease resembles exonuclease VII of *E. coli* (see Section 5-3) in that it is single-strand specific and initiates hydrolysis of such DNA from both 3′ and 5′ termini, yielding oligonucleotides averaging four in length. These oligonucleotides are released from both termini at equal rates with the same size distribution (117). Although not demonstrably active against intact native DNA, the enzyme can initiate hydrolysis at single-strand breaks, creating gaps that are 30 to 40 nucleotides long. Consistent with the limited degradation of incised duplex DNA, if a pyrimidine dimer is situated adjacent to such a break, the enzyme can excise the dimer (Fig. 6-21) (117). Not unexpectedly however, dimers are not preferentially removed from UV-irradiated DNA that is *randomly* incised with pancreatic DNase (Fig. 6-21). The pH optimum for the human placental enzyme is 8.0. Unlike exonuclease VII of *E. coli*, this enzyme has a requirement for divalent cation, which is satisfied by either Mg^{2+}, Co^{2+} or Mn^{2+} (117).

DNase VIII is another exonuclease from human placenta that is specific for single-strand DNA. It degrades this substrate exclusively

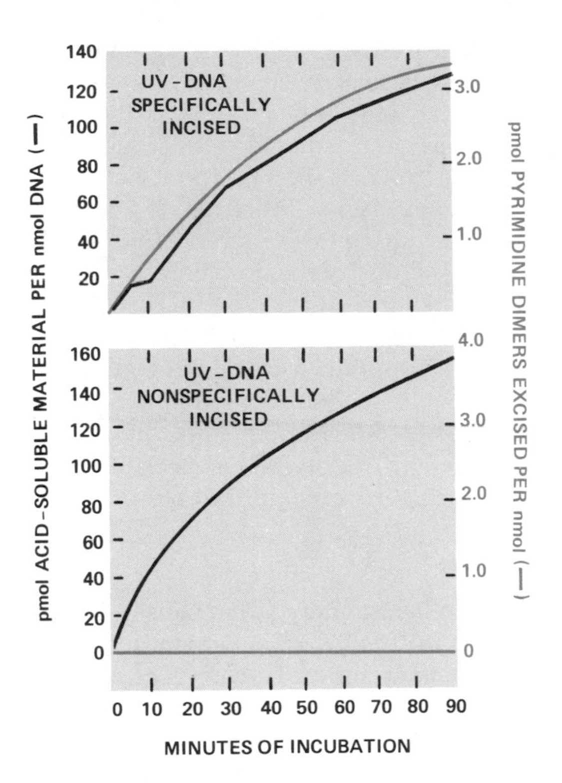

FIGURE 6-21
Kinetics of degradation and release of thymine dimers from UV-irradiated DNA by "correxonuclease" from human placenta. When UV-irradiated DNA is *specifically* incised at pyrimidine dimer sites by the *M. luteus* PD DNA glycosylase–AP endonuclease, dimers are preferentially excised by the exonuclease activity (top figure). When the DNA is *nonspecifically* incised with pancreatic DNase, essentially no dimers are lost from the DNA (bottom figure). (From J. Doniger and L. Grossman, ref. 117.)

from the 5′ terminus, liberating short oligonucleotides (118). It can also attack duplex DNA with incompletely hydrogen-bonded 5′ termini, and presumably as part of this function it catalyzes the excision of thymine dimers from specifically incised UV-irradiated duplex DNA (118). DNase VIII has a pH optimum of 9.5 and a Mg^{2+} optimum at 0.2 *M*. Both of these properties distinguish it from the placental correxonuclease. In addition, correxonuclease reaction products from DNA have an average size of four nucleotides, whereas the predominant reaction products of DNase VIII activity on single-strand DNA are dinucleotides (118). Another property that distinguishes these two exonucleases is their differential ability to attack nicked duplex DNA. DNase VIII cannot hydrolyze this substrate unless there is significant localized denaturation at the site of the nick, whereas correxonuclease can. Finally, DNase VIII apparently does not possess 3′ exonuclease activity (118).

Yet another enzyme with both 3′ → 5′ and 5′ → 3′ exonuclease activity (DNase V) has been isolated from Novikoff hepatoma cells (119) and from HeLa cells (120). The former enzyme was initially designated as stimulatory factor IV because it binds to homogeneous DNA polymerase β, resulting in an increase in the molecular weight, specific activity and stability of the polymerase (121). Purified DNase V requires a divalent cation and has optimal activity at 5 to 10 mM Mg^{2+}. The pH optimum is 8.8 in Tris-HCl or glycine-NaOH buffers, but the enzyme is active over a broad range of pH. On synthetic homopolymers exonuclease V shows a high degree of preference for double-strand substrates; however, the enzyme is unable to hydrolyze UV-irradiated poly(dA)·poly(dT), suggesting that it might have a limited role if any in the excision of bulky base damage (119).

3′ → 5′ Exonucleases

There is no evidence that in mammalian cells incision of DNA at pyrimidine dimer sites occurs by a mechanism that leaves 3′-terminal deoxyribose-phosphate moieties. However, as indicated previously (see Section 3-5), 3′-acting AP endonucleases have been demonstrated in extracts of mammalian cells. Thus it is possible that 3′-deoxyribose-phosphate moieties are generated in vivo following the incision of DNA containing apurinic or apyrimidinic sites. Such terminal deoxyribose-phosphate groups could be excised from DNA by 5′ AP endonucleases, a class of enzymes discussed fully in Section 3-5. Additionally, or alternatively, any 3′-acting exonuclease that can hydrolyze duplex DNA at nicks might be involved in the repair of 3′-terminal lesions. Two 3′ → 5′ exonucleases have been partially purified and characterized from mammalian cells and tissues.

DNase III

DNase III has been purified about 700-fold from normal rabbit bone marrow (113, 114, 122). The enzyme has a molecular weight of about 52,000, a pH optimum at 8.5 and a requirement for either Mg^{2+} or Mn^{2+}. An exonucleolytic mode of action is inferred from the sedimentation of enzyme-treated DNA in alkaline sucrose, which shows that after 2 percent of the DNA is digested into acid-soluble products, the remaining material shows no detectable decrease in sedimentation coefficient, i.e., no endonucleolytic activity is present (122). The hydrolysis of both single- and double-strand polymers is initiated preferentially from the 3′ end. Denatured DNA is degraded four times more rapidly than native DNA, with no apparent base specificity. The products of 3′ exonucleolytic degradation are 5′ mononucleotides as well as dinucleotides (122).

DNase VII is an exonuclease that has been purified 6000-fold from human placenta (123). The enzyme has an apparent molecular weight of 43,000, requires Mg^{2+} for activity and has a pH optimum of 7.8. The enzyme hydrolyzes single-strand and nicked duplex DNA at the same rate, proceeding in a $3' \rightarrow 5'$ direction and liberating 5′ mononucleotides.

Finally, recall that the so-called correxonuclease of human placenta can degrade single-strand DNA in vitro in both the $3' \rightarrow 5'$ and the $5' \rightarrow 3'$ direction (117), and DNase V can degrade duplex DNA bidirectionally (119).

Mammalian Cell DNA Polymerases

The two best-studied DNA polymerases in mammalian cells are nuclear enzymes called DNA polymerases α and β (110). Two other DNA polymerases designated γ (believed to be a mitochondrial enzyme) and δ have also been described, but these will not be discussed here since there is no indication that they are involved in the repair synthesis of nuclear DNA.

DNA Polymerase α

DNA polymerase α is the enzyme most clearly associated with chromosomal replication. In growing cells it accounts for more than 85 percent of the total DNA polymerase activity, but it only accounts for about 5 percent of the activity in quiescent cells. Thus, the activity of this enzyme undergoes a marked increase in actively proliferating cells (110). It has now been quite firmly established by the use of monoclonal antibodies that the enzyme from human KB cells has a nuclear location (124, 125), although extraction procedures during enzyme purification frequently result in massive leakage into the cytoplasm (126).

The exact molecular weight and physical composition of mammalian DNA polymerase α are not certain. The enzyme is very sensitive to proteolysis during purification and has a tendency to aggregate, especially at low ionic strength (110). Based on studies of the enzyme isolated from a variety of sources, catalytic activity appears to be associated with polypeptides of molecular weight 120,000 to 200,000 (110). It is likely that the polypeptides of higher molecular weight are the true catalytic subunits; smaller subunits are possibly the products of proteolysis (110). The preferred template-primer for DNA polymerase α is duplex DNA with gaps of 20 to 70 nucleotides (127). Nicked DNA is ineffective as a substrate (127). The polymerization mechanism is apparently the same as that for prokaryote enzymes, and both pyrophosphorolysis and pyrophosphate exchange have been demonstrated (110). DNA polymerase α as currently purified does not exhibit any exonuclease activities (128).

DNA Polymerase β

DNA polymerase β is also found in the nucleus. The purified enzyme from a variety of sources has a molecular weight of 40,000. The enzyme is less sensitive than polymerase α to SH blocking agents, and it is stimulated by salt. Like DNA polymerase α, its template-primer preference is for gapped DNA (129). It too has no associated exonuclease activity.

The principal properties of DNA polymerases α, β and γ from animal cells are summarized in Table 6-4.

TABLE 6-4
Distinctive features of animal DNA polymerases α, β and γ

	α	β	γ
Location	Nucleus	Nucleus	Mitochondria
Proposed principal function	Replication, nDNA	Repair	Replication, mtDNA
Mass (kDa)	120–220	30–50	150–300
S value	6–8	3–4	7–9
Subunits (kDa)	50–70; 155		
Aggregation in low salt	Yes	Yes	Yes
Isoelectric point	Acidic	Basic	Acidic
pH optimum	7.2	8.5	8.0
K_m for dNTPs (μM)	10	10	0.5
Preferred cation	Mg^{2+}	Mg^{2+} or Mn^{2+}	Mg^{2+} or Mn^{2+}
Template-primer			
Preferred	Gapped DNA	Gapped DNA	Ribohomopolymers
Ribo template-deoxy primer	No	Yes	Yes
Deoxy template-ribo primer	Yes	No	No
Nuclease activity	No	No	Uncertain
Relative activity			
Growing cells	(100)	10	2
Resting cells	0–5	(100)	10
Inhibitory effects			
Sulfhydryl (e.g., N-ethylmaleimide)	Strong	Weak	Strong
Salt (e.g., 0.2 *M* NaCl)	Yes	Stimulates	Stimulates
Phosphate (e.g. 0.1 *M*)	No	Yes	Stimulates
Heat (45°, 10 min)*	0	70–100	70
Dideoxy NTPs	Weak	Strong	Strong
Arabinosyl NTPs	Strong	Weak	Weak
Aphidicolin	Yes	No	No

*Percent of activity lost.
From A. Kornberg, ref. 110.

The Role of DNA Polymerases α and β in Repair Synthesis of DNA

The relative role of DNA polymerases α and β in repair synthesis of DNA is controversial (74, 130–139). A major limitation is the lack of available mutants totally defective in one or the other of these enzymes, necessitating the use of less direct experimental approaches to the issue. Perhaps the most physiological of these are experiments on terminally differentiated cells such as neurons or skeletal muscle cells, which are purported to contain little or no DNA polymerase α (140–143). Such cells do show repair synthesis following UV radiation and other forms of DNA damage, suggesting that an enzyme(s) other than DNA polymerase α is involved in excision repair. However, it is not clear that the residual polymerase α that may be present in these cells is inadequate to carry out this synthesis. In addition, the absence of DNA polymerase α in terminally differentiated cells does not necessarily imply that the enzyme is not involved in the repair synthesis of DNA in cells that do possess adequate levels of activity.

Aphidicolin (Fig. 6-22) is a tetracyclic diterpinoid obtained from *Cephalosporium aphidicola* that selectively inhibits DNA polymerase α (144) (Fig. 6-23). This selectivity offers the possibility that the inhibitor might be useful for establishing whether DNA polymerase α or β plays an exclusive role in repair synthesis during excision repair in mammalian cells. However, the results of such studies are equivocal. For example, UV-irradiated HeLa cells incubated in the presence of aphidicolin apparently carry out only repair synthesis of DNA, suggesting that this synthesis is a function of DNA polymerase β activity (131) (Fig. 6-24). In addition, in mitotic HeLa cells in which no replicative DNA synthesis is occurring, aphidicolin does not reduce the level of repair synthesis detected in UV-irradiated cells, leading to the conclusion that DNA polymerase α is certainly not *essential* for repair synthesis (139). On the other hand, other studies show that this compound inhibits *both* semiconservative DNA synthesis in unirradiated cells and repair synthesis in UV-irradiated cells, implying that DNA polymerase α is responsible for the latter phenomenon (132). In an attempt to measure the effect of aphidicolin specifically on DNA *repair* synthesis, a semi-in-vitro system consisting of isolated nuclei from G1 or G2 (i.e., non-S-phase) cells was utilized (132). The nuclei were incubated in the presence of hydroxyurea and cytosine arabinoside to inhibit any residual semiconservative DNA synthesis, and in addition ATP (an essential component for semiconservative DNA synthesis in isolated nuclei) was omitted. Aphidicolin inhibited the incorporation of [^{3}H]dTMP into the nuclei of UV-irradiated cells by more than 90 percent. In view of the apparent specificity of aphidicolin for DNA polymerase α, these experiments support the contention that this enzyme is required for repair

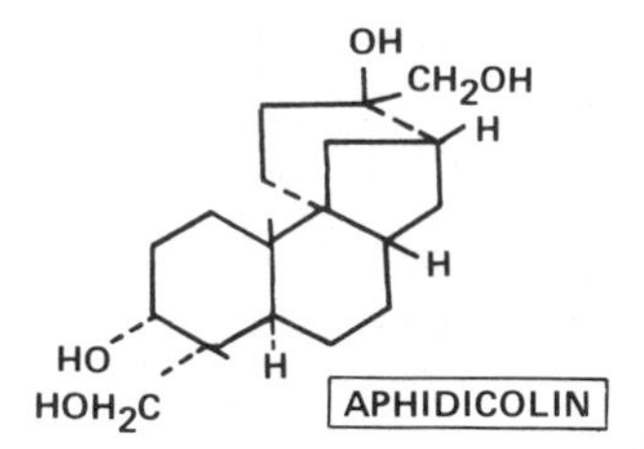

FIGURE 6-22
Structure of aphidicolin.

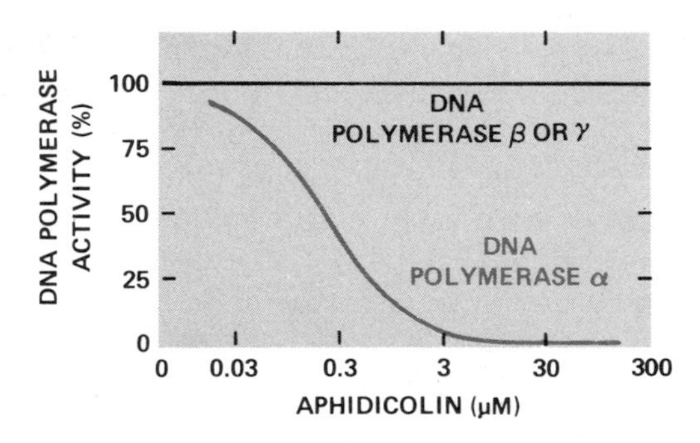

FIGURE 6-23
DNA polymerase α activity in vitro is selectively inhibited by aphidicolin. (From G. Pedrali-Noy and S. Spadari, ref. 133.)

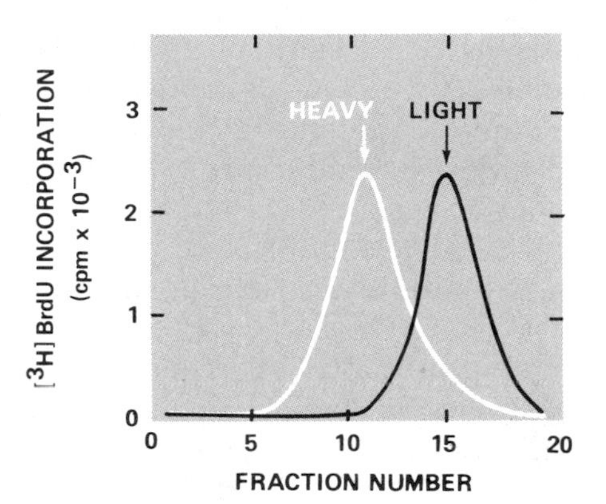

FIGURE 6-24
When *unirradiated* HeLa cells are incubated in the presence of [^{3}H]BrdU and the DNA is sedimented to equilibrium in CsCl, essentially all radioactivity is in the position of hybrid (heavy) DNA due to semiconservative DNA synthesis (see Fig. 5-2). When UV-irradiated cells are incubated in the presence of aphidicolin and subjected to equilibrium sedimentation analysis of DNA, the radioactivity is detected at parental (light) density (see Fig. 5-2). Thus in these experiments aphidicolin did not inhibit repair synthesis of DNA. (From G. Pedrali-Noy and S. Spadari, ref. 133.)

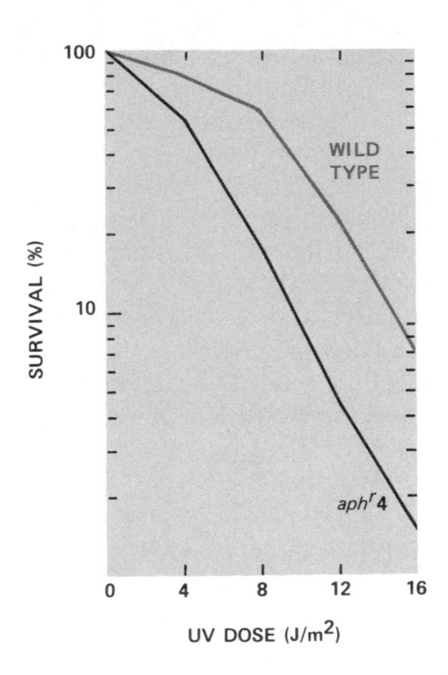

FIGURE 6-25
The aphidicolin-resistant mutant *aph*^{r}-4 is abnormally sensitive to killing by UV radiation, suggesting an involvement of DNA polymerase α in the repair of DNA damage caused by UV light. (After C-C. Chang et al., ref. 145.)

synthesis. A possible (though not necessarily exclusive) role for DNA polymerase α in DNA repair is also suggested by the isolation of mutant Chinese hamster ovary cells that are resistant to aphidicolin (145, 146). One such mutant contains aphidicolin-resistant DNA polymerase α activity in vitro and is UV sensitive (Fig. 6-25) (146).

Some experiments have led to the conclusion that *both* DNA polymerases α and β are involved in repair synthesis, but that each enzyme functions selectively in the repair of different kinds of DNA damage (134, 135). For example, aphidicolin or cytosine arabinoside was used as a specific inhibitor of DNA polymerase α and dideoxythymidine triphosphate as a selective inhibitor of DNA polymerase β. Based on the presumed selectivity of these inhibitors of DNA synthesis, it was concluded that the latter enzyme is primarily responsible for repair synthesis induced by agents such as bleomycin or neocarzinostatin, whereas polymerase α plays a more prominent role in repair synthesis associated with the repair of alkylation damage (134, 135).

There are indications that at low doses of DNA damage to human fibroblasts in culture by a variety of agents, repair synthesis of DNA is largely mediated by an enzyme other than DNA polymerase α, presumably DNA polymerase β (136). With increasing amounts of DNA damage there is increased repair synthesis and a greater participation in this process by DNA polymerase α. At high doses of damage the fraction of repair synthesis mediated by DNA polymerase α reaches a maximum, which varies as a function of the particular DNA damaging agent (136). With agents such as UV radiation and

N-acetoxy-2-acetyl-2-aminofluorene, which produce bulky base damage, DNA polymerase α (defined by aphidicolin inhibition) contributes to about 80 percent of the total repair synthesis. However, this enzyme appears to be significantly less involved in repair synthesis associated with damage caused by bleomycin.

One of the problems associated with the interpretation of these and other experiments on aphidicolin-treated cells is that the amount of DNA synthesis (and perhaps the amount of DNA polymerase) required for DNA repair is very small relative to that observed during semiconservative replication of DNA. Thus, even though the compound inhibits more than 98 percent of the total DNA synthesis in undamaged cells, the residual synthesis may still be catalyzed by DNA polymerase α and may be sufficient for DNA repair. Similarly, in damaged cells one can never be certain that the DNA synthesis inhibited by aphidicolin is exclusively or even wholly repair synthesis, since very small levels of residual replicative synthesis are difficult to distinguish from repair synthesis. It has also been argued that some of the controversy surrounding the use of inhibitors of DNA synthesis such as cytosine arabinoside and aphidicolin can be reconciled by considering the particular cells or the state of the cell used experimentally (74). Little or no inhibition of repair synthesis is observed in exponentially growing cells; however, in contact-inhibited cells repair synthesis is reduced to 20 to 30 percent of normal over a wide concentration range of both agents. In actively growing cells inhibition of repair synthesis is observed in the additional presence of hydroxyurea, suggesting that in cells actively involved in semiconservative DNA synthesis, the levels of DNA polymerase α and dCTP (the presumed targets for the inhibitors) are too high for the inhibitors to have any noticeable effect on repair synthesis.

The ability of DNA polymerases α and β from HeLa cells to carry out DNA synthesis in vitro at incisions made by various endonucleases that recognize sites of base loss in DNA is instructive (120). When partially depurinated PM2 DNA is cleaved with *E. coli* endonuclease III (a 3′ AP endonuclease; see Section 3-4) to produce incisions containing 3′-deoxyribose and 5′-phosphomonoester termini, the resulting incisions do not support detectable DNA synthesis by either DNA polymerases α or β. Thus, like *E. coli* DNA polymerase I (see Fig. 3-34), these enzymes apparently cannot utilize 3′-terminal deoxyribose moieties as primers for DNA synthesis. In contrast, when depurinated DNA is cleaved with a 5′ AP endonuclease from HeLa cells, the resulting 3′-hydroxyl nucleotide termini prime some strand displacement DNA synthesis by DNA polymerase β, but not by DNA polymerase α (120).

The inability of depurinated DNA incised by *E. coli* endonuclease III to serve as a primer for HeLa DNA polymerase β is relieved by removal of the 3′-apurinic residue by incubation with a 5′ AP endonuclease from HeLa cells. Such treatment leaves one-nucleotide gaps containing 3′-hydroxyl nucleotide termini. However, this substrate still does not support DNA synthesis by DNA polymerase α (120). On the other hand, when such gapped DNA is incubated with DNA polymerase β in the presence of dGTP and DNA ligase, complete repair of a significant fraction of the gaps occurs. Hence DNA polymerase β would seem to be ideally suited for a role in very short patch repair in eukaryotic cells, because unlike DNA polymerase α (which does not recognize gaps smaller than 10 nucleotides), it will initiate DNA synthesis from nicks or from one-nucleotide gaps (127). In addition, the β enzyme can completely resynthesize gaps up to 50 nucleotides, whereas the α enzyme does not completely fill large gaps (127). This difference in gap filling is consistent with the observation that DNA polymerase β catalyzes highly distributive DNA synthesis, dissociating from the template after the incorporation of each mononucleotide. DNA polymerase α, in contrast, is a processive enzyme that only incorporates 10 to 20 nucleotides with each enzyme-template interaction (128, 147).

The formation of excision gaps during DNA repair that are larger than a *single* nucleotide requires the action of an exonuclease following DNA incision (Fig. 3-1). If DNA repair synthesis from a nick *precedes* exonucleolytic degradation, then in the case of DNA polymerase β a limited incorporation (about 15 to 20 nucleotides) can occur by strand displacement during DNA synthesis in vitro (120) (Fig. 6-2). As indicated in Section 6-1, repair synthesis of UV-irradiated human fibroblasts in which DNA polymerase α is inhibited by cytosine arabinoside may indeed proceed by a strand displacement mechanism in vivo. The displaced strand could then be a substrate for any of the 5′ → 3′ single-strand-specific nucleases described above, e.g., DNase VIII or correxonuclease. Alternatively, if repair synthesis *follows* exonucleolytic degradation during excision repair, then the latter could be catalyzed by a double-strand-specific exonuclease such as DNase VII, DNase VIII, correxonuclease or DNase IV.

In summary, it is very likely that both classes of DNA polymerases are involved in the repair of DNA damaged by physical or chemical agents, and that if for any reason one polymerase is limiting, the other is able to replace it, as appears to be true for *E. coli*. Thus, mammalian DNA polymerases may respond to signals to replicate DNA with suitable template-primers, irrespective of whether the polymerization is concerned ultimately with maintenance or with propagation of the genome. In addition, it is likely that different DNA polymerases have different relative affinities for the substrates produced by endonucleases and/or exonucleases during the preceding events in excision repair of various types of DNA damage.

Mammalian DNA Ligases

Mammalian polynucleotide ligase was first purified from extracts of rabbit bone marrow (148). This enzyme, estimated to have a molecular weight of about 95,000, is dependent for its activity on a double-strand DNA substrate with adjacent 3′-OH and 5′-P termini, ATP and Mg^{2+}. NAD cannot be substituted for ATP. The rabbit enzyme catalyzes an ATP-dependent pyrophosphate exchange in the absence of DNA. ATP is also hydrolyzed by the enzyme in the absence of DNA. This reaction is stimulated by the presence of DNA nicked with pancreatic DNase I, suggesting that an enzyme-adenylate complex is a reaction intermediate. In studies with an enzyme purified from calf thymus, a covalently linked enzyme-adenylate complex has indeed been isolated (149). The complex catalyzes the joining of adjacent 3′-OH and 5′-P termini in DNA in the absence of added ATP, with release of the adenylate residue. Further, following incubation of ligase-[^{14}C]adenylate complex with pyrophosphate, radioactivity is recovered as ATP (149). Thus, the mammalian polynucleotide ligase apparently functions in the same way as the enzymes from prokaryote sources (see Fig. 5-23). About 60 percent of the mammalian ligase activity is in the nuclear fraction of cells, 35 percent is in the cytoplasm and 5 percent is in a crude mitochondrial fraction (149). However, in view of the well-known tendency of nuclear enzymes to leach into the cytoplasm during aqueous extraction of nuclei, these distributions may have no functional significance.

Polynucleotide ligase activity has been isolated from a number of other mammalian cell sources, including extracts of chicken, hamster, mouse, monkey and human cells grown in culture (150). After infection of mouse embryo, monkey kidney and HeLa cells with polyoma virus, SV40 and vaccinia virus respectively, the enzyme activity in extracts of these cell types increases, suggesting that virus infection induces an activity that is indistinguishable (at least in terms of cofactor requirement) from that present in uninfected cells (150).

Two distinct forms of DNA ligase with estimated molecular weights of approximately 95,000 and approximately 190,000 respectively have been identified from EUC cells in culture (151, 152). The enzyme from fresh crude extract of these cells elutes from a Sephadex G-100 column as a single peak of molecular weight 190,000. When this peak of activity is pooled and rechromatographed on Sephadex G-100, about 40 percent of the activity elutes as the 95,000 molecular weight fraction. Furthermore, if fresh crude extract is allowed to stand at 0°C for 20 days, initial gel filtration yields both molecular weight forms. This could arise from artifactual aggregation and disaggregation of the enzyme during aging or during its handling. However, other investigations have provided more per-

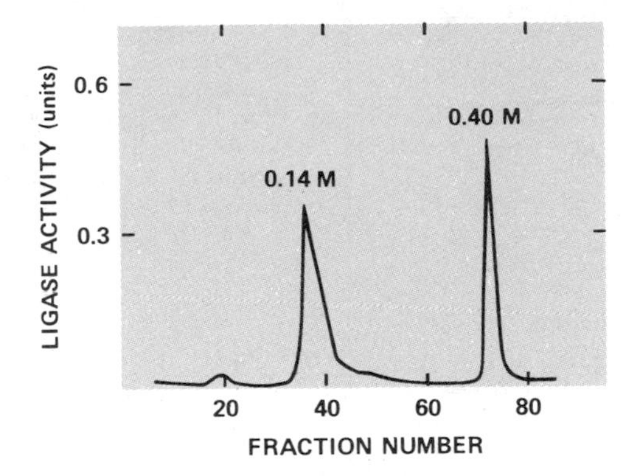

FIGURE 6-26
Separation by hydroxylapatite chromatography of two peaks of DNA ligase activity from human placenta by stepwise elution with phosphate buffer at the indicated concentrations. (From S. Söderhäll and T. Lindahl, ref. 154.)

suasive evidence of two distinct polynucleotide ligase activities. Thus, two species of enzyme from calf thymus can be separated by hydroxylapatite chromatography (153) (Fig. 6-26). Calculations of molecular weight based on the sedimentation coefficients of the two activities yield values of 175,000 and 85,000. Both activities require Mg^{2+} and ATP as cofactors, but a number of properties distinguish the two activities. DNA ligase I (the high molecular weight fraction) is more stable to both heat and storage in the cold than is DNA ligase II (153). In addition, DNA ligase I has a broad pH optimum between 7.4 and 8.0, whereas DNA ligase II is maximally active at pH 7.8. Finally, at pH 7.0, ligase I is more active in 2-(N-morpholino)-ethanesulfonic acid–KOH buffer than in Tris-HCl buffer, whereas the reverse is true of DNA ligase II (153).

Both DNA ligases from calf thymus have been purified, and an antiserum has been prepared against DNA ligase I (154). The antiserum specifically inhibits this enzyme and not DNA ligase II at equivalent protein concentrations (Fig. 6-27). The antiserum is also specific when tested against a mixture of DNA ligases I and II. In addition to its failure to inhibit DNA ligase II activity, the DNA ligase I antiserum does not bind to that enzyme, suggesting that the two ligases are antigenically quite distinct. Examination of extracts of calf spleen, calf liver, rabbit spleen, human placenta and mouse ascites cells has confirmed the presence of the two DNA ligase forms (154). In all cases (including calf thymus extracts), DNA ligase I activity is about 10 times higher than that of DNA ligase II.

Dissociation of DNA ligase I into a lower molecular weight form in

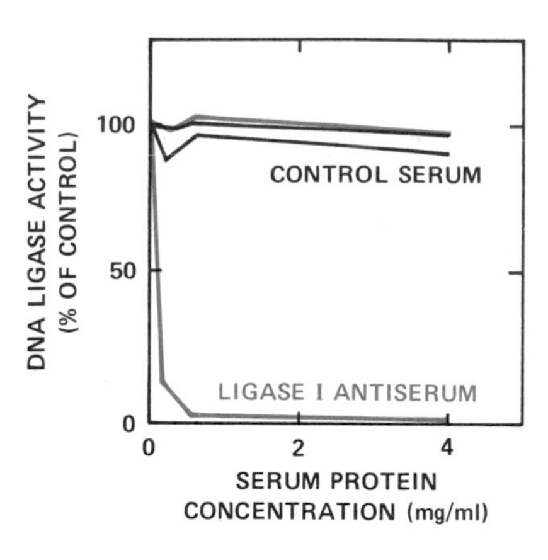

FIGURE 6-27
DNA ligase I antiserum causes inhibition of DNA ligase I activity, but DNA ligase II activity is unaffected (blue lines). In the presence of control serum, normal levels of both DNA ligases are observed (black lines). (From S. Söderhäll and T. Lindahl, ref. 154.)

vitro has been confirmed (154), thus providing an explanation for the two forms of the enzyme initially observed in EUC cell extracts. This form of DNA ligase can, however, be distinguished from true DNA ligase II. It has the same heat sensitivity and pH optimum as DNA ligase I, and it is also inhibited by DNA ligase I antiserum (154). In addition, attempts to generate true DNA ligase II by proteolysis of DNA ligase I have not been successful (154).

DNA ligase activity has been measured at different stages of cell culture growth and after different times following UV irradiation of monkey kidney cells (155). The level of total enzyme activity is about twice as high in exponentially growing as in confluent cultures of cells. In addition, UV radiation results in an increase in ligase activity in stationary-phase cultures equivalent to that observed in cultures of unirradiated exponentially growing cells (Fig. 6-28). This increase in activity is apparently not the result of inactivation of inhibitors, but it has not been established whether the increase requires de novo protein synthesis. It has also not been established whether *both* DNA ligases I and II are "induced" under these conditions. This increase in ligase activity may be related to the increased synthesis of poly(ADP-ribose) associated with nicking of DNA during excision repair (see Section 6-4).

Mammalian DNA Topoisomerases and DNA Repair

Type I and type II DNA topoisomerases are present in eukaryotic cells, including mammalian cells (155–158). Eukaryotic type I topoisomerases show distinct differences from those present in *E. coli* and in *M. luteus*, however (110). The eukaryotic enzymes do not have an absolute requirement for Mg^{2+}. In addition, the bacterial enzymes act only on negative superhelical turns in DNA and lose catalytic

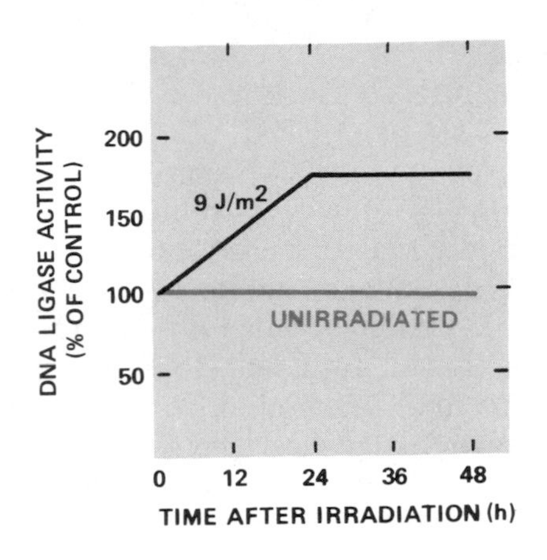

FIGURE 6-28
In monkey kidney cells exposed to UV radiation there is an increased level of DNA ligase activity relative to that of unirradiated cells. (From M. Mezzina and S. Nocentini, ref. 155.)

efficiency as the degree of superhelicity decreases, whereas the eukaryotic enzymes can work on DNA with either negative or positive superhelicity (110). Finally, whereas the prokaryotic nicking and closing enzymes form transient nicks by covalent linkage at the 5′ phosphoryl group, the eukaryotic enzymes have been found linked to the 3′ phosphoryl group.

A role for DNA gyrase in excision repair of DNA in eukaryotic cells is suggested by observations on the sensitivity of repair to inhibitors of this enzyme. The accumulation of DNA strand breaks in UV-irradiated HeLa cells undergoing excision repair is reduced by treatment of cells with novobiocin, suggesting that DNA gyrase may be involved in incision or preincision steps of excision repair (159). In addition, novobiocin results in a time-dependent reversible inhibition of excision of thymine-containing pyrimidine dimers and of repair synthesis of DNA (160).

The specificity of these effects of novobiocin is questionable, however (161). This antibiotic has also been shown to be an effective inhibitor of partially purified DNA polymerase α from monkey cells (162) and rat brain (163) and of DNA polymerases I and II from *S. cerevisiae* (164). There are also indications that the inhibition of semiconservative DNA synthesis observed in novobiocin-treated cells may not be related to the relaxing of supercoils in DNA, and in *E. coli* novobiocin appears to inhibit a number of other ATPase-dependent reactions (165). At concentrations similar to those required to inhibit DNA gyrase, novobiocin has been shown to inhibit nick translation by *E. coli* DNA polymerase I and DNA ligation by phage T4 DNA ligase (165). All these observations suggest that the inhibitory effects of novobiocin in complex cellular systems may not be useful indicators of the specific involvement of DNA gyrase in DNA repair.

6-6 Mismatch Repair in Mammalian Cells

Evidence in support of a mismatch excision repair system in mammalian cells is not as extensive as for *E. coli*. However, studies using heteroduplex DNA derived from animal viruses do suggest the existence of such a phenomenon in higher cells. For example, during the transformation of African green monkey kidney cells with heterozygous DNA prepared from pairs of SV40 *ts* mutants, some pairs generate molecules that are infectious under restrictive conditions (166). This suggests that the infectivity of SV40 heterozygotes may be dependent on a cell-mediated enzymatic activity that can correct presumed mismatches in the DNA.

More direct studies have measured the frequency of conversion of selected heteroduplexes into homoduplexes following transfection of mouse embryo cells with polyoma viral DNA (167). Heteroduplex DNA molecules were generated from two distinct polyoma variants called TS-A and CR, which differ genotypically at four sites in the genome (Fig. 6-29). One site (A) is a mutation that governs the ability of the variant strains to produce infectious virus at 38.5°C. The CR strain will form plaques at this temperature, but the TS-A strain will not. Thus correction of the mismatch at this site in the heteroduplexes can be selected phenotypically. The other three genotypic differences are phenotypically silent mutations, but they could be detected in viral progeny DNA by restriction endonuclease analysis. The results of these experiments were more consistent with the existence of a cellular mismatch repair system than with effects of DNA replication or recombination (167). In addition, it was noted that those markers separated by stretches of DNA about 600 nucleotides or more in length (all except B and C) segregated independently of one another, suggesting that the excision tracts operative in the putative excision repair system usually do not exceed about 600 nucleotides. By the same reasoning, markers separated by less than 90 nucleotides (B and C) appeared to be contained within a single excision tract (167).

Mismatch repair in *E. coli* indicates that the strand selectivity instructed by specific methylation events allows for the discrimination between the "correct" and "incorrect" base in the mismatch, thereby providing a mechanism for the removal of potentially mutagenic mispairs in DNA (see Section 5-7). However, one can envisage selective advantages to the operation of mismatch repair systems that are *not* strand directed (i.e., that are random with respect to the members of the mispairing), the biological significance of which would be the *generation* of nucleotide sequence divergence. An aspect of mammalian cell DNA metabolism well suited to such consid-

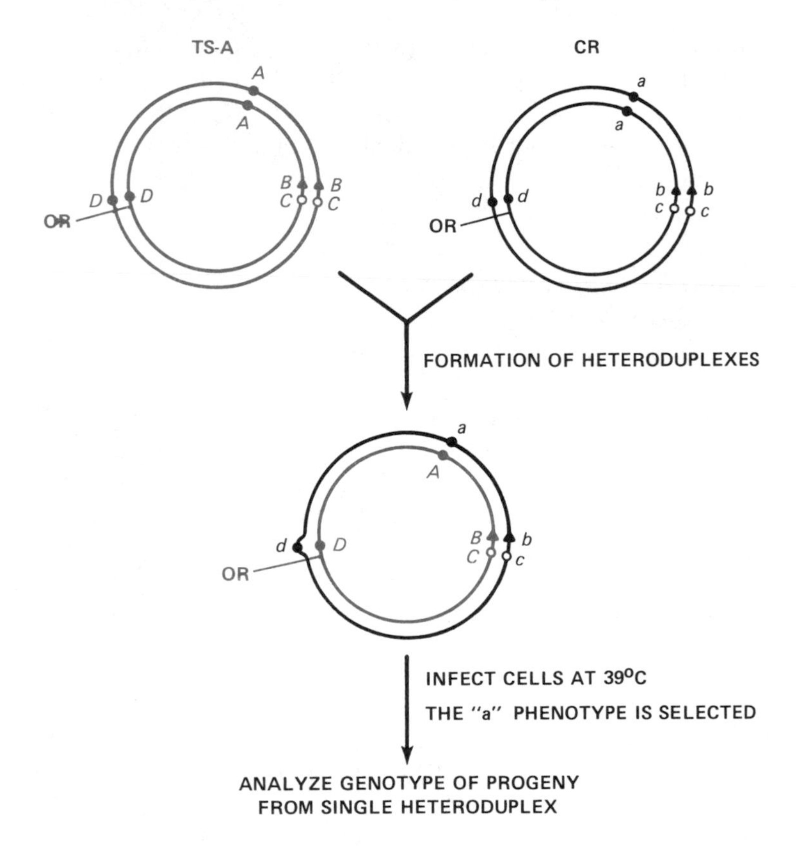

FIGURE 6-29
Two variants of polyoma, TS-A and CR, differ genotypically at four distinct sites (A/a, B/b, C/c, and D/d), the approximate positions of which are indicated relative to the origin of DNA replication (OR). The upper and lower case letters do not indicate wild-type and mutant alleles. The repair of the four sites of mispairing in the heteroduplex molecules generated was monitored by analysis of progeny DNA molecules obtained from a *single plaque*. Molecules that are a/a are able to plaque at 38.5°C and hence could be selected phenotypically. The genotype of the other three markers was then established by restriction enzyme analysis. (From L. K. Miller et al., ref. 167.)

erations relates to the molecular mechanism for generating antibody diversity. Examination of the established nucleotide sequence of a number of mouse germ line V_H genes and collation of the sum of the single base pair mismatches that would be formed in different heteroduplex combinations of these genes reveals a correlation between the "hottest" region of sequence divergence among the germ line V_H genes examined and a particular hypervariable region of mouse immunoglobulin (168). This observation is certainly consistent with the notion that a mechanism for generating antibody diversity involves a nonselective mode of mismatch repair of DNA that permits the fixation of mutations rather than mitigating against them. It is therefore quite likely that in mammalian cells (if not in other eukaryotes and even prokaryotes) the unraveling of the biochemistry of mismatch repair may reveal both selective (e.g., strand directed) and random modes for this repair process.

6-7 General Concluding Comments on Excision Repair of DNA in Prokaryotes and Eukaryotes

Although many important details remain to be solved, the basic biochemical pathways for the excision repair of base damage in DNA are well established. An area in which there is still limited information is whether or not the coupled incision-excision mechanism illustrated by the uvrABC endonuclease of *E. coli* is general in prokaryotic cells, and particularly whether the same general mechanism for excision of bulky base damage operates in eukaryotic cells. If so, why are so many genes required for DNA incision in yeast and apparently also in human cells, and what is the role of those genes that seem to be *involved* in, but not absolutely *required* for, the incision of damaged DNA? The application of gene cloning techniques has had a major impact on this topic, and it seems safe to assume that the mechanism of action in vitro of at least the uvrABC complex of *E. coli* will be understood in considerable detail in the near future. Similarly, the successful cloning of *RAD* genes from *S. cerevisiae* offers the prospect of solving the biochemistry of DNA incision in a eukaryotic organism. This solution could provide significant insights into the molecular biology of DNA incision in other eukaryotes, including humans. The observation that mutant cells defective in the incision of DNA are abnormally sensitive to killing by a large number of different chemical and physical agents attests to the biological importance of the nucleotide excision type of pathway in the repair of DNA damage. As indicated previously (see Chapter 3), DNA glycosylases are likely to be limited in number in a given cell type and thus can only account for the repair of a relatively small number of specific forms of base damage.

With respect to excision repair that is mediated by DNA glycosylases–AP endonucleases and hence requires *independent* postincision exonucleolytic degradation, there appears to be significant redundancy of enzymes for both excision of base damage and resynthesis of DNA. The evidence in *E. coli* suggests that the multiple enzymes for these functions serve "backup" rather than strictly alternative roles. Thus, under normal conditions DNA polymerase I of *E. coli* probably carries out all excision-resynthesis by a coordinate rather than a sequential mechanism. There are indications that this enzyme is regulated, but it remains to be firmly established what factor(s) determines the relatively homogeneous size of the small repair patches in constitutive cells. Nor is it understood what determines the size of the long repair patches in induced *E. coli*.

In addition to the enzymology of excision repair, an aspect of this area of DNA biochemistry that still requires much clarification in eukaryotes is how the various enzymatic events associated with excision repair take place in chromatin and in chromosomes. Are there nucleosome unfolding events that are unique to the repair of damaged DNA, or do eukaryotes utilize a single basic mechanism for providing access of enzymes to DNA during replication, transcription, recombination and repair? Also, what if any is the role of ADP-ribosylation of proteins in DNA repair? It is hoped that further attention will be focused on these important questions.

Suggestions for Further Reading

General aspects of excision repair in mammalian cells, refs. 38, 50, 63, 74; poly(ADP) ribose and DNA repair, refs. 98, 103; mammalian DNA ligases, ref. 154.

References

1. LaBelle, M., and Linn, S. 1982. *In vivo* excision of pyrimidine dimers is mediated by a DNA N-glycosylase in *Micrococcus luteus* but not in human fibroblasts. *Photochem. Photobiol.* 36:319.
2. Williams, J. I., and Cleaver, J. E. 1978. Excision repair of ultraviolet damage in mammalian cells. Evidence for two steps in the excision of pyrimidine dimers. *Biophys. J.* 22:265.
3. Williams, J. I., and Cleaver, J. E. 1979. Removal of T4 endonuclease V–sensitive sites from SV40 DNA after exposure to ultraviolet light. *Biochim. Biophys. Acta* 562:429.
4. Kantor, G. J., and Setlow, R. B. 1981. Rate and extent of DNA repair in non-dividing human diploid fibroblasts. *Cancer Res.* 41:819.
5. Paterson, M. C., Lohman, P. H. M., and Sluyter, M. L. 1973. Use of a UV endonuclease from *Micrococcus luteus* to monitor the progress of DNA repair in UV-irradiated human cells. *Mutation Res.* 19:245.
6. Ehmann, U. K., and Friedberg, E. C. 1980. An investigation of the effect of radioactive labeling of DNA on excision repair in UV-irradiated human fibroblasts. *Biophys. J.* 31:285.
7. Mitchell, D. L., Nairn, R. S., Alvillar, J. A., and Clarkson, J. M. 1982. Loss of thymine dimers from mammalian cell DNA. The kinetics for antibody binding sites are not the same as that for T4 endonuclease V sites. *Biochim. Biophys. Acta* 697:270.

8. Cornelis, J. J., Rommelaere, J., Urbain, J., and Errera, M. 1977. A sensitive method for measuring pyrimidine dimers *in situ*. *Photochem. Photobiol.* 26:241.

9. Cornelis, J. J. 1978. The influence of inhibitors on dimer removal and repair of single-strand breaks in normal and bromodeoxyuridine substituted DNA of HeLa cells. *Biochim. Biophys. Acta* 521:134.

10. Cornelis, J. J. 1978. Characterization of excision repair of pyrimidine dimers in eukaryotic cells as assayed with antidimer sera. *Nucleic Acids Res.* 5:4273.

11. Amacher, D. E., Elliott, J. A., and Lieberman, M. W. 1977. Differences in removal of acetylaminofluorene and pyrimidine dimers from the DNA of cultured mammalian cells. *Proc. Natl. Acad. Sci.* (USA) 74:1553.

12. Ehmann, U. K., Cook, K. H., and Friedberg, E. C. 1978. The kinetics of thymine dimer excision in ultraviolet-irradiated human cells. *Biophys. J.* 22:249.

13. Konze-Thomas, B., Levinson, J. W., Maher, V. M., and McCormick, J. J. 1979. Correlation among the rates of dimer excision, DNA repair replication, and recovery of human cells from potentially lethal damage induced by ultraviolet radiation. *Biophys. J.* 28:315.

14. Inoue, M. and Takebe, H. 1978. DNA repair capacity and rate of excision repair in UV-irradiated mammalian cells. *Japan J. Genet.* 58:285.

15. Kantor, G. J., Petty, R. S., Warner, C., Phillips, D. J. H., and Hull, D. R. 1980. Repair of radiation induced DNA damage in nondividing populations of human diploid fibroblasts. *Biophys. J.* 30:399.

16. Duncan, J., Slor, H., Cook, K., and Friedberg, E. C. 1975. Thymine dimer excision by extracts of human cells. In *Molecular mechanisms for repair of DNA*. P. C. Hanawalt and R. B. Setlow, eds., p. 643. New York: Plenum.

17. Regan, J. D., Trosko, J. E., and Carrier, W. L. 1966. Evidence for excision of ultraviolet-induced pyrimidine dimers from the DNA of human cells *in vitro*. *Biophys. J.* 8:319.

18. Isomura, K., Nikaido, C., Horikama, M., and Sugahara, T. 1973. Repair of DNA damage in ultraviolet-sensitive cells isolated from HeLa S3 cells. *Radiat. Res.* 53:143.

19. Setlow, R. B., Regan, J. D., German, J., and Carrier, W. L. 1969. Evidence that xeroderma pigmentosum cells do not perform the first step in the repair of ultraviolet damage to their DNA. *Proc. Natl. Acad. Sci.* (USA) 64:1035.

20. Ben-Hur, E., and Ben-Ishai, R. 1971. DNA repair and ultraviolet light irradiated HeLa cells and its reversible inhibition by hydroxyurea. *Photochem. Photobiol.* 13:337.

21. Smith, C. A., and Hanawalt, P. C. 1976. Repair replication in cultured normal and transformed human fibroblasts. *Biochim. Biophys. Acta* 447:121.

22. Edenberg, H. J., and Hanawalt, P. C. 1973. The timecourse of DNA repair replication in ultraviolet irradiated HeLa cells. *Biochim. Biophys. Acta* 324:206.

23. Conner, W. G., and Norman, A. 1971. Unscheduled DNA synthesis in human leucocytes. *Mutation Res.* 13:393.

24. Clarkson, J. M., and Evans, H. J. 1972. Unscheduled DNA synthesis in human leucocytes after exposure to UV light, γ-rays, and chemical mutagens. *Mutation Res.* 14:413.

25. Smets, L. A. 1969. Ultraviolet-light-enhanced precursor incorporation and the repair of DNA damage. *Int. J. Radiat. Biol.* 16:407.

26. Stefanini, M., Dalpra, L., Zei, G., Giorgi, R., Falaschi, A., and Nuzzo, F. 1976. Incorporation of [^{3}H]thymidine stimulated by ultraviolet radiation into human fibroblast cultures. *Mutation Res.* 34:313.

27. Evans, R. G., and Norman, A. 1968. Radiation stimulated incorporation of thymidine into the DNA of human lymphocytes. *Nature* 217:455.

28. Burk, P. D., Lutzner, M. A., Clarke, D. D., and Robbins, J. H. 1971. Ultraviolet-stimulated thymidine incorporation in xeroderma pigmentosum lymphocytes. *J. Lab. Clin. Med.* 77:759.

29. Bowman, P. D., Meed, R. L., and Daniel, C. W. 1976. Decreased unscheduled DNA synthesis in non-dividing aged WI38 cells. *Mech. Aging Dev.* 5:251.

30. Rasmussen, R. E. 1968. DNA synthesis in ultra-violet irradiated cultured human cells. Ph.D. thesis. San Francisco: University of California.

31. Speigler, P., and Norman, A. 1970. Temperature dependence of unscheduled DNA synthesis in human lymphocytes. *Radiat. Res.* 43:187.

32. Williams, J. I. 1977. Excision repair and DNA synthesis after ultraviolet light exposure in African green monkey kidney CV-1 cells and simian virus 40 (SV40). Ph.D. thesis. San Francisco: University of California.

33. Reynolds, R. J., Cook, K. H., and Friedberg, E. C. 1981. Measurement of thymine-containing pyrimidine dimers by one-dimensional thin-layer chromatography. In *DNA repair—a laboratory manual of research procedures*. E. C. Friedberg and P. C. Hanawalt, eds., p. 11. New York: Dekker.

34. Cleaver, J. E. 1981. Sensitivity of excision repair in normal human, xeroderma pigmentosum variant and Cockayne's syndrome fibroblasts to inhibition by cytosine arabinoside. *J. Cell. Physiol.* 108:163.

35. Cleaver, J. E. 1983. Structure of repaired sites in human DNA synthesized in the presence of inhibitors of DNA polymerases alpha and beta in human fibroblasts. *Biochim. Biophys. Acta* 729:301.

36. Ben-Ishai, R., and Peleg, L. 1975. Excision in primary cultures of mouse embryo cells and its decline in progressive passages and established cell lines. In *Molecular mechanisms for repair of DNA*. P. C. Hanawalt and R. B. Setlow, eds., p. 607. New York: Plenum.

37. Peleg, L., Raz, E., and Ben-Ishai, R. 1977. Changing capacity for DNA excision repair in mouse embryonic cells *in vitro*. *Exp. Cell. Res.* 104:301.

38. Hanawalt, P. C., Cooper, P. K., Ganesan, A. K., and Smith, C. A. 1979. DNA repair in bacteria and mammalian cells. *Ann. Rev. Biochem.* 48:783.

39. Regan, J. D., and Setlow, R. B. 1974. Two forms of repair in DNA of human cells damaged by chemical carcinogens and mutagens. *Cancer Res.* 34:3318.

40. Painter, R. B., and Young, B. R. 1972. Repair replication in mammalian cells after x-irradiation. *Mutation Res.* 14:225.

41. Setlow, R. B., Faulcon, F. M., and Regan, J. D. 1976. Defective repair of gamma-ray-induced DNA damage in xeroderma pigmentosum cells. *Int. J. Radiat. Biol.* 29:125.

42. Walker, I. G., and Th'ng, J. P. H. 1982. Excision-repair patch size in DNA from human KB cells treated with UV-light, or methyl methanesulfonate. *Mutation Res.* 105:277.

43. Th'ng, J. P., and Walker, I. G. 1983. DNA repair patch size measurements with nucleosomal DNA. *Carcinogenesis* 4:975.

44. Edenberg, H., and Hanawalt, P. C. 1972. Size of repair patches in the DNA of ultraviolet-irradiated HeLa cells. *Biochim. Biophys. Acta* 272:361.

45. Smith, C. A., and Hanawalt, P. C. 1978. Phage T4 endonuclease V stimulates DNA repair replication in isolated nuclei from ultraviolet-irradiated human cells, including xeroderma pigmentosum fibroblasts. *Proc. Natl. Acad. Sci. (USA)* 75:2598.

46. Kaye, J., Smith, C. A., and Hanawalt, P. C. 1980. DNA repair in human cells containing photoadducts of 8-methoxypsoralen or angelicin. *Cancer Res.* 40:696.

47. Phillips, D. H., Hanawalt, P. C., Miller, J. A., and Miller, E. C. 1981. The *in vivo* formation and repair of DNA adducts from 1′-hydroxysafrole. *J. Supramol. Struct. Cell Biochem.* 16:83.

48. Francis, A. A., Blevins, R. D., Carrier, W. L., Smith, D. P., and Regan, J. D. 1979. Inhibition of DNA repair in ultra-violet-irradiated human cells by hydroxyurea. *Biochim. Biophys. Acta* 563:385.

49. Rosenstein, B. S., Setlow, R. B., and Ahmed, F. E. 1980. Use of the dye Hoechst 33258 in a modification of the bromodeoxyuridine photolysis technique for the analysis of DNA repair. *Photochem. Photobiol.* 31:215.

50. Hanawalt, P. C., Cooper, P. K., and Smith, C. A. 1981. Repair replication schemes in bacteria and human cells. *Prog. Nucleic Acid Res. Mol. Biol.* 26:181.

51. Gottesfeld, J. M., Garrard, W. T., Bagi, G., Wilson, R. F., and Bonner, J. 1979. Partial purification of the template-active fraction of chromatin: a preliminary report. *Proc. Natl. Acad. Sci.* (USA) 71:2193.

52. Mathis, D., Oudet, P., and Chambon, P. 1980. Structure of transcribing chromatin. *Prog. Nucleic Acid Res. Mol. Biol.* 24:1.

53. Weintraub, H., and Groudine, M. 1976. Chromosomal subunits in active genes have an altered conformation. *Science* 193:848.

54. Igo-Kemenes, T., Hörz, W., and Zachau, H. G. 1982. Chromatin. *Ann. Rev. Biochem.* 51:89.

55. Bodell, W. J. 1977. Nonuniform distribution of DNA repair in chromatin after treatment with methylmethane sulfonate. *Nucleic Acids Res.* 4:2619.

56. Cleaver, J. E. 1977. Nucleosome structure controls rate of excision repair in DNA of human cells. *Nature* 270:451.

57. Smerdon, M. J., Tlsty, T. D., and Lieberman, M. W. 1978. Distribution of ultraviolet-induced DNA repair synthesis in nuclease sensitive and resistant regions of human chromatin. *Biochemistry* 17:2377.

58. Tlsty, T. D., and Lieberman, M. W. 1978. The distribution of DNA repair synthesis in chromatin and its rearrangement following damage with N-acetoxy-2-acetylaminofluorene. *Nucleic Acids Res.* 6:3261.

59. Bodell, W. J., and Banerjee, M. R. 1979. The influence of chromatin structure on the distribution of DNA repair synthesis studied by nuclease digestion. *Nucleic Acids Res.* 6:359.

60. Williams, J. I., and Friedberg, E. C. 1979. Deoxyribonucleic acid excision repair in chromatin after ultraviolet irradiation of human fibroblasts in culture. *Biochemistry* 18:3965.

61. Smerdon, M. J., Kastan, M. B., and Lieberman, M. W. 1979. Distribution of repair-incorporated nucleotides and nucleosome rearrangement in the chromatin of normal and xeroderma pigmentosum human fibroblasts. *Biochemistry* 18:3732.

62. Oleson, F. B., Mitchell, B. L., Dipple, A., and Lieberman, M. W. 1979. Distribution of DNA damage in chromatin and its relation to repair in human cells treated with 7-bromomethylbenz(a)anthracene. *Nucleic Acids Res.* 7:1343.

63. Lieberman, M. W., Smerdon, M. J., Tlsty, T. D., and Oleson, F. B. 1979. The role of chromatin structure in DNA repair in human cells damaged with chemical carcinogens and ultraviolet radiation. In *Environmental carcinogenesis*. P. Emmelot and E. Kriek, eds., p. 345. Amsterdam: Elsevier/North-Holland Biomedical Press.

64. Cleaver, J. E. 1977. DNA repair processes and their impairment in some human diseases. In *Progress in genetic toxicology*. Vol. 2. D. Scott, B. A. Bridges and F. Sobels, eds., p. 29. Amsterdam: Elsevier.

65. Smerdon, M. J., Watkins, J. F., and Lieberman, M. W. 1982. Effect of histone H1 removal on the distribution of ultraviolet-induced deoxyribonucleic acid repair synthesis within chromatin. *Biochemistry* 21:3879.

66. Zolan, M. E., Smith, C. A., Calvin, N. M., and Hanawalt, P. C. 1982. Rearrangement of mammalian chromatin structure following excision repair. *Nature* 299:462.

67. Ramanathan, R., Rajalakshmi, S., Sarma, D. S. R., and Farber, E. 1976. Nonrandom nature of *in vivo* methylation by dimethylnitrosamine and the subsequent removal of methylated products from rat liver chromatin DNA. *Cancer Res.* 36:2073.

68. Wiesenhahn, G. P., Hyde, J. E., and Hearst, J. E. 1979. The photoaddition of trimethylpsoralen to *Drosophila melanogaster* nuclei: a probe for chromatin substructure. *Biochemistry* 16:925.

69. Cech, T., and Pardue, M. L. 1977. Cross-linking of DNA with trimethylpsoralen is a probe for chromatin structure. *Cell* 11:631.

70. Hiss, E. A., and Preston, R. J. 1977. The effect of cytosine arabinoside on the frequency of single-strand breaks in DNA of mammalian cells following irradiation or chemical treatment. *Biochim. Biophys. Acta* 478:1.

71. Johnson, R. T., and Collins, A. R. S. 1978. Reversal of the changes in DNA and chromosome structure which follow the inhibition of UV-induced repair in human cells. *Biochem. Biophys. Res. Comm.* 80:361.

72. Dunn, W. C., and Regan, J. D. 1979. Inhibition of DNA excision repair in human cells by arabinofuranosyl cytosine: effect on normal and xeroderma pigmentosum cells. *Mol. Pharmacol.* 15:367.

73. Hashem, N., Bootsma, D., Keijzer, W., Greene, A., Coriell, L., Thomas, G., and Cleaver, J. E. 1980. Clinical characteristics, DNA repair, and complementation groups in xeroderma pigmentosum patients from Egypt. *Cancer Res.* 40:13.

74. Smith, C. A. Analysis of repair synthesis in the presence of inhibitors. In *DNA repair and its inhibition: towards an analysis of mechanism*, vol. 13, Nucleic Acids Symposium Series, A. R. S. Collins, C. S. Downs and R. J. Johnson, eds. London: IRL Press Ltd., in press.

75. Bodell, W. J., Kaufmann, W. K., and Cleaver, J. E. 1982. Enzyme digestion of intermediates of excision repair in human cells irradiated with ultraviolet light. *Biochemistry* 21:6767.

76. Weintraub, H. 1976. Cooperative alignment of Nu bodies during chromosome replication in the presence of chloramphenicol. *Cell* 9:419.

77. Bodell, W. J., and Banerjee, M. R. 1976. Reduced DNA repair in mouse satellite DNA after treatment with methylmethanesulfonate, and N-methyl-N-nitrosourea. *Nucleic Acids Res.* 3:1689.

78. Berliner, J., Hunes, S. W., Aoki, C. T., and Norman, A. 1975. The sites of unscheduled DNA synthesis within irradiated human lymphocytes. *Radiat. Res.* 63:544.

79. Harris, C. C., Conner, R. J., Jackson, F. E., and Lieberman, M. W. 1974. Intranuclear distribution of DNA repair synthesis induced by chemical carcinogens or ultraviolet light in human diploid fibroblasts. *Cancer Res.* 34:3461.

80. Lieberman, M. W., and Poirier, M. C. 1974. Distribution of deoxyribonucleic acid repair synthesis among repetitive and unique sequences in the human diploid genome. *Biochemistry* 13:3018.

81. Lieberman, M. W., and Poirier, M. C. 1974. Intragenomal distribution of DNA repair synthesis: repair in satellite and mainband DNA in cultured mouse cells. *Proc. Natl. Acad. Sci.* (USA) 71:2461.

82. Shoyab, M. 1981. Organization of UV-induced repair replicated DNA in the cellular genome of cultured human and mouse cells. *Exp. Cell Res.* 121:9.

83. Zolan, M., Cortopassi, G. A., Smith, C. A., and Hanawalt, P. C. 1982. Deficient repair of chemical adducts in α DNA of monkey cells. *Cell* 28:613.

84. Leadon, S. A., Zolan, M. D., and Hanawalt, P. C. 1983. Restricted repair of aflatoxin B1–induced damage in α DNA of monkey cells. *Nucleic Acids Res.* 11:5675.

85. Zolan, M. E. 1983. Ph.D. thesis. Stanford University.

86. Christman, J. K., Price, P., Pedrinan, L., and Acs, G. 1977. Correlation between hypomethylation of DNA and expression of globin genes in Friend erythroleukemia cells. *Eur. J. Biochem.* 81:53.

87. Compere, S. J., and Palmiter, R. D. 1981. DNA methylation controls the inducibility of the mouse metallothionein-1 gene in lymphoid cells. *Cell* 25:233.

88. Groudine, M., Eisenman, R., and Weintraub, H. 1981. Chromatin structure of endogenous retroviral genes and activation by an inhibitor of DNA methylation. *Nature* 292:311.

89. Many, T. N., and Cedar, H. 1981. Active gene sequences are undermethylated. *Proc. Natl. Acad. Sci.* (USA) 78:4246.

90. Razin, A., and Friedman, J. 1981. DNA methylation and its possible biological roles. *Prog. Nucl. Acids Res. Mol. Biol.* 25:33.

91. Razin, A., and Riggs, A. 1980. DNA methylation and gene function. *Science* 210:604.

92. Bird, A. P. 1978. Use of restriction enzymes to study eukaryotic DNA methylation: II. The symmetry of methylated sites supports semi-conservative copying of the methylation pattern. *J. Mol. Biol.* 118:49.

93. Holliday, R., and Pugh, J. E. 1975. DNA modification mechanisms and gene activity during development. *Science* 187:226.

94. Holliday, R. 1980. Possible relationships between DNA damage, DNA modification and carcinogenesis. In *Progress in environmental mutagenesis*. M. Alečević, ed., p. 93. Amsterdam: Elsevier/North-Holland Biomedical Press.

95. Kastan, M. B., Gowans, B. J., and Lieberman, M. W. 1982. Methylation of deoxycytidine incorporated by excision-repair synthesis of DNA. *Cell* 30:509.

96. Wilson, V. L., and Jones, P. A. 1983. Inhibition of DNA methylation by chemical carcinogens *in vitro*. *Cell* 32:239.

97. Hayaishi, O., and Ueda, K. 1977. Poly (ADP-ribose) and ADP-ribosylation of proteins. *Ann. Rev. Biochem.* 46:95.

98. Shall, S. 1981. $(ADP\text{-ribose})_n$, a new component in DNA repair. In *Chromosome damage and repair*. E. Seeberg and K. Kleppe, eds., p. 477. New York: Plenum.

99. Parnell, M. R., Stone, P. R., and Whish, W. J. D. 1980. ADP-ribosylation of nuclear proteins. *Biochem. Soc. Trans.* 8:215.

100. Halldorsson, H., Gray, D. A., and Shall, S. 1978. Poly (ADP-ribose) polymerase activity in nucleotide permeable cells. *FEBS Lett.* 85:349.

101. Skidmore, C. J., Davies, M. I., Goodwin, P. M., Halldorsson, H., Lewis, P., Shall, S., and Zia'ee, A-A. 1979. The involvement of poly (ADP-ribose) polymerase in

the degradation of NAD caused by γ-radiation and N-methyl-N-nitrosourea. *Eur. J. Biochem.* 101:135.

102. Durkacz, B. W., Omidiji, O., Gray, D. A., and Shall, S. 1980. (ADP-ribose)$_n$ participates in excision repair. *Nature* 283:593.

103. Shall, S., Durkacz, B. W., Gray, D. A., Irwin, J., Lewis, P. J., Perera, M., and Tavassoli, M. 1982. (ADP-ribose)$_n$ participates in DNA excision repair. In *Mechanisms of chemical carcinogenesis.* C. C. Harris and P. A. Cerutti, eds., p. 389. New York: Alan R. Liss.

104. Durkacz, B. W., Nduka, N., Omidiji, O., Shall, S., and Zia'ee, A-A. 1980. ADP-ribose and DNA repair. In *Novel ADP-ribosylations of regulatory enzymes and proteins.* M. Smulson and T. Sugimura, eds., p. 207. Amsterdam: North Holland/Elsevier.

105. Nduka, N., Skidmore, C. J., and Shall, S. 1980. The enhancement of cytotoxicity of N-methyl-N-nitrosourea and of γ-radiation by inhibitors of poly (ADP-ribose) polymerase. *Eur. J. Biochem.* 105:525.

106. Hori, T-A. 1981. High incidence of sister chromatid exchanges and chromatid interchanges in the conditions of lowered activity of poly (ADP-ribose) polymerase. *Biochem. Biophys. Res. Comm.* 102:38.

107. Creissen, D., and Shall, S. 1982. Regulation of DNA ligase activity by poly (ADP-ribose). *Nature* 196:271.

108. Morgan, W. F., and Cleaver, J. E. 1983. Effect of 3-aminobenzamide on the rate of ligation during repair of alkylated DNA. *Cancer Res.* 43:3104.

109. Cleaver, J. E., Bodell, W. J., Morgan, W. F., and Zelle, B. 1983. Differences in the regulation of poly (ADP-ribose) of repair of DNA damage from alkylating agents and ultraviolet light according to cell type. *J. Biol. Chem.* 258:9059.

110. Kornberg, A. 1980. DNA replication. San Francisco: Freeman.

111. Lindahl, T., Gally, J. A., and Edelman, G. M. 1969. Deoxyribonuclease IV: a new exonuclease from mammalian tissues. *Proc. Natl. Acad. Sci.* (USA) 62:597.

112. Lindahl, T. 1971. The action pattern of mammalian deoxyribonuclease IV. *Eur. J. Biochem.* 18:415.

113. Lindahl, T. 1971. Excision of pyrimidine dimers from ultraviolet-irradiated DNA by exonucleases from mammalian cells. *Eur. J. Biochem.* 18:407.

114. Lindahl, T. 1972. Mammalian deoxyribonucleases acting on damaged DNA. In *Molecular and cellular repair processes.* R. F. Beers, R. M. Herriott and R. C. Tilghman, eds., p. 3. Baltimore: Johns Hopkins University Press.

115. Cook, K. H., and Friedberg, E. C. 1978. Multiple thymine dimer excising nuclease activities in extracts of human KB cells. *Biochemistry* 17:850.

116. Cook, K. H., and Friedberg, E. C. 1978. Partial purification and characterization of three thymine dimer excising activities from human KB cells. In *DNA repair mechanisms.* P. C. Hanawalt, E. C. Friedberg and C. F. Fox, eds., p. 301. New York: Academic.

117. Doniger, J., and Grossman, L. 1976. Human correxonuclease. Purification and properties of a DNA repair exonuclease from placenta. *J. Biol. Chem.* 251:4579.

118. Pedrini, A. M., and Grossman, L. 1983. Purification and characterization of DNase VIII, a $5' \rightarrow 3'$ directed exonuclease from human placental nuclei. *J. Biol. Chem.* 258:1536.

119. Mosbaugh, D. W., and Meyer, R. R. 1980. Interaction of mammalian deoxyribonuclease V, a double strand $3' \rightarrow 5'$ and $5' \rightarrow 3'$ exonuclease, with deoxyribonucleic acid polymerase-β from the Novikoff hepatoma. *J. Biol. Chem.* 255:10239.

120. Mosbaugh, D. W., and Linn, S. 1983. Excision repair and DNA synthesis with a combination of HeLa DNA polymerase β and DNase V. *J. Biol. Chem.* 258:108.

121. Mosbaugh, D. W., Stalker, D. M., Probst, G. S., and Meyer, R. R. 1977. Novikoff hepatoma deoxyribonucleic acid polymerase. Identification of a stimulatory protein bound to the β-polymerase. *Biochemistry* 16:1512.

122. Lindahl, T., Gally, J. A., and Edelman, G. M. 1969. Properties of deoxyribonuclease III from mammalian tissues. *J. Biol. Chem.* 244:5014.

123. Hollis, G. F., and Grossman, L. 1981. Purification and characterization of DNase VII, a $3' \rightarrow 5'$ directed exonuclease from human placenta. *J. Biol. Chem.* 256:8074.

124. Tanaka, S., Hu, S-Z., Wang, T. S-F., and Korn, D. 1982. Preparation and preliminary characterization of monoclonal antibodies against human DNA polymerase α. *J. Biol. Chem.* 257:8386.

125. Bensch, K., Tanaka, S., Hu, S-Z., Wang, T. S-F., and Korn, D. 1982. Intracellular localization of human DNA polymerase α with monoclonal antibodies. *J. Biol. Chem.* 257:8391.

126. Eichler, D. C., Fisher, P. A., and Korn, D. 1977. Effect of calcium on the recovery and distribution of DNA polymerase α from cultured human cells. *J. Biol. Chem.* 252:4011.

127. Wang, T. S-F., and Korn, D. 1980. Reactivity of KB cell deoxyribonucleic acid polymerases α and β with nicked and gapped deoxyribonucleic acid. *Biochemistry* 19:1782.

128. Fisher, P. A., Wang, T. S-F., and Korn, D. 1979. Enzymological characterization of DNA polymerase α. Basic catalytic properties, processivity, and gap utilization of the homogeneous enzyme from human KB cells. *J. Biol. Chem.* 254:6128.

129. Wang, T. S-F., and Korn, D. 1982. Specificity of the catalytic interaction of human DNA polymerase β with nucleic acid substrates. *Biochemistry* 21:1597.

130. Bertazzoni, U., Stefanini, M., Pedrali-Noy, G., Guilotto, E., Nuzzo, F., Falaschi, A. and Spadari, S. 1976. Variations of DNA polymerases-α and -β during prolonged stimulation of human lymphocytes. *Proc. Natl. Acad. Sci.* (USA) 73:785.

131. Pedrali-Noy, G., and Spadari, S. 1979. Effect of aphidicolin on viral and human DNA polymerases. *Biochem. Biophys. Res. Comm.* 88:1194.

132. Hanoaka, F., Kato, H., Ikegami, S., Ohashi, M., and Yamada, M-A. 1979. Aphidicolin does inhibit repair replication in HeLa cells. *Biochem. Biophys. Res. Comm.* 87:575.

133. Pedrali-Noy, G., and Spadari, S. 1980. Aphidicolin allows a rapid and simple evaluation of DNA repair synthesis in damaged human cells. *Mutation Res.* 70:389.

134. Miller, M. R., and Chinault, D. N. 1982. Evidence that DNA polymerases α and β participate differentially in DNA repair synthesis induced by different agents. *J. Biol. Chem.* 257:46.

135. Miller, M. R., and Chinault, D. N. 1982. The role of DNA polymerases α, β and γ in DNA repair synthesis induced in hamster and human cells by different DNA damaging events. *J. Biol. Chem.* 257:10204.

136. Dresler, S. L., and Lieberman, M. W. 1983. Identification of DNA polymerases involved in DNA excision repair in diploid human fibroblasts. *J. Biol. Chem.* 258:9990.

137. Mattern, M. R., Paone, R. F., and Day, III R.S. 1982. Eukaryotic DNA repair is blocked at different steps by inhibitors of DNA topoisomerases and DNA polymerases α and β. *Biochim. Biophys. Acta* 697:6.

138. Seki, S., Ohashi, M., Ogura, H., and Oda, T. 1982. Possible involvement of DNA polymerases α and β in bleomycin-induced unscheduled DNA synthesis in permeable HeLa cells. *Biochem. Biophys. Res. Comm.* 104:1502.

139. Guilotto, E., and Mondello, C. 1981. Aphidicolin does not inhibit the repair synthesis of mitotic chromosomes. *Biochem. Biophys. Res. Comm.* 99:1287.

140. Stockdale, F. E. 1971. DNA synthesis in differentiating skeletal muscle cells: initiation by ultraviolet light. *Science* 171:1145.

141. Stockdale, F. E., and O'Neill, M. C. 1972. Repair DNA synthesis in differentiated embryonic muscle cells. *J. Cell. Biol.* 52:589.

142. Wicha, M., and Stockdale, F. E. 1972. DNA-dependent DNA polymerases in differentiating embryonic muscle cells. *Biochem. Biophys. Res. Comm.* 48:1079.

143. Hubscher, U., Kuenzle, C. C., Limacher, W., Schoner, R., and Spadari, S. 1978. Functions of DNA polymerases α, β and γ in neurons during development. *Cold Spring Harbor Symp. Quant. Biol.* 43:625.

144. Ikegami, S., Taguchi, T., Ohashi, M., Oguro, M., Nagano, H., and Mano, Y. 1978. Aphidicolin prevents mitotic cell division by interfering with the activity of DNA polymerase α. *Nature* 275:458.

145. Chang, C-C., Boezi, J. A., Warren, S. T., Sabourin, C. L. K., Liu, P. K., Glatzer, L., and Trosko, J. E. 1981. Isolation and characterization of a UV-sensitive hypermutable aphidicolin-resistant Chinese hamster cell line. *Somat. Cell Genet.* 7:235.

146. Liu, P. K., Chang, C-C., Trosko, J. E., Dube, D. K., Martin, G. M., and Loeb, L. A. 1983. Mammalian mutator mutant with an aphidicolin-resistant DNA polymerase α. *Proc. Natl. Acad. Sci.* (USA) 80:797.

147. Hockensmith, J. W., and Bambara, R. A. 1981. Kinetic characteristics which distinguish two forms of calf thymus DNA polymerase α. *Biochemistry* 20:227.

148. Lindahl, T., and Edelman, G. M. 1968. Polynucleotide ligase from myeloid and lymphoid tissues. *Proc. Natl. Acad. Sci.* (USA) 61:680.

149. Söderhäll, S., and Lindahl, T. 1973. Mammalian deoxyribonucleic acid ligase. *J. Biol. Chem.* 248:672.

150. Sambrook, J., and Shatkin, A. J. 1969. Polynucleotide ligase activity in cells infected with simian virus 40, polyoma virus or vaccinia virus. *J. Virology* 4:719.

151. Spadari, S., Ciarrocchi, G., and Falaschi, A. 1971. Purification and properties of a polynucleotide ligase from human cell cultures. *Eur. J. Biochem.* 22:75.

152. Pedrali-Noy, G. C. F., Spadari, S., Ciarrocchi, G., Pedrini, A. M., and Falaschi, A. 1973. Two forms of DNA ligase of human cells. *Eur. J. Biochem.* 39:343.

153. Söderhäll, S., and Lindahl, T. 1973. Two DNA ligase activities from calf thymus. *Biochem. Biophys. Res. Comm.* 53:910.

154. Söderhäll, S., and Lindahl, T. 1975. Mammalian DNA ligases. Serological evidence for two separate enzymes. *J. Biol. Chem.* 250:8438.

155. Mezzina, M., and Nocentini, S. 1978. DNA ligase activity in UV-irradiated monkey kidney cells. *Nucleic Acids Res.* 5:4317.

156. Hsieh, T-S., and Brutlag, D. 1980. ATP-dependent DNA topoisomerase from *D. melanogaster* reversibly catenates duplex DNA rings. *Cell* 21:115.

157. Baldi, M. I., Benedetti, P., Mattoccia, E., and Tocchini-Valentini, G. P. 1980. *In vitro* catenation and decatenation of DNA and a novel eukaryotic ATP-dependent topoisomerase. *Cell* 20:461.

158. Miller, K. G., Liu, L. F., and Englund, P. T. 1981. A homogeneous type II DNA topoisomerase from HeLa cell nuclei. *J. Biol. Chem.* 256:9334.

159. Collins, A., and Johnson, R. 1979. Novobiocin: an inhibitor of the repair of UV-induced but not X-ray-induced damage in mammalian cells. *Nucleic Acids Res.* 7:1311.

160. Mattern, M. R., and Scudiero, D. A. 1981. Dependence of mammalian DNA synthesis on DNA supercoiling. III. Characterization of the inhibition of replicative and repair-type DNA synthesis by novobiocin and naladixic acid. *Biochim. Biophys. Acta* 653:248.

161. Snyder, R. D., Van Houten, B., and Regan, J. D. 1982. Studies on the inhibition of repair of ultraviolet- and methylmethane sulfonate–induced damage in the DNA of human fibroblasts by novobiocin. *Nucleic Acids Res.* 10:6207.

162. Edenberg, H. J. 1980. Novobiocin inhibition of simian virus 40 DNA replication. *Nature* 286:529.

163. Sung, S. C. 1974. Effect of novobiocin on DNA-dependent DNA polymerases from developing rat brain. *Biochim. Biophys. Acta* 361:115.

164. Nakayawa, K., and Sugino, A. 1980. Novobiocin and naladixic acid target proteins in yeast. *Biochem. Biophys. Res. Comm.* 96:306.

165. Cleaver, J. E. 1982. Specificity and completeness of inhibition of DNA repair by novobiocin and aphidicolin. *Carcinogenesis* 3:1171.

166. Wilson, J. H. 1977. Genetic analysis of host range mutant viruses suggests an uncoating defect in simian virus 40–resistant monkey cells. *Proc. Natl. Acad. Sci.* (USA) 74:3503.

167. Miller, L. K., Cooke, B. E., and Fried, M. 1976. Fate of mismatched base-pair regions in polyoma heteroduplex DNA during infection of mouse cells. *Proc. Natl. Acad. Sci.* (USA) 73:3073.

168. Radman, M. 1983. Diversification and conservation of genes by mismatch repair: a case for immunoglobulin genes. In *Cellular responses to DNA damage.* E. C. Friedberg and B. A. Bridges, eds., p. 287. New York: Alan R. Liss.

169. Smith, C. A. 1978. Removal of T4 endonuclease V sensitive sites and repair replication in confluent human diploid fibroblasts. In *DNA repair mechanisms.* P. C. Hanawalt, E. C. Friedberg and C. F. Fox, eds., p. 311. New York: Academic.

170. Smerdon, M. J., and Lieberman, M. W. 1978. Distribution of UV-induced DNA repair synthesis in human chromatin. In *DNA repair mechanisms.* P. C. Hanawalt, E. C. Friedberg and C. F. Fox, eds., p. 327. New York: Academic.

171. Cook, K. H., Friedberg, E. C., and Cleaver, J. E. 1975. Excision of thymine dimers from specifically incised DNA by extracts of xeroderma pigmentosum cells. *Nature* 256:235.

7

DNA Damage Tolerance in Prokaryotic Cells

7-1 Introduction

DNA repair in the strict sense used in this book (see Section 2-1) appears to be largely confined to nonreplicating regions of the genome and is also conceivably restricted in regions actively involved in other aspects of metabolism, such as transcription and recombination, that alter the "resting" topology of DNA. The next two chapters deal with the *tolerance* of base damage to DNA and primarily involve a consideration of the cellular responses to damage at or near *replication forks.* Such responses do not necessarily exclude DNA repair; indeed, there are suggestions that excision repair of lesions associated with replication forks may occur in cells (1, 2), but the experimental evidence in support of this notion is limited and insufficient data are available to mechanistically distinguish such putative repair from the excision repair pathways presented in earlier chapters. This issue notwithstanding, the responses discussed in this chapter are not considered to be DNA repair phenomena. Nonetheless, in the literature descriptive terms that include the word repair are frequently used to describe these responses, and these will be noted in the appropriate sections for purposes of clarification.

The cellular responses to be considered have important biological consequences. With respect to individual members of a population of dividing cells, these responses provide potential mechanisms for survival in the face of replicative arrest, possibly allowing for DNA repair to occur once template DNA containing base damage has been replicated. In addition some damage tolerance mechanisms are associated with a significant increase in mutation frequency, thus providing the potential for genetic diversification within a population of affected cells.

Once again the principal biological system that will be considered is *E. coli*, and the specific type of base damage most often referred to is the (presumably now familiar) pyrimidine dimer. But it should be remembered that cells probably respond very similarly in general if not specifically to a variety of base damages that interfere with normal semiconservative DNA synthesis. In particular, DNA damaged by UV radiation contains (6-4) pyrimidine adducts (see Section 1-5), in addition to cyclobutyl pyrimidine dimers; however, the effect of the former lesions on semiconservative DNA replication has not been systematically explored.

7-2 Inhibition of DNA Synthesis by Damage to DNA Templates

Bulky base adducts inhibit DNA replication both in vivo and in vitro (3–10). Figure 7-1 shows the incorporation of radiolabeled thymidine into the DNA of *E. coli* cells as a function of UV radiation dose. Note that a strain defective in the excision repair of pyrimidine dimers is much more sensitive to inhibition of DNA synthesis than are wild-type strains and that photoreactivation of pyrimidine dimers partially reverses this effect (3). Hence, clearly pyrimidine dimers can be implicated in the inhibition of DNA synthesis in vivo. Although such studies do not indicate *how* pyrimidine dimers inhibit DNA synthesis in *E. coli*, at least two formal possibilities, which are not mutually exclusive, can be considered. First, dimers may inhibit the *initiation* of replication by some mechanism that is UV dose dependent, perhaps as a consequence of some kind of structural alteration at the origin of replication. Alternatively (or in addition) the *progression* of replication forks may be arrested when dimers are encountered in template strands. The latter explanation appears to be the most likely in *E. coli* and presumably in other prokaryotes. Thus, for example,

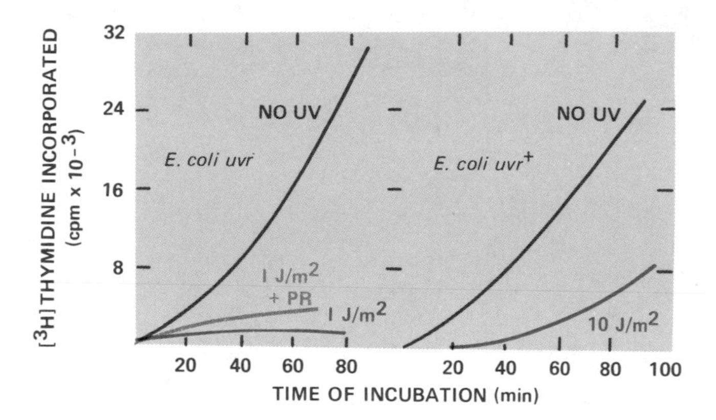

FIGURE 7-1
When *E. coli* cells are UV-irradiated, semiconservative DNA synthesis is inhibited, as evidenced by the reduced rate of incorporation of [^{3}H]thymidine (both panels). Cells that are defective in excision repair of bulky base damage such as pyrimidine dimers (left panel) are much more sensitive to inhibition of DNA synthesis by UV radiation than are cells that are excision proficient (right panel). (Note the different doses of UV radiation used in the two panels.) If cells are exposed to enzymatic photoreactivation (PR) after irradiation, some recovery of the rate of thymidine incorporation is observed (left panel). (After R. B. Setlow et al., ref. 3.)

pyrimidine dimers in the single-strand DNA of phage ϕX174 constitute blocks to the progression of replication forks following infection of *E. coli* (4, 5). Figure 7-2 shows contour tracings from electron micrographs of unirradiated and UV-irradiated ϕX174 DNA that demonstrate this phenomenon. Studies of DNA synthesis in vitro, using either purified *E. coli* enzymes or cell extracts as a source of DNA polymerase and UV-irradiated single-strand ϕX174 DNA as a template, also show that pyrimidine dimers block replication (9).

As indicated in Section 2-2, a particularly sensitive technique for measuring the effect of base damage on DNA synthesis in vitro involves the use of DNA templates of defined nucleotide sequence (6, 8). Single-strand ϕX174 DNA damaged in a variety of ways, including exposure to UV radiation or to various chemicals such as N-acetoxy-2-acetylaminofluorene (AAF) or the 7,8-hydroxy-9,10-epoxy derivative of benzo(a)pyrene, can be annealed with primers obtained by restriction enzyme digestion of ϕX174 DNA. DNA synthesis can then be effected in vitro with the use of any one of a variety of DNA polymerases. The products of the polymerization reactions can be sequenced and the *sites of termination of DNA synthesis* compared

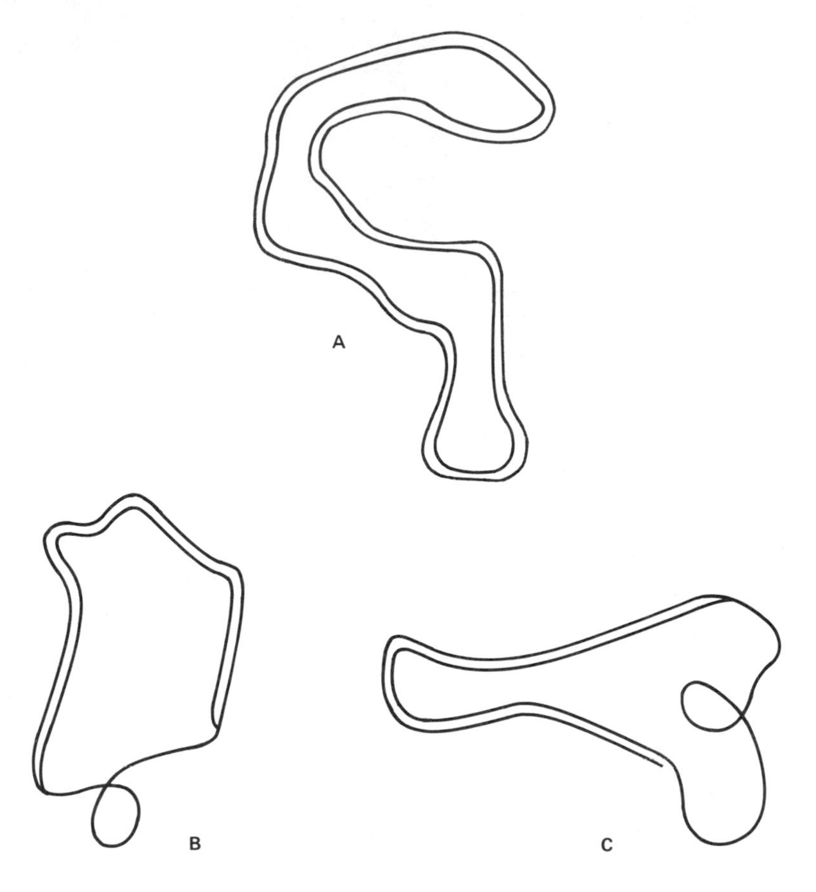

FIGURE 7-2
Contour tracings of replicating ϕX174 DNA molecules observed with electron microscopy following exposure of phage-infected *E. coli* cells to unirradiated or UV-irradiated conditions. Unirradiated molecules show complete replication to the RFI form (A); UV-irradiated molecules containing an average of one pyrimidine dimer per molecule show single-strand regions, indicating sites of replicative arrest (B, C). (From R. M. Benbow et al., ref. 4.)

with the known nucleotide sequence of the template. With both AAF- and UV radiation-treated DNA, synthesis arrests one nucleotide before the site of base damage; N-acetyl-N-(guan-8-yl)-2-aminofluorene in the former case and pyrimidine dimers in the latter (6, 8) (see Fig. 2-1).

UV radiation (and other DNA damaging agents) also causes inhibition of DNA synthesis in mammalian cells (10). However, in such cells our understanding of the biochemistry and molecular biology of DNA replication is far less complete and the exact mechanism(s) of inhibition of synthesis is less clear. In addition, the *E. coli* chromosome has a unique origin of replication from which it is replicated bidirectionally (11). Thus, there are only two replication forks in the single chromosome typically present in an *E. coli* cell. (However, cells in the exponential phase of growth may contain as many as four copies of the genome and in enriched medium each chromosome may contain multiple replication forks.) Eukaryotic chromosomes, on the other hand, contain multiple tandem replicons, each of which is about 200- to 600-fold smaller than the single *E. coli* replicon (10). The complexities that arise from this fundamental distinction create problems in the interpretation of experiments with mammalian cells, which are considered in greater detail in Chapter 8.

7-3 Resumption of DNA Synthesis in *E. coli*

DNA Synthesis with Daughter-Strand Discontinuities

The arrest of DNA replication in *E. coli* at sites of pyrimidine dimers in template strands is not permanent. This can be inferred from a careful examination of the data shown in Fig. 7-1, for example. Based on the known number of dimers produced in DNA as a function of UV radiation dose, the residual DNA synthesis observed in the excision-defective strains suggests that some DNA replication must proceed past these lesions. There are at least two mechanisms whereby cells could resume DNA synthesis on templates containing replicative blocks and in so doing enhance their potential for survival. One is by reinitiating DNA synthesis some distance downstream from the blocks (essentially circumventing them), thereby creating *gaps* or *discontinuities* in the daughter DNA strands, which subsequently become filled in by some mechanism. The second is by continuing DNA synthesis past the template lesion in *a continuous mode* after an initial arrest.

A model for the former type of cellular response to DNA damage was first suggested by Paul Howard-Flanders and his colleagues (12) and is frequently referred to as *postreplication repair* of DNA (13–15). The essential features of this model are shown in Fig. 7-3; a more refined model of this phenomenon is presented below in "Strand Transfer and Complete Strand Exchange." (see Fig. 7-18). Note that the model does not postulate any biochemical events that lead to the actual removal of pyrimidine dimers from DNA. For this reason, I prefer not to use the term "repair" in this context. It has been pointed out that the events described in the model are associated with the

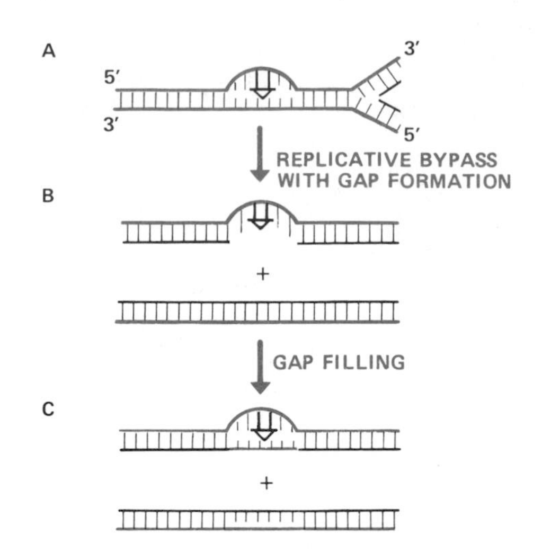

FIGURE 7-3
Simplified model of the events associated with discontinuous replication of UV-irradiated DNA with gap filling. A single pyrimidine dimer is shown in one of the template DNA strands (A). Replicative bypass of the template damage results in one normal sister duplex DNA molecule and one containing a gap opposite the dimer. A nonreciprocal recombinational event fills the gap in this duplex (B), leaving a temporary gap in the isopolar strand of the other sister duplex. The latter gap is not shown in the figure but its filling by repair synthesis using the normal complementary strand as a template is indicated by the colored region in this parental DNA strand (C).

repair of gaps in daughter DNA strands and hence the term "daughter strand gap repair" is sometimes used (16). However, consistent with the restrictive biochemical definition of DNA repair employed throughout this text, less specific phrases such as *replication of damaged DNA with gap formation* or *discontinuous DNA synthesis with gap formation* are used here.

The experiments that led to the model shown in Fig. 7-3 provided the first indication that the survival of UV-irradiated cells can be effected by cellular responses other than excision repair of lesions such as pyrimidine dimers. The search for an alternative response to DNA damage was prompted by the observation that cells defective in the *uvr*-dependent excision repair mode that also carried a mutation in the *recA* gene (known to be required for generalized recombination; see "The *RecA* Gene of *E. coli*: Its Role in Recombinational Events," below) were killed by a dose of UV radiation that produced an average of about 1 pyrimidine dimer per *E. coli* genome equivalent (17) (Fig. 7-4). One plausible interpretation of these results is that *recA*$^+$ cells utilize a *DNA damage tolerance* mechanism that facilitates *survival without excision repair* of the dimers.

This putative excision-independent cellular response to DNA damage was investigated by examination of DNA synthesis in UV-irradiated *E. coli*. When excision-defective (*uvr*) cells were pulse labeled with [^{3}H]thymidine shortly after exposure to UV radiation, the average size of the newly synthesized daughter strands was significantly smaller than that observed in unirradiated cells (12, 18) (Fig. 7-5). However, with increasing time after irradiation, the daughter strands became longer, ultimately reaching the size of those syn-

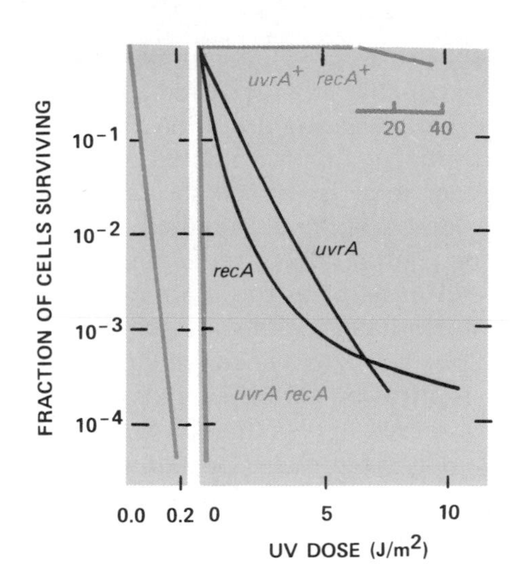

FIGURE 7-4
E. coli cells that are proficient in both excision repair (*uvrA*$^+$) and in discontinuous DNA synthesis with gap filling (*recA*$^+$) are resistant to killing by UV radiation. Cells that are defective in both of these processes (*uvrArecA*) are extremely UV sensitive. The survival curve of the *uvrArecA* double mutant is reproduced on the left with an expanded UV dose scale. From this curve it can be estimated that at the dose of UV radiation that results in a probability of 100 percent killing (D_0), there is on the average about one pyrimidine dimer per *E. coli* genome equivalent. (After P. Howard-Flanders and R. P. Boyce, ref. 17.)

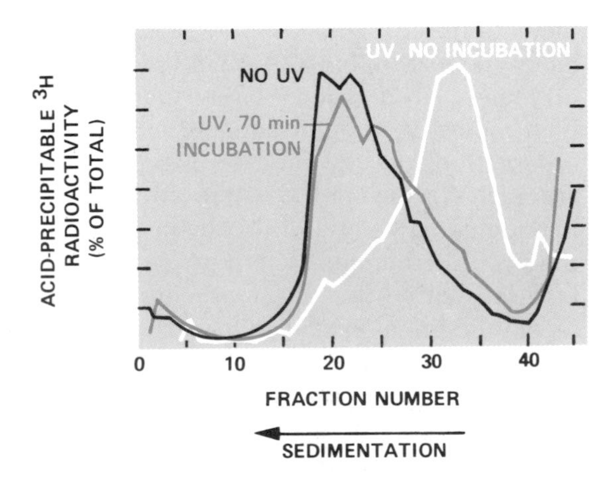

FIGURE 7-5
DNA synthesized immediately after UV irradiation of *E. coli* has a lower molecular weight than normal. *E. coli uvr* cells were exposed to UV radiation and pulse labeled briefly with [^{3}H]thymidine. Cells were analyzed by sedimentation in alkaline sucrose gradients either immediately after labeling (no incubation) or following incubation for 70 minutes. Immediately following the pulse label, the newly synthesized DNA is of low molecular weight and sediments near the top of the gradient. However, with time the newly synthesized DNA approaches the size of the unirradiated control (no UV). (From W. D. Rupp and P. Howard-Flanders, ref. 12.)

thesized in unirradiated cells (12, 18) (Fig. 7-5). A number of explanations for this observation were considered (12). A reduced rate of DNA synthesis in UV-irradiated cells was ruled out, because when the radiolabeling periods for irradiated or unirradiated cells were adjusted so that equal amounts of total isotope incorporation occurred in both cell cultures, the same result was obtained. A second possible explanation was that the shorter DNA strands reflected the onset of new rounds of DNA synthesis from the normal origin of replication. However, when UV-irradiated cells were incubated in unlabeled medium for a period long enough to ensure completion of any newly initiated rounds of replication and then pulse labeled, shorter DNA strands were still observed. It was therefore concluded that the DNA synthesized on damaged templates contained *discontinuities.*

This conclusion led to the specific suggestion that after UV irradiation, DNA replication continues at a normal rate along the chromosome until a dimer is reached, at which point replication is delayed before resuming at some point beyond the dimer, thereby generating a discontinuity or gap in the daughter strand (12) (Fig. 7-3). The increase in sedimentation rate of the daughter strands during subsequent incubation indicates that the defects (or gaps) in these strands ultimately disappear. A likely mechanism of the reconstitution of high molecular weight daughter-strand DNA was suggested by the observation that the process of replication generates two *identical sister DNA molecules,* thus providing for the possibility of gap filling in daughter DNA strands by recombinational events. As indicated in the simplified model shown in Fig. 7-3, a daughter-strand gap could be filled with a homologous region of undamaged DNA derived by recombinational exchange from the isopolar parental strand present on the sister DNA duplex. The resulting discontinuity in the parental strand could then be eliminated by repair synthesis using the undamaged region of the complementary daughter strand as a template. The final products of these events would be two intact sister duplex DNA molecules, one of which still contained pyrimidine dimers in its parental strands.

Evidence in Support of the Model

Experiments with *E. coli* have yielded results that are consistent with the general model shown in Fig. 7-3. The predictions of the model that have been most extensively examined are the presence of gaps in newly replicated DNA, the specific location of these gaps with respect to pyrimidine dimers in template DNA strands and the reconstitution of high molecular weight DNA by recombinational and/or repair synthesis events (10, 16).

Gaps in Newly Synthesized DNA. The size of newly synthesized DNA fragments in UV-irradiated cells approximates the average interdimer distance in the template strands (19, 20), a result expected if replication arrested at a dimer and then resumed with discontinuities. Similarly, if the number of "arresting" lesions (e.g., dimers) in the template DNA was reduced by excision repair (i.e., when the cells are *uvr*$^+$) or by enzymatic photoreactivation prior to replication, the formation of correspondingly longer daughter-strand fragments was observed during a brief labeling pulse (21). In addition, under such conditions shorter post-UV incubation times were required to convert newly synthesized low molecular weight DNA into high molecular weight products (21).

Further evidence in support of discontinuities or gaps in the DNA synthesized in UV-irradiated bacteria and an indication of their position relative to that of pyrimidine dimers in the template strands came from conjugational crosses between *Flac*$^+$ donors and *lac* recipients (18). Such conjugations between unirradiated cells are normally accompanied by the transfer of one of the DNA strands of the sex factor to the recipient cell and its replication in the recipient during transfer (22, 23). The same process is presumed to occur when the *Flac*$^+$ episome is UV-irradiated. If discontinuities were generated in the daughter strands synthesized during the transfer of the episome, they could be opposite pyrimidine dimers, or displaced between them (Fig. 7-6). An attempt to distinguish between these two possible configurations was made by examination of the susceptibility of the transferred DNA to excision repair and to enzymatic photoreactivation (18). The former process depends on the presence of an intact complementary DNA strand and presumably could not take place if dimers were opposite gaps. However, dimers opposite gaps should still be sensitive to enzymatic photoreactivation (Fig. 7-6).

In order to minimize recombination events between the transferred episome and the recipient chromosome (which could obviously complicate the interpretation of such experiments), conjugations were carried out using recombination-defective (*recA*) recipient cells (18). Under these experimental conditions the yield of lac$^+$ colonies decreased rapidly with increasing UV radiation dose relative to that observed in wild-type recipients (Fig. 7-7). In contrast, there was no significant difference in the yield of lac$^+$ colonies when recipient strains were compared that were excision proficient or excision deficient; i.e., excision repair was apparently not operative on the UV-irradiated *Flac*$^+$ episome (Fig. 7-7). The damaged episomes were, however, sensitive to photoreactivation in the recipient by exposure to visible light (Fig. 7-8). The lack of detectable excision repair of the transferred episome together with the apparent repair of

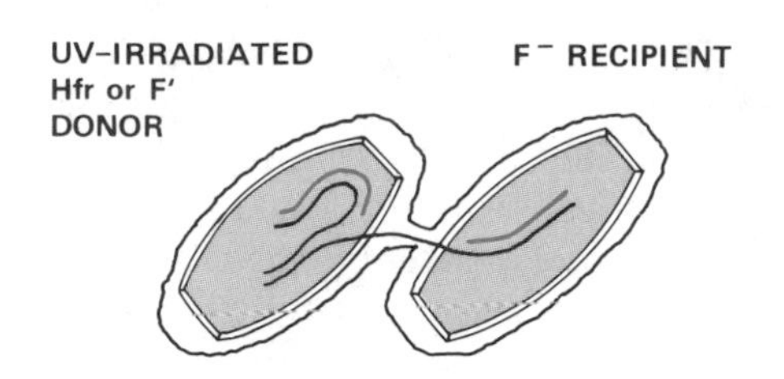

SUBSTRATE	REPAIR BY EXCISION	REPAIR BY PHOTOREACTIVATION
	–	+
	+	+

FIGURE 7-6
DNA transferred from a UV-irradiated Hfr or F′ donor to an unirradiated F$^-$ recipient contains pyrimidine dimers (not shown here for the sake of clarity). When this DNA is replicated in the recipient, discontinuous DNA synthesis may generate gaps opposite dimers or at sites removed from dimers (lower figure). Dimers situated opposite gaps should not be amenable to excision repair because this process requires a duplex configuration of the DNA. However, these dimers should be repairable by enzymatic photoreactivation, albeit possibly at a reduced rate. On the other hand if the gaps in the daughter strands are not opposite dimers in template strands, the dimers should be repairable by both excision repair and by enzymatic photoreactivation. (See Figs. 7-7 and 7-8.) (After P. Howard-Flanders et al., ref. 18.)

dimers by photoreactivation is consistent with the presence of gaps in the strand synthesized during DNA transfer and the location of these gaps opposite pyrimidine dimers (18).

Structures suggestive of postreplicative gaps in DNA have been visualized with electron microscopy of DNA using direct aqueous spreading or "staining" with T4 gene *32* protein (a protein that specifically binds to single-strand regions of DNA) (24). Measurements from such electron micrographs have yielded gap sizes ranging from 1500 to 40,000 nucleotides (24). Independent estimates of average gap sizes of about 1000 nucleotides have been obtained by the isolation of duplex DNA molecules containing single-strand regions (presumed to be postreplicative gaps) with chromatography on benzoylated-naphthoylated DEAE-cellulose (25), a chromatographic matrix that discriminates between single- and double-strand DNA.

Evidence for Recombinational Events. Gap filling in nascent DNA of UV-irradiated *E. coli* appears to be associated with recombina-

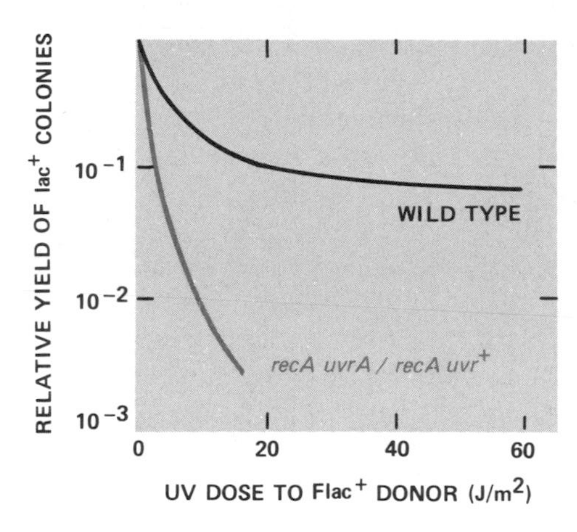

FIGURE 7-7
The relative yield of lac⁺ colonies after an excision-defective *Flac*⁺ strain was mated to mutant *lac* recipients. When the recipient is defective in the *recA* gene, the yield of lac⁺ colonies is significantly reduced relative to that observed in a wild-type recipient. However, there is no detectable difference in the yield of lac⁺ colonies in recipients that are either excision repair proficient (*recA uvr*⁺) or excision repair deficient (*recA uvrA*). (After P. Howard-Flanders et al., ref. 18.)

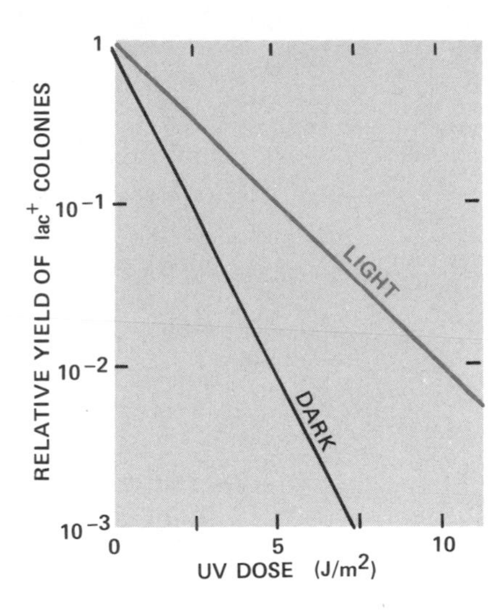

FIGURE 7-8
Flac⁺ DNA transferred from a UV-irradiated donor to an unirradiated recipient shows enhanced survival if following transfer the cells are exposed to photoreactivating light. This result is consistent with (but certainly does not prove) the presence of replicative gaps opposite pyrimidine dimers (see Fig. 7-6). (From P. Howard-Flanders et al., ref. 18.)

tional events. Thus, cells carrying a mutation in the *recA* gene and as a consequence severely deficient in generalized recombination fail to convert short nascent DNA strands into high molecular weight products (21). In addition, UV-irradiated DNA that has undergone replication is highly recombinogenic (18, 26–28). For example, in the conjugation experiments described above in which UV-irradiated *Flac*⁺ episomes were transferred into mutant *lac* unirradiated cells, when the recipient was *recA*⁺, the yield of recombinant progeny that resulted from genetic exchanges between donor and recipient DNA was significantly greater than in crosses with unirradiated *Flac*⁺ episomes (18). Experiments with phage λ also suggest that photoproducts in phage DNA stimulate genetic exchanges between homologous phage chromosomes and that such DNA must be replicated in order for this to occur (28).

Direct evidence for exchanges between DNA strands during post-replicative gap filling comes from the use of appropriate density labels (26). *E. coli* cells were grown for several generations in medium containing the heavy isotopes ^{13}C and ^{15}N. The cells were then exposed to various doses of UV light and grown for less than one generation in a medium without the density markers (light medium) containing [^{3}H]thymidine, which created hybrid density DNA with the *radiolabel uniquely in the light strands* (see Fig. 7-9). If exchanges occur between sister DNA duplexes, light (^{3}H-labeled) material should become covalently associated with heavy (nonradioactive) DNA (Fig. 7-9). Heavy DNA from UV-irradiated cells replicated in the presence of [^{3}H]thymidine does indeed contain strands of intermediate density, as demonstrated by isopycnic sedimentation of denatured extracted DNA in CsCl (26). This DNA was resolved into light and heavy components after shearing to a molecular weight of less than 5×10^5, suggesting that the exchanges involved segments of at least this size.

The hybrid molecules did not arise from labeling of the *ends* of the heavy parental DNA strands, since the cells were grown in nonradioactive medium prior to UV irradiation for a period sufficient to complete any unfinished cycles of replication. The hybrid molecules also could not have arisen from degradation of DNA and reincorporation of density-labeled material into progeny strands, since this should not result in *discrete segments* of heavy nonradioactive DNA joined to radioactive light DNA (26). An estimate of the number of exchanges in these experiments showed that a DNA segment containing both a heavy nonradioactive and a light radioactive region was formed for every 1.7 pyrimidine dimers measured; i.e., there was about one genetic exchange for each dimer (26).

That value correlates well with the one obtained using the BU photolysis technique discussed in Section 5-2 (29–31). The existence of 5-BU in *parental* DNA has been observed after replication of UV-irradiated excision-defective cells in the presence of this thymine analogue. Quantitative photolysis of the DNA by irradiation at 313 nm indicated that these regions were about 1.5×10^4 nucleotides in length and that one such region was present for every 1.2 pyrimidine dimers in the parental DNA. These regions were presumed to have arisen either by repair synthesis across gaps created by recombinative transfer of parental DNA to fill postreplicative daughter-strand gaps (Fig. 7-3) or by the recombinational events themselves (30, 31) (Fig. 7-3).

Further evidence for DNA strand exchanges during the semiconservative replication of damaged DNA comes from analysis of the distribution of pyrimidine dimers in the progeny of UV-irradiated cells. The model shown in Fig. 7-3 predicts that during successive rounds of DNA replication, pyrimidine dimers would remain confined to the original parental strands. However, probing for the presence of dimers in DNA with the T4 PD DNA glycosylase–AP endonuclease (see Section 3-4) showed that these lesions were

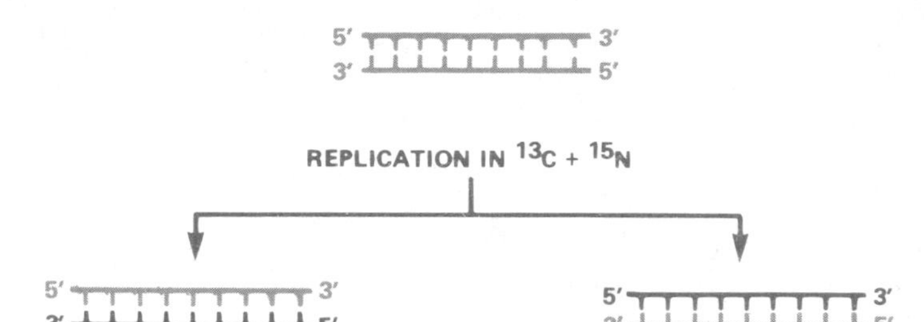

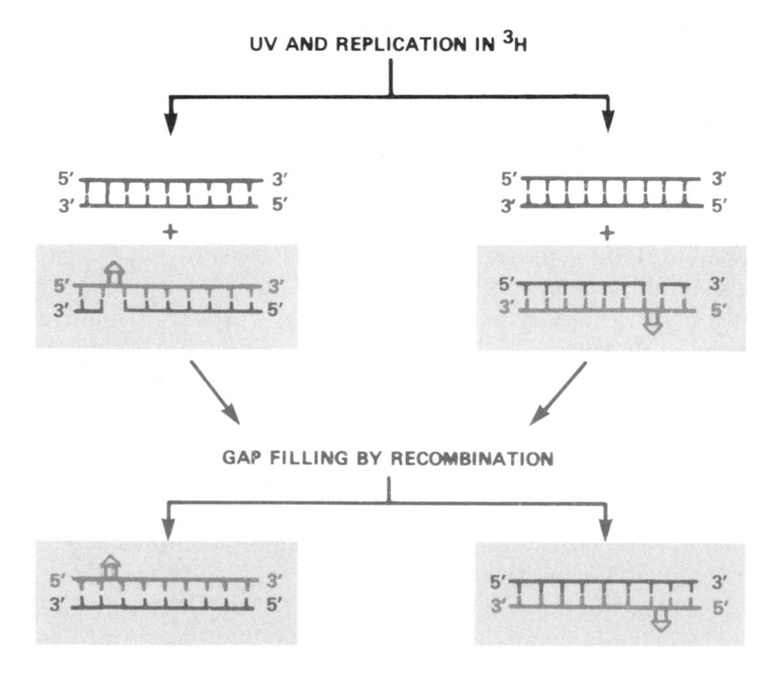

FIGURE 7-9
Demonstration of recombinational events during replication of UV-irradiated *E. coli*. Cells were density labeled in one DNA strand by allowing replication to take place in the presence of the heavy isotopes ^{13}C and ^{15}N. These cells were then UV-irradiated and allowed to replicate in the presence of [^{3}H]thymidine, thus generating radiolabeled daughter strands with putative gaps. If these gaps are filled by recombinational events as shown in the figure (see also Fig. 7-3), then light radioactive DNA should become covalently associated with heavy nonradioactive DNA, which was shown to occur.

equally distributed between parental and progeny strands (32, 33). Thus dimers apparently become randomly transferred into nascent daughter strands, an observation consistent with the notion that strand exchanges do accompany the replicative response to the presence of pyrimidine dimers in DNA. This observation also suggests that the strand exchanges are not as selective as suggested by the simplified model shown in Fig. 7-3; i.e., they may be *reciprocal* and are not necessarily confined specifically to regions that span gaps.

Genetic Functions Required for the Discontinuous Synthesis of DNA with Gap Filling

A number of genes have been implicated in the tolerance of DNA damage associated with discontinuous synthesis and gap filling of daughter-strand DNA (16). However, the precise role of most of these genes is not clear, nor has it been established whether or not they function in a single or in multiple biochemical pathways. There is also considerable uncertainty concerning the absolute distinction between those biochemical functions specifically required for *discontinuous replication with gap formation* and for *continuous* (*so-called translesion*) *DNA synthesis* during semiconservative replication following DNA damage. Some of this uncertainty stems directly from the limitations of the techniques used to study these events. For example, replicative responses to template DNA damage are frequently investigated by measuring changes in the molecular weight of newly synthesized DNA. Sedimentation of pulse-labeled DNA in alkaline sucrose density gradients is a particularly convenient procedure for this purpose (Fig. 7-5). However, the demonstration of the conversion of small newly synthesized DNA fragments into larger ones does not necessarily imply that the conversion is associated with the formation of gaps that are subsequently filled by recombinational events. Thus, stalling of the replication complex at a pyrimidine dimer and subsequent resumption of fork progression (as is postulated during translesion semiconservative DNA synthesis; see "DNA Damage Tolerance by Continuous Synthesis on Damaged Templates: Translesion DNA Synthesis," below) cannot be distinguished from discontinuous DNA synthesis just by sedimentation analysis (Fig. 7-10). Furthermore, intermediate reactions and the formation of discrete reaction products such as gaps, gap filling by the exchange of DNA sequences between isologous strands of sister DNA molecules and repair synthesis of DNA have not been specifically demonstrated in many studies (16).

Another difficulty in clearly distinguishing the discontinuous–gap filling mode of DNA replication from the continuous-translesion mode is that most studies have deliberately utilized strains of *E. coli* that are defective in the *uvrA*, *uvrB* or *uvrC* functions in order to examine replicative responses to DNA damage in the absence of competing excision events. However, as will be noted in Section 7-9, these genes may play a role in some of the replicative responses to DNA damage. Hence the molecular events that ensue during the

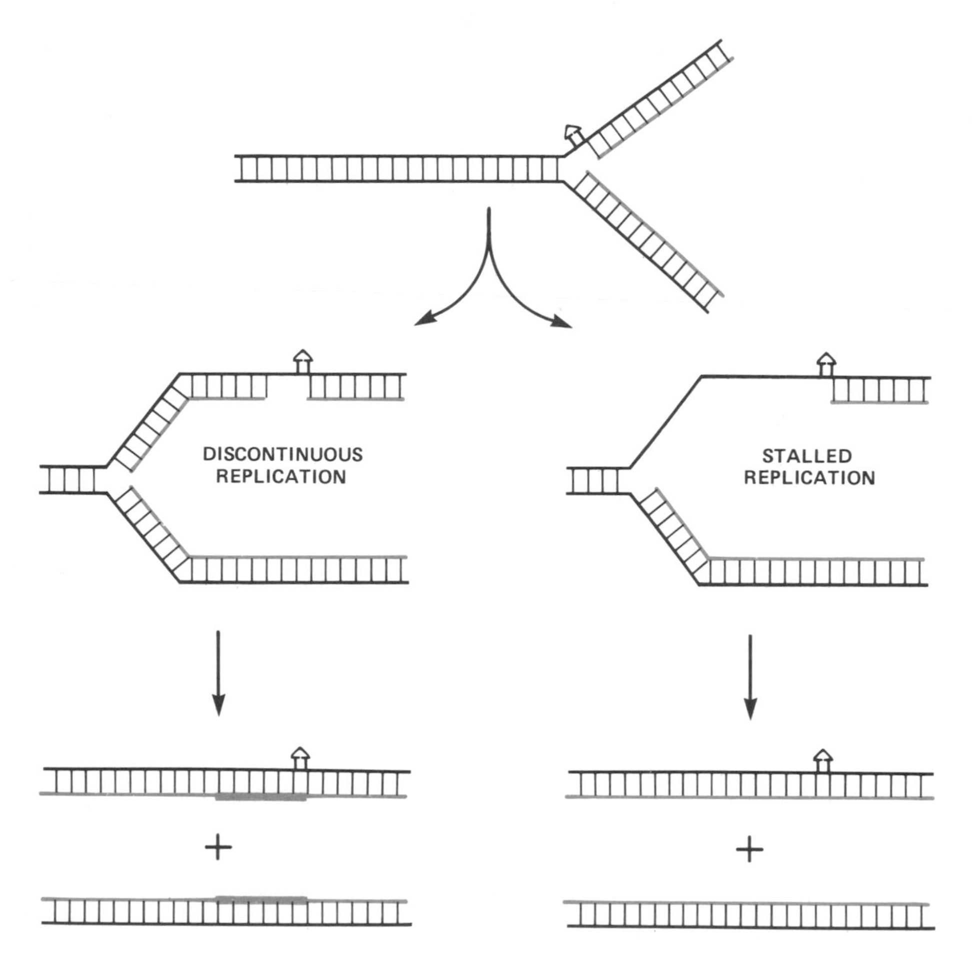

FIGURE 7-10
Both discontinuous replication and stalled replication generate small nascent DNA molecules that subsequently are converted into high molecular weight ones. Thus it can be very difficult if not impossible to differentiate these processes using the sedimentation velocity of radiolabeled newly synthesized DNA. The thicker lines in the lower left diagram indicate regions of gap filling by recombinational exchange and by nonsemiconservative DNA synthesis (see Fig. 7-3).

replication of damaged DNA may differ depending on whether cells have functional *uvrA*, *uvrB* or *uvrC* genes or not. Indeed, the *uvrA* and *uvrB* genes have been implicated in the discontinuous–gap filling mode of DNA synthesis by the demonstration that *uvrA* or *uvrB* derivatives of strains carrying *recA* (34), *recB* (20) or *recF* (35) mutations carry out less complete repair of daughter-strand gaps than do the corresponding uvr^+ strains (16).

Because of these manifold complexities and uncertainties, the only genetic function discussed in depth here is that of the *recA* gene, since the requirement of it for recombinational events is undisputed. Other biochemical functions of the *recA* gene that may be pertinent to *translesion* DNA synthesis are discussed later in the chapter.

The RecA *Gene of* E. coli: *Its Role in Recombinational Events.* The *recA* gene (which maps at 58 minutes on the *E. coli* genetic map, Fig. 7-11) and its translational product, recA protein, have been the subject of intensive investigations that have been facilitated by the over-expression of this protein in induced cells (see Section 7-6) and in cells transformed with plasmids containing the cloned *recA* gene. RecA protein is amazingly versatile and carries out a number of apparently discrete biochemical functions. Of these, the one that concerns us here is its role in exchanges between DNA molecules, i.e., genetic recombinations, with a particular emphasis on the re-

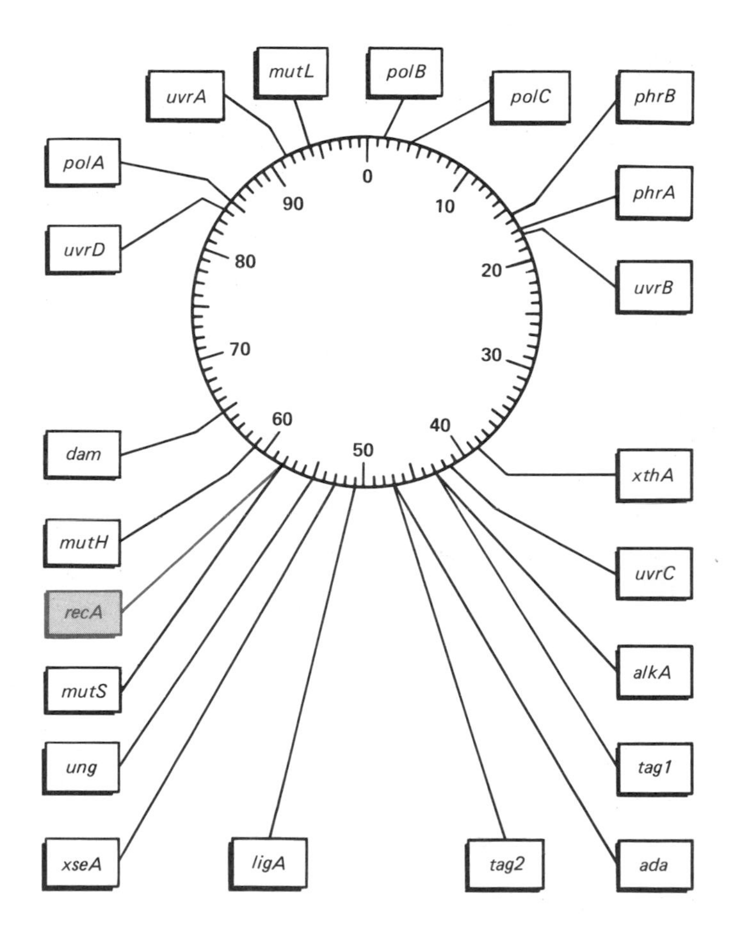

FIGURE 7-11
The location of the *recA* gene on the *E. coli* genetic map.

combinational events associated with discontinuous DNA synthesis and gap filling.

THE HOMOLOGOUS PAIRING OF DNA MOLECULES. A feature that is critical to normal recombination between DNA molecules is the alignment (or pairing) of the DNA such that exchanges between regions of sequence homology can occur. RecA protein (molecular weight, 37,800) (36) interacts with single-strand and duplex DNA in vitro in a number of different ways that are consistent with the promotion of homologous pairing of DNA molecules (37–55). In catalytic amounts recA protein promotes the renaturation of denatured DNA (*pairing of complementary DNA strands*) in an ATP-dependent reaction (40). In stoichiometric amounts sufficient to cover all single-strand DNA present, pairing can also be initiated *between duplex DNA molecules* and *homologous single-strand DNA molecules*, resulting in the displacement of a portion of one of the strands of the duplex to form a D loop (41–45) (Figs. 7-12 and 7-13). For this reaction less recA protein is required in the presence of helix-destabilizing proteins

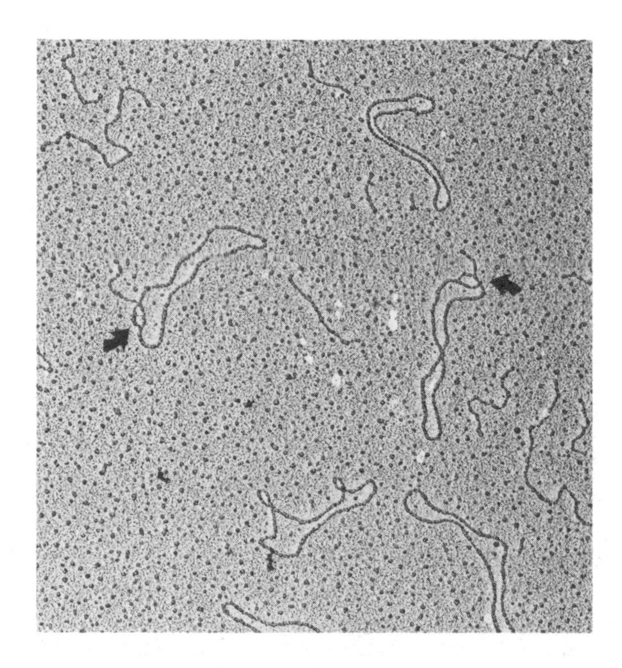

FIGURE 7-12
DNA molecules showing D-loops (arrows) generated during pairing between ϕX174 RF1 and homologous single strands (also see Fig. 7-13). The DNAs were incubated in the presence of recA protein and ATP. (Courtesy of Dr. Kevin McEntee.)

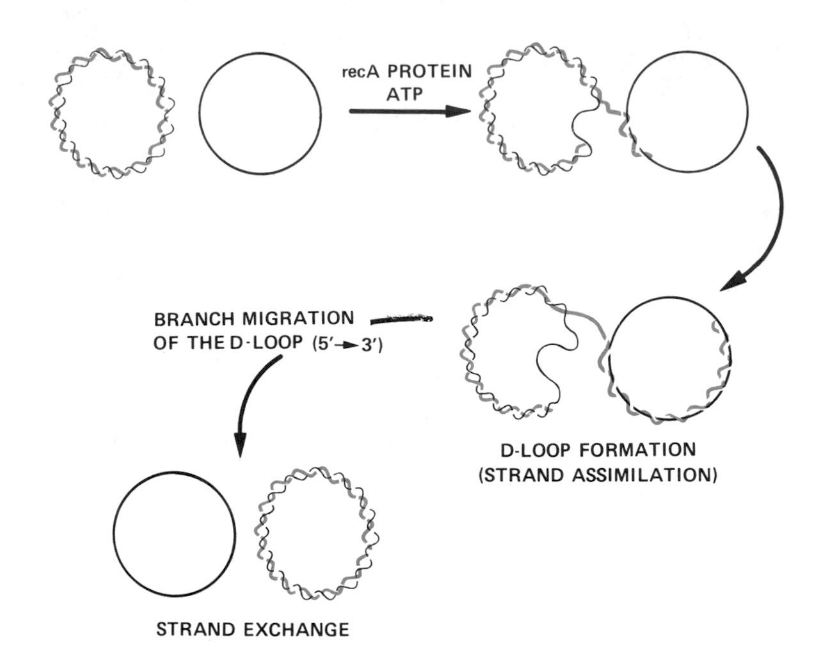

FIGURE 7-13
In the presence of recA protein and ATP a nicked circular duplex DNA will undergo strand exchange with a complementary intact circular single-strand DNA molecule. The displaced single-strand region of the duplex molecule forms a displacement (D) loop during this process of strand transfer and assimilation. Branch migration in the 5′ → 3′ direction results in a complete exchange, so that one strand of the original duplex molecule is now hydrogen bonded to the original single-strand circle. (After K. McKentee and G. M. Weinstock, ref. 49.)

(such as single-strand DNA binding protein of *E. coli*, SSB), which also bind to single-strand DNA (46–48). Like the renaturation of denatured DNA, D-loop formation is also ATP dependent and is highly specific since recA protein purified from a cold-sensitive mutant fails to promote D-loop formation at the restrictive temperature. The effect of single-strand DNA in stimulating the partial unwinding of duplex DNA is not dependent on homology between the two DNA species (43). This is consistent with the notion that during generalized recombination there must be a molecular mechanism that allows for the *search for homology* before such regions actually pair (37).

The role of the ATPase function of recA protein (49, 50) in recombination is not clear. RecA protein binds very strongly to DNA in the presence of the nonhydrolyzable ATP analogue, ATP[γ-S], whereas the presence of ADP promotes dissociation from DNA. The hydrolysis of ATP may thus play a role in a cycle of binding and detachment of the protein from DNA during recombination (37, 46, 51).

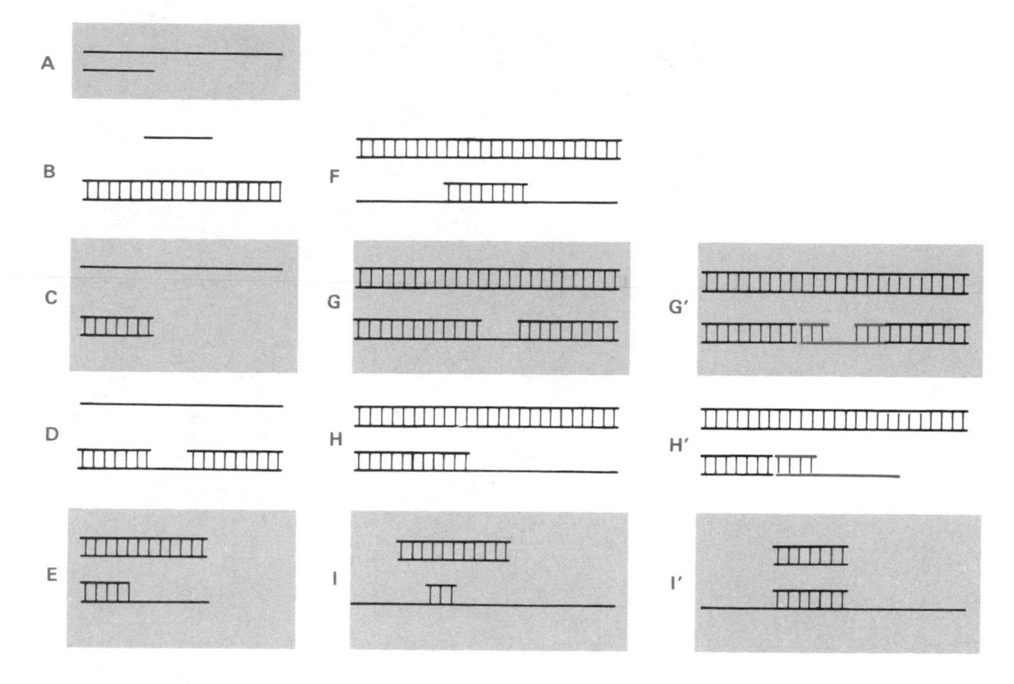

FIGURE 7-14
Twelve different combinations of DNA molecules are shown that have been examined for homologous pairing and the formation of stable joint molecules. In each case the two DNA molecules schematically represented are homologous. Single lines represent single-strand DNA and double lines joined by vertical bars represent duplex DNA. Structures A through I can all form stable joint molecules, whereas structures G′, H′ and I′ do not. Structures G′ and H′ are identical to structures G and H respectively except that the single-strand regions and the ends flanking the gaps are in regions of nonhomology with respect to the other DNA duplex. Structure I′ has no complement to the single-strand region. (From C. M. Radding, ref. 37.)

The process of D-loop formation in vitro provides the basis for potentially understanding how two DNA molecules might initiate homologous pairing during genetic recombination in vivo. DNA molecules paired in a region of homology are sometimes referred to as *joint molecules* (45), and the formation of stable joint molecules can be catalyzed by recA protein from a variety of structural variants, provided that one of the DNA molecules is at least partially single strand and that a free end exists in one of them (37). Figure 7-14 diagrammatically illustrates a number of such examples. Example B leads to typical D-loop structures and example C yields joint molecules in which the single strand extensively displaces one strand of the duplex, resulting in the formation of heteroduplex regions that can be up to thousands of base pairs long. Example E (and presumably example H) creates so-called Holliday structures (Fig. 7-15), in which *reciprocal strand exchanges* also result in long heteroduplex joint molecules. Example G (Fig. 7-14) simulates conformations that

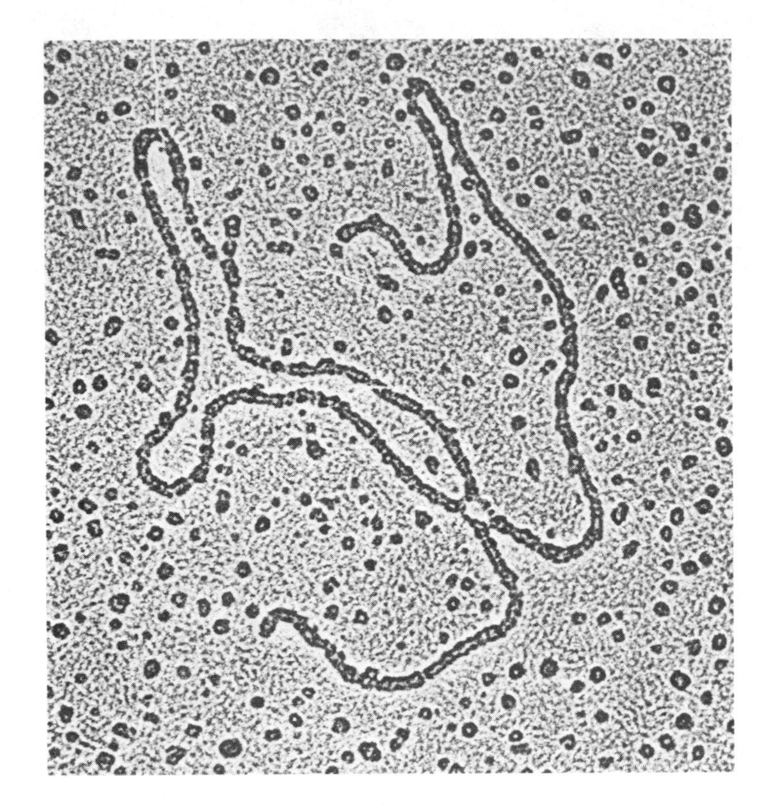

FIGURE 7-15
Reciprocal strand exchanges between two duplex DNA molecules result in so-called Holliday structures, named after Robin Holliday, who first proposed the occurrence of these intermediates during recombination (see also Fig. 7-18, structures [4]-[6]). (Courtesy of Dr. C. M. Radding.)

may be generated by discontinuous DNA synthesis on damaged DNA templates. Indeed, when circular duplex plasmid DNA molecules containing gaps were incubated with homologous intact DNA in the presence of recA protein, Mg^{2+} and ATP, joint DNA molecules were detected (38, 44) (Fig. 7-16).

The pairing of gapped DNA (such as that which might form during discontinuous semiconservative DNA replication) with intact homologous duplex DNA by recA protein results in joint molecules that are stable when deproteinized by detergents and proteins, but unstable in alkali (45, 52, 53). The paired molecules are thus apparently linked by noncovalent bonds and show a thermal stability approaching that of duplex DNA. However, little is known about the detailed structure in the pairing regions of these joint molecules. As suggested above, some models for genetic recombination between two DNA duplexes postulate that one of them contains a nick or gap;

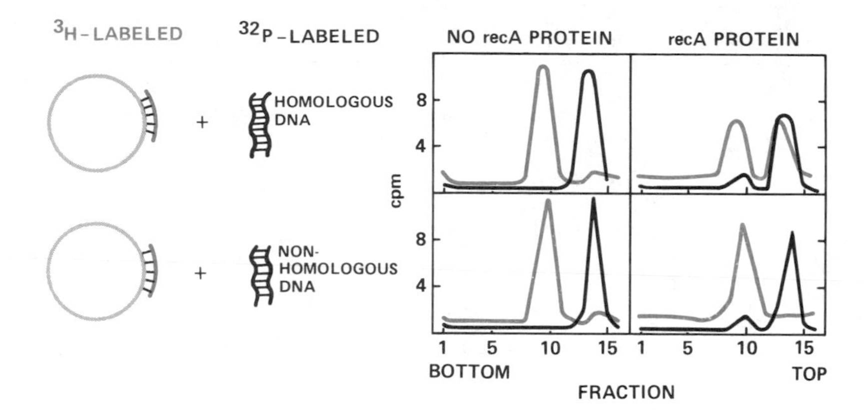

FIGURE 7-16
The formation of joint molecules between duplex gapped DNA and homologous intact duplex DNA. Duplex gapped DNA was generated by hybridizing a restriction fragment (^{3}H labeled) of ϕX174 DNA to a complementary intact circular single-strand molecule. This gapped DNA was incubated with duplex linear intact ^{32}P-labeled homologous (top) or nonhomologous (bottom) DNA, with or without recA protein. Following incubation, the DNA was sedimented in neutral sucrose gradients and ^{3}H and ^{32}P radioactivity were monitored in individual fractions. In the presence of recA protein joint molecules (detected as peaks containing radioactivity from both isotopes) formed between the gapped DNA (^{3}H) and the homologous DNA (^{32}P) (top right panel). No joint molecules were formed in the absence of recA protein (left panels) or when nonhomologous DNA (right bottom panel) was used. (After S. C. West et al., ref. 44.)

i.e., there is a region of unwound single-strand DNA with a free end that acts as a template in the search for complementary sequences in the second duplex (Fig. 7-17) (56). Once homology is established, the free end is assimilated to form a heteroduplex, with the displaced strand of the invaded DNA duplex giving rise to the D-loops observed in vitro (Figs. 7-12 and 7-13) (45, 46).

The role of single-strand regions in genetic recombination is currently quite complex. For example, whereas structures G and H in Fig. 7-14 form joint molecules in the presence of recA protein, structures G′ and H′ (in which the DNA ends and the single-strand domains are in *regions of nonhomology*) do not (37). There is also evidence that in reactions involving *partially* single-strand DNA (e.g., structures E through I in Fig. 7-14) pairing must take place first between the single-strand region and its complement in the duplex, since structure I forms stable joint molecules whereas structure I′ does not (37).

In addition to these complexities, the results of some experiments cannot be readily accommodated by models in which pairing between two duplexes requires the initial unwinding of a free end (45). For example, the introduction of interstrand cross-links in DNA im-

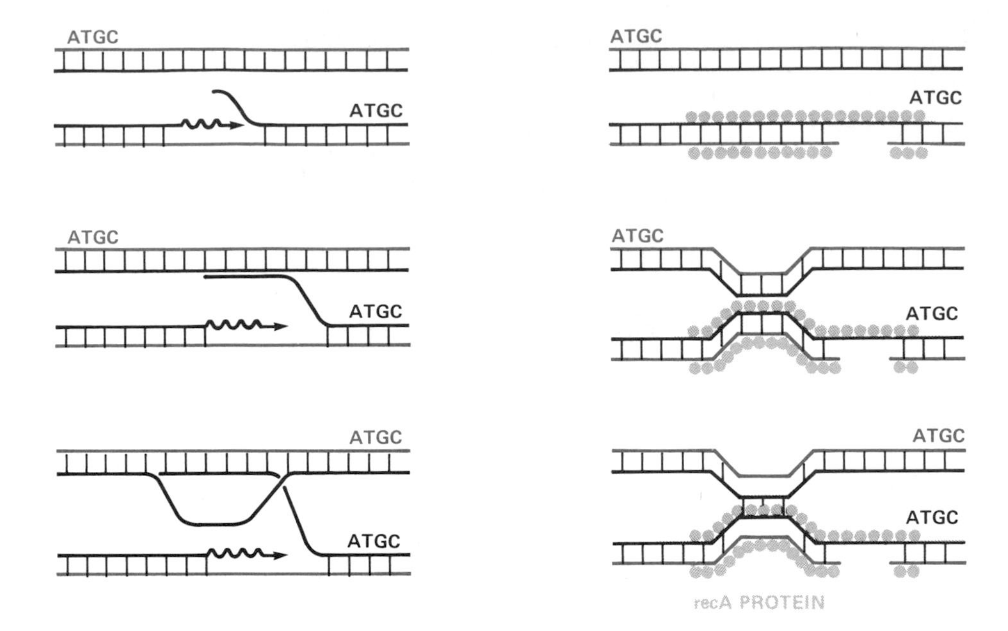

FIGURE 7-17
Two models proposed for the early steps in generalized genetic recombination are shown. The model on the left proposes the displacement of a single strand containing a free end of one DNA duplex by replication (wavy line). The free single strand searches for homology in the other (intact) DNA duplex, and when such homology is found (indicated diagrammatically by alignment of the sequences ATGC) a three-strand recombinational intermediate is formed, resulting in D-loop formation. The model on the right shows the initial steps postulated in four-strand homologous pairing before strand separation. According to this model, recA protein (indicated by the filled circles) binds cooperatively to single-strand DNA at the site of a gap and then loads the adjoining duplex DNA. The loaded DNA molecules interact nonspecifically while searching for homology. When homology between two DNA molecules is established (indicated again by the aligned ATGC sequences), formation of homologous contacts makes possible hydrogen bonding between them. This site of initial pairing through hydrogen bonding may or may not (as shown here) include the gap. (After S. C. West et al., ref. 45.)

paired strand separation, yet had no effect on the yield of joint molecules (53). Additionally, electron micrographs of joint molecules show that the joint is often distant from the single-strand gap (52). Thus, although recA protein may indeed process some DNA substrates via D-loop formation involving strand transfer, it may process others between two DNA duplex molecules, one of which contains single-strand gaps but not necessarily unwound single-strand regions with free ends (45). According to the latter model, recA protein binds initially at the gap but spreads along the adjacent duplex (Fig. 7-17). Hence the requirement for gaps may be purely to provide sites for the immediate binding of recA protein. The initial contact between the two DNA molecules may therefore be nonproductive, but when associated with recA protein such contact leads to a search for homology with the ultimate formation of joint molecules held together by hydrogen bonds (45).

Strand Transfer and Complete Strand Exchange

The formation of *synapses* or *stable joint molecules* by which DNA molecules are placed in homologous register can be distinguished in vitro from the phenomenon of *strand transfer*. This reaction may immediately succeed homologous pairing during recombination in vivo, resulting in the formation of long heteroduplex joints by a concerted and polar mechanism (44, 57–59). In interactions between a completely single-strand DNA and a duplex DNA molecule this reaction is observed as *branch migration of the D loop* in the $5' \rightarrow 3'$ direction (Fig. 7-13). Branch migration has also been noted in DNA containing base damage. For instance, in the presence of SSB protein and ATP, recA protein promoted the complete exchange of strands between circular single-strand DNA containing pyrimidine dimers and a homologous linear duplex, converting the dimer-containing single-strand DNA into a circular duplex form, thus providing a direct in vitro model for gap filling by recombination during discontinuous replication of DNA containing bulky base damage (60).

The recognition of sequence homology and the formation of long, stable heteroduplex joints, the two experimentally observable parameters of recombination mediated in vitro by recA protein, may be related mechanistically. RecA protein may function as a special type of DNA helicase that can displace a strand from a DNA duplex in a unidirectional way and replace it with another strand in the same direction. According to such a model, recA protein may begin to unwind duplex DNA in response to *any* available single-strand region but can only complete this action (thereby resulting in a heteroduplex joint) when a homologous strand is aligned in sequence register and perhaps when a free end makes the process topologically feasible (37). Of course, in vivo other enzymes, including nucleases and DNA ligase, must be involved in order to resolve intermediates, complete the strand exchanges and restore covalent integrity to all DNA molecules.

Based on the information gained from the studies just summarized, we can think of the process of discontinuous semiconservative DNA synthesis with gap filling in more detailed terms than the simple model originally presented (Fig. 7-3). A dimer in parental DNA results in the formation of a postreplicative gap opposite a stretch of single-strand DNA (Fig. 7-18). RecA protein binds to the single-strand region and aligns it with a homologous region of the sister duplex. When homologous pairing is achieved, an enzyme nicks the duplex to generate a free end in the undamaged parental strand. The free end now switches with the isologous newly synthesized daughter strand, producing a crossed-strand exchange and giving rise to a so-called Holliday structure. The gaps in the heteroduplexes are then repaired by repair synthesis and DNA ligation and eventually the Holliday structure is resolved by cutting and rejoining of strands.

The Role of SSB Protein. In addition to the proteins already mentioned, there is evidence that accessory DNA binding proteins such as SSB protein are involved in recombination. In the presence of this protein recA protein interacts with single-strand DNA to form a complex whose stability is dependent on the presence of SSB protein (61). The SSB protein prevents dissociation of recA protein from single-strand DNA, rendering the binding of recA protein to single-strand DNA irreversible. Under these conditions, the homologous pairing phase of the strand exchange reaction is accelerated to the point that it is no longer rate limiting (47). Similarly, as indicated previously, the rate and extent of strand transfer between full-length linear duplex and single-strand circular DNA to form heteroduplex structures were enhanced considerably when SSB protein was added after incubation of the DNA with recA protein (47, 48) (Fig. 7-19).

Evidence of a requirement for SSB protein in the recombinational responses associated with the tolerance of DNA damage also comes from studies with living cells. Conditional lethal mutants defective in SSB protein are abnormally sensitive to UV radiation at permissive temperatures (62). Certain strains of *E. coli* suppress the conditional lethality caused by the mutant allele *ssb-1* (63). However, at elevated temperatures these strains arc still extremely UV sensitive (Fig. 7-20) and degrade their DNA extensively after UV irradiation (62). Both DNA degradation and UV sensitivity (Fig. 7-20) persist under conditions in which recA protein is copiously expressed by the transduction of an operator-constitutive *recA* mutant allele (which allows for constitutive expression of the normally inducible *recA* gene; see Section 7-6). Thus recA protein cannot function efficiently in the recombinational events associated with the tolerance of DNA damage unless normal SSB protein is also present (64).

FIGURE 7-18
A detailed model for the mechanism of discontinuous semiconservative DNA synthesis with gap filling. See the text for a full description of the model. (After P. Howard-Flanders, ref. 84.)

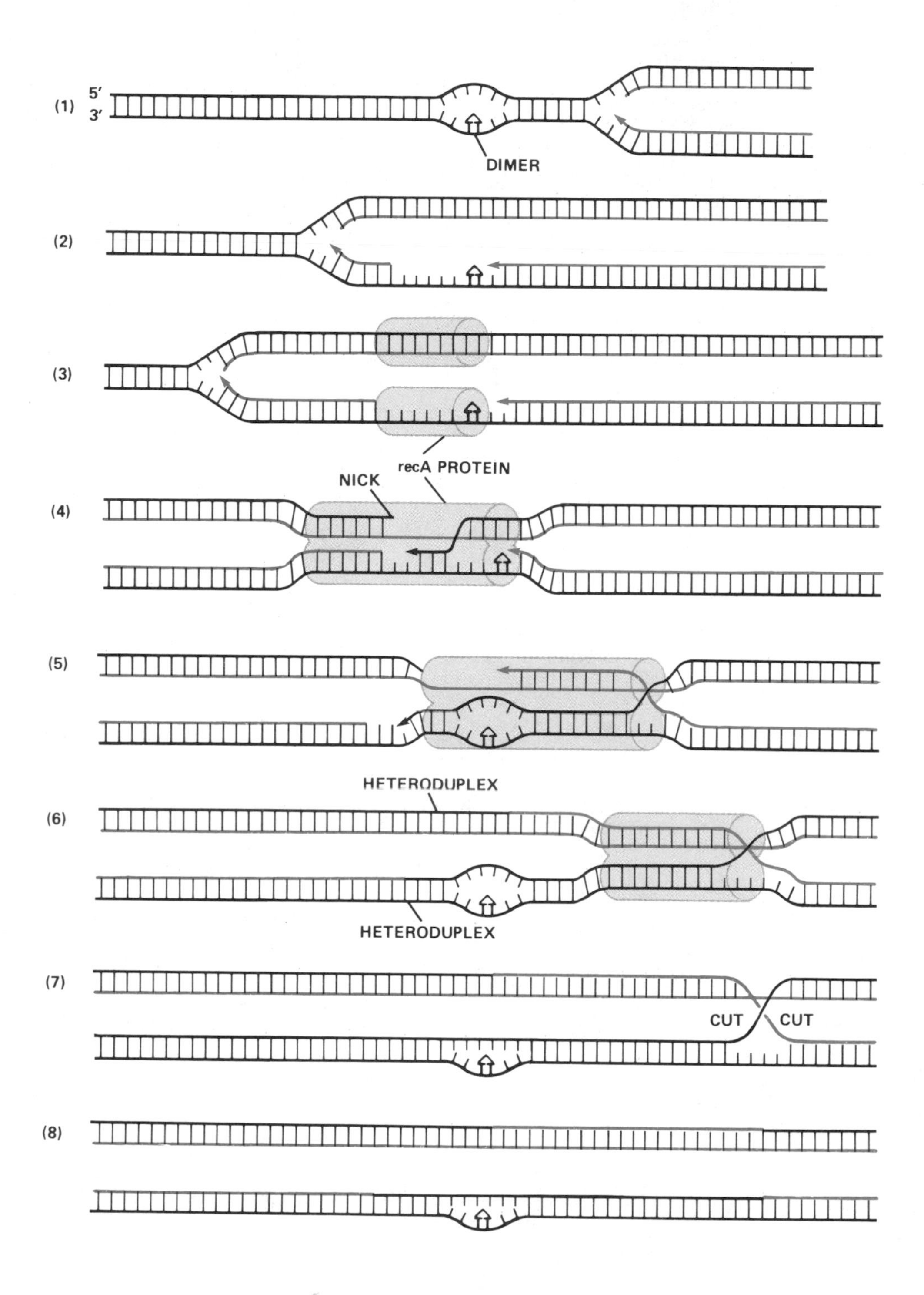
(1)
5′
3′
DIMER
(2)
(3)
recA PROTEIN
NICK
(4)
(5)
HETERODUPLEX
(6)
HETERODUPLEX
(7)
CUT
CUT
(8)

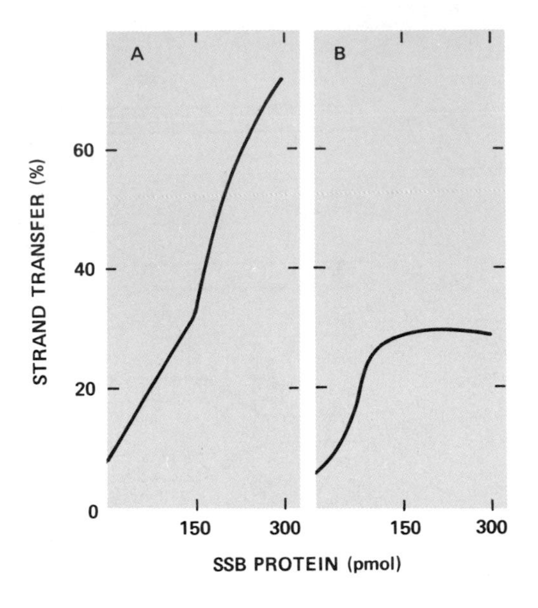

FIGURE 7-19
Effect of increasing single strand binding (SSB) protein concentration on recA protein-mediated DNA strand transfer. (A) SSB protein added five minutes after recA protein; (B) SSB protein added five minutes before recA protein. DNA strand transfer was measured by neutral sucrose gradient sedimentation as the percent of ^{32}P from a small linear duplex ϕX174 restriction fragment transferred to unlabeled single-strand ϕX174 DNA. (After S. C. West et al., ref. 48.)

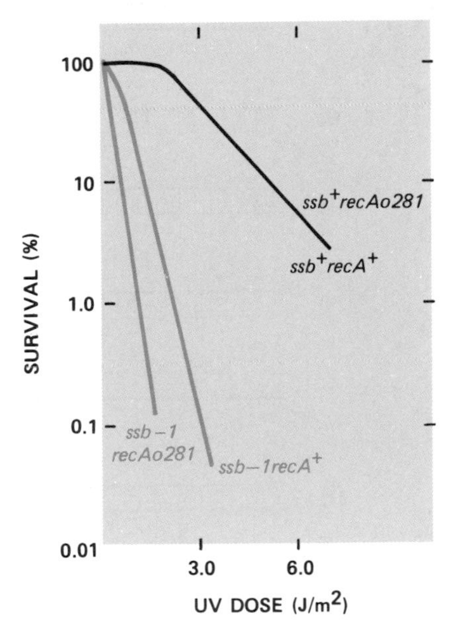

FIGURE 7-20
A strain that is defective in SSB protein at restrictive temperatures (*ssb-1 recA*$^+$) is sensitive to killing by ultraviolet radiation relative to a *ssb*$^+$ *recA*$^+$ strain. This sensitivity is not alleviated by transduction of a *recA* gene containing an operator-constitutive mutation (*recAo281*) that allows for constitutive expression of recA protein. In fact, the double mutant (*ssb-1 recAo281*) is slightly *more* UV sensitive. (From H. B. Lieberman and E. M. Witkin, ref. 64.)

The Interaction of RecA Protein with DNA. The recombination exchanges that occur during the filling of gaps generated by the tolerance of DNA damage during discontinuous DNA replication (and during generalized recombination) clearly involve specific interactions of recA protein with DNA. We are just beginning to gain some

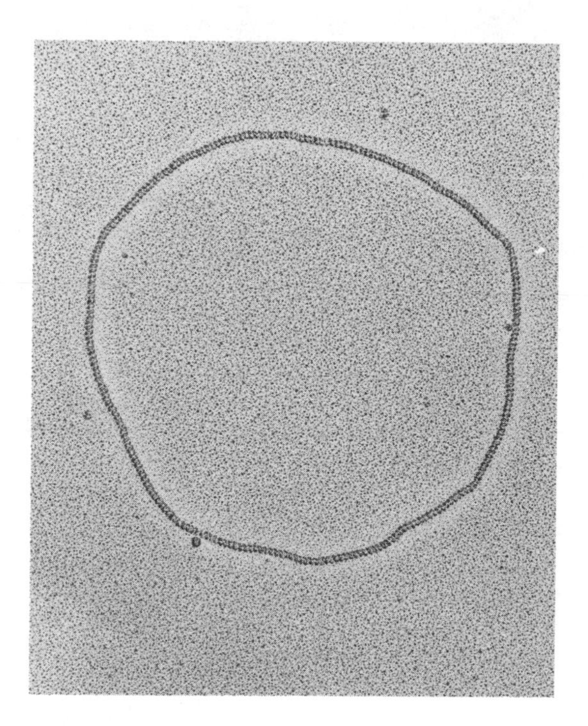

FIGURE 7-21
Binding of recA protein to plasmid DNA. A nicked circular plasmid DNA molecule (a derivative of pBR322) is completely covered with recA protein. Its contour length is stretched by 50 percent relative to that observed without bound recA protein. There are 268 turns of recA-DNA complex in the molecule, visible as striations on the shadowed specimen. (Also see Figs. 7-22 and 7-23) (Courtesy of Dr. Andrzej Stasiak.)

insight into the physical chemistry of this process. In the presence of ATP[γS], recA protein binds cooperatively to gapped duplex DNA, forming thick, rigid rods readily observed in the electron microscope (54). The formation of these structures is dependent on the presence of gaps in the DNA duplex, suggesting that single-strand regions in duplex molecules serve as the initial binding sites from which recA proteins migrate (54). Under certain conditions helical fibers can also be observed on covalent circular duplexes, which can be completely covered by recA protein (55) (Fig. 7-21). The protein forms a right-handed helix around DNA with about 18.6 base pairs per turn (Fig. 7-22) (39, 55, 65). Thus the DNA within this complex is stretched from 3.4 to 5.1 Å per base pair (Fig. 7-23), a 50 percent elongation similar to that caused by the intercalation of ethidium bromide (39, 55, 65). The stoichiometry of binding deduced from electron-scattering estimates of protein mass per nucleotide is three base pairs per protein monomer and six protein monomers per helical turn (55, 65) (Figs. 7-22 and 7-23).

Helical fibers are also observed in crystals of pure recA protein in the absence of any nucleotide (66). An examination of hexagonal crystals by X-ray diffraction shows the protein monomers to be oriented head to tail in a helix with six monomers per turn, similar to the helical protein–DNA fibers (55) (Fig. 7-22). On the other hand, crystals formed in the presence of ADP are tetragonal rather than hexagonal (55). It is suggested that under these conditions the monomeric protein units are once again arranged head to tail in a helical fiber, but this time with four monomers per helical turn (55) (Fig. 7-22).

The observations on the structure of recA protein helical fibers have led to a model for the involvement of recA protein in the homologous pairing between an intact DNA duplex and the single-strand regions in a gapped duplex molecule (Figs. 7-18 and 7-24). The model suggests that recA protein monomers form a single helical sheath of protein surrounding the locally paired single-strand DNA

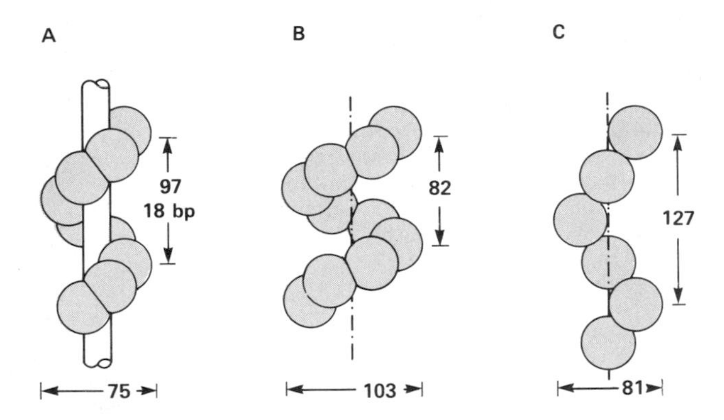

FIGURE 7-22
Diagrammatic representations of the helical fibers of recA protein. (A) RecA protein fiber condensed on a DNA duplex as visualized by electron microscopy (see Fig. 7-21). The protein is shown condensed in a spiral with six monomers per turn, a pitch of nearly 100 Å per turn and a diameter of 75 Å. This pitch extends the DNA by about 50 percent, resulting in approximately 18 bp per helical turn of DNA (see Fig. 7-23). (B) Helical fiber in the crystal form of pure protein without DNA. Examination by X-ray diffraction shows the protein monomers to be oriented head to tail in a helix with six monomers per turn and a pitch of approximately 82 Å. (C) X-ray diffraction of crystals formed in the presence of ADP suggests that the protein monomers may be arranged head to tail in a helical fiber with four monomers per turn and a pitch of 127 Å. (From P. Howard-Flanders and S. C. West, ref. 55.)

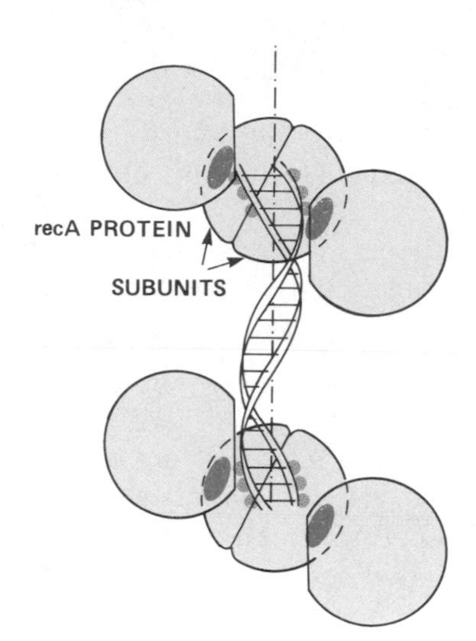

FIGURE 7-23
A sectional view of a recA protein fiber surrounding a DNA duplex. Each monomer is thought possibly to have binding sites for two to three bases on each strand (three are shown here with small colored circles), as well as two protein binding sites (colored ovoids) and a site for ATP (not shown). The DNA chain is shown to be extended to approximately 18 bp per helical turn (see Fig. 7-22). (From P. Howard-Flanders and S. C. West, ref. 55.)

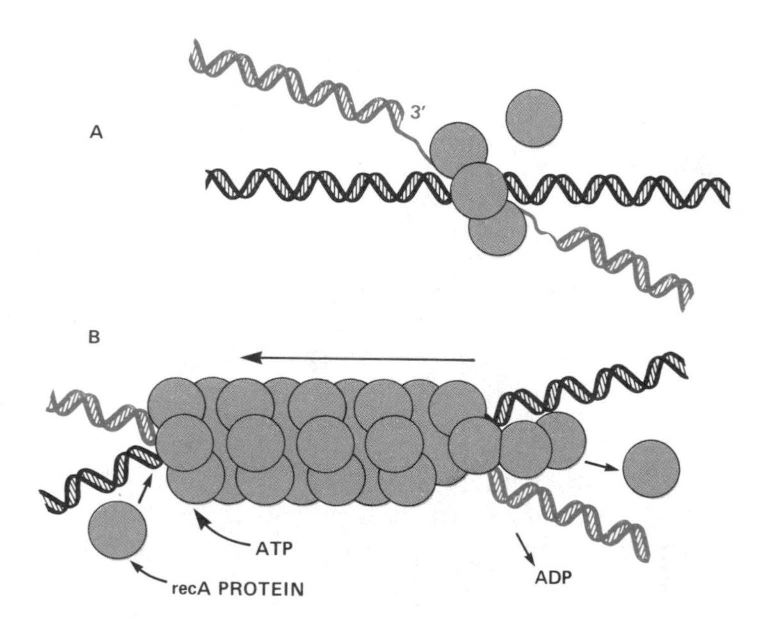

FIGURE 7-24
Diagrammatic representation of recA protein bound to DNA during genetic recombination. A few monomers binding head to tail on a single-strand gap may promote initial contacts between single-strand and duplex regions of DNA (A). Once homologous contacts between the single-strand and the duplex are made, more recA monomers bind head to tail, extending the complex toward the left (arrow) (B). ATP hydrolysis may occur at the tail end (right), where recA protein could undergo allosteric changes that lead to its dissociation. (From P. Howard-Flanders and S. C. West, ref. 55.)

and DNA duplex. Once the sheath is formed, further binding of protein monomers to the head end extends the sheath in the $3' \rightarrow 5'$ direction of the single strand, presumably extending along the single-strand gap and beyond onto the two duplexes. The hydrolysis of ATP to ADP is associated with an allosteric change in the recA protein at the tail end of the sheath, forcing the paired strands into a heteroduplex configuration (39, 55). Further studies will undoubtedly shed light on the accuracy of this model.

Other Functions Required for Discontinuous Synthesis of Damaged DNA. Aside from the *recA* and *ssb-1* genes, a number of other genetic functions in *E. coli* have been implicated in the discontinuous synthesis of DNA in UV-irradiated cells and the associated filling of postreplicative gaps. The role of some of these, such as the *recB* (20, 35, 77) (and presumably the *recC*) gene is readily understood, since they have been clearly implicated in genetic recombination (67). However, the role of the *uvrD* (20), *polA* and *polC* (68–72), *dnaG* (73), *dnaB* (74), *lexA* (34, 75, 76), *recF* (35, 76, 77) and *recL* (78) genes, all of which have been suggested to be involved in or required for "postreplication repair", is not clear (16). Certainly *lexA* or *recF* mutants are not defective in generalized recombination in the absence of other mutations (79, 80).

In Sections 4-2 and 5-3 the *lexA* gene (Fig. 7-25) was referred to as an essential component in the regulation of certain inducible responses following DNA damage in *E. coli*. The details of this induction phenomenon are addressed in Section 7-6; however, it is germane to note here that one of the several genes under *lexA* control is the *recA* gene and that some mutations in the *lexA* gene result in failure to induce expression of recA protein (81–84). Thus, defects in postreplicative gap filling of DNA in *lexA* mutants may reflect a requirement for recA protein for recombinational or other events essential for the discontinuous synthesis of damaged DNA. For example, it is well known that recA protein is required to prevent degradation of DNA following UV irradiation, and there is evidence that protection from such degradation may be afforded by a mechanism(s) under *lexA* control (85). However, since the *lexA* gene does not appear to be required for *generalized* recombination (except in cells defective in the *recB* or *recC* genes) (79, 80), the requirement for recA protein in the recombinational events associated with gap filling and in generalized recombination may be distinct.

The role of the three DNA polymerases of *E. coli* in discontinuous DNA synthesis with gap formation is also currently vague, although there are indications that DNA polymerase III is required for normal postreplicative gap closure (68–72).

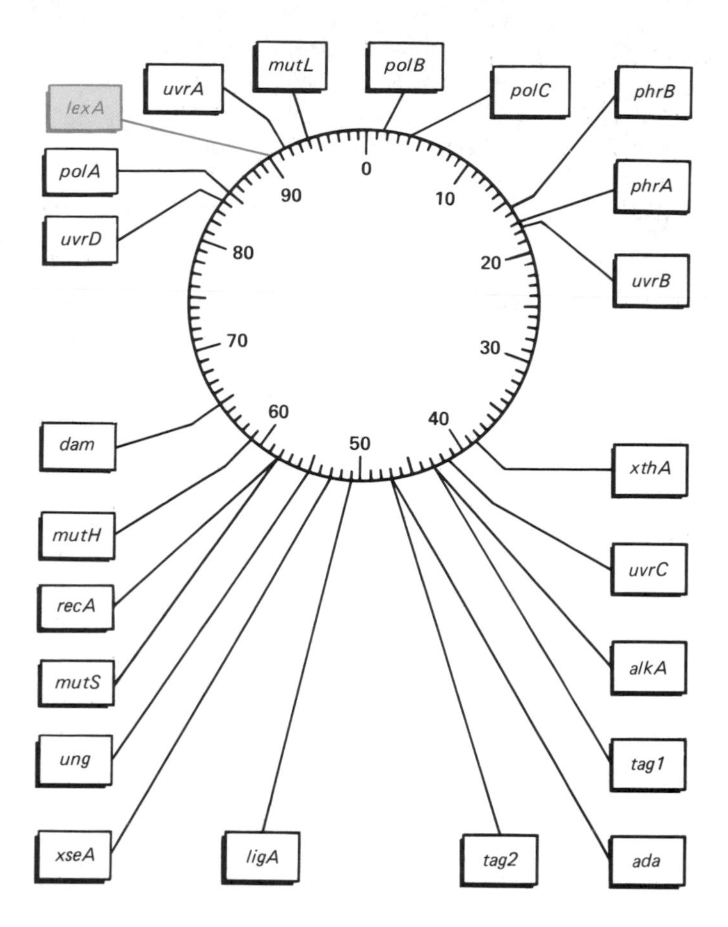

FIGURE 7-25
The location of the *lexA* gene is at approximately 90 minutes on the *E. coli* genetic map.

DNA Damage Tolerance by Continuous Synthesis on Damaged Templates: Translesion DNA Synthesis

The discontinuous mode of DNA synthesis just discussed appears to be the predominant cellular response to DNA damage in excision-defective cells during the first hour after UV irradiation of *E. coli* (12, 26). However, there are indications (less well substantiated experimentally) that *E. coli* may also be able to effect semiconservative synthesis of damaged DNA *without creating gaps*. This phenomenon is referred to as *translesion DNA synthesis* or as *transdimer synthesis* when the template damage is known to be pyrimidine dimers. However, as indicated in the introduction to this chapter, since UV-irradiated DNA contains other photoproducts, such as the (6-4) lesion described in Section 1-5, that might also act as replicative blocks, it might be more accurate to use the more general term *translesion synthesis* rather than *transdimer synthesis*, and this nomenclature will be adopted here.

During early investigations on discontinuous synthesis of DNA it was noted that if cells were pulse labeled with [^{3}H]thymidine later than an hour after irradiation, a significant fraction of the newly synthesized DNA sedimented in alkaline sucrose gradients at the position of DNA from unirradiated cells, rather than at the position of the smaller DNA observed at earlier times after irradiation (12). This is consistent with the completion of discontinuous synthesis and gap filling, so that DNA replication presumably could then occur on damage-free templates. However, it will be recalled that enzymatic probing for pyrimidine dimers in DNA after discontinuous synthesis showed these lesions to be distributed between *both* template and daughter strands. Although this redistribution should result in a progressive dilution of dimers on every chromosome with each successive round of DNA replication, it was calculated from the doses of radiation used in experiments that there should not be *any* dimer-free templates available for DNA replication without gaps within 60 minutes of irradiation (10).

Another explanation for the synthesis of high molecular weight DNA at later times after irradiation is that gap filling by recombinational events becomes more efficient with time, so that eventually short DNA intermediate structures simply are not detected (10). A more likely explanation is that specific alterations in the damaged DNA, in the DNA replication complex itself or in both lead to the *insertion of nucleotides opposite photoproducts* in template strands; i.e., translesion DNA synthesis occurs. Such DNA synthesis need not necessarily generate dimer-free daughter strands if at the same time randomization of dimer-containing fragments by recombinational events occurs; hence this explanation is not inconsistent with the presence of dimers in daughter DNA strands (10).

Translesion DNA Synthesis Is an Inducible Response to DNA Damage

The available evidence suggests that if translesion synthesis of DNA does occur in UV-irradiated *E. coli*, it is an induced cellular response that is a component of a spectrum of inducible responses collectively referred to as the *SOS responses* (81–84, 86). Subsets of these responses are also frequently referred to in the literature as "*error-prone DNA repair*" or "*inducible DNA repair*" (81). By analogy with the phenomenon of so-called *postreplication repair* discussed in Section 7-3 ("DNA Synthesis with Daughter Strand Discontinuities") these terms were coined to reflect the observation that these cellular responses are associated with *enhanced survival* of cells following their exposure to UV radiation. At the risk of being tedious, it is reiterated here that there is no evidence that induced cellular responses are associated with the repair of lesions in a strict biochemical sense, although this possibility has not been rigorously excluded.

Since these induced responses are typically associated with the replication of damaged DNA, they are also frequently treated in the literature as components of postreplication repair, and attempts have been made to distinguish between *error-free postreplication repair* (to connote the discontinuous mode of DNA replication associated with recombinational gap filling) and *error-prone postreplication repair* (to describe translesion DNA synthesis) (81). However, as indicated in Section 7-3, distinct postreplicative events are not readily identifiable since both of these responses to DNA damage may be dependent on many of the same gene functions. Thus, conceivably, cells that are incapable of repairing DNA damage by its excision may

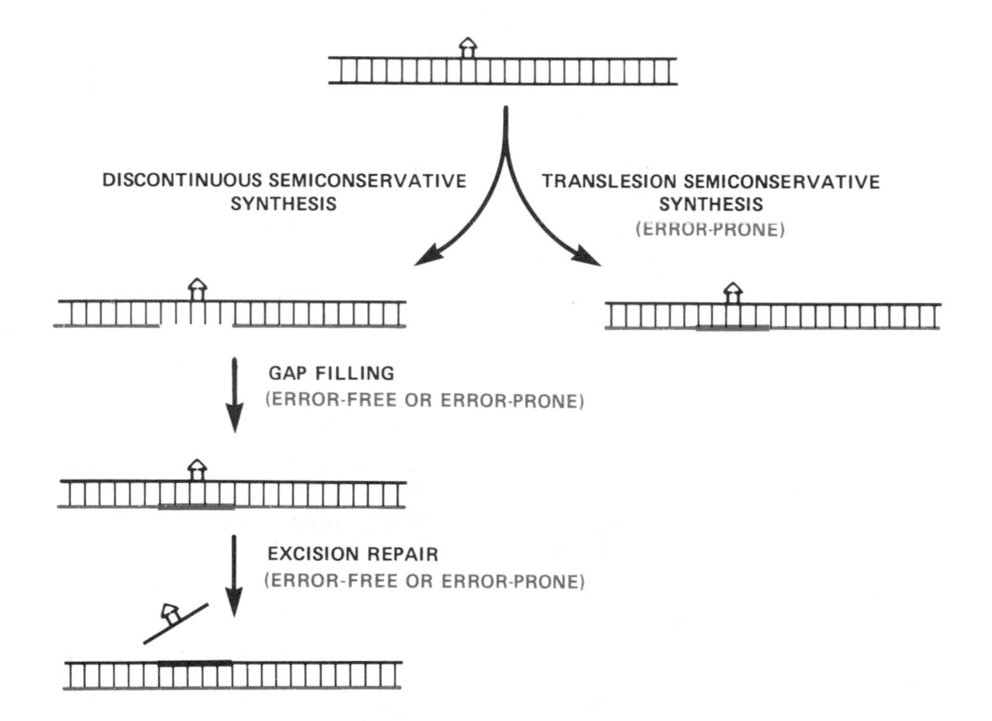

FIGURE 7-26
Mutations can arise from error-prone DNA synthesis occurring in a variety of possible situations. Translesion (continuous) semiconservative DNA synthesis may be error-prone. Alternatively or additionally, the filling of gaps during discontinuous semiconservative replication might involve error-prone *repair synthesis* or such synthesis might be prone to error during the excision of base damage, particularly at sites of closely spaced lesions on opposite DNA strands (see Fig. 7-39).

tolerate template damage by discontinuous semiconservative DNA synthesis *or* by translesion DNA synthesis. At the present time we have no idea whether these events are mutually exclusive in a given cell or in a population of cells. To add yet further difficulty to this panoply of complexities, the generation of mutations by translesion synthesis (the feature that led to the phrase "error-prone") may result from *semiconservative DNA synthesis* or from *nonsemiconservative DNA synthesis*. The latter could take the form of repair synthesis across gaps generated during discontinuous DNA replication or across gaps generated during excision repair of bulky base damage (Fig. 7-26).

Since the continuous semiconservative synthesis of DNA (translesion DNA synthesis) on damaged templates is one of the inducible responses to DNA damage collectively known as the SOS response, the succeeding sections of this chapter are devoted to a historical consideration of this cellular response to DNA damage and to its broader physiological implications. In Section 7-9 we shall return to a consideration of the translesion synthesis of DNA as a component of the SOS phenomenon and as a specific mechanism for tolerating base damage in DNA.

7-4 The SOS Phenomenon in *E. coli:* Historical and General Considerations

The idea that damage to DNA or that the physiological consequences of such damage might initiate a regulatory signal causing the simultaneous derepression of a number of genes in *E. coli*, the products of some or all of which enhance the survival of the cell or of its normal resident phages, was first formally promulgated by Miroslav Radman at a conference on mutagenesis held in Rochester, New York, in 1973 (86). The *international distress signal* (SOS), the term he appropriated to describe this phenomenon, is frequently misinterpreted to imply a last-ditch attempt by the cell to survive the lethal effects of DNA damage after other cellular responses have failed. However, a direct quote from the written proceedings of the 1973 meeting indicates that he coined the phrase SOS primarily to emphasize the dramatic nature of this cellular response to a distress signal; i.e., DNA damage.

> The principal idea is that *E. coli* possesses a DNA repair system which is repressed under normal physiological conditions but which can be induced by a variety of DNA lesions. Because of its "response" to DNA-damaging treatments we call this hypothetical repair "SOS repair." The 'danger' signal which induces SOS repair is probably a temporary blockage of the normal DNA replication and possibly just the presence of DNA lesions in the cell. (86)

The experimental evidence that prompted this hypothesis is rooted in a number of disparate areas of the molecular biology of *E. coli* and its phages, including prophage induction, phage mutagenesis and survival, host mutagenesis and survival, cell filamentation, cellular responses to DNA damage (including excision repair and DNA damage tolerance) and genetic recombination. A detailed consideration of all these phenomena is beyond the scope of this book. However, it is instructive to briefly review some of these areas in the context in which they most obviously contributed to the SOS hypothesis, before discussing the phenomenon itself and its regulation.

Studies carried out over 30 years ago on the molecular biology of lysogenization by bacteriophage λ included pedigree analysis of individual bacteria from a lysogenic strain of *B. megaterium*, in order to prove that lysogeny represents a hereditary potentiality of the bacterium to generate phage (87). It was noted that the proportion of bacteria that liberated phage changed substantially from one experiment to another, which attracted attention to environmental variables. It was subsequently demonstrated that exposure of lysogens to relatively small doses of UV light caused phage induction and lysis of cells (88). It was also shown that induction of the prophage requires conditions in which protein synthesis can occur. Since then it has been established that treatments other than exposure to UV light can induce λ lysogens. These include exposure to ionizing radiation, mitomycin C, nitrogen mustard, hydrogen peroxide and organic peroxides (89, 90). Induction also follows treatment of lysogens with the base analogues 6-azauracil (91) or 5-fluorouracil (90) and thymine starvation of thymine auxotrophs (90). In addition temperature elevation in certain mutants unable to synthesize DNA at high temperatures results in induction (92). All these perturbations share in common that they *arrest DNA replication*.

The classic studies of Francois Jacob and Eli Wollman and their coworkers established the well-known operator-repressor model for the maintenance of prophage and the regulation of lysogenic induction (93). This model proposed that the normal prophage state is maintained by the synthesis of a *repressor* that acts on a sensitive *operator*, thereby preventing expression of the structural genes required for phage production. Following induction by UV radiation or by any one of the other treatments mentioned above that inhibit DNA replication, the repressor is inactivated, structural genes are expressed that lead to phage production and cell lysis results in the liberation of normal temperate phage. This model has been supported by a large number of experiments that are beyond the scope of the present discussion.

By the late 1960s, striking parallels between the biology of phage λ induction and the SOS response in *E. coli* began to emerge from a number of independent observations. For example, Evelyn Witkin noted similarities between prophage induction in *E. coli* K-12 (λ) and filament formation (septation) in *E. coli* B (94) (Table 7-1). She pointed out that both phenomena are mass effects, occurring in virtually every member of a population of cells exposed to low doses of UV radiation. In addition, both effects can be induced by a variety of agents that damage DNA. She proposed that certain bacterial functions (including the synthesis of a protein that inhibits septum formation) may be governed by repressors similar enough to λ repressor to respond to the same inducer produced when DNA synthesis is interrupted (94). Very shortly after this suggestion it was independently observed that an *E. coli* mutant thermosensitive for prophage induction in the *absence* of DNA damage (now referred to as the *tif* mutation; see below) *formed filaments in response to DNA damage* when cured of prophage. The conditions for filament formation in the cured strain were very similar to those for induction of prophage in the lysogenic state (95). As will be seen in Section 7-6, both filamentation and prophage induction are components of the SOS phenomenon in *E. coli* (81).

Another interesting relation between λ induction and cellular responses to DNA damage emerged from a different sphere of interest: *genetic recombination* in *E. coli*. In the mid-1960s John Clark and his coworkers isolated a mutant defective in generalized recombination; the now well-known mutant *recA* strain (96). As indicated previously, this mutant is unusually sensitive to UV radiation, and Clark and his colleagues observed that induction of λ prophage by UV was defective in this strain (97) (Fig. 7-27). Subsequently it was shown that the failure to induce λ in the *recA* mutant is associated with persistence of λ repressor (98).

At the same time as a role for the *recA* gene in λ induction was established, independent studies provided evidence for its involvement in *UV mutagenesis*. It was known that when uvr^+ or *uvr* strains of *E. coli* are exposed to the same low UV fluence of radiation, neither significant killing nor mutagenesis is detected in the former group. Mutant (*uvr*) strains, however, show a high frequency of UV-induced

TABLE 7-1
Similarities between filament formation and lysogenic induction in *E. coli*

1.	Both are mass effects, occurring in virtually every member of a population
2.	Both can be initiated by a variety of DNA damaging agents
3.	Both occur occasionally in untreated cultures and in "old" cultures
4.	Both are reduced by photoreactivation after exposure to UV light at 254 nm
5.	Both are dependent on active protein synthesis

After E. M. Witkin, ref. 94.

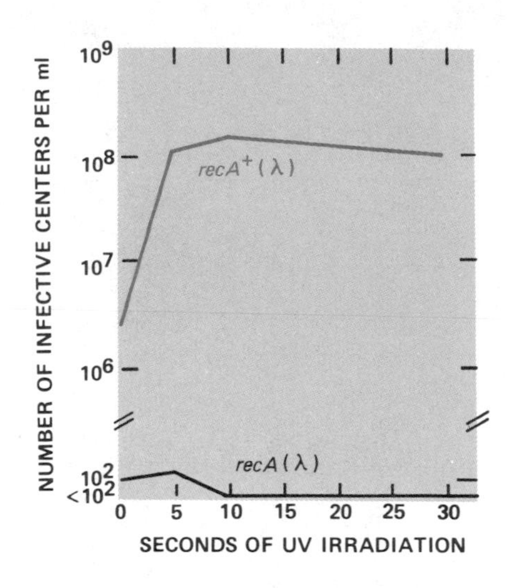

FIGURE 7-27
E. coli cells that are defective in the *recA* gene do not induce λ prophage following exposure to UV radiation. (From K. Brooks and A. J. Clark, ref. 97.)

mutations among the survivors (99–102) (Table 7-2). Since the highly mutable population is defective in excision repair, an explanation for the mutagenesis was sought in terms of other cellular responses to DNA damage. It will be recalled that at about this time the *recA* gene was also implicated as an essential function for so-called postreplicational repair of UV-irradiated DNA in excision-defective strains of *E. coli* (21). Furthermore, strains of *E. coli* defective at a locus originally called *exr* but subsequently shown to be identical to *lexA* were known to be totally nonmutable by UV radiation (80, 102) (Table 7-3), indicating that the *lexA* gene is required for an error-prone response to DNA damage. In view of the observations implicating the *recA* gene in "postreplication repair," Witkin was prompted to examine the effect of inactivation of this gene on UV mutability, and she showed that like *lexA* mutants, *recA* mutants are also UV nonmutable (103) (Table 7-3).

TABLE 7-2
Frequency of mutations into streptomycin resistance and into prototrophy following UV irradiation in uvr^+ and *uvr* strains of *E. coli*

Genotype	UV Dose (J/m^2)	Survival (%)	Mutations/10^7 Survivors	
			Str^r	Prototrophs
uvr^+	2	100.0	<0.05	<0.5
uvr	2	28.6	39.8	275.0

From E. M. Witkin, ref. 102.

TABLE 7-3
Mutagenesis in *recA* and *lexA* mutants of *E. coli*

Genotype	Mutation Scored	UV Dose (J/m^2)	Survival (%)	Mutations/10^8 Survivors Induced by UV
recA$^+$	Str^r	2.0	100	4.0
recA	Str^r	2.0	0.9	<0.008
lexA$^+$	Str^r	2.5	100	0.43
lexA	Str^r	2.5	11	<0.05

After D. W. Mount et al., ref. 80, and E. M. Witkin, ref. 103.

To summarize at this point, one can trace a historical thread that begins with studies in the early 1950s on the biology of phage λ induction and that by the late 1960s had related the *recA* gene both to this phenomenon (known to be an *inducible* function requiring synthesis of a protein repressor) and to the phenomenon of host-cell mutagenesis produced by UV radiation. In addition, by the late 1960s it was established that host-cell mutagenesis has a requirement for a functional *lexA* gene. Finally, it had been noted that there were striking similarities between the events associated with induction of phage λ in lysogenic strains and with induction of filamentation in nonlysogenic strains. To complete the historical framework for the SOS hypothesis we must return to the 1950s to consider other seminal observations on bacteriophage λ.

In 1953, Jean Weigle observed that the survival of UV-irradiated phage λ was enhanced when a *further dose of radiation was given to the phage-bacterial complexes* (Fig. 7-28) *or to the bacteria*, before or after adsorption of the irradiated phage (104). If irradiation of the bacteria was administered before phage infection, the cells retained their reactivating ability for many days when kept in buffer at 4°C. Weigle termed this phenomenon *UV restoration* and noted that among the reactivated phages a large proportion were *mutant*. He also showed that photoreactivation of the host cells prior to phage infection eliminated these effects, suggesting that pyrimidine dimers in host DNA were necessary for this phenomenon. The phenomenon is now generally referred to as *Weigle* or *W reactivation* because terms such as *UV restoration* or *UV reactivation* (an alternative phrase coined to describe it) are readily confused with *host-cell reactivation* and with *multiplicity reactivation*, which describe entirely different phenomena (105). (*Host-cell reactivation* describes the effect of constitutive host responses to DNA damage in phage genomes that do not encode their own DNA repair functions. *Multiplicity reactivation* describes the enhanced survival of phages exposed to DNA

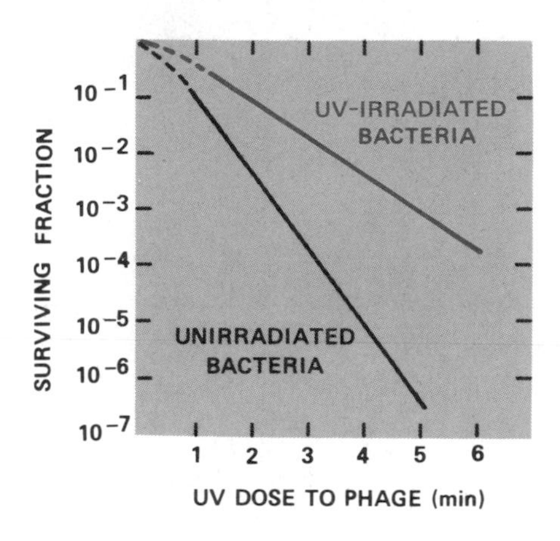

FIGURE 7-28
When UV-irradiated phage λ is plated on *E. coli* cells that were previously irradiated, the phage survival is greater than on unirradiated bacteria. This phenomenon was first described by Jean Weigle and is referred to as W (Weigle) reactivation of the phage. (From J. J. Weigle, ref. 104.)

damage following infection of cells with *multiple* phage genomes that reconstitute one or more undamaged genomes by recombinational events.) In addition, W reactivation of UV-irradiated phage can be effected by many types of DNA damage to the host (the same spectrum of perturbations that effect induction of lysogenic bacteria or that induce filamentation); hence the designation *UV* reactivation is too limited (105).

Numerous studies were carried out in efforts to explain the biology of W reactivation. A critical observation relevant to the present discussion was that strains defective in the *recA* gene are *defective in both the enhanced survival and the enhanced mutagenesis of irradiated bacteriophage* λ (106–108). Additionally, the *lexA* gene of *E. coli* is required for these responses and W reactivation also requires active protein and RNA synthesis (109). By the early 1970s a number of investigators had commented on the striking similarities between the genetic and physiological requirements for W reactivation, UV mutagenesis and prophage induction of bacteriophage λ (105). Experiments were carried out to test this correlation and attention was directed to earlier observations that were now recognized to support these correlations (105). Specifically:

1. The genetic locus mutated in *E. coli* strains that are thermosensitive for phage λ induction in the *absence* of DNA damage is called *tif* (*t*hermal *i*nduction of λ and thermal *f*ilamentation). Such strains support W reactivation and mutagenesis of UV-irradiated phage λ following exposure to elevated temperatures (110, 111). In fact, these responses are quantitatively indistinguishable from those observed using UV radiation as the inducing stimulus at temperatures *nonpermissive* for thermoinduction (110, 111) (Table 7-4).

TABLE 7-4
Mutagenesis of UV-irradiated phage λ when plated on *E. coli* tif^{+} and *tif* strains

Strain	Preincubation Temperature (°C)	Frequency of Clear Plaque Mutants ($\times 10^{-4}$)	
		Unirradiated cells	UV-irradiated cells
tif^{+}	30	3.6	25.0
	42	4.0	20.0
tif	30	6.0	23.0
	42	20.0	36.0

From M. Castellazzi et al., ref. 110.

The *tif* mutant also demonstrates enhanced *bacterial* mutagenesis following exposure to low doses of UV radiation at elevated temperatures (112) (Fig. 7-29).

2. It had been previously observed that in wild-type *E. coli* the induction of mutations by UV irradiation of excision-defective strains obeys "*two-hit kinetics,*" i.e., the slope of the mutation frequency curve shows a *quadratic* rather than a linear relation to UV radiation dose (81) (Fig. 7-29). This has been interpreted as a requirement for *two separate events for mutagenesis,* both of which are dependent on UV irradiation of the cell (i.e., DNA damage and induction of a mutagenic cellular response to that damage) (113, 114). Consistent with this interpretation, the induction of mutations at raised temperatures following UV irradiation of the *tif* mutant shows a *linear* rather than a dose-squared relation (112) (Fig. 7-29), suggesting that the requirement for one of the two UV-dependent events (i.e., induction of the functions required for mutation) is met by elevation of temperature.

3. Both UV mutagenesis and λ induction are abolished by inhibition of protein synthesis in the *tif* mutant (112).

4. If prophage induction and W reactivation are strongly correlated, one would expect that any inducing signal for the former should also induce the latter under the appropriate experimental conditions. This prediction has been borne out (105); even *indirect* induction of both W reactivation and mutation in phage λ results when an F^{-} recipient of *E. coli* infected with UV-irradiated phage receives a sex factor from a UV-irradiated donor host (115).

It is hoped that this rather brief historical account of some of the experimental observations that contributed to the formulation of the

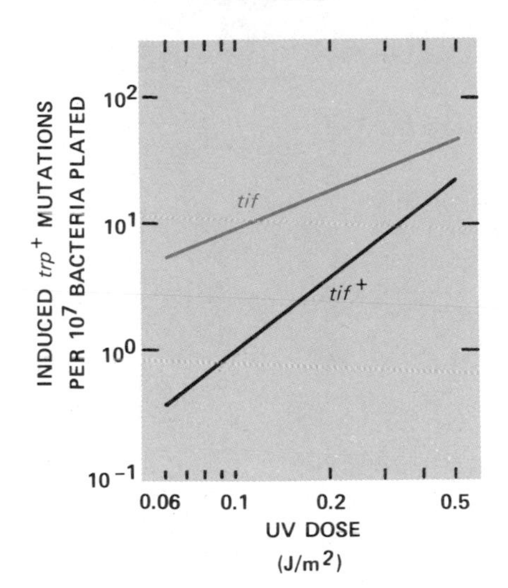

FIGURE 7-29
When wild-type (*tif*$^+$) cells are UV-irradiated and then incubated at either high or low temperatures, the slope of the curve relating mutation frequency to UV dose fits a theoretical "two-hit" curve. In contrast, *tif* mutants generate mutations at the *trp* locus with a linear dependency on UV dose; i.e., the slope fits a theoretical "one-hit" curve. (After E. M. Witkin, ref. 81.)

SOS hypothesis provides a framework for understanding the cellular and molecular biology of this intriguing phenomenon. Despite the prominent emphasis on the physiology of phage λ induction as an experimental model, heuristically it was recognized that it was unlikely that cells would have evolved a response to DNA damage specifically for the benefit of this phage, and so it was assumed that the enhanced survival operates to the benefit of the host cell as well (105). However, it was not immediately obvious how to test this experimentally, since insult to the cellular genome by an agent such as UV radiation is at the same time an inducing signal and a potentially lethal treatment. More recently, it has been shown that *preinduction* of the SOS system by small nonlethal doses of UV radiation enhances the survival of cells exposed to a subsequent dose of more extensive radiation (116).

7-5 The SOS Phenotype

Aside from the phenotype of enhanced survival and enhanced mutagenesis, the SOS phenomenon includes a number of other responses to DNA damage, all of which are inducible and all of which require functional *recA* and *lexA* genes for their expression (81–84). These include prophage induction, cell division delay (sometimes leading to filamentous growth), inhibition of the degradation of host DNA, aberrant reinitiation of DNA replication, long patch repair synthesis of DNA during excision repair and shutoff of cellular respiration. A

TABLE 7-5

Pleiotropic effects of *recA* and *lexA* mutations in *E. coli* as interpreted with the SOS hypothesis

Phenotype	*recA*$^{+}$ *lexA*$^{+}$	*recA*	*lexA*	Possible Inducible *recA*$^{+}$*lexA*$^{+}$-Dependent (SOS) Function(s)
UV sensitivity	+	+ + +	+ +	SOS "repair" activity
X-ray sensitivity	+	+ + +	+ + +	Growth medium-dependent single-strand break "repair" activity
UV inducibility of λ prophage	+	–	±	Lytic phage growth
DNA degradation after UV irradiation	+	+ + +	+ + +	Exonuclease V inhibitor
Mutagenic W reactivation of λ phage	+	–	–	SOS "repair" activity
UV mutagenesis (bacterial)	+	–	–	SOS "repair" activity
Aberrant reinitiation of DNA replication at chromosomal origin after arrest	+	–	?	Reinitiation protein(s)
Delayed cell division after UV irradiation (filamentous growth when extreme)	+ (*lon*$^{+}$) + + + (*lon*)	–	–	Septum inhibitor
Synthesis of recA protein after treatment with UV radiation, nalidixic acid, etc.	+	–	–	RecA protein
Respiration shutoff after UV irradiation	+	–	–	Respiration control protein(s)
Chloramphenicol-sensitive postreplication "repair" pathway	+	–	–	SOS "repair"
Long patch excision repair pathway	+	–	–	SOS "repair"
Thermal inducibility of SOS functions in *tif-1* mutant strains	+	–	–	All of the above
Recombination ability	+	–	+	None

After E. M. Witkin, ref. 81.

more complete list of these responses is provided in Table 7-5. It is thus apparent that a variety of treatments that damage DNA or interrupt its synthesis generate a regulatory signal, the *SOS signal.* This initiates a complex induction process culminating in the derepression of a group of metabolically diverse but coordinately regulated functions, *SOS functions,* all of which presumably promote the survival of the damaged cell and/or that of its resident phages (81–84, 105). As will be noted presently, the genes involved in the SOS response are not linked in a single operon but are scattered around the genome, and hence they are collectively referred to as the *SOS regulon.*

7-6 Regulation of the SOS Phenomenon

As indicated in Section 7-4, in order to explain the similar requirements for induction of prophage λ in lysogenic strains and of filamentous growth after UV irradiation of nonlysogenic *E. coli,* it was proposed that a bacterial gene coding a septum-inhibiting protein is regulated by a repressor similar enough to λ repressor to respond to the same inducer and that the common inducer is synthesized when DNA replication is interrupted (94). As additional SOS functions were identified and their common requirements for the *recA* and *lexA* genes were recognized this hypothesis was evolved further, and it was proposed that *all* SOS functions have repressors that respond to the same inducer, and that the *recA* and *lexA* gene products are necessary for induction of all of these functions. However, in order to understand the biochemistry of the SOS phenomenon it was clearly necessary to identify and characterize the products encoded by the *recA* and *lexA* genes.

The *RecA* Gene Product

Certain experimental conditions (which later were recognized as inducing stimuli for the SOS response) lead to markedly increased synthesis of a protein of molecular weight 40,000, which was initially called protein X, the synthesis of which is defective in *recA* and *lexA* mutants (117–119). It was soon established that protein X is in fact recA protein (120, 121). In addition, the *tif* mutation was mapped to the position of *recA* (122), indicating that the *tif* gene product is an altered form of recA protein. The observations that induction of SOS functions (either by DNA damage or, in the case of *tif,* simply by raising the temperature) is accompanied by increased

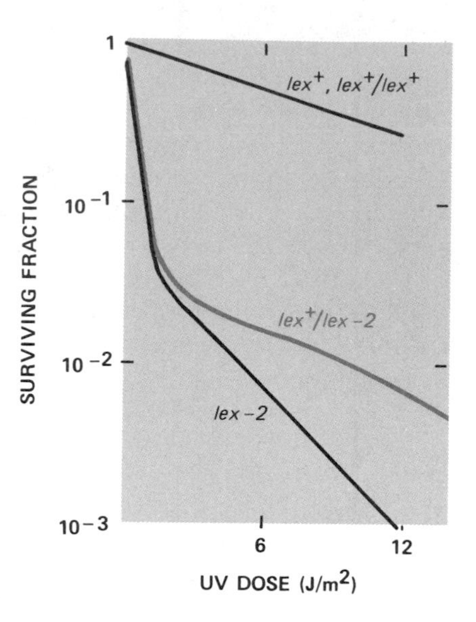

FIGURE 7-30
Mutations in the *lexA* gene are typically dominant. A mutant (*lex-2*) is abnormally sensitive to UV radiation relative to *lexA*$^+$ or *lexA*$^+$/*lexA*$^+$ strains. Following introduction of an episome carrying the *lexA*$^+$ gene, the UV resistance of the *lexA*$^+$/*lex-2* strain is not enhanced. The shallower slope of the survival curve at higher UV doses suggests that a small fraction (about 5 percent) of the *lexA-2* cells are UV resistant. These may be either *lexA*$^+$/*lexA*$^+$ homozygous segregants or haploid *lexA*$^+$ segregants. (From D. W. Mount et al., ref. 80.)

amounts of recA protein in the cells suggest that SOS induction might result in specific derepression of the *recA* gene. If so, how is expression of this gene regulated? The answer to this question came primarily from further study of the *lexA* gene. It was observed that most *lexA mutants are genetically dominant,* i.e., cells carrying both the wild-type and the mutant gene display the mutant phenotype (Fig. 7-30) and are *noninducible* by DNA damage or arrested DNA synthesis (80). These mutants are designated *lexA* or, more recently, as *lexA*(*Ind*) (for noninducible) (82).

Dominance in a mutant gene is often an indication that the gene has a regulatory function. If so, it should be possible to isolate mutants in which the regulatory function is recessive rather than dominant. In the case of the *lexA* gene this was accomplished by first isolating mutant *lexA* derivatives of *tif* strains in which the inhibition of cell division associated with SOS induction (which manifests itself as filamentation of the cells) was suppressed coordinately with suppression of phage λ induction, W reactivation and mutagenesis (123). The availability of these *lexA tif sfi* (*s*uppressor of *fi*lamentation) mutants facilitated the identification of *lexA* mutants with an inactive product in which *spontaneous SOS induction* (measured by the frequency of backward mutation in *his* mutants) occurs at both low and high temperatures (124) (Table 7-6). Such mutants cannot be lysogenized by phage λ because λ repressor is spontaneously inactivated (see Fig. 7-33). Hence this mutation is sometimes called *spr* (*sp*ontaneous *r*epressor inactivation) (124) or *lexA*(*Def*) (for de-

TABLE 7-6
Spontaneous mutator effect in *lexA* recessive mutants of *E. coli*

Genotype	Temperature (°C)	His$^+$ colonies per 10^8 viable cells
lexA$^+$ spr$^+$ tif$^+$ sfi$^+$	30	12
	40	25
lexA$^+$ spr$^+$ tif sfi	30	17
	40	145
lexA spr$^+$ tif sfi	30	6
	40	12
lexA spr tif sfi	30	145
	40	347

From D. W. Mount, ref. 124.

fective induction of λ) (82). Independent studies also showed that the expression of *lexA* is autoregulated. Thus, for example, the *lexA* gene is repressed in cells containing high levels of lexA protein (82).

All these observations led to the model for the regulation of the SOS response shown in Fig. 7-31 (82–84). The model postulates that in uninduced cells, synthesis of recA protein is repressed by the product of the *lexA* gene; i.e., *lexA protein is a repressor* that binds to the operator of the *recA* gene, to its own operator and to the operators of all other genes that are SOS inducible. The model further postulates that DNA damage and/or arrested DNA synthesis are inducing stimuli that somehow lead to the activation of preexisting (constitutive) recA protein to a form that destroys the lexA repressor, thereby derepressing its own synthesis and the synthesis of any other genes under the control of the *lexA* gene. Activated recA protein is also able to destroy λ repressor, thereby facilitating induction of prophage.

The model predicts that various alleles of the *recA* and *lexA* genes should encode products with particular biochemical alterations (Table 7-7), and this prediction has been substantiated by direct examination of a number of the relevant proteins. For example, the *recA* gene carrying the *tif* mutation encodes an altered recA protein, activation of which occurs spontaneously at elevated temperatures. Dominant *lexA*(*Ind*) mutations result in a form of lexA protein that is insensitive to activated recA protein. Similarly, the *spr* mutant encodes a form of lexA protein that is defective as a repressor; hence the designation *lexA*(*Def*). Both the wild-type *recA* and the wild-type *lexA* gene products have been purified, and their detailed biochemical characterization generally supports this model of the regulation of the SOS response (82–84).

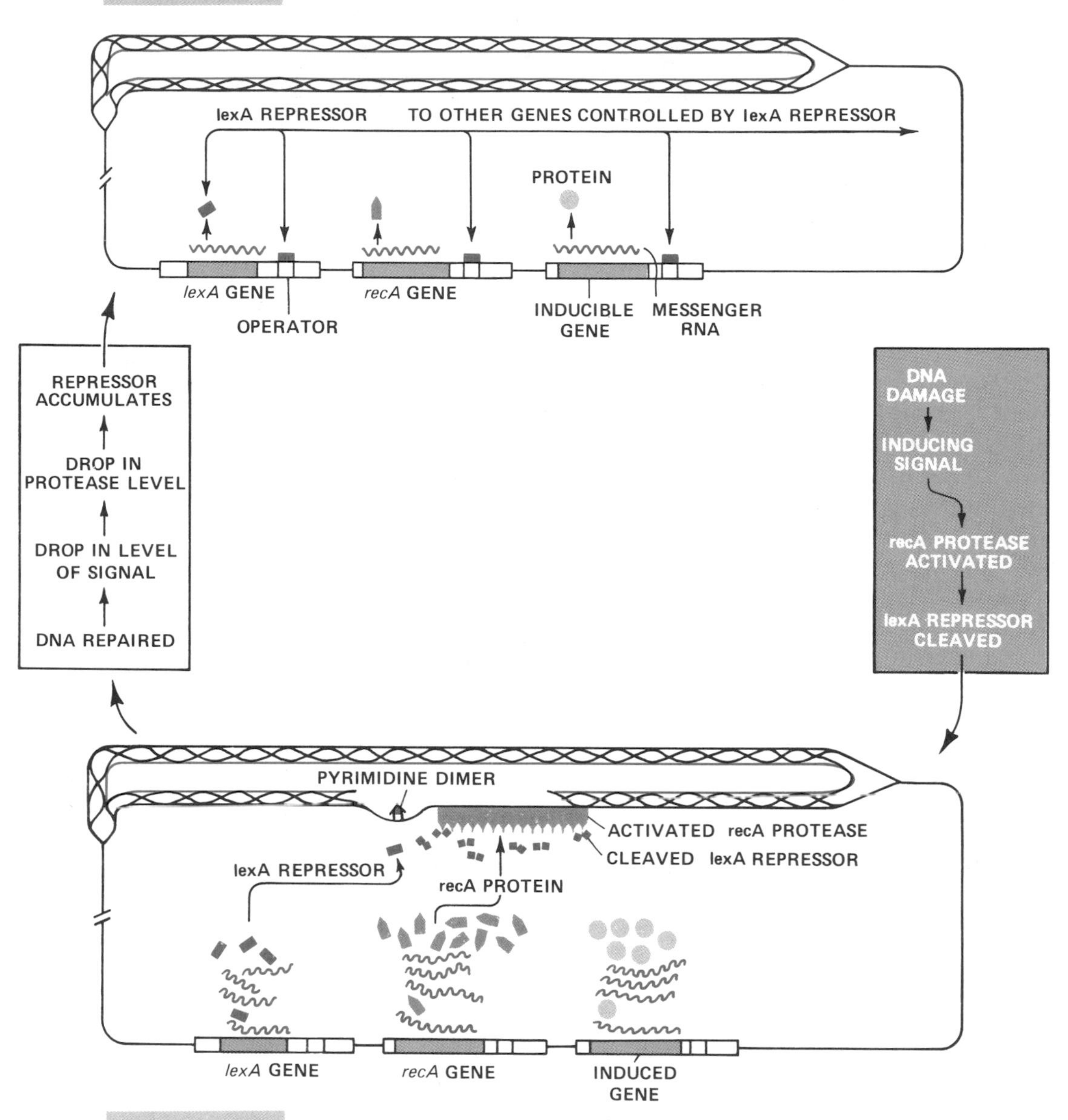
UNINDUCED STATE
lexA REPRESSOR
TO OTHER GENES CONTROLLED BY lexA REPRESSOR
PROTEIN
lexA GENE
recA GENE
OPERATOR
INDUCIBLE GENE
MESSENGER RNA
REPRESSOR ACCUMULATES
DROP IN PROTEASE LEVEL
DROP IN LEVEL OF SIGNAL
DNA REPAIRED
DNA DAMAGE
INDUCING SIGNAL
recA PROTEASE ACTIVATED
lexA REPRESSOR CLEAVED
PYRIMIDINE DIMER
ACTIVATED recA PROTEASE
CLEAVED lexA REPRESSOR
lexA REPRESSOR
recA PROTEIN
lexA GENE
recA GENE
INDUCED GENE
INDUCED STATE

FIGURE 7-31
Diagrammatic representation of the mechanism by which the *lexA-recA* regulon is regulated. In the *uninduced* state (top) lexA repressor protein constitutively expressed in low amounts is bound to the *lexA* operator and to the operators of the *recA* gene and other genes under *lexA* control. These genes are still able to express small amounts of the proteins they encode; thus there is some recA protein constitutively present in uninduced cells. Following DNA damage (e.g., the presence of a pyrimidine dimer near a replication fork after induction by UV radiation), existing recA protein is activated to a form that is required for the generation of an active protease that cleaves lexA repressor, perhaps by binding to the single-strand DNA in the gaps created by discontinuous DNA synthesis past the dimers (bottom figure). In the *induced* state (bottom) derepression of the *recA* gene results in the production of large amounts of recA protein so that more protease can be generated to continue degrading lexA repressor protein. Other genes under *lexA* control are also derepressed, although not necessarily with identical kinetics. When the inducing signal disappears (perhaps by repair of the single-strand gap), the level of active protease drops, lexA repressor accumulates and genes under *lexA* control are once again repressed. (After P. Howard-Flanders, ref. 84, and J. W. Little and D. W. Mount, ref. 82.)

TABLE 7-7
Alleles of *recA* and *lexA*

Allele*	Phenotype	Biochemical Alteration
recA alleles		
recA	Recombination deficient; cannot induce SOS functions; very sensitive to DNA damage	Defective protein
recA441 (*tif-1*)	Spontaneous SOS expression at 42°C without DNA damage	Protease more easily activated
recA430 (*lexB30*)	Recombination proficient; cannot induce λ or SOS	Specific defect in protease, binds to single-strand DNA
recA142	Recombination deficient; λ shows spontaneous induction	More defective in DNA-binding activities than in protease
recAo	Cis-dominant constitutive *recA* expression	Operator mutant
lexA alleles		
lexA(Ind) (*lexA*)	Dominant to *lexA*$^+$; no induction of SOS functions; sensitive to DNA damage	Protease-resistant repressor
lexA(ts) (*tsl*)	Recessive; spontaneous filamentation and *recA* expression at 42°C	Thermosensitive repressor
lexA(Def) (*spr*)	Recessive; constitutive for *recA* and other *lexA* target genes; constitutive for most SOS functions in combination with *recA441*	Defective repressor

*The designations in parentheses are those previously used in the literature. The various alleles of *lexA* have been renamed to conform more closely to the current understanding of their function. Ind, ts and Def stand for noninducible, thermosensitive and defective, respectively.
From J. W. Little and D. W. Mount, ref. 82.

RecA Protein Is a Protease or Is Required for Proteolysis of Repressors

In Section 7-4 evidence was presented that recA protein interacts with DNA in ways that are consistent with its role(s) in genetic recombination. Several lines of evidence also indicate that constitutive recA protein must be *activated* to function as a protease during induction of SOS genes (82–84). For example, a number of SOS functions (including prophage induction) are *not* efficiently induced merely by the presence of high levels of wild-type recA protein generated by *recA* operator-constitutive mutations (that allow constitutive expression of the *recA* gene) or by the transformation of cells with multicopy plasmids containing the cloned *recA* gene (125, 126). Furthermore, in a merodiploid strain carrying both the $lexA^+$ gene and the dominant *lexA*(*Ind*) mutant gene (the latter codes for a protease-resistant repressor; see Table 7-7), rapid degradation of the product of the $lexA^+$ but not of the *lexA*(*Ind*) gene product is observed under inducing conditions (127). Since the *lexA*(*Ind*) repressor remains intact, little if any amplification of recA protein can occur in this strain, implying that *constitutive* recA protein must have become activated. Thus a new modified form of recA protein is not synthesized as a result of SOS induction; rather some alteration of native recA protein takes place after its synthesis that results in its activation.

Further support for the notion that recA protein is activated to a protease under appropriate conditions comes from an examination of the nucleotide sequence of the cloned *recA* gene (36, 128). The predicted amino acid sequence in the amino-terminal region of the protein reveals striking homologies with the active sites of several serine proteases (36). Additionally, three cysteine and two histidine residues are present in this region. These amino acids are found in the active site of most known *E. coli* proteases (129).

RecA Protein Inactivates λ Repressor

Lambda repressor exists in solution in equilibrium between monomers and dimers ($K_d = 10^{-8}$ *M*). At the concentration found in a lysogen (about 10^{-7} *M*) most of the repressor is in the form of dimers. Only the dimerized form binds to the phage operators with high affinity and specificity (82). The monomer has two functional and physical domains (Fig. 7-32). The amino-terminal domain determines recognition of operator sites, and a carboxy-terminal domain provides most of the contacts important for dimerization (82). The cleavage site for recA protease lies between these two domains in a "hinge" region that is sensitive to many proteases (Fig. 7-32). Since only dimers bind tightly to operators, separation of the two domains

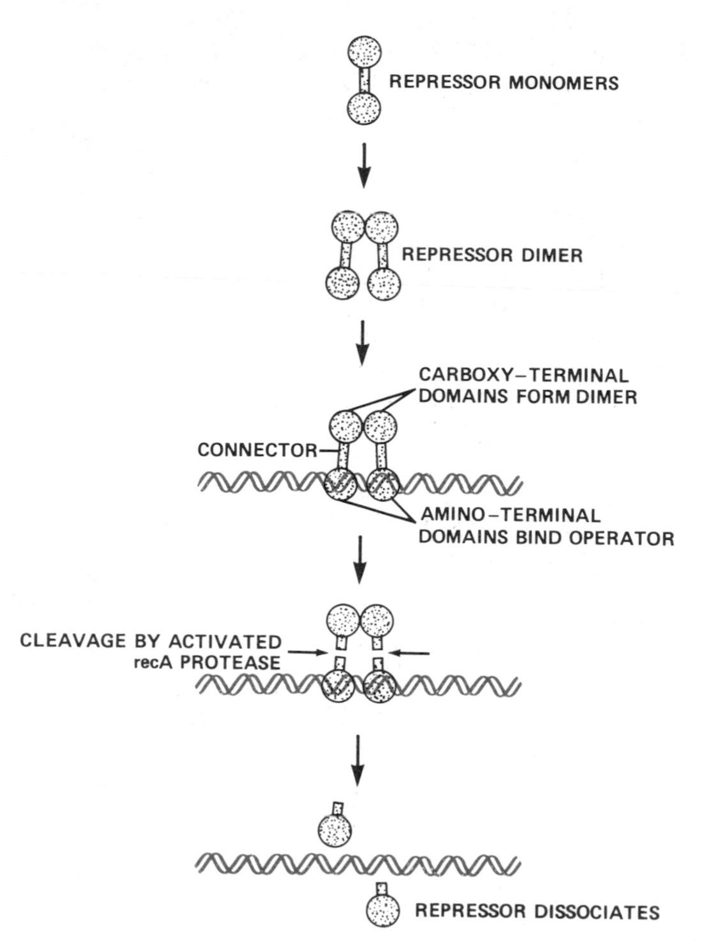

FIGURE 7-32
Mechanism of inactivation of phage λ repressor. The repressor protein exists in a monomeric form but dimerizes through interactions at its carboxy-terminal domains. In the dimerized state repressor binds to the operator in DNA through interactions involving the amino-terminal domains of the proteins. Cleavage by activated recA protease (or possibly some other protease for the activation of which recA protein is essential) occurs in the connector regions of the proteins between the ends. (After B. Lewin, ref. 191.)

by proteolytic cleavage prevents repressor dimerization and thereby inactivates the protein (82).

That λ repressor is indeed cleaved proteolytically during prophage induction was established with experiments in which the fate of the repressor was followed in extracts of induced cells (130–132). As induction proceeds, repressor monomer protein progressively disappears and concurrently a protein fragment approximately one half the size of the monomer appears, suggesting a precursor-product relationship between the two. Repressor synthesized by a noninducible (*ind*$^-$) phage is not cleaved in vivo and cleavage is also not observed in a mutant *recA* host. Furthermore, extracts of the *spr* mutant (which constitutively expresses phage induction and other SOS genes) contain a proteolytic activity that cleaves wild-type phage λ repressor in vitro in an ATP-dependent reaction (Fig. 7-33). Under conditions of in vitro cleavage of λ repressor (133, 134), two fragments of approximately equal size are recovered (Fig. 7-34), in contrast to the single fragment detected in vivo, which suggests that λ repressor is inactivated by a single cleavage event near the middle of the repressor monomer and that one of the fragments is degraded in the cell.

Cleavage of λ repressor directed by highly purified recA protein is dependent on the presence of both ATP and polynucleotide (133, 134). The ATP analogue ATP[γS] substitutes effectively for ATP in vitro and in fact supports a severalfold higher rate of cleavage than does ATP. In addition, a variety of single-strand polynucleotides support the cleavage of repressor by recA protein, including oligonucleotides as short as six nucleotides in length, DNA restriction fragments, circular DNA molecules and polyribonucleotides and polydeoxyribonucleotides (134). Native DNA is much less effective, particularly in reactions at high salt concentrations (134). Cleavage of λ repressor is catalytic and the rate depends on the ratio of recA monomer to nucleotide added to the reaction mixture, reaching a maximum at an approximate ratio of six nucleotides to one monomer (134).

How does recA protein become activated to a protease? The purified protein binds ATP[γS] tightly in the presence of DNA, probably

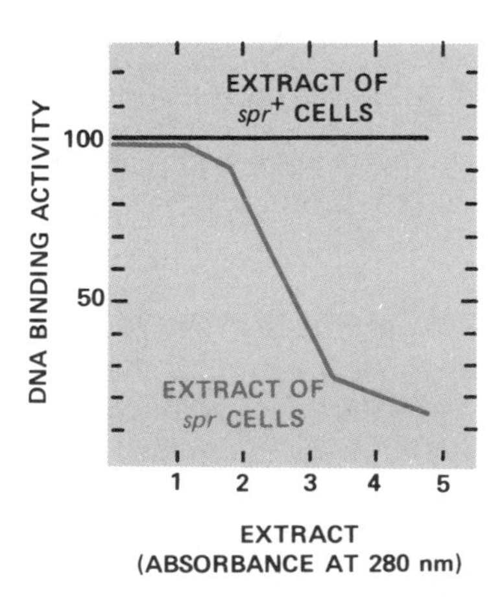

FIGURE 7-33
Extracts of wild-type (*spr*$^+$) cells do not contain activated protease, and hence when they are incubated with purified λ repressor in the presence of ATP, no inactivation of repressor (measured by binding to ^{32}P-labeled λ DNA) is observed. However, extracts of mutant *spr* cells contain activated protease constitutively, and these extracts degrade λ repressor in vitro. (From J. W. Roberts et al., ref. 131.)

in a ternary complex of all three components (134). An equivalent complex may form with natural triphosphates, and binding of recA protein into this initial complex *without* hydrolysis of the triphosphate may invoke its proteolytic activity against λ repressor. This model is consistent with the observation made above that ATP[γS] stimulates proteolytic activity of recA protein more efficiently than does ATP, yet is far less sensitive to hydrolysis (133). Furthermore, less than one ATP[γS] molecule is hydrolyzed for each repressor monomer cleaved, indicating that the hydrolysis of ATP[γS] is not directly coupled to the cleavage event (133). In vivo, recA protein may become activated by interaction with single-strand regions of DNA or with small oligonucleotides generated by the degradation of damaged DNA.

Most studies on the *recA* gene product in vitro have been carried out using recA (tif) protein derived from the *spr tif sfi* strain, in which expression of the protein is conveniently constitutive. However, in vivo the mechanism of activation of recA and of recA (tif) proteins may not be identical, since differences in activation of these two proteins have been observed in vitro (135, 136). For example, short

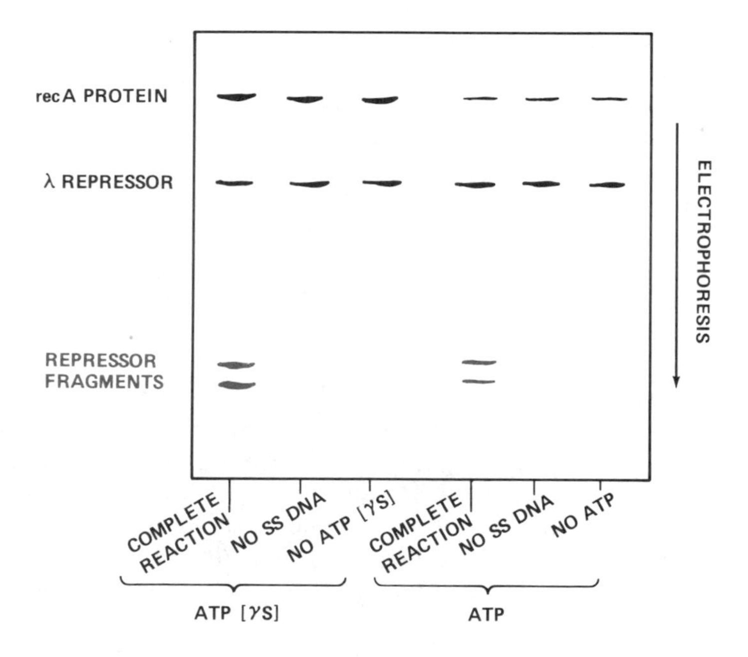

FIGURE 7-34
Cleavage of λ repressor protein in vitro has a requirement for recA protein, single strand DNA and ATP or ATP[γS]. Following gel electrophoresis of incubation mixtures, recA protein λ repressor and degraded λ repressor can be observed in the appropriate lanes. (From N. L. Craig and J. W. Roberts, ref. 134.)

oligonucleotides such as poly(dT)$_{\overline{12}}$ or poly(dA)$_{\overline{14}}$ stimulate neither the protease nor the ATPase activities of wild-type recA protein (135). In contrast, these short oligonucleotides do activate recA (tif) protein to cleave λ repressor without stimulating its ATPase activity (135). Moreover, both the ATPase and protease activities of recA (tif) protein are stimulated by poly(rU) and poly(rC), whereas wild-type recA protein does not respond to these ribopolymers (135). Thus, in vivo, activation of recA (tif) protein may occur more readily by short single-strand regions or gaps in replicating DNA. Wild-type protein may require oligonucleotides generated by the degradation of damaged DNA. This may explain why *tif* mutants can be induced in the absence of DNA damage to the cell (135).

RecA Protease Also Inactivates LexA Repressor

Based on the information garnered from studies on the inactivation of λ repressor, the model of SOS regulation discussed above would predict that in the presence of activated recA protease lexA protein (the putative repressor of the *recA* gene and of other SOS genes) is cleaved. A number of experiments support this prediction. For example, lexA protein can be detected as an intact entity of the appropriate molecular weight after infection with *lexA*$^+$ transducing phages of cells containing little or no active recA protease (127). However, after infection of cells containing high levels of recA (tif) protein (which is more active as a protease than recA protein), intact lexA protein cannot be recovered (127). Analogously, when *tif* cells are infected with *lexA*$^+$ transducing phage, repression of *recA* does not occur, whereas in cells containing less active forms of recA protein SOS functions are repressed following such infection (127).

LexA protein has a molecular weight of 22,700 (137, 138). The amino terminus of the protein is the portion that binds to DNA (139). In vitro, at concentrations of 10^{-5} to 10^{-6} *M*, lexA protein is a dimer (140), and there is evidence that, like the phage repressor, lexA protein requires an intact carboxyl end to dimerize efficiently (141, 142).

Cleavage of the lexA polypeptide has been demonstrated in vitro using maxicell extracts (see Section 4-2) containing radiolabeled lexA protein (143). The reaction is dependent on the addition of both recA protein and ATP or ATP[γS]. The 22,700 molecular-weight protein gradually disappears during the reaction, and concurrently two smaller species can be identified. LexA (Ind) protein is largely resistant to cleavage in this reaction (143), thus providing a satisfying biochemical verification of the dominant phenotype of *lexA* (Ind) mutants. Wild-type lexA protein has been purified to better than 95 percent homogeneity from cells harboring a recombinant plasmid carrying the *lexA* gene under the control of the *E. coli lac* promoter

(144). The amino acid composition of the purified protein and its amino-terminal sequence are in agreement with that predicted from the known nucleotide sequence of the *lexA* gene (144). LexA protein is cleaved into two polypeptides during incubation with purified recA protein in the presence of ATP and single-strand DNA; the site of this cleavage is between Ala^{84} and Gly^{85} of the polypeptide (144) (Fig. 7-35).

In the discussions presented here on the derepression of the SOS regulon it has been assumed that the proteolytic degradation of both the lexA and the λ repressor protein is effected *directly* by recA protein acting as a protease. Although a significant body of experimental data is consistent with this view, very recent studies (139) suggest that while recA protein is clearly *required* for repressor degradation, the catalytically active protease may be constituted by the *repressor proteins themselves* rather than by recA protein. Thus further studies may be required to establish the precise role of recA protein in repressor cleavage during SOS induction.

Purified LexA Protein Represses Both the *recA* and the *lexA* Gene

In vitro lexA protein binds to regulatory sequences in both the cloned *recA* and *lexA* genes (141, 143, 145). With these binding reactions 20 bp palindromic *operator sequences* (so-called SOS boxes) have been identified in these genes (Fig. 7-36) that are protected by lexA protein against nucleases and against certain chemical alkylations (82). The *recA* gene has a single such box, and the *lexA* gene has a tandem pair (Fig. 7-36). There is considerable homology between the binding sites in these two genes with sequences in the 5′ noncoding regions of other SOS genes (82). No two operator sites are identical in their nucleotide sequence, although every site so far examined has a number of bases in common (142) (Fig. 7-36). *LexA* operator sites in the regulatory regions of different genes occur at different positions relative to the promoter; in some genes the operator is near the start point of transcription whereas in others it is farther upstream (142).

Repressor activity of the lexA protein has been demonstrated in vitro by showing specific inhibition of transcription from both the *lexA* and the *recA* promoter and from other cloned genes that are part of the SOS regulon (145, 146). These observations support the contention that during normal growth of *E. coli* the genes of the SOS regulon are repressed by lexA protein, which is maintained at a constant level by the autoregulatory repression that the protein exerts on its own synthesis (Fig. 7-31). This autoregulation provides a

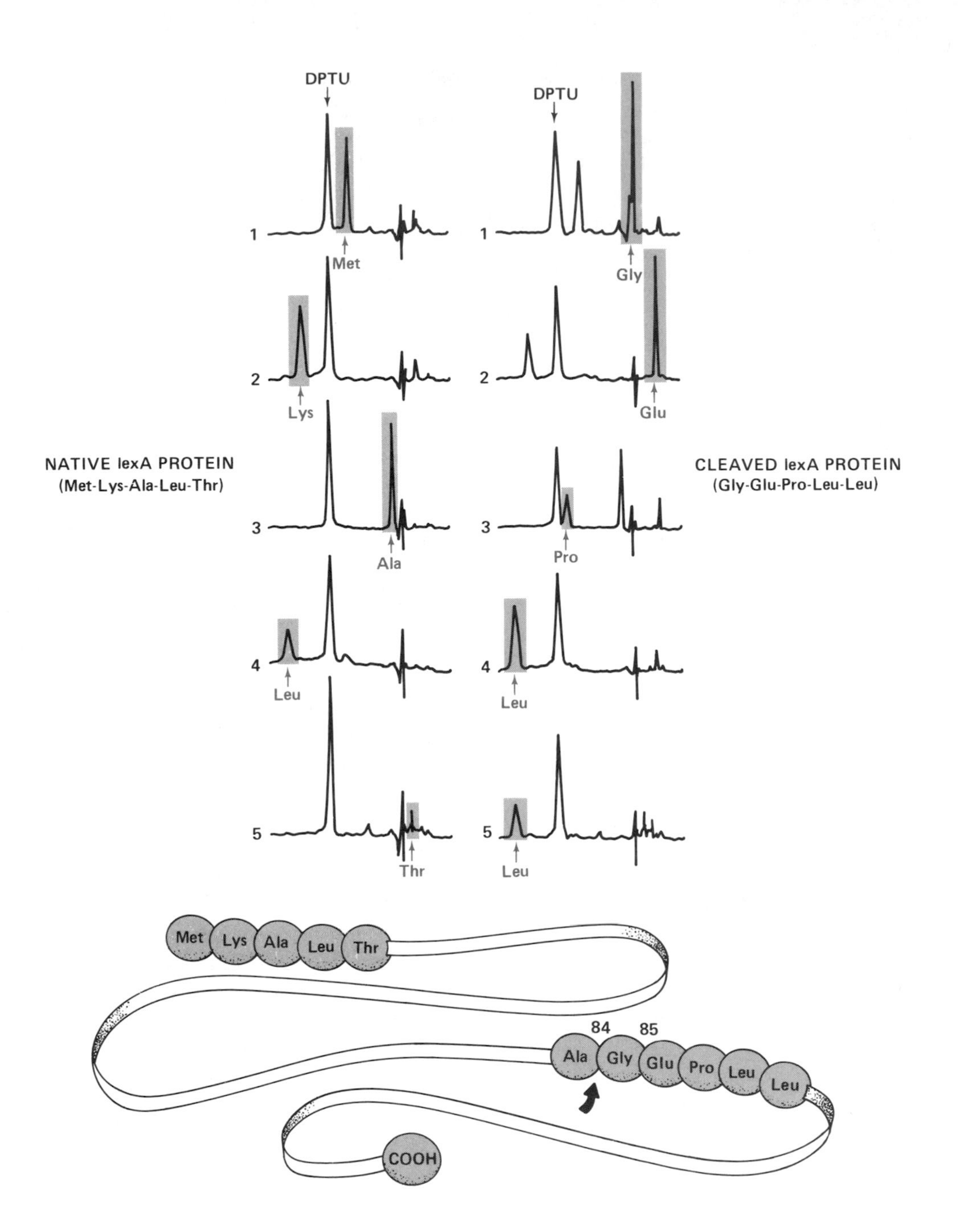

DPTU
Met
Lys
Ala
Leu
Thr
DPTU
Gly
Glu
Pro
Leu
Leu
1
2
3
4
5
NATIVE lexA PROTEIN
(Met-Lys-Ala-Leu-Thr)
CLEAVED lexA PROTEIN
(Gly-Glu-Pro-Leu-Leu)
Met
Lys
Ala
Leu
Thr
84
85
Ala
Gly
Glu
Pro
Leu
Leu
COOH

FIGURE 7-35
The site of cleavage of lexA protein by activated recA protease is between Ala[84] and Gly[85] of the peptide (lower figure). LexA protein was incubated with recA protein in the presence of ATP[γS] and single-strand DNA. After incubation the reaction mixture was dialyzed, lyophilized and subjected to partial degradation by the Edman procedure. Phenylthiohydantoin derivatives were determined by high-performance liquid chromatography and were detected by absorbance at 269 nm (right panel). Diphenylthiourea (DPTU) is a by-product of the Edman degradation. In this way the amino acid sequence at the amino terminus of the cleavage product (and hence the site of cleavage) was determined. The left panel shows confirmation of the previously established amino-terminal sequence of native lexA protein. (After T. Horii et al., ref. 144.)

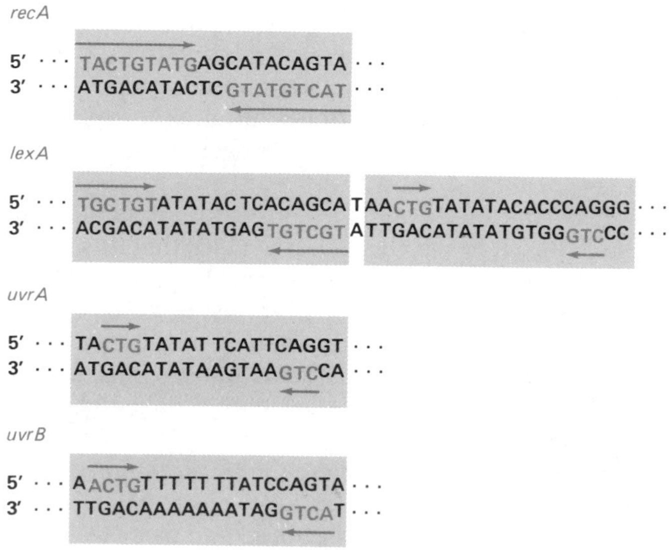

---

recA

5′ · · · TACTGTATGAGCATACAGTA · · ·
3′ · · · ATGACATACTCGTATGTCAT · · ·

lexA

5′ · · · TGCTGTATATACTCACAGCA TAACTGTATATACACCCAGGG · · ·
3′ · · · ACGACATATATGAGTGTCGT ATTGACATATATGTGGGTCCC · · ·

uvrA

5′ · · · TACTGTATAT TCATTCAGGT · · ·
3′ · · · ATGACATATAAGTAAGTCCA · · ·

uvrB

5′ · · · AACTGT TT TT TTATCCAGTA · · ·
3′ · · · TTGACAAAAAAATAGGTCAT · · ·

FIGURE 7-36
Operator regions near the beginning of the *recA*, *lexA*, *uvrA* and *uvrB* genes have similar base sequences, about 20 base pairs long, that are binding sites for the lexA repressor. All the binding sites (sometimes referred to as *SOS boxes*) include inverted repeat sequences (arrows). The *lexA* gene has two nearly identical SOS boxes. (From P. Howard-Flanders, ref. 84.)

quick return to a steady-state level of repressor following removal of the inducing signal (82, 83, 141, 147).

The type of biochemical investigations just described are beginning to yield interesting clues as to how finer regulation of the SOS response may occur (82, 83, 141, 142, 147, 148). For example, recA protease cleaves lexA protein more efficiently than it does λ repressor. This may favor survival of the phage by ensuring that lysogenic induction does not result until levels of DNA damage exceed the capacity of the other inducible responses. Another interesting observation is that the operator site (SOS box) in the *recA* gene binds lexA protein more strongly than do the operator sites in the *lexA*, *uvrB* and *sfiA* genes (141, 142, 147, 148). These observations suggest that different SOS genes with variations in operator nucleotide sequence may be derepressed after different amounts of lexA protein have been inactivated (82, 83, 141, 142, 147, 148). Due to the lower affinity of lexA protein for the binding site in the *lexA* gene relative to that for the *recA* gene, the *lexA* promoter should become derepressed *before* the *recA* promoter becomes proportionately derepressed. This relative increase in lexA protein synthesis may prolong the period during which its level is dropping, thereby permitting finer discrimination for switching on part but not all of the SOS regulon. Lower repressor affinity of the *lexA* operator relative to that of other operators might be expected on theoretical grounds alone, since the gene presumably must synthesize sufficient product to ensure repression of the rest of the regulon before it switches off its own expression (141, 142, 147, 148).

At least two genes in the *recA-lexA* regulon (*lexA* and colE1) have two operator sites. It has been suggested (142) that this may allow for cooperative interactions between bound lexA protein molecules, and, in fact, cooperativity has been observed in vitro (141, 142). Such cooperativity might serve as a regulatory mechanism for derepression of these genes, since cooperative interactions would help to maintain repression until *all* lexA protein in the cell had been eliminated by proteolysis (142).

At least 80 percent of pulse-labeled lexA protein is cleaved within three minutes after exposure of cells to low levels of UV radiation. Thus activation of the protease is prompt and cleavage of this repressor is a very early step in SOS induction (82, 148). In contrast, it takes about 20 minutes for cleavage of λ repressor to be more than 90 percent complete (130). These differences correlate with the observation that lexA repressor is a much better substrate for recA protease than is λ repressor (143). Another observation consistent with the rapid inactivation of lexA protein by SOS-inducing treatments is that both recA and lexA mRNA synthesis occurs within five minutes after an inducing stimulus (82, 149).

The kinetics of the return to normal growth conditions after induction has also been investigated. Between 30 and 60 minutes after nonlethal doses of UV irradiation, the rate of recA protein synthesis declines (150). The rate of synthesis of recA mRNA also diminishes after about 30 minutes and that of lexA mRNA about 20 minutes later, consistent with the weaker affinity of lexA repressor for the *lexA* operator relative to that for the *recA* operator.

7-7 Repertory of SOS Genes

How many different genes in *E. coli* are controlled by the *recA-lexA* regulatory mechanism? One way to approach this question is to identify genes that are induced by DNA damage and then determine whether their induction is dependent on the *recA* and *lexA* functions. This has been achieved by the use of a specialized transducing phage that can act as a single-step operon fusion vector. The phage, designated Mud, is a derivative of the temperate bacteriophage Mu that integrates into the bacterial chromosome with no appreciable site specificity (151–153). The phage carries the lactose structural genes but no promoter capable of initiating their transcription (Fig. 7-37). However, when the phage integrates within a bacterial transcriptional unit, the *lac* genes can be expressed as a result of continued transcription into the phage genome. Such an insertion creates an operon fusion in which the synthesis of β-galactosidase is under the control of a cellular regulatory locus (151, 152). Loci involved in the SOS response can thus be identified by the generation of random

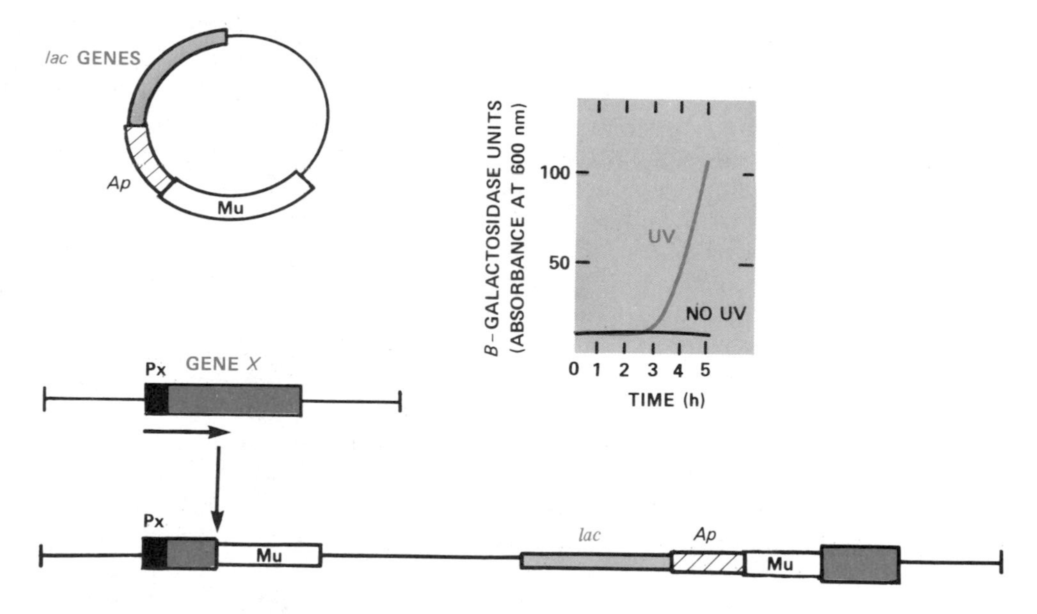

FIGURE 7-37
Diagrammatic representation of a derivative of phage Mu (Mud, or *lac* Ap^r) which contains the *lac* genes without a promoter and also the ampicillin-resistance (Ap) gene. By random insertion into *E. coli* gene *X* the phage can bring the expression of the *lac* genes under the control of the *X* promoter (Px). If expression of gene *X* is induced by DNA damage, this can be monitored by observing enhanced synthesis of β-galactosidase in the induced cells (inset).

TABLE 7-8
DNA damage-inducible genes in *E. coli*

Gene	Map Location	Function
recA	58′	General recombination; protease (?)
lexA	91′	SOS repressor
uvrA	92′	Excision repair
uvrB	17′	Excision repair
uvrC	41.5′	Excision repair
uvrD	84′	Excision repair (?); mismatch repair
umuC,D	25′	Mutagenesis
sfiA	22′	Cell division inhibitor (?)
himA	38′	Site-specific recombination
dinA	2′	Unknown
dinB	~8′	Unknown
dinD	80–85′	Unknown
dinF	91′	Unknown

From J. W. Little and D. W. Mount, ref. 82.

gene fusions and then direct screening for cells in which Mud has inserted, so that the *lac* genes are expressed in response to DNA damage. In certain cases the *function* of such loci can be deduced because insertions that occur within a gene or proximal to its transcriptional unit abolish the function of that gene. With this experimental approach a number of so-called *din* (*d*amage *in*ducible) genes have been identified in *E. coli*. Two of these (obviously not detectable by the experiments described above since they are *required* for induction) are *lexA* and *recA*. Six others have been established as the *uvrA*, *uvrB*, *uvrC*, *sfiA*, *umuC,D* and *himA* genes (151–157) (Table 7-8). It has also been independently established that the *uvrC* (158) and *uvrD* (159) genes are inducible by DNA damage (Table 7-8), presumably under control of the *recA-lexA* regulon. The observation that all three genes of *E. coli* required for excision repair of bulky base damage (*uvrA*, *uvrB* and *uvrC*) are inducible suggests that the sensitivity to UV radiation of *recA* mutants must be interpreted both in terms of reduced excision repair and in terms of defective replicative bypass with gap filling. The nature and function of the *dinA*, *dinB*, *dinD* and *dinF* genes (Table 7-8) remain to be established.

The *umuC,D* gene is required for UV and chemical mutagenesis in *E. coli* since *umu* mutants are nonmutable by these agents (160–163). The *himA* gene is required for site-specific recombination such as the integration of phage λ DNA into the *E. coli* chromosome. LexA protein is a repressor of all these loci, and the *recA* gene is required for their induction.

7-8 Signal for SOS Induction

It has not been precisely established how the numerous perturbations known to induce the SOS regulon actually signal the activation of constitutive recA protein to an active form required for degradation of repressor (82). As noted previously, many of the agents that induce SOS genes cause base damage to DNA. However, DNA damage can have a number of *secondary consequences*, including replicative arrest, alterations in nucleoid structure or altered superhelicity of DNA, to mention just a few. In addition, replicative arrest may be associated with particular alterations of DNA structure due to the accumulation of replicative intermediates. Finally, the degradation of DNA associated with DNA damage (and possibly with arrested replication) could generate oligonucleotides and other small molecules, as well as alterations in the pool sizes of various DNA and nucleotide precursors (82).

The observation that cleavage of the lexA repressor is a rapid process argues against the involvement of systems that regulate relatively slow cellular processes (such as cell division or the timing of chromosomal replication) for the activation of recA protein (82). However, the direct analysis of rapid changes in induced cells is experimentally difficult because proof that a particular alteration represents an inducing signal requires the ability to effect such an alteration at will and to perturb both the metabolism of the signal and the nature of its target (82).

The notion that DNA degradation products may serve as an inducing signal has received considerable attention. Products of the degradation of newly synthesized DNA by exonuclease V (the products of the *recB* and *recC* genes of *E. coli*) result in the induction of the SOS response (164, 165). However, exonuclease V is apparently not absolutely required for induction since recA protein can be induced by UV irradiation of mutants (*recB*) that are defective in this enzyme (150). Further evidence in support of the role of DNA degradation in the induction process comes from studies with nalidixic acid. This compound is an effective inducer of some (but not all) SOS functions (119) and also results in the selective degradation of newly replicated DNA by binding to DNA gyrase subunit A. Furthermore, treatment of cells with novobiocin, which leads to the inhibition of DNA replication by inactivation of the B subunit of DNA gyrase *without* causing DNA degradation (166), does not result in induction of SOS functions (164).

To test the hypothesis that DNA degradation products can serve as an inducing signal, attempts have been made to derepress the *recA* gene in vivo by providing exogenous DNA fragments in a variety of ways, for example, by infecting cells with phages containing unmodified DNA, which should be degraded in the cells. However, infection with either unmodified phage λ or phage T4 does not result in efficient induction of recA protein synthesis (150, 167). Another experimental system has utilized permeabilized cells for delivering exogenous oligonucleotides (168). This system has revealed a remarkable specificity for prophage ϕ80 induction by particular oligonucleotides (Fig. 7-38). The prophage is efficiently induced by the addition of two classes of guanine-containing deoxyoligonucleotides. One class contains specific deoxydinucleotides having the sequence A (or G or I)-G; the other contains oligodeoxyguanylates with a 5′ phosphate and a chain length of 6 to 18 (168). It is difficult to evaluate the general significance of these findings because it is not known whether the molecular mechanism of the induction of phages ϕ80 and λ are identical. However, the induction of phage ϕ80 in the permeabilized cell system does require an active *recA* gene.

The phenomenon of *indirect induction* provides a useful model with which to examine the nature of the inducing signal, since damage to the host cell is avoided. Following conjugational transfer of

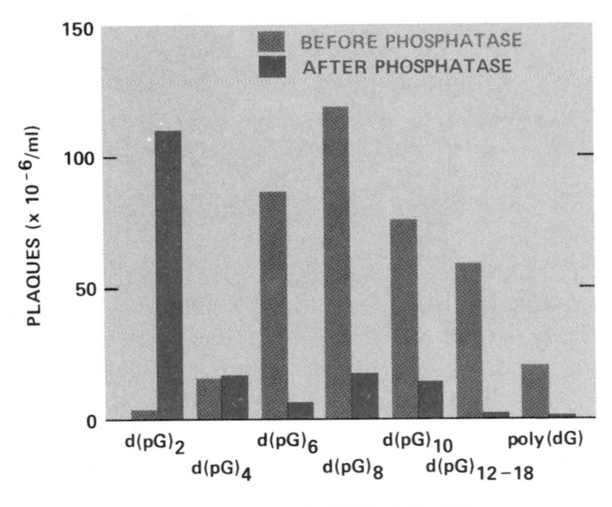

FIGURE 7-38
Induction of prophage ϕ80 can be effected by a number of oligodeoxyguanylyl phosphates introduced into plasmolyzed cells. Most of the oligomers require a 5′ terminal phosphate since inducing activity is lost after treatment with alkaline phosphatase. An exception is the dinucleotide d(pG-G), which only becomes activated when *converted* into d(G-G) by phosphatase treatment. (From M. Oishi et al., ref. 168.)

UV-damaged DNA from an Hfr or F′ *lac* strain, or after infection of cells by UV-damaged phage P1, certain SOS functions are induced in the recipient cell (169). This has also been shown to occur with self-transmissible replicons such as F, R and ColI. The part of the UV-damaged replicon whose introduction into the cell elicits λ prophage induction appears to be a small DNA fragment encompassing the origin of replication of the F sex factor (*oriF*) (169). The simplest general explanation for this type of induction is that abortive replication of the incoming damaged DNA generates an inducing signal that activates recA protein (169).

The general conclusion from these studies and from other studies of a similar nature is that intermediates in DNA metabolism are likely to be important in the molecular mechanism(s) of activation of constitutive recA protein to a form that is either directly or indirectly required for proteolysis of protein repressors. However, whether or not this activation involves a unique biochemical pathway, what the nature of this pathway is and whether induction involves the appearance of a metabolite(s) not present in uninduced cells remain to be established.

7-9 Mechanism of Some SOS Responses

Another aspect of the SOS response that suffers from a paucity of detailed understanding is the precise mechanism by which the various SOS phenotypes manifest themselves. A notable exception is the induction of phage λ, a topic that is outside the scope of this text and so will not be considered here. However, before concluding this discussion on the SOS response, it is appropriate to return to the topic that served as the introduction to this subject much earlier in the chapter: that is, the replication of DNA past base damage in template strands (translesion DNA synthesis; Section 7-3) and its possible relation to one of the most biologically significant of the SOS phenotypes: mutagenesis.

Error-Prone DNA Synthesis

Mutations are a very important component of the SOS response to DNA damage. However, their spectrum in both prokaryotic and eukaryotic cells is very diverse. In addition, the molecular biology of mutagenesis in both SOS-induced and SOS-uninduced cells is a vast field that deserves consideration in a comprehensive text of its own; thus a systematic treatment of mutagenesis is not included here. For the purposes of the present discussion, however, it should be noted that the great majority of mutations that develop in UV-irradiated *E. coli* (and presumably in cells treated with other agents that cause bulky base damage in DNA) are dependent on the operation of the *recA-lexA* inducible regulon just discussed. Despite the fact that the biology of this important phenotypic response to DNA damage has been studied in detail, the exact mechanism(s) whereby UV-induced mutations arise in *E. coli* is still not understood in molecular or biochemical terms. Not the least of the many problems associated with studying mutagenesis in these terms is the fact that it is a very rare event not obviously amenable to biochemical analysis. Furthermore, it usually occurs some time *after* DNA containing base damage has been replicated (either semiconservatively and/or nonsemiconservatively). Thus, the kinetics of the crucial biochemical events are hard to predict. Finally, the biochemistry of replicational fidelity is still not fully understood in *undamaged* prokaryotic cells, let alone cells exposed to DNA damage.

What is known about UV irradiation-induced mutagenesis in *E. coli*? Although pyrimidine dimers may be required for the *induction* of error-prone cellular responses to UV radiation damage in DNA, the widely held notion that those photoproducts are the *principal* source of template damage that leads to mutations has been challenged, since the distribution of potential (6-4) photoproducts in DNA (see Section 1-5) correlates as well if not better with that of certain UV radiation-induced mutations than does the distribution of potential pyrimidine dimers (170, 171).

It was indicated in Section 7-3 that cells apparently can circumvent noninstructional photoproducts in DNA by a discontinuous mode of semiconservative DNA synthesis associated with gap filling by *recA*-dependent recombinational events. If all unexcised pyrimidine dimers and/or (6-4) photoproducts result in daughter-strand gaps during DNA replication, then error-prone filling of gaps is a tenable explanation of UV-induced mutations, at least in cells defective in the *uvrA*, *uvrB* or *uvrC* genes (mutant *uvr* strains) (81, 172). Two specific mechanisms for error-prone gap filling can be envisaged. One is error-prone repair synthesis utilizing a DNA polymerase capable of patching gaps despite the presence of a photoproduct opposite a portion of the gap. This would qualify as a form of *nonsemiconservative translesion DNA synthesis*. Alternatively, the recombinational repair of gaps may be inaccurate, although this seems unlikely since induced mutagenesis can occur in strains of *E. coli* deficient in all the major pathways of recombination (81, 172).

There is evidence that supports the involvement of DNA polymerase III in error-prone *nonsemiconservative* translesion synthesis in vivo. This comes from a study on UV mutagenesis in an *E. coli uvrA polC ts* mutant, in which mutation fixation was measured by loss of the ability to remove potential mutations by photoreactivation (loss of photoreversibility) (173). Although loss of photoreversibility occurs normally during incubation at the permissive temperature, it ceases immediately after transfer to the restrictive temperature. This suggests that the mutational event requires active DNA polymerase III (173). However, in spite of the *polC* mutation, DNA replication and daughter-strand gap repair continue for some time at the non-permissive temperature in the particular strain employed in this study (173). A possible explanation for this apparent discrepancy is that *non*semiconservative translesion synthesis catalyzed by DNA polymerase III is involved in the repair of only a few daughter-strand gaps, a fraction too small to be detected biochemically under these conditions, but large enough to account for mutagenesis (173).

A possible mechanism for the role of DNA polymerase III in error-prone *nonsemiconservative* DNA synthesis is the expression in induced cells of a component that associates with the polymerase III holoenzyme and alters its fidelity. This possibility is supported by the properties of the holoenzyme from two strains of *E. coli* carrying mutations called *dnaQ49* and *mutD5* (174) that confer greatly increased mutation rates. These are both located at five minutes on the *E. coli* genetic map and are probably mutations in the same gene. Of particular interest is the observation that partially purified holoenzyme from *dnaQ49* has enhanced thermolability and salt sensitivity compared with the wild-type enzyme (174). The product of the *dnaQ* gene, known to be a protein of molecular weight 25,000 (175), was compared with purified DNA polymerase III holoenzyme by two-dimensional gel electrophoresis (176). The *dnaQ* product comigrated with the ϵ subunit, a 25 kd protein of the DNA polymerase III "core enzyme" (see Table 5-3). Interestingly, the $3' \rightarrow 5'$ exonuclease proofreading function of DNA polymerase III from the *mutD* and *dnaQ* mutants is defective. Thus in vitro the mutant enzymes demonstrate reduced excision of a terminal mispaired base from a copolymer substrate, relative to that observed with wild-type enzyme, as well as reduced turnover of dTTP into dTMP during replication of G4 DNA (174).

A third possible mechanism for *recA-lexA*-dependent mutagenesis is by error-prone *excision repair* of pyrimidine dimers (81, 172). Indeed, in *uvr*$^+$ bacteria, fixation of most mutations in relation to pyrimidine dimers (as measured by loss of their photoreversibility) occurs in the *absence* of DNA replication and with a time course similar to that for excision repair (177). In fact, it has been estimated that at least 75 percent of the UV-induced mutations in *uvr*$^+$ strains arise during excision repair (177). Since the excision repair of pyrimidine dimers is largely error-free in mutant *recA* or *lexA* cells (172), how does induction of the SOS phenomenon alter this process? An aspect of excision repair in induced cells that was discussed in Section 5-3 is its association with very long patches of repair synthesis of DNA; so-called long patch excision repair. As indicated in that chapter, this mode of excision repair is dependent on the *recA* and *lexA* genes, it requires protein synthesis and it is induced by exposure of cells to UV radiation. All these observations suggest that long patch excision repair is one of the phenotypic manifestations of the SOS response.

A specific model has been proposed that relates long patch excision repair to mutagenesis. The model considers the threat to cell viability posed by the simultaneous excision repair of *very closely spaced pyrimidine dimers on opposite strands of DNA*. According to this model, excision repair initiated at one lesion is inhibited at the second in order to avoid a lethal double-strand break in the genome and requires *transdimer repair synthesis* for its completion, a mode of repair synthesis that results in the long patches observed experimentally (Fig. 7-39) (16). Direct experimental support for this postulated mechanism of UV-induced mutagenesis in uvr^+ cells is lacking. The hypothesis would predict that the proportion of available sites repaired by the long patch excision repair mode should increase nonlinearly with UV dose since it requires the presence of *two* dimers. However, a dose-squared relation is not observed experimentally (178). In addition, long patch excision repair occurs to the same extent in *umuC,D* mutants as in wild-type cells (178). The *umuC,D* gene is required for SOS-induced mutagenesis since *umuC,D* mutants are UV nonmutable (160, 163).

The final postulated mechanism for *recA-lexA*-dependent UV mutagenesis to be considered here takes us back to Section 7-3, concerning the tolerance of DNA base damage, i.e., to the model that induction of the SOS response results in a modification of one or more components of the *E. coli* replisome such that *semiconservative* DNA synthesis occurs across template lesions such as pyrimidine dimers and/or (6-4) photoproducts. Once again the experimental evidence in support of this phenomenon is limited and in the main inferential rather than direct. For example, the spectrum of base changes resulting from forward mutations induced by UV light in the *lacI* gene of *E. coli* has been examined in detail (179). A genetic analysis of over 4000 nonsense mutations facilitated assignment of many of them to specific codons and allowed deduction of the precise base changes that occurred. A significant percentage of the UV-induced nonsense mutations were found to be tandem double nucleotide changes. Nearly all of these occurred at the position occupied by two adjacent pyrimidines in the wild-type codon, as expected if incorrect nucleotides had been inserted during replication past pyrimidine dimers and/or (6-4) photoproducts (170) (Table 7-9). In addition, an

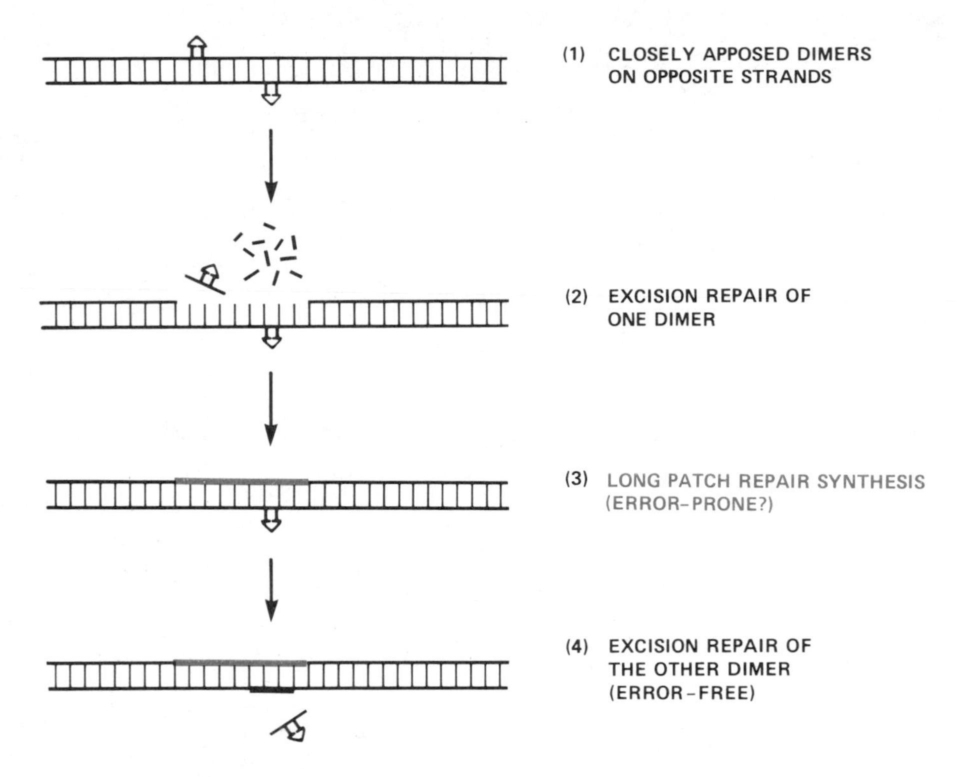

FIGURE 7-39
Excision repair of closely spaced lesions on opposite strands. It is proposed that excision of one dimer involves degradation of a long tract of DNA, resulting in long patch repair synthesis. This process may require induction of SOS genes and may be error-prone. During its operation the excision of the second lesion is inhibited. Once long patch repair is completed the other lesion can be repaired by typical excision repair. The relative sizes of the patches shown here are not drawn to scale.

analysis of a number of UV-induced mutations in the *cI* gene of phage λ shows a remarkable correlation with the distribution of (6-4) photoproducts in DNA (180).

Targeted v. Untargeted Mutagenesis

The correlation of mutations with nucleotide sequence alterations in daughter strands directly opposite base damage in template DNA is referred to as *targeted mutagenesis*. Arguments have been marshalled in support of the notion that all mutations measured with the *lacI* system are of this type. Figure 7-40 shows a comparison of the specificities for three carcinogens known to be dependent on the SOS system for mutagenic activity. Each of the mutagens has a different

TABLE 7-9
Distribution of UV-induced base damage in the *lacI* gene of *E. coli*

Dinucleotide	Site†	Sequence	(6-4) Products (%)	Pyrimidine Dimers (%)	Nonsense Mutants	Base Substitution
TC	652	*TC*	0.86	0.82	43	C→T
	658	T*TC*	0.49	0.36	80	C→T
	672	T*TC*C	0.60	0.86	1	C→A
	678	T*TC*C	0.83	—	4	C→A
	689	C*TC*C	0.57	0.58	39	C→T
	706	TTT*TC*	0.98	0.91	60	C→T
	732	CC*CTC*	1.08	—	0	C→A
CC	418	*CC**	0.01	0.35	7	C→T
	484	*CC*	0.16	0.25	20	C→T
	688	CT*CC*	0.15	0.53	9	C→T
TT	636	T*TT*	0.08	2.08	1	T→A

†First base of the dinucleotide; C* denotes 5-methylcytosine.
After W. A. Haseltine, ref. 170.

specificity, suggesting that mutations dependent on the *E. coli* SOS system are probably mainly targeted to sites of template damage. If mutations were exclusively or largely untargeted, one would expect similar or even identical mutational spectra with these agents (181).

However, there is evidence that UV mutagenesis in bacteriophages for which *E. coli* is the natural host may also be *nontargeted*. Indeed, the frequency of mutations in both phage λ and the single-strand phage M13 is increased in *unirradiated* phages when SOS proficient host cells are previously induced (182). This observation is consistent with the idea that the relaxed fidelity of DNA replication associated with induction of the SOS system not only facilitates translesion synthesis but also creates an increased probability of replicational errors opposite *normal* template DNA or opposite cryptic lesions such as sites of base loss (apurinic or apyrimidinic sites). The contribution of untargeted replicative errors to total UV mutagenesis may be higher for phage genomes than for bacterial genomes because phage DNA undergoes more rounds of replication and because postreplicative mismatch correction of untargeted mutations may be less efficient in phage DNA (182). Aside from the specific postulated relation between SOS-dependent translesion synthesis of DNA and both targeted and untargeted mutagenesis, the observation that in some experimental systems (e.g., M13 infection of SOS-induced cells) mutations are *preferentially untargeted* suggests that UV pho-

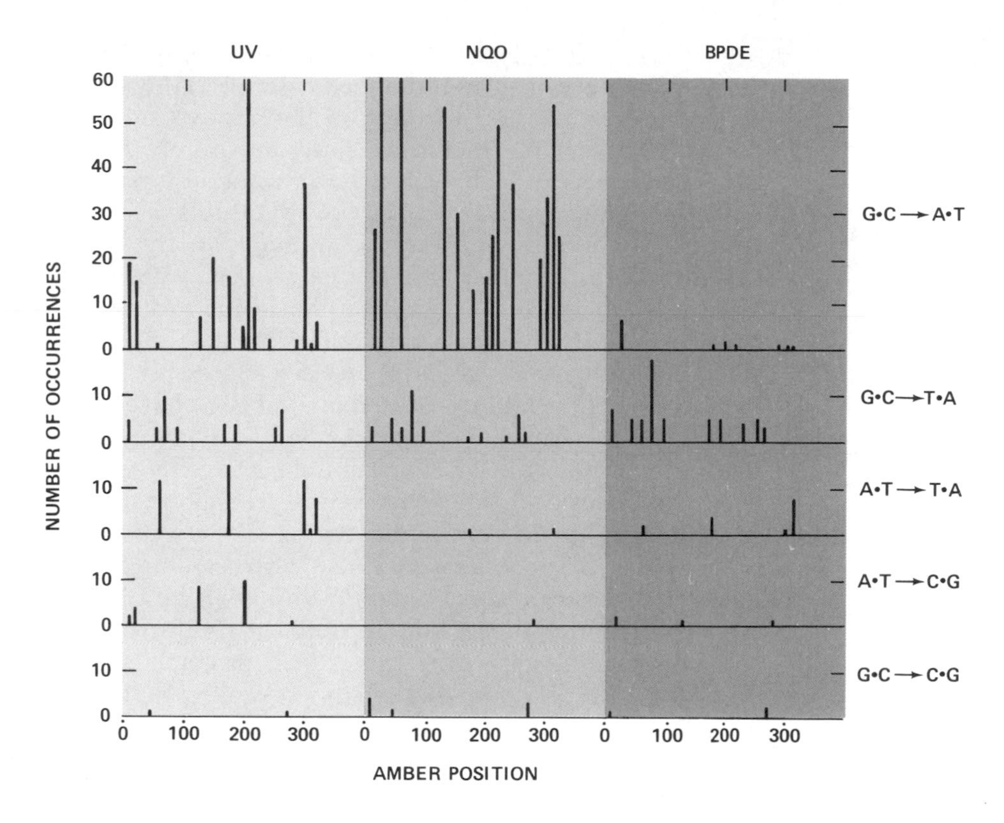

FIGURE 7-40
UV radiation, 4-nitroquinoline-1-oxide (NQO) and benzo(a)pyrene diolepoxide (BPDE) cause nonsense codon mutations at sites in different positions of the *lacI* gene of *E. coli*. This pattern suggests that these mutations are largely (if not exclusively) targeted to sites of base damage, since if they were mainly untargeted one would expect the sites of mutation to be very similar for all three types of DNA damage. (From J. H. Miller, ref. 181.)

toproducts such as pyrimidine dimers may primarily act as *inducing* rather than as targeting lesions for an SOS process that can lead to both targeted and untargeted mutations (182).

Possible Mechanisms of Semiconservative Translesion DNA Synthesis

Aside from the mutagenic nature of semiconservative translesion DNA synthesis, what other evidence supports the existence of this phenomenon as a mechanism for the tolerance of base damage? Some studies have examined the *extent* of DNA synthesis on UV-irradiated templates in induced and uninduced cells. As indicated in Section

7-2, UV-irradiated ϕX174 DNA is apparently only replicated to the first pyrimidine dimer in uninduced cells (4, 5) (Fig. 7-2). However, in UV-induced cells an increase in the relative amount of duplex ϕX174 DNA results, including some apparently fully replicated DNA molecules (5). Such experiments do not fully exclude the possibility that multiple initiations of DNA synthesis occur at sites other than the origin of replication, nor do they directly demonstrate the complete covalent integrity of the newly synthesized DNA. In fact, attempts to demonstrate complete translesion synthesis of UV-irradiated ϕX174 DNA in vitro with unfractionated or partially purified extracts of induced *E. coli* have not been successful. Nonetheless, in vitro studies of the replication of damaged DNA have been providing some interesting insights into how SOS induction might promote translesion DNA synthesis. For example, a large increase in the turnover of nucleoside triphosphates into free monophosphate is observed during the in vitro replication of UV-irradiated ϕX174 DNA by *E. coli* DNA polymerase I (9). It has been suggested that this nucleotide turnover may be due to idling by the DNA polymerase (i.e., incorporation of nucleotides opposite noninstructional lesions and subsequent excision by the $3' \rightarrow 5'$ proofreading exonuclease function of the polymerase), thereby preventing replication past the lesions. If a similar phenomenon occurs in vivo, a dampening of the $3' \rightarrow 5'$ exonuclease function of DNA polymerase III in induced cells might facilitate error-prone translesion DNA synthesis (9).

Inhibition of the $3' \rightarrow 5'$ exonuclease activity of *E. coli* DNA polymerase I or use of eukaryote polymerases lacking $3' \rightarrow 5'$ exonuclease activity (9) does enhance DNA synthesis on UV-irradiated templates in vitro but does not completely alleviate inhibition of DNA synthesis (Fig. 7-41). Although these results indicate that inhibition or absence of 3′-terminal proofreading of newly replicated DNA in and of itself does not allow complete translesion DNA synthesis, it is difficult to conceive of any mechanism of such synthesis that would not *require* the inhibition of this function (9). Thus, while the absence of proofreading is not sufficient to allow translesion synthesis in vitro, its involvement in a more complex system associated with the induction of SOS functions in vivo cannot be excluded.

Very recently it has been shown (183) that replication in vitro of UV-irradiated circular single-strand M13 DNA by *E. coli* RNA polymerase plus *E. coli* DNA polymerase III holoenzyme in the presence of single-strand DNA binding protein yields full-length as well as partially replicated products. A similar result was obtained with G4 DNA primed with *E. coli* DNA primase and ϕX174 DNA primed with a synthetic oligonucleotide. The fraction of full-length DNA detected was several orders of magnitude greater than that expected if pyrimidine dimers in template DNA constituted absolute blocks to DNA replication. Substitution of DNA polymerase I for DNA polymerase

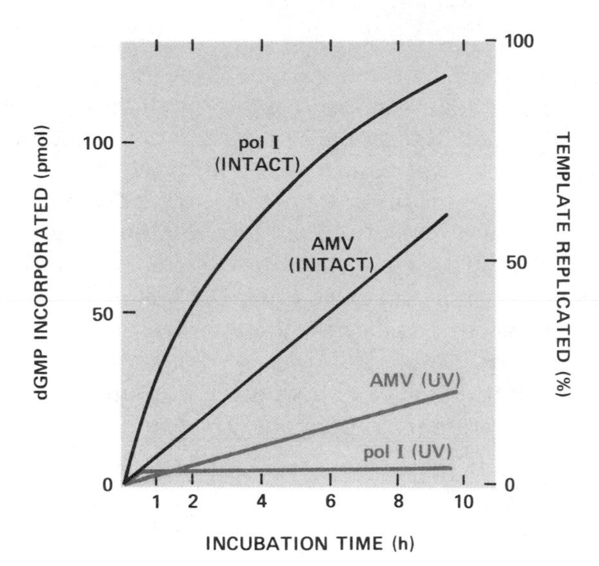

FIGURE 7-41
Extent of DNA synthesis by avian myeloblastosis virus (AMV) DNA polymerase (reverse transcriptase) or by *E. coli* DNA polymerase I on intact or UV-irradiated primed ϕX174 phage DNA in vitro. If the $3' \rightarrow 5'$ exonuclease activity of *E. coli* DNA polymerase I were to continually remove incorrect nucleotides inserted opposite template damage, then replication past these lesions would be effectively inhibited, as shown in the figure. On the other hand, an enzyme with no $3' \rightarrow 5'$ exonuclease (such as the AMV enzyme) might allow some bypass DNA synthesis to occur, as shown here. (From G. Villani et al., ref. 9.)

III holoenzyme also resulted in apparent translesion DNA synthesis in vitro, though at a reduced efficiency (183). The conditions for translesion synthesis in these experiments resemble those presumably operating constitutively in *E. coli*. Thus it is suggested (183) that transdimer DNA synthesis may occur in both uninduced and induced cells and that the nucleotide sequence and/or conformation of the resulting DNA products in these two states may differ.

The discussions earlier in this section suggested a role of the *uvrA*, *uvrB* and *uvrC* genes in UV mutagenesis exclusively in terms of a possible error-prone excision repair mode of DNA. However, some experiments suggest that these functions may also be required for translesion semiconservative synthesis of DNA. When previously uninduced *uvr*$^+$ cells are exposed to UV radiation, excision of pyrimidine dimers occurs to the level where few if any can be detected in the acid-precipitable DNA. In addition, semiconservative DNA synthesis in such cells is inhibited for well over an hour after irradiation (184). But if *E. coli* cells are SOS induced under conditions permissive for protein synthesis (e.g., by thymine starvation), a significant *inhibition of thymine dimer excision* results, while semiconservative DNA synthesis *resumes* within 45 to 60 minutes (184) (Fig. 7-42). Separation of replicated from nonreplicated DNA in these induced cells shows that both fractions contain the same concentration of thymine-containing pyrimidine dimers, suggesting that replication past such lesions occurred (185, 186). This dimer tolerance is dependent on the *recA* and *lexA* genes as well as on the *uvrA* and *uvrB* functions (187, 188).

In contradistinction to the dimers present in unreplicated DNA in these cells, those present in replicated DNA are *insensitve to attack by the M. luteus PD DNA glycosylase–AP endonuclease* (186). The spectrum of structural modifications of DNA containing dimers that might render these photoproducts insensitive to the *M. luteus* (and presumably phage T4) enzyme is not known. However, an intriguing possibility is that the 5′ glycosyl bond of the dimerized pyrimidines is already hydrolyzed, and that this event somehow facilitates the transdimer synthesis of DNA (189). Such a model of transdimer synthesis obviously requires the existence of a PD DNA glycosylase activity in *E. coli*, preliminary evidence for which was discussed in

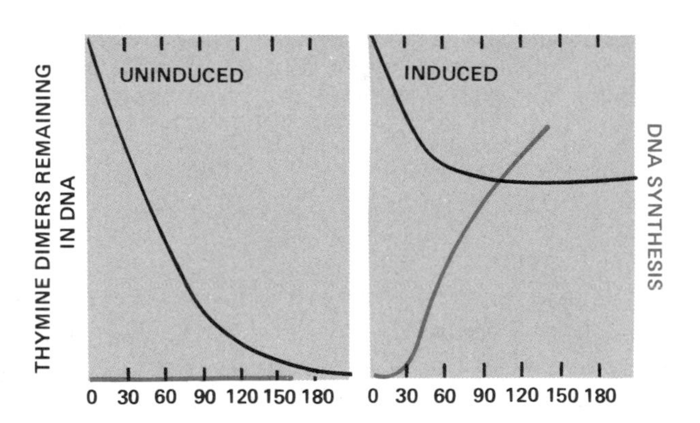

FIGURE 7-42
Diagrammatic representation of the kinetics of thymine dimer excision and of semiconservative DNA synthesis in uninduced and induced *E. coli*. If UV-irradiated cells are not able to induce SOS functions (left), resumption of semiconservative DNA synthesis is delayed. However, if cells are induced (by thymine starvation for example) prior to UV irradiation (right), excision of thymine-containing pyrimidine dimers is inhibited but semiconservative DNA synthesis resumes. This suggests that at least some of this DNA synthesis uses template strands that still contain pyrimidine dimers.

Section 3-4. If this activity is dependent on one or more of the *E. coli* *uvr* genes, it could explain the *uvr* dependence of putative semiconservative transdimer DNA synthesis (187–189).

Finally, a novel form of DNA polymerase I activity has been isolated from induced *E. coli* cells. This enzyme is referred to as DNA polymerase I* and, like the constitutive enzyme, it is insensitive to N-ethylmaleimide, is inhibited by antibody to DNA polymerase I and is not detected in a *polA1* strain (190). DNA polymerase I* activity sediments through sucrose gradients as a broad peak with an S value significantly larger than that of the constitutive enzyme. The fidelity of the induced enzyme species is relatively low with a variety of synthetic templates, although the enzyme appears to have normal levels of $3' \rightarrow 5'$ exonuclease activity (190).

7-10 Summary

The mechanism(s) by which unexcised DNA damage is tolerated in *E. coli* is clearly complex, and much remains to be done to clearly distinguish those biochemical events that are constitutive from those that are induced. This distinction is difficult to make and experimentally is particularly complicated by the fact that both constitutive and induced modes of DNA damage tolerance are dependent on the *recA* gene, the product of which is involved in both recombination and repressor inactivation. It is evident too that the plethora of genes that are derepressed in response to the various inducing stimuli that turn on the SOS regulon are under complex regulatory control, and there is still much to be learned about the details of this regulation. For example, it is not known whether all the SOS target genes have been identified, nor whether other substrates exist for activated recA protein. This issue is particularly germane if activated recA protein is not really a protease, for under such circumstances one could envisage that activated recA protein could play a variety of roles. During derepression of the *recA-lexA* regulon activated recA protein may be required for repressor inactivation; however, the requirement of it for mutagenesis may involve some other role.

It is still not known precisely what the inducing signals are that activate recA protein. Nor is it clear what the relations are between various SOS target genes and the biological end points that classically define the SOS response. The sequence analysis of a number of genes in the SOS system, including the *recA* and *lexA* genes, has provided valuable insights into repressor-operator interactions, and the biochemistry of purified recA and lexA protein is under active study. It will certainly be of great interest to learn in greater detail how individual repressor-operator interactions at each SOS gene affect the expression of these genes and the expression of the overall SOS response.

The molecular mechanisms of the expression of various SOS phenotypes are still poorly understood. Among these the particular response that has obvious biological relevance is that of mutagenesis. Does this arise by a single mechanism such as translesion semiconservative DNA synthesis or are other pathways such as error-prone excision repair and error-prone gap filling involved? Finally, a major area of interest is whether a comparable regulatory mechanism exists in higher organisms. Genetic evidence suggests the existence of several damage-inducible functions in yeast, and evidence exists for inducible repair and mutagenesis in mammalian cells. This aspect of cellular responses to DNA damage is considered in greater detail in the next chapter.

Suggestions for Further Reading

DNA synthesis on damaged templates, refs. 10, 16, 38, 39; recA protein, refs. 49, 50; recombination, ref. 67; the SOS response, refs. 81–84, 105, 141, 142, 147, 148, 192; mutagenesis, refs. 81, 163, 171, 172, 192.

References

1. Clark, A. J., and Volkert, M. R. 1978. A new classification of pathways repairing pyrimidine dimer damage in DNA. In *DNA repair mechanisms*. P. C. Hanawalt, E. C. Friedberg and C. F. Fox, eds., p. 57. New York: Academic.
2. Moustacchi, E., Ehmann, U. K., and Friedberg, E. C. 1979. Defective recovery of semiconservative DNA synthesis in xeroderma pigmentosum cells following split-dose ultraviolet irradiation. *Mutation Res.* 62:159.
3. Setlow, R. B., Swenson, P. A., and Carrier, W. L. 1963. Thymine dimers and inhibition of DNA synthesis by ultraviolet irradiation of cells. *Science* 142:1464.
4. Benbow, R. M., Zuccarelli, A. J., and Sinsheimer, R. L. 1974. A role for single-strand breaks in bacteriophage ϕX174 genetic recombination. *J. Mol. Biol.* 88:629.
5. Caillet-Fauquet, P., Defais, M., and Radman, M. 1977. Molecular mechanisms of induced mutagenesis. Replication in vivo of bacteriophage ϕX174 single-stranded, ultraviolet light-irradiated DNA in intact and irradiated host cells. *J. Mol. Biol.* 177:95.
6. Moore, P., and Strauss, B. S. 1979. Sites of inhibition of *in vitro* DNA synthesis in carcinogen- and UV-treated ϕX174 DNA. *Nature* 278:664.
7. Hsu, W-T., Lin, E. J. S., Harvey, R. G., and Weiss, S. B. 1977. Mechanism of phage ϕX174 DNA inactivation by benzo[a]pyrene-7,8-dihydrodiol-9,10-epoxide. *Proc. Natl. Acad. Sci.* (USA) 74:3335.

8. Moore, P. D., Bose, K. K., Rabkin, S. D., and Strauss, B. S. 1981. Sites of termination of *in vitro* DNA synthesis on ultraviolet- and N-acetylaminofluorene-treated ϕX174 templates by prokaryotic and eukaryotic DNA polymerases. *Proc. Natl. Acad. Sci.* (USA) 78:110.

9. Villani, G., Boiteux, S., and Radman, M. 1978. Mechanisms of ultraviolet-induced mutagenesis: extent and fidelity of *in vitro* DNA synthesis on irradiated templates. *Proc. Natl. Acad. Sci.* (USA) 75:3037.

10. Hall, J. D., and Mount, D. W. 1981. Mechanisms of DNA replication and mutagenesis in ultraviolet-irradiated bacteria and mammalian cells. *Prog. Nucleic Acid Res. Mol. Biol.* 25:53.

11. Gefter, M. L. 1975. DNA replication. *Ann. Rev. Biochem.* 44:45.

12. Rupp, W. D., and Howard-Flanders, P. 1968. Discontinuities in the DNA synthesized in an excision-defective strain of *Escherichia coli* following ultraviolet irradiation. *J. Mol. Biol.* 31:291.

13. Hanawalt, P. C., and Setlow, R. B., eds. 1975. *Molecular mechanisms for repair of DNA*. New York: Plenum.

14. Hanawalt, P. C., Friedberg, E. C., and Fox, C. F., eds. 1978. *DNA repair mechanisms*. New York: Academic.

15. Friedberg, E. C., and Bridges, B. A., eds. 1983. *Cellular responses to DNA damage*. New York: Alan R. Liss.

16. Hanawalt, P. C., Cooper, P. K., Ganesan, A. K., and Smith, C. A. 1979. DNA repair in bacteria and mammalian cells. *Ann. Rev. Biochem.* 48:783.

17. Howard-Flanders, P., and Boyce, R. P. 1966. DNA repair and genetic recombination: studies on mutants of *Escherichia coli* defective in these processes. *Radiat. Res.* (supp.) 6:156.

18. Howard-Flanders, P., Rupp, W. D., Wilkins, B. M., and Cole, R. S. 1968. DNA replication and recombination after UV-irradiation. *Cold Spring Harbor Symp. Quant. Biol.* 33:195.

19. Sedgwick, S. G. 1975. Genetic and kinetic evidence for different types of post-replication repair in *Escherichia coli*. *J. Bacteriol.* 123:154.

20. Youngs, D. A., and Smith, K. C. 1976. Genetic control of multiple pathways of postreplicational repair in *uvrB* strains of *Escherichia coli* K-12. *J. Bacteriol.* 125:102.

21. Smith, K. C., and Meun, D. H. C. 1970. Repair of radiation-induced damage in *Escherichia coli*. I. Effect of *rec* mutations on postreplication repair of damage due to ultraviolet radiation. *J. Mol. Biol.* 51:459.

22. Ohki, M., and Tomizawa, J. I. 1968. Asymmetric transfer of DNA strands in bacterial conjugation. *Cold Spring Harbor Symp. Quant. Biol.* 33:651.

23. Rupp, W. D., and Ihler, G. 1968. Strand selection during bacterial mating. *Cold Spring Harbor Symp. Quant. Biol.* 33:647.

24. Johnson, R. C., and McNeill, W. F. 1978. Electron microscopy of UV-induced postreplication repair of daughter strand gaps. In *DNA repair mechanisms*. P. C. Hanawalt, E. C. Friedberg and C. F. Fox, eds., p. 95. New York: Academic.

25. Iyer, V. N., and Rupp, W. D. 1971. Usefulness of benzoylated naphthoylated DEAE-cellulose to distinguish and fractionate double-stranded DNA bearing different extents of single-stranded regions. *Biochim. Biophys. Acta* 228:117.

26. Rupp, W. D., Wilde, C. E. III, Reno, D. L., and Howard-Flanders, P. 1971. Exchanges between DNA strands in ultraviolet-irradiated *Escherichia coli*. *J. Mol. Biol.* 61:25.

27. Wilkins, B. M., and Howard-Flanders, P. 1968. The genetic properties of DNA transferred from ultraviolet-irradiated *Hfr* cells of *Escherichia coli* K-12 during mating. *Genetics* 60:243.

28. Lin, P-F., and Howard-Flanders, P. 1976. Genetic exchanges caused by ultraviolet photoproducts in phage λ DNA molecules: the role of DNA replication. *Molec. Gen. Genet.* 146:107.

29. Setlow, R. B., and Regan, J. D. 1981. Measurement of repair synthesis by photolysis of bromouracil. In *DNA repair—a laboratory manual of research procedures.* E. C. Friedberg and P. C. Hanawalt, eds., p. 307. New York: Dekker.

30. Ley, R. D. 1973. Postreplication repair in an excision-defective mutant of *Escherichia coli:* ultraviolet light–induced incorporation of bromodeoxyuridine into parental DNA. *Photochem. Photobiol.* 18:87.

31. Ley, R. D. 1975. Ultraviolet-light induced incorporation of bromodeoxyuridine into parental DNA of an excision-defective mutant of *Escherichia coli.* In *Molecular mechanisms for repair of DNA.* P. C. Hanawalt and R. B. Setlow, eds., p. 313. New York: Plenum.

32. Ganesan, A. K. 1974. Persistence of pyrimidine dimers during post-replication repair in ultraviolet light-irradiated *Escherichia coli* K12. *J. Mol. Biol.* 87:103.

33. Ganesan, A. K. 1975. Distribution of pyrimidine dimers during postreplication repair in UV-irradiated excision-deficient cells of *Escherichia coli* K12. In *Molecular mechanisms for repair of DNA.* P. C. Hanawalt and R. B. Setlow, eds., p. 317. New York: Plenum.

34. Ganesan, A. K., Seawell, P. C., and Mount, D. W. 1978. Effect of tsl (thermosensitive suppressor of *lex*) mutation on postreplication repair in *Escherichia coli* K-12. *J. Bacteriol.* 135:935.

35. Rothman, R. H., and Clark, A. J. 1977. The dependence of postreplication repair on *uvr*B in a *rec*F mutant of *Escherichia coli* K-12. *Molec. Gen. Genet.* 155:279.

36. Sancar, A., Stachelek, C., Konigsberg, W., and Rupp, W. D. 1980. Sequences of the *rec*A gene and protein. *Proc. Natl. Acad. Sci.* (USA) 77:2611.

37. Radding, C. M. 1981. Recombination activities of *E. coli* recA protein. *Cell* 25:3.

38. Howard-Flanders, P., Cassuto, E., and West, S. C. 1981. Regulatory and enzymatic functions of recA protein in recombination and postreplication repair. In *Chromosome damage and repair.* E. Seeberg and K. Kleppe, eds., p. 169. New York: Plenum.

39. Howard-Flanders, P. 1983. Workshop summary: recombination repair. In *Cellular responses to DNA damage.* E. C. Friedberg and B. A. Bridges, eds., p. 577. New York: Alan R. Liss.

40. Weinstock, G. M., McEntee, K., and Lehman, I. R. 1979. ATP-dependent renaturation of DNA catalyzed by the recA protein of *Escherichia coli. Proc. Natl. Acad. Sci.* (USA) 76:126.

41. Shibata, T., DasGupta, C., Cunningham, R. P., and Radding, C. M. 1979. Purified *Escherichia coli* recA protein catalyzes homologous pairing of superhelical DNA and single stranded fragments. *Proc. Natl. Acad. Sci.* (USA) 76:1638.

42. McEntee, K., Weinstock, G. M., and Lehman, I. R. 1979. Initiation of general recombination catalyzed *in vitro* by the recA protein of *Escherichia coli. Proc. Natl. Acad. Sci.* (USA) 76:2615.

43. Cunningham, R. P., Shibata, T., DasGupta, C., and Radding, C. M. 1979. Single strands induce recA protein to unwind duplex DNA for homologous pairing. *Nature* 281:191.

44. West, S. C., Cassuto, E., and Howard-Flanders, P. 1981. RecA protein promotes homologous-pairing and strand-exchange reactions between duplex DNA molecules. *Proc. Natl. Acad. Sci.* (USA) 78:2100.

45. West, S. C., Cassuto, E., and Howard-Flanders, P. 1981. Homologous pairing can occur before DNA strand separation in general genetic recombination. *Nature* 290:29.

46. McEntee, K., Weinstock, G. M., and Lehman, I. R. 1980. RecA protein-catalyzed strand assimilation: stimulation by *Escherichia coli* single-stranded DNA-binding proteins. *Proc. Natl. Acad. Sci.* (USA) 77:857.

47. Cox, M. M., Soltis, D. A., Livneh, Z., and Lehman, I. R. 1983. On the role of single-stranded DNA binding protein in recA protein–promoted DNA strand exchange. *J. Biol. Chem.* 258:2577.

48. West, S. C., Cassuto, E., and Howard-Flanders, P. 1982. Role of SSB protein in RecA promoted branch migration reaction. *Mol. Gen. Genet.* 186:333.

49. McEntee, K., and Weinstock, G. M. 1981. The recA enzyme of *Escherichia coli* and recombination assays. In *The enzymes.* 3rd ed. P. D. Boyer, ed., Nucleic acids, vol. XIV, part A, p. 445. New York: Academic.

50. Weinstock, G. M. 1982. Enzymatic activities of the recA protein of *Escherichia coli. Biochimie* 64:611.

51. McEntee, K., Weinstock, G. M., and Lehman, I. R. 1981. Binding of the recA protein of *Escherichia coli* to single- and double-stranded DNA. *J. Biol. Chem.* 256:8835.

52. Cunningham, R. P., DasGupta, C., Shibata, T., and Radding, C. M. 1980. Homologous pairing in genetic recombination: recA protein makes joint molecules of gapped circular DNA and closed circular DNA. *Cell* 20:223.

53. Cassuto, E., West, S. C., Mursalim, J., Conlon, S., and Howard-Flanders, P. 1980. Initiation of genetic recombination: homologous pairing between duplex DNA molecules promoted by recA protein. *Proc. Natl. Acad. Sci.* (USA) 77:3962.

54. West, S. C., Cassuto, E., Mursalim, J., and Howard-Flanders, P. 1980. Recognition of duplex DNA containing single-stranded regions by recA protein. *Proc. Natl. Acad. Sci.* (USA) 77:2569.

55. Howard-Flanders, P., and West, S. C. 1983. Enzymatic mechanism of post-replication repair. In *Cellular responses to DNA damage.* E. C. Friedberg and B. A. Bridges, eds., p. 399. New York: Alan R. Liss.

56. Meselson, M. S., and Radding, C. M. 1975. A general model for genetic recombination. *Proc. Natl. Acad. Sci.* (USA) 72:358.

57. DasGupta, C., Shibata, T., Cunningham, R. P., and Radding, C. M. 1980. The topology of homologous pairing promoted by recA protein. *Cell* 24:213.

58. Kalin, R., Cunningham, R. P., DasGupta, C., and Radding, C. M. 1981. Polarity of heteroduplex formation promoted by *Escherichia coli* recA protein. *Proc. Natl. Acad. Sci.* (USA) 78:4786.

59. Cox, M. M., and Lehman, I. R. 1981. Directionality and polarity in recA protein-promoted branch migration. *Proc. Natl. Acad. Sci.* (USA) 78:6018.

60. Livneh, Z., and Lehman, I. R. 1982. Recombinational bypass of pyrimidine dimers promoted by the recA protein of *Escherichia coli. Proc. Natl. Acad. Sci.* (USA) 79:3171.

61. Cox, M. M., and Lehman, I. R. 1982. RecA protein–promoted DNA strand exchange. *J. Biol. Chem.* 257:8523.

62. Glassberg, J., Meyer, R. R., and Kornberg, A. 1979. Mutant single-strand binding protein of *Escherichia coli*: genetic and physiological characterization. *J. Bacteriol.* 140:14.

63. Lieberman, H. B., and Witkin, E. M. 1981. Variable expression of the *ssb*-1 allele in different strains of *Escherichia coli* K12 and B: differential suppression of its effects on DNA replication, DNA repair and ultraviolet mutagenesis. *Mol. Gen. Genet.* 183:348–355.

64. Lieberman, H. B., and Witkin, E. M. 1983. DNA degradation, UV sensitivity, and SOS-mediated mutagenesis in strains of *Escherichia coli* deficient in single-strand DNA binding protein: effects of mutations and treatments that alter levels of exonuclease V or recA protein. *Molec. Gen. Genet.* 190:92.

65. DiCapua, E., Engel, E., Stasiak, A., and Koller, T. 1982. Characterization of complexes between recA protein and duplex DNA by electron microscopy. *J. Mol. Biol.* 157:87.

66. McKay, D. B., Steitz, T. A., Weber, I. T., West, S. C., and Howard-Flanders, P. 1980. Crystallization of monomeric recA protein. *J. Biol. Chem.* 255:6662.

67. Radding, C. M. 1978. Genetic recombination: strand transfer and mismatch repair. *Ann. Rev. Biochem.* 47:480.

68. Tait, R. C., Harris, A. L., and Smith, D. W. 1974. DNA repair in *Escherichia coli* mutants deficient in DNA polymerase I, II and/or III. *Proc. Natl. Acad. Sci.* (USA) 71:675.

69. Sedgwick, S. G., and Bridges, B. A. 1974. Requirement for either DNA polymerase I or DNA polymerase III in postreplication repair in excision-proficient *Escherichia coli*. *Nature* 249:348.

70. Johnson, R. C. 1978. Reduction of postreplication DNA repair in two *Escherichia coli* mutants with temperature-sensitive polymerase III activity: implications for the postreplication repair pathway. *J. Bacteriol.* 136:125.

71. Sedgwick, S. G. 1975. Genetic and kinetic evidence for different types of postreplication repair in *Escherichia coli* B. *J. Bacteriol.* 123:154.

72. Barfknecht, T. R., and Smith, K. C. 1978. The involvement of DNA polymerase I in the postreplication repair of ultraviolet radiation induced damage in *Escherichia coli* K-12. *Molec. Gen. Genet.* 167:37.

73. Johnson, R. C. 1976. DNA post-replication repair gap filling in temperature-sensitive *dan*G, *dna*C, *dna*A mutants of *Escherichia coli*. *Biochem. Biophys. Res. Comm.* 70:791.

74. Johnson, R. C. 1975. Post-replication repair gap filling in an *Escherichia coli* strain deficient in *dna*B gene product. In *Molecular mechanisms for repair of DNA*. P. C. Hanawalt and R. B. Setlow, eds., p. 325. New York: Plenum.

75. Youngs, D. A., and Smith, K. C. 1973. Evidence for the control by *exr*A and *pol*A genes of two branches of the *uvr* gene–dependent excision repair pathway in *Escherichia coli* K-12. *J. Bacteriol.* 116:175.

76. Wang, T-C. V., and Smith, K. C. 1981. Effect of *recB*21, *uvrD*3, *lexA*101 and *recF*143 mutations on ultraviolet radiation sensitivity and genetic recombination in Δ*uvr*B strains of *Escherichia coli* K-12. *Molec. Gen. Genet.* 183:37.

77. Wang, T. V., and Smith, K. C. 1983. Mechanism for *recF*-dependent and *recB*-dependent pathways of postreplication repair in UV-irradiated *Escherichia coli uvrB*. *J. Bacteriol.* 156:1093.

78. Rothman, R. H., and Clark, A. J. 1977. Defective excision and postreplication repair of UV-damaged DNA in a *rec*L mutant strain of *E. coli* K-12. *Molec. Gen. Genet.* 155:267.

79. Horii, Z-I., and Clark, A. J. 1973. Genetic analysis of the *rec*F pathway to genetic recombination in *Escherichia coli* K-12: isolation and characterization of mutants. *J. Mol. Biol.* 80:327.

80. Mount, D. W., Low, K. B., and Edmiston, S. J. 1972. Dominant mutations (*lex*) in *Escherichia coli* K-12 which affect radiation sensitivity and frequency of ultraviolet light–induced mutations. *J. Bacteriol.* 112:886.

81. Witkin, E. M. 1976. Ultraviolet mutagenesis and inducible DNA repair in *Escherichia coli. Bact. Rev.* 40:869.

82. Little, J. W., and Mount, D. W. 1982. The SOS regulatory system of *Escherichia coli. Cell* 29:11.

83. Sedgwick, S. G., and Yarranton, G. T. 1982. How cells in distress use SOS. *Nature* 296:606.

84. Howard-Flanders, P. 1981. Inducible repair of DNA. *Sci. Am.* 245:72.

85. Volkert, M. R., George, D. L., and Witkin, E. M. 1976. Partial suppression of the *lex*A phenotype by mutations (*rnm*) which restore ultraviolet resistance but not ultraviolet mutability to *Escherichia coli* B/r *uvr*A *lex*A. *Mutation Res.* 36:17.

86. Radman, M. 1974. Phenomenology of an inducible mutagenic DNA repair pathway in *Escherichia coli:* SOS repair hypothesis. In *Molecular and environmental aspects of mutagenesis.* L. Prakash, F. Sherman, M. Miller, C. Lawrence and H. W. Tabor, eds., p. 128. Springfield: Thomas.

87. Lwoff, A., and Gutmann, A. 1950. Recherches sur un *Bacillus mégathérium* lysogène. *Ann. Inst. Pasteur* 78:711.

88. Lwoff, A., Siminovitch, L., and Kjelgaard, N. 1950. Induction de la production de bacteriophages chez une bactérie lysogène. *Ann. Inst. Pasteur* 79:815.

89. Jacob, F., and Wollman, E. L. 1959. Lysogeny. In *The viruses.* F. N. Burnet and W. M. Stanley, eds., vol. 2, p. 319. New York: Academic.

90. Korn, D., and Weissbach, A. 1962. Thymineless induction of *Escherichia coli* K-12 (λ). *Biochim Biophys Acta* 61:775.

91. Zgaga, V., and Militić, B. 1965. Superinfection with homologous phage of *Escherichia coli* K-12 (λ) treated with 6-azauracil. *Virology* 27:205.

92. Goldthwait, D., and Jacob, F. 1964. Sur la mechanisme de l'induction du developpement du prophage chez les bactéries lysogènes. *Comp. Rend. Acad. Sci. Paris* 259:661.

93. Hayes, W. 1968. *The genetics of bacteria and their viruses.* 2nd ed. London: Blackwell Scientific Publications.

94. Witkin, E. M. 1967. The radiation sensitivity of *Escherichia coli* B: a hypothesis relating filament formation and prophage induction. *Proc. Natl. Acad. Sci.* (USA) 57:1275.

95. Kirby, E. P., Jacob, F. and Goldthwait, D. A. 1967. Prophage induction and filament formation in a mutant strain of *Escherichia coli. Proc. Natl. Acad. Sci.* (USA) 58:1903.

96. Clark, A. J., and Margulies, A. D. 1965. Isolation and characterization of recombination-deficient mutants of *Escherichia coli* K-12. *Proc. Natl. Acad. Sci.* (USA) 53:451.

97. Brooks, K., and Clark, A. J. 1967. Behavior of λ bacteriophage in a recombination deficient strain of *Escherichia coli. J. Virology* 1:283.

98. Hertman, I., and Luria, S. E. 1967. Transduction studies on the role of a *rec*$^+$ gene in the ultraviolet induction of prophage lambda. *J. Mol. Biol.* 23:117.

99. Bridges, B. A., and Munson, R. J. 1966. Excision-repair of DNA damage in an auxotrophic strain of *E. coli*. *Biochem. Biophys. Res. Comm.* 22:268.

100. Hill, R. F. 1965. Ultraviolet-induced lethality and reversion to prototrophy in *Escherichia coli* strains with normal and reduced dark repair ability. *Photochem. Photobiol.* 4:563.

101. Witkin, E. M. 1966. Radiation-induced mutations and their repair. *Science* 152:1345.

102. Witkin, E. M. 1967. Mutation-proof and mutation-prone modes of survival in derivatives of *Escherichia coli* B differing in sensitivity to ultraviolet light. *Brookhaven Symp. Biol.* 20:17.

103. Witkin, E. M. 1969. The mutability toward ultraviolet light of recombination-deficient strains of *Escherichia coli*. *Mutation Res.* 8:9.

104. Weigle, J. J. 1953. Induction of mutations in a bacterial virus. *Proc. Natl. Acad. Sci.* (USA) 39:628.

105. Radman, M. 1975. SOS repair hypothesis: phenomenology of an inducible repair which is accompanied by mutagenesis. In *Molecular mechanisms for repair of DNA*. P. C. Hanawalt and R. B. Setlow, eds., p. 355. New York: Plenum.

106. Miura, A., and Tomizawa, J-I. 1968. Studies on radiation-sensitive mutants of *E. coli*. III. Participation of the *rec* system in induction of mutation by ultraviolet irradiation. *Molec. Gen. Genet.* 103:1.

107. Ogawa, H., Shimada, K., and Tomizawa, J-I. 1968. Studies on radiation-sensitive mutants of *E. coli*. I. Mutants defective in the repair synthesis. *Molec. Gen. Genet.* 101:227.

108. Kneser, H. 1968. Relationship between K-reactivation and UV-reactivation of bacteriophage λ. *Virology* 36:303.

109. Defais, M., Fauquet, P., Radman, M., and Errera, M. 1971. Ultraviolet reactivation and ultraviolet mutagenesis of λ in different genetic systems. *Virology* 43:495.

110. Castellazzi, M., George, J., and Buttin, G. 1972. Prophage induction and cell division in *E. coli*. I. Further characterization of the thermosensitive mutation *tif*-1 whose expression mimics the effect of UV-irradiation. *Molec. Gen. Genet.* 119:139.

111. Castellazzi, M., George, J., and Buttin, G. 1972. Prophage induction and cell division in *E. coli*. II. Linked (*rec*A *zab*) and unlinked (*lex*) suppressors of *tif*-1-mediated induction and filamentation. *Mol. Gen. Genet.* 119:153.

112. Witkin, E. M. 1974. Thermal enhancement of ultraviolet mutability in a *tif*-1 *uvr*A derivative of *Escherichia coli* B/r: evidence that ultraviolet mutagenesis depends upon an inducible function. *Proc. Natl. Acad. Sci.* (USA) 71:1930.

113. Doudney, C. O. 1975. The two-lesion hypothesis for UV-induced mutation in relation to recovery of capacity for DNA replication. In *Molecular mechanisms for the repair of DNA*. P. C. Hanawalt and R. B. Setlow, eds., p. 389. New York: Plenum.

114. Witkin, E. M., and George, D. L. 1973. Ultraviolet mutagenesis in *pol*A and *uvr*A *pol*A derivatives of *Escherichia coli* B/r: evidence for an inducible error-prone repair system. *Genetics* 73(suppl.):91.

115. George, J., Devoret, R., and Radman, M. 1974. Indirect ultraviolet-reactivation of phage λ. *Proc. Natl. Acad. Sci.* (USA) 71:144.

116. Cooper, P. K. 1981. Inducible excision repair in *Escherichia coli*. In *Chromosome damage and repair*. E. Seeberg and K. Kleppe, eds., p. 139. New York: Plenum.

117. Gudas, L. J., and Pardee, A. B. 1975. Model for the regulation of *Escherichia coli* DNA repair functions. *Proc. Natl. Acad. Sci.* (USA) 72:2330.

118. Gudas, L. J., and Pardee, A. B. 1976. DNA synthesis inhibition and the induction of protein X in *Escherichia coli*. *J. Mol. Biol.* 101:459.

119. Gudas, L. J. 1976. The induction of protein X in DNA repair and cell division mutants of *Escherichia coli*. *J. Mol. Biol.* 104:567.

120. McEntee, K. 1977. Protein X is the product of the *rec*A gene of *Escherichia coli*. *Proc. Natl. Acad. Sci.* (USA) 74:5275.

121. Little, J. W., and Kleid, D. G. 1977. *Escherichia coli* protein X is the *rec*A gene product. *J. Biol. Chem.* 252:6251.

122. Moody, E. E. M., Low, K. B., and Mount, D. W. 1973. Properties of strains of *Escherichia coli* K-12 carrying mutant *lex* and *rec* alleles. *Mol. Gen. Genet.* 121:197.

123. George, J., Castellazzi, M., and Buttin, G. 1975. Prophage induction and cell division of *E. coli*. III. Mutations *sfi*A and *sfi*B restore division in *tif* and *lon* strains and permit the expression of mutator properties of *tif*. *Mol. Gen. Genet.* 140:309.

124. Mount, D. W. 1977. A mutant of *Escherichia coli* showing constitutive expression of the lysogenic induction and error-prone DNA repair pathways. *Proc. Natl. Acad. Sci.* (USA) 74:300.

125. Uhlin, B. E., and Clark, A. J. 1981. Overproduction of the *Escherichia coli* recA protein without stimulation of its proteolytic activity. *J. Bacteriol.* 148:386.

126. Volkert, M. R., Margossian, L. J., and Clark, A. J. 1981. Evidence that *rnm*B is the operator of the *Escherichia coli* *rec*A gene. *Proc. Natl. Acad. Sci.* (USA) 78:1786.

127. Little, J. W., Edmiston, S. H., Pacelli, L. Z., and Mount, D. W. 1980. Cleavage of the *Escherichia coli* lexA protein by the recA protease. *Proc. Natl. Acad. Sci.* (USA) 77:3225.

128. Horii, T., Ogawa, T., and Ogawa, H. 1980. Organization of the *rec*A gene of *Escherichia coli*. *Proc. Natl. Acad. Sci.* (USA) 77:313.

129. Mount, D. W. 1980. The genetics of protein degradation in bacteria. *Ann. Rev. Genet.* 14:279.

130. Roberts, J. W., and Roberts, C. W. 1975. Proteolytic cleavage of bacteriophage lambda repressor in induction. *Proc. Natl. Acad. Sci.* (USA) 72:147.

131. Roberts, J. W., Roberts, C. W., and Mount, D. W. 1977. Inactivation and proteolytic cleavage of phage λ repressor *in vitro* in an ATP-dependent reaction. *Proc. Natl. Acad. Sci.* (USA) 74:2283.

132. Roberts, J. W., Roberts, C. W., and Craig, N. L. 1978. *Escherichia coli* *rec*A gene product inactivates phage λ repressor. *Proc. Natl. Acad. Sci.* (USA) 75:4714.

133. Craig, N. L., and Roberts, J. W. 1981. Function of nucleoside triphosphate and polynucleotide in *Escherichia coli* recA protein–directed cleavage of phage λ repressor. *J. Biol. Chem.* 256:8039.

134. Craig, N. L., and Roberts, J. W. 1980. *E. coli* recA protein–directed cleavage of phage λ repressor requires polynucleotide. *Nature* 283:26.

135. McEntee, K., and Weinstock, G. M. 1981. *Tif*-1 mutation alters polynucleotide recognition by the recA protein of *Escherichia coli*. *Proc. Natl. Acad. Sci.* (USA) 78:6061.

136. Phizicky, E. M., and Roberts, J. W. 1981. Induction of SOS functions: regulation of proteolytic activity of *E. coli* recA protein by interaction with DNA and nucleoside triphosphate. *Cell* 25:259.

137. Horii, T., Ogawa, T., and Ogawa, H. 1981. Nucleotide sequence of the *lex*A gene of *E. coli*. *Cell* 23:689.

138. Markham, B. E., Little, J. W., and Mount, D. W. 1981. Nucleotide sequence of the *lex*A gene of *Escherichia coli* K-12. *Nucleic Acids Res.* 9:4149.

139. Little, J. W. 1984. Autodigestion of lexA and phage λ repression. *Proc. Natl. Acad. Sci.* (USA) 81:1375.

140. Brent, R., and Ptashne, M. 1981. Mechanism of action of the *lex*A gene product. *Proc. Natl. Acad. Sci.* (USA) 78:4204.

141. Brent, R. 1982. Regulation and autoregulation by *lex*A protein. *Biochimie* 64:565.

142. Brent, R. 1983. Regulation of the *E. coli* SOS response by the *lex*A gene product. In *Cellular responses to DNA damage*. E. C. Friedberg and B. A. Bridges, eds., p. 361. New York: Alan R. Liss.

143. Little, J. W., Mount, D. W., and Yanisch-Perron, C. R. 1981. Purified lexA protein is a repressor of the *rec*A and *lex*A genes. *Proc. Natl. Acad. Sci.* (USA) 78:4199.

144. Horii, T., Ogawa, T., Nakatani, T., Hase, T., Matsubara, H., and Ogawa, H. 1981. Regulation of SOS functions: purification of *E. coli* lexA protein and determination of its specific site cleaved by the recA protein. *Cell* 27:515.

145. Sancar, A., Sancar, G. B., Rupp, W. D., Little, J. W., and Mount, D. W. 1982. LexA protein inhibits transcription of the *E. coli uvr*A gene *in vitro*. *Nature* 298:96.

146. Sancar, G. B., Sancar, A., Little, J. W., and Rupp, W. D. 1982. The *uvr*B gene of *Escherichia coli* has both *lex*A-repressed and *lex*A-independent promoters. *Cell* 28:523.

147. Little, J. W. 1983. Variations in the *in vivo* stability of *lex*A repressor during the SOS regulatory cycle. In *Cellular responses to DNA damage*. E. C. Friedberg and B. A. Bridges, eds., p. 369. New York: Alan R. Liss.

148. Little, J.W. 1982. Control of the SOS regulatory system by the level of recA protease. *Biochimie* 64:585.

149. McPartland, A., Green, L., and Echols, H. 1980. Control of *rec*A gene RNA in *E. coli*: regulatory and signal genes. *Cell* 20:731.

150. Little, J. W., and Hanawalt, P. C. 1977. Induction of protein X in *Escherichia coli*. *Mol. Gen. Genet.* 150:237.

151. Kenyon, C. J., and Walker, G. C. 1980. DNA-damaging agents stimulate gene expression at specific loci in *Escherichia coli*. *Proc. Natl. Acad. Sci.* (USA) 77:2819.

152. Walker, G. C., Elledge, S. J., Kenyon, C. J., Krueger, J. H., and Perry, K. L. 1982. Mutagenesis and other responses induced by DNA damage in *Escherichia coli*. *Biochimie* 64:607.

153. Kenyon, C. J., and Walker, G. C. 1981. Expression of the *E. coli uvr*A gene is inducible. *Nature* 289:808.

154. Fogliano, M., and Schendel, P. F. 1981. Evidence for the inducibility of the *uvr*B operon. *Nature* 289:196.

155. Bagg, A., Kenyon, C. J., and Walker, G. C. 1981. Inducibility of a gene product required for UV and chemical mutagenesis in *Escherichia coli*. *Proc. Natl. Acad. Sci.* (USA) 78:5749.

156. Huisman, O., and d'Ari, R. 1981. An inducible DNA replication–cell division coupling mechanism in *E. coli*. *Nature* 290:797.

157. Miller, H. I., Kirk, M., and Echols, H. 1981. SOS induction and autoregulation of the *him*A gene for site-specific recombination in *E. coli*. *Proc. Natl. Acad. Sci.* (USA) 78:6754.

158. Backendorf, C. M., van den Berg, E. A., Brandsma, J. A., Kartašova, T., van Sluis, C. A., and van de Putte, P. 1983. *In vivo* regulation of the *uvr* and *ssb* genes in *Escherichia coli*. In *Cellular responses to DNA damage*. E. C. Friedberg and B. A. Bridges, eds., p. 161. New York: Alan R. Liss.

159. Kumura, K., Oeda, K., Akiyama, M., Horiuchi, T., and Sekiguchi, M. 1983. The *uvr*D gene of *E. coli*. Molecular cloning and expression. In *Cellular responses to DNA damage*. E. C. Friedberg and B. A. Bridges, eds., p. 51. New York: Alan R. Liss.

160. Kato, J., and Shinoura, Y. 1977. Isolation and characterization of mutants of *Escherichia coli* deficient in induction of mutation by ultraviolet light. *Mol. Gen. Genet.* 156:121.

161. Steinborn, G. 1978. *Uvm* mutants of *Escherichia coli* K-12 deficient in UV mutagenesis. *Mol. Gen. Genet.* 165:87.

162. Elledge, S. J., and Walker, G. C. 1983. Proteins required for ultraviolet light and chemical mutagenesis in *Escherichia coli*: identification of the products of the *umu*C locus of *Escherichia coli*. *J. Mol. Biol.* 164:175.

163. Elledge, S. J., Perry, K. L., Krueger, J. H., Mitchell, B. B., and Walker, G. C. 1983. Cellular components required for mutagenesis. In *Cellular responses to DNA damage*. E. C. Friedberg and B. A. Bridges, eds., p. 353. New York: Alan R. Liss.

164. Smith, C. L., and Oishi, M. 1978. Early events and mechanisms in the induction of bactorial SOS functions: analysis of the phage repressor inactivation process *in vivo*. *Proc. Natl. Acad. Sci.* (USA) 75:1657.

165. Smith, C. L., and Oishi, M. 1978. Inactivation of phage repressor in a permeable cell system: role of *rec*BC DNase in induction. *Proc. Natl. Acad. Sci.* (USA) 75:3569.

166. Higgins, N. P., Peebles, C. L., Sugino, A., and Cozzarelli, N. R. 1978. Purification of subunits of *Escherichia coli* DNA gyrase and reconstitution of enzymatic activity. *Proc. Natl. Acad. Sci.* (USA) 75:1773.

167. Dharmalingam, K., and Goldberg, E. B. 1980. Restriction *in vivo*. V. Induction of SOS functions in *Escherichia coli* by restricted T4 phage DNA, and alleviation of restriction by SOS functions. *Mol. Gen. Genet.* 178:51.

168. Oishi, M., Irbe, R. M., and Morin, L. M. E. 1981. Molecular mechanisms for the induction of "SOS" functions. *Prog. Nuc. Acid. Res. Mol. Biol.* 26:281.

169. Devoret, R. 1981. Inducible error-prone repair and induction of prophage lambda in *Escherichia coli*. *Prog. Nuc. Acid. Res. Mol. Biol.* 26:251.

170. Haseltine, W. A. 1983. Site specificity of ultraviolet light induced mutagenesis. In *Cellular responses to DNA damage*. E. C. Friedberg and B. A. Bridges, eds., p. 3. New York: Alan R. Liss.

171. Haseltine, W. A. 1983. Ultraviolet light repair and mutagenesis revisited. *Cell* 33:13.

172. Witkin, E. M. 1975. Relationships among repair, mutagenesis, and survival: overview. In *Molecular mechanisms for repair of DNA*. P. C. Hanawalt and R. B. Setlow, eds., p. 347. New York: Plenum.

173. Bridges, B. A., Mottershead, R. P., and Sedgwick, S. G. 1976. Mutagenic DNA repair in *Escherichia coli*. III. Requirement for a function of DNA polymerase III in ultraviolet-light mutagenesis. *Mol. Gen. Genet.* 144:53.

174. Echols, H., Lu, C., and Burgers, P. M. J. 1983. Mutator strains of *Escherichia coli*, *mut*D and *dna*Q, with defective exonucleolytic editing by DNA polymerase III holoenzyme. *Proc. Natl. Acad. Sci.* (USA) 80:2189.

175. Horiuchi, T., Maki, H., Maruyama, M., and Sekiguchi, M. 1981. Identification of the *dna*Q gene product and location of the structural gene for RNase H of *Escherichia coli* by cloning of the genes. *Proc. Natl. Acad. Sci.* (USA) 78:3770.

176. Scheuermann, R., Tam, S., Burgers, P. M. J., Lu, C., and Echols, H. 1983. Identification of the ϵ subunit of *Escherichia coli* DNA polymerase III holoenzyme as the *dna*Q gene product: a fidelity subunit for DNA replication. *Proc. Natl. Acad. Sci.* (USA) 80:7085.

177. Green, M. H. P., Bridges, B. A., Eyfjord, J. E., and Muriel, W. J. 1977. An error-free excision-dependent mode of post-replication repair of DNA in *Escherichia coli*. *Coll. Internat. du CNRS* no. 256, p. 227. Paris CNRS.

178. Cooper, P. K. 1982. Characterization of long patch excision repair of DNA in ultraviolet-irradiated *Escherichia coli*: an inducible function under rec-lex control. *Mol. Gen. Genet.* 185:189.

179. Coulondre, C., and Miller, J. H. 1977. Genetic studies of the *lac* repressor. IV. Mutagenic specificity in the *lac*I gene of *Escherichia coli*. *J. Mol. Biol.* 117:577.

180. Hutchinson, F., Skopek, T. R., and Wood, R. D. 1983. Mechanism of mutagenesis for lambda phage. In *Cellular responses to DNA damage*. E. C. Friedberg and B. A. Bridges, eds., p. 501. New York: Alan R. Liss.

181. Miller, J. H. 1982. Carcinogens induce targeted mutations in *Escherichia coli*. *Cell* 31:5.

182. Brandenburger, A., Godson, G. N., Radman, M., Glickman, B. W., Van Sluis, C. A., and Doubleday, O. P. 1981. Radiation-induced base substitution mutagenesis in single-stranded DNA phage M13. *Nature* 294:180.

183. Livneh, Z. Personal communication.

184. Sedliaková, M., Slezáriková, V., Măsek, F., and Brozmanová, J. 1978. UV-inducible repair: influence on survival, dimer excision, DNA replication and breakdown in *Escherichia coli* B/r *hcr*$^+$ cells. *Mol. Gen. Genet.* 160:81.

185. Sedliaková, M., Brozmanová, J., and Măsek, F. 1975. Persistence of thymine dimers in the replicated DNA of *Escherichia coli* B/r *hcr*$^+$. *FEBS Lett.* 60:161.

186. Sedliaková, M., Brozmanová, J., Măsek, F., and Kleibl, K. 1981. Evidence that dimers remaining in preinduced *Escherichia coli* B/r *hcr*$^+$ become insensitive after DNA replication to the extract from *Micrococcus luteus*. *Biophys J.* 36:429.

187. Sedliaková, M., Brozmanová, J., Măsek, F., and Slezáriková, V. 1977. Interaction of restoration processes in UV irradiated *Escherichia coli* cells. *Photochem. Photobiol.* 25:259.

188. Sedliaková, M., Slezáriková, V., and Pĭrsel, M. 1978. UV-inducible repair II: its role in various defective mutants of *Escherichia coli* K-12. *Mol. Gen. Genet.* 167:209.

189. Bonura, T., Radany, E. H., McMillan, S., Love, J. D., Schultz, R. A., Edenberg, H. J., and Friedberg, E. C. 1982. Pyrimidine dimer–DNA glycosylases: studies on bacteriophage T4-infected and on uninfected *Escherichia coli*. *Biochimie* 64:643.

190. Lackey, D., Krauss, S. W., and Linn, S. 1982. Isolation of an altered form of DNA polymerase I from *Escherichia coli* cells induced for *rec*A/*lex*A functions. *Proc. Natl. Acad. Sci.* (USA) 79:330.

191. Lewin, B. 1983. *Genes*. New York: Wiley.

192. Walker, G. C. 1984. Mutagenesis and inducible responses to deoxyribonucleic acid damage in *Escherichia coli*. *Microbiol. Rev.* 48:60.

8

DNA Damage Tolerance in Eukaryotic Cells

8-1 Introduction

As indicated in Section 7-2, lesions such as pyrimidine dimers also represent blocks to semiconservative DNA synthesis in eukaryotic cells, including mammalian cells (1–5). However, current understanding of the tolerance by eukaryotes of such damage is far less complete than in the case of the prokaryote *E. coli*. In no small measure this paucity of information stems from the greater complexity of DNA replication in eukaryotic cells (1–5). In addition, the genetic framework that was so crucial to the establishment of inducible cellular responses to DNA damage in *E. coli* has not yet been constructed for any eukaryotic system, although attempts are being made to do so with genetically accessible organisms such as *S. cerevisiae, D. melanogaster* and mammalian viruses (6–10). Studies on putative inducible functions are further complicated by the fact that agents that *selectively* inhibit protein synthesis in prokaryotic cells (such as chloramphenicol) and thereby provide biochemical evidence for the activation of genes are not available for use in higher organisms. Hence even biochemical approaches to the regulation of

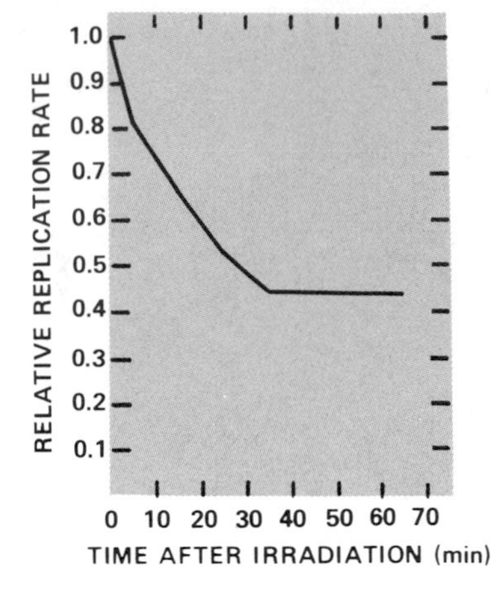

FIGURE 8-1
Inhibition of DNA synthesis in Chinese hamster ovary cells as a function of time after UV irradiation. (From J. Doniger, ref. 26.)

gene expression in eukaryotic cells exposed to DNA damage are controversial. Finally, a significant problem associated with studies on the cellular responses to DNA damage in eukaryotic cells comes from the tendency to extrapolate molecular mechanisms and biochemical pathways directly from *E. coli* and other prokaryotic systems. Although searches for parallels between *E. coli* and humans are certainly appropriate and often useful, it is important to remember that related biological phenomena in organisms at different levels of organization may differ profoundly and that experimental approaches that are productive in one organism may be fruitless in another.

8-2 Inhibition of DNA Synthesis in Mammalian Cells

It is generally observed that UV irradiation of mammalian cells reduces the rate of DNA synthesis (Fig. 8-1), presumably due to the presence of photoproducts such as pyrimidine dimers. However, the difficulties associated with enzymatic photoreactivation in mammalian cells in vivo (see Section 2-2) have precluded a definitive investigation of the photoreversibility of this inhibition as has been achieved with *E. coli* (see Section 7-2). Nonetheless, evidence for the involvement of pyrimidine dimers in DNA synthesis inhibition comes from studies on other eukaryotic cells in which enzymatic photoreactivation can be demonstrated in vivo. For example, in chick fibroblasts (11), marsupial cells (12) and frog cells (13), photoreactivation of dimers is accompanied by a significant recovery of the

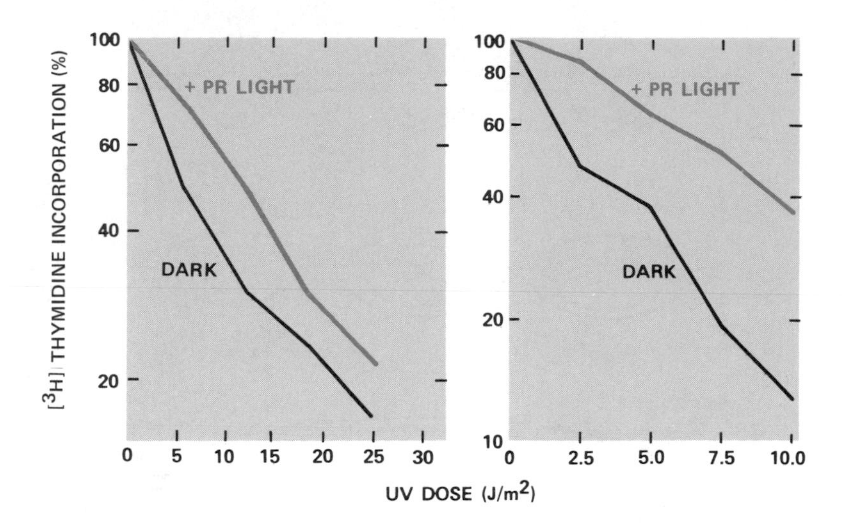

FIGURE 8-2
Exposure of chick cells (left) or frog cells (right) to photoreactivating light after UV irradiation decreases the extent of inhibition of DNA synthesis measured by the incorporation of [^{3}H]thymidine into DNA. (From A. R. Lehmann and S. Stevens, ref. 11, and B. S. Rosenstein and R. B. Setlow, ref. 13.)

rate of DNA synthesis (Fig. 8-2). This recovery is usually not complete, however, and it is therefore possible that other photoproducts, such as (6-4) photoproducts (see Section 1-5), play a role in the inhibition of DNA synthesis, or that dimers in replication forks are less photoreactivable than those in nonreplicating DNA (3).

A major problem in deciphering the response of mammalian cells to DNA damage at replication forks is that such cells normally replicate their DNA in multiple relatively small replicons (Fig. 8-3), the exact size and number of which is not firmly established (14–17) and which may vary from cell to cell and in different stages of cell development. The existence of multiple small replicons presents problems in the interpretation of experiments aimed at investigating phenomena such as discontinuous DNA synthesis with gap formation and translesion DNA synthesis. These problems will be explicated more specifically later in Section 8-2.

Inhibition of DNA synthesis in UV-irradiated mammalian cells may result from any one or more of a number of possible perturbations of normal DNA replication (1). First, *initiation* of synthesis in certain replicons may be inhibited by photoproducts present in those replicons. The initiation of DNA replication is one of the most sensitive points of regulation in cells subjected to DNA damage. Low doses of X rays, UV light or other DNA damaging agents that result

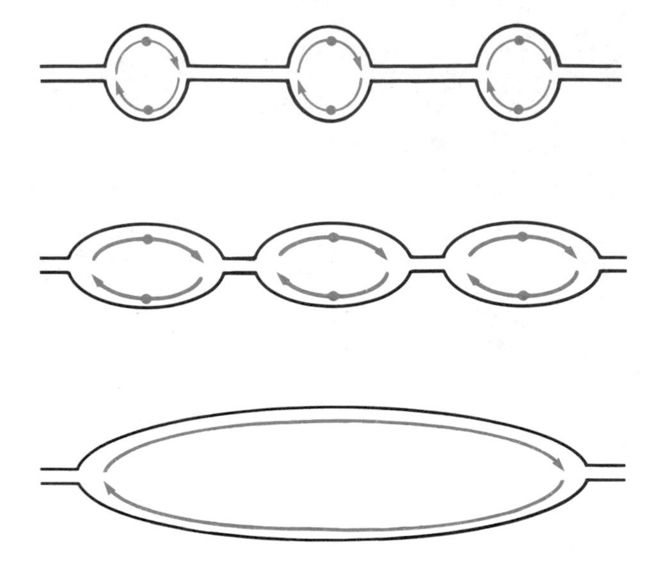

FIGURE 8-3
Schematic representation of DNA replication in a cluster of three replicons in a mammalian chromosome. Replication is initiated simultaneously at all three origins (the blue circles) and proceeds bidirectionally until newly replicated strands from adjacent replicons meet, allowing joining of these strands into DNA lengths greater than that of single replicons. Arrows on the newly replicated strands show their polarity. (After J. D. Hall and D. W. Mount, ref. 1.)

in low frequencies of direct or enzymatically produced (during repair) incisions in DNA rapidly inhibit as many as 50 percent of the replicon initiations that should occur within an hour of exposure to damage (18–20). It has even been suggested that the presence of a lesion in a replicon *domain* (consisting of multiple replicons under coordinate regulation) might inhibit initiation of DNA synthesis in all of the replicons in that domain (21–25). As will be seen in Section 9-3, in at least one human disease (ataxia telangiectasia) arrested initiation of replication is less sensitive to DNA damage than in normal human cells. The abnormal response in this disease suggests that the inhibition of initiation normally seen is not a passive consequence of DNA damage but is actively mediated by specific gene products (18).

Second, *cessation or slowing of replication* may result from regulatory disruptions effected by damage in preinitiated replicons or replicon domains. Inhibition of DNA synthesis may also result from *direct blockage of replication fork progression* by dimers and/or other lesions, and such blocks may be permanent or temporary in different replicons. Finally, enzymes utilized in normal semiconservative replication might be sequestered for DNA repair in such a way that replication is inhibited while certain repair processes are active. The relative influence of any or all of these parameters may vary with UV dose, time after irradiation, culture conditions and cell type (1–5).

It is instructive to address the evidence for and against some of these explanations for the inhibition of DNA synthesis in mammalian cells. This is certainly a useful way of gaining some perspective on the experimental difficulties and interpretive complexities associated with studies on cells exposed to DNA damage. In these discussions emphasis is placed on UV radiation damage, and for convenience this will be considered to be pyrimidine dimers. However, many relevant studies have employed other DNA damaging agents with essentially the same results.

Effects of DNA Damage on Initiation of DNA Replication

When Chinese hamster ovary cells are pulse labeled with [^{3}H]thymidine, the distances between replication forks that were initiated from the same origin can be measured with fiber autoradiography (26) (Fig. 8-4). A recently replicated region appears as a tandem pair of labeled segments separated by an unlabeled section (synthesized prior to the addition of label) that contains the origin of replication (Fig. 8-4). The length of the labeled regions increases as a function of the pulse times with radiolabel, which makes possible measurement of the *rate of replication fork progression*. The relative proportion of very short unlabeled regions to longer ones is also an indication of whether new replicons have been initiated normally. The results of such experiments indicate no differences between irradiated and unirradiated cells (26) (Fig. 8-5). However, it has been pointed out (1) that such studies do not provide any indication of what *fraction of the total number of replicons* in irradiated cells are functional compared with that in unirradiated cells. In addition this type of experiment only scores those replicons in which initiation proceeds bidirectionally. Thus, faulty initiation in which one fork fails to progress from the origin would not be detected (1).

The ability of human cells to initiate DNA synthesis after UV irradiation has also been inferred by measuring the *total amount of DNA synthesized* per unit time and the *size of the newly synthesized DNA* (27). In irradiated cells both of these parameters initially decrease and then recover with time (Fig. 8-6). The recovery of normal-size labeled DNA strands is generally interpreted as an indication that previously initiated replication forks that were stalled at pyrim-

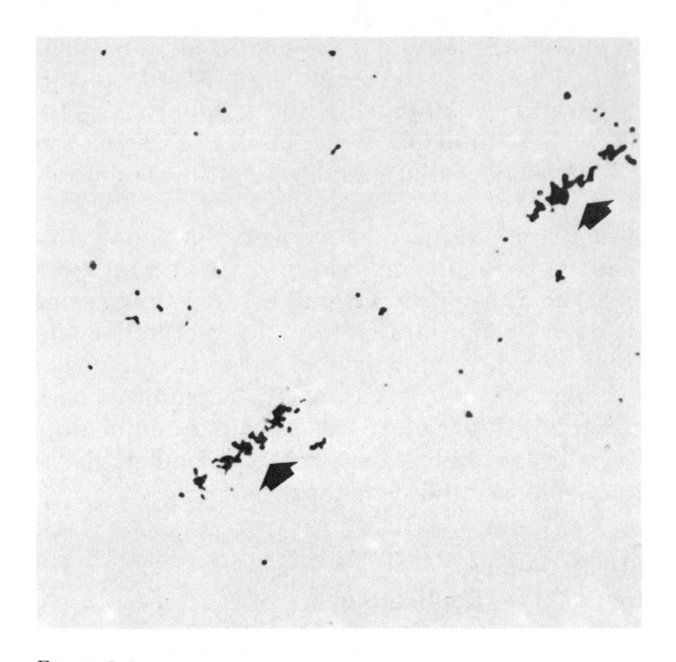

FIGURE 8-4
DNA fiber autoradiography. The linear arrays of autoradiographic grains (arrows) represent segments of DNA that were replicated during a pulse with [^{3}H]thymidine. A bidirectionally replicating region appears as a tandem pair of labeled segments, each of which is the product of a single replication fork, separated by an unlabeled section that was synthesized prior to the pulse and from which replication initiated at a single origin. The mean lengths of the unlabeled segments can be plotted as a function of the pulse times, and from the resulting slope the rate of replication fork progression can be calculated (see Fig. 8-5). In addition the relative proportion of very short unlabeled regions to longer ones in UV-irradiated cells provides a qualitative indication of whether inhibition of replicon initiation is occuring. (Courtesy of Dr. Jay Doniger.)

idine dimers eventually overcame this block by one or more mechanisms (27). An observation more germane to the present discussion on initiation of replication is that when the recovery of fork progression is apparently complete, the *total* amount of DNA synthesis per unit time is still *reduced* relative to that of unirradiated cells (Fig. 8-6). This suggests that the inhibition of DNA synthesis reflects the *failure of some replicons to be initiated*. The same interpretation has been given to the results of similar experiments, in which cells were exposed to ionizing radiation or to chemical damage to DNA. In such cells there is a decrease in the number of small newly labeled DNA molecules relative to the number in untreated cells, whereas the

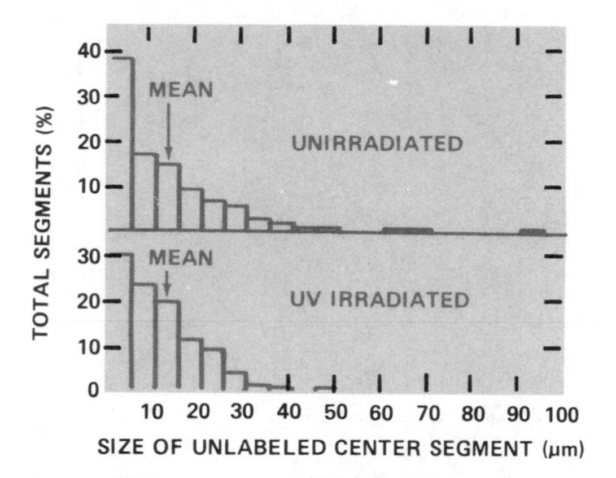

FIGURE 8-5
The distribution of unlabeled replicon segments according to their length (see Fig. 8-4) in unirradiated and UV-irradiated (5 J/m^2) Chinese hamster ovary cells is the same. Only replicons found in tandem sets were measured, since these unequivocally represent bidirectionally replicating replicons. (From J. Doniger, ref. 26.)

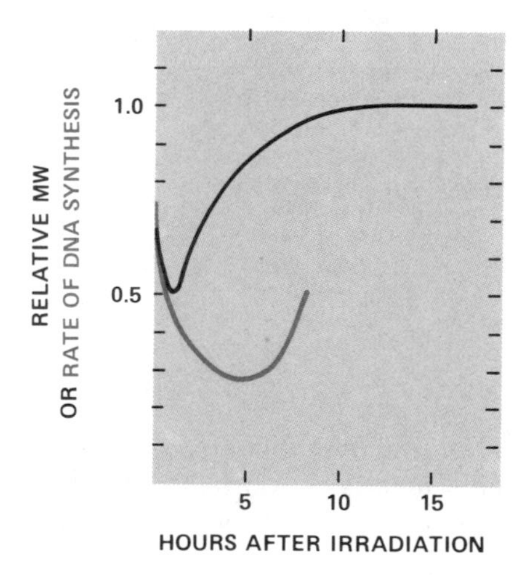

FIGURE 8-6
The relative molecular weight of newly synthesized DNA can be measured by the sedimentation of pulse-labeled cells in alkaline sucrose gradients. The overall rate of DNA synthesis is measured by the incorporation of [^{3}H]thymidine. The observation that the molecular weight of the newly synthesized daughter strands (black curve) in UV-irradiated cells reaches that seen in unirradiated cells at a time when the rate of DNA synthesis (blue curve) has not yet recovered, suggests that some replicons failed to initiate replication in the irradiated cells. (After S. D. Park and J. E. Cleaver, ref. 27.)

quantity of large pulse-labeled molecules is hardly affected (Fig. 8-7). This suggests that growing DNA strands continue to be elongated but new chains cannot be initiated, leading to the relative paucity of low molecular weight DNA (21–25, 28, 29).

As indicated above it has been proposed that agents (such as ionizing radiation) that cause DNA strand breaks inhibit initiation of DNA

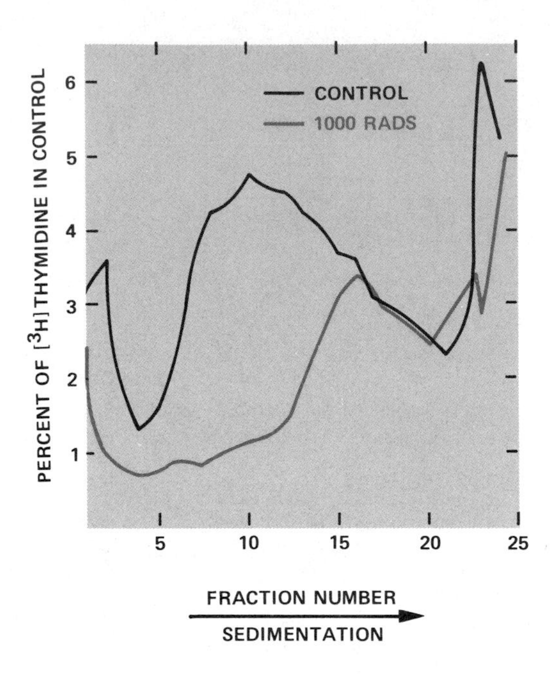

FIGURE 8-7
Alkaline sucrose density gradient profiles of DNA from unirradiated (control) or X-irradiated (1000 rads) Chinese hamster ovary cells. In the X-irradiated cells there is a decrease in the relative amount of small DNA molecules (fractions 1–10) that reflect regions of DNA in which replication was initiated soon after the irradiation. However, the relative quantity of *large* molecules (fractions 15–20) that reflect the progression of replicons that were initiated *prior* to the irradiation is not significantly decreased. This suggests that *growing* DNA strands continue to be elongated but new chains cannot be initiated. (From R. B. Painter and B. R. Young, ref. 22.)

synthesis in *clusters of replicons* extending over domains as large as 10^9 in molecular weight (21–25). This idea is supported by the fact that normal human fibroblasts exposed to UV radiation under conditions permissive for excision repair (which of course results in enzyme-catalyzed DNA strand breaks) also show a relative paucity of short nascent DNA strands, where xeroderma pigmentosum cells, which are defective in excision repair (see Section 9-2), do not show this putative inhibition of replicon initiation (27, 30) (Fig. 8-8). However, it has been pointed out (4) that after ionizing radiation, the rejoining of the vast majority of DNA strand breaks in mammalian cells is very rapid, whereas the rates of DNA synthesis are slowest *after* most of the strand breaks have been rejoined (31). In addition,

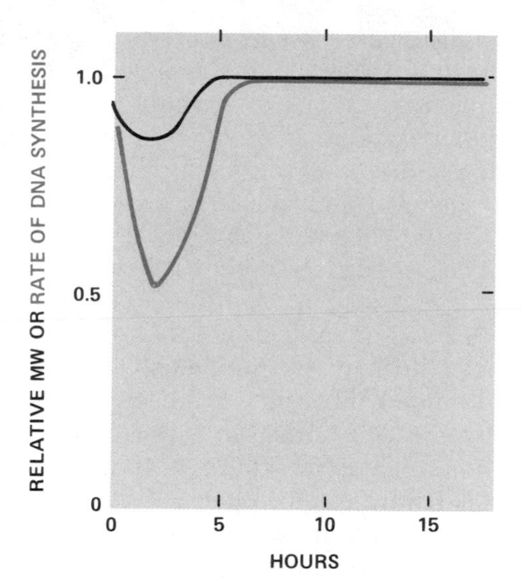

FIGURE 8-8
When xeroderma pigmentosum (XP) cells are exposed to UV radiation, the overall rate of recovery of DNA synthesis (blue curve) coincides with the rate of recovery of normal-size nascent DNA fragments (black curve). This suggests that the XP cells do not suffer inhibition of replicon initiation as do normal cells (see Fig. 8-6) and that the inhibition in normal cells is due to strand breaks produced during excision repair, since XP cells are defective in enzyme-induced incision of DNA following UV irradiation. (After S. D. Park and J. E. Cleaver, ref. 27.)

if X-irradiated cells are incubated in the presence of fluorodeoxyuridine (an inhibitor of DNA synthesis), this has no effect on the rejoining of strand breaks (32). Yet even after four hours in fluorodeoxyuridine, on removal of the inhibitor a depression of DNA synthesis rates is still observed (4, 31). Finally, after UV irradiation, the inhibition of DNA synthesis in xeroderma pigmentosum cells is at least as great as in normal cells, despite the fact that these cells do not carry out significant enzyme-catalyzed strand breakage of DNA (5). Thus a direct relation between DNA strand breaks and inhibition of replicon initiation is still equivocal (4, 5).

Effects of DNA Damage on DNA Elongation

Initiation of replication is organized at the level of replicon clusters approximately 10^9 daltons in size, which present large targets for DNA damage. Another possible explanation for the reduced overall rate of DNA synthesis after UV irradiation of mammalian cells is reduction in the *rate of DNA chain elongation following normal initiation* of synthesis, due to *arrest of replication* at sites of damage in template strands. Fork progression is a more locally regulated process involving individual replicons and therefore presents a smaller target for damage, and so effects tend to be observed only at relatively high levels of base damage to DNA (18). The arrest of replication fork progression might be permanent for some or all replicons, or it might be overcome by either a continuous or a discontinuous mode of DNA synthesis, as described in *E. coli* (see Section 7-3). It has proved difficult to distinguish between these various alternatives, and much controversy has been generated over the interpretation of experiments aimed at investigating them (1–5). For example, an autoradiographic study of human HeLa cells (33) showed that immediately after irradiation with fluences greater than 5 J/m^2, fibers accumulate with lengths approximating interdimer distances in the parental DNA. This suggests that replication forks are indeed blocked at sites of pyrimidine dimers. On the other hand, in the fiber autoradiographic experiments with Chinese hamster ovary cells described in the previous subsection, the *rate of fork progression* was also examined by measuring the lengths of the labeled regions as a function of labeling time. These experiments led to the interpretation that after low UV fluences some replicons cease to synthesize DNA within 30 minutes after irradiation, but those that continue to synthesize DNA do so with normal rates of fork progression (26). Thus if DNA synthesis is blocked at sites of dimers, it apparently can rapidly resume past the site of blockage.

Attempts have been made to establish the relative contributions of inhibition of initiation of replication and of DNA chain growth to the overall depression in DNA synthesis observed in mammalian cells subjected to DNA damage. A general conclusion that emerged from some studies (34) is that within the first hour after irradiation DNA replication is affected in a way that depends on the *average number of lesions per replicon*. Hence, at low numbers of lesions per replicon inhibition of initiation is the predominant response, whereas at higher numbers of lesions per replicon blockage of DNA chain growth is also observed. Furthermore, after irradiation of human fibroblasts with a dose of UV light that blocks chain growth, the rate at which cells recover their ability to synthesize increasingly more and larger-size DNA is a function both of *replicon size* and of *excision repair capacity*. Cells with small replicons recover more rapidly than cells with large replicons (Fig. 8-9) and excision-repair-defective cells recover less rapidly than excision-competent cells (34). The curve relating the recovery rate (k) to replicon size corresponds to the relation $k \alpha$ (replicon size)$^{-1}$ (Fig. 8-9). Since for cells

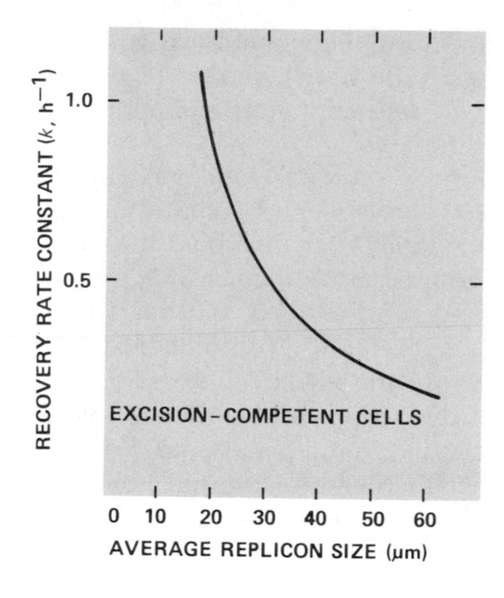

FIGURE 8-9
The relationship between the rate constant for the recovery of DNA replication and the average replicon size shows that in cells that are competent for excision repair, the smaller the replicon size, the more rapid the recovery. This is consistent with the observation that the smaller the replicon size the greater the number of initiation sites for DNA synthesis per cell. (From J. E. Cleaver et al., ref. 34.)

with similar DNA contents replicon size is roughly inversely proportional to the number of replicon initiation sites per cell, the larger the number of replicon initiation sites, the faster the rate of recovery of DNA synthesis.

Daughter-Strand Gap Formation during Semiconservative Synthesis in Mammalian Cells

Aside from the parameters of excision repair capacity and number of replicon initiation sites, the rate of recovery of DNA synthesis in cells containing template base damage is presumably also influenced by factors affecting the *bypass* of lesions that arrest fork progression or slow it down. Hence considerable time and effort have been devoted to the investigation of a possible discontinuous mode of DNA synthesis with daughter-strand gap formation in mammalian cells. An experimental approach analogous to that described for *E. coli* (see Section 7-3) reveals a cellular response that in general resembles the discontinuous mode of DNA synthesis defined in prokaryotes (35–38). However, the interpretation of the radioactivity profiles from alkaline sucrose sedimentation of pulse-labeled DNA from mammalian cells is much more controversial than for bacterial DNA (2). First, there are technical problems. For example, the DNA molecules that are isolated from gently lysed mammalian cells sediment anomalously because of their very large size. This problem has been addressed by intentionally reducing their size immediately prior to sedimentation by shearing or by exposing them to radiation to cause strand breaks (36, 38–42). But even under these conditions sedimentation profiles tend to be broad and calculations of the molecular weight of DNA fragments are subject to significant error, particularly with respect to smaller molecules in the upper portion of the gradients, where the relation between sedimentation and molecular weight is difficult to establish accurately.

As indicated earlier in this section, a more fundamental problem arises from the fact that eukaryotic chromosomes contain multiple tandem replicons, which are considerably smaller than the single bacterial replicon (33, 43, 44). Mammalian replicons range in size from 8 to 560 million daltons, with an average of 200 million daltons, whereas the *E. coli* replicon is about 2800 million daltons in size (2). The multiple mammalian replicons are arranged in clusters and have unique internal initiation sites from which replication is mainly bidirectional (2). As a consequence, the overall pattern of DNA replication in mammalian chromosomes involves the joining of daughter strands at several different levels (Fig. 8-3). For example, low molecular weight replicative intermediates must be joined within individual replicons. In addition, daughter strands in adjacent replicons must be joined. The latter process causes particular interpretive problems with respect to DNA synthesis in damaged cells, because depending on the amount of damage incurred in the DNA, the distance between lesions may approximate or exceed the size of many of the replicons. Furthermore, the rate at which daughter strands in normal adjacent replicons join may be similar to that at which daughter-strand discontinuities caused by damage are eliminated (Fig. 8-10). Finally, discontinuities in daughter strands may result not only from gaps left opposite lesions in template strands but also from the inhibition of replicon initiation, strand elongation or both. When strand elongation is inhibited, the discontinuities occur between *adjacent* active replicons. When initiation is inhibited, the discontinuities may also occur between nonadjacent active replicons separated by one or more inactive replicons (2) (Fig. 8-11). In summary, with so many possible explanations for discontinuities during DNA replication, it is often difficult to interpret experiments that demonstrate the phenomenon.

Other indicators of discontinuous DNA synthesis with gap filling in mammalian cells have been sought. It will be recalled that in bacteria the average size of the newly replicated DNA in alkaline sucrose gradients is roughly the same as the average distance between adjacent pyrimidine dimers on the template DNA strand (see Section 7-3). Similar comparisons have been made in mammalian cells. For a given mammalian cell type such a correspondence has been observed in some studies (26, 36, 45, 46), but in others it has been found that the average size of the newly synthesized strands *exceeds* that of the interdimer distance (47, 48), suggesting that no gaps are formed.

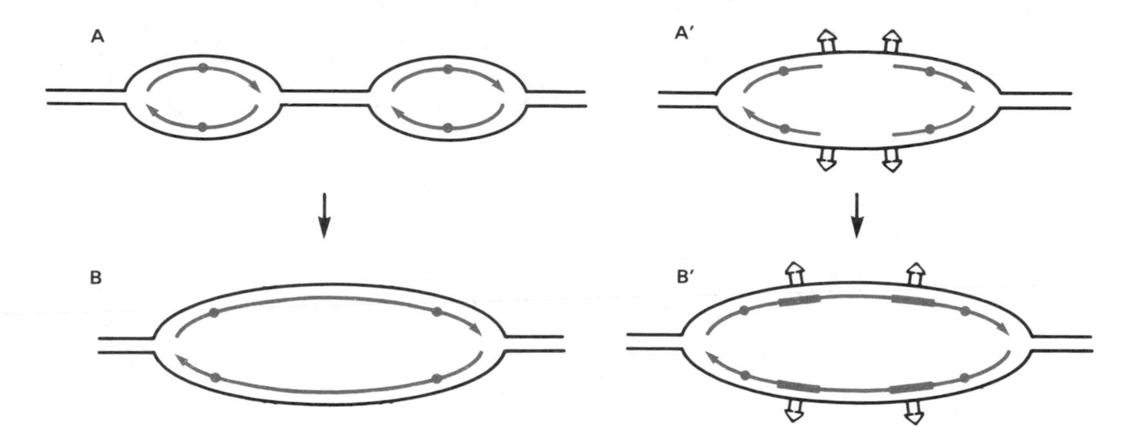

FIGURE 8-10
Diagrammatic representation of how the complexity of DNA replication in mammalian cells can result in difficulty in interpreting the results of sedimentation velocity analyses of UV-irradiated cells. The size of nascent DNA in unirradiated cells (A) may approximate that caused by replicative arrest in UV-irradiated cells (A′), and the joining of adjacent replicons (B) may be indistinguishable from gap filling opposite sites of dimers (B′).

If DNA synthesis in mammalian cells leaves gaps opposite pyrimidine dimers, as has been postulated in the classical model of discontinuous replicative bypass in *E. coli*, it should be possible to directly detect such gaps in the nascent strands opposite the lesions. For example, daughter-strand gaps should be sensitive to degradation by single-strand-specific nucleases, which would generate double-strand breaks in the DNA (Fig. 8-12). Such breaks have been observed, and the size of the double-strand fragments produced approximates the average interdimer distance (3, 49, 50), arguing in favor of gaps opposite dimers. However, a distinction between these gaps and other structures in replicating DNA that might also generate regions sensitive to single-strand-specific nucleases is problematic. More direct evidence for the presence of gaps in DNA *specifically located opposite pyrimidine dimers* would be the formation of double-strand breaks by the treatment of newly replicated DNA from UV-irradiated cells with the PD DNA glycosylase–AP endonuclease from *M. luteus* or phage T4, since these enzymes should specifically attack dimers present in intact parental strands (Fig. 8-12). Generally, such experiments have yielded negative results (46, 51, 52).

In a different experimental approach to the detection of putative daughter-strand gaps, cells were pulse labeled with radioactive thymidine soon after UV irradiation (Fig. 8-13). As expected, DNA isolated immediately after a radioactive pulse sedimented as small molecules relative to the DNA isolated from unirradiated cells. However,

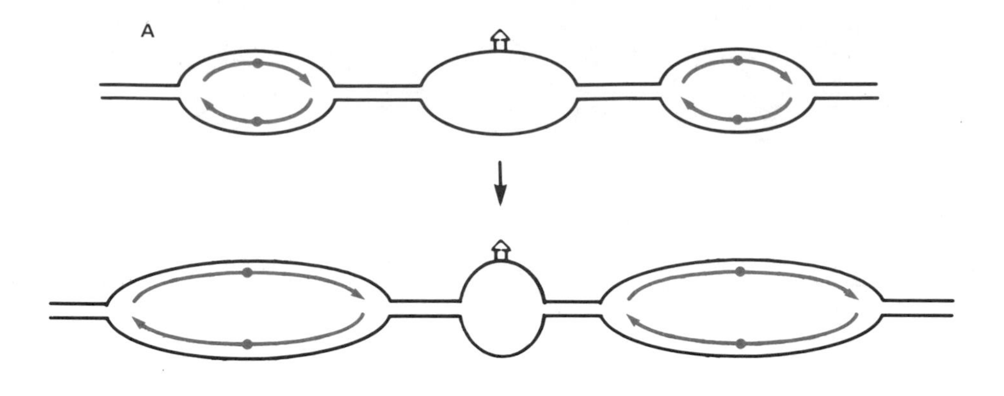

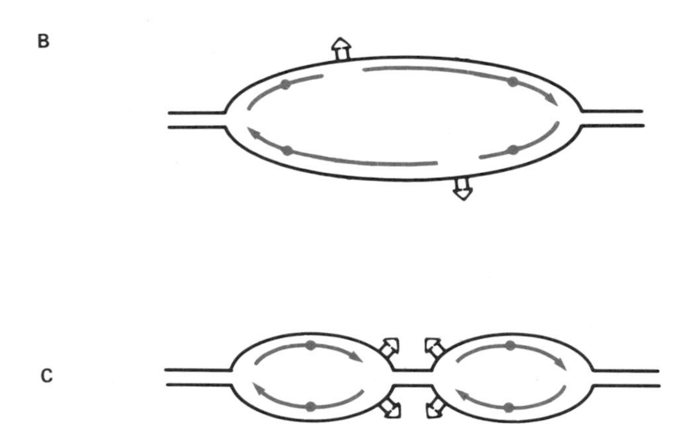

FIGURE 8-11
Gaps in newly synthesized UV-irradiated DNA may result from multiple processes. (A) Discontinuities result from failure of initiation of DNA synthesis in a replicon (containing a pyrimidine dimer) flanked by two normal replicons; (B) gaps arise from interrupted DNA synthesis; (C) discontinuities result from arrested replication.

if following the pulse the cells were incubated for some time in BrdU, the DNA sedimented as longer fragments, suggesting that putative gaps were filled in by de novo DNA synthesis (53) (Fig. 8-13). Consistent with this interpretation, when the longer DNA was exposed to 313 nm light, photolysis of the incorporated 5-BU (Section 5-2) resulted in the genesis of smaller molecules similar in size to those obtained immediately after pulse labeling. An alternative expla-

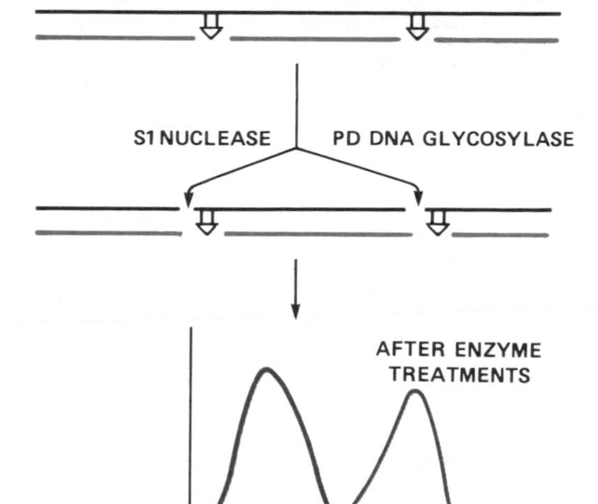

FIGURE 8-12
A number of enzyme probes have been used in attempts to demonstrate the presence of replicative gaps opposite pyrimidine dimers in DNA. Thus treatment of the DNA with either S1 nuclease (which attacks the single-strand region opposite the gap) or with a dimer-specific enzyme probe such as the phage T4 PD DNA glycosylase–AP endonuclease (Section 3-4) is expected to result in double-strand breaks in the DNA that can be detected by a reduced rate of sedimentation in neutral sucrose gradients.

nation for these results, however, is that following temporary blockage at dimers, replication continues *without* covalent interruption; that is, *end addition occurs*, and this is the source of the incorporated BrdU. Attempts were made to distinguish between these possibilities by calculating the length of the photolabile segments. The calculations revealed lengths of approximately 400 to 1600 bases (53). Since the incubation period in the presence of BrdU was sufficient to allow replication of entire replicons, it was concluded that the relatively small BU-substituted regions represented filled gaps rather than continuous DNA synthesis across sites of base damage in template strands. One weak point in this conclusion (54) is that in UV-irradiated cells pool sizes may be altered such that BU substitution occurs at abnormally low levels. Thus the lengths of DNA synthesized might actually be much longer and could indeed result from end addition onto pulse-labeled molecules rather than from gap filling.

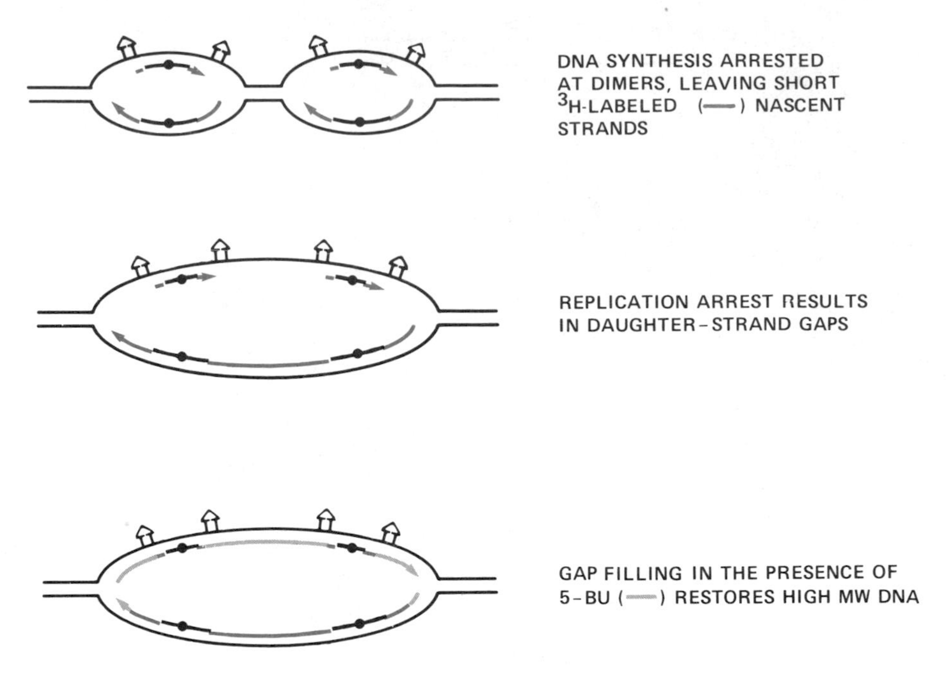

FIGURE 8-13
When cells are UV-irradiated and pulse labeled with [^{3}H]thymidine, short replicative intermediates are detected due to arrested replication (presumably at sites of pyrimidine dimers in template strands). When the cells are incubated further in the presence of unlabeled medium containing BrdU, high molecular weight DNA containing photolabile 5-BU is detected suggesting that postreplicative gaps are filled in by nonsemiconservative DNA synthesis.

Gap Filling by Recombinational Events in Mammalian Cells

It will be recalled that the evidence from *E. coli* suggests that gaps created during replication of irradiated DNA are filled by recombination with undamaged homologous DNA from sister chromosomes (see Section 7-3). Experiments designed to detect UV-induced strand exchanges using density-labeled DNA have been inconclusive in mammalian cells (55, 56). An alternative approach has been to probe DNA after UV irradiation for the presence of pyrimidine dimers in *daughter* strands, using the sensitivity of radiolabeled newly synthesized DNA molecules to the *M. luteus* or T4 enzymes as a specific assay for the presence of dimers. If no DNA molecules were replicating at the time of UV irradiation, then all the dimers should be confined to *parental* strands following post-UV DNA synthesis in the presence of [^{3}H]thymidine (Fig. 8-14). The presence of dimers associated with daughter (radiolabeled) DNA would thus be evidence for recombinational exchanges (Fig. 8-14). Some experiments of this

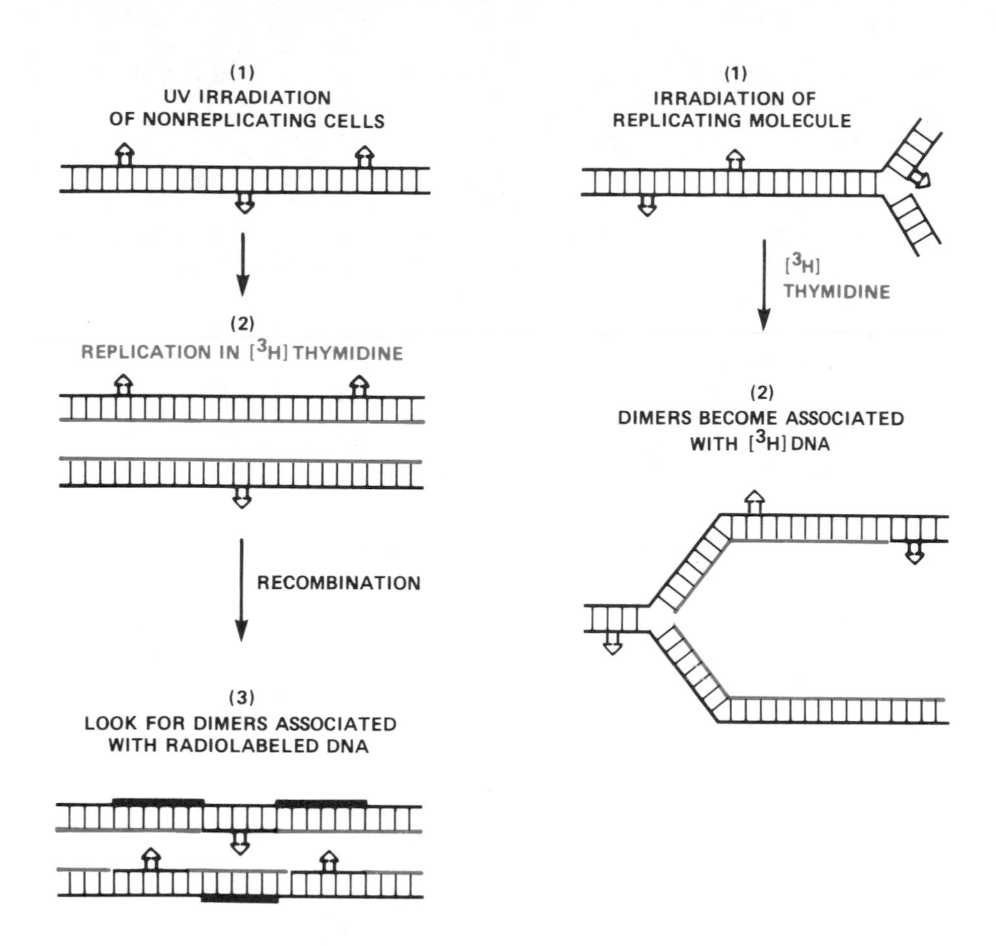

FIGURE 8-14
Recombinational exchange of DNA segments containing pyrimidine dimers must be distinguished from simple end addition. At the left *nonreplicating* DNA is exposed to UV radiation and the DNA is replicated in the presence of [^{3}H]thymidine. The covalent association of pyrimidine dimers with radiolabeled DNA is due to recombinational exchanges. (Thick lines indicate regions of possible repair synthesis of gaps created by recombinational exchange.) However, a covalent association of dimers with radiolabeled DNA can also result if dimers are present in DNA daughter strands generated *prior* to incorporation of [^{3}H]thymidine (right).

type show that the transfer of DNA segments containing pyrimidine dimers to daughter strands does occur, but at a level significantly lower than that observed in *E. coli* (52, 57–59). However, the interpretation of these results is complicated by the fact that if any DNA molecules were in the process of semiconservative DNA synthesis at the time of UV irradiation, then end addition to daughter strands already initiated and containing dimers could result in the covalent association of radiolabeled DNA with DNA containing dimers (Fig. 8-14).

This argument does not necessarily eliminate a recombinational mechanism for gap filling in mammalian cells, since modes of strand exchange are possible in which dimer transfer does not occur. For example, very short regions relative to the average distance between pyrimidine dimers might be exchanged, thereby reducing the probability of dimer transfer to newly synthesized DNA (1). Nonetheless, an obvious way of avoiding the irradiation of daughter strands whose synthesis was initiated prior to irradiation is to use cells in the G0 or G1 phase of the cell cycle. Such experiments have been carried out. In one study (60) three dimers per 10^9 daltons of daughter-strand DNA were detected after UV irradiation of a number of cell types, and in another study (61) as many as 12 dimers per 10^9 daltons of daughter-strand DNA were detected. This translates into a transfer of 3 to 20 percent of the dimers from parental to daughter DNA. Pyrimidine dimers in daughter-strand DNA were also detected in experiments in which the DNA was density labeled *prior* to irradiation. Following UV irradiation, heavy DNA (containing any daughter strands that were elongated by synthesis adjacent to dimer sites) was separated from normal-density DNA by isopycnic sedimentation. The normal DNA still contained as many as 20 dimers per 10^9 daltons of daughter DNA (61).

Although the role of recombination in the cellular responses to DNA damage in mammalian cells is still controversial, genetic recombination stimulated by exposure to UV radiation has been observed in virus-infected mammalian cells. Thus, the formation of recombinant animal viruses increases after infection of cells with pairs of genetically distinguishable UV-irradiated herpes simplex or SV40 viruses (62–64). In addition, multiplicity reactivation (a process in which enhanced survival of multiply infecting viruses is effected by genetic exchanges between damaged viral genomes) of UV-irradiated herpes simplex virus, human adenovirus and SV40 virus has been observed in human and monkey cells (63, 65, 66). If multiplicity reactivation of mammalian cell viruses occurs by a recombinational mechanism as it does in phage-infected bacteria (67), these observations provide some support for the notion that damage in *host-cell* DNA might also be tolerated by mechanisms involving recombinational strand exchanges.

Continuous Replicative Bypass of DNA Damage in Mammalian Cells: Translesion Synthesis during DNA Replication

In many experiments in which mammalian cells are exposed to UV radiation, the newly synthesized DNA is not smaller or is only slightly smaller than that in undamaged cells (1), suggesting either that if gaps are formed, they are sealed very rapidly, or that a continuous (*translesion*) DNA synthesis mode operates. The most persuasive evidence for the latter phenomenon comes from UV irradiation of SV40-infected monkey cells, using a temperature-sensitive SV40 mutant blocked in the initiation of DNA replication at nonpermissive temperatures in order to synchronize viral DNA replication (68). In this experimental system the size of the newly synthesized viral DNA fragments after irradiation closely approximates the average interdimer distance in the template strands (68, 69), which is consistent with replication of SV40 molecules bidirectionally from the origin until a pyrimidine dimer is encountered. The ratio of radiolabel incorporated into SV40 DNA sequences near the origin of replication to that in sequences farther removed from the origin is higher for irradiated than for unirradiated viruses, suggesting that DNA synthesis does not reinitiate beyond dimers. Nonetheless, when pulse-labeled cultures are subsequently incubated in medium containing unlabeled DNA precursors, labeled viral molecules that were initially small become progressively longer, as though replication past dimers occurs in a continuous mode after temporary arrest (68, 69). This conclusion is strengthened by the observation that sites sensitive to the phage T4 PD DNA glycosylase–AP endonuclease enzyme probe are not detected in nascent SV40 DNA strands; that is, there is no evidence for a recombinational mode of filling gaps created by discontinuous synthesis (68, 69).

A specific model for continuous replicative bypass of template base damage involves the so-called *copy-choice mechanism* (56, 70) (Fig. 8-15). This model postulates that when replication is blocked in one nascent DNA strand of a replication fork, replication of the other nascent strand can continue. Subsequently the nascent strands are partially displaced and pair, allowing the strand initially blocked at the dimer in the template strand to copy the required DNA sequence from the complementary nascent strand. The nascent strands then pair with the complementary respective parental strands and the normal mechanism of semiconservative DNA synthesis resumes until the next block is reached.

In summary, the investigation of the tolerance of DNA base damage in replicating mammalian cells is fraught with experimental difficulties and interpretative complexities. The weight of the evidence presently available suggests that DNA synthesis in UV-irradiated mammalian cells, although initially blocked at sites of UV damage, eventually continues past these lesions. Whether replication resumes by a continuous or discontinuous mode has not been adequately established, and it should be borne in mind that these modes need not be mutually exclusive. Although several models for continuous modes of replication have been proposed, none has been well documented experimentally (1). Thus it is not clear exactly how mammalian cells manage to tolerate unrepaired base damage in DNA, and this area of molecular biology requires considerable further study.

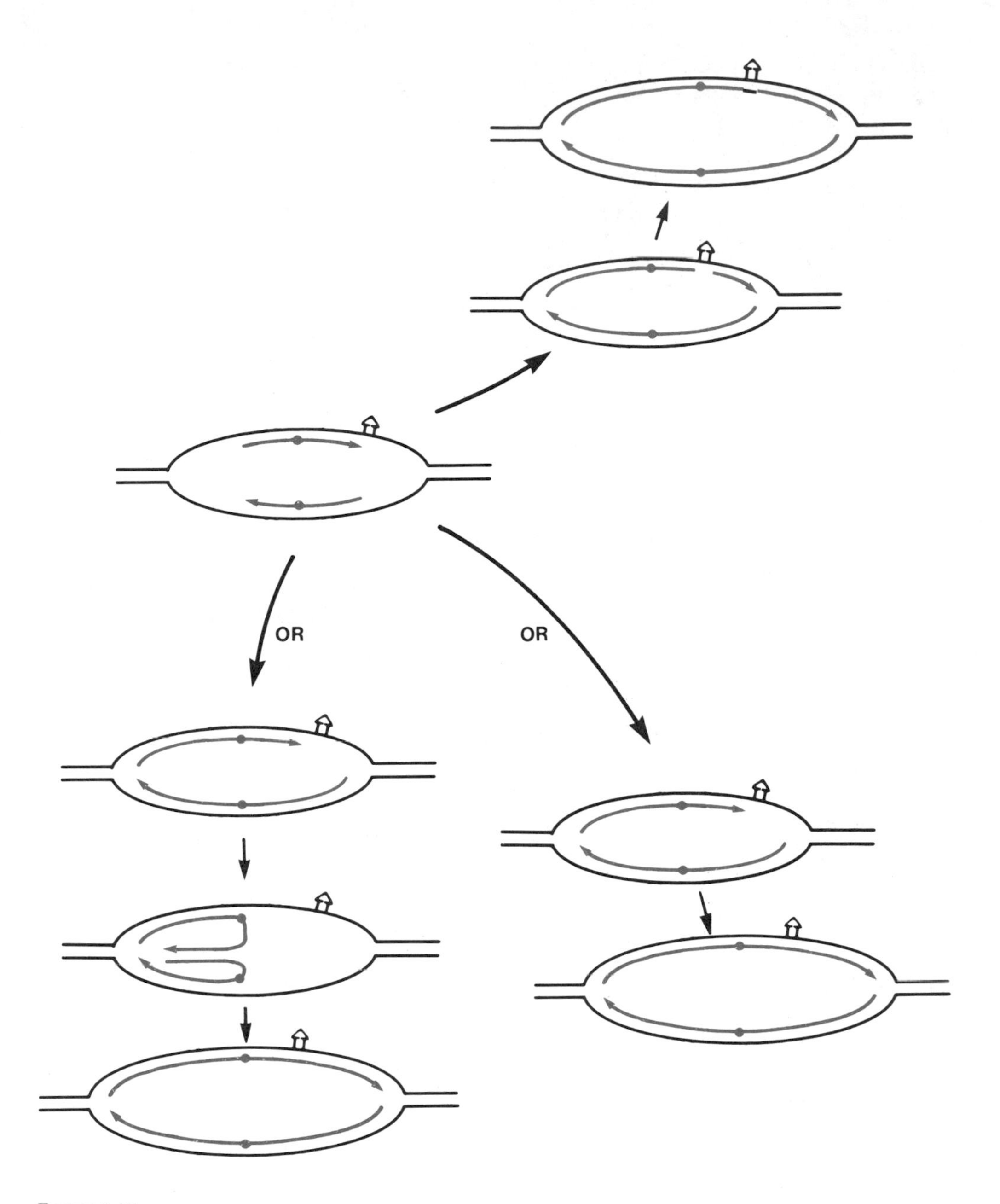

FIGURE 8-15
A summary of models for DNA replication in mammalian cells on templates containing bulky base damage such as pyrimidine dimers. The model at the upper right shows replicative bypass with gap filling (either by de novo synthesis or recombinational exchange). The model at the lower right shows temporary arrest at a dimer with subsequent transdimer synthesis. The model at the lower left shows a copy-choice mechanism in which the top nascent strand uses its complementary nascent sister strand rather than the parental strand as a template.

8-3 Inducible Responses to DNA Damage in Mammalian Cells?

The discovery of the SOS response in *E. coli* and the indications that inducible functions may be associated with the tolerance of DNA damage in living cells encouraged a search for analogous phenomena in eukaryotic cells. A number of observations suggest that mammalian cells respond to physiological perturbations that act as inducing signals in *E. coli*. However, unequivocal evidence for the derepression of previously repressed genetic functions is lacking, and there is certainly no indication that eukaryotic cells respond to inducing signals by expressing a battery of genes under the control of a common regulatory system, as is true in *E. coli* and other prokaryotic cells (see Chapter 7).

Enhanced Survival of UV-Irradiated Viruses in Host Cells Subjected to DNA Damage

Attempts have been made to demonstrate the phenomenon of Weigle (W) reactivation in mammalian cells using a variety of UV-irradiated DNA viruses and "inducing" host cells with a variety of different DNA damaging agents (6). Infection of UV-irradiated herpes virus (71–78), adenovirus (79) or single-strand DNA virus (71) in rodent, monkey, or human cells treated with UV radiation (71, 72, 77, 79–82), X rays (83), aflatoxin B_1, N-acetoxy-2-acetylaminofluorene, monofunctional alkylating agents, hydroxyurea or cycloheximide (72, 74, 82) is associated with *enhanced survival of the viruses*, although similar experiments with vaccinia and polio viruses (79) have not yielded this result. The enhanced reactivation of UV-irradiated herpes simplex virus in CV-1 monkey kidney cells is shown in Fig. 8-16.

The interpretation of these experiments must be viewed with caution. For example, unirradiated and UV-irradiated viruses are typically assayed by plaque formation on monolayers of UV-irradiated cells, an experimental protocol that creates potential problems because only one in 10^2 to 10^3 virus particles yields a plaque in the monolayer. It has been suggested that the enhanced survival (or reactivation) of the viruses may simply reflect the fact that damaged host cells are better recipients for viral adsorption and infection (84). However, studies with *transfecting SV40 DNA* show levels of enhanced reactivation that are quite comparable to those observed with the intact virus (81).

In general, the capacity of mammalian cells to support plaque formation by *unirradiated* viruses decreases as a function of UV

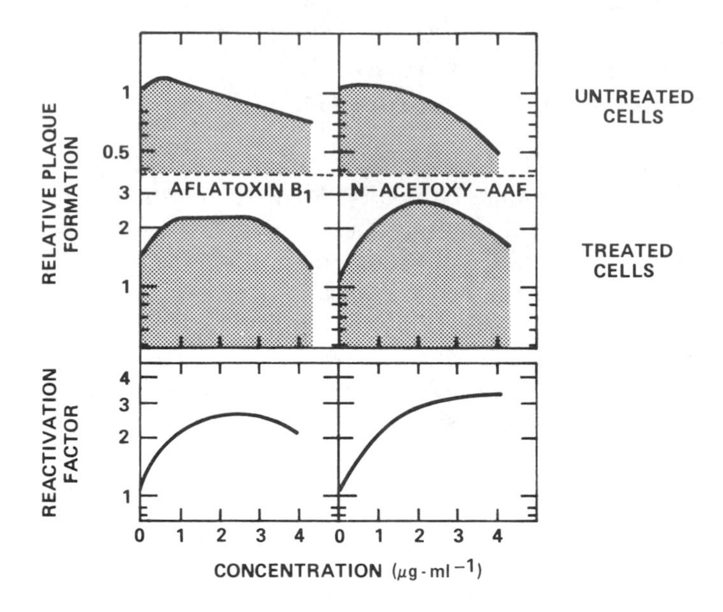

FIGURE 8-16
Effect of different concentrations of aflatoxin B_1 or of N-acetoxy-AAF on plaque formation by untreated (top figures) or treated (middle figures) CV-1 monkey kidney cells following infection with UV-irradiated herpes simplex virus. The reactivation factor (bottom figures) is the ratio of the plaque formation on treated and untreated cells. (From C. D. Lytle et al., ref. 72.)

exposure of the cells; however, the relative yield of plaques by *irradiated* virus is increased by low levels of UV exposure to the cells. The extent of the reactivation can be quantitated in terms of the so-called *UV reactivation factor*, obtained from the normalized ratio of the survival of unirradiated virus to irradiated virus (Fig. 8-16). In the majority of such experiments the reactivation factor is modest and certainly far less than that observed with phage λ after infection of *E. coli* (6) (see Section 7-4 and Fig. 7-28). Interestingly, the highest reactivation factor is found when infection following host cell irradiation (or damage by some other mechanism) is delayed for an optimum time corresponding to one or two cell generations, which in terms of cell division cycle time is roughly equivalent to the peak induction time in *E. coli* infected with UV-irradiated phage λ (6). Additionally, the UV-enhanced reactivation of herpes virus and of Kilham rat virus is inhibited in the presence of cycloheximide (71), suggesting a requirement for active protein synthesis. However, cycloheximide does not block protein synthesis selectively in mammalian cells; DNA synthesis is also arrested by this agent.

Little is known about the molecular mechanism of enhanced reactivation of mammalian viruses. In general, it is associated with treatments that cause breaks or gaps in cellular DNA or that inhibit DNA synthesis (85). Its expression in human cells is apparently not dependent on excision repair, since the phenomenon can also be demonstrated in XP cells that are excision repair defective (see Section 9-2) (86).

Enhanced Mutagenesis of Viruses in Host Cells Subjected to DNA Damage

A feature of W reactivation in *E. coli* is that it is highly mutagenic for UV-irradiated phages and moderately mutagenic for undamaged phages (87–89; see Section 7-4). Since UV radiation-enhanced reactivation of animal viruses resembles W reactivation in bacteria in some respects, it is relevant to ask whether or not the former phenomenon is also mutagenic. Some studies give an affirmative answer to this question. For example, the rate of reversion to wild-type of UV-irradiated SV40 temperature-sensitive mutants grown in UV-irradiated monkey kidney cells is increased relative to that in unirradiated cells (69, 90, 91) (Table 8-1). The extent of mutagenesis increases with increasing UV dose to the *virus*; however, no obvious correlation has been observed between the reversion frequency and the UV dose to the *cells*. This dependence on the state of the host cell for mutagenesis of SV40 virus has not been observed in an adenovirus–human fibroblast system (92).

TABLE 8-1
Frequency of reversion of SV40 *tsA58* mutants at 33°C

UV Irradiation (J/m^2)		
Virus	Cells	Reversion Frequency
0	0	$<5 \times 10^{-5}$
0	5	$<5 \times 10^{-5}$
0	10	$<10 \times 10^{-5}$
0	15	$<6 \times 10^{-5}$
1200	0	$<1 \times 10^{-4}$
1200	5	8×10^{-4}
1200	10	15×10^{-4}
1200	15	13.5×10^{-4}

From A. Sarasin and A. Benoit, ref. 90.

The multiplicity of infection with virus appears to be an important determinant of the extent of mutagenesis observed in mammalian cells (6, 85). Thus, the frequency of forward mutation into IrdU-resistant thymidine kinase mutants of UV-irradiated herpes simplex virus is not enhanced in irradiated relative to unirradiated cells under conditions of a low multiplicity of infection but is increased when the multiplicity of infection is greater than 10 (75). Under the latter conditions *multiplicity reactivation* of the virus is also observed, suggesting that enhanced mutagenesis (at least of herpes virus) may be better correlated with multiplicity reactivation than with enhanced reactivation due to some induced cellular response (77, 85).

The frequency of mutations resulting from UV irradiation of SV40 *ts* mutants, *without irradiation of the host cells,* is also apparently influenced by the multiplicity of infection (90). Thus, under circumstances in which the host cells are irradiated, one must consider the possibility that the enhanced viral mutagenesis observed has an explanation unrelated to induced error-prone responses that are directly comparable to W reactivation of phage λ in *E. coli*. For instance, it has been shown that UV irradiation of SV40 virus is highly recombinogenic (8). A similar effect of UV radiation on the *host* genome might enhance viral mutation frequencies by some recombinational-dependent mechanism.

The role of inducible host-cell functions in viral mutagenesis has also been explored using parvoviruses as models (93). These viruses contain a linear single-strand DNA genome that is converted into a duplex replicative form in infected cells. The single-strand configuration of these genomes offers a number of distinct advantages for studies on putative inducible functions in mammalian cells. For example, premutagenic lesions in the viral genome cannot be removed by conventional excision repair prior to replication. Nor can the effect of such lesions on replicative functions be mitigated by recombination with sister DNA molecules. Enhanced mutagenesis of UV-irradiated virus was observed in both unirradiated and UV-irradiated host cells (Fig. 8-17). Interestingly, the level of mutagenesis among the descendants of *both unirradiated and UV-irradiated* parvovirus was enhanced at low multiplicities of infection when host (human embryonic kidney) cells were irradiated prior to infection (93) (Fig. 8-17). The frequency of mutations was only slightly higher for irradiated relative to undamaged viruses, suggesting that the "induced" cellular mutator activity is mainly untargeted on the viral genome (Fig. 8-17).

Although these results bear a striking superficial resemblance to those obtained following infection of SOS-competent strains of *E. coli* with single-strand phages (87, 94, 95; see Section 7-9), there are

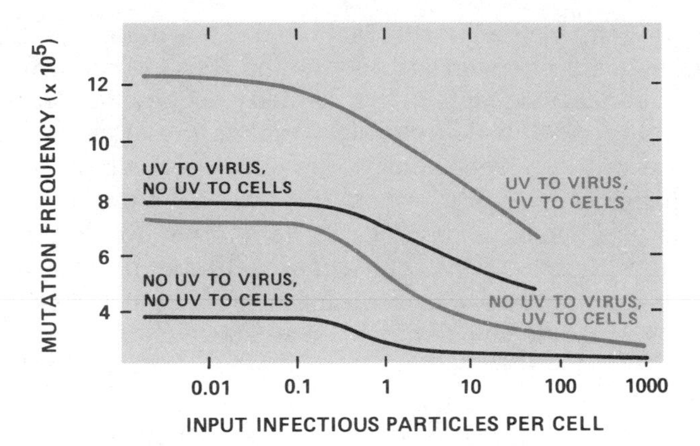

FIGURE 8-17
When UV-irradiated parvovirus H-1 is introduced into previously UV-irradiated cells, a modest enhancement of viral mutagenesis is observed relative to that in unirradiated cells (upper two curves). However, in both UV-irradiated cells (blue lines) and in unirradiated cells (black lines), the frequency of mutation of UV-irradiated virus is not much greater than that of unirradiated virus, suggesting that a significant fraction of the viral mutations may not be targeted to sites of photoproducts in the viral DNA. (From J. J. Cornelis et al., ref. 93.)

important differences. The most notable of these is that UV irradiation of parvovirus H-1 induces a significant frequency of mutations *without* prior host-cell irradiation (93) (Fig. 8-17). In contrast, the exposure of ϕX174 or M13 to UV light results in a much smaller or even undetectable level of viral mutagenesis in unirradiated bacteria.

Enhanced mutagenesis of herpes simplex virus, which contains a duplex DNA genome, has also been observed in the human embryonic kidney cell line used to propagate the parvovirus H-1 (96). However, in contrast to the results obtained with the single-strand virus, there is no evidence for mutagenesis of *unirradiated* herpes virus in irradiated host cells. In this regard it is of interest that the polymerases used for DNA replication have different sources for the two viruses. Herpes simplex virus encodes its own DNA polymerase (97), whereas H-1 virus employs a host cell DNA polymerase (98). Presumably the utilization of different DNA polymerases by the two viruses, together with the different structure of their respective genomes, gives rise to the differences in mutagenesis observed (96).

In addition to the increase in mutation frequency in parvovirus H-1 found in the studies mentioned above, there is a modest increase in mutation frequency of *undamaged herpes simplex virus* following infection of UV-irradiated hosts (Table 8-2). Thus in mammalian cell viruses untargeted mutagenesis may be the result of an error-prone damage tolerance mechanism that is activated by host-cell DNA damage, an observation that closely parallels the untargeted mutagenesis observed in phage-bacterial systems (see Section 7-9).

Finally, irrespective of whether the level of mutagenesis is increased in previously *irradiated hosts relative to that in unirradiated hosts*, most studies agree that irradiation of virus alone is sufficient to produce significant mutagenesis (Fig. 8-17; Table 8-2). Two possibilities have been advanced to account for these results: either there is a constitutive mutagenic component for viral UV mutagenesis in mammalian cells, or the damaged viral DNA alone is sufficient to trigger an error-prone cellular response (6). The latter, if true, would be not unlike the phenomenon of indirect induction of UV mutagenesis in *E. coli* (99).

Translesion Synthesis of Viral DNA

If UV irradiation of host cells or the simple introduction of UV-irradiated viral DNA into unirradiated cells induces "error-prone" enhanced reactivation and mutagenesis of viruses, one might anticipate that these phenomena are related to *translesion synthesis of DNA* during semiconservative DNA replication. The semiconservative replication of UV-irradiated SV40 DNA was discussed in Section 8-2 (see page 477). A possible relation between an inducing stimulus and enhanced DNA synthesis has been observed in another virus-host system, once again raising the intriguing possibility of a translesion DNA synthesis mode. UV irradiation of the single-strand genome of the minute virus of mice (MVM) prior to infection of unirradiated cells results in a dose-dependent inhibition of formation of the replicative forms (RFI and RFII) (Fig. 8-18), attributed to the presence of absolute blocks that prevent elongation of the newly synthesized complementary strand (100, 101). However, exposure of the host cells to UV light *prior to infection* with UV-

TABLE 8-2
Mutagenesis in the thymidine kinase (TK) gene of herpes simplex virus

UV Radiation to Cells	TK Mutant Frequency	
	Irradiated virus	Unirradiated virus
–	4.8×10^{-4}	1.5×10^{-4}
+	13.0×10^{-4}	3.5×10^{-4}

From U. B. Das Gupta and W. C. Summers, ref. 77.

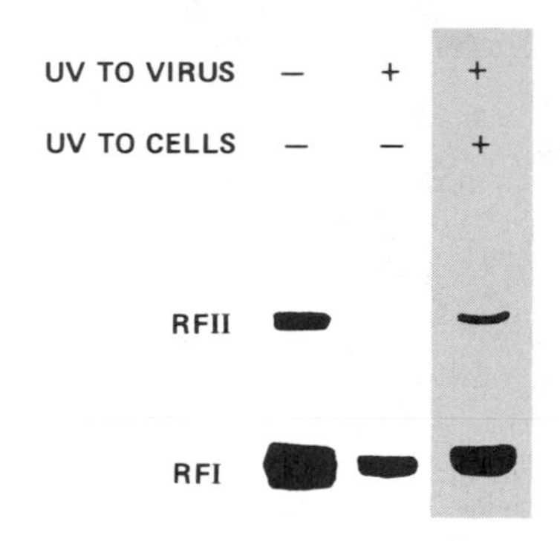

FIGURE 8-18
UV irradiation of minute virus of mice (MVM) prior to infection of *unirradiated* mouse cells (middle lane) results in a decrease in the total amount of RF molecules (both RFI and RFII) synthesized from input single-strand DNA. However, if the cells were exposed to a sublethal dose of UV radiation prior to infection with *UV-irradiated* virus (right lane), a higher level of total viral DNA synthesis is achieved. (From J-M. Vos and J. Rommelaere, ref. 101.)

irradiated MVM enhances the fraction of input viral DNA that is converted to the replicative forms (Fig. 8-18). This enhancement apparently requires de novo protein synthesis during the interval between cell irradiation and virus infection (100, 101).

Provirus Induction

Integrated DNA from DNA or RNA viruses can be activated to produce viral particles by treatment of cells with a variety of DNA-damaging mutagens and carcinogens (102–105) (Fig. 8-19). This process bears at least a superficial resemblance to prophage induction in lysogenic bacteria, one of the best-characterized SOS responses (see Section 7-4). An interesting parallelism in dose-response curves has been demonstrated for *UV induction of SV40 provirus* and *enhanced reactivation of irradiated herpes simplex virus* (78) (Fig. 8-19) and for the induction of SV40 provirus and the formation of *sister-chromatid exchanges* (see Section 9-4) following exposure to ethylmethanesulfonate or mitomycin C (106). It remains to be established whether these correlations signify a common regulation of the expression of the three functions.

Split-Dose Experiments: Possible Evidence for Inducible DNA Synthesis in Mammalian Cells

A number of experiments designed to investigate inducible responses to DNA damage in mammalian cells have utilized *split-dose protocols* in which the first, so-called *tickling dose* is relatively small and is intended to be an inducing stimulus. In order to examine the influence of putative induced functions on DNA damage, a second, larger *DNA damaging dose* is applied after a suitable interval has elapsed to allow putative induction to occur.

When Chinese hamster ovary cells are exposed to small doses of

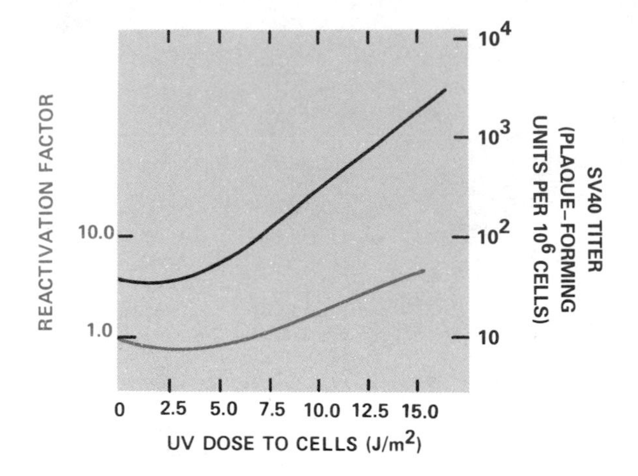

FIGURE 8-19
Hamster kidney cells already stably transformed with SV40 virus were UV-irradiated and infected with UV-irradiated or unirradiated herpes simplex virus. The reactivation factor is defined as the ratio of the titer of UV-irradiated herpes simplex virus to that of unirradiated virus at a particular UV dose to the cells, divided by a value similarly determined for unirradiated cells. Values greater than 1.0 indicate reactivation, i.e., enhanced survival of herpes simplex virus (see Fig. 8-16). UV irradiation of these cells also results in induction of SV40 provirus. The dose-response curves for enhanced reactivation of irradiated herpes virus and for induction of SV40 provirus show a striking parallelism. This parallelism is reminiscent of that between prophage induction and W reactivation of phage λ in *E. coli*, both of which are inducible responses to DNA damage. (From G. B. Zamansky et al., ref. 78.)

UV light several hours prior to a larger dose and are subsequently pulse labeled and analyzed in alkaline sucrose gradients, the DNA synthesized after exposure to the larger dose of UV light is predictably smaller than that observed for DNA from unirradiated cells. However, the DNA increases in size when the cells are subsequently incubated in unlabeled medium (107) (Fig. 8-20). This apparently enhanced rate of DNA chain elongation is not observed in cells incubated with cycloheximide between the two UV treatments (107), suggesting the need for de novo protein synthesis.

These results have been interpreted in terms of a possible inducible mechanism for enhanced DNA synthesis that facilitates replication past dimers. However, other studies (108, 109) have shown that an initial low UV dose to mammalian cells alters the pattern of DNA synthesis, causing an abnormal distribution of the size of newly

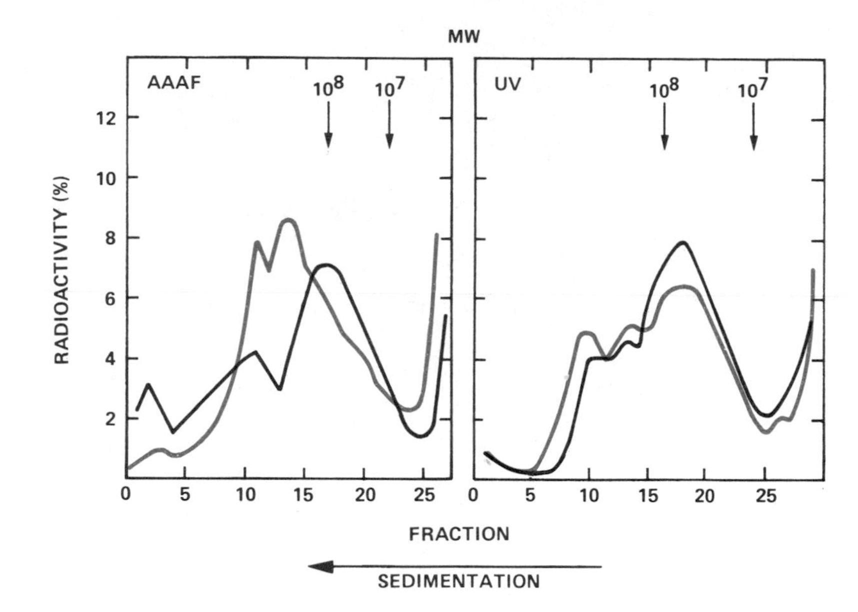

FIGURE 8-20
Sedimentation after single or split doses of N-acetoxy-N-2-acetylaminofluorene (AAAF) or UV. The panel on the left shows the sedimentation profile of newly synthesized DNA in cells treated with 2.5 μM AAAF (blue line) or without AAAF (black line) before treatment with 5.0 μM AAAF two hours later. Similarly, cells from the experiment shown in the panel on the right were either given 2.5 J/m^2 of UV radiation (blue line) or no UV (black line) prior to exposure to 5 J/m^2 of UV light two hours later. In both cases the DNA from cells treated with the split-dose protocol attains a higher molecular weight than that from cells treated with an equivalent single dose. This suggests that the first DNA damaging treatment may induce the enhanced tolerance of the damage produced by the second treatment. The arrow under the abscissa indicates the direction of sedimentation. (After S. M. D'Ambrosia and R. B. Setlow, ref. 107.)

synthesized DNA molecules at the time of a second UV dose. It has therefore been argued that the apparent larger size of pulse-labeled DNA extracted from cells exposed twice to UV light might result from the *preferential labeling of molecules of longer size* during the pulse. This criticism has been countered by the demonstration that if the first UV dose is given in the G2 phase of the cell cycle, enhanced elongation of newly replicated DNA still occurs in cells irradiated with a second UV dose prior to the S phase (110). Since DNA replication obviously cannot occur between the two doses under these experimental conditions, this enhancement cannot be attributed to altered patterns of DNA synthesis after the first UV exposure.

Measurement of the *overall rate of DNA synthesis* through autoradiography of cells radiolabeled with [^{3}H]thymidine also shows enhancement by split-dose protocols. For example, when human fibro-

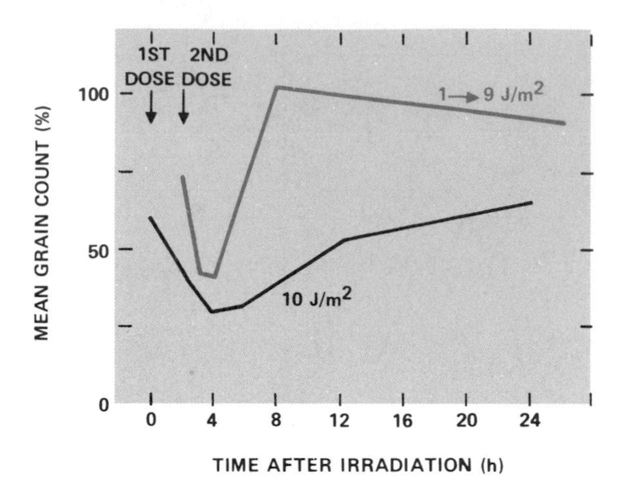

FIGURE 8-21
Effect of split dose UV irradiation to normal human fibroblasts on semiconservative DNA synthesis measured by grain counts from autoradiographic preparations. The second UV dose (9 J/m^2) was given two hours after the initial UV dose (1 J/m^2). Split dose irradiation results in less severe inhibition of DNA synthesis than a single equivalent total dose of 10 J/m^2 and in a more rapid and complete recovery of DNA synthesis. (From E. Moustacchi et al., ref. 111.)

blasts in culture are exposed to a UV dose of 10 J/m^2, an inhibition of the rate of DNA synthesis occurs that is followed by a recovery some hours later (Fig. 8-21) (111, 112). When the kinetics of this inhibition and recovery are compared with those observed in cells exposed to a split dose of 1 J/m^2 followed two hours later by 9 J/m^2, a significant enhancement in both the rate and the extent of recovery is observed (Fig. 8-21). The extent of the initial inhibition of synthesis in cells given the split-dose irradiation relative to that in unirradiated cells is also reduced (111, 112) (Fig. 8-21).

Split-dose protocols have also been used to examine effects on mutagenesis. Chinese hamster ovary cells were irradiated with X rays at times between 0 and 17 hours before being irradiated with UV light (113). X rays do not produce ouabain-resistant mutations in Chinese hamster ovary cells, but UV light does (114–116). Thus the X rays served as a nonmutagenic inducer of a putative error-prone cellular replicative response that should have yielded more UV-induced ouabain-resistant mutants. No synergism was observed between the two radiations for the production of mutants resistant to either ouabain or 6-thioguanine. Negative results were also obtained in a different study with V79 cells derived from Chinese hamsters (117). However, in the latter study fractionation of UV exposure did

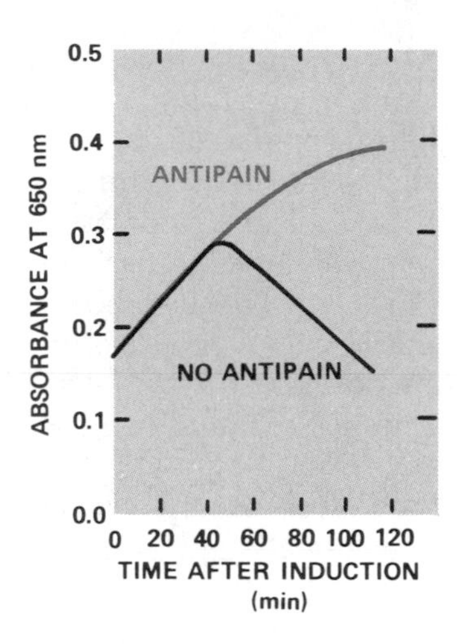

FIGURE 8-22
The protease inhibitor antipain blocks the thermal induction of phage λ in a *tif* mutant of *E. coli*. Induction was measured by a drop in the absorbance of the culture at 650 nm. (From M. S. Meyn et al., ref. 118.)

increase the *colony-forming ability* of the cells, and exposure to cycloheximide for the period between the two UV applications mitigated this effect.

The Role of Proteases in the Cellular Reponses to DNA Damage

There is good evidence that the expression of the SOS phenomenon in *E. coli* is associated with proteolytic degradation of specific gene repressors by the activation of recA protein (see Section 7-6). It might reasonably be anticipated therefore that general inhibition of protease activity in *E. coli* would interfere with the expression of SOS functions, and indeed this is so. The protease inhibitor *antipain* has no effect on *E. coli* cell growth, overall RNA or protein synthesis, or induction of β-galactosidase (which does *not* involve proteolysis of the lac repressor) (118). However, this compound drastically decreases UV mutagenesis in *E. coli*, blocks the expression of thermally induced mutator activity and filamentous growth in the *tif* mutant at 42°C and prevents thermal induction of λ prophage in a *tif* lysogen (Fig. 8-22) (118).

These observations prompted an examination of the effects of protease inhibitors on the cellular responses to DNA damage in mammalian cells. Antipain has no effect on the frequency of 8-azaguanine–resistant, 6-thioguanine–resistant or ouabain-resistant mutants in V79 Chinese hamster cells treated with UV light, N-methyl-N′-nitro-N-nitrosoguanidine (MNNG), or 3-methylcholanthrene (119, 120). Antipain also does not inhibit the enhanced reactivation of herpes simplex virus by UV light. On the other hand, formation of sister chromatid exchanges (see Section 9-4) resulting from the treatment of cells with the tumor promoter 12-O-tetradecanoylphorbol-13-acetate (TPA) (121) or with X rays (122) and the formation of chromosomal aberrations in cells treated with MNNG (121–124) are inhibited by antipain treatment (Table 8-3). In addition, neoplastic transformation of mouse 10T½ cells (a mouse embryo-derived fibroblast cell line) following exposure to X rays and TPA is reduced by treatment with either antipain or *leupeptin*, also a protease inhibitor (125), and induction of endogenous dormant C-type retrovirus by infection of cells with irradiated herpes simplex virus is inhibited by antipain (6).

Other evidence suggestive of a role of proteases in the expression of gene functions following DNA damage in mammalian cells stems from studies on the expression of plasminogen activator. This compound is a protease that has received considerable attention because of evidence relating the appearance of it to neoplastic transformation (126). A quantitative relation between plasminogen activity and neoplastic transformation in vitro has been shown for a number of cell types (127), and its activity is also enhanced in many normal and transformed cells exposed to TPA (127, 128). As stated above, TPA-enhanced neoplastic transformation of cells in culture is inhibited by protease inhibitors. It is thus interesting that antipain (but not leupeptin or soybean trypsin inhibitor) inhibits the induction of plasminogen activator by TPA in 10T½ cells (129). Activator synthesis is also enhanced by UV irradiation or chemical treatment of either chicken, hamster, rat, mouse or human embryonic fibroblasts (130), but this effect is not observed in UV-irradiated normal *adult* human fibroblasts (130). However, plasminogen activator is induced in the fibroblasts of patients suffering from xeroderma pigmentosum (131) (see Section 9-2) (Fig. 8-23). The most striking effect is observed in excision repair-deficient (classical) XP cells (131). In these cells induction occurs in a narrow UV dose range and results in up to a

TABLE 8-3
Inhibition of MNNG-induced chromosomal aberrations by antipain

Treatment	Quadriradials	Triradials	Breaks	Other Exchange-type Aberrations
Control	0	0	0	0
Antipain (1 m*M*)	1	0	0	0
MNNG (0.2 μg/ml)	11	8	7	12
MNNG (0.2 μg/ml) + antipain (1 m*M*)	2	2	0	0
MNNG (0.5 μg/ml)	11	17	0	4
MNNG (0.5 μg/ml) + antipain (1 m*M*)	0	0	0	0

From A. Kinsella and M. Radman, ref. 121.

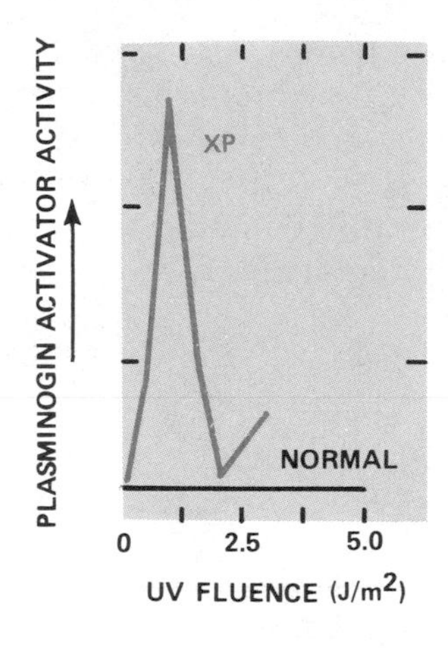

FIGURE 8-23
Expression of plasminogen activator is markedly increased in UV-irradiated xeroderma pigmentosum (XP) cells. (From R. Miskin and R. Ben-Ishai, ref. 131.)

20-fold increase in the levels of plasminogen activator. The UV doses required for this effect are 1/10th those required for the expression of comparable levels of activator in XP heterozygotes (proficient in excision repair) or in normal embryonic cells (131). Thus unrepaired DNA damage may be involved in the inducing signal for synthesis of plasminogen activator in human cells. Indeed, a comparison of the expression of plasminogen activator with UV-enhanced mutagenesis and induced viral reactivation in XP cells shows that all these functions exhibit a similar inverse relation between residual DNA repair capacity and the inducing dose of UV radiation (131). Furthermore, synthesis of plasminogen activator in XP cells after UV exposure is completely inhibited by cycloheximide, suggesting a requirement for protein synthesis (131).

Other Examples of Possible Induction of Gene Expression in Mammalian Cells Exposed to DNA Damage

1. A glycosylated nuclear membrane protein of molecular weight 35,000 called XM1 is expressed and turned over at a high rate in B lymphocyte-derived cells of various species (132). Splenic or thymic T cells or fibroblasts do not express this protein. On polyclonal stimulation of B lymphocyte cell lines or of proliferating B lymphocyte-derived cell lines the synthesis of XM1 is turned off, but it is expressed when these cells are exposed to DNA damaging agents such as mitomycin C, UV light or γ radiation and after replicative arrest by hydroxyurea (132).

2. When permeabilized Friend erythroleukemia cells previously treated briefly with *dimethylsulfoxide* (DMSO) are exposed to cell-

free extracts from UV-irradiated (but not unirradiated) cells, a small but significant number of cells show evidence of erythroid differentiation (133). A similar activity is detected in extracts of cells treated with mitomycin C, but not in cells treated with DMSO. The induction of the factor(s) stimulating erythroid differentiation is not specific to Friend cells, since similar inducing activity is observed in extracts of UV-irradiated nonerythroid cells.

Evidence for DNA Damage Inducible Functions in Lower Eukaryotes

S. cerevisiae

A number of lower eukaryotes, particularly the yeast *S. cerevisiae*, show promise of providing more definitive examples of induction of gene expression by DNA damage. For example, there is evidence that inducible mechanisms are involved in mitotic recombination and in mutagenesis in *S. cerevisiae*. Specifically, recombination between unirradiated chromosomes is induced by UV or X irradiation of haploid cells that are subsequently mated to unirradiated heteroallelic diploids (Fig. 8-24) (7, 134). The recombination events selected do not involve the direct participation of the irradiated chromosome, which appears to function exclusively as an inducer. Photoreactivation of the UV-irradiated haploid cells results in a decrease in recombinants, suggesting that lesions in the DNA (presumably pyrimidine dimers) are responsible for the induction of the recombinational ability.

As far as *mutagenesis in yeast* is concerned, it will be recalled from the previous chapter (Section 7-4) that in wild-type *E. coli* cells, the kinetics of UV mutagenesis show a dose-squared relation, whereas in *tif* cells the mutational response has a linear relation to UV dose (see Fig. 7-29). The response in wild-type cells has been interpreted as evidence for two independent events required for mutation: induction of the SOS response and production of mutagenic DNA damage. Such an interpretation is consistent with the observation that *tif* cells do not require DNA damage for induction and hence are dependent only on a single UV dose-related event; that is, production of mutagenic DNA damage.

In haploid repair-competent strains of *S. cerevisiae*, mutation kinetics are biphasic: linear at low doses but squared at high doses (135) (Fig. 8-25). These curves can be fitted mathematically by a general equation for repair-mediated mutagenesis if it is assumed that there exists a significant constitutive level of error-prone repair (to account for the linear kinetics) plus an inducible component (to account for the dose-squared response) (135). If protein synthesis is blocked in yeast cells for three days after UV irradiation, mutation

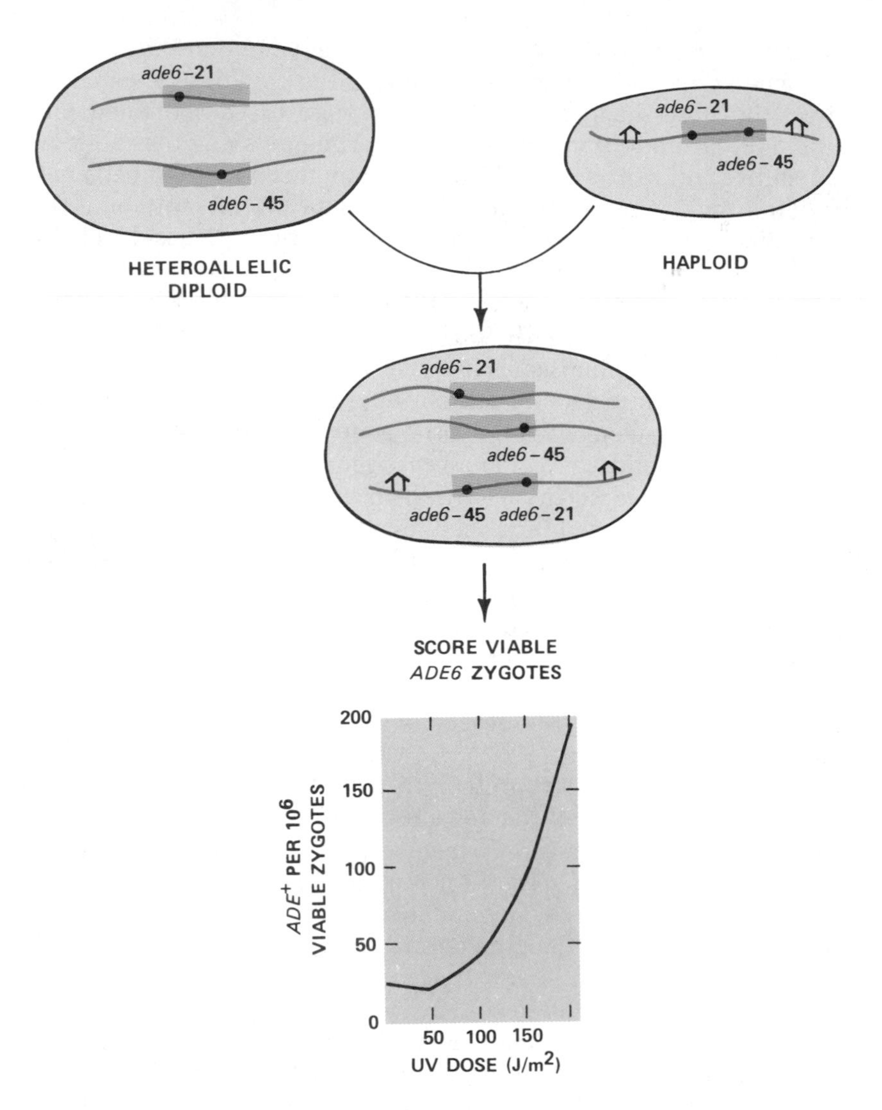

FIGURE 8-24
Induction of recombination following UV irradiation in yeast. An unirradiated diploid strain heteroallelic for the *ade6* locus was mated to a UV-irradiated haploid strain carrying both mutant alleles. Wild-type (*ADE6*) recombinants can only derive from recombination between the heteroallelic chromosomes from the diploid strain. This recombination is apparently induced by the presence of the irradiated chromosome derived from the haploid and shows a UV dose dependency. (After F. Fabre and H. Roman, ref. 134.)

frequencies in the higher dose range are reduced and the resulting frequencies fit a linear extrapolation of the low-dose kinetics; in effect the dose-squared component appears to be abolished (Fig. 8-25) (135). In addition, split-dose experiments show that both UV mutability and resistance to killing are enhanced in yeast cells held in buffer for 24 hours after an initial dose of UV radiation and that both effects are blocked by inhibition of protein synthesis during the liquid holding period (135).

Perhaps the most persuasive evidence for regulated expression of eukaryotic genes in response to DNA damage comes from studies on *din* (*d*amage-*in*ducible) genes in yeast (10). A pool of yeast genes was cloned into plasmid vectors containing the *E. coli lacZ* gene so as to bring the coding element of the latter gene under the control of yeast promoters. Plasmids in which a yeast gene was properly fused produced β-galactosidase in yeast transformants. Screening for transformants that expressed increased levels of β-galactosidase after exposure to DNA damage has yielded a number of *din-lacZ* fusions (10). Some of them are induced after treatment with low levels of UV or γ radiation, methylmethanesulfonate, 4-nitroquinoline-1-oxide or mitomycin C or after thymine starvation. One of the fusions is induced by UV but not by γ radiation or methylmethanesulfonate. Induction is reflected by increased mRNA synthesis after DNA damage (10).

Possible modulations of proteolytic activities that may be required for the induction of cellular responses to DNA damage are largely unexplored in *S. cerevisiae*. However, following exposure to UV radiation, proteinase B activity increases by about threefold in *RAD* yeast cells and there is a moderate increase in carboxypeptidase Y and aminopeptidase I (leucine aminopeptidase) (136, 137). The observed increase in proteinase B activity is inhibited by cycloheximide treatment and is influenced by the repair competency of the cell (136, 137). Thus yeast mutants defective in the excision repair pathway (see Section 4-3) still show induction, but those defective in a pathway required for mutation in yeast (*rad6*-1 and *pso2*-1 mutants) fail to show an increase in proteinase activity, suggesting that a proteolytic event may be involved in the molecular mechanism(s) of UV mutagenesis in yeast (136, 137).

Ustilago maydis

Inhibition of protein synthesis by cycloheximide inhibits radiation-induced allelic recombination in the fungus *U. maydis* (138). It has also been shown that maximum survival of UV-irradiated *U. maydis* requires a two- to three-hour period of postirradiation RNA and protein synthesis (139). Split-dose irradiation shows that this require-

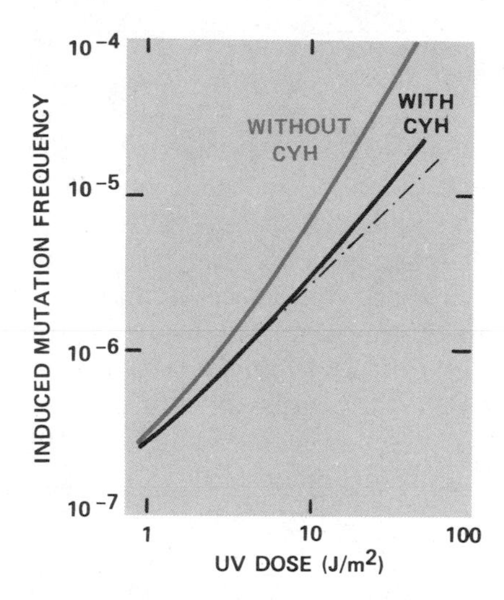

FIGURE 8-25
The kinetics of induction of mutations from *lys* to *LYS* in yeast are biphasic (linear-quadratic) in the absence of cycloheximide (CYH). In the presence of CYH the quadratic component is largely abolished. The dotted line shows an extrapolation that yields an idealized curve with a slope of 1. (From F. Eckhardt et al., ref. 135.)

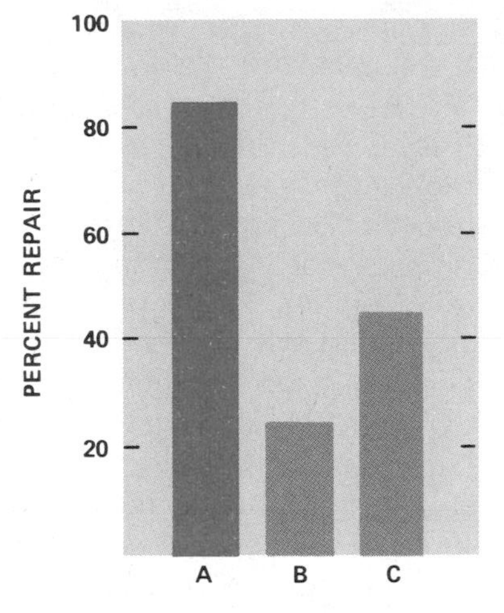

FIGURE 8-26
Excision repair in *N. crassa* is increased when cells in phosphate buffer are irradiated at a low dose of UV and subsequently with a higher dose (C) relative to that observed with a single large dose (B). The amount of excision repair in cells given a single large dose and incubated in complete medium (A) is shown for comparison. (From T. I. Baker, ref. 140.)

ment correlates with the development of a radioresistant state of the cell that is induced by an initial small dose of UV radiation (139). Once induced, the radio-resistant state precludes the need for further RNA and protein synthesis for proficient repair of DNA damage caused by a second UV dose. Such radioresistance is retained for up to 75 hours and may represent the expression of inducible responses to DNA damage (139).

Neurospora crassa

Wild-type *N. crassa* removes pyrimidine dimers from its DNA by excision repair during postirradiation incubation (140).This excision repair process proceeds much more rapidly in nutritive medium than in phosphate buffer. Cycloheximide added immediately after a damaging UV dose inhibits repair in a growth medium but does not affect repair in phosphate buffer. Furthermore, a low dose of UV radiation administered prior to the damaging dose increases excision repair activity in phosphate buffer (Fig. 8-26) (140).

8-4 Summary

It is obvious that the mechanisms whereby DNA damage at or near replication forks is tolerated are far less defined in eukaryotic cells than in the prokaryote *E. coli*. In some measure this reflects the difficulty of interpreting the types of experiments that have been relatively successful in bacteria, because of the greater complexity associated with the molecular biology of DNA replication in eukaryotic cells. The obvious challenge here is to devise alternative experimental approaches to study DNA synthesis in both normal eukaryotes and those exposed to DNA damage. However, a far more serious limitation to the study of cellular responses to DNA damage in higher organisms is that mentioned in the introduction to this chapter: the lack of an established genetic framework for most eukaryotic cells and certainly for mammalian cells. Indications that this situation is changing give cause for optimism, however. Certainly the yeast *S. cerevisiae* is becoming an increasingly informative model eukaryotic system. The fruit fly *D. melanogaster* has also served classical genetics well for many years and although it was not discussed in this chapter, attempts to exploit the genetic potential of this organism as an experimental model for studying eukaryotic responses to DNA damage are beginning.

The available evidence suggests that mammalian cells can tolerate base damage that is not removed by DNA repair. A number of perturbations of semiconservative DNA synthesis are observed when replication occurs or is attempted in cells that have sustained DNA damage. Two distinct aspects of regulation have been identified: initiation and replication fork progression. In general, DNA base damage (and/or its attempted repair) inhibits initiation of replication in relatively low amounts and in large amounts also inhibits fork progression. But many important questions remain unanswered. Do eukaryotic cells perform replicative bypass of template DNA damage with gap formation? Does translesion DNA synthesis occur in such cells? Alternatively, do some, if not all, eukaryotic cells possess heretofore undiscovered ways of negotiating replicative blocks that are not observed in prokaryotic models? Do eukaryotic cells respond to DNA damage and/or replicative arrest by derepressing specific genes and, if so, are these genes organized into a single regulon comparable to the SOS regulon in *E. coli*? Precise answers to these and other questions are still forthcoming. One hopes that clarity will be brought to the present state of relative confusion by the application of at least two important experimental principles given in the introduction to this chapter. First, the choice of experimental models is critical. Undeniably we wish to understand the response to DNA damage in human cells. But human cells may not be the best cell type with which to approach this area of molecular biology at the present time. Second, although it may seem intuitively obvious, one should not forget that experimental designs and specific experimental techniques that are particularly useful in the study of bacteria cannot always be successfully extrapolated to the study of higher organisms.

Suggestions for Further Reading

References 1–6, 14, 18, 44, 71, 84, 85.

References

1. Hall, J. D., and Mount, D. W. 1981. Mechanisms of DNA replication and mutagenesis in ultraviolet-irradiated bacteria and mammalian cells. *Prog. Nucleic Acid Res. Mol. Biol.* 25:53.
2. Hanawalt, P. C., Cooper, P. K., Ganesan, A. K., and Smith, C. A. 1979. DNA repair in bacteria and mammalian cells. *Ann. Rev. Biochem.* 48:783.
3. Meneghini, R., Menck, C. F. M., and Schumacher, R. I. 1981. Mechanisms of tolerance to DNA lesions in mammalian cells. *Quart. Rev. Biophys.* 14:381.
4. Lehmann, A. R., and Karran, P. 1981. DNA repair. *Int. Rev. Cytol.* 72:101.
5. Lehmann, A. R. 1981. DNA replication in mammalian cells damaged by mutagens. In *Chromosome damage and repair.* E. Seeberg and K. Kleppe, eds., p. 383. New York: Plenum.
6. Radman, M. 1980. Is there SOS induction in mammalian cells? *Photochem. Photobiol.* 32:823.
7. Fabre, F. 1983. Mitotic transmission of induced recombinational ability in yeast. In *Cellular responses to DNA damage.* E. C. Friedberg and B. A. Bridges, eds., p. 379. New York: Alan R. Liss.
8. Gentil, A., Margot, A., and Sarasin, A. 1983. Effect of UV-irradiation on genetic recombination of simian virus 40 mutants. In *Cellular responses to DNA damage.* E. C. Friedberg and B. A. Bridges, eds., p. 385. New York: Alan R. Liss.
9. Boyd, J. B., Harris, P. V., Presley, J. M., and Narachi, M. 1983. *Drosophila melanogaster:* a model eukaryote for the study of DNA repair. In *Cellular responses to DNA damage.* E. C. Friedberg and B. A. Bridges, eds., p. 107. New York: Alan R. Liss.
10. Ruby, S. W., Szostak, J., and Murray, A. W. 1983. Cloning regulated yeast genes from a pool of *lacZ* fusions. *Methods in Enzymol.* 101:253.
11. Lehmann, A. R., and Stevens, S. 1975. Postreplication repair of DNA in chick cells: studies using photoreactivation. *Biochim. Biophys. Acta* 402:179.
12. Krishnan, D., and Painter, R. B. 1973. Photoreactivation and repair replication in rat kangaroo cells. *Mutation Res.* 17:213.
13. Rosenstein, B. S., and Setlow, R. B. 1980. DNA repair after UV irradiation of ICR 2A frog cells: pyrimidine dimers are long acting blocks to nascent DNA synthesis. *Biophys J.* 31:195.

14. Edenberg, H., and Huberman, J. 1975. Eukaryotic chromosome replication. *Ann. Rev. Genet.* 9:245.

15. Kapp, L. N., and Painter, R. B. 1978. Replicon sizes in mammalian cells as measured by an X-ray plus bromodeoxyuridine photolysis method. *Biophys. J.* 24:739.

16. Hand, R. 1978. Eukaryotic DNA: organization of the genome for replication. *Cell* 15:317.

17. Ockey, C. H. 1979. Quantitative replicon analysis of DNA synthesis in cancer-prone conditions and the defects in Bloom's syndrome. *J. Cell Sci.* 40:125.

18. Cleaver, J. E. 1984. DNA repair and replication. In *Biochemical mechanisms of genetic disorders*. P. J. Benke, ed. New York: Dekker (in press).

19. Painter, R. B. 1977. Rapid test to detect agents that damage human DNA. *Nature* 265:650.

20. Painter, R. B. 1978. DNA synthesis inhibition in HeLa cells as a simple test for agents that damage human DNA. *J. Environ. Path. Toxicol.* 2:65.

21. Painter, R. B., and Rasmussen, R. E. 1964. Organization of the deoxyribonucleic acid replicating system in mammalian cells as revealed by the use of X-radiation and bromouracil deoxyriboside. *Nature* 201:172.

22. Painter, R. B., and Young, B. R. 1976. Formation of nascent DNA molecules during inhibition of replicon initiation in mammalian cells. *Biochim. Biophys. Acta* 418:146.

23. Povirk, L. F., and Painter, R. B. 1976. The effect of 313 nanometer light on initiation of replicons in mammalian cell DNA containing bromodeoxyuridine. *Biochim. Biophys. Acta* 432:267.

24. Painter, R. B. 1977. Inhibition of initiation of HeLa replicons by methyl methane sulfonate. *Mutation Res.* 42:299.

25. Painter, R. B. 1978. Inhibition of DNA replicon initiation by 4-nitroquinoline-1-oxide, adriamycin and ethyleneimine. *Cancer Res.* 38:4445.

26. Doniger, J. 1978. DNA replication in UV light irradiated Chinese hamster cells: the nature of replicon inhibition and post-replication repair. *J. Mol. Biol.* 120:433.

27. Park, S. D., and Cleaver, J. E. 1979. Recovery of DNA synthesis after ultraviolet irradiation of xeroderma pigmentosum cells depends on excision repair and is blocked by caffeine. *Nucleic Acids Res.* 6:1151.

28. Makino, F., and Okada, S. 1975. Effects of ionizing radiations on DNA replication in cultured mammalian cells. *Radiat. Res.* 62:37.

29. Walters, R. A., and Hildebrand, C. E. 1975. Evidence that X-irradiation inhibits DNA replicon initiation in Chinese hamster cells. *Biochem. Biophys. Res. Comm.* 65:265.

30. Kaufmann, W. K., Cleaver, J. E., and Painter, R. B. 1980. Ultraviolet radiation inhibits replicon initiation of S phase human cells. *Biochim. Biophys. Acta* 608:191.

31. Tolmach, L. J., and Jones, R. W. 1977. Dependence of the rate of DNA synthesis in X-irradiated HeLa cells on dose and time after exposure. *Radiat. Res.* 69:117.

32. Ormerod, M. G., and Stevens, U. 1971. The rejoining of X-ray-induced strand breaks in the DNA of a murine lymphoma cell (L5178Y). *Biochim. Biophys. Acta* 232:72.

33. Edenberg, H. J. 1976. Inhibition of DNA replication by UV light. *Biophys J.* 16:849.

34. Cleaver, J. E., Kaufmann, W. K., Kapp, L. N., and Park, S. D. 1983. Replicon size and excision repair as factors in the inhibition and recovery of DNA synthesis from ultraviolet damage. *Biochim. Biophys. Acta* 739:207.

35. Cleaver, J. E., and Thomas, G. H. 1969. Single-strand interruptions in DNA and the effect of caffeine in Chinese hamster cells irradiated with UV-light. *Biochem. Biophys. Res. Comm.* 36:203.

36. Lehmann, A. R. 1972. Postreplication repair of DNA in UV-irradiated mammalian cells. *J. Mol. Biol.* 66:319.

37. Lehmann, A. R., and Kirk-Bell, S. 1972. Postreplication repair of DNA in UV-irradiated mammalian cells. No gaps in DNA synthesized late after UV-irradiation. *Eur. J. Biochem.* 31:438.

38. Ormerod, M. G. 1976. Radiation-induced strand breaks in the DNA of mammalian cells. In *Biology of radiation carcinogenesis*. J. M. Yuhas, R. W. Tennant and J. D. Regan, eds., p. 67. New York: Raven Press.

39. Regan, J. D., Setlow, R. B., and Ley, R. D. 1971. Normal and defective repair of damaged DNA in human cells: a sensitive assay utilizing the photolysis of bromodeoxyuridine. *Proc. Natl. Acad. Sci.* (USA) 68:708.

40. Lehmann, A. R., and Ormerod, M. G. 1971. The replication of DNA in murine lymphoma cells (L5178Y). II. Size of replicating units. *Biochim. Biophys. Acta* 272:191.

41. Elkind, M. M., and Kamper, C. 1970. Two forms of repair of DNA in mammalian cells following irradiation. *Biophys. J.* 10:237.

42. Lett, J. T., Klucis, E. S., and Sun, C. 1970. On the size of the DNA in the mammalian chromosome. Structural subunits. *Biophys. J.* 10:277.

43. Huberman, J. A., and Riggs, A. D. 1968. On the mechanism of DNA replication in mammalian chromosomes. *J. Mol. Biol.* 32:327.

44. Sheinin, R., Humbert, J., and Pearlmann, R. E. 1978. Some aspects of eukaryotic DNA replication. *Ann. Rev. Biochem.* 47:277.

45. Buhl, S. N., Stillman, R. M., Setlow, R. B., and Regan, J. D. 1972. DNA chain elongation and joining in normal human and xeroderma pigmentosum cells after UV-irradiation. *Biophys. J.* 12:1183.

46. Clarkson, J. M., and Hewitt, R. R. 1976. Significance of dimers to the size of newly-synthesized DNA in UV-irradiated Chinese hamster ovary cells. *Biophys. J.* 16:1155.

47. Bowden, G. T., Giesselbach, B., and Fusenig, N. E. 1978. Postreplication repair of DNA in ultraviolet light–irradiated normal and malignantly transformed mouse epidermal cell cultures. *Cancer Res.* 38:2709.

48. Park, S. D., and Cleaver, J. E. 1979. Postreplication repair: questions of its definition and possible alterations in XP cell strains. *Proc. Natl. Acad. Sci.* (USA) 76:3927.

49. Cordeiro-Stone, M., Schumacher, R. I., and Meneghini, R. 1979. Structure of replication fork in UV light irradiated human cells. *Biophys. J.* 27:287.

50. Meneghini, R., Cordeiro-Stone, M., and Schumacher, R. I. 1981. Size and frequency of gaps in newly synthesized DNA of xeroderma pigmentosum human cells irradiated with UV light. *Biophys. J.* 33:81.

51. Meneghini, R., and Hanawalt, P. C. 1975. Postreplication repair in human cells: on the presence of gaps opposite dimers and recombination. In *Molecular mechanisms for repair of DNA*. P. C. Hanawalt and R. B. Setlow, eds., p. 639. New York: Plenum.

52. Meneghini, R., and Hanawalt, P. C. 1976. T4 endonuclease sensitive sites in DNA from UV irradiated human cells. *Biochim. Biophys. Acta* 425:428.

53. Buhl, S. N., Setlow, R. B., and Regan, J. D. 1972. Steps in DNA chain elongation and joining after ultra-violet irradiation of human cells. *Int. J. Radiat. Biol.* 22:417.

54. Painter, R. B. 1974. DNA damage and repair in eukaryotic cells. *Genetics* 78:139.

55. Rommelaere, J., and Miller-Faures, A. 1975. Detection by density equilibrium centrifugation of recombinant-like DNA molecules in somatic mammalian cells. *J. Mol. Biol.* 98:195.

56. Fujiwara, Y., and Tatsumi, M. 1976. Replicative bypass repair of UV damage to DNA of mammalian cells: caffeine sensitive and caffeine resistant mechanisms. *Mutation Res.* 37:91.

57. Fujiwara, Y., and Tatsumi, M. 1977. Low levels of DNA exchanges in normal human and xeroderma pigmentosum cells after UV irradiation. *Mutation Res.* 43:279.

58. Waters, R., and Regan, J. D. 1976. Recombination of UV induced pyrimidine dimers in human fibroblasts. *Biochem. Biophys. Res. Comm.* 72:803.

59. Lehmann, A. R., and Kirk-Bell, S. 1978. Pyrimidine dimer sites associated with the daughter DNA strands in UV-irradiated human fibroblasts. *Photochem. Photobiol.* 27:297.

60. Fornace, A. J., Jr. 1983. Recombination of parent and daughter strand DNA after UV-irradiation in mammalian cells. *Nature* 304:552.

61. Meneghini, R., and Menck, C. F. M. 1978. Pyrimidine dimers in DNA strands of mammalian cells synthesized after UV-irradiation. In *DNA repair mechanisms*. P. C. Hanawalt, E. C. Friedberg and C. F. Fox, eds., p. 493. New York: Academic.

62. Das Gupta, U. B., and Summers, W. C. 1980. Genetic recombination of *Herpes simplex* virus, the role of the host cell and UV-irradiation of the virus. *Mol. Gen. Genet.* 178:617.

63. Hall, J. D., Featherston, J. D., and Almy, R. 1980. Evidence for repair of ultraviolet light–damaged herpes virus in human fibroblasts by a recombination mechanism. *Virology* 105:490.

64. Dubbs, D. R., Rachmeler, M., and Kit, S. 1974. Recombination between temperature-sensitive mutants of simian virus 40. *Virology* 57:161.

65. Yamamoto, H., and Shimojo, H. 1971. Multiplicity reactivation of human adenovirus type 12 and simian virus 40 irradiated by ultraviolet light. *Virology* 45:529.

66. Selsky, C. A., Henson, P., Weichselbaum, R. R., and Little, J. B. 1979. Defective reactivation of ultraviolet light–irradiated herpes virus by a Bloom's syndrome fibroblast line. *Cancer Res.* 39:3392.

67. Hayes, W. 1968. *The genetics of bacteria and their viruses*. Oxford: Blackwell.

68. Sarasin, A. R., and Hanawalt, P. C. 1980. Replication of ultraviolet-irradiated simian virus 40 in monkey kidney cells. *J. Mol. Biol.* 138:299.

69. Mezzina, M., Gentil, A., and Sarasin, A. 1981. Simian virus 40 as a probe for studying inducible repair functions in mammalian cells. *J. Supramol. Struct. Cell Biochem.* 17:121.

70. Higgins, N. P., Kato, K., and Strauss, B. 1976. A model for replication repair in mammalian cells. *J. Mol. Biol.* 101:417.

71. Lytle, C. D. 1978. Radiation-enhanced virus reactivation in mammalian cells. *Natl. Cancer Inst. Monograph* 50:145.

72. Lytle, C. D., Coppey, J., and Taylor, W. D. 1978. Enhanced survival of ultraviolet-irradiated herpes simplex virus in carcinogen-pretreated cells. *Nature* 272:60.

73. Lytle, C. D., Goddard, J. G., and Buchta, F. L. 1978. Protease inhibitors prevent UV-enhanced virus reactivation in *E. coli* but not in monkey kidney cells. In *DNA repair mechanisms*. P. C. Hanawalt, E. C. Friedberg and C. F. Fox, eds., p. 559. New York: Academic.

74. Lytle, C. D., and Goddard, J. G. 1979. Enhanced virus reactivation in mammalian cells: effects of metabolic inhibitors. *Photochem. Photobiol.* 29:959.

75. Lytle, C. D., Goddard, J. G., and Lin, C. 1980. Repair and mutagenesis of herpes simplex virus in UV-irradiated monkey cells. *Mutation Res.* 70:139.

76. Lytle, C. D., Iacangelo, A. L., Lin, C. H., and Goddard, J. G. 1981. UV-enhanced reactivation in mammalian cells: increase by caffeine. *Photochem. Photobiol.* 33:123.

77. Das Gupta, U. B., and Summers, W. C. 1978. Ultraviolet reactivation of herpes simplex virus is mutagenic and inducible in mammalian cells. *Proc. Natl. Acad. Sci.* (USA) 75:2378.

78. Zamansky, G. B., Kleinman, L. F., Black, P. H., and Kaplan, J. C. 1980. Reactivation of herpes simplex virus in a cell line inducible for simian virus 40 synthesis. *Mutation Res.* 70:1.

79. Bockstahler, L. E., and Lytle, C. D. 1977. Radiation enhanced reactivation of nuclear replicating mammalian viruses. *Photochem. Photobiol.* 25:477.

80. Bockstahler, L. E., and Lytle, C. D. 1970. UV light enhanced reactivation of a mammalian virus. *Biochem. Biophys. Res. Comm.* 41:184.

81. Cornelis, J. J., Lupker, J. H., and van der Eb, A. J. 1980. UV-reactivation, virus production and mutagenesis of SV40 in UV-irradiated monkey kidney cells. *Mutation Res.* 71:139.

82. Sarasin, A. R., and Hanawalt, P. C. 1978. Carcinogens enhance survival of UV-irradiated simian virus 40 in treated monkey cells: induction of a recovery pathway? *Proc. Natl. Acad. Sci.* (USA) 75:346.

83. Bockstahler, L. E., and Lytle, C. D. 1971. X-ray enhanced reactivation of ultraviolet-irradiated human virus. *J. Virology* 8:601.

84. Cleaver, J. E. 1978. DNA repair and its coupling to DNA replication in eukaryotic cells. *Biochim. Biophys. Acta* 516:489.

85. Bockstahler, L. E. 1981. Induction and enhanced reactivation of mammalian viruses by light. *Prog. Nucleic Acid Res. Mol. Biol.* 26:303.

86. Lytle, C. D., Day, R. S. III, Hellman, K. B., and Bockstahler, L. E. 1976. Infection of UV-irradiated normal human and XP fibroblasts by herpes simplex virus: studies on capacity and Weigle reactivation. *Mutation Res.* 36:257.

87. Bleichrodt, J. F., and Verheij, W. S. D. 1974. Mutagenesis by ultraviolet radiation in bacteriophage ϕX174: on the mutation stimulating process induced by ultraviolet radiation in the host bacterium. *Mol. Gen. Genet.* 135:10.

88. Caillet-Fauquet, P., and Defais, M. 1977. Kinetics of induction of error-prone repair of bacteriophage λ by temperature shift in an *Escherichia coli dna*B mutant. *Mol. Gen. Genet.* 155:321.

89. Ichikawa-Ryo, H., and Kondo, S. 1975. Indirect mutagenesis in phage lambda by ultraviolet preirradiation of host bacteria. *J. Mol. Biol.* 97:77.

90. Sarasin, A., and Benoit, A. 1980. Induction of an error-prone mode of DNA repair in UV-irradiated monkey kidney cells. *Mutation Res.* 70:71.

91. Gentil, A., Margot, A., and Sarasin, A. 1982. Enhanced reactivation and mutagenesis after transfection of carcinogen-treated monkey kidney cells with UV-irradiated simian virus 40 (SV40) DNA. *Biochimie* 64:693.

92. Day, R. S. III, and Ziolkowski, C. 1978. Studies on UV-induced viral reversion, Cockayne's syndrome, and MNNG damage using adenovirus 5. In *DNA repair mechanisms.* P. C. Hanawalt, E. C. Friedberg and C. F. Fox, eds., p. 535. New York: Academic.

93. Cornelis, J. J., Su, Z. Z., and Rommelaere, J. 1982. Direct and indirect effects of ultraviolet light on the mutagenesis of parvovirus H-1 in human cells. *The EMBO J.* 1:693.

94. Brandenberger, A., Godson, G. N., Radman, M., Glickman, B. W., Van Sluis, C. A., and Doubleday, O. P. 1981. Radiation-induced base substitution mutagenesis in single-stranded DNA phage M13. *Nature* 294:180.

95. Yatagai, F., Kitayama, S., and Matsuyama, A. 1981. Weigle reactivation and Weigle mutagenesis in phage ϕX174 by various types of radiation. *Mutation Res.* 91:3.

96. Lytle, C. D., and Knott, D. C. 1982. Enhanced mutagenesis parallels enhanced reactivation of herpes virus in a human cell line. *The EMBO J.* 1:701.

97. Purifoy, D. J. M., and Powell, K. L. 1977. Herpes simplex virus DNA polymerase as the site of phosphonoacetate sensitivity: temperature-sensitive mutants. *J. Virology* 24:470.

98. Rhode, S. L. III. 1974. Replication process of the parvovirus H-1. III. Factors affecting H-1 RF DNA synthesis. *J. Virology* 14:791.

99. George, J., Devoret, R., and Radman, M. 1974. Indirect ultraviolet-reactivation of phage λ. *Proc. Natl. Acad. Sci.* (USA) 71:144.

100. Rommelaere, J., and Ward, D. C. 1982. Effect of UV-irradiation on DNA replication of the parvovirus minute-virus-of-mice in mouse fibroblasts. *Nucleic Acid Res.* 10:2577.

101. Vos, J-M., and Rommelaere, J. Personal communication.

102. Fogel, M., and Sachs, L. 1970. Induction of virus synthesis in polyoma transformed cells by ultraviolet light and mitomycin C. *Virology* 40:174.

103. Hirsch, M. S., and Black, P. H. 1974. Activation of mammalian leukemia viruses. *Adv. Virus Res.* 19:265.

104. Lowy, D. R., Rowe, W. P., Teich, N., and Hartley, J. W. 1971. Murine leukemia virus: high-frequency activation *in vitro* by 5-iododeoxyuridine and 5-bromodeoxyuridine. *Science* 174:155.

105. Rothschild, H., and Black, P. H. 1970. Analysis of SV40-induced transformation of hamster kidney tissue *in vitro.* VII. Induction of SV40 virus from transformed hamster cell clones by various agents. *Virology* 42:251.

106. Kaplan, J. C., Zamansky, G. B., Black, P. H., and Latt, S. A. 1978. Parallel induction of sister chromatid exchanges and infectious virus from SV40-transformed cells by alkylating agents. *Nature* 271:662.

107. D'Ambrosio, S. M., and Setlow, R. B. 1976. Enhancement of postreplication repair in Chinese hamster calls. *Proc. Natl. Acad. Sci.* (USA) 73:2396.

108. Painter, R. B. 1978. Does ultraviolet light enhance postreplication repair in mammalian cells? *Nature* 275:243.

109. Painter, R. B. 1980. Response of Chinese hamster ovary cells to DNA damage after a conditioning exposure to ultraviolet light. *Biochim. Biophys. Acta* 609:257.

110. D'Ambrosio, S. M., Aebersold, P. M., and Setlow, R. B. 1978. Enhancement of postreplication repair in ultraviolet-light-irradiated Chinese hamster cells by irradiation in G2 or S-phase. *Biophys. J.* 23:71.

111. Moustacchi, E., Ehmann, U. K., and Friedberg, E. C. 1979. Defective recovery of semiconservative DNA synthesis in xeroderma pigmentosum cells following split dose ultraviolet-irradiation. *Mutation Res.* 62:159.

112. Friedberg, E. C., Moustacchi, E., Paul, B. R., and Ehmann, U. K. 1980. Possible evidence for inducible repair in human cells in culture. *Proc. Conference on Structural Pathology in DNA and the Biology of Ageing,* p. 80. Bonn: *Deutsche Forschungsgemeinschaft.*

113. Cleaver, J. E. 1978. Absence of interaction between X-rays and UV light in inducing ouabain- and thioguanine-resistant mutants in Chinese hamster cells. *Mutation Res.* 52:247.

114. Arlett, C. F., Turnbull, D., Harcourt, S. A., Lehmann, A. R., and Colella, C. M. 1975. A comparison of the 8-azaguanine and ouabain-resistance systems for the selection of induced mutant Chinese hamster cells. *Mutation Res.* 33:261.

115. Friedrich, U., and Coffino, P. 1977. Mutagenesis in S49 mouse lymphoma cells: induction of resistance to ouabain, 6-thioguanine and dibutyryl cyclic AMP. *Proc. Natl. Acad. Sci.* (USA) 74:679.

116. Cleaver, J. E. 1977. Induction of thioguanine- and ouabain-resistant mutants and single-strand breaks in the DNA of Chinese hamster cells by ^{3}H-thymidine. *Genetics* 87:129.

117. Chang, C-C., D'Ambrosio, S. M., Schultz, R. A., Trosko, J. E., and Setlow, R. B. 1978. Modification of UV-induced mutation frequencies in Chinese hamster cells by dose fractionation, cyclohexamide and caffeine treatments. *Mutation Res.* 52:231.

118. Meyn, M. S., Rossman, T., and Troll, W. 1977. A protease inhibitor blocks SOS functions in *Escherichia coli:* antipain prevents λ repressor inactivation, ultraviolet mutagenesis, and filamentous growth. *Proc. Natl. Acad. Sci.* (USA) 74:1152.

119. Kuroki, T., and Drevon, C. 1979. Inhibition of chemical transformation in C3H/10T ½ cells by protease inhibitors. *Cancer Res.* 39:2755.

120. Kinsella, A. R., and Radman, M. 1980. Inhibition of carcinogen-induced chromosomal aberrations by an anticarcinogenic protease inhibitor. *Proc. Natl. Acad. Sci.* (USA) 77:3544.

121. Kinsella, A. R., and Radman, M. 1978. Tumor promoter induces sister chromatid exchanges: relevance to mechanisms of carcinogenesis. *Proc. Natl. Acad. Sci.* (USA) 75:6149.

122. Nagasawa, H., and Little, J. B. 1979. Effect of tumor promoters, protease inhibitors, and repair processes on X-ray-induced sister chromatid exchanges in mouse cells. *Proc. Natl. Acad. Sci.* (USA) 76:1943.

123. Radman, M., Villani, G., Boiteux, S., Kinsella, A. R., Glickman, B. W., and Spadari, S. 1979. Replicational fidelity: mechanisms of mutation avoidance and mutation fixation. *Cold Spring Harbor Symp. Quant. Biol.* 43:937.

124. Radman, M., and Kinsella, A. R. 1980. Chromosomal events in carcinogenic initiation and promotion: implications for carcinogenicity testing and cancer

prevention strategies. In *Molecular and cellular aspects of carcinogen screening tests*. R. Montesano, H. Bartsch and L. Tomatis, eds., p. 75. *IARC Sci. Publ.* no. 27, Lyon.

125. Kennedy, A. R., and Little, J. B. 1978. Protease inhibitors suppress radiation-induced malignant transformation *in vitro*. *Nature* 276:825.

126. Quigley, J. 1979. Proteolytic enzymes of normal and malignant cells. In *Surfaces of normal and malignant cells*. R. O. Hynes, ed., p. 247. Chichester, U.K.: Wiley.

127. Goldfarb, R. H., and Quigley, J. P. 1978. Synergistic effect of tumor virus transformation and tumor promoter treatment on the production of plasminogen activator by chick embryo fibroblasts. *Cancer Res.* 38:4601.

128. Wigler, M., and Weinstein, I. B. 1976. Tumor promoter induces plasminogen activator. *Nature* 259:232.

129. Long, S. D., Quigley, J. P., Troll, W., and Kennedy, A. R. 1981. Protease inhibitor antipain suppresses 12-O-tetradecanoyl-phorbol-13-acetate induction of plasminogen activator in transformable mouse embryo fibroblasts. *Carcinogenesis* 2:933.

130. Miskin, R., and Reich, E. 1980. Plasminogen activator: induction of synthesis by DNA damage. *Cell* 19:217.

131. Miskin, R., and Ben-Ishai, R. 1981. Induction of plasminogen activator by UV light in normal and xeroderma pigmentosum fibroblasts. *Proc. Natl. Acad. Sci.* (USA) 78:6236.

132. Rahmsdorf, H. J., Mallick, U., Ponta, H., and Herrlich, P. 1982. A B-lymphocyte specific high turnover protein: constitutive expression in resting B cells and induction of synthesis in proliferating cells by inhibition of replication. *Cell* 29:459.

133. Nomura, S., and Oishi, M. 1982. An intracellular factor which affects erythroid differentiation in mouse Friend cells. *Biochimie* 64:763.

134. Fabre, F., and Roman, H. 1977. Genetic evidence for inducibility of recombination competence in yeast. *Proc. Natl. Acad. Sci.* (USA) 74:1667.

135. Eckhardt, F., Moustacchi, E., and Haynes, R. H. 1978. On the inducibility of error-prone repair in yeast. In *DNA repair mechanisms*. P. C. Hanawalt, E. C. Friedberg and C. F. Fox, eds., p. 421. New York: Academic.

136. Schwencke, J., and Moustacchi, E. 1982. Proteolytic activities in yeast after UV irradiation. I. Variation in proteinase levels in repair proficient Rad^+ strains. *Mol. Gen. Genet.* 185:290.

137. Schwencke, J., and Moustacchi, E. 1982. Proteolytic activities in yeast after UV irradiation. II. Variation in proteinase levels in mutants blocked in DNA-repair pathways. *Mol. Gen. Genet.* 185:296.

138. Holliday, R. 1975. Further evidence for an inducible recombination repair system in *Ustilago maydis*. *Mutation Res.* 29:149.

139. Lee, M. G., and Yarranton, G. T. 1982. Inducible DNA repair in *Ustilago maydis*. *Mol. Gen. Genet.* 185:245.

140. Baker, T. I. 1983. Inducible nucleotide excision repair in *Neurospora*. *Mol. Gen. Genet.* 190:295.

9

DNA Damage and Human Disease

9-1 Introduction

The study of DNA repair in human cells was limited for many years by the lack of available mutant cells defective in their response to DNA damage. In 1968 James Cleaver observed that fibroblasts in culture derived from the skin of a human patient with the disease *xeroderma pigmentosum* (XP) were defective in excision repair following exposure to UV radiation (1). This was the first indication of a DNA repair defect associated with a human disease. Since then, an enormous amount of attention has been focused on the response of XP cells to DNA damage by a variety of chemical and physical agents (2–16). Cleaver's observations also provided an impetus to examine the phenotypic response to DNA damaging agents of cells from a number of other hereditary human diseases, particularly those associated with chromosomal abnormalities or with an abnormally raised incidence of neoplasia (7, 11, 12, 14, 17–20). In addition, the observation that human subjects suffering from XP are highly susceptible to malignant neoplasms caused by the well-characterized mutagen UV radiation has prompted further exploration of the relation between DNA damage, mutagenesis and neoplastic transformation at a variety of levels. This chapter deals primarily with a number of *selected* hereditary diseases in which there is evidence for defects in some aspect of DNA metabolism following DNA damage by physical and/or chemical agents. It also includes a more general discussion of some components of the complex relation between DNA damage and cancer in humans.

Despite extensive studies during the past 15 years, there is still no precise understanding of the molecular defect(s) in DNA repair in XP. However, as will be seen in the course of the chapter, there is little doubt that the most prevalent form of the disease, so-called *classical XP*, is characterized by defective or deficient excision repair of DNA. The role of defective DNA repair in the other human diseases discussed here is more controversial. In this regard it should be noted that in general, the demonstration of abnormal sensitivity of cells to killing by physical and/or chemical agents known to interact with DNA does not necessarily imply a defect in *excision repair* in the disease in question. The persistence of damage to DNA may, for example, result from a dysregulation of DNA replication. Indeed, as will be seen in Section 9-3, this has been suggested as the primary basis for the sensitivity of ataxia telangiectasia (AT) cells to ionizing radiation. In addition, some cell types may sustain greater levels of DNA damage than normal cells following an equivalent insult. A general mechanism by which this could occur is by the operation of one or more membrane defects that allow access of higher levels of certain chemicals to the nucleus. Another possible mechanism could involve qualitative and/or quantitative disturbances in the metabolic activation of unreactive polar compounds to forms that are more reactive with DNA (see Section 1-3). For example, there is evidence that some agents have a particular avidity for mitochondria because of their marked lipophilicity (21). Such agents might cause selective or preferential damage to mitochondrial rather than nuclear DNA, interfering with normal mitochondrial functions and thereby altering cell survival. Finally, there is evidence that some human diseases may be characterized by the abnormal production of metabolic products that damage DNA in the absence of exogenous damaging agents and also create heightened sensitivity to DNA damage in their presence (18). This phenomenon is discussed in relation to the human disease called Bloom's syndrome (Section 9-3).

9-2 Xeroderma Pigmentosum

This disease is clinically characterized chiefly by the early onset of severe photosensitivity of the exposed regions of the skin (Fig. 9-1) and eyes, a very high incidence of skin cancers and frequent neurologic abnormalities (2–16). A particular syndrome of neurological

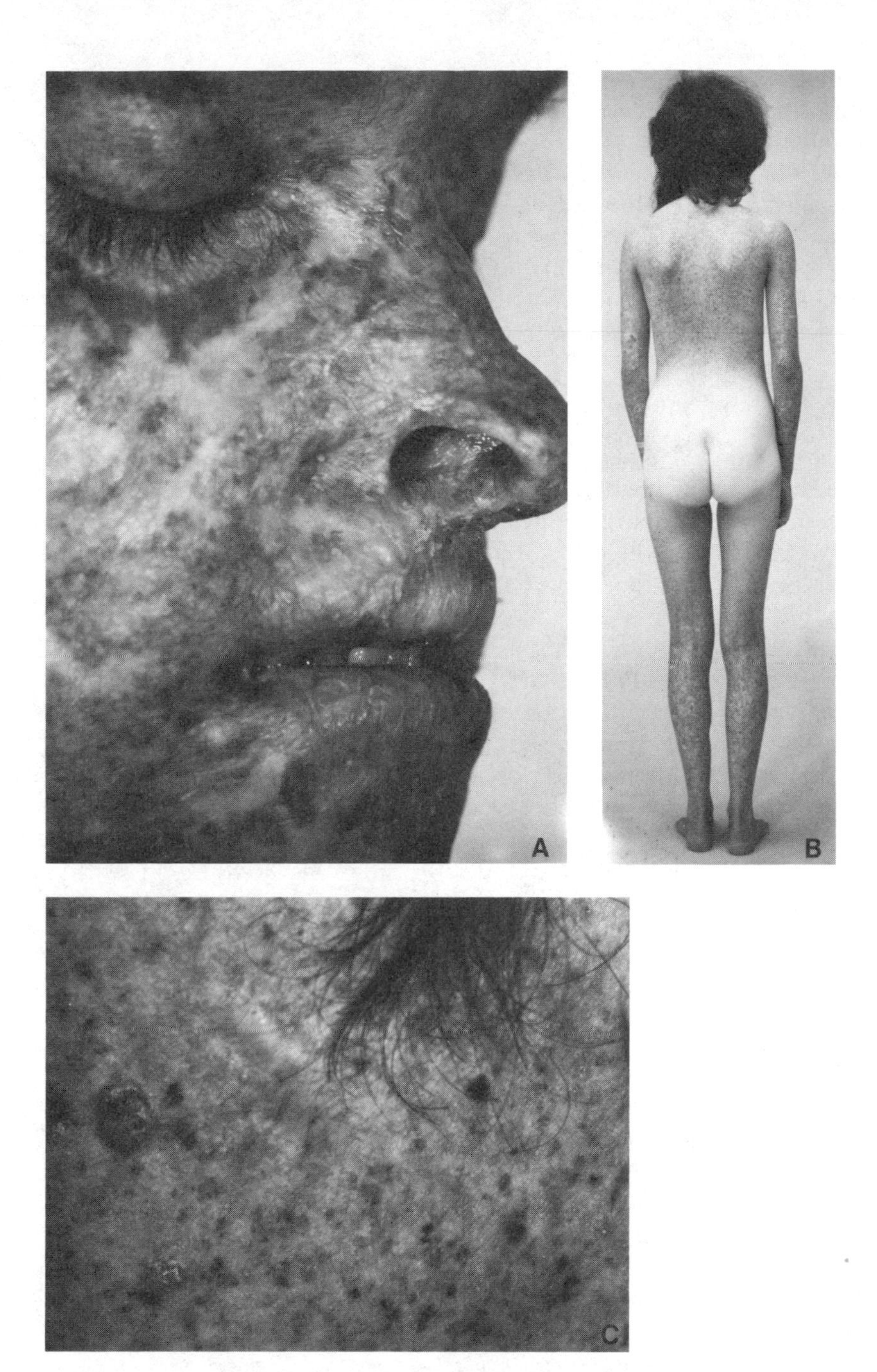

FIGURE 9-1
Individuals with xeroderma pigmentosum show severe skin disorders in regions normally exposed to actinic radiation (B). The most prominent of these disorders are extreme dryness (xerosis) and areas of alternating hyper- and hypopigmentation (A, C). The large lesion shown on the left in C is a basal cell carcinoma of the skin. (Courtesy of Dr. K. H. Kraemer.)

defects in association with typical skin symptoms and signs was first described by De Sanctis and Cacchione and is referred to as the *De Sanctis–Cacchione syndrome* (22). It should be noted, however, that neurologic abnormalities frequently accompany the dermatological manifestations of XP and not all such cases qualify as examples of this syndrome. XP has a worldwide distribution, but the incidence varies from about 1 in 200,000 in Europe and the U.S. to as high as 1 in 40,000 in Japan (3, 23). In all known cases the disease is inherited through an autosomal recessive mode (2–16).

The Genetic Complexity of XP

A large array of phenotypic alterations have been described at both the cellular and the biochemical level in XP cells exposed to chemical or radiation damage (Table 9-1), many of which are difficult to accommodate on the basis of a single molecular defect. Perhaps the first level of complexity that merits discussion is the apparent genetic heterogeneity of the disease. Cells from the initial cases of XP that were documented in terms of their defect in excision repair were exposed to UV radiation and examined by measurement of repair synthesis and/or pyrimidine dimer excising capability as a function of post-UV incubation time (24, 25). Both parameters of excision repair were found to be markedly defective. (Figs. 9-2, 9-3, 9-4). However, as increasing numbers of cases were reported and studied, a large variability in the repair defect as measured by unscheduled DNA synthesis (UDS) was noted (26). This suggested the possibility of genetic heterogeneity, a postulate that was examined by fusing cells from different XP patients and comparing the levels of UDS in

TABLE 9-1
Some phenotypic abnormalities in xeroderma pigmentosum cells in culture

1. Increased sensitivity to killing by UV radiation and UV-mimetic chemicals.
2. Defective host-cell reactivation of UV-irradiated viruses.
3. Some cells show increased sensitivity to killing by ionizing radiation.
4. Defective elimination of thymine-containing dimers from acid-insoluble DNA.
5. Failure to lose sites sensitive to PD DNA glycosylase enzyme probes.
6. Defective repair synthesis of DNA.
7. Increased sister chromatid exchanges following exposure to UV radiation.
8. Enhanced mutagenesis after treatment with mutagens.
9. Enhanced susceptibility to neoplastic transformation.
10. Defective AP endonuclease activity (group D).
11. Lower levels of DNA photolyase activity.

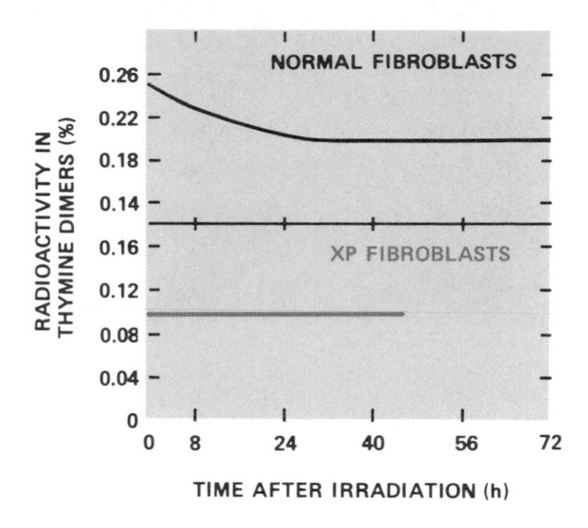

FIGURE 9-2
Normal human fibroblasts in culture show a significant loss of thymine-containing pyrimidine dimers from the acid-insoluble fraction of radiolabeled DNA during post-irradiation incubation. No detectable loss of dimers is observed from the DNA of XP patients belonging to complementation group A. (From U. K. Ehmann et al., ref. 90.)

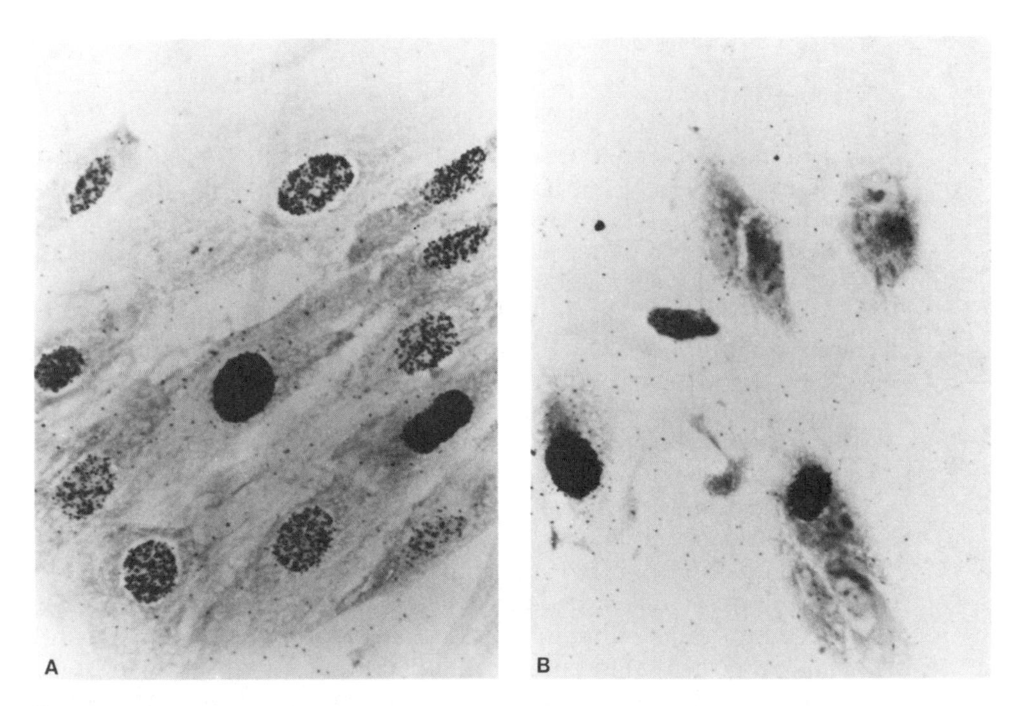

FIGURE 9-3
Normal human fibroblasts (A) in culture show autoradiographic grains in S phase cells and in all cells not in the S phase of the cell cycle. The latter reflects repair synthesis or so-called unscheduled DNA synthesis (UDS) (see Section 5-2) of DNA following exposure to ultraviolet radiation. The cells that are intensely labeled are cells in the S phase of the cycle and are carrying out semiconservative DNA synthesis. Fibroblasts from patients with xeroderma pigmentosum (B) show no detectable UDS in non-S-phase cells, although labeling of S-phase cells is normal. (Courtesy of Dr. J. E. Cleaver.)

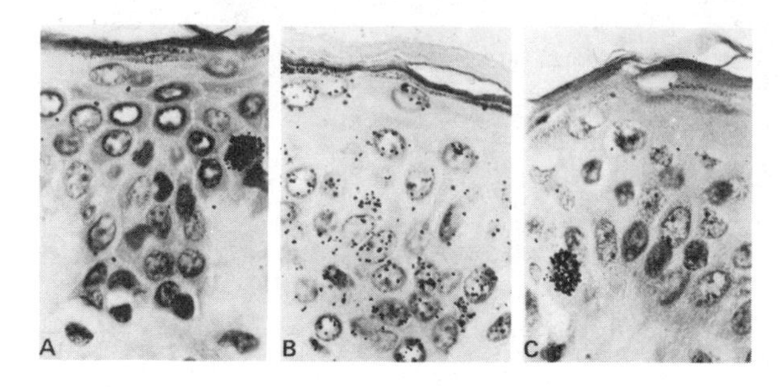

FIGURE 9-4
Defective unscheduled DNA synthesis (UDS) of DNA can be demonstrated in the epidermal cells of living XP individuals following injection of [^{3}H]thymidine into a UV-irradiated area of skin. The histological section on the left (A) is from an unirradiated normal subject, and some cells show semiconservative DNA synthesis. The center section (B) is from a normal UV-irradiated subject and shows UDS in most cells. No UDS is observed in the section from an XP patient who was exposed to UV radiation (C). (From J. H. Epstein et al., ref. 277).

heterodikaryons and unfused monokaryons (27). It was observed that cells from different XP patients can complement one another, restoring levels of UDS to normal (see Section 4-3) (Figs. 4-21 and 9-5).

Systematic complementation analysis by cell fusion has led to the identification of eight *complementation groups*, designated A through H (14, 16, 28). In addition, approximately 10 percent of all cases of clinically typical XP do not have an apparent defect in the excision repair of DNA damage caused by UV radiation (29–31). These cases are referred to as the *XP variant form* to distinguish them from the so-called *classical XP cases* (31, 32). A recent worldwide compilation of XP cases and the complementation groups into which they fall is shown in Table 9-2. Patients from groups A and C constitute the majority of cases; however, it is interesting that group C apparently is rare in Japan (23). If the results of these complementation analyses reflect classical Mendelian genetics and also reflect strictly *intergenic* rather than *intragenic* (*interallelic*) complementation, it appears that in normal human fibroblasts at least eight genes are involved in the excision repair of bulky base damage to DNA.

Intragenic and intergenic complementation may in some cases be distinguishable, because in the former case no completely normal gene product should be synthesized and the efficiency of complementation is expected to be relatively low. This general principle has been applied by fusing XP cells with each other or with normal fibroblasts and selecting heterokaryons with varying numbers of wild-type alleles (33). One might anticipate that under such condi-

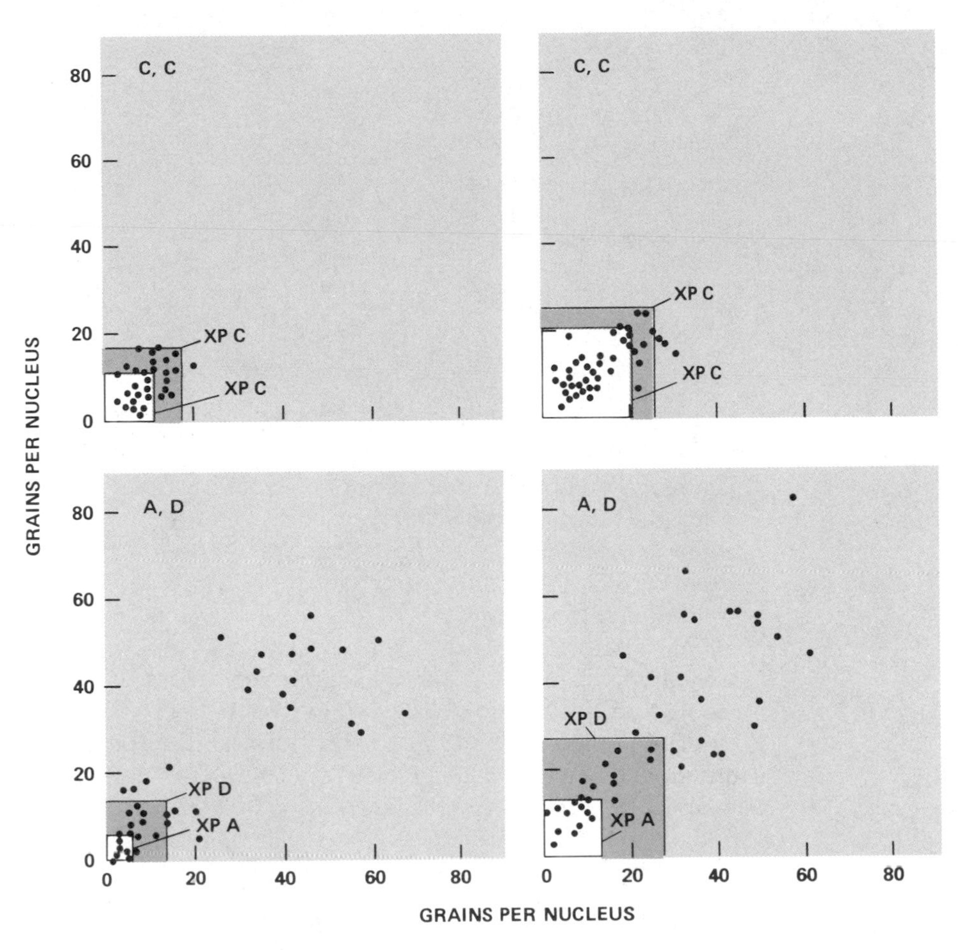

FIGURE 9-5
When cells from two XP group C individuals are fused, there is no complementation of the excision repair defect and levels of UDS (measured as the number of autoradiographic grains per nucleus) remain low in heterokaryons exposed to either UV radiation (top left) or 4-nitroquinoline-1-oxide (top right). However, when cells from two *different* complementation groups (A and D) are fused, increased levels of UDS are observed in the heterokaryons exposed to UV radiation (bottom left) or to 4-nitroquinoline-1-oxide (bottom right). (After B. Zelle, ref. 278.)

tions, allelic mutants restricted to *intragenic* complementation would not restore DNA repair capacity to the heterokaryons as efficiently as fusions between XP and normal cells. What is observed is that XP group A, C and D cells complement one another at least as well as each is complemented by normal human cells, suggesting that the defects in these three complementation groups are at differ-

TABLE 9-2
Relative frequencies of various XP complementation groups among different populations

	Complementation Group (No. of Patients)								
Population	A	B	C	D	E	F	G	H	Variant
North America	3	1	5	5	0	0	0	0	2
Europe	10	0	14	8	2	0	2	1	5
Japan	21	0	1	1	0	1	0	0	9
Egypt	7	0	12	0	0	0	0	0	5

From J. E. Cleaver, ref. 16.

ent genetic loci (33). This conclusion is supported by differences in the kinetics of complementation with these three cell types. For example, XP group A nuclei achieve maximum complementation within six hours, even in the presence of the protein synthesis inhibitor cycloheximide. XP group C cells, on the other hand, require closer to twenty-four hours to attain this state, and XP group D cells require even longer (33).

DNA repair synthesis in the nuclei of XP group A cells is complemented very rapidly by normal cells (within 20 to 30 minutes), suggesting that the XP group A gene product is freely diffusable in and out of the cell nucleus (33). In addition, when protein synthesis is inhibited, a single normal cell can almost fully complement as many as three XP group A cells, suggesting that the XP group A gene product is present in excess. This observation, together with the independent observations of a rapid turnover of XP group A factor (33), could also account for the rapid complementation by normal cells. In contrast to the results obtained with XP group A cells, the cytoplasm of normal fibroblasts appears to have only a small surplus of the XP C and D gene products, since these XP cells are complemented poorly in the absence of protein synthesis. There are also indications that the XP group C and D gene products do not readily diffuse out of the nucleus of normal cells and that they may interact with each other (33).

The Biochemical Defects in XP Cells

As indicated in Section 4-3, the available evidence suggests that much, if not all, of the genetic complexity in XP concerns the *incision* of damaged DNA during excision repair rather than later events such as repair synthesis or DNA ligation. In general, the various parameters by which excision repair capacity is measured in human cells correlate well within any given complementation group, but there are significant exceptions (14, 16) (Table 9-3). For example,

TABLE 9-3
Comparative parameters of DNA repair in xeroderma pigmentosum cells

Complementation Group	Repair Synthesis* (% of Normal)	Loss of ESS† in 32 h (% of Normal)	Relative UV Sensitivity
A	0–5	0	+ + + +
B‡	3–7	10	+ + +
C	10–15	20	+ +
D	10–40	20	+ + + +
E	40–60	50	+
F	0–10	60	+
G	10–15	0	+ + + +
H‡	30	?	?

*The repair synthesis values quoted represent the average of a number of different reported studies.

†ESS = endonuclease-sensitive sites (sites in DNA sensitive to pyrimidine dimer-specific enzyme probes).

‡Both complementation groups B and H are represented by only a single case and both have been diagnosed as examples of Cockayne's syndrome together with XP.

most cells in group A have very low rates of loss of sites sensitive to dimer-specific enzyme probes (Fig. 4-24, Table 9-3) and of thymine dimers (measured directly) from their DNA (Fig. 9-2), have low levels of UDS (Figs. 9-3, 9-4; Table 9-3) and are very sensitive to killing by UV radiation (14, 16) (Fig. 9-6). Similarly, most group C cells show residual UDS (Table 9-3) (16, 34) and, in at least one reported study,

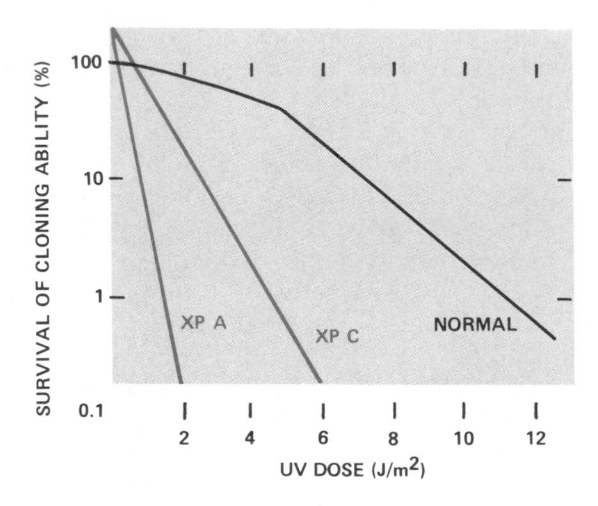

FIGURE 9-6
XP cells from complementation groups A and C are considerably more sensitive to killing by UV radiation than normal human fibroblasts are. (From V. M. Maher and J. J. McCormick, ref. 62.)

an increased number of (presumably) enzyme-mediated DNA strand breaks during post-UV incubation in vivo compared with group A cells (16, 34). Their sensitivity to cell killing is also reduced relative to that of group A cells (16, 34) (Fig. 9-6). On the other hand, most group D cells are as UV sensitive as most group A cells (35) and they show a rate of loss of specific enzyme-sensitive sites (pyrimidine dimers) no more rapid than that of group C cells (36) (Fig. 4-24, Table 9-3). Yet many group D cells demonstrate significant residual levels of UDS (Table 9-3) (37, 38). Finally, although group F cells show a pronounced residual capacity for loss of enzyme-sensitive sites, UDS is disproportionately low (Table 9-3) (39). Of the various parameters mentioned, UDS is perhaps the least reliable indicator of excision repair, since it is sensitive to physiological perturbations in the pool size of thymidine in the cells. In addition, repair synthesis reflects not only the number of sites in DNA repaired but also the size of the repair patch at any given site. Thus, longer average repair patch sizes at a reduced number of sites of repair may not be reflected by a reduction in total repair synthesis.

Why So Many Complementation Groups?

The apparent requirement for so many genetic functions for the incision of UV-irradiated DNA in human cells is not totally surprising. As was pointed out in Section 4-3, incision of DNA following UV irradiation of *E. coli* is dependent on three gene products, all of which are required to constitute a damage-specific DNA incising activity (40). In the lower eukaryote yeast *S. cerevisiae* at least five *RAD* genes are required for the incision of UV-irradiated DNA in vivo (41) (see Section 4-3). A model that attempts to explain the requirement for multiple gene products for the incision of damaged DNA is fully elaborated in Section 4-3 and essentially suggests that these products constitute a large *multiprotein complex* that is required both for the specific cutting of DNA associated with the excision of damaged bases and for whatever conformational changes in the chromosome may be necessary for access of the repair enzyme to sites of base damage in DNA. An alternative possibility is that the human DNA incising activity may be a smaller entity but may be part of a *multienzyme complex* required for *all* the biochemical events in excision repair, including repair synthesis and DNA ligation. Inactivation of any single component of the complex may render the entire complex inoperative, thereby creating the phenotype of incision defectiveness. This model is somewhat unlikely, since measurements of DNA polymerase and of DNA ligase activity in a number of XP and normal cell lines have not demonstrated any significant differences (42–44).

The notion of a large, multicomponent protein complex required either for preincision and/or incision events or for incision, excision, resynthesis and DNA ligation has not been substantiated by any direct biochemical evidence, because such an activity has not been isolated from human cells. However, as indicated in Section 4-3, studies with extracts of calf thymus suggest the existence of a very large protein that catalyzes incision at pyrimidine dimers in UV-irradiated DNA (45). There is also interesting *indirect* evidence for a large excision repair complex in human cells. For example, it has been shown that both repair synthesis and excision of thymine-containing pyrimidine dimers from the DNA of UV-irradiated normal human cells are inhibited by exposure of the cells to the monofunctional alkylating agent methylmethanesulfonate (MMS) (46–48). Since XP cells are not abnormally sensitive to MMS (14), it is likely that the repair pathways for excision of pyrimidine dimers and of monofunctional alkylation base damage are distinct in normal human cells (46–48). Thus the inhibition of excision repair of pyrimidine dimers by alkylation treatment may reflect the inactivation of an *alkylation-sensitive protein target*. A model formally analogous to target theory in radiobiology, postulating that one "hit" in a sensitive target is sufficient to inactivate it, has facilitated calculations of the approximate size of a protein in which one alkylation site per molecule would result following exposure of human cells to 1 to 2 m*M* MMS. These calculations yield a target size of approximately 10^6 daltons (46–48).

Perhaps the most plausible explanation for the genetic complexity of excision repair in human cells is the possible requirement for significant chromosomal structural modification as a *preincision event* (see Section 4-3). However, aside from the demonstration that extracts of XP cells from complementation groups A, C and G are indeed defective in the excision of thymine-containing pyrimidine dimers from chromatin but not from deproteinized chromatin or naked DNA (49–51) (see Section 4-3), once again there is a regrettable paucity of experimental data. When XP cells in culture are treated with proteases in an effort to "relax" chromatin structure, no increase in UDS is observed after exposure of the treated cells to UV radiation (52). In a different approach to this issue XP cells were treated with sodium butyrate, an agent known to increase the transcriptional activity of cells in culture by altering the extent of histone acetylation (53). This agent promotes increased UDS in both normal and XP irradiated cells; however, the relative difference between the two cell populations is maintained (54).

A specific aspect of chromosomal protein modification that has received particular attention with respect to DNA repair in mammalian cells is ADP-ribosylation of chromosomal proteins by poly(ADP-ribose) polymerase (see Section 6-4). Any perturbation of chromatin that causes DNA strand breakage results in increased poly(ADP-ribose) polymerase activity and decreased NAD content, indicative of increased poly(ADP-ribose) synthesis (55, 56). XP cells from complementation groups A, B, C and D fail to show these responses following UV irradiation (57, 58). These results have led to the suggestion that in normal human cells, enzyme-catalyzed incision of DNA stimulates the synthesis of poly(ADP-ribose), which somehow produces alterations in chromatin conformation that facilitate base damage excision, repair synthesis and DNA ligation (57). Since XP cells are defective or deficient in the incision of DNA, the sequence of biochemical events leading to chromatin modification fails to occur, further complicating the defect in excision repair (57).

XP Cells Are Also Defective in the Repair of UV-Mimetic Chemical Damage

A large number of chemical agents promote increased killing of XP relative to normal human cells (14, 16) (Table 9-4). In addition, the repair capacity of XP cells following treatment with a variety of chemicals is defective in a number of instances (14, 16). Such chemicals are generally bulky adducts of the type expected to cause significant helix distortion, analogous to that produced by pyrimidine dimers. Thus, as is true of the *E. coli* DNA incising activity (see Section 4-2), it appears that the analogous activity in human cells is not specific for pyrimidine dimers in DNA. On the other hand, studies in human cells on combined DNA damage produced by both UV radiation and chemicals (59–61) suggest the existence of a branched excision repair pathway in which some steps in the repair of chemical and UV damage are common, whereas others are not. A defect in a step common to both pathways could also explain the sensitivity of XP cells to both chemicals and UV radiation.

It is obvious from the preceding discussion that an understanding of the molecular basis for the defect(s) in excision repair of damage to DNA in XP cells requires further investigation. It is hoped that the numerous complexities cited above will be explained when the biochemistry and enzymology of excision repair in normal human cells are better understood. Irrespective of the precise nature of the defect(s) in DNA excision repair in XP cells, a number of the phenotypes of these cells related above follow logically from this observation. For example, all classical XP cells have an increased sensitivity to UV radiation and to UV-mimetic chemicals and a decreased ability to reactivate DNA viruses (Fig. 9-7) exposed to these forms of DNA damage (14, 16). In addition, as mentioned previously,

TABLE 9-4
Some chemicals causing increased killing of XP fibroblasts relative to that of normal cells

4-Nitroquinoline-1-oxide
4-Hydroxyaminoquinoline-1-oxide
2-Methyl-4-nitroquinoline-1-oxide
3-Methyl-4-nitropyridine-1-oxide
Benz(a)anthracene-5,6-epoxide
N-hydroxy and N-acetoxy ester derivatives of 4-acetylaminobiphenyl-2-acetylaminofluorene, 2-acetylaminophenanthrene, 4-acetylaminostilbene
Activated aflatoxin B_1
Activated sterigmatocystin
Activated (hydroxy and 5,6-epoxide) derivates of benz(a)anthracene, dibenz(a,h)anthracene, chyrsene, 7,12-dimethylbenz(a)anthracene and benzo(a)pyrene
Nitrogen mustard
Mitomycin C
Methylmethanesulfonate
Ethylmethanesulfonate
Bisulfan
8-Methoxypsoralen and long-wavelength UV
Decarbamoyl mitomycin C

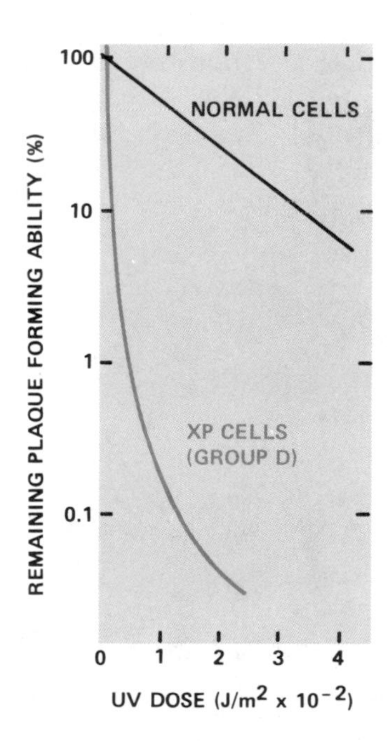

FIGURE 9-7
The plaque-forming ability of UV-irradiated adenovirus 2 is markedly reduced when the virus is plated on XP cells (group D cells here) compared with when it is plated on normal human fibroblasts. (From R. S. Day III et al., ref. 279.)

XP cells show decreased levels of repair synthesis when examined with the techniques used to measure this parameter. They also fail to lose sites in their DNA sensitive to dimer-specific endonuclease probes and are defective or deficient in the loss of thymine dimers from the acid-insoluble fraction of their DNA (14, 16).

Other Defects in XP Cells

XP cells in culture demonstrate an increased frequency of mutation at a number of different genetic loci following exposure to UV or UV-mimetic chemicals (62–65) (Fig. 9-8). Although the karyotype of the vast majority of XP individuals is normal, cells in culture have an abnormal incidence of chromosomal aberrations following exposure to UV or to certain chemicals (66–68). The level of spontaneously induced sister chromatid exchanges (SCEs) (see Section 9-4) is normal, but following exposure of XP cells to DNA damage, a significant increase in SCEs occurs (69–72). For example, a survey of complementation groups A, B, C, D and E showed an increase in SCEs after UV irradiation in all but group E (71). Finally, it has been shown that cells from an XP individual are more susceptible to neoplastic transformation by UV radiation than normal cells are (73).

Collectively, these biochemical and cellular defects in XP cells are consistent with the observation of marked sensitivity to sunlight in living XP individuals, and the prevalence of skin cancer in this disease provides important indirect evidence that DNA damage may result in mutation and neoplastic transformation of human cells. However, there are a number of reported observations on XP cells and individuals that are not conveniently reconcilable with a defect in the excision repair of pyrimidine dimers or UV-mimetic chemical damage, and these merit brief discussion.

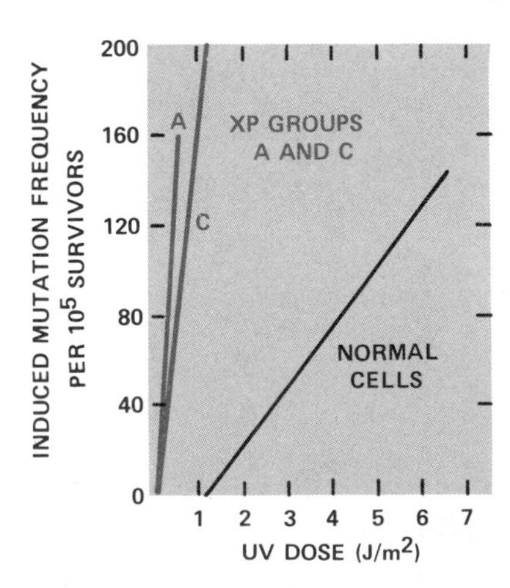

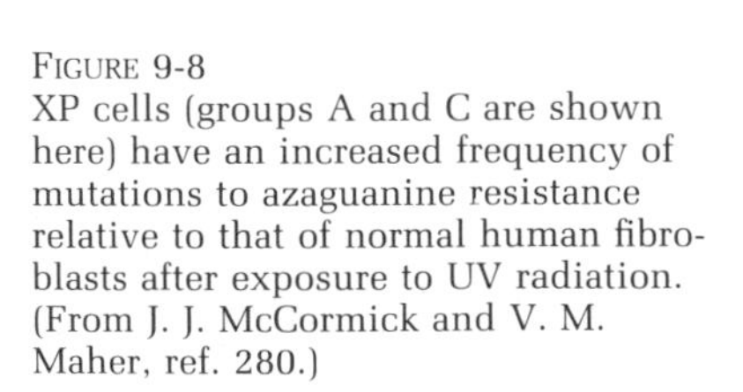

FIGURE 9-8
XP cells (groups A and C are shown here) have an increased frequency of mutations to azaguanine resistance relative to that of normal human fibroblasts after exposure to UV radiation. (From J. J. McCormick and V. M. Maher, ref. 280.)

1. It has been reported that extracts of normal human cells have two species of AP endonuclease activity (see Section 3-5) that can be resolved by phosphocellulose chromatography: one with a high K_m for apurinic duplex DNA and the other with a low K_m for this substrate (74). Cells from complementation group D have about one-sixth the normal level of AP endonuclease activity and are apparently lacking the low K_m species (74, 75). In view of these results it might be anticipated that XP cells from complementation group D would be relatively deficient in the repair of viral DNA containing apurinic sites. In fact, a significantly *higher* level of viral infectious centers is observed in XP group D cells compared with that in normal human fibroblasts (Fig. 9-9) (76). These results suggest the existence of an alternative (and presumably more efficient) pathway for the repair of apurinic sites in the DNA of normal human cells. A possible candidate for such a pathway is the DNA purine insertase activity purported to be present in extracts of normal human fibroblasts (see Section 2-5).

2. Prior to 1974, there was no documented evidence of either enzymatic photoreactivation in living cells from placental mammals or of photoreactivating enzyme in extracts of such cells. Since then a protein with the properties of DNA photolyase has been purified from extracts of human peripheral blood leukocytes (77) (see Section 2-2). This activity has also been identified in a number of other placental mammalian sources (78, 79), and it has been observed that the levels of photoreactivating enzyme in extracts of fibroblasts from XP individuals are lower than in normal cell extracts (80, 81). These differences are independent of the age of the individual and of the

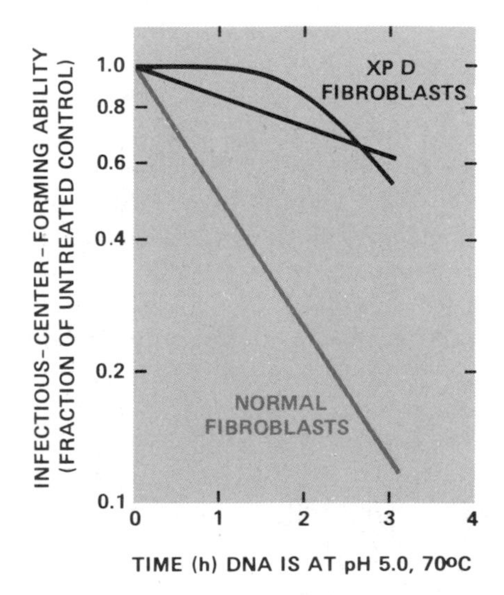

FIGURE 9-9
The yield of infectious virus from heat-acid depurinated SV40 DNA is greater in XP group D cells than in normal human fibroblasts. (From R. D. Kudrna et al., ref. 76.)

cell passage number. In addition, there is no evidence that the enzyme from XP cells has different properties or cofactor requirements than the enzyme from normal cells, and no enzyme inhibitor has been found in XP cell extracts (80).

3. Some XP cell strains, particularly a well-studied line from complementation group G, are abnormally sensitive to killing by ionization as well as by UV radiation (68, 82).

4. It is not obvious why defective DNA repair should result in neurological defects. In addition, although only a limited number of cases have been adequately studied, the available evidence indicates that organs other than the skin are not particularly susceptible to neoplastic transformation in XP individuals (83). This is surprising, since if defective DNA repair is the underlying basis for the predilection for UV-induced skin cancer, one might expect that cancer in other organs would also be increased. This has prompted speculation about other possible pathogenic mechanisms, such as impaired immune responsiveness (84), to account for the high incidence of skin cancer in the disease. In this regard studies on XP have provided some evidence of impaired cellular immunity (85), and in addition there are indications that immunosuppression correlates with skin cancer in normal subjects (86).

The XP Variant Form

As indicated previously, shortly after the discovery of defective DNA repair in classical XP cell lines, a group of clinically typical XP patients was found to have normal levels of UDS and repair synthesis as measured by other procedures (29–32). Cells from these patients had only a slightly increased sensitivity to UV radiation (Fig. 9-10) and a modest decrease in the ability to reactivate UV-irradiated adenovirus (87, 88). The assumption that pyrimidine dimer excision is normal in these cells is supported by a normal rate of loss of sites sensitive to pyrimidine dimer-specific enzyme probes (89). Also, the kinetics of the loss of thymine dimers from the acid-insoluble fraction of UV-irradiated cells is normal (90). The term *XP variant* was coined to describe this form of the disease, characterized by typical clinical features but with no apparent defect in nucleotide excision repair following exposure to UV radiation (29–32). However, there may be a subtle abnormality in excision repair associated with later steps in the process, such as repair synthesis and/or DNA ligation. For example, most XP variant cell lines have an enhanced sensitivity to inhibition of the completion of excision repair in the presence of cytosine arabinoside, an inhibitor of DNA synthesis (91). This is reflected in the persistence in XP variant cells of a slight but reproducibly larger number of DNA incisions (strand breaks) that pre-

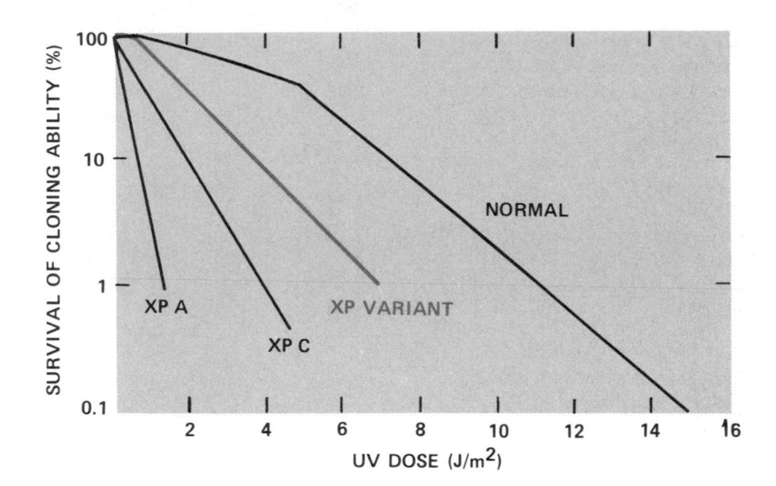

FIGURE 9-10
XP variant fibroblasts have a modest sensitivity to killing by UV radiation relative to that of normal human cells; in fact many XP variant cells show no detectable increase in UV sensitivity. The survival of classical XP cells (complementation groups A and C) is shown for comparison. (From V. M. Maher and J. J. McCormick, ref. 62.)

sumably initiated excision repair relative to the number in normal cells (91).

XP variant cells also have a defect in the replication of damaged DNA. The weight-average molecular weight of newly synthesized DNA (labeled with a pulse of [^{3}H]thymidine) in unirradiated or UV-irradiated *normal skin fibroblasts* is about 8×10^7 daltons. Following a chase in nonradioactive thymidine, the molecular weight reaches about 1.5×10^8 daltons (the maximal weight-average molecular weight measurable under the experimental conditions generally used) (89). The molecular weight distribution of the DNA of *unirradiated XP variant cells* is identical to that of normal cells. However, following irradiation, XP variant DNA pulse labeled for the same time as normal cells has a lower molecular weight (about 3×10^7 daltons) and consequently requires a longer chase time to reach the maximal detectable molecular weight (89) (Fig. 9-11). In the presence of caffeine, the restoration of high molecular weight DNA is completely inhibited in XP variant cells within the time of the usual chase, whereas normal cells are hardly affected at all by this compound (Fig. 9-12). Since the process of conversion of low molecular weight DNA synthesized on damaged templates into high molecular weight DNA constitutes a particular cellular response to DNA damage whereby lesions that are not repaired may be tolerated (so-called postreplication repair) (see Chapters 7 and 8), it has been

FIGURE 9-11
Increase in the molecular weight of newly synthesized DNA. Both normal human and XP variant fibroblasts were UV-irradiated, pulse labeled for one hour with [^{3}H]thymidine and then chased for various times in medium with unlabeled thymidine. Weight-average molecular weights were calculated from molecular weight distributions obtained from alkaline sucrose gradient profiles. The weight-average molecular weight of the DNA from the XP variant patients is smaller than that of normal subjects immediately after irradiation of the cells, and hence takes longer to reach normal high molecular weight values. (After A. R. Lehmann et al., ref. 89.)

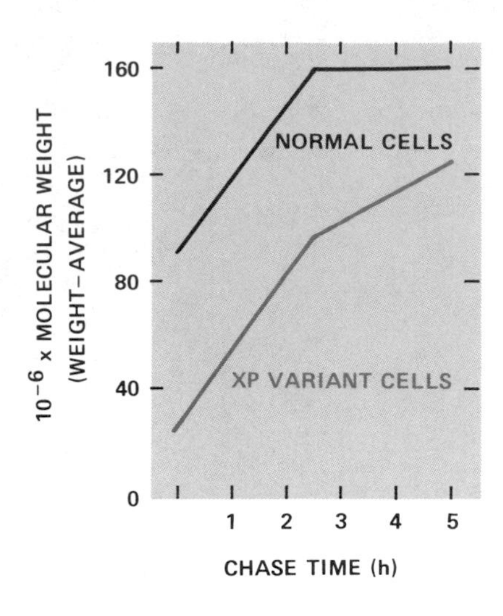

FIGURE 9-12
When normal human fibroblasts (top panel) exposed to UV radiation are pulse labeled with [^{3}H]thymidine and then chased with unlabeled thymidine, low molecular weight DNA (not shown) is converted into a high molecular weight form (also see Fig. 9-11). This conversion is unaffected by the presence of caffeine (blue line). However, in XP variant cells (bottom panel), in the presence of caffeine the conversion of low molecular weight DNA into a high molecular weight form is inhibited. (After A. R. Lehmann et al., ref. 89.)

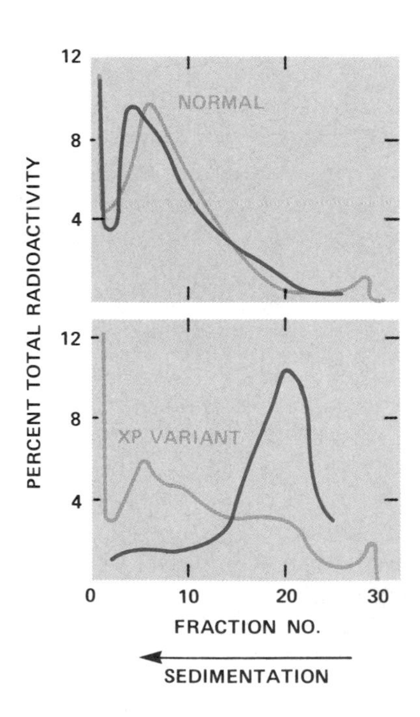

suggested that XP variant cells are defective in this response (89, 92, 93).

The type of pulse-chase experiments just described are open to multiple interpretations, a problem discussed in detail in Section 8-2. For example, as indicated above, when semiconservative DNA synthesis is examined in UV-irradiated XP variant and normal fibroblasts, the *absolute size* of the DNA fragments synthesized during brief pulses of [^{3}H]thymidine given one to two hours after irradiation is decreased relative to that in unirradiated cells, with the XP variant showing a greater decrease than normal cells (Fig. 9-11). Thus, an alternative model postulates that pyrimidine dimers (and presumably other sites of base damage in DNA) act as all-or-nothing blocks to the progress of replication forks and that XP variant cells have an increased probability (relative to that of normal cells) of replicative arrest at sites of DNA damage (94–97).

Classical XP cells also show a reduced ability to convert low molecular weight DNA into a higher molecular weight form following exposure to UV radiation, although quantitatively the effect is not as marked as that observed in XP variant cells (92, 93) (Fig. 9-13). This raises the interesting question of whether XP variant and classical XP cells share a *common defect* in the response to base damage at replication forks. Some studies suggest that they may. For instance, when normal human cells are exposed to UV radiation, there is an initial decrease in the rate of semiconservative DNA synthesis followed by a dose-dependent recovery toward normal rates (Fig. 9-14) (98).

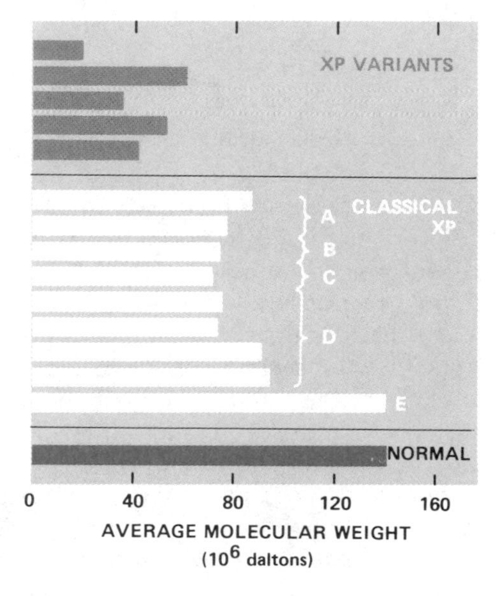

FIGURE 9-13
UV-irradiated cells from normal, XP variant or classical XP individuals were pulse labeled with [^{3}H]thymidine and then chased in medium containing unlabeled thymidine in the presence of caffeine. The cells were lysed on top of alkaline sucrose gradients and the weight-average molecular weights of the DNA were determined after centrifugation (see Fig. 9-12). The letters in the middle panel denote the complementation groups to which the various cell strains belong. Note that caffeine inhibits the restoration of high molecular weight DNA not only in XP variant cells but also in a number of the classical XP cells. In general this inhibition is quantitatively greater in XP variant relative to classical XP cells. (From A. R. Lehmann et al., ref. 92.)

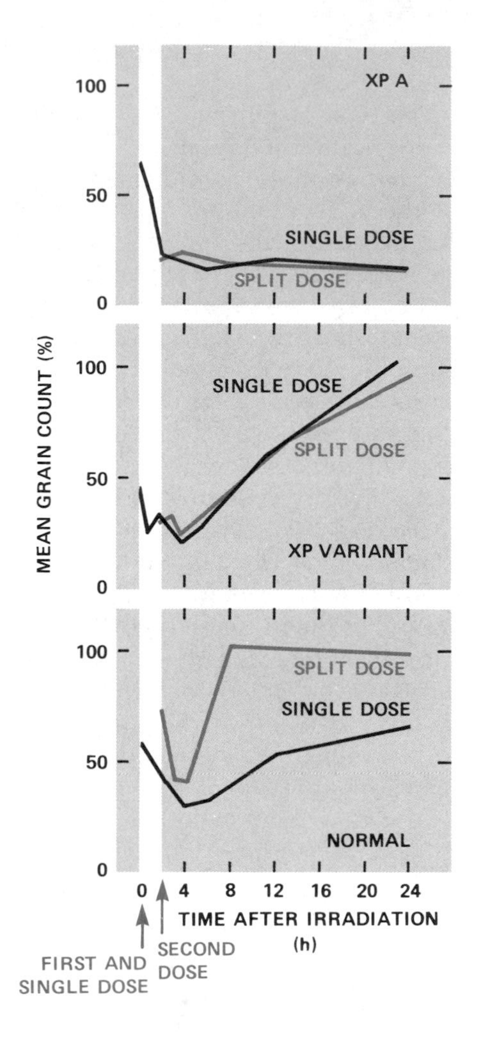

FIGURE 9-14
When normal human fibroblasts are exposed to a split dose of UV radiation (bottom panel), the rate and extent of recovery of DNA synthesis is enhanced compared with that of cells that received an equivalent single dose of radiation. However, both XP variant (middle panel) and classical XP cells (top panel) are defective in this enhanced recovery. (After E. Moustacchi et al., ref. 99.)

However, when the UV dose is fractionated by a two-hour period, that is, 1 J/m^2 of UV radiation is followed by 9 J/m^2 of irradiation instead of a single dose of 10 J/m^2, the rate of recovery from maximal inhibition of DNA synthesis is increased in normal cells (99) (Fig. 9-14). This type of experiment was described in Section 8-3 as possible evidence for an inducible response to DNA damage in human cells, based on the argument that the first dose induces a particular cellular response that deals with the second (DNA damaging) dose. Irrespective of this interpretation, the point to be made here is that *both* classical XP and XP variant cells fail to show this split-dose-dependent enhanced recovery phenomenon (99) (Fig. 9-14). These observations suggest that both cell types may share a defect in the excision repair of those dimers that bear a particular relation to replication forks (Fig. 9-15). This defect in excision repair may result

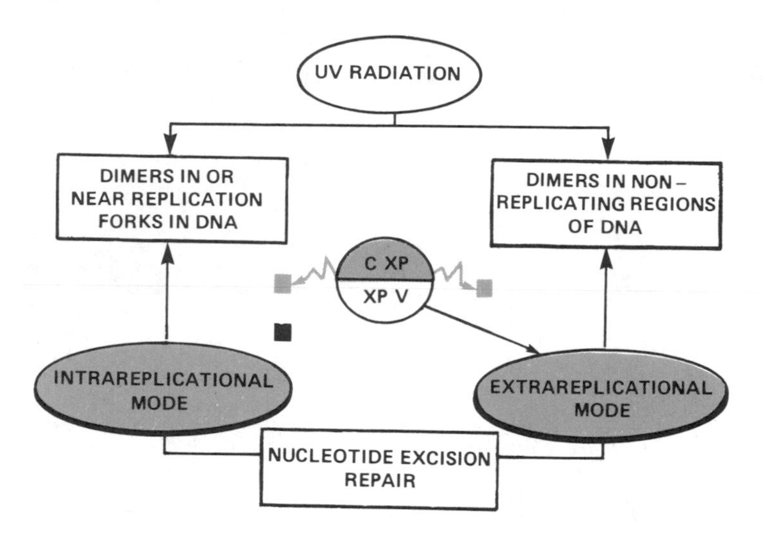

FIGURE 9-15
A model for the excision repair defects in classical XP (C XP) and XP variant (XP V) cells. The model proposes that lesions in or near replication forks in DNA are normally repaired by an "intrareplicational" excision repair mode that is defective in both XP cell types. Dimers in nonreplicating regions of DNA are excised by an "extrareplicational" mode that is defective in classical XP cells but not in XP variant cells.

in mutations, perhaps through the operation of an error-prone replicative mode for bypassing DNA damage. An attractive feature of this model is that it provides an explanation for the observations that the clinical manifestations of XP (including the predisposition to skin cancer) in both the variant and the classical forms are indistinguishable and that both cell types show enhanced mutagenesis in culture (Figs. 9-8, 9-16). It is, of course, necessary to postulate that classical XP cells *but not XP variant cells* are additionally defective in the excision repair of lesions not specifically related to replication forks (Fig. 9-15).

9-3 Other Human Diseases with Evidence for Altered Cellular Responses to DNA Damage

The remainder of this chapter deals with a number of other human diseases in which the evidence for a primary defect in the repair of DNA damage is less well substantiated than in XP, and in some of the examples to be discussed there are good indications of a defect in some aspect of DNA metabolism other than DNA repair.

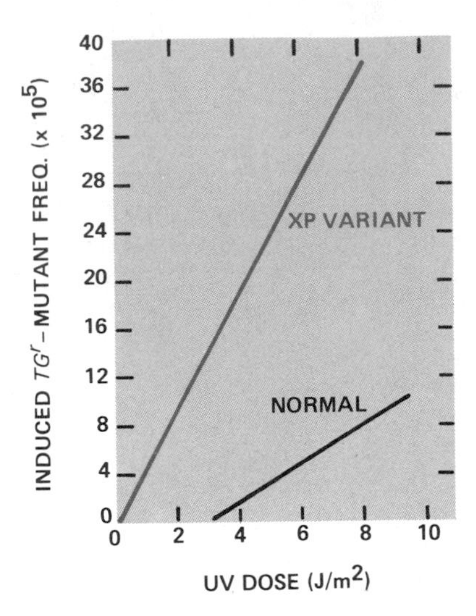

FIGURE 9-16
XP variant cells are more mutable to thioguanine resistance through UV radiation than normal human fibroblasts are. (From B. C. Myhr et al., ref. 283.)

Ataxia Telangiectasia (AT) (Louis-Bar Syndrome)

AT is another genetic disorder with an autosomal recessive mode of inheritance (100). The disease occurs with an incidence of about 1 per 40,000 live births and affects many systems of the body, particularly the nervous system, the immune system and the skin (7, 11, 12, 14, 17, 19). The most characteristic neurological disorder is a profound cerebellar ataxia resulting in a staggering gait, severe muscular uncoordination and progressive mental retardation (101, 102) (Fig. 9-17). The term telangiectasia describes the marked dilation of small blood vessels, most easily observed in the eye and the skin (Fig. 9-17). A significant feature of the disease is an immune dysfunction that renders affected individuals susceptible to intercurrent infection (103). This dysfunction also manifests itself with an increased incidence of neoplasms of the lymphoreticular system. Approximately 10 percent of AT individuals develop neoplasms, 88 percent of which are lymphoreticular, including Hodgkin's disease and non-Hodgkin's lymphomas (104, 105). Almost all of the neoplasms occur before AT individuals are 20 years of age, in contrast to the median age in the general population of about 55 years. Heterozygous relatives of AT patients are more likely to develop neoplasms than unrelated persons, and it has been estimated that the presence of the AT gene in the population may be responsible for about 5 percent of all persons dying of cancer before the age of 45 years (7).

Disordered DNA Metabolism in AT Cells

Cells with polyploid or endoreduplicated chromosomes are 10 to 20 times more prevalent in AT than in normal individuals (106). In

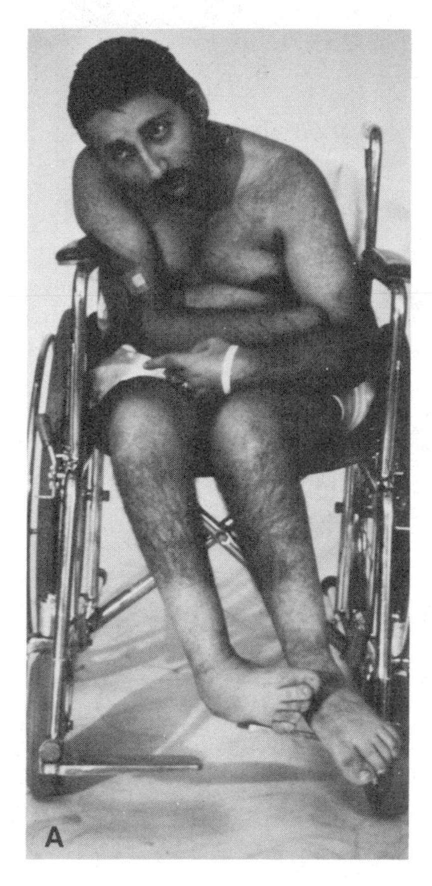

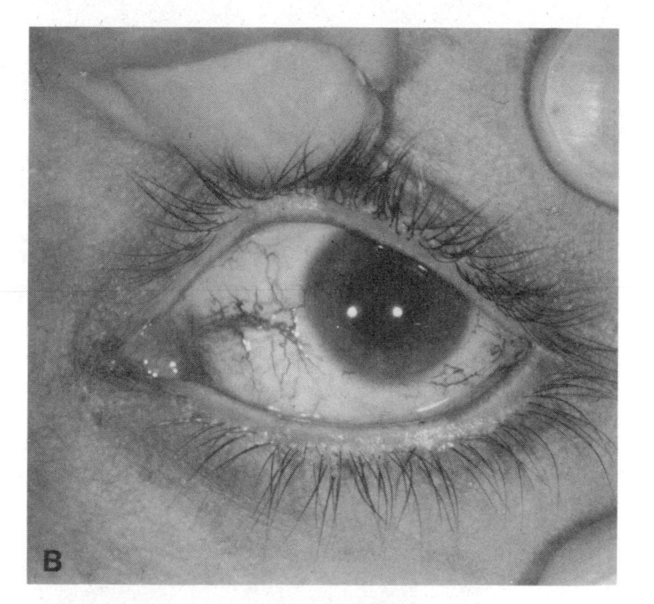

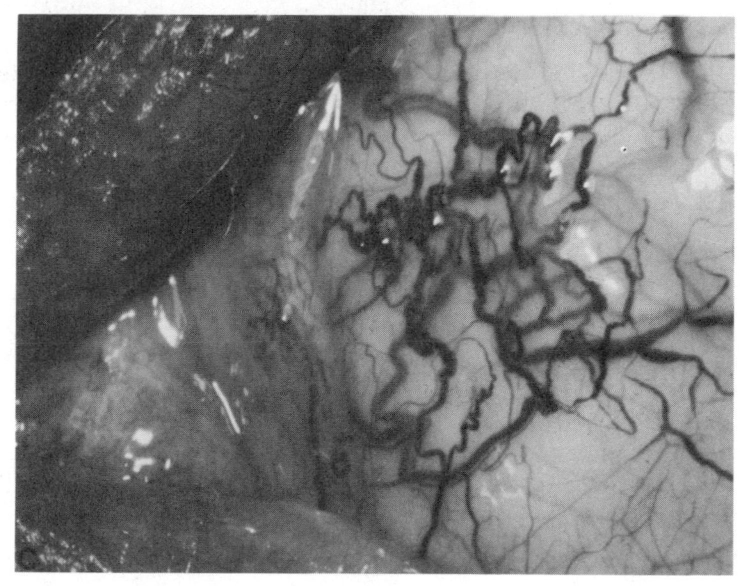

FIGURE 9-17
Patients suffering from ataxia telangiectasia (AT) have mental impairment and cerebellar ataxia (A) as well as abnormal dilation of small blood vessels. The last abnormality is easily observed in the conjunctiva of the eye (B, C). (Courtesy of Dr. K. H. Kraemer.)

addition, cells from AT individuals have a high frequency of spontaneous chromosomal abnormalities and X irradiation markedly increases the number of chromosomal breaks in such cells (106–112). Faulty DNA repair was first suspected when several AT patients with tumors demonstrated severe or fatal reactions to ionizing radiation therapy (113–116). This prompted an examination of the response of AT fibroblasts in culture to ionizing radiation, and it was observed that AT cells have a significantly increased sensitivity to ionizing radiation in all cases (117). There is virtually no overlap in the survival of normal and AT cells exposed to ionizing radiation (118–121), and this extreme radiosensitivity is one of the most outstanding features that is common to all AT cell lines so far examined (122) (Fig. 9-18). AT cells are not abnormally sensitive to UV radiation or to chemicals classified as UV mimetic [e.g., N-acetoxy-N-2-acetylaminofluorene or 4-nitroquinoline-1-oxide (4-NQO)] but are sensitive to X-ray-mimetic chemicals such as bleomycin (123–125).

Studies on the biochemistry of DNA repair in irradiated AT cells have revealed no detectable defects in the kinetics of rejoining of single- or double-strand breaks in DNA (110, 126–128). However, some (but not all) AT cell lines show a decreased rate and extent of repair synthesis of DNA after exposure to ionizing radiation (17, 19) (Fig. 9-19). In addition, when DNA from irradiated normal and AT cells is incubated with an extract of *M. luteus* containing an enzyme activity that apparently recognizes an unidentified form of base damage caused by ionizing radiation, sites sensitive to the enzyme disappear from the DNA of some AT cells much more slowly than from the DNA of normal cells, suggesting that AT cells are defective in the excision of some form of ionizing radiation damage (129, 130) (Fig. 9-20).

Other experiments have provided preliminary evidence suggestive of a DNA repair defect in AT cells (131–133). For example, crude extracts of normal human cells apparently contain an enzyme(s) that enhances the priming activity of γ-irradiated colE1 DNA for purified DNA polymerase I of *E. coli*. Exactly what effects this enhancement of DNA synthesis is not known, but it has been speculated that sites of base damage in γ-irradiated DNA may be attacked by endonucleases that leave 3′ moieties that are ineffective primers for DNA synthesis. This is very reminiscent of the effect of 3′-terminal ap-

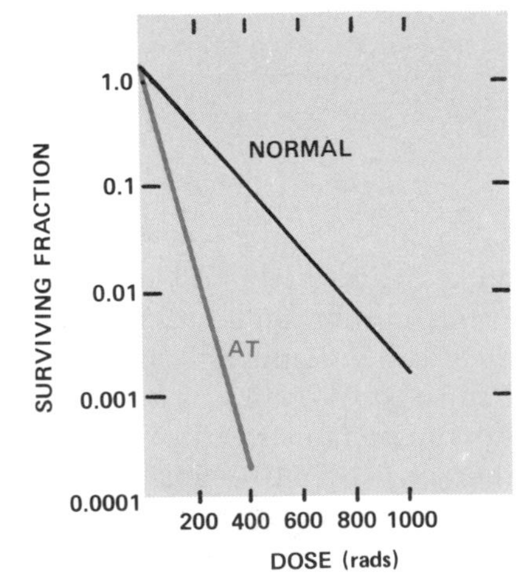

FIGURE 9-18
Fibroblasts from patients with ataxia telangiectasia (AT) are distinctly more sensitive to killing by X-rays than are cells from normal individuals. (After A. M. R. Taylor et al., ref. 117.)

urinic or apyrimidinic sites on DNA synthesis catalyzed by *E. coli* DNA polymerase I (see Section 5-4). The removal of these moieties by activities present in human cell extracts may then facilitate enhanced DNA synthesis. Extracts of a number of AT cell lines contain considerably lower amounts of this putative priming-enhancer activ-

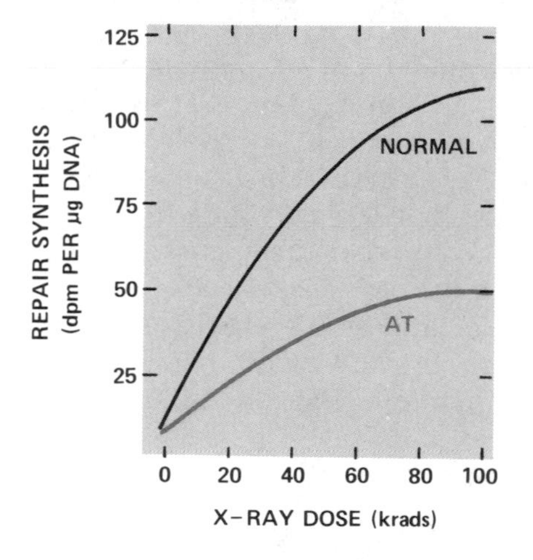

FIGURE 9-19
Cells from some ataxia telangiectasia (AT) patients show a decreased rate and extent of repair synthesis after exposure to ionizing radiation. Repair synthesis was measured by the incorporation of [^{3}H]thymidine under conditions in which semiconservative DNA synthesis was inhibited. (After M. C. Paterson et al., ref. 129.)

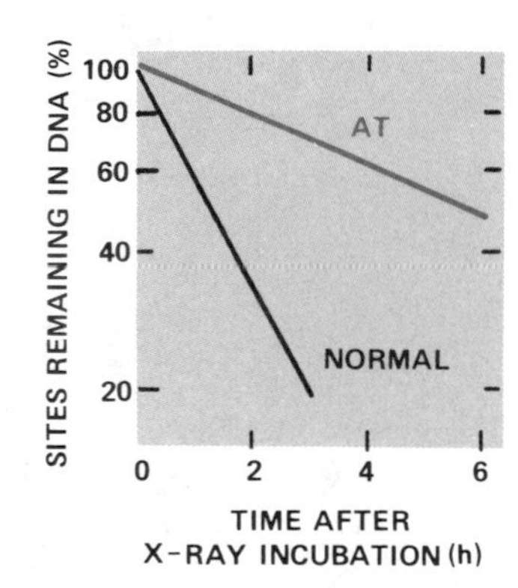

FIGURE 9-20
Loss of enzyme-sensitive sites (ESS) (see Section 5-2) from the DNA of normal and AT cells. The DNA in fibroblasts from ataxia telangiectasia and normal human subjects was radiolabeled with [^{3}H]thymidine, then the fibroblasts were exposed to X irradiation and incubated for the times indicated. The DNA was extracted and incubated with an extract of *M. luteus*, following which the DNA was sedimented on alkaline sucrose gradients. Sites of DNA damage produced by the X irradiation and which were not repaired during the postirradiation incubation are recognized by an enzyme activity in the *M. luteus* extract, resulting in nicking of the DNA and hence a reduced rate of sedimentation. Based on the sedimentation rate, the fraction of ESS lost from the DNA during postirradiation incubation can be calculated. Normal cells lose such sites more rapidly and more extensively than AT cells do. (From M. C. Paterson et al., ref. 129.)

ity, whereas the activity of AT heterozygotes is indistinguishable from that of normal cells (131–133).

More recently attention has been focused on a defect in semiconservative DNA synthesis that appears to be *common to all AT cell lines* exposed to ionizing radiation or to bleomycin. When normal human fibroblasts are exposed to ionizing radiation, there is a dose-dependent inhibition of the rate of DNA synthesis, as measured by the uptake of radiolabeled thymidine (134–138) (Fig. 9-21). AT cells, on the other hand, are significantly more resistant to radiation- and bleomycin-induced DNA synthesis inhibition (134–140) (Fig. 9-21). Based on the results of additional experiments in which the molecular weight distribution of newly synthesized DNA in alkaline sucrose gradients was examined, it appears that *replicon initiation* is inhibited by X irradiation or by bleomycin to a much lesser extent in AT cells than in normal cells (135). It has been suggested (see Section 8-2) that in normal cells DNA damage induces a change in *clusters of replicons* (the domain of a cluster having a molecular weight of about 10^9 daltons) that precludes the initiation of DNA synthesis in any

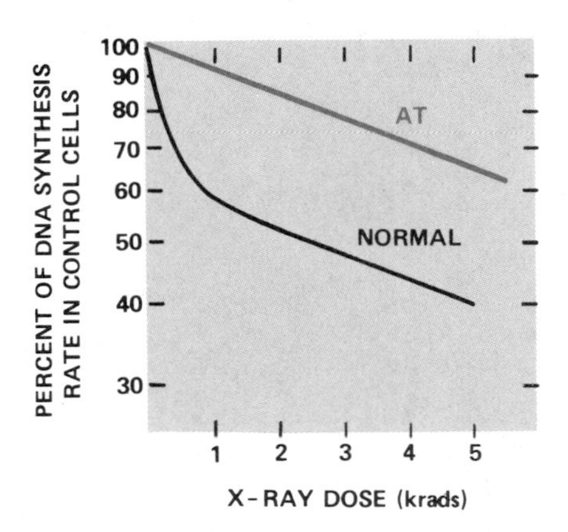

FIGURE 9-21
When normal human fibroblasts are exposed to X irradiation (or bleomycin), there is a dose-dependent decrease in the rate of semiconservative DNA synthesis. AT cells fail to show the same degree of inhibition of DNA synthesis. Note that normal cells show an initial rapid inhibition component sensitive to low doses of X-rays, followed by a less severe inhibition component sensitive to larger doses of X-rays. The former component is thought to possibly reflect inhibition of replicon initiation and is sensitive to low doses of X-rays because the target constitutes a relatively large replicon cluster domain. The less severe component of DNA synthesis inhibition may reflect inhibition of replication fork progression, a smaller target for DNA damage. The observation that inhibition of fork progression occurs to the same extent in both normal and AT cells (indicated by the parallel slopes of the second component curves), suggests that AT cells are primarily insensitive to inhibition of replicon initiation following X irradiation. (From R. B. Painter and B. R. Young, ref. 135.)

replicon in that cluster until the DNA damage is repaired (141, 142). Thus AT cells may owe their radiosensitivity to a failure to manifest a DNA damage-induced delay that allows normal cells to repair such damage rather than to their inability to repair DNA damage per se (135). However, it has been argued (143) that it is the very process of excision repair that causes arrest of replicon initiation in normal cells and that the radio resistance of DNA synthesis in AT cells is a reflection of defective excision repair.

The biochemical basis for the radio-resistant DNA synthesis is not known. However, this phenomenon can be partly mimicked by treating normal cells with caffeine or by incubating them in hypertonic medium after irradiation (144). Because both of these perturbations are known to alter chromatin structure, it is suggested that the radio-resistant DNA synthesis in AT cells may be due to an intrinsic difference in chromatin structure between AT and normal cells (144). No such differences have yet been identified.

Other DNA defects have been observed in AT cells. For example, cocultivation of plasma and lymphocytes from AT patients with those of normal individuals results in a significant increase in chromosomal damage in the normal cells (145). Tissue culture medium used to cultivatc AT skin fibroblasts also significantly increases chromosomal breakage in phytohemagglutinin-stimulated normal lymphocytes (145) (Table 9-5). These observations suggest that AT cells may produce clastogenic factors (factors that cause chromosomal abnormalities; see Section 1-2). The possible role of such factors in the potentiation of DNA damage caused by exogenous agents is discussed in greater detail below in the section on Bloom's syndrome.

AT cells are not hypermutable by ionizing radiation (146) (Fig. 9-22). Thus the increased incidence of lymphoreticular neoplasms observed in AT individuals may not be pathogenetically related to mutations in somatic cells but rather, as was suggested in Section 9-2

TABLE 9-5
Effect of conditioned medium derived from skin fibroblast cultures on chromosome damage in normal lymphocytes

Source of Medium	Chromosome Breaks/Cell
AT cells	0.060
AT cells	0.095
Normal cells	0.015
Normal cells	0.030

After M. Shaham et al., ref. 145.

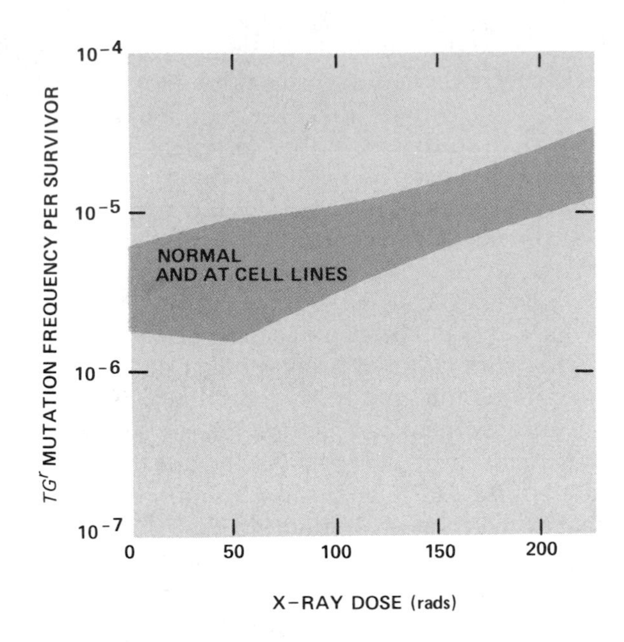

FIGURE 9-22
The mutation frequency to thioguanine resistance of a series of normal and AT cell lines exposed to ionizing radiation is defined by the shaded area. There is no significant difference between the two groups. (After C. F. Arlett, ref. 146.)

for skin cancer in XP patients (84), may result from the immune deficiency that is associated with AT.

Genetic Heterogeneity in AT

The radio-resistant DNA synthesis that characterizes AT cells has provided the basis for an assay to detect complementation of the defect in cell hybrids (Fig. 9-23). Such studies have identified a minimum of five complementation groups, indicating a considerable genetic complexity (147, 148). This is not surprising, considering the complexity of replicon initiation in eukaryotic cells and the fact that AT is about 10 times more frequent in humans than XP is (7).

Bloom's Syndrome (BS)

The syndrome originally described by Bloom is characterized by a strikingly low birth weight and by stunted growth (149–152). Light-induced telangiectasia develops on the skin of the face in typical "butterfly" lesions (Fig. 9-24), the head is elongated and BS patients are highly susceptible to infections (149–152). The incidence of BS

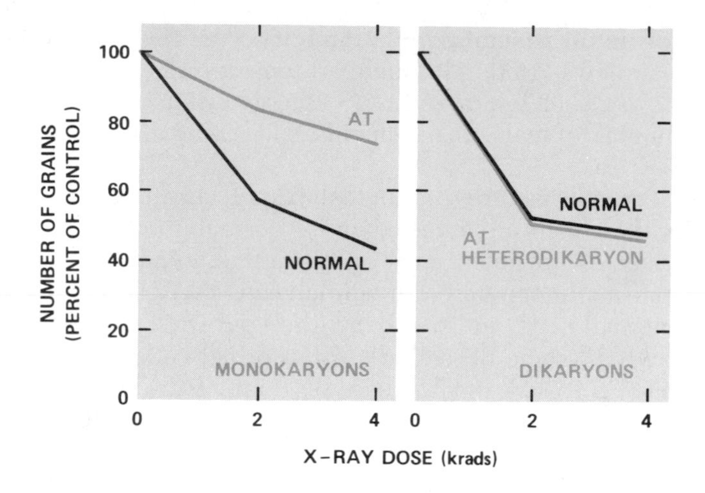

FIGURE 9-23
AT cells can be grouped into different complementation groups by monitoring the ability of fused cells to carry out normal inhibition of DNA synthesis after exposure to X irradiation. Monokaryons from AT cells show *reduced* inhibition of DNA synthesis relative to that of normal cells (left panel) (see Fig. 9-21). However, heterodikaryons formed by fusion of AT cells from different complementation groups show a normal response to X irradiation (right panel). (From N. G. J. Jaspers and D. Bootsma, ref. 148.)

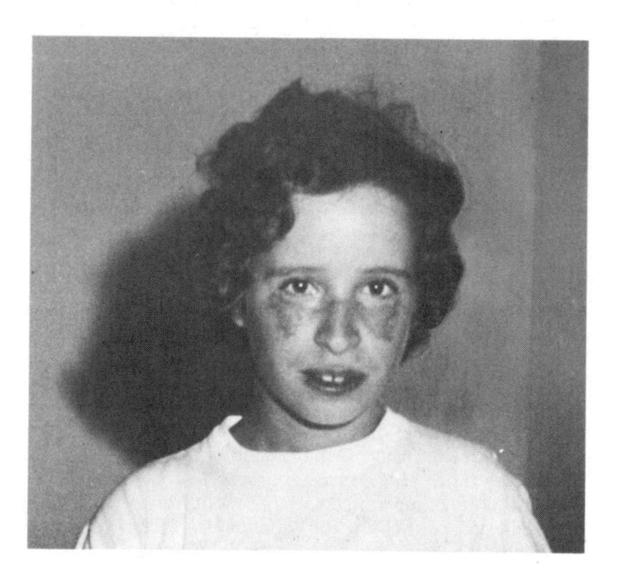

FIGURE 9-24
Patients with Bloom's syndrome typically have light-induced capillary dilatation of the skin of the face in a butterfly distribution. (Courtesy of Dr. James German.)

is very low in the general population but occurs in about 1 in 58,000 Ashkenazic Jews (153). Although evidence points to an autosomal recessive mode of inheritance, BS is diagnosed in a significantly higher number of males than females, suggesting a sex linkage (151, 152).

Several notable features distinguish BS cells in culture. Generation times for these cells are longer than for normal fibroblasts, and plating efficiencies are lower (154). Cells undergoing DNA synthesis incorporate about one-half as much labeled DNA precursor per unit time as normal cells, and newly synthesized DNA strands elongate more slowly in some BS cells than in normal cells (154–156) (Fig. 9-25). Levels of DNA polymerases α, β and γ are normal in these cells, however (157). BS cells in culture also have high numbers of spontaneous chromosomal breaks and rearrangements (158, 159). The distinguishing chromosome rearrangements in this syndrome are sister chromatid exchanges (SCEs) (see Section 9-4), which occur with a frequency of between 10 and 15 times that in normal cells (159–161). Disproportionate numbers of particular chromosomal abnormalities referred to as quadriradial figures, so called because the

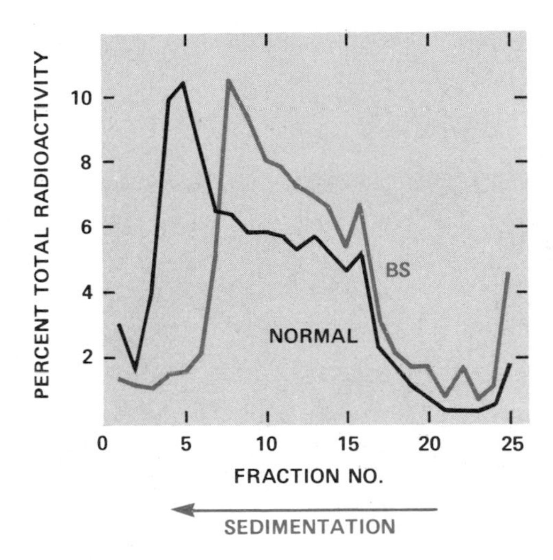

FIGURE 9-25
When cells from normal and Bloom's syndrome (BS) patients are exposed to UV radiation and then pulsed with [^{3}H]thymidine to label nascent DNA, the labeled DNA from some BS cells sediments more slowly in alkaline sucrose gradients than does DNA from normal cells. This observation suggests that BS cells have a defect in semiconservative DNA synthesis on damaged DNA templates. (From F. Gianelli et al., ref. 154.)

chromosome is apparently extended in four directions (Fig. 9-26), are also present in BS cells (160, 161).

The aforementioned characteristics of sunlight sensitivity and high numbers of chromosomal aberrations, together with the observation that one out of nine BS patients develops malignancies at an average age of 20, have prompted speculation that the primary defect in BS may be in DNA repair (14). So far the evidence for this is marginal. Although one study has reported that two BS fibroblast lines are abnormally sensitive to UV irradiation at 254 nm (154), other studies have found no such differences (14). A number of BS lines examined show increased DNA strand breaks following exposure to near-UV light (313 nm) (162, 163). In addition, one BS cell line is not able to reactivate UV-irradiated herpes simplex virus as efficiently as two normal control cell lines or two other BS cell lines that have been examined (164). However, all other measures of DNA repair in UV-irradiated BS cells have so far proved to be normal.

The sensitivity of BS cells to chemical mutagens has also been tested. Treatment of these cells with ethylmethanesulfonate (EMS) results not only in increased killing of them relative to that of normal cells (165) but also in other types of measurable damage. For example, when peripheral blood lymphocytes from BS patients are treated with EMS, the number of SCEs is increased relative to that in cells from normal donors (166) (as stated above, the *basal* level of this chromosomal aberration in normal individuals is approximately 1/10th to 1/15th that of BS patients). Furthermore, as indicated in Section 1-2, cocultivation of BS cells with other cell types affects the frequency of chromosomal abnormalities in all the cells (167, 168). The propagation of normal lymphocytes in conditioned medium derived from BS fibroblast cultures results in a significant increase in the frequency of SCEs in the normal lymphocytes (see Table 1-4). Conversely, cocultivation of BS skin fibroblasts with Chinese hamster ovary cells or with normal human cells reduces the frequency of SCEs in the former population. The putative clastogenic factor be-

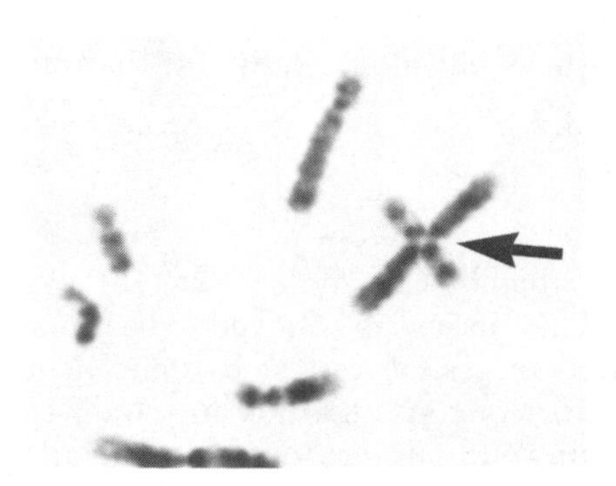

FIGURE 9-26
Cells from Bloom syndrome patients frequently show quadriradial chromosomes (indicated by the arrow), producing a typical cruciform appearance. (Courtesy of Dr. James German.)

TABLE 9-6
Aberrations per mitosis induced in normal human lymphocytes by concentrated ultrafiltrates of BS fibroblast culture media

Source of Ultrafiltrate	Superoxide Dismutase	Aberrations/Mitosis
BS	−	0.22
BS	+	0.10
BS	−	0.37
BS	+	0.03

From I. Emerit and P. Cerutti, ref. 168.

lieved to be responsible for these chromosomal rearrangements has been partially characterized from highly concentrated medium in which BS cells were grown (168). It has a molecular weight between 1000 and 10,000, the same size range as that of clastogenic factors identified in the serum of patients with systemic lupus erythematosis (169) and in certain strains of NZB mice with high frequencies of spontaneous chromosomal aberrations (170). Addition of bovine superoxide dismutase to cultures of phytohemagglutinin-stimulated lymphocytes from normal individuals suppresses the clastogenic activity present in concentrated ultrafiltrates of BS media (168) (Table 9-6). This observation has led to the suggestion that BS cells may be deficient in the detoxification of reactive oxygen species that lead to DNA damage (168) (see Section 1-2).

The formation and characterization of clastogenic factors from human cells and the study of their relation to DNA damage merit further study. If such factors can be isolated and their potential for producing DNA damage clarified, it may become reasonable to characterize human diseases such as AT and BS in terms of a *relative* deficiency in DNA repair capacity caused by a raised background of spontaneous DNA damage, rather than in terms of an intrinsic defect in DNA repair. This model is supported by the observation that the spontaneous mutation rate in BS fibroblasts is elevated relative to that observed in normal cells (171).

Cockayne Syndrome (CS)

The rare syndrome originally described by Cockayne (172, 173) may include a DNA repair deficiency. This disease is believed to be transmitted genetically in an autosomal recessive way, as evidenced by the familial incidence and the history of consanguinity in some parents of affected children (174). Individuals are characteristically dwarfed (Fig. 9-27), with growth and development arrested at ages of from several months to a few years, although levels of growth hormone are normal (175). The other prevalent symptoms of the disease include prominent photosensitivity, deafness, optic atrophy, intracranial calcifications, mental deficiency, large ears and nose, sunken eyes, long arms and legs proportional to body size, type II lipoproteinemia and a general appearance of premature aging (Fig. 9-27) (172–176).

The unusual sensitivity of CS individuals to sunlight has led to the suspicion that they may be defective in some form of repair of UV damage to DNA. Skin fibroblasts from these individuals are distinctly more sensitive to killing by UV radiation at 254 nm than are normal cells (176–182) (Fig. 9-28). Reports on the ability of CS cells to reactivate UV-irradiated adenovirus (host-cell reactivation) are conflicting. In one study no abnormality in the host-cell reactivating capacity of CS cells was observed (183). Another study, however, suggests that only approximately 20 percent of CS cells are able to repair UV-irradiated virus normally, and that the remaining cells have only about 30 percent of the normal ability to reactivate the damaged viruses (184).

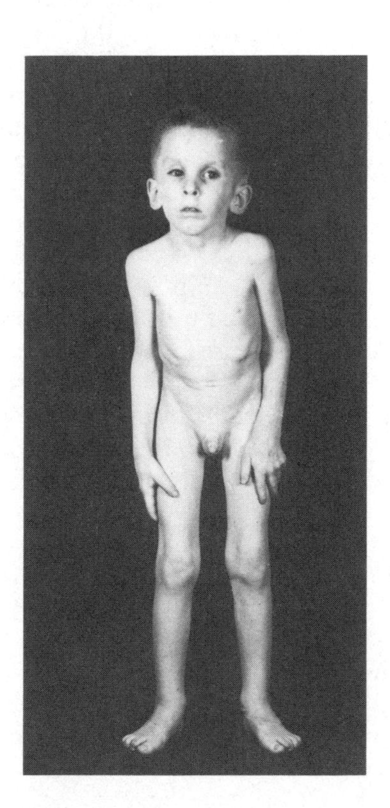

FIGURE 9-27
Individuals with Cockayne's syndrome have arrested growth and development, giving rise to a dwarfed appearance. The sunken eyes and disproportionately long arms and legs are also very characteristic. (Courtesy of Dr. James German.)

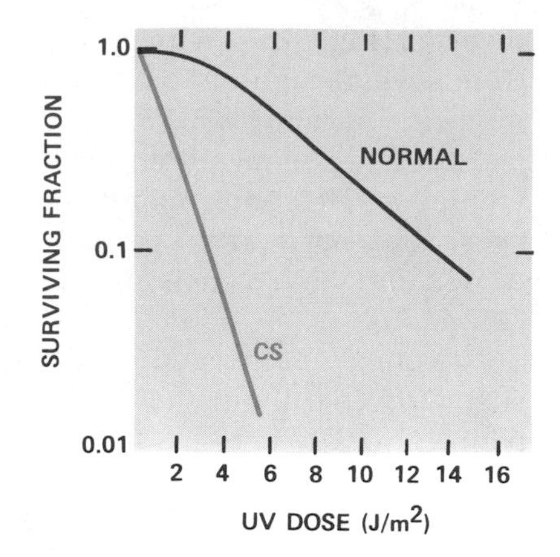

FIGURE 9-28
Cells from individuals with Cockayne's syndrome (CS) are abnormally sensitive to killing by UV radiation. (From K. Tanaka et al., ref. 189.)

The sensitivity of CS cells to killing by UV light is paralleled by their sensitivity to certain UV-mimetic chemicals. Thus the survival of CS cells treated with either N-acetoxy-N-2-acetylaminofluorene or with 4-NQO is significantly lower than that of normal cells so treated (178–180). However, according to most studies, CS cells survive as well as normal cells when treated with agents that are X-ray mimetic (e.g., monofunctional alkylating agents) (176, 178–180).

Several parameters of DNA repair have been studied in CS. Cells show a normal decrease in molecular weight of DNA during post-irradiation incubation, suggesting that DNA incisions are made at pyrimidine dimer and (6-4) photoproduct sites (14). Excision of thymine-containing pyrimidine dimers also occurs at normal levels and UDS is normal in CS cells (185, 186). Caffeine does not alter the survival of UV-irradiated CS cells and no defect in any DNA polymerase has been observed in them (178). A number of CS cell lines show an increased number of SCEs induced by UV radiation; however, there is no close correlation between the levels of exchange and of cell killing by UV radiation (182).

As mentioned above in the discussion of the XP variant form, when normal mammalian cells are exposed to UV radiation, there is an immediate and dose-dependent depression in the overall rate of DNA synthesis in the culture, with a progressive recovery to normal rates of DNA synthesis in the succeeding five to eight hours (Fig. 9-14). CS cells, like those from XP patients, show a pronounced defect in this recovery phenomenon (Fig. 9-29) (182, 186–189). At first glance it is tempting to speculate that like XP cells, CS cells may be specifically defective in the repair of lesions that block DNA replication. However, these cells are abnormally sensitive to the lethal effects of UV radiation even when they are not actively synthesizing DNA (188). In growing CS cells RNA synthesis is also

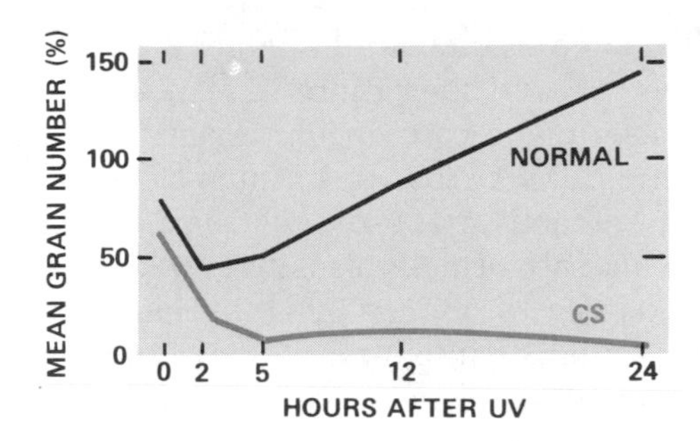

FIGURE 9-29
When cells from Cockayne's syndrome (CS) patients are exposed to UV radiation, the recovery toward normal rates of semiconservative DNA synthesis following initial inhibition is severely delayed relative to that in normal cells. Semiconservative DNA synthesis was measured by autoradiography of [^{3}H]-labeled cells. (From A. R. Lehmann et al., ref. 186.)

inhibited by UV radiation, but this aspect of nucleic acid metabolism recovers more rapidly after irradiation than DNA synthesis does (188). These observations have led to the interesting suggestion that immediately following UV irradiation of cells, there is normally a rapid excision of photoproducts from actively transcribing regions, thereby preventing any significant inhibition of RNA synthesis. This putative preferential DNA repair is thought to be defective in CS cells but undetectable by conventional measurements because quantitatively (as was suggested for XP variant cells) it represents a negligible proportion of total excision repair (188).

Recovery of normal rates of semiconservative DNA synthesis after UV irradiation results after fusion of CS cells from different individuals, and on this basis three genetic complementation groups have so far been identified in the disease (189, 190). Two of the complementation groups are pure CS and the third overlaps with xeroderma pigmentosum complementation group B (see Section 9-2).

Fanconi's Anemia (FA)

FA is characterized clinically by pancytopenia (depression of all cellular elements in the blood, i.e., red cells, white cells and platelets) and diverse congenital abnormalities (191). Arguments have been presented in favor of an autosomal recessive character, but the limited number of documented cases and the high ratio of affected to unaffected siblings points to caution in assuming recessive transmission (192, 193). Several reports have noted a high incidence of leukemia and other malignant neoplasms in individuals with the disease and in asymptomatic blood relatives (107, 193–196), and it has been estimated that the risk to an FA heterozygote individual of death from cancer is three times that of the general white population in the U.S. (195). If the conclusions of these reports are correct, the FA genotype may account for as much as 1 percent of all cancer deaths and 5 percent of deaths due to acute leukemia.

Cells from individuals with FA have an increased number of spontaneous chromosomal aberrations (14) and aberrations in response to treatment with a variety of mutagenic and carcinogenic agents, particularly those that cause interstrand DNA cross-links, such as nitrogen mustard, mitomycin C and photoactivated psoralen (see Section 1-3) (68, 196–200). FA cells are also abnormally sensitive to killing by these agents, but not to DNA damage produced by UV or ionizing radiation (201). A defect in the repair of interstrand cross-links in DNA has not been directly demonstrated. However, the persistence of reversibly denaturable DNA (an expression of interstrand DNA cross-links; see Section 1-3) after treatment of cells with mitomycin C suggests that cross-links are removed abnormally slowly from the DNA of FA cells (202). It has also been reported that cells from FA individuals have lower NAD levels than cells from normal donors (203). It will be recalled from the discussion in Section 6-4 that NAD in the nuclei of eukaryotic cells is principally metabolized to poly(ADP-ribose), which is believed to play a role in DNA excision repair, possibly by activation of DNA ligase activity. However FA cells show no detectable defect in DNA strand joining.

Retinoblastoma

Retinoblastoma is a rare malignant eye tumor that appears spontaneously in otherwise apparently normal individuals and in genetically predisposed individuals (204–206). A small percentage of the latter group show a deletion in the long arm of chromosome 13, which is referred to as *D-deletion-type retinoblastoma* (205–207), whereas in the remainder of this group, no deletion is evident, which is referred to as *hereditary-type retinoblastoma* (14). However, it is possible that so-called hereditary-type retinoblastoma involves a deletion so small that it is undetectable by cytological means. Those who survive hereditary-type retinoblastoma have a higher than normal probability of developing malignancies of other types, primarily osteosarcomas (205, 208, 209). The incidence of malignant tumor formation from radiation administered for therapy is also greater than normal in these individuals (14).

Fibroblasts from retinoblastoma individuals have been examined for sensitivity to X-rays. Cells from both D-deletion and hereditary forms of the disease are apparently slightly more sensitive to cell killing than normal cells (210, 211) (Table 9-7), but some studies have failed to confirm this observation (212). X-ray survival is normal, however, in cells from individuals who spontaneously develop retinoblastoma and in whom there is no family history of the disease (211). No specific DNA repair defect has been demonstrated in retinoblastoma cells.

TABLE 9-7
X-ray sensitivity of various cell strains

Strain	Cloning Efficiency (%)	D_0 ± SEM*	Mean D_0 of Group
SPORADIC RETINOBLASTOMA			
HL21	13.6–14.7	167 ± 11	
RbMe	12.8–34.2	167 ± 7	
RbCH	2.8–9.2	164 ± 7	
HL23B	11.3–17.0	144 ± 7	151 ± 5
AG 1947	11.5–22.8	142 ± 6	
AG 1946	5.5–21.0	141 ± 7	
AG 1979	0.93–2.5	135 ± 8	
HEREDITARY RETINOBLASTOMA			
AG 1980	2.0–8.1	140 ± 8	
AG 1131	3.0–16.5	132 ± 4	
AG 1879	1.4–8.0	128 ± 9	
AG 1123	1.2–12.4	121 ± 4	123 ± 6
AG 1408	10.4–12.9	119 ± 5	
AG 1880	4.3–20.5	98 ± 6	
D-DELETION RETINOBLASTOMA			
AG 1142	0.7–6.5	95 ± 5	
LKR	21.8–44.0	139 ± 8	
NORMAL			
Li 106	1.0–6.3	149 ± 7	
AG 1518	11.6–25.9	152 ± 12	
GM 1381	0.6–1.7	147 ± 9	147 ± 3
EX 25	3.2–12.7	140 ± 4	

*SEM = standard error of the mean. The D_0 is the calculated dose of radiation (in rads) that would result in an average of one lethal event per cell.
From R. R. Weichselbaum et al., ref. 211.

Additional Human Diseases of Possible Interest

The correlation between sensitivity to sunlight, predisposition to malignant neoplasms of the skin and defective DNA repair in XP cells has stimulated a broad search for DNA repair defects in other diseases that are characterized by light sensitivity or by a high incidence of cancer. The search has yielded negative results for diseases such as Rothmund-Thompson disease, lupus erythematosis, psoriasis, basal cell nevus syndrome and malignant melanoma (14). Similarly, the association of neurological defects in AT with increased sensitivity to ionizing radiation and to certain chemicals such as

N-methyl-N′-nitro-N-nitrosoguanidine (MNNG) has prompted investigators to examine cells from individuals with other primary neurological disorders. Initial case studies suggest that individuals with Huntington's disease, Friedreich's ataxia, familial dysautonomia and oligopontocerebellar atrophy may be abnormally sensitive to ionizing radiation (213). In a more comprehensive study it has been shown that cell strains from six patients with Huntington's disease and from four patients with familial dysautonomia are abnormally sensitive to killing by MNNG (213).

9-4 DNA Damage and Cancer: General Considerations

The diseases discussed above constitute specific examples of a possible pathogenetic relation between abnormal DNA metabolism in response to DNA damage and neoplastic transformation. However, the relation between DNA damage and carcinogenesis may be more general, and the remainder of the chapter is devoted to a limited consideration of some aspects of this possible relation.

One of the most striking properties of the neoplastic state is its *heretible nature;* that is, the neoplastic phenotype is typically observed in both daughter cells at each division. This is also true of the preneoplastic state, an arbitrarily designated period between the time that a cell undergoes some critical initiating molecular event(s) and the time of frank expression of the transformed phenotype by the descendants of that cell. Not unexpectedly, the heretible nature of both the preneoplastic and the neoplastic state has focused speculation that the pathogenesis of cancer involves permanent alterations in gene expression, perhaps accompanied by permanent alterations in gene structure, i.e., mutation. These loosely knit concepts are embodied in the general theory of carcinogenesis called the *somatic mutation theory* (214, 215), which dates back to observations by T. Boveri (216) of altered nuclear morphology in cancer cells. The somatic mutation theory may be stated as follows. Agents that initiate neoplatic transformation (carcinogens) do so by interacting with the DNA of cells, causing damage. If the cellular response(s) to this damage involves the misrepair of lesions, or if DNA repair fails to occur and the predominant response is an error-prone mechanism, mutations may arise in the descendants of the affected cells. These mutations, affecting specific genes (or sets of genes) that regulate critical aspects of cellular growth, may express themselves phenotypically as neoplastic transformation.

Rigorous proof of the somatic mutation theory of cancer obviously requires the demonstration in a single experimental system of at least the following:

1. That exposure of target cells to known carcinogens results in defined DNA damage
2. That the cellular response(s) to such damage results in mutations
3. That these mutations are expressed in the descendants of the damaged cells and that their localization to specific genes is always correlated with cellular transformation.

Neoplastic transformation is peculiar to complex multicellular organisms and is of special interest in mammals, particularly humans. However, such organisms are not well suited to experimental approaches of the type just described. As a corollary, prokaryotes such as *E. coli* and even lower eukaryotes such as the yeast *S. cerevisiae*, in which such studies are potentially feasible, do not undergo neoplastic transformation. Because of these experimental constraints, evidence for the somatic mutation theory is mainly indirect and comes from correlative studies that have examined selected aspects of the postulated relation between DNA damage, DNA repair, mutagenesis and neoplastic transformation in both prokaryotic and eukaryotic systems.

Epidemiological studies have provided a wealth of evidence suggesting a direct correlation between exposure to specific environmental agents and a number of human cancers (217). Perhaps the most celebrated recent example is the documented association between lung cancer and cigarette smoking in both men and women (218). Table 9-8 lists other associations between various forms of human cancer and exposure to specific environmental agents. Another fruitful source of information concerning the relation between environmental agents and neoplasia has been the various animal experimental models extensively used for well over half a century. In recent years, neoplastic transformation has been reproducibly achieved in mammalian cells in tissue culture. Of particular interest is the successful transformation of human fibroblasts in culture following treatment with ionizing radiation (219, 220), a variety of chemicals (221, 222) or UV radiation at 254 nm (223–226). All these agents have been extensively characterized with respect to DNA damage both in vivo and in vitro (see Section 1-3).

Tissue culture systems are potentially more useful for detailed biochemical and biological studies than are whole animals. However, a number of limitations in these systems should be kept in mind in

TABLE 9-8
Some occupational and environmental cancers in humans

Agent	Occupation	Site of Cancer
Ionizing radiations		
Radon	Certain underground miners (uranium, fluorspar, hematite)	Bronchus
X-rays, radium	Radiologists, radiographers	Skin
Radium	Luminous-dial printers	Bone
Ultraviolet light (sunlight)	Farmers, sailors	Skin
Polycyclic hydrocarbons in soot, tar, oil	Chimney sweepers, manufacturers of coal gas, many other groups of exposed industrial workers	Scrotum, skin, bronchus
2-Naphthylamine, 1-naphthylamine	Chemical workers, rubber workers, manufacturers of coal gas	Bladder
Benzidine; 4-aminobiphenyl	Chemical workers	Bladder
Asbestos	Asbestos workers, shipyard and insulation workers	Bronchus, pleura and peritoneum
Arsenic	Sheep-dip manufacturers, gold miners, some vineyard workers and ore smelters	Skin and bronchus
Bis(chloromethyl) ether	Makers of ion-exchange resins	Bronchus
Benzene	Workers with glues, varnishes, etc.	Marrow (leukemia)
Mustard gas	Poison gas makers	Bronchus, larynx, nasal sinuses
Vinyl chloride	PVC manufacturers	Liver (angiosarcoma)
Chrome ores	Chromate manufacturers	Bronchus
Nickel ore	Nickel refiners	Bronchus, nasal sinuses
Isopropyl oil	Isopropylene manufacturers	Nasal sinuses
Use of kangri and dhoti		Skin of abdomen and thigh
Chewing betel, tobacco, lime		Mouth
Reverse smoking		Palate
Smoking		Mouth, pharynx, larynx, bronchus, esophagus, bladder
Alcoholic drinks		Mouth, pharynx, larynx, esophagus
Aspect of sexual intercourse (virus?)		Cervix uteri
Infectious mononucleosis (?)		Hodgkin's disease
Aflatoxin		Liver
Shistosomiasis		Bladder

After R. Doll, ref. 217.

evaluating the significance of such studies. First, most tissue culture systems consist of *fibroblasts* in culture. In the living animal such cells are transformed into *sarcomas*, whereas the great majority of human malignant tumors arise from epithelial cells and hence are *carcinomas*. It is possible that the molecular events associated with neoplastic transformation are not identical in these two cell types. In addition, there is considerable controversy as to whether the various biological end points used to measure transformation in tissue culture, such as the formation of multicellular foci during growth on solid surfaces and anchorage-independent growth (for example growth in agar), really are synonymous with neoplasia in vivo in all cases.

An alternative approach to investigating the role of DNA damage in the pathogenesis of cancer is to perturb the DNA repair capacity of a cellular or biological system and measure the effect on neoplastic transformation after exposure to a DNA damaging agent. If interference with DNA repair increases the carcinogenicity of a given agent, it is reasonable to conclude that the interaction of this agent with DNA (rather than with other cellular components) is critical to the neoplastic process. Studies of this type have been carried out using the Amazon molly *Poecilia formosa* as an experimental model (227–230). This fish is gynogenetic (nonsexually reproducing) and can be cloned in a laboratory setting to produce a genetically identical population of animals. Thus the transplantation of cells or tissues from one member of a cloned population to another can be carried out with impunity. Another useful feature of the species is that it does not possess an excision repair system for pyrimidine dimers in DNA, although the repair of such lesions can be effected by enzymatic photoreactivation (see Section 2-2). Target cells from a given fish can therefore be exposed to UV radiation at 254 nm and either maintained in the dark or in photoreactivating light prior to inoculation into recipient hosts. When cells from the thyroid gland or the liver were subjected to this experimental protocol, an increased incidence of tumors in recipients bearing cells maintained under *non-DNA repair conditions* was observed, thereby implicating DNA damage (pyrimidine dimers) as a specific etiological factor in the neoplastic transformation (Table 9-9).

Less direct studies have led to the same general conclusion. For example, caffeine is an inhibitor of excision repair and of so-called postreplication repair in many cells (231–233) and potentiates mutagenesis in these cells. Exposure of experimental mice to UV radiation after painting one ear of each animal with a solution of caffeine results in about one-half the tumor incidence in the unpainted ears relative to that in the caffeine-sensitized ears (234). In a different study using Syrian hamster embryo cells, the UV action spectra for

TABLE 9-9
Correlation between pyrimidine dimers in DNA and tumors in *P. formosa*

Treatment	Fraction of Fish with Granulomas	Granulomas per Fish	Fraction of Fish with Thyroid Cancer by	
			Gross pathology	Histology
UV	51/63	1.8	34/34	29/29
2.5 min PR + UV	49/57	2.0	26/26	22/22
5.0 min PR + UV	37/42	1.9	48/50	22/23
UV + 5.0 min PR	15/43	0.4	1/43	0/6

From R. W. Hart and R. B. Setlow, ref. 230.

neoplastic transformation, cell killing and pyrimidine dimer formation correlated closely (235, 236), again suggesting a causal relation between DNA damage and neoplasia. Finally, the demonstration of the photoreactivation of UV-induced ouabain-resistant mutations in haploid frog cells in vitro adds further inferential evidence that pyrimidine dimers are mutagenic lesions in vertebrates (237).

Use of Mutagen Screening Tests as an Indicator of the Relation between DNA Damaging Agents and Carcinogenicity

Further indirect support for the somatic mutation theory of cancer comes from studies demonstrating that many carcinogens are mutagens. In these studies a number of specific biological systems have been utilized, some of which have come to enjoy widespread popularity in mutagen screening tests. The tests described in this chapter may indeed be sensitive and practically useful indicators of potential carcinogens in the environment. However, specifically with respect to the somatic mutation theory of cancer, the observation that a given agent results in mutations of selected genetic loci in prokaryotes on the one hand, and in neoplastic transformation of animal cells on the other, does not necessarily imply a pathogenetic relation between these events. The mutagen screening tests described here should thus perhaps be critically evaluated not so much from the perspective of their contribution to our understanding of the pathogenesis of cancer as from the perspective of their utilitarian value as rapid, simple and relatively inexpensive predictors of *potential* carcinogens.

The *Salmonella* mutagen test system utilizes strains of *S. typhimurium* extensively characterized with respect to mutations in the histidine operon (238, 239). The test, developed by Ames and his colleagues (240), scores reversion of histidine auxotrophs to prototrophy following the application of a suspected mutagen to agar plates on which the tester strain (about 10^9 cells per plate) is seeded (Fig. 9-30). Since its original inception, considerable refinements in the test have greatly enhanced its sensitivity and efficiency. For example, mutations in genes coding for error-free DNA repair functions such as excision repair have been introduced into tester strains. In addition, strains have been selected for increased permeability of their cell wall to bulky chemicals that otherwise could not gain access to the interior of the bacteria. Use has also been made of certain plasmids that confer increased sensitivity for the detection of mutagens by strains bearing them. Finally, the incorporation into tester plates of a fraction of homogenized mammalian liver provides the enzymes necessary for the metabolism of a number of compounds that require specific activation from weakly polar hydrophobic states to more reactive electrophilic forms that can bind covalently to DNA (see Section 1-3).

The *Salmonella* test has been evaluated with about 300 suspected carcinogens and noncarcinogens (defined by animal tests). Carcinogenicity and mutagenicity correlated in close to 90 percent of the compounds tested; however, the correlation was significantly lower for certain classes of carcinogens such as hydrazines and heavily chlorinated chemicals (241). The test scored equally well with respect to noncarcinogenic chemicals (241). The most obvious concerns are *false positive* and *false negative* results. The former are difficult to evaluate since the "falsity" of any positive result is defined by the failure to directly demonstrate carcinogenicity of the agent in question. Thus it has been suggested (241) that many false positive results may really reflect limitations in carcinogenicity testing or the testing of very weak carcinogens. Such suggestions are not without foundation. A number of agents initially claimed to be noncarcinogens in animal test systems have been shown to be mutagenic with the *Salmonella* test, prompting further carcinogenicity testing, which yielded positive results. A specific example is the chemical furylfuramide (AF-2), a food additive used in Japan until it was identified as a highly potent mutagen with the *Salmonella* test (242).

A more serious issue is that of *false negative* results. Since the assay depends on the reversion of mutant to wild-type enzyme activity, mutations in the relevant *Salmonella* gene may fail to score as such if critical codons in the gene are not "hit." In addition, it might be anticipated that liver extracts may not be efficient activators of all precarcinogens tested in vitro, and some bulky compounds may not gain access to the interior of cells despite increased permeability of the cell membrane of the tester strains.

Prophage Induction Test (Inductest)

This mutagen screening test is based on the long-established observation that certain agents causing DNA damage in *E. coli* strains that are lysogenic for bacteriophage λ result in the induction of the prophage, with subsequent lysis of the indicator bacteria, giving rise to readily identifiable and quantifiable plaques (243) (Fig. 9-30) (see Section 7-4). To enhance its sensitivity, the test incorporates many of the same perturbations described for the *Salmonella* test, that is, deficiency in error-free DNA excision repair, increased membrane permeability of the *E. coli* strains used for establishing lysogeny and the presence of liver cell extract in the test system. A notable advantage of this test is that in contrast to the *Salmonella* one it is relatively insensitive to the toxic effect of the agents tested, since viability of the bacterial strain is not a prerequisite. Lysogenized strains of *E. coli* have been constructed in which the expression of the structural gene for galactokinase is under control of the λ repressor (244). Thus, any agent that promotes induction of the phage simultaneously allows expression of the enzyme, which can be readily quantitatively assayed.

DNA Repair Test

Since DNA damage is common to both mutagenesis and to DNA repair, some tests of potential carcinogens rely on the simple detection of a DNA repair response in living cells. One such test utilizes the demonstration of DNA repair synthesis in mammalian (including human) cells by autoradiography (UDS) (245, 246) (see Section 5-2). This is a technically very simple procedure and has the distinct advantage of utilizing the cells of direct interest as targets for carcinogens. As an alternative to autoradiography, the repair response to DNA damage can be monitored by measuring the uptake of radiolabeled DNA precursors (e.g., [^{3}H]thymidine) by liquid scintillation spectrometry. G0 cells such as peripheral blood lymphocytes, in which there is essentially no detectable semiconservative DNA synthesis, are well suited to this purpose (247).

It is highly likely that agents that cause DNA damage and elicit a DNA repair response are potentially mutagenic. Hence the probability of false positives in a DNA repair test should be no greater than in any of the tests that measure mutations directly. However, it is entirely possible that some carcinogens cause DNA damage that is

AMES (*SALMONELLA*) TEST

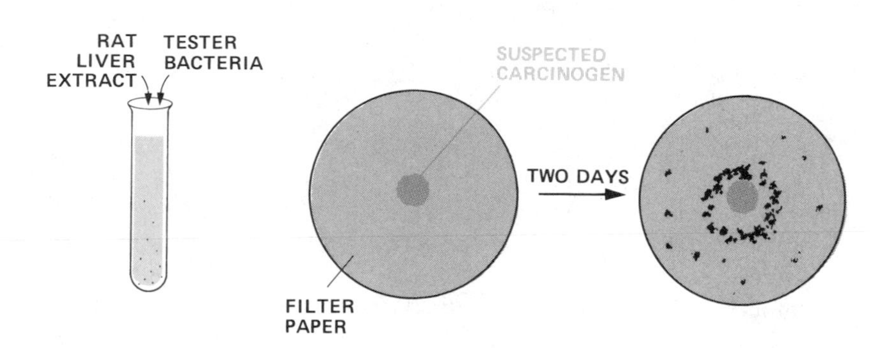

INDUCTEST

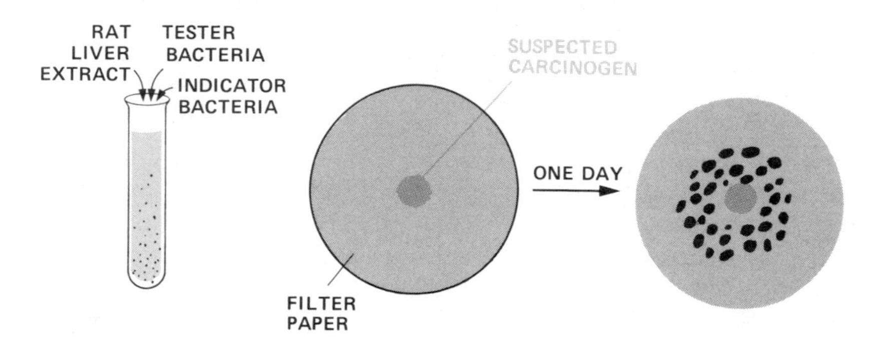

FIGURE 9-30
Mutagens can be detected very simply and quickly by either the *Salmonella* (Ames) test (top) or the phage λ inductest (bottom). In the *Salmonella* test, tester bacteria (his^-) are mixed with a rat liver extract to activate certain precarcinogens to more reactive ultimate carcinogens, and the mixture is plated on an agar plate without added histidine. A dose of the suspected carcinogen to be tested for mutagenicity is placed on a filter paper disk on the plate. After two days most of the tester bacteria are dead because of the lack of available histidine. However, mutations caused by the suspected carcinogen result in reversion of some bacteria to his^+ and these propagate as visible colonies.

In the inductest lysogenic tester bacteria carrying λ prophage are mixed with rat liver extract and with indicator bacteria (to detect lysis). The mixture is plated and the plates overlaid with a filter disk as in the *Salmonella* test. After 24 hours most of the plate is covered by a thick layer of indicator bacteria. However, where the chemical being tested has diffused from the disk, the DNA damage it causes leads to induction of mature phage, resulting in plaque formation due to cell lysis. (After R. Devoret, ref. 281.)

not subject to excision repair (on which the demonstration of UDS depends) and yield mutations promoted by error-prone responses to DNA damage. Obviously this could produce false negative results. In addition, UDS is a very crude quantitative indicator of DNA repair, since levels of repair synthesis are highly dependent on the pool sizes of thymidine in cells and these can fluctuate as a function of the physiological state of the cells tested.

Sister Chromatid Exchange (SCE) as an Indicator of Mutagenicity and Carcinogenicity

The morphological demonstration of the reciprocal exchange of DNA between homologous regions of sister chromatids was first reported over 25 years ago (248). Initial techniques utilized autoradiography of cells that were replicated for a single cycle in the presence of [^{3}H]thymidine (249). Since the advent of BrdU substitution in DNA for imparting differential staining characteristics to metaphase chromosomes, SCEs can be readily demonstrated without the use of autoradiography. Thus, when cells are grown in the presence of BrdU for two rounds of DNA replication, one of the sister chromatids in each chromosome is *unifilarly substituted* (i.e., only one of the polynucleotide strands contains the thymine analogue) and the other is *bifilarly substituted* (250) (Fig. 9-31). Alternatively, cells can be grown in the presence of BrdU for a single round of DNA synthesis followed by a second cycle in its absence, in which case one of the sister chromatids is unifilarly substituted and the other unsubstituted (249). In either event, the subsequent staining of metaphase spreads of cells with any one of a variety of dyes results in the differential staining of the two sister chromatids of each chromosome and the simple identification of SCEs (249). The use of both the fluorescent dye Hoechst 33258 and of Giemsa stain, for example, yields so-called harlequin chromosomes (Fig. 9-32) (249).

The spontaneous level of SCEs in mammalian cells is not known, since both [^{3}H]thymidine and BrdU themselves induce SCEs at some low level. Nonetheless, significant increments in the frequency of SCEs are observed when cells are exposed to a variety of agents that interact with and damage DNA (251). In addition, as indicated in Section 9-3, the spontaneous level of SCEs is increased by at least an order of magnitude in cells from individuals suffering from Bloom's syndrome (159–161). Cells from these individuals also show a marked increase (relative to normal cells) of SCEs following exposure to ethylmethanesulfonate, a mutagenic chemical to which Bloom's syndrome cells are abnormally sensitive (165). Similarly, in cells from individuals with XP, the spontaneous level of SCEs is normal, but increased exchanges are observed following exposure of XP cells to UV radiation or certain monofunctional alkylating agents (70-72, 252).

These observations have led to the suggestion that quantitation of SCEs in human cells may be a sensitive indicator of DNA damage and of mutagenic and carcinogenic agents. Tests with known mutagens

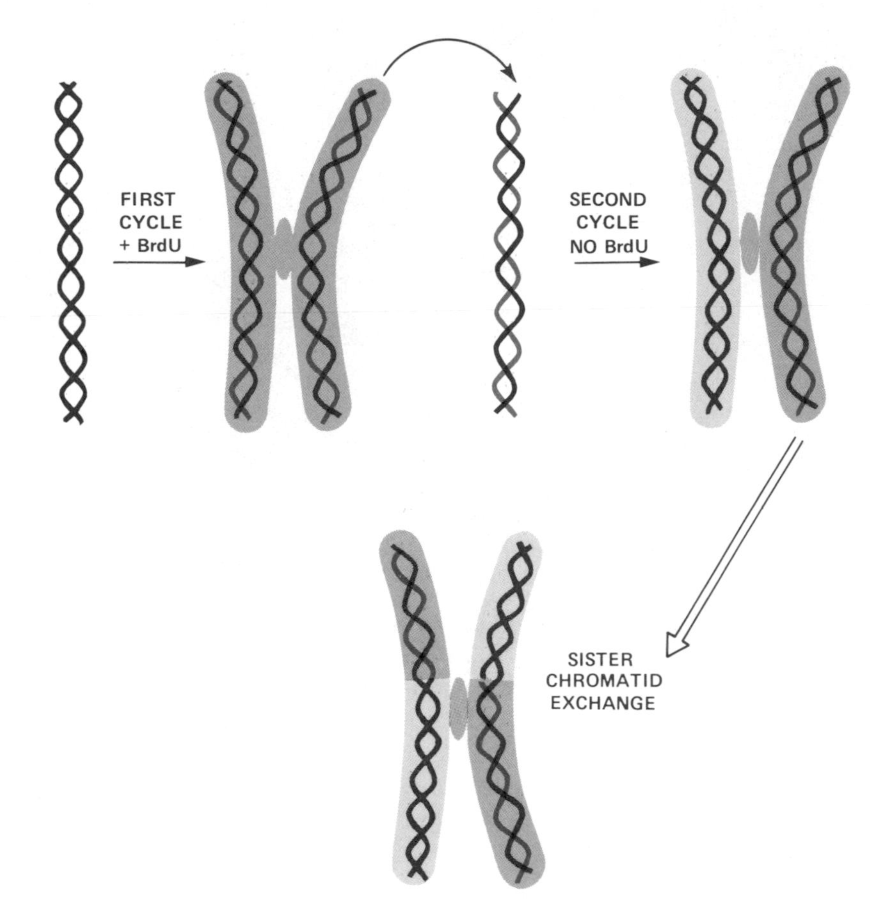

FIGURE 9-31
Sister chromatid exchanges can be visualized by allowing cells to replicate in BrdU. After one cycle of replication in BrdU each chromatid is unifilarly labeled with a density marker (5-BU). After a second cycle of replication without BrdU, each daughter chromosome (only one of which is shown in the figure) will contain one chromatid that is unfilarly substituted and one chromatid that is unsubstituted. This imparts differential staining properties to the sister chromatids and facilitates the visualization of exchanges between them (see Fig. 9-32).

and carcinogens on Chinese hamster ovary cells have shown an increase in induced SCEs with all direct-acting agents (253), and liver cell extracts have been incorporated into cell cultures in order to effect metabolic activation of indirect-acting chemicals so that these can also be evaluated (254). In a study in which SCEs and mutation at the *HGPRT* (hypoxanthine-guanine phosphoribosyltransferase) locus of Chinese hamster ovary cells were compared, a linear relation was observed after treatment of the cells with either ethylmethane-

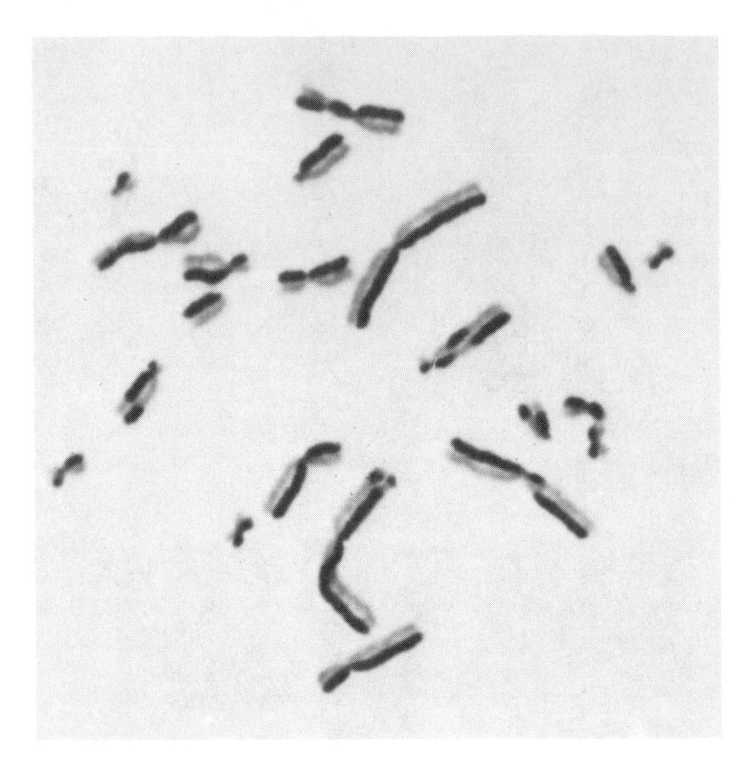

FIGURE 9-32
Sister chromatid exchanges can be seen under the light microscope as so-called harlequin chromosomes, produced by exchanges between stained and unstained sister chromatids (see Fig. 9-31). (Courtesy of Dr. Sheldon Wolff.)

sulfonate, N-ethyl-N-nitrosourea, mitomycin C or proflavine sulfate (255) (Fig. 9-33).

Is an increase in SCEs an accurate predictor of DNA damage and/or the action of mutagenic and carcinogenic compounds? An unequivocal answer to this question cannot be given as yet. Part of the difficulty is that correlations between SCEs and exposure to physical and chemical mutagens have not been as extensive as with other systems, e.g., the *Salmonella* test. However, a significant aspect of this uncertainty relates directly to our current ignorance in molecular terms of how SCEs are generated and what their biological significance is.

Correlations between DNA Damage, DNA Repair, Mutagenesis and Transformation in a Single Experimental System

Studies on mouse 10T½ cells (a mouse embryo-derived fibroblast cell line) in culture represent one of the few attempts to correlate all the various biological end points considered in this chapter. When

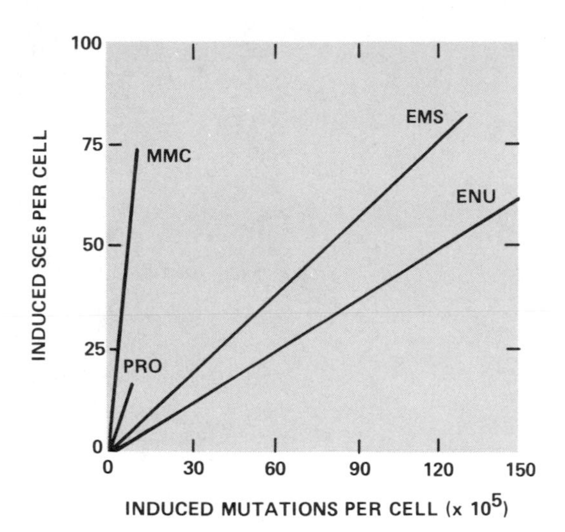

FIGURE 9-33
The relation between induced sister chromatid exchanges (SCEs) and induced mutations following treatment of Chinese hamster ovary cells with mitomycin C (MMC), proflavine sulfate (PRO), ethylmethanesulfonate (EMS) or N-ethyl-N-nitrosourea (ENU). (From A. V. Carrano et al., ref. 255.)

these cells are exposed to DNA damage and then held under conditions that preclude cell division, an enhancement of survival is observed. This phenomenon is called recovery from *potentially lethal damage* (PLD) (256–258). The recovery apparently reflects the action of cellular DNA repair mechanisms, because cells from XP patients failed to demonstrate this response to UV irradiation (259), and cells from patients with AT failed to demonstrate the response to X irradiation (259). The recovery from PLD after UV irradiation of 10T½ cells was paralleled by a significant decline in the frequency of SCEs, chromosomal aberrations and neoplastic transformation (258) (Fig. 9-34). In addition, both mutation to ouabain resistance and neoplastic transformation increased exponentially with increasing UV radiation exposure (258). These results are consistent with the notion that all the biological end points measured are caused by the same DNA lesions.

The results obtained following exposure of 10T½ cells to X-rays were more complex. After this type of DNA damage the repair of PLD was maximal by three to six hours and remained at this level for at least another six hours (258). The frequency of transformants also increased to a maximum at about three hours but then declined progressively and was undetectable in cells held under nongrowth conditions for 24 to 48 hours (258). Chromosome aberrations *de-*

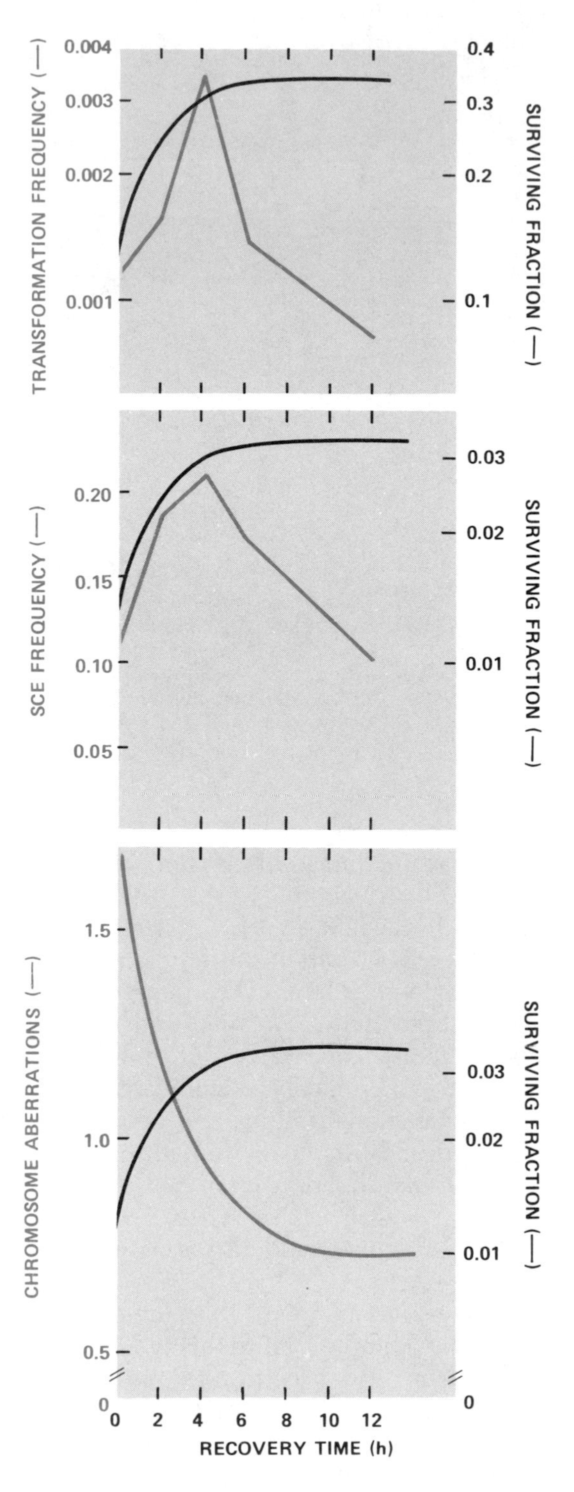

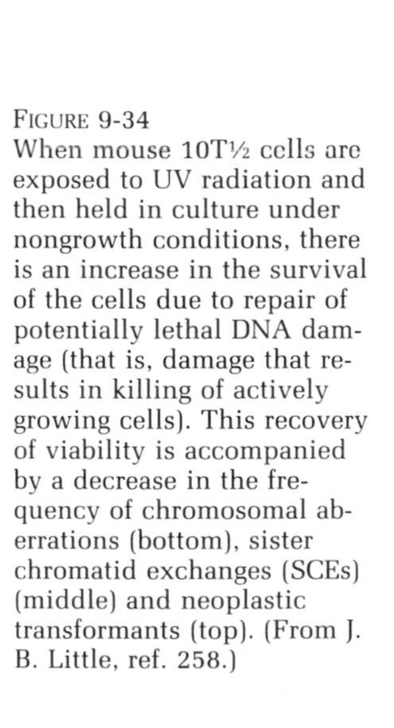

FIGURE 9-34
When mouse 10T½ cells are exposed to UV radiation and then held in culture under nongrowth conditions, there is an increase in the survival of the cells due to repair of potentially lethal DNA damage (that is, damage that results in killing of actively growing cells). This recovery of viability is accompanied by a decrease in the frequency of chromosomal aberrations (bottom), sister chromatid exchanges (SCEs) (middle) and neoplastic transformants (top). (From J. B. Little, ref. 258.)

clined during the first four to six hours of postirradiation holding, while the kinetics of induction of SCEs very closely paralleled that of transformation (258). Thus, unlike the findings with UV radiation, changes in the frequency of transformation and of SCEs during recovery from X-ray damage to DNA did not correlate with the frequency of survival and of chromosome aberrations (258). In order to explain these results, it has been postulated that X irradiation of 10T½ cells leads to the production of lethal lesions, the repair of which is associated with an increase in survival and a decrease in the frequency of chromosomal aberrations. However, this DNA repair mode may be *error-prone*, leading to new lesions in the DNA that subsequently express as increased mutations, SCEs and cell transformation. A slower-acting DNA repair process that removes these "secondary" lesions over the course of six to 24 hours of postirradiation holding could account for the subsequent decline in these biological end points (258). In support of the possible operation of error-prone cellular responses to DNA damage, it should be noted that the increase in SCEs induced by X irradiation of 10T½ cells can be suppressed by protease inhibitors such as antipain and leupeptin (260) (See Section 8-3). In addition, proteases have been shown to suppress X-ray-induced neoplastic transformation of cells (261).

Arguments against the Somatic Mutation Theory

A series of provocative arguments have challenged the concept that all cancers arise from DNA damage that becomes fixed as mutations in somatic cells. For example, according to the somatic mutation theory one would expect humans with inherited defects in DNA repair to have a significantly increased incidence of the common human cancers (e.g., lung, colon and breast cancers). As indicated in Section 9-2, XP individuals are extremely sensitive to sunlight and do indeed suffer an age-specific incidence of skin cancer much greater than that in the general population (83). However, even though the DNA repair defect is presumably present in all the cells of the body, the limited available evidence indicates that these individuals do not suffer an abnormal incidence of internal cancers, suggesting that the commonly fatal cancers in humans, for example those arising in the intestine and lung, may not be caused by mutations arising from DNA damage (83). In addition, cancers in the general population affecting organs such as the liver, stomach and esophagus are strongly associated with the dietary intake of known or suspected carcinogens in many underdeveloped countries. However, these neoplasms constitute a small fraction of the total cancers in the United States and Europe, where occupational cancers probably account for less than 5 percent of all cancer deaths (83). Without discounting the evidence that certain human cancers are unquestionably related to specific environmental agents known to promote DNA damage and mutagenesis, the question arises whether such agents and mechanisms account for *all* human cancers and, if not, what alternative explanations can be considered (83).

The existence of transposable genetic elements, which can translocate to different chromosomal sites and thereby modulate gene expression, was first described in maize by Barbara McClintock over a quarter of a century ago. Since then, similar elements have been found at other levels of biological organization. These include transposons, insertion sequences and the mutation-inducing bacteriophage mu in bacteria; copia sequences in *Drosophila,* and elements called Ty sequences in the yeast *S. cerevisiae* (262, 263). Rearrangements of these elements and larger-scale rearrangements, some of which are visible at the light-microscopic level (chromosomal rearrangements), have been implicated in the pathogenesis of cancer (83, 264). Chromosomal rearrangements, principally deletions, translocations and the gain or loss of entire chromosomes, are associated with a variety of human tumors (264, 265).

It is increasingly apparent that many (perhaps all) of these rearrangements are quite specific, sometimes to the extent that a particular rearrangement may be diagnostic for a particular type of human cancer. For example, about 90 percent of chronic myeloid leukemias show a translocation from chromosome 22 to chromosome 9, creating the characteristically shortened 22 known as the Philadelphia chromosome (265). Individuals with ataxia telangiectasia are often found to be harboring an expanding clone of lymphocytes with a translocation affecting chromosome 14, and the lymphatic leukemias that commonly occur in these individuals are thought to develop from these clones (108, 266, 267). Other examples of specific chromosomal rearrangements associated with human tumors are provided in Table 9-10. Chromosomal rearrangements are also associated with tumors in rodents. These are frequently characteristic of the type of tumor and the particular compound or treatment used to induce carcinogenesis (268). Finally, chromosomal rearrangements, particularly ploidy changes, often accompany neoplastic transformation of cells in culture (268–270). The demonstration of chromosomal rearrangements associated with these examples of neoplasia is certainly consistent with the hypothesis that such rearrangements are pathogenetically related to transformation. It has even been argued that the failure to find such an association in all cancers simply reflects the limitations of existing technology.

Genetic transposition has been clearly associated with the regulation of gene expression in a number of biological systems (83, 264). Well-studied examples include changes in the flagellar antigen of

TABLE 9-10
Consistent chromosomal rearrangements in human neoplasms

Malignancy	Chromosomal Rearrangement	Patients with Rearrangements (%)	Comments
Chronic myelogenous leukemia	t(9;22)(q34;q11)	85	Variant translocations occur in 7 percent of patients.
Acute myeloid leukemia (M2)	t(8;21)(q22;q22)	10	Variant translocations may occur. Unique loss of sex chromosomes in 30 to 80 percent of both male and female patients with t(8;21).
Acute promyeloid leukemia (M3)	t(15;17)(q22;q21)	60	Marked variation in frequency according to geographic area; this variation may be related to methodology. Variant translocations have been reported. t(15;17) has never been observed in other leukemias.
Acute myelomonocytic leukemia (M4)	inv(16)(p13q22) del(16)(q22)	25	These rearrangements are subtle; therefore the frequency may change when more individuals have been analyzed or when previously reported individuals are reviewed.
Acute nonlymphocytic leukemia	+8 −7	25 12	
Acute lymphocytic leukemia	t(4;11)(q21;q23) t(9;22)(q34;q11)	5 12	
Acute lymphocytic leukemia (B cell)	14q+	100	Donor chromosome is variable.
Burkitt's lymphoma	t(8;14)(q24;q32)	90	Present in African and non-African tumors and EBV-positive and negative tumors. Variant translocations may occur: t(8;22), t(2;8).
Acute monocytic leukemia Acute myelomonocytic leukemia	t(9;11)(p22;q23)	—	
Non-Hodgkin's lymphoma	t(14;18)(q32;q21)	—	Observed in follicular lymphomas, small cleaved, mixed or large cell.
	t(11;14)(q13;q32)	—	Observed in large cell lymphomas.
Neuroblastoma	del(1)(p31p36)	—	
Small cell lung carcinoma	del(3)(p14p23)	—	
Ovarian carcinoma	t(6;14)(q21;q24)	—	
Retinoblastoma	del(13q)	—	Band q14 is consistently deleted.
Wilms' tumor	del(11p)	—	Band p13 is consistently deleted.
Ewing's sarcoma	t(11;22)(q24;q12)	—	

Based on data from J. D. Rowley, ref. 282 and revised by M. LeBeau and J. D. Rowley.

Salmonella, specific surface antigens of certain trypanosomes, mating-type switching in yeast and the generation of antibody diversity by transpositions that occur during B cell differentiation (83). No experimental evidence directly linking transposition of smaller DNA elements (equivalent to transposons and insertion sequences in bacteria) to neoplastic transformation in humans has yet been obtained. However, in recent years it has been shown that certain DNA sequences present in both normal and neoplastic human cells cause neoplastic transformation of NIH 3T3 cells in culture, following transfection of the cells (271–275). The obvious interpretation of these results is that the transfected DNA becomes integrated into the host genome, thus providing the possibility for quiescent oncogenes to come under the regulatory control of promoters that facilitate their expression. This transfection assay has provided the basis for cloning human cellular oncogenes and for comparing the nucleotide sequence of such genes cloned from tumor cells and from their non-neoplastic normal counterparts. Surprisingly, such studies on a human bladder oncogene did not reveal any gross DNA sequence rearrangement. Instead, the oncogene differs from the naturally occurring normal gene only in a single base substitution, the effect of which is to substitute the amino acid valine for glycine in the 12th position of a protein called p21 (276). The examination of other cellular oncogenes by nucleotide sequencing is of obvious interest.

It should be evident from the discussion in this chapter that the precise role of DNA damage and the cellular responses to DNA damage in the pathogenesis of cancer are far from certain. However, from the mass of data accumulated in the literature (only a small fraction of which has been treated here) it seems reasonably well established that many forms of neoplastic transformation are a consequence of DNA damage. It is not at all unlikely that, depending on the precise nature of the damage, its location with respect to active replicons, the nature of the cellular response to the damage and perhaps other variables of DNA structure and function, permanent alterations in gene expression may arise by mutation, by one of several modes of gene rearrangement or by other mechanisms, thereby leading ultimately to the phenotypic states of the cell that are currently defined as neoplastic transformation.

Suggestions for Further Reading

Human diseases with defective cellular responses to DNA damage, refs. 3, 5–19, 102, 130, 151, 188, 201; DNA damage and cancer in general, refs. 83, 230, 241, 258, 264, 274–276.

References

1. Cleaver, J. E. 1968. Defective repair replication of DNA in xeroderma pigmentosum. *Nature* 218:652.

2. Reed, W. B., Landing, B., Sugarman, G., Cleaver, J. E., and Melnyk, J. 1969. Xeroderma pigmentosum. Clinical and laboratory investigation of its basic defect. *JAMA* 207:2073.

3. Robbins, J. H., Kraemer, K. H., Lutzner, M. A., Festoff, B. W., and Coon, H. G. 1974. Xeroderma pigmentosum. An inherited disease with sun sensitivity, multiple cutaneous neoplasms and abnormal DNA repair. *Ann. Int. Med.* 80:221.

4. Cleaver, J. E., Bootsma, D., and Friedberg, E. C. 1975. Human diseases with genetically altered DNA repair processes. *Genetics* 79:215.

5. Cleaver, J. E., and Bootsma, D. 1975. Xeroderma pigmentosum: biochemical and genetic characteristics. *Ann. Rev. Genet.* 9:19.

6. Bootsma, D. 1977. Defective DNA repair and cancer. In *Research in photobiology*. A. Castellani, ed., p. 455. New York: Plenum.

7. Kraemer, K. H. 1977. Progressive degenerative diseases associated with defective DNA repair: xeroderma pigmentosum and ataxia telangiectasia. In *DNA repair processes*. W. W. Nichols and D. G. Murphy, eds., p. 37. Miami: Symposia Specialists.

8. Bootsma, D. 1978. Xeroderma pigmentosum. In *DNA repair mechanisms*. P. C. Hanawalt, E. C. Friedberg and C. F. Fox, eds., p. 589. New York: Academic.

9. Robbins, J. H. 1978. Significance of repair of human DNA: evidence from studies of xeroderma pigmentosum. *J. Natl. Cancer Inst.* 61:645.

10. Friedberg, E. C. 1978. Xeroderma pigmentosum: recent studies on the DNA repair defects. *Arch. Pathol. Lab. Med.* 102:3.

11. Arlett, C. F., and Lehmann, A. R. 1978. Human disorders showing increased sensitivity to the induction of genetic damage. *Ann. Rev. Genet.* 12:95.

12. Setlow, R. B. 1978. Repair deficient human disorders and cancer. *Nature* 271:713.

13. Cleaver, J. E. 1978. Xeroderma pigmentosum: genetic and environmental influences in skin carcinogenesis. *Int. J. Derm.* 17:435.

14. Friedberg, E. C., Ehmann, U. K., and Williams, J. I. 1979. Human diseases associated with defective DNA repair. *Adv. Rad. Biol.* 8:85.

15. Friedberg, E. C. 1981. Xeroderma pigmentosum—a human model of defective DNA repair. In *Chromosome damage and repair*. E. Seeberg and K. Kleppe, eds., p. 313. New York: Plenum.

16. Cleaver, J. E. Xeroderma pigmentosum. 1983. In *The metabolic basis of inherited disease*, 5th ed. J. B. Stanbury, J. B. Wyngaarden, D. S. Fredrickson, J. L. Goldstein and M. S. Brown, eds., p. 1227. New York: McGraw-Hill.

17. Paterson, M. C. 1978. Ataxia telangiectasia: a model inherited disease linking deficient DNA repair with radiosensitivity and cancer proneness. In *DNA repair mechanisms*. P. C. Hanawalt, E. C. Friedberg and C. F. Fox, eds., p. 637. New York: Academic.

18. German, J. 1978. DNA repair defects and human disease. In *DNA repair mechanisms*. P. C. Hanawalt, E. C. Friedberg and C. F. Fox, eds., p. 625. New York: Academic.

19. Paterson, M. C. 1979. Ataxia telangiectasia: an inherited human disorder involving hypersensitivity to ionizing radiation and related DNA-damaging chemicals. *Ann. Rev. Genet.* 13:291.

20. Lehmann, A. R., and Mayne, L. 1981. The response of Cockayne syndrome cells to UV-irradiation. In *Chromosome damage and repair.* E. Seeberg and K. Kleppe, eds., p. 367. New York: Plenum.

21. Backer, J. M., and Weinstein, I. B. 1982. Interaction of benzo(a)pyrene and its dihydrodiol-epoxide derivative with nuclear and mitochondrial DNA in C3H10T ½ cell cultures. *Cancer Res.* 42:2764.

22. DeSanctis, C., and Cacchione, A. 1932. L'idiozia xerodermica. *Riv. Sper. Frentiatr.* 56:269.

23. Takebe, H., Miki, Y., Kozuka, T., Furuyama, J-I., Tanaka, K., Sasaki, M. S., Fujiwara, Y. and Akiba, H. 1977. DNA repair characteristics and skin cancers of xeroderma pigmentosum patients in Japan. *Cancer Res.* 37:490.

24. Cleaver, J. E. 1969. Xeroderma pigmentosum: a human disease in which an initial stage of DNA repair is defective. *Proc. Natl. Acad. Sci.* (USA) 63:428.

25. Setlow, R. B., Regan, J. D., German, J., and Carrier, W. L. 1969. Evidence that xeroderma pigmentosum cells do not perform the first step in the repair of ultraviolet damage to their DNA. *Proc. Natl. Acad. Sci.* (USA) 64:1035.

26. Bootsma, D., Mulder, M. P., Pot, E., and Cohen, J. A. 1970. Different inherited levels of DNA repair replication in xeroderma pigmentosum cell strains after exposure to ultraviolet irradiation. *Mutation Res.* 9:507.

27. De Weerd-Kastelein, E. A., Keijzer, W., and Bootsma, D. 1972. Genetic heterogeneity of xeroderma pigmentosum demonstrated by somatic cell hybridization. *Nature New Biol.* 238:80.

28. Moshell, A. N., Ganges, M. B., Lutzner, M. A., Coon, H. G., Barrett, S. F., Dupuy, J-M., and Robbins, J. H. 1983. A new patient with both xeroderma pigmentosum and Cockayne syndrome establishes the new xeroderma pigmentosum complementation group H. In *Cellular responses to DNA damage.* E. C. Friedberg and B. A. Bridges, eds., p. 209. New York: Alan R. Liss.

29. Burk, P. G., Lutzner, M. A., Clarke, D. D., and Robbins, J. H. 1971. Ultraviolet-stimulated thymidine incorporation in xeroderma pigmentosum lymphocytes. *J. Lab. Clin. Med.* 77:759.

30. Robbins, J. H., and Kraemer, K. H. 1972. Abnormal rate and duration of ultraviolet-induced thymidine incorporation into lymphocytes from patients with xeroderma pigmentosum and associated neurological complications. *Mutation Res.* 15:92.

31. Cleaver, J. E. 1972. Xeroderma pigmentosum: variants with normal DNA repair and normal sensitivity to ultraviolet light. *J. Invest. Dermatol.* 58:124.

32. Cleaver, J. E., Greene, A. E., Coriell, L. L., and Mulivor, R. A. 1981. Xeroderma pigmentosum variants. *Cytogenet. Cell Genet.* 31:188.

33. Giannelli, F., Pawsey, S. A., and Avery, J. A. 1982. Differences in patterns of complementation of the more common groups of xeroderma pigmentosum: possible implications. *Cell* 29:451.

34. Hashem, N., Bootsma, D., Keijzer, W., Greene, A., Coriell, L., Thomas, G., and Cleaver, J. E. 1980. Clinical characteristics, DNA repair and complementation groups in xeroderma pigmentosum patients from Egypt. *Cancer Res.* 40:13.

35. Andrews, A., Barrett, S. F., and Robbins, J. H. 1978. Xeroderma pigmentosum neurological abnormalities correlate with colony-forming ability after ultraviolet radiation. *Proc. Natl. Acad. Sci.* (USA) 75:1984.

36. Zelle, B., and Lohman, P. H. M. 1979. Repair of UV-endonuclease-susceptible sites in the 7 complementation groups of xeroderma pigmentosum A through G. *Mutation Res.* 62:363.

37. Kraemer, K. H., de Weerd-Kastelein, E. A., Robbins, J. H., Keijzer, W., Barrett, S. F., Petinga, R. A., and Bootsma, D. 1975. Five complementation groups in xeroderma pigmentosum. *Mutation Res.* 33:327.

38. Kraemer, K. H., Coon, H. G., Petinga, R. A., Barrett, S. F., Rahe, A. E., and Robbins, J. H. 1975. Genetic heterogeneity in xeroderma pigmentosum: complementation groups and their relationship to DNA repair rates. *Proc. Natl. Acad. Sci.* (USA) 72:59.

39. Zelle, B., Berends, F., and Lohman, P. M. H. 1980. Repair of damage by ultraviolet radiation in xeroderma pigmentosum cell strains of complementation groups E and F. *Mutation Res.* 73:157.

40. Sancar, A., and Rupp, W. D. 1983. A novel repair enzyme: UVRABC excision nuclease of *Escherichia coli* cuts a DNA strand on both sides of the damaged region. *Cell* 33:249.

41. Friedberg, E. C., Naumovski, L., Yang, E., Pure, G., Schultz, R. A., and Love, J. D. 1983. Approaching the biochemistry of excision repair in eukaryotic cells: the use of cloned genes from *Saccharomyces cerevisiae*. In *Cellular responses to DNA damage*. E. C. Friedberg and B. A. Bridges, eds., p. 63. New York: Alan R. Liss.

42. Pedrini, A. M., Dalpra, L., Ciarrocchi, G., Pedrali Noy, G. C. F., Spadari, S., Nuzzo, F., and Falaschi, A. 1974. Levels of some enzymes acting on DNA in xeroderma pigmentosum. *Nucleic Acids Res.* 1:193.

43. Parker, V. P., and Lieberman, M. W. 1977. Levels of DNA polymerases α, β and γ in control- and repair-deficient human diploid fibroblasts. *Nucleic Acids Res.* 4:2029.

44. Bertazzoni, U., Stefanini, M., Pedrali Noy, G., Nuzzo, F., and Falaschi, A. 1977. Levels of DNA polymerase α and β in normal and xeroderma pigmentosum fibroblasts. *Nucleic Acids Res.* 4:141.

45. Waldstein, E. A., Peller, S., and Setlow, R. B. 1979. UV-endonuclease from calf thymus with specificity toward pyrimidine dimers in DNA. *Proc. Natl. Acad. Sci.* (USA) 76:3746.

46. Gruenert, D. C., and Cleaver, J. E. 1981. Repair of ultraviolet damage in human cells also exposed to agents that cause strand breaks, crosslinks, monoadducts, and alkylations. *Chem.-Biol. Interact.* 33:163.

47. Park, S. D., Choi, K. H., Hong, S. W., and Cleaver, J. E. 1981. Inhibition of excision-repair of ultraviolet damage in human cells by exposure to methylmethane sulfonate. *Mutation Res.* 82:365.

48. Cleaver, J. E. 1982. Inactivation of ultraviolet repair in normal and xeroderma pigmentosum cells by methylmethane sulfonate. *Cancer Res.* 42:860.

49. Mortelmans, K., Friedberg, E. C., Slor, H., Thomas, G., and Cleaver, J. E. 1976. Evidence for a defect in thymine dimer excision in extracts of xeroderma pigmentosum cells. *Proc. Natl. Acad. Sci.* (USA) 73:2757.

50. Friedberg, E. C., Rudé, J. M., Cook, K. H., Ehmann, U. K., Mortelmans, K., Cleaver, J. E., and Slor, H. 1977. Excision repair in mammalian cells and the current status of xeroderma pigmentosum. In *DNA repair processes*. W. W. Nichols and D. G. Murphy, eds., p. 21. Miami: Symposia Specialists.

51. Fujiwara, Y., and Kano, Y. 1983. Characteristics of thymine dimer excision from xeroderma pigmentosum chromatin. In *Cellular responses to DNA damage*. E. C. Friedberg and B. A. Bridges, eds., p. 215. New York: Alan R. Liss.

52. Hays, M. D., Schenk, R. U., and Vincent, R. A., Jr. 1981. Xeroderma pigmentosum cells treated with proteases to relax chromatin structure do not exhibit increased unscheduled DNA synthesis. *Mutation Res.* 91:147.

53. Sealy, L., and Chalkley, R. 1978. The effect of sodium butyrate on histone modification. *Cell* 14:115.

54. Williams, J. I., and Friedberg, E. C. 1982. Increased levels of unscheduled DNA synthesis in UV-irradiated human fibroblasts pretreated with sodium butyrate. *Photochem. Photobiol.* 36:423.

55. Shall, S., Durkacz, B., Ellis, D., Irwin, J., Lewis, P., and Perera, M. 1981. (ADP-ribose)n, a new component in DNA repair. In *Chromosome damage and repair*. E. Seeberg and K. Kleppe, eds., p. 477. New York: Plenum.

56. Shall, S., Durkacz, B. W., Gray, D. A., Irwin, J., Lewis, P. J., Perera, M., and Tavassoli, M. 1982. (ADP-ribose)n participates in DNA excision repair. In *Mechanisms of chemical carcinogenesis*. C. C. Harris and P. A. Cerutti, eds., p. 389. New York: Alan R. Liss.

57. Berger, N. A., Sikorski, G. W., Petzold, S. J., and Kurohara, K. K. 1980. Defective poly (adenosine diphosphoribose) synthesis in xeroderma pigmentosum. *Biochemistry* 19:289.

58. McCurry, L. S., and Jacobson, M. K. 1981. Poly (ADP-ribose) synthesis following DNA damage in cells heterozygous or homozygous for the xeroderma pigmentosum genotype. *J. Biol. Chem.* 256:551.

59. Ahmed, F. E., and Setlow, R. B. 1980. DNA excision in repair proficient and deficient human cells treated with a combination of ultraviolet radiation and acridine mustard (ICR-170) or 4-nitroquinoline-1-oxide. *Chem.-Biol. Interact.* 29:31.

60. Ahmed, F. E., and Setlow, R. B. 1979. DNA repair in xeroderma pigmentosum cells treated with combinations of ultraviolet radiation and N-acetoxy-2-acetylaminofluorene. *Cancer Res.* 39:471.

61. Setlow, R. B., and Grist, E. 1981. Excision repair of bulky lesions in the DNA of mammalian cells. In *Chromosome damage and repair*. E. Seeberg and K. Kleppe, eds., p. 131. New York: Plenum.

62. Maher, V. M., and McCormick, J. J. 1976. Effect of DNA repair on the cytotoxicity and mutagenicity of UV-irradiation and of chemical carcinogens in normal and xeroderma pigmentosum cells. In *Biology of radiation carcinogenesis*. J. M. Yuhas, R. W. Tennant and J. D. Regan, eds., p. 129. New York: Raven Press.

63. Maher, V. M., McCormick, J. J., Grover, P. L., and Sims, P. 1977. Effect of DNA repair on the cytotoxicity and mutagenicity of polycyclic hydrocarbon derivatives in normal and xeroderma pigmentosum human fibroblasts. *Mutation Res.* 43:117.

64. Maher, V. M., Dorney, D. J., Mendrala, A. L., Konze-Thomas, B., and McCormick, J. J. 1979. DNA excision-repair processes in human cells can eliminate the cytotoxic and mutagenic consequences of ultraviolet irradiation. *Mutation Res.* 62:311.

65. Glover, T. W., Chang, C. C., Trosko, J. E., and Li, S. S. 1979. Ultraviolet light induction of diphtheria toxin-resistant mutants of normal and xeroderma pigmentosum human fibroblasts. *Proc. Natl. Acad. Sci.* (USA) 76:3982.

66. Parrington, J. M., Delhanty, J. D. A., and Baden, H. P. 1971. Unscheduled DNA synthesis, UV-induced chromosome aberrations and SV40 transformation in cultured cells from xeroderma pigmentosum. *Ann. Human Genet.* 35:149.

67. Sasaki, M. S. 1973. DNA repair capacity and susceptibility to chromosome breakage in xeroderma pigmentosum cells. *Mutation Res.* 20:291.

68. Sasaki, M. S., Toda, K., and Ozawa, A. 1977. Role of DNA repair in the susceptibility to chromosome breakage and cell killing in cultured human fibroblasts. In *Biochemistry of cutaneous epidermal differentiation.* M. Seiji and I. A. Bernstein, eds., p. 167. University of Tokyo Press.

69. Wolff, S., Bodycote, J., Thomas, G. H., and Cleaver, J. E. 1975. Sister chromatid exchanges in xeroderma pigmentosum cells that are defective in DNA excision repair or post-replication repair. *Genetics* 81:349.

70. Wolff, S., Rodin, B., and Cleaver, J. E. 1977. Sister chromatid exchanges induced by mutagenic carcinogens in normal and xeroderma pigmentosum cells. *Nature* 265:347.

71. De Weerd-Kastelein, E. A., Keijzer, W., Rainaldi, G., and Bootsma, D. 1977. Induction of sister chromatid exchanges in xeroderma pigmentosum cells after exposure to ultraviolet light. *Mutation Res.* 45:253.

72. Schönwald, A. D., and Passarge, E. 1977. UV-light induced sister chromatid exchanges in xeroderma pigmentosum lymphocytes. *Human Genet.* 36:213.

73. Maher, V. M., Rowan, L. A., Silinskas, K. C., Kateley, S. A., and McCormick, J. J. 1982. Frequency of UV-induced neoplastic transformation of diploid human fibroblasts is higher in xeroderma pigmentosum cells than in normal cells. *Proc. Natl. Acad. Sci.* (USA) 79:2613.

74. Kuhnlein, U., Penhoet, E. E., and Linn, S. 1976. An altered apurinic DNA endonuclease activity in group A and group D xeroderma pigmentosum fibroblasts. *Proc. Natl. Acad. Sci.* (USA) 73:1169.

75. Kuhnlein, U., Lee, B., Penhoet, E. E., and Linn, S. 1978. Xeroderma pigmentosum fibroblasts of the D group lack an apurinic DNA endonuclease species with a low apparent K_m. *Nucleic Acids Res.* 5:951.

76. Kudrna, R. D., Smith, J., Linn, S., and Penhoet, E. E. 1979. Survival of apurinic SV40 DNA in the D complementation group of xeroderma pigmentosum. *Mutation Res.* 62:173.

77. Sutherland, B. M. 1974. Photoreactivating enzyme from human leucocytes. *Nature* 248:109.

78. Sutherland, B. M., Runge, P., and Sutherland, J. C. 1974. DNA photoreactivating enzyme from placental mammals. Origin and characterization. *Biochemistry* 13:4710.

79. Harm, H. 1980. Damage and repair in mammalian cells after exposure to non-ionizing radiations. III. Ultraviolet and visible light irradiation of cells of placental mammals, including humans, and determination of photorepairable damage *in vitro*. *Mutation Res.* 69:167.

80. Sutherland, B. M., Rice, M., and Wagner, E. K. 1975. Xeroderma pigmentosum cells contain low levels of photoreactivating enzyme. *Proc. Natl. Acad. Sci.* (USA) 12:103.

81. Sutherland, B. M. 1977. Human photoreactivating enzymes. In *Research in photobiology*. A. Castellani, ed., p. 307. New York: Plenum.

82. Arlett, C. F., Harcourt, S. A., Lehmann, A. R., Stevens, S., Ferguson-Smith, M. A., and Mosley, W. N. 1980. Studies on a new case of xeroderma pigmentosum (XP3BR) from complementation group G with cellular sensitivity to ionizing radiation. *Carcinogenesis* 1:745.

83. Cairns, J. 1981. The origin of human cancers. *Nature* 289:353.

84. Bridges, B. 1981. How important are somatic mutations and immune control in skin cancer? Reflections on xeroderma pigmentosum. *Carcinogenesis* 2:471.

85. Dupuy, J. M., and Lafforet, D. 1974. A defect of cellular immunity in xeroderma pigmentosum. *Clin. Immunol. Immunopathol.* 3:52.

86. Kinlen, L. 1979. Collaborative United Kingdom/Australian study of cancer in patients treated with immunosuppressive drugs. *Br. Med. J.* 2:1461.

87. Day, R. S. III. 1974. Studies on repair of adenovirus 2 by human fibroblasts using normal, xeroderma pigmentosum, and xeroderma pigmentosum heterozygous strains. *Cancer Res.* 34:1965.

88. Day, R. S. III. 1975. Xeroderma pigmentosum variants have decreased repair of ultraviolet-damaged DNA. *Nature* 253:748.

89. Lehmann, A. R., Kirk-Bell, S., Arlett, C. F., Paterson, M. C., Lohman, P. H. M., deWeerd-Kastelein, E. A., and Bootsma, D. 1975. Xeroderma pigmentosum cells with normal levels of excision repair have a defect in DNA synthesis after UV-irradiation. *Proc. Natl. Acad. Sci.* (USA) 72:219.

90. Ehmann, U. K., Cook, K. H., and Friedberg, E. C. 1978. The kinetics of thymine dimer excision in ultraviolet irradiated human cells. *Biophys. J.* 22:249.

91. Cleaver, J. E. 1981. Sensitivity of excision repair in normal human, xeroderma pigmentosum variant and Cockayne's syndrome fibroblasts to inhibition by cytosine arabinoside. *J. Cell. Physiol.* 108:163.

92. Lehmann, A. R., Kirk-Bell, S., and Jaspers, N. G. J. 1977. Post-replication repair in normal and abnormal human fibroblasts. In *DNA repair processes*. W. W. Nichols and D. G. Murphy, eds., p. 203. Miami: Symposium Specialists.

93. Lehmann, A. R., Kirk-Bell, S., and Arlett, C. F. 1977. Post-replication repair in human fibroblasts. In *Research in photobiology*. A. Castellani, ed., p. 293. New York: Plenum.

94. Park, S. D., and Cleaver, J. E. 1979. Recovery of DNA synthesis after ultraviolet irradiation of xeroderma pigmentosum cells depends on excision repair and is blocked by caffeine. *Nucleic Acids Res.* 6:1151.

95. Park, S. D., and Cleaver, J. E. 1979. Post-replication repair: questions of its definition and possible alteration in xeroderma pigmentosum cell strains. *Proc. Natl. Acad. Sci.* (USA) 76:3929.

96. Cleaver, J. E. 1981. Inhibition of DNA replication by hydroxyurea and caffeine in an ultraviolet-irradiated human fibroblast cell line. *Mutation Res.* 82:159.

97. Kauffman, W. K., and Cleaver, J. E. 1981. Mechanisms of inhibition of DNA replication by ultraviolet light in normal human and xeroderma pigmentosum fibroblasts. *J. Mol. Biol.* 149:171.

98. Rudé, J., and Friedberg, E. C. 1977. Semiconservative deoxyribonucleic acid synthesis in unirradiated and ultraviolet-irradiated xeroderma pigmentosum and normal human skin fibroblasts. *Mutation Res.* 42:433.

99. Moustacchi, E., Ehmann, U. K., and Friedberg, E. C. 1979. Defective recovery of semiconservative DNA synthesis in xeroderma pigmentosum cells following split-dose ultraviolet-irradiation. *Mutation Res.* 62:159.

100. Tadjoedin, M. K., and Fraser, E. C. 1965. Heredity of ataxia telangiectasia (Louis-Bar syndrome). *Am. J. Dis. Child.* 110:64.

101. Sedgewick, R. P., and Boder, E. 1972. Ataxia telangiectasia. In *Handbook of clinical neurology*. P. J. Vinken and G. W. Bruyn, eds., p. 267. Amsterdam: North-Holland.

102. McFarlin, D. E., Strober, W., and Waldmann, T. A. 1972. Ataxia telangiectasia. *Medicine* 51:281.

103. Boder, E., and Sedgewick, R. P. 1958. Ataxia telangiectasia. A familial syndrome of progressive cerebellar ataxia, oculocutaneous telangiectasia and frequent pulmonary infection. *Pediatrics* 21:526.

104. Kersey, J. H., Spector, B. D., and Good, R. 1973. Primary immunodeficiency diseases and cancer: the immunodeficiency-cancer registry. *Int. J. Cancer* 12:333.

105. Spector, B. D. 1977. Immunodeficiency-cancer registry: 1975 update. In *Progress in cancer research and therapy*. J. J. Mulvihill, R. W. Miller and J. F. Fraumeni, eds., vol. 3, p. 339. New York: Raven Press.

106. Bochkov, N. P., Lopukhin, Y. M., Kuleshov, N. P., and Kovalchuk, L. V. 1974. Cytogenetic study of patients with ataxia-telangiectasia. *Humangenetik* 24:115.

107. German, J. 1972. Genes which increase chromosomal instability in somatic cells and predispose to cancer. In *Progress in medical genetics*. A. Steinberg and A. Bearn, eds., p. 61. New York: Grune & Stratton.

108. Cohen, M. M., Shahan, M., Dagan, J., Shmueli, E., and Kohn, G. 1975. Cytogenetic investigation in families with ataxia telangiectasia. *Cytogenet. Cell Genet.* 15:338.

109. Webb, T., Harnden, D. G., and Harding, M. 1977. The chromosome analysis and susceptibility to transformation by simian virus 40 of fibroblasts from ataxia-telangiectasia. *Cancer Res.* 37:997.

110. Taylor, A. M. R., Metcalfe, J. A., Oxford, J. M., and Harnden, D. G. 1976. Is chromatid-type damage in ataxia telangiectasia after irradiation at G_0 a consequence of defective repair? *Nature* 260:441.

111. Rary, J. M., Bender, M. A., and Kelly, T. E. 1975. A 14/14 marker chromosome lymphocyte clone in ataxia telangiectasia. *J. Hered.* 66:33.

112. Oxford, J. M., Harnden, D. G., Parrington, J. M., and Delhanty, J. D. A. 1975. Specific chromosome aberrations in ataxia telangiectasia. *J. Med. Genet.* 12:251.

113. Levin, S., and Perlov, S. 1971. Ataxia telangiectasia in Israel with observations on its relationship to malignant disease. *Isr. J. Med. Sci.* 7:1535.

114. Morgan, J. L., Holcomb, T. M., and Morrissey, R. W. 1968. Radiation reactions in ataxia telangiectasia. *Am. J. Dis. Child.* 116:557.

115. Gotoff, S. P., Amirmokri, E., and Liebner, E. 1967. Ataxia telangiectasia, untoward response to x-irradiation and tuberous sclerosis. *Am. J. Dis. Child.* 114:617.

116. Cunliffe, P. N., Mann, J. R., Cameron, A. H., Roberts, K. D., and Ward, H. W. C. 1975. Radiosensitivity in ataxia telangiectasia. *Br. J. Radiol.* 48:374.

117. Taylor, A. M. R., Harnden, D. G., Arlett, C. F., Harcourt, S. A., Lehmann, A. R., Stevens, S., and Bridges, B. A. 1975. Ataxia telangiectasia: a human mutation with abnormal radiation sensitivity. *Nature* 258:427.

118. Paterson, M. C., Bech-Hansen, N. T., and Smith, P. J. 1981. Heritable radiosensitive and DNA repair deficient disorders in man. In *Chromosome damage and repair*. E. Seeberg and K. Kleppe, eds., p. 335. New York: Plenum.

119. Chen, P. C., Lavin, M. F., Kidson, C., and Moss, D. 1978. Identification of ataxia telangiectasia heterozygotes, a cancer prone population. *Nature* 274:484.

120. Paterson, M. C., Anderson, A. K., Smith, B. P., and Smith, P. J. 1979. Enhanced radiosensitivity of cultured fibroblasts from ataxia telangiectasia heterozygotes manifested by defective colony-forming ability and reduced DNA repair replication after hypoxic γ-irradiation. *Cancer Res.* 39:3725.

121. Arlett, C. F., and Harcourt, S. A. 1980. Survey of radiosensitivity in a variety of human cell strains. *Cancer Res.* 40:926.

122. Cox, R., Hosking, G. P., and Wilson, J. 1978. Ataxia telangiectasia: the evaluation of radiosensitivity in cultured skin fibroblasts as a diagnostic test. *Arch. Dis. Child.* 53:386.

123. Lehmann, A. R., and Stevens, S. 1979. The response of ataxia telangiectasia cells to bleomycin. *Nucleic Acids Res.* 6:1953.

124. Taylor, A. M. R., Rosney, C. M., and Campbell, J. B. 1979. Unusual sensitivity of ataxia telangiectasia cells to bleomycin. *Cancer Res.* 39:1046.

125. Cohen, M. M., Simpson, S. J., and Pazos, L. 1981. Specificity of bleomycin-induced cytotoxic effects on ataxia telangiectasia lymphoid cell lines. *Cancer Res.* 41:1817.

126. Vincent, R. A., Jr., Sheridan, R. B., and Huang, P. C. 1975. DNA strand breakage repair in ataxia telangiectasia fibroblast-like cells. *Mutation Res.* 33:357.

127. Sheridan, R. B. III, and Huang, P. C. 1977. Single strand breakage and repair in eukaryotic DNA as assayed by S1 nuclease. *Nucleic Acids Res.* 4:299.

128. Lehmann, A. R., and Stevens, S. 1977. The production and repair of double strand breaks in cells from normal humans and from patients with ataxia telangiectasia. *Biochim. Biophys. Acta* 474:49.

129. Paterson, M. C., Smith, B. P., Lohman, P. H. M., Anderson, A. K., and Fishman, L. 1976. Defective excision repair of γ-ray-damaged DNA in human (ataxia telangiectasia) fibroblasts. *Nature* 260:444.

130. Paterson, M. C., Smith, B. P., Knight, P. A., and Anderson, A. K. 1977. Ataxia telangiectasia: an inherited human disease involving radiosensitivity, malignancy, and defective DNA repair. In *Research in photobiology*. A. Castellani, ed., p. 207. New York: Plenum Press.

131. Inoue, T., Hirano, K., Yokoiyama, A., Kada, T., and Kato, H. 1977. DNA repair enzymes in ataxia telangiectasia and Bloom's syndrome fibroblasts. *Biochim. Biophys. Acta* 479:497.

132. Inoue, T., Yokoiyama, A., and Kada, T. 1981. DNA repair enzyme deficiency and *in vitro* complementation of the enzyme activity in cell-free extracts from ataxia telangiectasia fibroblasts. *Biochim. Biophys. Acta* 655:49.

133. Edwards, M. H., Taylor, A. M. R., and Duckworth, G. 1980. An enzyme activity in normal and ataxia telangiectasia cell lines which is involved in the repair of γ-irradiation-induced DNA damage. *Biochem. J.* 188:677.

134. Houldsworth, J., and Lavin, M. F. 1980. Effect of ionizing radiation on DNA synthesis in ataxia telangiectasia cells. *Nucleic Acids Res.* 9:3709.

135. Painter, R. B., and Young, B. R. 1980. Radiosensitivity in ataxia telangiectasia: a new explanation. *Proc. Natl. Acad. Sci.* (USA) 77:7315.

136. Ford, M. D., and Lavin, M. F. 1981. Ataxia telangiectasia: an anomaly in DNA replication after irradiation. *Nucleic Acids Res.* 9:1395.

137. DeWit, J., Jaspers, N. G. J., and Bootsma, D. 1981. The rate of DNA synthesis in normal human and ataxia telangiectasia cells after exposure to x-irradiation. *Mutation Res.* 80:221.

138. Edwards, M. J., and Taylor, A. M. R. 1980. Unusual levels of (ADP-ribose)n and DNA synthesis in ataxia telangiectasia cells following γ-ray irradiation. *Nature* 287:745.

139. Edwards, M. J., Taylor, A. M. R., and Flude, E. J. 1981. Bleomycin induced inhibition of DNA synthesis in ataxia telangiectasia cell lines. *Biochem. Biophys. Res. Comm.* 102:610.

140. Painter, R. B. 1981. Radioresistant DNA synthesis: an intrinsic feature of ataxia telangiectasia. *Mutation Res.* 84:183.

141. Povirk, L. F., and Painter, R. B. 1976. The effect of 313 nanometer light on initiation of replicons in mammalian cell DNA containing bromodeoxyuridine. *Biochim. Biophys. Acta* 432:267.

142. Painter, R. B. 1978. Inhibition of DNA replicon initiation by 4-nitroquinoline-1-oxide, adriamycin, and ethyleneimine. *Cancer Res.* 38:4445.

143. Smith, P. J., and Paterson, M. C. 1980. Gamma-ray induced inhibition of DNA synthesis in ataxia telangiectasia fibroblasts is a function of excision repair capacity. *Biochem. Biophys. Res. Comm.* 97:897.

144. Painter, R. B. 1982. Structural changes in chromatin as the basis for radiosensitivity in ataxia telangiectasia. *Cytogenet. Cell Genet.* 33:139.

145. Shaham, M., Becker, Y., and Cohen, M. M. 1980. A diffusable clastogenic factor in ataxia telangiectasia. *Cytogenet. Cell Genet.* 27:155.

146. Arlett, C. F. 1980. Mutagenesis in repair-deficient human cell strains. In *Progress in environmental mutagenesis.* M. Alečević, ed., p. 161. Amsterdam: Elsevier/North Holland Biomedical Press.

147. Murnane, J. P., and Painter, R. B. 1982. Complementation of the defects in DNA synthesis in irradiated and unirradiated ataxia telangiectasia cells. *Proc. Natl. Acad. Sci.* (USA) 79:1960.

148. Jaspers, N. G. J., and Bootsma, D. 1982. Genetic heterogeneity in ataxia telangiectasia studied by cell fusion. *Proc. Natl. Acad. Sci.* (USA) 79:2641.

149. Bloom, D. 1954. Congenital telangiectatic erythema resembling lupus erythematosus in dwarfs. *Amer. J. Dis. Child.* 88:754.

150. Bloom, D. 1966. The syndrome of congenital telangiectatic erythema and stunted growth: observations and studies. *J. Pediat.* 58:103.

151. German, J. 1969. Chromosomal breakage syndromes. *Birth Defects* 5:117.

152. German, J. 1969. Bloom's syndrome. I. Genetical and clinical observations in the first twenty-seven patients. *Am. J. Hum. Genet.* 21:196.

153. German, J., Bloom, D., Passarge, E., Fried, K., Goodman, R. M., Katzenellenbogen, I., Laron, Z., Legum, C., Levin, S., and Wahrman, J. 1977. Bloom's syndrome. VI. The disorder in Israel and an estimation of the gene frequency in Ashkenazim. *Am. J. Hum. Genet.* 29:553.

154. Gianelli, F., Benson, P. F., Pawsey, S. A., and Polani, P. E. 1977. Ultraviolet light sensitivity and delayed DNA-chain maturation in Bloom's syndrome fibroblasts. *Nature* 265:466.

155. Hand, R., and German, J. 1975. A retarded rate of DNA chain growth in Bloom's syndrome. *Proc. Natl. Acad. Sci.* (USA) 72:758.

156. Henson, P., Selsky, C. A., and Little, J. B. 1981. Excision of ultraviolet damage and the effect of irradiation on DNA synthesis in a strain of Bloom's syndrome fibroblasts. *Cancer Res.* 41:760.

157. Parker, V. P., and Lieberman, M. W. 1977. Levels of DNA polymerases α, β and γ in control and repair-deficient human diploid fibroblasts. *Nucleic Acids Res.* 4:2029.

158. German, J., Archibald, R., and Bloom, D. 1965. Chromosomal breakage in a rare and probably genetically determined syndrome of man. *Science* 148:506.

159. German, J., Schonberg, S., Louis, E., and Chaganti, R. S. K. 1977. Bloom's syndrome. IV. Sister-chromatid exchanges in lymphocytes. *Am. J. Hum. Genet.* 29:248.

160. Chaganti, R. S. K., Schonberg, S., and German, J. 1974. A manyfold increase in sister chromatid exchanges in Bloom's syndrome lymphocytes. *Proc. Natl. Acad. Sci.* (USA) 71:4508.

161. German, J., Crippa, L. P., and Bloom, D. 1974. Bloom's syndrome. III. Analysis of the chromosome aberration characteristic of this disorder. *Chromosoma* 48:361.

162. Zbinden, I., and Cerutti, P. 1981. Near-ultraviolet sensitivity of skin fibroblasts of patients with Bloom's syndrome. *Biochem. Biophys. Res. Comm.* 98:579.

163. Hirschi, M., Netrawali, M. S., Remsen, J. F., and Cerutti, P. A. 1981. Formation of DNA single-strand breaks by near-ultraviolet and γ-rays in normal and Bloom's syndrome skin fibroblasts. *Cancer Res.* 41:2003.

164. Selsky, C. A., Henson, P., Weichselbaum, R. R., and Little, J. B. 1979. Defective reactivation of ultraviolet light–irradiated herpes virus by a Bloom's syndrome fibroblast strain. *Cancer Res.* 39:3392.

165. Arlett, C. F., and Harcourt, S. A. 1978. Cell killing and mutagenesis in repair-defective human cells. In *DNA repair mechanisms*. P. C. Hanawalt, E. C. Friedberg and C. F. Fox, eds., p. 633. New York: Academic.

166. Krepinsky, A. B., Heddle, J. A., and German, J. 1979. Sensitivity of Bloom's syndrome lymphocytes to ethyl methanesulfonate. *Hum. Genet.* 50:151.

167. Tice, R., Windler, G., and Ray, J. 1978. Effect of cocultivation on sister chromatid exchange frequencies in Bloom's syndrome and normal fibroblast cells. *Nature* 273:538.

168. Emerit, I., and Cerutti, P. 1981. Clastogenic activity from Bloom syndrome fibroblast cultures. *Proc. Natl. Acad. Sci.* (USA) 78:1868.

169. Emerit, I., Michelson, A., Levy, A., Camus, J., and Emerit, J. 1980. Chromosome breaking agent of low molecular weight in human systemic lupus erythematosus. Protector effect of superoxide dismutase. *Hum. Genet.* 55:341.

170. Emerit, I., Levy, A., and deVaux Saint-Cyr, C. 1980. Chromosome damaging agent of low molecular weight in the serum of New Zealand black mice. *Cytogenet. Cell. Genet.* 26:41.

171. Warren, S. T., Schultz, R. A., Chang, C-C., Wade, M. H., and Trosko, J. E. 1981. Elevated spontaneous mutation rate in Bloom syndrome fibroblasts. *Proc. Natl. Acad. Sci.* (USA) 78:3133.

172. Cockayne, E. A. 1936. Dwarfism with retinal atrophy and deafness. *Arch. Dis. Child.* 11:1.

173. Cockayne, E. A. 1946. Dwarfism with retinal atrophy and deafness. *Arch. Dis. Child.* 21:52.

174. Guzzetta, F. 1972. Cockayne-Neill-Dingwall syndrome. In *Handbook of clinical neurology*. P. J. Vinken and G. W. Bruyn, eds., vol. 13, p. 431. Amsterdam: North Holland.

175. Fujimoto, W., Greene, M., and Seegmiller, J. 1969. Cockayne's syndrome: report of a case with hyperlipoproteinemia, hyperinsulinemia, renal disease, and normal growth hormone. *J. Pediat.* 75:881.

176. Schmickel, R. D., Chu, E. H. Y., Trosko, J. E., and Chang, C-C. 1977. Cockayne syndrome: a cellular sensitivity to ultraviolet light. *Pediat.* 60:135.

177. Andrews, A. D., Yoder, F. W., Barrett, S. F., and Robbins, J. H. 1976. Cockayne's syndrome fibroblasts have decreased colony forming ability but normal rates of unscheduled DNA synthesis after ultraviolet irradiation. *Clin. Res.* 24:624A.

178. Wade, M. H., and Chu, E. H. Y. 1978. Effects of DNA damaging agents on cultured fibroblasts derived from patients with Cockayne syndrome. In *DNA repair mechanisms*. P. C. Hanawalt, E. C. Friedberg and C. F. Fox, eds., p. 667. New York: Academic.

179. Arlett, C. F., and Harcourt, S. A. 1978. Cell killing and mutagenesis in repair defective human cells. In *DNA repair mechanisms*. P. C. Hanawalt, E. C. Friedberg and C. F. Fox, eds., p. 633. New York: Academic.

180. Wade, M. H., and Chu, E. H. Y. 1979. Effects of DNA damaging agents on cultured fibroblasts derived from patients with Cockayne syndrome. *Mutation Res.* 59:49.

181. Deschavanne, P. J., Diatloff-Zito, C., Maciera-Coelho, A., and Malaise, E-P. 1981. Unusual sensitivity of two Cockayne's syndrome cell strains to both UV and γ irradiation. *Mutation Res.* 91:403.

182. Marshall, R. R., Arlett, C. F., Harcourt, S. A., and Broughton, B. A. 1980. Increased sensitivity of cell strains from Cockayne's syndrome to sister chromatid–exchange induction and cell killing by UV light. *Mutation Res.* 69:107.

183. Hoar, D. I., and Davis, F. 1979. Host-cell reactivation of UV-irradiated adenovirus in Cockayne syndrome. *Mutation Res.* 62:401.

184. Day, R. S. III, and Ziolkowski, C. 1978. Studies on UV-induced viral reversion, Cockayne's syndrome and MNNG damage using adenovirus 5. In *DNA repair mechanisms*. P. C. Hanawalt, E. C. Friedberg and C. F. Fox, eds., p. 535. New York: Academic.

185. Ahmed, F. E., and Setlow, R. B. 1978. Excision repair in ataxia telangiectasia, Fanconi's anemia, Cockayne syndrome, and Bloom's syndrome, after treatment with ultraviolet radiation and N-acetoxy-2-acetylaminofluorene. *Biochim. Biophys. Acta* 521:805.

186. Lehmann, A. R., Kirk-Bell, S., and Mayne, L. 1979. Abnormal kinetics of DNA synthesis in ultraviolet light–irradiated cells from patients with Cockayne's syndrome. *Cancer Res.* 39:4237.

187. Ikenaga, M., Inoue, M., Kozuka, T., and Sugita, T. 1981. The recovery of colony-forming ability and the rate of semi-conservative DNA synthesis in ultraviolet-irradiated Cockayne and normal human cells. *Mutation Res.* 91:87.

188. Lehmann, A. R., and Mayne, L. 1981. The response of Cockayne syndrome cells to UV-irradiation. In *Chromosome damage and repair*. E. Seeberg and K. Kleppe, eds., p. 367. New York: Plenum.

189. Tanaka, K., Kawai, K., Kumahara, Y., Ikenaga, M., and Okada, Y. 1981. Genetic complementation groups in Cockayne syndrome. *Somat. Cell Genet.* 7:445.

190. Lehmann, A. R. 1982. Three complementation groups in Cockayne's syndrome. *Mutation Res.* 106:347.

191. Fanconi, G. 1967. Familial constitutional panmyelocytopathy, Fanconi's anemia (F.A.). I. Clinical aspects. *Seminars Hemat.* 4:233.

192. Reinhold, J. D. L., Neumark, E., Lightwood, R., and Carter, C. O. 1952. Familial hypoplastic anemia with congenital abnormalities (Fanconi's syndrome). *Blood* 7:915.

193. Swift, M. R., and Hirschhorn, K. 1966. Fanconi's anemia. Inherited susceptibility to chromosome breakage in various tissues. *Ann. Int. Med.* 65:496.

194. Garriga, S., and Crosby, W. H. 1959. The incidence of leukemia in families of patients with hypoplasia of the marrow. *Blood* 14:1008.

195. Swift, M. 1971. Fanconi's anemia in the genetics of neoplasia. *Nature* 230:370.

196. Schroeder, T. M., and Kurth, R. 1971. Spontaneous chromosomal breakage and high incidence of leukemia in inherited disease. *Blood* 37:96.

197. Sasaki, M. S., and Tonomura, A. 1973. A high susceptibility of Fanconi's anemia to chromosome breakage by DNA cross-linking agents. *Cancer Res.* 33:1829.

198. Auerbach, A. D., and Wolman, S. R. 1976. Susceptibility of Fanconi's anemia fibroblasts to chromosome damage by carcinogens. *Nature* 261:494.

199. Kato, H., and Stich, H. F. 1976. Sister chromatid exchanges in ageing and repair-deficient human fibroblasts. *Nature* 260:447.

200. Auerbach, A. D., and Wolman, S. R. 1978. Carcinogen-induced chromosome breakage in Fanconi's anemia heterozygous cells. *Nature* 271:69.

201. Sasaki, M. S. 1978. Fanconi's anemia. A condition possibly associated with defective DNA repair. In *DNA repair mechanisms*. P. C. Hanawalt, E. C. Friedberg and C. F. Fox, eds., p. 675. New York: Academic.

202. Fujiwara, Y., Tatsumi, M., and Sasaki, M. S. 1977. Cross-link repair in human cells with its possible defects in Fanconi's anemia cells. *J. Mol. Biol.* 113:635.

203. Berger, N. A., Berger, S. J., and Catino, D. M. 1982. Abnormal NAD^+ levels in cells from patients with Fanconi's anemia. *Nature* 299:271.

204. Knudson, A. G., Jr. 1971. Mutation and cancer: statistical study of retinoblastoma. *Proc. Natl. Acad. Sci.* (USA) 68:820.

205. Knudson, A. G., Jr., Meadows, A. T., Nichols, W. W., and Hill, R. 1976. Chromosomal deletion and retinoblastomas. *N. Eng. J. Med.* 295:1120.

206. Bonaite-Pellie, C., Briar-Guillemot, M. L., Feingold, J., and Frezal, J. 1976. Mutation theory of carcinogenesis in retinoblastoma. *J. Natl. Cancer Inst.* 57:269.

207. Francke, U. 1976. Retinoblastoma and chromosome 13. *Cytogenet. Cell Genet.* 16:131.

208. Jensen, R. D., and Miller, R. W. 1971. Retinoblastoma: epidemiologic considerations. *N. Eng. J. Med.* 285:307.

209. Kitchen, F. D., and Ellsworth, R. M. 1974. Pleiotropic effects of the gene for retinoblastoma. *J. Med. Genet.* 11:244.

210. Weichselbaum, R. R., Nove, J., and Little, J. B. 1977. Skin fibroblasts from a D-deletion type retinoblastoma patient are abnormally X-ray sensitive. *Nature* 266:726.

211. Weichselbaum, R. R., Nove, J., and Little, J. B. 1978. X-ray sensitivity of diploid fibroblasts from patients with hereditary or sporadic retinoblastoma. *Proc. Natl. Acad. Sci.* (USA) 75:3962.

212. Ejima, Y., Sasaki, M. S., Utsumi, H., Kaneko, A., and Tanooka, H. 1982. Radiosensitivity of fibroblasts from patients with retinoblastoma and chromosome-13 anomalies. *Mutation Res.* 103:177.

213. Scudiero, D. A., Meyer, S. A., Clatterbuck, B. E., Tarone, R. E., and Robbins, J. H. 1981. Hypersensitivity to N′-methyl-N′-nitro-N-nitrosoguanidine in fibroblasts from patients with Huntington disease, familial dysautonomia, and other primary neuronal degenerations. *Proc. Natl. Acad. Sci.* (USA) 78:6451.

214. Florey, H., ed. 1962. *General pathology*, 3rd ed. London: Lloyd-Luke.

215. Trosko, J. E., and Chang, C-C. 1978. The role of mutagenesis in carcinogenesis. *Photochem. Photobiol. Rev.* 3:135.

216. Boveri, T. 1914. *Zur Frage der Entstehung Maligne Tumoren*. Jena: Gustav Fisher.

217. Doll, R. 1977. Introduction. In *Origins of human cancer*. H. H. Hiatt, J. D. Watson and J. A. Winsten, eds., Book A, p. 1. Cold Spring Harbor Laboratory.

218. Hammond, E. C., Garfinkel, L., Seidman, H., and Lew, E. A. 1977. Some recent findings concerning cigarette smoking. In *Origins of human cancer*. Book A. H. H. Hiatt, J. D. Watson and J. A. Winsten, eds., p. 101. Cold Spring Harbor Laboratory.

219. Namba, M., Nishitani, K., and Kimoto, T. 1978. Carcinogenesis in tissue culture: neoplastic transformation of a normal human diploid cell strain W1-38 with Co-60 gamma rays. *Jap. J. Exp. Med.* 48:303.

220. Borek, C. 1980. X-ray-induced *in vitro* neoplastic transformation of human diploid cells. *Nature* 283:776.

221. Kakunaga, T. 1978. Neoplastic transformation of human diploid fibroblast cells by chemical carcinogens. *Proc. Natl. Acad. Sci.* (USA) 75:1334.

222. Milo, G. E., Jr., and DiPaolo, J. A. 1978. Neoplastic transformation of human diploid cells *in vitro* after chemical carcinogen treatment. *Nature* 275:130.

223. Sutherland, B. M. 1978. Photoreactivation: evaluation of pyrimidine dimers in ultraviolet radiation–induced cell transformation. *Natl. Cancer Inst. Monogr.* 50:129.

224. Sutherland, B. M., Cimino, J. S., Delihas, N., Shih, A. G., and Oliver, R. P. 1980. Ultraviolet light–induced transformation of human cells to anchorage-independent growth. *Cancer Res.* 40:1934.

225. Sutherland, B. M., Delihas, N. C., Oliver, R. P., and Sutherland, J. C. 1981. Action spectra for ultraviolet-induced transformation of human cells to anchorage-independent growth. *Cancer Res.* 41:2211.

226. Milo, G. E., Weisbrode, S. A., Zimmerman, R., and McCloskey, J. A. 1981. Ultraviolet radiation–induced neoplastic transformation of normal human cells, *in vitro*. *Chem.-Biol. Interact.* 36:45.

227. Hart, R. W., Setlow, R. B., and Woodhead, A. D. 1977. Evidence that pyrimidine dimers in DNA can give rise to tumors. *Proc. Natl. Acad. Sci.* (USA) 74:5574.

228. Woodhead, A. D., Setlow, R. B., and Hart, R. W. 1977. Tissue changes resulting from the injection of γ-irradiated cells into the gynogenetic teleost, *Poecilia formosa*. *Cancer Res.* 37:4261.

229. Woodhead, A. D., and Scully, P. M. 1977. A comparative study of the pretumerous thyroid gland of the gynogenetic teleost, *Poecilia formosa*, and that of other *poeciliid* fishes. *Cancer Res.* 37:3751.

230. Hart, R. W., and Setlow, R. B. 1975. Direct evidence that pyrimidine dimers in DNA result in neoplastic transformation. In *Molecular mechanisms for repair of DNA*. P. C. Hanawalt and R. B. Setlow, eds., p. 719. New York: Plenum.

231. Lieb, M. 1961. Enhancement of ultraviolet-induced mutation in bacteria by caffeine. *Z. Vererbungslehre* 92:416.

232. Roberts, J. J., Friedlos, F., and Belka, E. S. 1978. DNA template breakage and decreased excision of hydrocarbon derived adducts from Chinese hamster cell DNA following caffeine-induced inhibition of post-replication repair. In *DNA repair mechanisms*. P. C. Hanawalt, E. C. Friedberg and C. F. Fox, eds., p. 527. New York: Academic.

233. Trosko, J. E., and Chu, E. H. 1973. Inhibition of repair of UV-damaged DNA by caffeine and mutation induction of Chinese hamster cells. *Chem.-Biol. Interact.* 6:317.

234. Zajdela, F., and Laterjet, R. 1974. The inhibitory effect of caffeine on the induction of cutaneous tumors in mice by ultraviolet rays. In *Excerpta Med. Int. Cong. Ser.* No. 351, vol. 3, p. 211. Amsterdam: Excerpta Medica.

235. DiPaolo, J. A., and Donovan, P. J. 1978. Transformation frequency of Syrian golden hamster cells and its modulation by ultraviolet irradiation. *National Cancer Inst. Monograph* 50:75.

236. Doniger, J., Jacobson, E. D., Krell, K., and DiPaolo, J. A. 1981. Ultraviolet light action spectra for neoplastic transformation and lethality of Syrian hamster embryo cells correlate with spectrums for pyrimidine dimer formation in cellular DNA. *Proc. Natl. Acad. Sci.* (USA) 78:2378.

237. Massey, H., Olejkowski, J. A., Hoess, R. H., and Freed, J. 1976. Photoreversal of UV-induction of ouabain resistance. *Inst. Cancer Res.* (*21st Scientific Report of the Fox Chace Cancer Center*, 1975–76), p. 140.

239. Ames, B. N. 1971. The detection of chemical mutagens with enteric bacteria. In *Chemical mutagens, principles and methods for their detection*. A. Hollaender, ed., vol. I, p. 267. New York: Plenum.

239. Ames, B. N. 1972. A bacterial system for detecting mutagens and carcinogens. In *Mutagenic effects of environmental contaminants*. H. E. Sutton and M. I. Harris, eds., p. 57. New York: Academic.

240. Maron, D. M., and Ames, B. N. 1983. Revised methods for the Salmonella mutagenicity test. *Mutation Res.* 113:173.

241. Ames, B. N. 1979. Identifying environmental chemicals causing mutations and cancer. *Science* 204:587.

242. Nagao, M., Sugimura, T., and Matsushima, T. 1978. Environmental mutagens and carcinogens. *Ann. Rev. Genet.* 12:117.

243. Moreau, P., Bailone, A., and Devoret, R. 1976. Prophage λ induction in *Escherichia coli* K12 *env*A *uvr*B: a highly sensitive test for potential carcinogens. *Proc. Natl. Acad. Sci.* (USA) 73:3700.

244. Levine, A., Moreau, P. L., Sedgwick, S. G., Devoret, R., Adhya, S., Gottesman, M., and Das, A. 1978. Expression of a bacterial gene turned on by a potent carcinogen. *Mutation Res.* 50:29.

245. Han, R. H. C., and Stich, A. F. 1975. DNA repair synthesis of cultured human cells as a rapid bioassay for chemical carcinogens. *Int. J. Cancer* 16:284.

246. Stich, H. F., Lam, P., Lo, L. W., Koropatnick, D. J., and San, R. H. C. 1975. The search for relevant short term bioassays for chemical carcinogens: the tribulation of a modern Sisyphus. *Canad. J. Genet. Cytol.* 17:471.

247. Agarwal, S. S., Brown, D. Q., Katz, E. J., and Loeb, L. A. 1977. DNA repair in human lymphocytes. In *Genetics of human cancer.* J. J. Mulvihill, R. W. Miller and J. F. Fraumeni, Jr., eds., p. 365. New York: Raven Press.

248. Taylor, J. H. 1958. Sister chromatid exchanges in tritium labeled chromosomes. *Genetics* 43:515.

249. Wolff, S. 1977. Sister chromatid exchange. *Ann. Rev. Genet.* 11:183.

250. Zakharov, A. F., and Egolina, N. A. 1972. Differential spiralisation along mammalian mitotic chromosomes. I. BUdR-revealed differentiation in Chinese hamster chromosomes. *Chromosoma* 38:341.

251. Perry, P., and Evans, H. J. 1975. Cytological detection of mutagen-carcinogen exposure by sister chromatid exchange. *Nature* 258:121.

252. Evans, H. J. 1982. Sister chromatid exchanges and disease states in man. In *Sister chromatid exchanges.* S. Wolff, ed., p. 183. New York: Wiley.

253. Stetka, D. G., and Wolff, S. 1976. Sister chromatid exchange as an assay for genetic damage induced by mutagen-carcinogens. I. *In vivo* test for compounds requiring metabolic activation. *Mutation Res.* 41:333.

254. Takehisa, S., and Wolff, S. 1977. Induction of sister chromatid exchanges in Chinese hamster cells by carcinogenic mutagens requiring metabolic activation. *Mutation Res.* 45:263.

255. Carrano, A. V., Thompson, L. H., Lindl, P. A., and Minkler, J. L. 1978. Sister chromatid exchanges as an indicator of mutagenesis. *Nature* 271:551.

256. Little, J. B. 1969. Repair of sublethal and potentially lethal radiation damage in plateau phase cultures of human cells. *Health Physics* 16:469.

257. Little, J. B. 1973. Factors influencing the repair of potentially lethal radiation damage in growth-inhibited cells. *Radiat. Res.* 56:320.

258. Little, J. B. 1981. Radiation transformation *in vitro:* implications for mechanisms of carcinogenesis. In *Advances in Modern Environmental Toxicology.* Vol. 1. N. Mishra, V. Dunkel and M. Mehlman, eds., p. 383. Princeton Junction, N.J.: Senate Press.

259. Weichselbaum, R. R., Nove, J., and Little, J. B. 1978. Deficient recovery from potentially lethal radiation damage in ataxia telangiectasia and xeroderma pigmentosum. *Nature* 271:261.

260. Nagasawa, H., and Little, J. B. 1979. Effect of tumor promoters and protease inhibitors on spontaneous and x-ray induced sister chromatid exchanges in mouse cells. *Proc. Natl. Acad. Sci.* (USA) 76:1943.

261. Kennedy, A. R., and Little, J. B. 1978. Protease inhibitors suppress radiation induced malignant transformation *in vitro. Nature* 276:825.

262. Calos, M. D., and Miller, J. H. 1980. Transposable elements. *Cell* 20:579.

263. Kleckner, N. 1981. Transposable elements in prokaryotes. *Ann. Rev. Genet.* 15:341.

264. Radman, M., Jeggo, P., and Wagner, R. 1982. Chromosomal rearrangement and carcinogenesis. *Mutation Res.* 98:249.

265. Rowley, J. D. 1973. A new consistent chromosomal abnormality in chronic myelogenous leukemia identified by quinacrine fluorescence and giemsa staining. *Nature* 243:290.

266. Rary, J. M., Bender, M. A., and Kelly, T. E. 1974. Cytogenetic studies of ataxia telangiectasia. *Am. J. Human Genet.* 26: 70A.

267. McCaw, B. K., Hecht, F., Harnden, D. G., and Teplitz, R. L. 1975. Somatic rearrangement of chromosome 14 in human lymphocytes. *Proc. Natl. Acad. Sci.* (USA) 72:2071.

268. Levan, A., Levan, G., and Mitelman, F. 1977. Chromosomes and cancer. *Hereditas* 86:15.

269. Saxholm, H. J., and Digernes, V. 1980. Progressive loss of DNA and lowering of the chromosomal mode in chemically transformed C3H/10T ½ cells during development of their oncogenic potential. *Cancer Res.* 40:4254.

270. Cowell, J. K., and Wigley, C. B. 1980. Changes in DNA content during *in vitro* transformation of mouse salivary gland epithelium. *J. Natl. Cancer Inst.* 64:1443.

271. Cooper, G. M., Okenquist, S., and Silverman, L. 1980. Transforming activity of DNA of chemically transformed and normal cells. *Nature* 284:418.

272. Shih, C., Shilo, B., Goldfarb, M. P., Dannenberg, A., and Weinberg, R. A. 1979. Passage of phenotypes of chemically transformed cells via transfection of DNA and chromatin. *Proc. Natl. Acad. Sci.* (USA) 76:5714.

273. Perucho, M., Goldfarb, M., Shimizu, K., Lama, C., Fogh, J., and Wigler, M. 1981. Human-tumor-derived cell lines contain common and different transforming genes. *Cell* 27:467.

274. Weinberg, R. A. 1982. Fewer and fewer oncogenes. *Cell* 30:3.

275. Bishop, J. M. 1983. Cancer genes come of age. *Cell* 32:1018.

276. Capon, D. J., Chen, E. Y., Levinson, A. D., Seeburg, P. H., and Goeddel, D. V. 1983. Complete nucleotide sequences of the T24 human bladder carcinoma oncogene and the normal homologue. *Nature* 302:33.

277. Epstein, J. H., Fukuyama, K., Reed, W. B., and Epstein, W. L. 1970. Defect in DNA synthesis in skin of patients with xeroderma pigmentosum *in vivo*. *Science* 168:1477.

278. Zelle, B. 1980. Ph.D. thesis, University of Rotterdam.

279. Day, R. S. III, Kraemer, K. H., and Robbins, J. H. 1975. Complementing xeroderma pigmentosum fibroblasts restore biological activity to UV-damaged DNA. *Mutation Res.* 28:251.

280. McCormick, J. J., and Maher, V. M. 1979. Mammalian cell mutagenesis as a biological consequence of DNA damage. In *DNA repair mechanisms*. P. C. Hanawalt, E. C. Friedberg and C. F. Fox, eds., p. 739. New York: Academic.

281. Devoret, R. 1979. Bacterial tests for potential carcinogens. *Sci. Am.* 241:40.

282. Rowley, J. D. 1980. Chromosome abnormalities in cancer. *Cancer Genet. Cytogenet.* 2:175.

283. Myhr, B. C., Turnbull, D., and DiPaolo, J. A. 1979. Ultraviolet mutagenesis of normal and xeroderma pigmentosum variant human fibroblasts. *Mutation Res.* 62:341.

Author Index

Subject Index